NOVA UNIVERSITY
CENTER FOR COMPUTER SCIENCE
3301 COLLEGE AVE.
FT. LAUDERDALE, FL 33314

Modern Digital Systems Design

Modern Digital Systems Design

John Y. Cheung
University of
Oklahoma—Norman

•

Jon G. Bredeson
University of
Alabama—Tuscaloosa

West Publishing Company

St. Paul   New York   Los Angeles   San Francisco

COVER IMAGE   Robert Llewellyn/Four By Five
COVER DESIGN   Frank Lamacchia/Viviani Productions, Inc.
TEXT DESIGN   Geri Davis/Quadrata, Inc.
EDITORIAL PRODUCTION SERVICES   Spectrum Publisher Services, Inc.
COMPOSITION   Syntax International

COPYRIGHT © 1990   By WEST PUBLISHING COMPANY
50 W. Kellogg Boulevard
P.O. Box 64526
St. Paul, MN 55164-1003

97 96 95 94 93 92 91 90   8 7 6 5 4 3 2 1 0

Library of Congress Cataloging-in-Publication Data

Cheung, John Y.
    Modern digital systems design/John Y. Cheung, Jon G. Bredeson.
      p. cm.
    Includes bibliographies and index.
    ISBN 0-314-47828-0
    1. Logic circuits. 2. Microprocessors. 3. Motorola 68010
(Microprocessor) 4. Intel 8085 (Microprocessor) I. Bredeson, Jon
G. II. Title.
TK7868.L6C465 1990
621.395—dc 19

88-29031
CIP

To our parents

Paul To-Kwong and Mei-Shin Cheung

Maurice and Myrtle Bredeson

Contents

[handwritten annotations:] DEC. POINT DEC — BINARY → HEX, → OCTAL, BIT #

[handwritten annotations:] SIGN + MAGNITUDE 1-S & 2-S COMPLEMENTS ADV. DISADV.]!

Fundamentals of Boolean Algebra

2

Design Techniques with Combinational Circuits

3

Advanced Design Techniques with Combinational Circuits

4

Sequential Circuit Design   5

Design Techniques with Logic Array Components

6

Fundamentals of Microprocessors

7

Memory and I/O

8

The 8085 Microprocessor 9

The 8085 Microprocessor- 10
Based System

The 68010 Microprocessor

11

The 68010 Microprocessor-Based System   12

Appendix A
The 8085 Instruction Set

Appendix B
The 68010 Instruction Set

Preface

The world of digital system design is infinitely variable and very exciting. The primary objectives of this book are to provide the student with basic digital system principles and practical design techniques. To achieve these goals, three aspects of system design are stressed: (1) comprehensive coverage of fundamental concepts; (2) practical design with real products; and (3) modular organization of text material. These three approaches are the major differences between this textbook and others in the digital system field.

Comprehensive coverage includes theory as well as application. For example, any complete textbook in digital system design must cover basic types of circuits: combinational, sequential, and microprocessor-based; in this book, equal emphasis is given to each of them. Regarding microprocessor-based circuits, many other texts discuss the methodology of making parts of a computer, which is helpful in understanding the architecture of a computer but is less practical to the system designer who must integrate the computer as part of the final product. The need today is not so much to learn how to design a better computer, but to learn how to use currently available microprocessors and supporting components as building blocks in system design. Hence, part of the comprehensive emphasis of this book is to discuss design techniques using microprocessors and other supporting chips as components in modern digital systems.

Practical design is another important emphasis. Far too often, an introductory text provides only a little of the design aspect and omits many details regarding real products. For example, most textbooks in the digital field cover inverters and other gates, but few point out the differences between inverters and other inverter-like gates such as buffers, drivers, line drivers, line receivers, and transceivers, like this book does. Numerous text examples use those currently available components and lab experiments to supplement the in-depth understanding of their behavior. These experiments further illustrate design concepts and provide concrete examples. Terminology used throughout this text is chosen to be as similar to industrial practice as possible.

Besides stressing comprehensive coverage and practical design, this book is also written in a modular fashion. That is, chapters dealing with fundamental concepts are separated from those dealing with design details. Many other texts that include microprocessors are limited in usage because they only discuss a single processor. To prevent this limitation, this text presents in-depth information on two popular microprocessors—an 8-bit and a 16-bit. This modular approach, treating different machines in separate sections, is advantageous to the student and the instructor. Two chapters are devoted to each machine at the end of the book and may be included or excluded at the discretion of the instructor without loss of continuity.

In looking at the overall chapter organization, the first half of the book (Chapters 1 through 6) covers basic concepts in combinational and sequential circuit design. Chapter 1 introduces the theory of binary numbers, presenting the positional numbering system for different bases, conversion techniques from one base to another, and theories of binary integer arithmetic for both signed and unsigned numbers. Binary codes for decimal representation are also introduced. Chapter 2 presents fundamentals of Boolean algebra—its definition, axioms, theorems, and functional representations. Techniques for minimization of Boolean functions by formulas, Karnaugh maps, and the variable-entered maps for both completely specified and incompletely specified functions are included. Chapter 3 covers design techniques using basic combinational components in the small-scale integration (SSI) category, and introduces real SSI products and their various subtleties. The design techniques are discussed and illustrated with real problems worked out all the way from concepts to components. Chapter 4 covers design techniques using complex combinational components in the medium-scale integration (MSI) category. Various circuit components in the MSI category are examined and design procedures are laid out in a top-down approach.

Chapter 5 introduces various sequential circuit components, including flip-flops, counters, clocks, and shift registers, with numerous application examples. Timing diagrams are stressed throughout the book from this point on. This chapter also presents fundamental concepts and design techniques of synchronous sequential circuits; procedures for both the synthesis and analysis of synchronous sequential circuits are clearly outlined in step form. Chapter 6 uses components in the large-scale integration (LSI) category, and principles of logic arrays and their design techniques are examined in detail.

The second half of the book (Chapters 7 through 12) covers microprocessor-based circuits. Chapters 7 and 8 are crucial in that they lay the foundation for computer organization as the student progresses into the very-large-scale integration (VLSI) component family. Chapter 7 first presents register transfer language, then a brief but intensive introduction to the concepts of system fundamentals. A generic computer system is examined in terms of its processor architecture as an illustration. Chapter 8 examines memory addressing modes, memory organization, and input/output interfacing.

Actual 8-bit and 16-bit microprocessors are then examined in detail. Chapter 9 explores software aspects of an 8-bit microprocessor, the Intel 8085: the internal organization, data formats, and addressing modes. Programming concepts of the 8085 are accompanied by plenty of examples. The implementation of commonly used language constructs are discussed as examples of assembly-language programming. Chapter 10 examines hardware aspects of the 8085: detailed electrical characteristics and their signals. Special emphasis is on timing diagrams as a way to explain the operation of the 8085. A review of the 8085-based system, together with several support chips, systematically shows how a microprocessor-based system is designed and furthers the student's understanding of 8085-based systems.

Chapters 11 and 12 examine the 16-bit microprocessor and may be studied independently of Chapters 9 and 10. Chapter 11 is dedicated to the software aspects of a 16-bit microprocessor, the Motorola 68010, including assembler directives and the instruction set. Numerous examples are given to clarify the use of some of the instructions. The implementations of many commonly used data structures are illustrated with assembly-program segments. Chapter 12 presents hardware aspects of the 68010, with much discussion centered on timing analysis and exception processing. Other topics include asynchronous bus operation, synchronous bus operation, direct memory access techniques, and 68010-based systems.

The chapter organization is ideal for a one-year sequence of courses in digital design, or for a two-semester sequence of courses (the first six chapters on combinational and sequential circuits can be used for the first semester while the rest of the book on microprocessor circuits can be used for the second semester). For those in a quarter system, an instructor could cover the first four chapters in the first quarter, then Chapters 5 through 8 in the second quarter, and Chapters 9 through 12 in the third quarter. Another option is to adapt the text to a one-semester course, using selected materials in different chapters based on the course level and emphasis of coverage. For example, a course in digital logic may cover selected material from the first seven chapters, ending with a brief introduction to microprocessor technology. Likewise, a one-semester course emphasizing the design aspects of microprocessors for those with a preliminary background in digital logic may begin with Chapter 6. Regardless of course use, the student will gain a comprehensive and practical perspective from which to design modern digital systems.

Acknowledgments

The authors thank the following reviewers for their excellent suggestions and valuable comments in the preparation of this publication: Martin Kaliski, California Polytechnic—San Luis Obispo; John Pavalt, Iowa State University; John Wallace, Georgia Institute of Technology; Manos Roumeliotis, West Virginia University; Charles Wright, Iowa State University; Nicolas Albertos, University of Tennessee; Veljko Malbasa, Washington State University; Aaron S. Collins, Clemson University; K. J. Pitts, San Jose State University; Barry Fagin, Dartmouth College; Shinji Nakamura, Dartmouth College; and Wayne McMurran, California Polytechnic—San Luis Obispo.

We also wish to express our sincere appreciation to all of our students who struggled through years of teaching as we refined the materials for this book. We are indebted to Advanced Micro Devices, Intel, Motorola, Signetics, and Texas Instruments for their generosity and cooperation in giving us permission to reprint some of their technical information. We also thank Marita Hardcastle and Kellean Kerr for proofing and typing the manuscript. Most of all, we owe the success of this book to West personnel Michael Slaughter (acquisitions editor) and Mary Garvey Verrill and Tamborah Moore (production editors) whose expertise and guidance throughout this project were indispensable. Special thanks are due in particular to Elise Oranges of Spectrum Publisher Services for her patience and hard work.

John Y. Cheung
Jon G. Bredeson

Number Theory

1

In this chapter the general number system is first presented. The formats for binary, octal, decimal, and hexadecimal number systems are next discussed in more detail, and conversions among the different number systems are also presented. We then examine the arithmetic operations for the binary number system for both unsigned and signed number representations. The last section of this chapter contains a brief discussion of the various codes used to represent decimal numbers and characters. We also present arithmetic operations for the decimal numbering system.

Most present-day digital systems and computer systems manipulate discrete elements of information, and these elements are represented in binary form. The reason for the binary form is that it is much easier to design electronic circuits that have only two stable states rather than three or more. Understanding computer systems requires a working knowledge of the binary number system as well as the octal and hexadecimal number systems, which are natural partners of the binary system. The octal and hexadecimal number systems are much easier to use than the binary number system, because the length of the numbers is shorter.

1.1

Number Systems

In general, a number system can be represented by a series of coefficients related to a particular base as follows:

$$N = (a_5 a_4 a_3 a_2 a_1 a_0 . a_{-1} a_{-2} a_{-3})$$

The coefficients a_j range in value from 0 to $r - 1$ in a base r system; and thus the previous number is actually taken as a shorthand notation to represent

$$N = (a_5 \times r^5) + (a_4 \times r^4) + (a_3 \times r^3) + (a_2 \times r^2) + (a_1 \times r^1)$$
$$+ (a_0 \times r^0) + (a_{-1} \times r^{-1}) + (a_{-2} \times r^{-2}) + (a_{-3} \times r^{-3}) \qquad (1\text{--}1)$$

Most people are already familiar with the decimal number system. The decimal number system is normally assumed in everyday use, and the powers of 10 are implied in most cases without further notation. The *base*, or *radix*, of the decimal number system is said to be 10. The base 10 was probably chosen because of the 10 fingers on human hands. For example, the decimal number 652.85_{10} is a shorthand notation for six groups of hundreds, five groups of tens, two single units, and eighty-five parts of a hundredth as follows:

$$652.85_{10} = (6 \times 10^2) + (5 \times 10^1) + (2 \times 10^0) + (8 \times 10^{-1}) + (5 \times 10^{-2})$$

For clarity, the subscript will be used to indicate the base of the number system.

Many other number systems are also possible, such as the binary (base 2) system, the ternary (base 3) system, the octal (base 8) system, and the hexadecimal (base 16) system. In the binary number system only two symbols are used, namely 0 and 1. For example, the binary number 110.11_2 represents

$$110.11_2 = (1 \times 2^2) + (1 \times 2^1) + (0 \times 2^0) + (1 \times 2^{-1}) + (1 \times 2^{-2})$$

The binary symbol is called a *bit*, which stands for a binary digit. It is customary to use the r decimal digits from 0 to $r - 1$ for a base r less than 10. The letters of the alphabet can be used in addition to the decimal digits for a base

Decimal (Base 10)	Binary (Base 2)	Ternary (Base 3)	Octal (Base 8)	Hexadecimal (Base 16)
00	0000	000	00	0
01	0001	001	01	1
02	0010	002	02	2
03	0011	010	03	3
04	0100	011	04	4
05	0101	012	05	5
06	0110	020	06	6
07	0111	021	07	7
08	1000	022	10	8
09	1001	100	11	9
10	1010	101	12	A
11	1011	102	13	B
12	1100	110	14	C
13	1101	111	15	D
14	1110	112	16	E
15	1111	120	17	F

Figure 1–1
Numbers with different bases

greater than 10. A ternary number system is constructed with base 3 using 0, 1, and 2 for digits. An octal number system is constructed with base 8 using 0, 1, ... , 7 for digits. In the hexadecimal number system the letters A, B, C, D, E, and F are used for the digits 10, 11, 12, 13, 14, and 15, respectively, to augment the 10 decimal digits. An example of a hexadecimal number is

$$(A6E2)_{16} = (A \times 16^3) + (6 \times 16^2) + (E \times 16^1) + (2 \times 16^0)$$
$$= (10 \times 16^3) + (6 \times 16^2) + (14 \times 16^1) + (2 \times 16^0)$$

The first 16 numbers in the decimal, binary, ternary, octal, and hexadecimal number systems are given in Figure 1–1.

1.2
Conversion between Number Systems with Different Bases

Any number can be converted from one base to another using (1–1) by expressing the original number in the form of a sum of powers in the original base but using the digits of the target number system to express the coefficients and the base of the original number, and by carrying out the resulting arithmetic in the target number system. For example, converting a number in base r to base 10 can be accomplished simply by forming the sum of the powers in r (the source base) with the coefficients in the base r number by means of decimal digits (the target base), and then performing the arithmetic in base 10. For example:

$$
\begin{aligned}
(1011.101)_2 &= (1 \times 2^3) + (0 \times 2^2) + (1 \times 2^1) + (1 \times 2^0) \\
&\quad + (1 \times 2^{-1}) + (0 \times 2^{-2}) + (1 \times 2^{-3}) \\
&= (11.625)_{10} \\
(36.75)_8 &= (3 \times 8^1) + (6 \times 8^0) + (7 \times 8^{-1}) + (5 \times 8^{-2}) \\
&= (30.953125)_{10} \\
(3A.F)_{16} &= (3 \times 16^1) + (A \times 16^0) + (F \times 16^{-1}) \\
&= (3 \times 16^1) + (10 \times 16^0) + (15 \times 16^{-1}) \\
&= (58.9375)_{10}
\end{aligned}
$$

This technique is always applicable for converting numbers from any base to any other base. However, it is not often convenient to carry out the arithmetic in the target number system if it is not the decimal number system. This difficulty can be overcome by rearranging (1–1) in a slightly more workable format. We will examine the number conversion procedure for the integer parts first and then for the fractional parts.

Integer Conversion

Let N be an integer number originally expressed in base s. We want to convert a number to target base r. In general, N consists of two parts:

$$
\begin{aligned}
N = N_I &\qquad\qquad\qquad + N_F \\
&= (a_n r^n + \cdots + a_1 r^1 + a_0 r^0) + (a_{-1} r^{-1} + a_{-2} r^{-2} + \cdots)
\end{aligned}
$$

where N_I and N_F are the integer and fractional parts, respectively. Let us consider the conversion of the integer part. There exists a unique set of integers, N_i's and a_i's, such that

$$
\begin{aligned}
N &= (r \times N_1) + a_0 \\
N_1 &= (r \times N_2) + a_1 \\
&\;\;\vdots \qquad\qquad \vdots \\
N_n &= (r \times 0) \;\; + a_n
\end{aligned}
\qquad (1\text{–}2)
$$

The arithmetic is performed in base r and $0 \le a_i < r$. The N_i's are the partial quotients and are obtained by dividing the original number repeatedly by the target base. The a_i's are the remainders of the divisions. Rearranging (1–2), we obtain:

$$
\begin{aligned}
N &= r(rN_2 + a_1) && + a_0 \\
&= r^2 N_2 && + a_1 r^1 + a_0 \\
&= r^2(rN_3 + a_2) && + a_1 r^1 + a_0 \qquad (1\text{-}3) \\
&\;\;\vdots && \;\;\vdots \\
&= a_n r^n + a_{n-1} r^{n-1} + \cdots + a_1 r^1 + a_0
\end{aligned}
$$

The above equation shows that the remainders a_i are exactly the desired coefficients in the target number system. This algorithm is particularly suited for converting numbers from decimal to another target base. The following examples will be done in base 10 arithmetic.

Example 1–1

Convert $(529)_{10}$ to a base 2 number.

$$
\begin{aligned}
529 &= (264 \times 2) + 1 \\
264 &= (132 \times 2) + 0 \\
132 &= (66 \times 2) + 0 \\
66 &= (33 \times 2) + 0 \\
33 &= (16 \times 2) + 1 \\
16 &= (8 \times 2) + 0 \\
8 &= (4 \times 2) + 0 \\
4 &= (2 \times 2) + 0 \\
2 &= (1 \times 2) + 0 \\
1 &= (0 \times 2) + 1
\end{aligned}
$$

$$(529)_{10} = (1\,0\,0\,0\,0\,1\,0\,0\,0\,1)_2$$

Check:

$$
\begin{aligned}
(1000010001)_2 &= (1 \times 2^9) + (1 \times 2^4) + (1 \times 2^0) \\
&= \quad 512 \quad + \quad 16 \quad + \quad 1 \\
&= (529)_{10}
\end{aligned}
$$

The original number is 529_{10}. Since the target base is 2, we will first divide 529_{10} by 2_{10}, yielding a quotient of 264_{10} and a remainder of 1. We next divide 264_{10} by 2_{10}, again yielding a quotient of 132 and a remainder of 0. This process continues until the quotient is zero. The coefficients of the original number in the target base are then the remainders in reverse order; i.e., the first remainder is the least significant digit and the last remainder is the most significant digit.

Example 1–2

Convert $(529)_{10}$ to a base 16 number.

$$529 = (33 \times 16) + 1$$
$$33 = (2 \times 16) + 1$$
$$2 = (0 \times 16) + 2$$

$$(529)_{10} = (2\ 1\ 1)_{16}$$

Check:

$$(211)_{16} = (2 \times 16^2) + (1 \times 16^1) + (1 \times 16^0)$$
$$= 512 + 16 + 1$$
$$= (529)_{10}$$

Example 1–3

Convert $(3311)_{10}$ to a base 16 number. In this case the letters are necessary.

$$3311 = (206 \times 16) + 15 \qquad \text{where: } 15 = F$$
$$206 = (12 \times 16) + 14 \qquad\qquad\quad 14 = E$$
$$12 = (0 \times 16) + 12 \qquad\qquad\quad 12 = C$$

Thus $(3311)_{10} = (CEF)_{16}$.

Check:

$$(CEF)_{16} = (C \times 16^2) + (E \times 16^1) + (F \times 16^0)$$
$$= (12 \times 16^2) + (14 \times 16^1) + (15 \times 16^0)$$
$$= 3072 + 224 + 15$$
$$= (3311)_{10}$$

We mentioned earlier that the conversion from a base r to a base s can be done directly in base s arithmetic. Most people are not proficient in performing arithmetic in bases other than 10; so, a conversion is possible from base r to base 10 and then to base s. All arithmetic is then handled in base 10.

Example 1–4 ▮▮▮▮▮▮▮▮▮▮▮▮▮▮▮▮▮▮▮

Convert $(2E4)_{16}$ to a base 8 number by direct conversion. Note that 10_{16} $(= 16_{10})$ in base 8 is 20_8.

$$(2E4)_{16} = [2_8 \times (20_8)^2] + [16_8 \times (20_8)^1] + [4_8 \times (20_8)^0]$$
$$= \quad 1000 \quad + \quad 340 \quad + \quad 4$$
$$= (1344)_8$$

(Note that all computations above are done in base 8.) We will now check the answer by first converting to decimal and then to the octal number system.

$$(2E4)_{16} = (2 \times 16^2) + (E \times 16^1) + (4 \times 16^0)$$
$$= \quad 512 \quad + \quad 224 \quad + \quad 4$$
$$= (740)_{10}$$

$$740 = (92 \times 8) + 4$$
$$92 = (11 \times 8) + 4$$
$$11 = (1 \times 8) + 3$$
$$1 = (0 \times 8) + 1$$

$$(2E4)_{16} = (740)_{10} = \quad (1\ 3\ 4\ 4)_8$$

The algorithm in (1–2) and (1–3) can thus be conveniently applied, particularly when the source number system is a decimal number system.

For converting numbers to decimal form from a nondecimal system, we can rewrite (1–1) in a recursive parenthetical format:

$$N = \quad a_n \times r^n \quad + \cdots \quad + a_1 \times r^1 + a_0$$
$$= \quad (a_n \times r^{n-1} + \cdots \quad + a_1) \times r + a_0$$
$$= \quad [(a_n \times r^{n-2} + \cdots + a_2) \times r \quad + a_1] \times r + a_0$$
$$\vdots$$
$$= \{\cdots [(a_n \times r \quad + a_{n-1}) \times r + a_{n-2}] \times r \cdots\} \times r + a_0 \quad (1\text{–}4)$$

The procedure to convert from one base to another is rather simple using the above equation. Start with the most significant digit (a_n), multiply it by the target base $(a_n \times r)$, add the product to the next significant digit $(a_n \times r + a_{n-1})$, multiply the resulting partial sum by the base $[(a_n \times r + a_{n-1}) \times r]$, and then add the resulting product to the next less significant digit $[(a_n \times r + a_{n-1}) \times r + a_{n-2}]$. This process is continued until the least significant digit has been added to the last partial result. Without further multiplication, the sum at this point is the desired result. This process is particularly suitable for the calculator, because all the operations are performed in a chain fashion.

Example 1-5

Convert $(2541)_8$ to a decimal number.

$$
\begin{aligned}
(2541)_8 &= [(2 \times 8 + 5) \times 8 + 4] \times 8 + 1 \\
&= [(16 \quad + 5) \times 8 + 4] \times 8 + 1 \\
&= (21 \qquad\qquad \times 8 + 4) \times 8 + 1 \\
&= (168 \qquad\qquad\qquad + 4) \times 8 + 1 \\
&= 172 \qquad\qquad\qquad\qquad \times 8 + 1 \\
&= 1376 \qquad\qquad\qquad\qquad\qquad + 1 \\
&= (1377)_{10}
\end{aligned}
$$

We urge readers to practice this technique on the calculator until they are proficient in this method of converting numbers from the other bases to the decimal number system.

Fractional Conversion

For the fractional part, N_F, we desire to find $a_{-1}, a_{-2}, \ldots$, for the base r number. Multiplying the fractional part by r gives

$$
rN_F = a_{-1} + a_{-2}r^{-1} + a_{-3}r^{-2} + \cdots
$$

Note that the integer part of rN_F is a_{-1}. Subtracting a_{-1} from both sides and multiplying by r again, we obtain

$$
r(rN_F - a_{-1}) = a_{-2} + a_{-3}r^{-1} + a_{-4}r^{-2} + \cdots
$$

Again, a_{-2} is obtained as the integer part of the resultant partial difference and then removed. This process is continued until the needed accuracy is obtained. Note that this process may not necessarily terminate after a finite number of steps, even though the number of fractional digits in the original number is finite.

Example 1-6

Convert $(0.625)_{10}$ to base 2.

$$
\begin{aligned}
2 \times (0.625) &= 1.25 & a_{-1} &= 1 \\
2 \times (0.25) &= 0.50 & a_{-2} &= 0 \\
2 \times (0.50) &= 1.00 & a_{-3} &= 1
\end{aligned}
$$

Hence, $(0.625)_{10} = (0.101)_2$. This fractional conversion process terminates in this case after three iterations.

Check:

$$(0.101)_2 = 1 \times 2^{-1} + 0 \times 2^{-2} + 1 \times 2^{-3}$$
$$= \quad 0.5 \quad + \quad 0 \quad + \quad 0.125$$
$$= (0.625)_{10}$$

Example 1–7

Convert $(87.68)_{10}$ to base 8.

$$87 = 10 \times 8 + 7 \qquad 8 \times (0.68) = 5.44 \qquad a_{-1} = 5$$
$$10 = 1 \times 8 + 2 \qquad 8 \times (0.44) = 3.52 \qquad a_{-2} = 3$$
$$1 = 0 \times 8 + 1 \qquad 8 \times (0.52) = 4.16 \qquad a_{-3} = 4$$
$$ \qquad 8 \times (0.16) = 1.28 \qquad a_{-3} = 1$$

Thus $(87.68)_{10} = (127.5341 \ldots)_8$. For this example, the conversion process continues indefinitely.

Conversion between Binary and Hexadecimal and between Binary and Octal Numbers

The conversion of binary numbers to octal and hexadecimal and vice versa is relatively easy because the octal and hexadecimal digits are related to the binary digits by powers of 2. These cases are also very interesting because of the many conversions that are performed between these bases in computer systems.

The conversion of a binary number to an octal number can be carried out by partitioning a binary number into groups of three to the left and right of the fractional point. (Note that it is not called a decimal point here, because this is not a decimal system.) This results from the fact that $2^3 = 8$, and three binary digits can represent any octal digit from 0 to 7. An example of the conversion follows:

$$(010\,110\,011\,111\,001 \,.\, 111\,010\,100)_2 = (26371.724)_8$$
$$2 \quad 6 \quad 3 \quad 7 \quad 1 \,.\, 7 \quad 2 \quad 4$$

One or more leading zeroes may be supplied to complete a group of three binary digits both at the beginning and at the end of the binary number as necessary. The conversion from base 8 to base 2 likewise can be obtained by using three binary digits for every octal digit as follows:

$$(621.53)_8 = (110\,010\,001 \,.\, 101\,011)_2$$
$$6 \quad 2 \quad 1 \,.\, 5 \quad 3$$

Similarly, converting from base 2 to base 16 involves partitioning the binary digits of the original number in groups of four from the fractional point, since

$2^4 = 16$. Two examples follow:

$$(1010\ 0110\ 1000\ .\ 1011\ 1000)_2 = (A68.B8)_{16}$$
$$A\quad 6\quad 8\ .\ B\quad 8$$
$$(8EF.A4)_{16} = (1000\ 1110\ 1111\ .\ 1010\ 0100)_2$$
$$8\quad E\quad F\ .\ A\quad 4$$

In octal-to-binary or hexadecimal-to-binary conversions, the groups of threes or fours are counted from the fractional point toward the left for the integer part and toward the right for the fractional part.

1.3

Binary Arithmetic for Unsigned Numbers

The ability to perform arithmetic operations in base 10 has been practiced by most of us for years, and the operations are committed to memory. To become proficient in other bases requires a considerable amount of practice. The primary arithmetic we will examine here is in base 2, because computer systems operate mostly with base 2. Note that arithmetic operations dealing with octal and hexadecimal numbers are very similar to those for binary numbers.

Addition and Subtraction

The addition, subtraction, and multiplication operations for base 2 numbers are shown in Figure 1–2. The most significant digit in the addition table or the subtraction table should be regarded as a carry or a borrow, respectively. The addition or subtraction operation is carried out in a bit-by-bit fashion starting from the rightmost digit, which is the least significant bit, with the carry or the borrow propagating from the least significant bit to the most significant bit.

Example 1–8

Add 1001_2 to 0101_2 and 1110_2 to 0111_2.

$$
\begin{array}{rr}
\text{Carry} \rightarrow & 1 \\
& 1001 \\
+ & 0101 \\
\hline
& 1110 \\
\end{array}
$$

$$
\begin{array}{rr}
\text{Carry} \rightarrow & 11 \\
& 1110 \\
+ & 0111 \\
\hline
& 11101 \\
\end{array}
$$

Addition table: sum = augend + addend

Addend

+	0	1
Augend 0	0, 0	0, 1
1	0, 1	1, 0

Carry, Sum

Subtraction table: difference = minuend − subtrahend

Subtrahend

−	0	1
Minuend 0	0, 0	1, 1
1	0, 1	0, 0

Borrow, Difference

Multiplication table: product = multiplicand × multiplier

Multiplier

×	0	1
Multiplicand 0	0	0
1	0	1

Product

Figure 1–2

The addition, subtraction, and multiplication operations

As can be seen from this example, if the word length is fixed, four in this case, then a carry-out from the most significant digit indicates an overflow condition in unsigned binary addition.

Example 1–9

Subtract 0010_2 from 1001_2 and 0110_2 from 0011_2.

$$
\begin{array}{r}
\text{Borrow} \rightarrow \quad 11 \\
1001 \\
-\ 0010 \\
\hline
0111
\end{array}
$$

$$
\begin{array}{r}
\text{Borrow} \rightarrow \quad 1 \\
0011 \\
-\ 0110 \\
\hline
11101
\end{array}
$$

This example again shows that a borrow-out from the most significant bit constitutes an underflow condition in unsigned subtraction. We can now generalize to state that any carry-out (or borrow-out) from the most significant bit indicates an overflow (or underflow) condition in unsigned binary addition or subtraction.

Multiplication and Division

Multiplication and division can be accomplished by the usual longhand procedure. For multiplication, each bit of the multiplier is used to multiply the entire multiplicand to form a partial product. The final product is then the sum of all the partial products with its position properly shifted.

Example 1–10

Multiply $(101)_2$ by $(110)_2$.

$$
\begin{array}{r}
101 \\
\times\ 110 \\
\hline
000 \\
101 \\
101 \\
\hline
11110
\end{array}
$$

Then $(101)_2 \times (110)_2 = (11110)_2$. Since it is not easy to add all the partial products together at the same time in a base other than 10, it is more practical to add them up one at a time.

For division, the divisor is subtracted from the dividend starting from the most significant bit, similar to the longhand division procedure in the decimal system. If the subtraction is successful, a 1 is entered into the corresponding bit of the quotient. If the subtraction is not successful, a 0 is entered into the corresponding quotient bit and the divisor is added back. In either case, the remainder is shifted to the left and the next bit of the dividend is added to the remainder. The process continues until all the bits of the dividend have been used. What is left in the remainder becomes the true remainder of the division process.

Example 1–11 ▬▬▬▬▬▬▬▬▬▬▬▬▬▬▬▬▬▬▬▬▬▬▬▬▬▬▬

Divide 1100111_2 by 110_2.

$$
\begin{array}{rl}
\ 10001 & \text{Quotient} \\
110\,|\,\overline{1100111} & \text{Divisor, dividend} \\
\ 110 & \text{Subtract divisor} \\
\ \overline{00} & \text{First partial remainder} \\
\ 00 & \text{Remainder insufficient for subtraction} \\
\ \overline{01} & \\
\ 00 & \text{Remainder insufficient for subtraction} \\
\ \overline{11} & \\
\ 00 & \text{Remainder insufficient for subtraction} \\
\ \overline{111} & \\
\ 110 & \text{Subtract divisor} \\
\ \overline{1} & \text{True remainder}
\end{array}
$$

Hence the division of 1100111_2 by 110_2 results in a quotient of 10001_2 and a remainder of 1_2.

1.4 ▬▬▬▬▬▬▬▬▬▬▬▬▬▬▬▬▬▬▬▬▬▬▬▬▬▬▬▬▬▬▬▬
Binary Arithmetic for Signed Numbers

The arithmetic for unsigned numbers is rather simple, because all quantities can be considered positive. The arithmetic for signed numbers is more complicated, because the sign must also be taken into consideration.

Representation of Negative Numbers

The normal process for handling negative numbers is to precede the magnitude of each negative number with a negative sign. Computing systems normally handle a negative number by using an additional bit for the sign. Hence, the most significant bit is usually reserved for the sign bit, with the sign bit being 0 for positive numbers and 1 for negative numbers. A number of different schemes exist for specifying negative numbers, such as the sign magnitude notation, the 1's complement notation, and the 2's complement notation. The most common one by far is the 2's complement notation, because it allows the same hardware to be used for both addition and subtraction.

The sign magnitude representation of a negative number is obtained from its positive counterpart by merely changing the sign bit from 0 to 1. The 1's complement representation is obtained by subtracting each bit of the positive counterpart from 1 and ignoring any individual borrow-out arising from the subtraction of each bit. The 2's complement representation is obtained by adding 1 to the 1's complement representation. These procedures can easily be extended for numbers in other bases. In the following examples we assume that the binary numbers have a total of 8 bits, with the most significant bit being the sign bit.

Example 1–12

Find the representation of positive and negative numbers assuming a word size of 8 bits.

	Sign Magnitude	1's Complement	2's Complement
Bit position	76543210	76543210	76543210
$+(90)_{10}$	01011010	01011010	01011010
$-(90)_{10}$	11011010	10100101	10100110

The three different representations will always be the same for positive numbers but will be different for negative numbers. The 1's complement gets its name because subtracting each bit from 1 and ignoring any borrow-out is the same as complementing each bit. The 1's complement of A, where

$$A = (a_7a_6a_5a_4a_3a_2a_1a_0)_2$$

can be expressed as

$$\text{1's compl of } A = (11111111)_2 - (a_7a_6a_5a_4a_3a_2a_1a_0)_2$$
$$= 2^8 - 1 - A$$

The 2's complement of A likewise can be expressed as

$$A_c = (\ 11111111)_2 - (a_7a_6a_5a_4a_3a_2a_1a_0) + 1$$
$$= (100000000)_2 - (a_7a_6a_5a_4a_3a_2a_1a_0)$$
$$= 2^8 - A$$

Therefore the 2's complement of A, which we will call A_c, is $(2^8 - A)$ for an 8-bit number or $(2^n - A)$ for an n-bit number. Note that if the fractional point is placed right after the sign bit, then the first term in the above expression is simply 2, hence the name. Also observe that the 2's complement of the 2's complement of A is A, since

$$(A_c)_c = 2^n - (2^n - A) = A$$

The range of representation for the sign magnitude, the 1's complement, and the 2's complement in an n-bit number are listed below:

$$\text{Sign magnitude:} \quad -2^{n-1} + 1 \leq N \leq 2^{n-1} - 1$$
$$\text{1's complement:} \quad -2^{n-1} + 1 \leq N \leq 2^{n-1} - 1$$
$$\text{2's complement:} \quad -2^{n-1} \quad\quad \leq N \leq 2^{n-1} - 1$$

The range for the 2's complement representation is one larger than the others. This is because, for both the sign magnitude and the 1's complement notations, there are two representations for zero, namely a plus zero and a minus zero. This, however, is not the case for the 2's complement notation, which has only one representation for zero.

Addition and Subtraction of Signed Numbers

Addition and subtraction of signed numbers represented in 1's complement and 2's complement notations can always be handled simply as an addition, regardless of whether the numbers are positive or negative. This is possible because a subtraction operation can always be converted to an addition operation by first complementing the subtrahend; that is,

$$A - B = A + (-B)$$

We can therefore consider the notion of a generalized addition operation that accounts for both addition and subtraction. Hence, in the 2's complement representation, subtraction is accomplished by taking the 2's complement of the number to be subtracted and then adding it to the minuend, regardless of which number is positive or negative.

For addition or subtraction of two numbers, A and B, there are two cases where A and B can be positive or negative in any combination. If A is negative, then A_c will be used. If B is negative, then B_c will be used. The machines always ignore the carry from the most significant bit when 2's complement addition and subtraction are performed. These two cases are examined separately below.

CASE 1
Both Numbers Have the Same Sign.
That is, $A \geq 0$ and $B \geq 0$ or $A < 0$ and $B < 0$.
  If both numbers are positive,

$$S = A + B$$

If both numbers are negative,

$$S = A_c + B_c$$
$$= (2^n - A) + (2^n - B)$$
$$= 2^n - (A + B) + 2^n \text{ (ignored)}$$

The result of adding two binary numbers when both have the same sign yields a sum that is always in the proper form. If both numbers are negative, an additional 2^n term results, indicated by a carry-out from the addition. This carry-out should be ignored and is not indicative of an overflow condition. However, since both numbers are of the same sign, the possibility of an overflow or underflow condition exists. An overflow condition occurs when the addition of two positive numbers results in a negative sum, while an underflow condition occurs when the addition of two negative numbers results in a positive sum.

Example 1–13

Perform the following 2's complement arithmetic.

$$01011100 + 00010011 = ?$$

The computation is as follows:

$$
\begin{array}{c}
\text{Carry bit} \rightarrow \quad 1 \\
\end{array}
$$

01011100	$+92_{10}$
$+\ 00010011$	$+\ +19_{10}$
01101111	111_{10}

If two positive numbers are added together, it is possible to have a result that is larger than $2^{n-1} - 1$, which is $(127)_{10} = (01111111)_2$ in this case. When this happens, the most significant bit would become 1. This represents a negative number internally and will cause an error if it goes undetected. This is termed an *overflow condition*, and most computers provide for the detection of an overflow.

Example 1–14

Perform the following 2's complement arithmetic.

$$01011100 + 01110011 = ?$$

The computation is as follows:

$$
\begin{array}{c}
\text{Carry bit} \rightarrow \quad 111 \\
\end{array}
$$

01011100	$+92_{10}$
$+\ 01110011$	$+\ +115_{10}$
OVERFLOW $\rightarrow$ $\quad 11001111$	207_{10}

The change in sign in the sum indicates that an overflow condition has occurred. In other words, the magnitude of the sum has exceeded the range that can be represented by the representation for signed numbers with the given word length.

Example 1–15

Perform the following 2's complement arithmetic.

$$10100011 + 11100011 = ?$$

The computation is

$$
\begin{array}{rr}
\text{Carry} \rightarrow \quad 11 \quad 11 & \\
10100011 & -93_{10} \\
+\ 11100011 & +\ -29_{10} \\
\hline
\cancel{1}10000110 & -122_{10}
\end{array}
$$

The result is obtained by ignoring the carry generated from the most significant bit. This is always the rule in 2's complement additions and subtractions when both members are of the same sign, as mentioned before.

Just as there are overflows when adding two positive numbers, similar situations may also arise with the addition of two negative numbers. If the resultant number is positive, an error has occurred because the range has been exceeded. This error condition is called *underflow* and is illustrated below.

Example 1–16

Perform the following 2's complement arithmetic.

$$10100011 + 10000011 = ?$$

The computation is

$$
\begin{array}{rr}
\text{Carry} \rightarrow \quad 11 & \\
10100011 & -93_{10} \\
+\ 10000011 & +\ -125_{10} \\
\hline
\text{UNDERFLOW} \rightarrow \quad 100100110 & -218_{10}
\end{array}
$$

For both the overflow and underflow cases, the programmer may opt to initiate corrective action.

Example 1–17 ◼◼◼◼◼◼◼◼◼◼◼◼◼◼◼◼◼◼◼◼◼◼◼◼◼◼◼◼◼◼◼◼◼◼

Perform the following 2's complement arithmetic.

$$10100011 - 00011101 = ?$$

The computation begins by finding the 2's complement of the number to be subtracted and then adding:

$$(00011101)_c = 11100011$$

$$
\begin{array}{rr}
\text{Carry} \rightarrow \quad 11 \quad 11 & 11 \\
10100011 & -93_{10} \\
+ \; 11100011 & + \; -29_{10} \\
\hline
\mathit{1}10000110 & -122_{10}
\end{array}
$$

The result is obtained by ignoring the carry generated from the most significant bit.

From the above examples it is readily seen that the overflow or underflow condition can be detected in general for both positive and negative numbers. Let c_0 be the least significant bit and c_{n-1} be the most significant bit of the sum. If the number is positive, a change in the resultant sign bit due to the carry-out c_{n-2} with no carry-out from the sign bit c_{n-1} signifies an overflow. Likewise, if the number is negative, there will always be a carry-out from the sign bit. But if there is no carry-out c_{n-2}, the sign bit will be positive, indicating an underflow condition. Hence the overflow or underflow can be represented in a general case as follows:

$$V = c_n \oplus c_{n-1}$$

where $\oplus$ represents the Exclusive OR function discussed later. The proof of this expression is left to the reader as an exercise. Many other expressions for overflow or underflow detection also exist.

CASE 2
Both Numbers Are of Opposite Signs.
That is, $A \geq 0$ and $B < 0$ or $A < 0$ and $B \geq 0$.

If $|A| \geq |B|$, then the result is positive:

$$S = A + B_c = A + 2^n - B = (A - B) + 2^n \text{ (ignored)}$$

If $|A| < |B|$, then the result is negative:

$$S = A + B_c = A + 2^n - B = 2^n - (B - A)$$

If the sum is positive, the 2^n factor should be ignored because the machine has a total of only n bits (note that bit $n + 1$ is the coefficient of 2^n) and therefore $S = 2^n + (A - B) = A - B + 2^n$, where the last 2^n is ignored. There are no underflow or overflow conditions when adding two numbers of different signs together.

Example 1–18

Perform the following 2's complement arithmetic.

$$01110101 + 10100100 = ?$$

The computation proceeds as follows:

$$
\begin{array}{lr}
\text{Carry} \rightarrow \quad 11 \quad 1 & \\
01110101 & +117_{10} \\
+\ 10100100 & +\ -92_{10} \\
\hline
\cancel{1}00011001 & +25_{10}
\end{array}
$$

In this example the sum is positive as indicated by the sign bit. The carry-out 2^n is ignored.

Example 1–19

Perform the following 2's complement arithmetic.

$$01011100 + 10000011 = ?$$

The computation proceeds as follows:

$$
\begin{array}{lr}
\text{Carry} \rightarrow & \\
01011100 & 92_{10} \\
+\ 10000011 & +\ -125_{10} \\
\hline
11011111 & -33_{10}
\end{array}
$$

The result for this case is negative, and the sum is represented as a 2's complement of $B - A$, which is the correct result.

1.5
Binary Codes

There are a number of cases where binary codes are used to store information representing decimal digits, letters, special characters on terminals, control characters, etc. The minimum number of bits to uniquely represent N different items is $n = \lfloor \log_2 N \rfloor$ where $\lfloor x \rfloor$ denotes the smallest integer greater than or equal to x. For the decimal digits 0 to 9, at least 4 bits would be needed to uniquely decode each character. A majority of the codes developed deal with the decimal system. Most of these codes can be represented in a weighted form:

$$N = \sum_{i=0}^{n-1} w_i b_i + B$$

where n is the number of bits in the code, B is the bias, w_i are the weights for each bit, and b_i are the bits themselves. However, the reader is cautioned that not all codes fit this scheme. Figure 1–3 gives several commonly used codes to represent the decimal digits.

We have seen that the decimal numbering system is a positional numbering system in base 10. Hence there are 10 decimal digits. Each decimal digit can be represented uniquely by a combination of four binary variables. Four binary variables can produce 16 unique codes. Since only 10 are required to represent the 10 decimal digits, there are many different possible ways to choose the 10 codes out of 16 to represent the decimal digits. A number of these schemes have been shown in Figure 1–3. Each scheme is chosen with some special properties in mind.

The BCD (binary-coded decimal) code is equivalent to the normal binary representation. The weights of each bit may be taken as the value of the given position. The weight assignment for the BCD code is thus 8, 4, 2, and 1. The BCD code is called a *weighted code*, because the addition of the weights for each position

Figure 1–3
Decimal digit codes

Decimal Digit	BCD 8421	Excess-3	2421	7421 or 2 out of 5	Gray $g_3 g_2 g_1 g_0$	Modified Gray $m_3 m_2 m_1 m_0$
0	0000	0011	0000	11000	0000	0010
1	0001	0100	0001	00011	0001	0110
2	0010	0101	0010	00101	0011	0111
3	0011	0110	0011	00110	0010	0101
4	0100	0111	0100	01001	0110	0100
5	0101	1000	1011	01010	0111	1100
6	0110	1001	1100	01100	0101	1101
7	0111	1010	1101	10001	0100	1111
8	1000	1011	1110	10010	1100	1110
9	1001	1100	1111	10100	1101	1010

of the code where it is 1 will result in the decimal digit; i.e., the decimal digit 7 is

$$7_{10} = (0 \times 8) + (1 \times 4) + (1 \times 2) + (1 \times 1)$$

The advantage of this code is that the representation is natural and is widely used. The disadvantage is that the discontinuity between 9 and 0 causes special precautionary steps to be taken in decimal arithmetic. This will be seen later. Unless explicitly specified, the term BCD will refer to the 8421 BCD representation in the remaining discussion.

The Excess-3 code is obtained from the BCD code by adding three to each position. By offsetting the decimal digits by three in the binary representation, the number of digits with the most significant bit being 0 is the same as those with the most significant bit being 1. This provides for faster carry propagation in addition operations. A glance down the most significant bit of the digit representations reveals that the first five digits have a 0 in this bit, while the last five digits have a 1. This balance in 0's and 1's provides fast carry generation and also simplifies the correction of invalid codes. Another advantage of the Excess-3 code is that the 1's complement of the code also yields a valid Excess-3 code. In fact, the 9's complement (discussed in the next section) of a digit represented in Excess-3 is obtained merely by taking the 1's complement of the code itself.

The Gray code is useful because there is only one bit change in the code for numbers differing by one. This property helps to reduce transmission errors in many types of digital systems. This is more evident when the codes are placed in a tabular form as shown in Figure 1–4. The Gray code, which is used mostly in communications, has the special property of only one bit change between successive digits. The modified Gray code is similar to the Gray code except that it is offset by three in much the same way as the Excess-3. The modified Gray code has the added advantage that there is also one bit change between 9 and 0. Both the Gray and modified Gray codes are used mostly in communications and are seldom used for decimal representations in arithmetic operations.

The 7421 BCD code is also called the 2-out-of-5 code because when a parity bit is added to the code, then there are always two bits that are one for any of the decimal digits. This provides additional error detection capability. The weights for the binary bits are 7, 4, 2, and 1, respectively; zero is treated as a special case. The 2-out-of-5 code has an additional bit in its code, which adds redundancy and allows for error detection.

Most computers have the ability to store the decimal digits internally in the BCD format and then perform arithmetic operations while maintaining the numbers in decimal format using the BCD code. In most cases, this method saves more time than converting the decimal numbers to binary, doing the arithmetic operations, and then converting the binary numbers back to decimal form using the BCD code. A more detailed discussion on the computational procedures is presented in the next section.

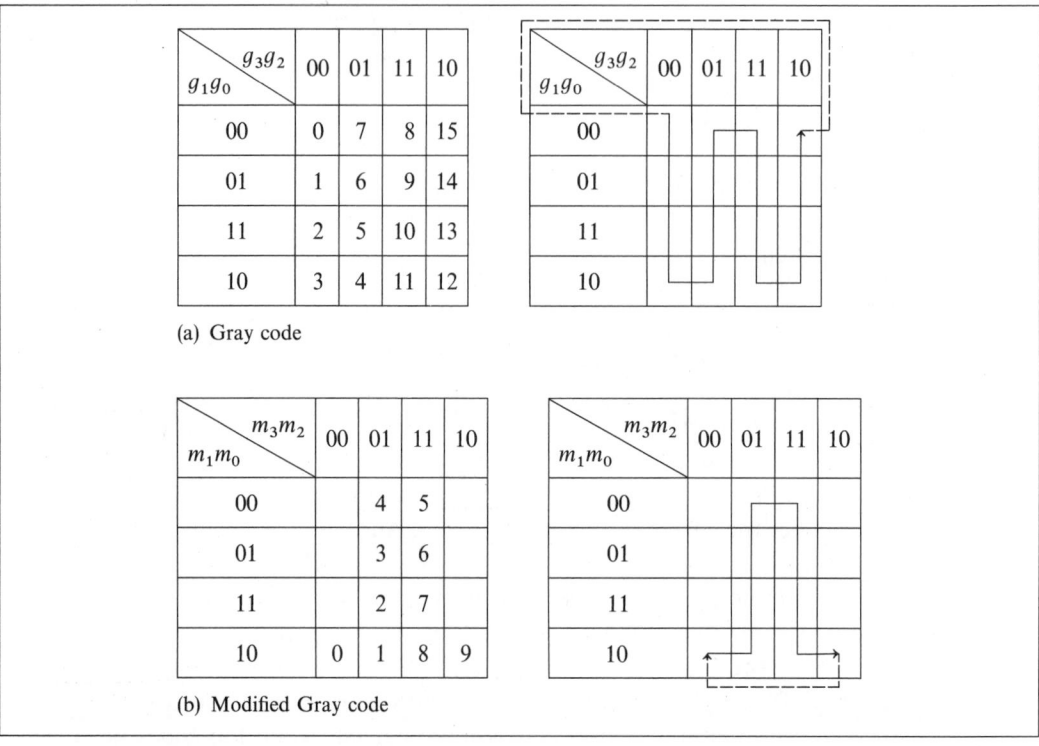

Figure 1–4

Graphic illustration of the (a) Gray and (b) modified Gray codes

charact

parity

The computer systems of today must have the capability of storing information that corresponds to letters, special characters, etc. This code is used to transfer files on all kinds of records, forms, and reservations. The two most commonly used codes are given in Figure 1–5. One code is known as ASCII (American Standard Code for Information Interchange) and the other as EBCDIC (Extended Binary-Coded Decimal Interchange Code). EBCDIC is used by International Business Machines (IBM) in many of its systems, while ASCII is used by many other computer manufacturers.

A number of codes add an additional bit called a *parity bit*. This bit is added to make the total number of 1 bits in the code even (even parity) or odd (odd parity). Any single bit change caused by an error will cause an even parity code to have an odd number of 1's, and likewise an odd parity code to have an even number of 1's. The digital systems can easily detect the change and perform what is called *error detection*. Note that an even number of bit errors cannot be detected by one parity bit, but an odd number of bit errors can be. The 7-bit ASCII

Character	7-Bit ASCII Code	8-Bit EBCDIC Code
A	100 0001	1100 0001
B	100 0010	1100 0010
C	100 0011	1100 0011
D	100 0100	1100 0100
E	100 0101	1100 0101
F	100 0110	1100 0110
G	100 0111	1100 0111
H	100 1000	1100 1000
I	100 1001	1100 1001
J	100 1010	1101 0001
K	100 1011	1101 0010
L	100 1100	1101 0011
M	100 1101	1101 0100
N	100 1110	1101 0101
O	100 1111	1101 0110
P	101 0000	1101 0111
Q	101 0001	1101 1000
R	101 0010	1101 1001
S	101 0011	1110 0010
T	101 0100	1110 0011
U	101 0101	1110 0100
V	101 0110	1110 0101
W	101 0111	1110 0110
X	101 1000	1110 0111
Y	101 1001	1110 1000
Z	101 1010	1110 1001
0	011 0000	1111 0000
1	011 0001	1111 0001
2	011 0010	1111 0010
3	011 0011	1111 0011
4	011 0100	1111 0100
5	011 0101	1111 0101
6	011 0110	1111 0110
7	011 0111	1111 0111
8	011 1000	1111 1000
9	011 1001	1111 1001
blank	010 0000	0100 0000
.	010 1110	0100 1011
(	010 1000	0100 1101
+	010 1011	0100 1110
$	010 0100	0101 1011
*	010 1010	0101 1100
)	010 1001	0101 1101
−	010 1101	0110 0000
/	010 1111	0110 0001
,	010 1100	0110 1011
=	011 1101	0111 1110

Figure 1-5 Alphanumeric codes

code is normally used with a parity bit in most systems, because this makes a total of 8 bits, often referred to as a *byte*.

1.6
Decimal Arithmetic

It is often desirable to perform arithmetic operations directly in the BCD code format without the burden of first converting the input decimal number to binary, performing the operation in binary, and then reconverting the binary result back to decimal for display and presentation. A number of codes can be used for representing the decimal digits, notably the 8-4-2-1 code and the Excess-3 code.

BCD Representation

A decimal number is represented digit by digit with its decimal equivalent. For example, the decimal number 123.456_{10} is represented as

$$123.456_{10} = 0001/0010/0011.0100/0101/0110$$

The slashes are included only for clarity and are not actually present in the machine representation. Addition or subtraction of these coded decimal numbers is done one digit at a time starting from the least significant bit as in the binary case.

Example 1–20

Represent the decimal number 8279 using the various codes.

Decimal	8	2	7	9
BCD	1000	0010	0111	1001
Excess-3	1011	0101	1010	1100
Gray	1100	0011	0100	1101
Modified Gray	1110	0111	1111	1010
7421 BCD	1001	0010	1000	1010

Negative decimal numbers can be represented by complements in the same way as negative binary numbers. In the decimal number system, there is a 9's complement analogous to the 1's complement of the binary number system, and a 10's complement analogous to the 2's complement. The 9's complement is that portion that must be added to the original number to make every digit a 9. The

Decimal Equivalent	BCD		Excess-3	
	True	9's Complement	True	9's Complement
0	0000	1001	0011	1100
1	0001	1000	0100	1011
2	0010	0111	0101	1010
3	0011	0110	0110	1001
4	0100	0101	0111	1000
5	0101	0100	1000	0111
6	0110	0011	1001	0110
7	0111	0010	1010	0101
8	1000	0001	1011	0100
9	1001	0000	1100	0011

Figure 1–6
The 9's complement for the decimal digits in the BCD and Excess-3 representations

10's complement is the 9's complement plus one on the least significant decimal digit. Figure 1–6 shows the 9's complement for both the BCD and the Excess-3 codes.

Example 1–21

Represent -4598_{10} in the various decimal representations.
In 9's complement representation:

Decimal	— 4 5 9 8
BCD	0101 / 0100 / 0000 / 0001
Excess-3	1000 / 0111 / 0011 / 0100

In 10's complement representation:

Decimal	— 4 5 9 8
BCD	0101 / 0100 / 0000 / 0010
Excess-3	1000 / 0111 / 0011 / 0101

BCD Addition/Subtraction

Decimal arithmetic can be performed either in BCD or in Excess-3 representation. The basic decimal addition operation is still performed by the full adders used for binary additions, but the result must be examined carefully to ensure that there are no resulting invalid codes. The BCD addition/subtraction is presented in this section, and the Excess-3 arithmetic will be presented in the next section.

When adding together two decimal digits in BCD and taking into account the possibility of a carry-in, the resultant sum can range between 0 and 19. Three possibilities can arise as exemplified in the three cases at the top of the next page.

Figure 1–7

The 20 possible results after adding two decimal digits and a possible carry-in in BCD representation

Group	Decimal Equivalent	Generated Carry-in	Code after Addition of Two BCD Digits	Correct Carry	Correct Code	Action to Be Taken
	0	0	0000	0	0000	
	1	0	0001	0	0001	
	2	0	0010	0	0010	
	3	0	0011	0	0011	
I	4	0	0100	0	0100	None
	5	0	0101	0	0101	
	6	0	0110	0	0110	
	7	0	0111	0	0111	
	8	0	1000	0	1000	
	9	0	1001	0	1001	
	10	0	1010	1	0000	
	11	0	1011	1	0001	
II	12	0	1100	1	0010	Add 0110_2
	13	0	1101	1	0011	
	14	0	1110	1	0100	
	15	0	1111	1	0101	
	16	1	0000	1	0110	
III	17	1	0001	1	0111	Add 0110_2
	18	1	0010	1	1000	
	19	1	0011	1	1001	

Observations

i. $0010 + 0111 = 0 / 1001$ Result is correct.

ii. $0101 + 0111 = 0 / 1100$ Resultant code is invalid and carry-out is missing.

iii. $1001 + 0111 = 1 / 0000$ Resultant code is incorrect.

 In the first case, the addition of two digits results in a single digit less than or equal to 9. The result of the addition is correct, and no modification is needed. In the second case, the result of the addition is greater than 10, but no carry is generated. This addition results in an invalid code, which requires a correction by adding six (0110) to the result. This correction is needed because the difference between nine ($0000 / 1001$) and ten ($0001 / 0000$) is one ($10 - 9 = 1$) in decimal but seven ($10000 - 01001 = 00110$) in binary. This discrepancy accounts for the addition of six in the correction. In the last case, though the addition generates a carry, the carry takes away 16; while in decimal, a carry takes away only 10. Hence this result also needs to be corrected by adding six to the result. Figure 1–7 shows which results must be corrected.

Example 1–22 ▪▬▬▬▬▬▬▬▬▬▬▬▬▬▬▬▬▬▬▬▬▬▬▬▬▬▬

Using BCD representation, add 197_{10} to 98_{10}.

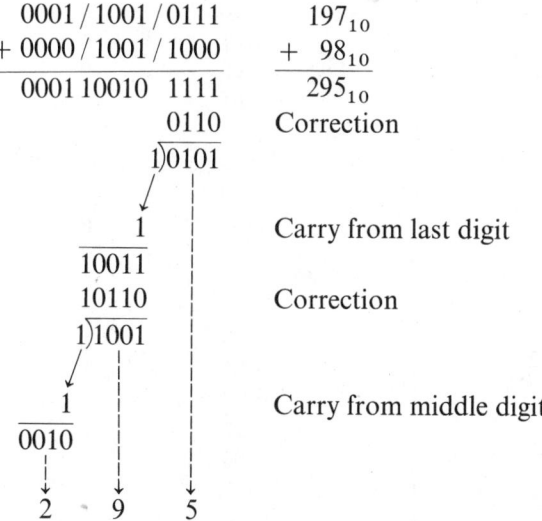

The steps for subtraction are identical except that the 9's complement of the subtrahend is taken before the addition is performed.

Excess-3 Addition/Subtraction

Addition and subtraction in the Excess-3 representation is similar to that in the BCD representation, except that a correction is always required. This is because there is always an extra offset of 0011 in the result when adding or subtracting in the Excess-3 representation. Let a and b be the BCD code of two decimal digits. Then their Excess-3 representations are $a + 3$ and $b + 3$, respectively. Therefore, the sum of the Excess-3 digits is $(a + b + 3) + 3$, which is always 0011 more than what the result should be. Hence a subtraction of 0011 is needed. This subtraction can be performed by adding 1101 and ignoring the generated carry. However, if there is a carry generated in the original calculation as shown in Figure 1–8, then this carry-out takes away 16, which is six more than it should. To remedy this situation, a correction is needed by adding 0011.

Figure 1–8

The 19 combinations after a decimal addition of two digits and a possible carry-in in Excess-3 representation

Group	Decimal Equivalent	Generated Carry-in	Code after Excess-3 Addition	Correct Carry	Correct Code	Action to Be Taken
	0	0	0110	0	0011	
	1	0	0111	0	0100	
	2	0	1000	0	0101	
	3	0	1001	0	0110	Subtract 0011_2
I	4	0	1010	0	0111	or add 1101_2
	5	0	1011	0	1000	and ignore carry
	6	0	1100	0	1001	
	7	0	1101	0	1010	
	8	0	1110	0	1011	
	9	0	1111	0	1100	
	10	1	0000	1	0011	
	11	1	0001	1	0100	
	12	1	0010	1	0101	
	13	1	0011	1	0110	
	14	1	0100	1	0111	
II	15	1	0101	1	1000	Add 0011_2
	16	1	0110	1	1001	
	17	1	0111	1	1010	
	18	1	1000	1	1011	
	19	1	1001	1	1100	

Example 1–23

Add 197_{10} to 98_{10} using Excess-3 representation.

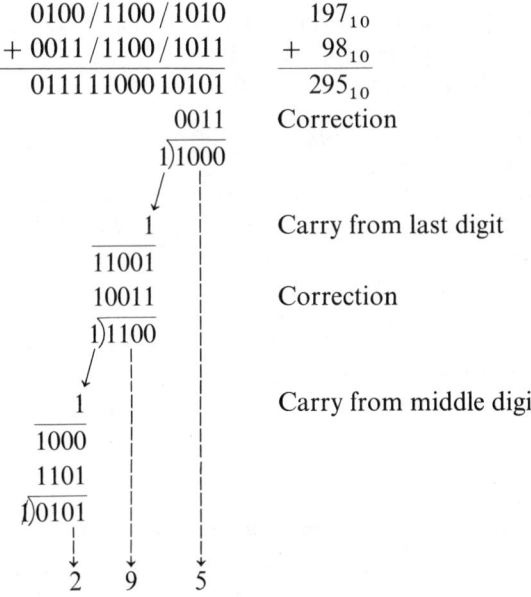

$$
\begin{array}{ll}
0100/1100/1010 & 197_{10} \\
+\ 0011/1100/1011 & +\ 98_{10} \\
\hline
0111\ 11000\ 10101 & 295_{10}
\end{array}
$$

0011   Correction

1)1000

1   Carry from last digit

11001

10011   Correction

1)1100

1   Carry from middle digit

1000

1101

1)0101

2   9   5

The correction of the most significant digit results in a carry that is always ignored.

References

1. Booth, Taylor L. *Introduction to Computer Engineering Hardware and Software Design.* John Wiley & Sons, New York, 1984.
2. Chu, Y. *Digital Computer Design Fundamentals.* McGraw-Hill, New York, 1962.
3. Dietmeyer, D. L. *Logic Design of Digital Systems,* 2d ed. Allyn & Bacon, Boston, 1978.
4. Fletcher, William I. *An Engineering Approach to Digital Design.* Prentice-Hall, Englewood Cliffs, N.J., 1980.
5. Flores, I. *The Logic of Computer Arithmetic.* Prentice-Hall, Englewood Cliffs, N.J., 1962.
6. Hayes, John P. *Digital System Design and Microprocessors.* McGraw-Hill, New York, 1984.
7. Hill, Frederick J., and Gerald R. Peterson. *Introduction to Switching Theory and Logical Design,* 3d ed. John Wiley & Sons, New York, 1981.
8. Humphrey, W. S., Jr. *Switching Circuits with Computer Applications.* McGraw-Hill, New York, 1958, chap. 4.
9. Kline, R. M. *Digital Computer Design.* Prentice-Hall, Englewood Cliffs, N.J., 1977.
10. Kohavi, Z. *Switching and Finite Automata Theory.* McGraw-Hill, New York, 1970.

11. Kostopoulos, G. K. *Digital Engineering.* John Wiley & Sons, New York, 1975.
12. Lewin, Morton H. *Logic Design and Computer Organization.* Addison-Wesley, Reading, Mass., 1983.
13. Mano, M. Morris. *Digital Logic and Computer Design.* Prentice-Hall, Englewood Cliffs, N.J., 1979.
14. Nagle, H. T. Jr., B. D. Carrol, and J. D. Irwin. *An Introduction to Computer Logic.* Prentice-Hall, Englewood Cliffs, N.J., 1975.
15. O'Connor, Patrick. *Digital and Microprocessor Technology.* Prentice-Hall, Englewood Cliffs, N.J., 1979.
16. Roth, Charles H., Jr. *Fundamentals of Logic Design,* 2d ed. West Publishing Co., St. Paul, Minn., 1979.
17. Taub, Herbert. *Digital Circuits and Microprocessors.* McGraw-Hill, New York, 1982.
18. Ware, W. H. *Digital Computer Technology and Design,* Vol. 1. John Wiley & Sons, New York, 1963.
19. Wiatrowski, Claude A., and Charles H. House. *Logic Circuits and Microcomputer Systems.* McGraw-Hill, New York, 1980.
20. Winkel, David, and Franklin Prosser. *The ART of Digital Design.* Prentice-Hall, Englewood Cliffs, N.J., 1980.

Problems

1. List the first 32 decimal digits in base 2 and base 4.

2. Convert the decimal number 360.25 to base 2, base 3, base 4, base 8, and base 16.

3. Convert the following binary numbers to decimal:
   (a) 1100110   (e) 1011
   (b) 11110000   (f) 10110
   (c) 0101010   (g) 101100
   (d) 10110010   (h) 1011000

4. Change the following binary numbers with fractions to decimal:
   (a) 1001.0101
   (b) 11110.11
   (c) 10110110.01

5. Convert the following numbers to their decimal equivalents:
   (a) $(120120)_3$ (e) $(A6B.F)_{16}$
   (b) $(123032)_4$ (f) $(7654.321)_8$
   (c) $(67512)_8$ (g) $(324.4)_5$
   (d) $(7.54)_8$ (h) $(1AB.2C)_{16}$

6. Complete the following table:

	Binary	Octal	Decimal	Hexadecimal
(a) 10_2				
(b) 10_8				
(c) 10_{10}				
(d) 10_{16}				

7. Why does the fractional conversion process from one base to another sometimes not terminate after a finite number of steps? Is there any relation between the numbers of fractional digits in the original number and in the converted number?

8. Perform the indicated binary arithmetic assuming that all numbers are 8 bits and unsigned:
   (a) 10111000 + 01000111
   (b) 01100110 − 11100100
   (c) 11100010 + 11011010
   (d) 00011001 − 01010100
   (e) 11001010 + 00010100
   (f) 11000100 − 00101010

9. Perform the indicated binary arithmetic assuming that all numbers are unsigned:
   (a) 10111000 × 0011
   (b) 01100110 × 10100
   (c) 11100010 × 1010
   (d) 00011001 / 0100
   (e) 11001010 / 10100
   (f) 11000100 / 1010

10. The following binary numbers are 8-bit numbers with the most significant bit being the sign bit. For each number, find the sign magnitude, the 1's complement, and the 2's complement representations for the number of the opposite sign.
   (a) 01101111   (e) 10011001
   (b) 00001000   (f) 01111101
   (c) 10011111   (g) 00100110
   (d) 01100110   (h) 11111111

11. Show that the 2's complement of any binary number can be obtained by complementing each bit to the left of the least significant 1 bit and leaving this 1 bit and any other less significant 0 bits the same.

12. The following numbers are 8-bit, 2's complement numbers. Find the decimal equivalents including the sign.
   (a) 01111111   (e) 11110000
   (b) 11111111   (f) 01111101
   (c) 10000000   (g) 01110010
   (d) 10101010   (h) 10101010

13. Perform the indicated arithmetic in 2's complement form for the following 8-bit numbers. Verify by converting the numbers and results to decimal.
   (a) 10001011 + 01101111
   (b) 10100110 + 01111011
   (c) 01110110 − 00101100
   (d) 10011101 − 10100011
   (e) 10000000 − 10000000

14. Show that an overflow flag (normally call V) is the Exclusive OR of the carry of the two most significant bits; that is, for an 8-bit machine,

$$V = C_7 \oplus C_6$$

where C_7 is the carry-out of the most significant bit and C_6 is the carry-in of the most significant bit. The operation of the Exclusive OR function is given below:

$\oplus$	0	1
0	0	1
1	1	0

15. Find other expressions for detecting overflow and underflow conditions.

16. Find the ASCII equivalent in binary form for your first and last name including the space between them.

17. Represent the decimal number 629 in (a) BCD code, (b) Excess-3 code, and (c) 2-out-of-5 code.

18. Develop a systematic means to assign a binary code to all 52 playing cards with a minimum number of bits.

19. How are overflow and underflow conditions detected when adding or subtracting 8-4-2-1 BCD numbers? Show that the subtraction of any two 2's complement numbers by finding the 2's complement of the subtrahend and adding gives the correct results.

20. Perform the following BCD arithmetic operations using the 8-4-2-1 representation assuming that all numbers are three-digit, unsigned decimal numbers.
    (a) 123 + 456
    (b) 123 − 456
    (c) 789 + 445
    (d) 789 − 445
    (e) 274 + 338
    (f) 932 − 677

21. Repeat Problem 20 assuming that all numbers are three-digit, signed decimal numbers in 10's complement format.

22. Repeat Problem 20 using the Excess-3 representation assuming that the given numbers are three-digit, unsigned decimal numbers.

23. Which of the codes presented in the chapter can be represented in the weighted-sum form? What are the values of the weights and the bias?

Fundamentals of Boolean Algebra

2

In this chapter we first introduce the defining postulates and theorems that constitute Boolean algebra. Boolean algebra forms the basis for the binary operations used in nearly all digital circuits, including all modern-day computer systems. We next examine the realization of binary logic and some commonly used binary operations. A major consideration of digital design is to simplify digital circuits. We present the minimization techniques by formulas, by Karnaugh maps, and by the variable entered maps. Karnaugh maps for two, three, and four variables are covered. The last section of this chapter deals with minimization techniques for incompletely specified functions.

In this chapter we introduce the basic mathematics necessary to analyze and synthesize digital systems. The formal mathematics is based on a two-value Boolean algebra sometimes called *switching algebra*. Boole formulated the laws of Boolean algebra, whereas Shannon showed that a two-value Boolean algebra could be used to investigate bistable electrical devices.

2.1

Postulates of Boolean Algebra

One of the most convenient sets of postulates for Boolean algebra was given by Huntington in 1904. These postulates define the basis of Boolean algebra.

I There exists a <u>set *S* of elements</u> subject to an equivalence relation. The equivalence relation, denoted "$=$" on a given set of elements, for example $\{0, 1, a, b, c\}$, satisfies the following properties:
  (a) Reflexive: For all a in S, then $a = a$.
  (b) Symmetric: If $a = b$, then $b = a$ for all a and b in S.
  (c) Transitive: If $a = b$ and $b = c$, then $a = c$ for all a, b, and c in S.
  (d) Substitutive: If $a = b$, then by substituting a for b in any expression, an equivalent relation will be obtained.

II (a) An operator "$+$" is defined such that $a + b$ is in S whenever a and b are in S.
  (b) An operator "$\cdot$" is defined such that $a \cdot b$ is in S whenever a and b are in S.

III (a) There exists an element 0 in S such that, for every a in S,

$$a + 0 = a$$

  (b) There exists an element 1 in S such that, for every a in S,

$$a \cdot 1 = a$$

IV The operators "$+$" and "$\cdot$" are commutative:
  (a) For all a and b in S, $a + b = b + a$.
  (b) For all a and b in S, $a \cdot b = b \cdot a$.

V The operators "$+$" and "$\cdot$" are distributive:
  (a) For all a, b, and c in S,

$$a + (b \cdot c) = (a + b) \cdot (a + c)$$

(b) For all a, b, and c in S,

$$a \cdot (b + c) = (a \cdot b) + (a \cdot c)$$

VI For every element a in S, there exists an element $\bar{a}$ (complement of a) such that

$$a \cdot \bar{a} = 0$$
$$a + \bar{a} = 1$$

VII There are at least two elements a and b in S such that

$$a \neq b$$

Postulate II is the closure property, Postulate IV is the commutative property, and Postulate V is the distributive property. These postulates are very similar to ordinary algebra except for Postulates V(a) and VI.

The simplest example of a Boolean algebra, sometimes called a switching algebra, consists of only two elements, 0 and 1. Therefore,

$$S = \{0, 1\}$$

and let the operations "+" and "·" be given by Figure 2–1.

Postulates I and II are satisfied because the set is defined and the results of the "+" and "·" operations are either 0 or 1. Postulate III is satisfied because 0 and 1 are defined and

$$0 + 0 = 0$$
$$0 + 1 = 1$$
$$1 \cdot 0 = 0$$
$$1 \cdot 1 = 1$$

Postulate IV is satisfied because the tables showing the "+" and "·" operations are symmetrical about the diagonal. Postulate V(a) can be shown by examining all combinations for a, b, and c in S as shown in Figure 2–2. This is called the *truth table approach*. A *truth table* is a table listing all possible combinations of

Figure 2–1

The Boolean "+" and "·" operations

+	0	1		·	0	1
0	0	1		0	0	0
1	1	1		1	0	1

	Left-Hand Side		Right-Hand Side		
$a \quad b \quad c$	$b \cdot c$	$a + (b \cdot c)$	$(a + b)$	$(a + c)$	$(a + b) \cdot (a + c)$
0 0 0	0	0	0	0	0
0 0 1	0	0	0	1	0
0 1 0	0	0	1	0	0
0 1 1	1	1	1	1	1
1 0 0	0	1	1	1	1
1 0 1	0	1	1	1	1
1 1 0	0	1	1	1	1
1 1 1	1	1	1	1	1

Figure 2–2
Proof of Huntington's Postulate V(a)

input and corresponding output values. The output set for the left-hand side of Postulate V(a) and that for the right-hand side are computed independently and compared. As seen from Figure 2–2, both sets produce identical outputs, indicating that under all possible combinations, the output of the left-hand side is identical to the output of the right-hand side. Postulate V(b) can also be shown with a similar table as given in Figure 2–3. Postulate VI is shown by proving that

$$\bar{0} = 1$$
$$\bar{1} = 0$$

Figure 2–3
Proof of Huntington's Postulate V(b)

	Left-Hand Side		Right-Hand Side		
$a \quad b \quad c$	$b + c$	$a \cdot (b + c)$	$(a \cdot b)$	$(a \cdot c)$	$(a \cdot b) + (a \cdot c)$
0 0 0	0	0	0	0	0
0 0 1	1	0	0	0	0
0 1 0	1	0	0	0	0
0 1 1	1	0	0	0	0
1 0 0	0	0	0	0	0
1 0 1	1	1	0	1	1
1 1 0	1	1	1	0	1
1 1 1	1	1	1	1	1

Let $a = 0$. Then

$$a \cdot \bar{a} = 0 \cdot \bar{0} = 0$$
$$a + \bar{a} = 0 + \bar{0} = 1 \Rightarrow \bar{0} = 1$$

Now let $a = 1$. Then

$$a \cdot \bar{a} = 1 \cdot \bar{1} = 0$$
$$a + \bar{a} = 1 + \bar{1} = 1 \Rightarrow \bar{1} = 0$$

Postulate VII is obvious because we have only two elements, 0 and 1.

Huntington's postulates are always given in pairs where one of the pairs can be obtained from the other by interchanging 0 and 1 and interchanging the "+" and "·" operators. When doing so, be careful not to alter the location of the parentheses, if any. For example:

$$a \cdot 1 = a$$
$$\downarrow$$
$$a + 0 = a$$

and
$$a + (b \cdot c) = (a + b) \cdot (a + c)$$
$$\downarrow \quad \downarrow \qquad \downarrow \quad \downarrow \quad \downarrow$$
$$a \cdot (b + c) = (a \cdot b) + (a \cdot c)$$

Therefore, any theorem that can be proved for a Boolean algebra has a dual that will also hold. Every step in the proof can be replaced by its dual and the result will still hold true. This is the *principle of duality*.

The use of the "·" symbol as a multiplication operation is usually omitted in ordinary algebra, and we will do that here for Boolean algebra as well, whenever it causes no confusion. Furthermore, the operator "·" will be performed before the operator "+" is performed, so that normally we will let

$$a + bc \Rightarrow a + (b \cdot c)$$

Should any ambiguity arise, the shortened notation will not be used.

2.2
Basic Theorems of Boolean Algebra

A number of theorems are necessary if we are to become proficient in manipulating Boolean expressions. The reader should become familiar with the following theorems as soon as possible. The proofs will list the postulates or other theorems that are invoked on the right-hand side. Alternatively, each theorem can be proved by the truth table technique.

Theorem 2-1 ──

The elements 0 and 1 are unique.

Proof Assume that two zero elements, 0_1 and 0_2, exist. Then, for any a_1 and a_2 in S,

$$a_1 + 0_1 = a_1 \qquad \text{P III(a)}$$

and

$$a_2 + 0_2 = a_2 \qquad \text{P III(a)}$$

Let

$$a_1 = 0_2$$
$$a_2 = 0_1$$

Then

$$0_2 + 0_1 = 0_2 \qquad \text{P I(d)}$$
$$0_1 + 0_2 = 0_1$$

Therefore,

$$0_2 = 0_2 + 0_1$$
$$= 0_1 + 0_2 \qquad \text{P IV(a)}$$
$$= 0_1$$

Hence,

$$0_2 = 0_1$$

This implies that 0 is unique. By the principle of duality it can easily be shown that 1 is also unique. In the following proofs, the dual relationship will be given at the end.

Theorem 2-2 ──

For every a in S,

$$a \cdot a = a$$
$$a + a = a$$

Proof

$$a + a = (a + a) \cdot 1 \qquad \text{P III(b)}$$
$$= (a + a) \cdot (a + \bar{a}) \qquad \text{P VI}$$
$$= a + a\bar{a} \qquad \text{P V(a)}$$
$$= a + 0 \qquad \text{P VI}$$

Hence,

$$a + a = a \qquad\qquad \text{P III(a)}$$
$$a \cdot a = a \qquad\qquad \text{Duality}$$

Theorem 2–3

For every a in S,

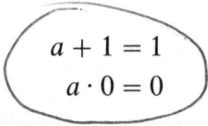

$$a + 1 = 1$$
$$a \cdot 0 = 0$$

Proof

$$
\begin{aligned}
a \cdot 0 &= 0 + a \cdot 0 \qquad && \text{P III(a)} \\
&= a\bar{a} + a \cdot 0 \qquad && \text{P VI} \\
&= a(\bar{a} + 0) \qquad && \text{P V(b)} \\
&= a \cdot \bar{a} \qquad && \text{P III(a)}
\end{aligned}
$$

Hence,

$$a \cdot 0 = 0 \qquad\qquad \text{P VI}$$
$$a + 1 = 1 \qquad\qquad \text{Duality}$$

Theorem 2–4

The elements 0 and 1 are distinct, and $\bar{0} = 1$.

Proof Let a be any element in S. Then

$$a \cdot 1 = a \qquad \text{P III(b)}$$
$$a \cdot 0 = 0 \qquad \text{T 2–3}$$

If $1 = 0$, then this implies that $a = 0$. Postulate VII stipulates that at least two elements exist. Hence we have a contradiction, since the only a that satisfies the above conditions is $a = 0$. The second part of the theorem is proved by writing

$$
\begin{aligned}
\bar{0} &= 0 + \bar{0} \qquad && \text{P III(a)} \\
&= 1 \qquad && \text{P VI}
\end{aligned}
$$

Theorem 2–5

For all a and b in S,

$$a + ab = a$$
$$a(a + b) = a$$

Proof

$$a(a + b) = (a + 0)(a + b) \qquad \text{P III(a)}$$
$$= a + 0 \cdot b \qquad \text{P V(a)}$$
$$= a + 0 \qquad \text{T 2–3}$$
$$= a \qquad \text{P III(a)}$$
$$a + ab = a \qquad \text{Duality}$$

Theorem 2–6 _____

For every a in S, the element $\bar{a}$ is unique.

Proof By contradiction, assume that two complements of a exist, namely, $\bar{a}_1$ and $\bar{a}_2$. Then

$$a + \bar{a}_1 = 1$$
$$a + \bar{a}_2 = 1$$
$$a \cdot \bar{a}_1 = 0$$
$$a \cdot \bar{a}_2 = 0$$

Therefore,

$$\bar{a}_2 = 1 \cdot \bar{a}_2 \qquad \text{P III(b)}$$
$$= (a + \bar{a}_1)\bar{a}_2 \qquad \text{Assumption}$$
$$= a\bar{a}_2 + \bar{a}_1\bar{a}_2 \qquad \text{P V(b)}$$
$$= 0 + \bar{a}_1\bar{a}_2 \qquad \text{Assumption}$$
$$= a\bar{a}_1 + \bar{a}_1\bar{a}_2 \qquad \text{Assumption}$$
$$= \bar{a}_1(a + \bar{a}_2) \qquad \text{P V(b)}$$
$$= \bar{a}_1 \cdot 1 \qquad \text{Assumption}$$
$$= \bar{a}_1 \qquad \text{P III(b)}$$

Theorem 2–7 _____

For every a in S,

$$a = \bar{\bar{a}}$$

Proof Let $x = \bar{\bar{a}}$. Then

$$\bar{a} \cdot x = 0 \qquad \text{P VI}$$
$$\bar{a} + x = 1 \qquad \text{P VI}$$
$$\bar{a} \cdot a = 0 \qquad \text{P VI}$$
$$\bar{a} + a = 1 \qquad \text{Duality}$$

Since both x and a satisfy the complement of $\bar{a}$ and the complement is unique, $x = a = \bar{\bar{a}}$.

Theorem 2–8

For all a, b, and c in S,

$$a[(a + b) + c] = [(a + b) + c]a = a$$

Proof The first two expressions are equal because of Postulate IV. Now,

$$
\begin{aligned}
a[(a + b) + c] &= a(a + b) + ac &&\text{P V(b)} \\
&= a + ac &&\text{T 2–5} \\
&= a &&\text{T 2–5}
\end{aligned}
$$

Theorem 2–9

For all a, b, and c in S,

$$a + (b + c) = (a + b) + c$$
$$a(bc) = (ab)c$$

Proof Let

$$
\begin{aligned}
Z &= [(a + b) + c][a + (b + c)] \\
&= [(a + b) + c]a + [(a + b) + c](b + c) &&\text{P V(b)} \\
&= a + [(a + b) + c](b + c) &&\text{T 2–8} \\
&= a + \{[(a + b) + c]b + [(a + b) + c]c\} &&\text{P V(b)} \\
&= a + (b + c) &&\text{T 2–8}
\end{aligned}
$$

Similarly,

$$
\begin{aligned}
Z &= [(a + b) + c][a + (b + c)] \\
&= (a + b)[a + (b + c)] + c[a + (b + c)] &&\text{P V(b)} \\
&= (a + b)[a + (b + c)] + c &&\text{T 2–8} \\
&= \{a[a + (b + c)] + b[a + (b + c)]\} + c &&\text{P V(b)} \\
&= (a + b) + c &&\text{T 2–8}
\end{aligned}
$$

Hence,

$$a + (b + c) = (a + b) + c$$

and

$$a(bc) = (ab)c \qquad\qquad \text{Duality}$$

This establishes the associative law from the postulates, without requiring another postulate. The rules $(a + b) + c = a + b + c$ and $a(bc) = abc$ can be extended to any number of products or sums.

Theorem 2–10

For any a and b in S,

$$a + \bar{a}b = a + b$$
$$a(\bar{a} + b) = ab$$

Proof

$$
\begin{aligned}
a(\bar{a} + b) &= a\bar{a} + ab & &\text{P V(b)} \\
&= 0 + ab & &\text{P VI} \\
&= ab & &\text{P III(a)} \\
a + \bar{a}b &= a + b & &\text{Duality}
\end{aligned}
$$

Theorem 2–11

For all a and b in S,

$$\overline{(a + b)} = \bar{a}\bar{b}$$
$$\overline{(ab)} = \bar{a} + \bar{b}$$

Proof The complement was shown to be unique, so if

$$(a + b) \cdot (\bar{a}\bar{b}) = 0$$

and

$$(a + b) + (\bar{a}\bar{b}) = 1$$

then this implies that

$$\overline{(a + b)} = \bar{a}\bar{b}$$

Now

$$
\begin{aligned}
(a + b)(\bar{a}\bar{b}) &= a \cdot \bar{a}\bar{b} + b \cdot \bar{a}\bar{b} & &\text{P V(b)} \\
&= 0 \cdot \bar{b} + b\bar{a}\bar{b} & &\text{P VI} \\
&= 0 + b\bar{a}\bar{b} & &\text{T 2–3} \\
&= b\bar{a}\bar{b} & &\text{P III} \\
&= \bar{a}b\bar{b} & &\text{P IV} \\
&= \bar{a} \cdot 0 & &\text{P VI} \\
&= 0 & &\text{T 2–3}
\end{aligned}
$$

$$(a + b) + (\overline{ab}) = [(a + b) + \overline{a}][(a + b) + \overline{b}] \qquad \text{P V(a)}$$
$$= (a + \overline{a} + b)(a + b + \overline{b}) \qquad \text{P V(a) and T 2-9}$$
$$= (1 + b)(a + 1) \qquad \text{P VI}$$
$$= 1 \cdot 1 \qquad \text{T 2-3}$$
$$= 1 \qquad \text{T 2-2}$$

Hence,

$$\overline{(a + b)} = \overline{a}\overline{b}$$

and
$$\overline{ab} = \overline{a} + \overline{b} \qquad \text{Duality}$$

The identities given in Theorem 2–11 represent two very important forms called *De Morgan's laws*. Based on Theorem 2–11, we can immediately conclude that the following two relationships hold:

$$a + b = \overline{\overline{a + b}} = \overline{\overline{a}\overline{b}}$$

and
$$ab = \overline{\overline{ab}} = \overline{\overline{a} + \overline{b}}$$

This just shows that there are many different forms of De Morgan's laws and it is important that the reader be able to recognize the various forms. De Morgan's laws are very useful for finding the complements of Boolean expressions. This will lead to increased ability in manipulating Boolean expressions. We will state without proof two corollaries to De Morgan's laws.

Corollary 1

The following are two nonidentities that are often overlooked. The complement operation and the "·" operation are not commutative, neither are the complement operation and the "+" operation.

$$\overline{ab} \neq \overline{a}\overline{b}$$

and
$$\overline{a + b} \neq \overline{a} + \overline{b}$$

Corollary 2

The two De Morgan's rules can be extended to any number of variables:

$$\overline{ab \cdots c} = \overline{a} + \overline{b} + \cdots + \overline{c}$$

and
$$\overline{a + b + \cdots + c} = \overline{a}\overline{b} \cdots \overline{c}$$

Theorem 2–12

For all a, b, and c in S,

$$ab + \overline{a}c + bc = ab + \overline{a}c$$

and
$$(a + b)(\overline{a} + c)(b + c) = (a + b)(\overline{a} + c)$$

Proof

$$ab + \bar{a}c + bc = ab + \bar{a}c + bc(a + \bar{a}) \qquad \text{P III(b) and P VI}$$
$$= ab + abc + \bar{a}c + \bar{a}bc \qquad \text{P V(b) and P IV}$$
$$= ab(1 + c) + \bar{a}c(1 + b) \qquad \text{P V(b)}$$
$$= ab + \bar{a}c \qquad \text{T 2–3 and P III}$$
$$(a + b)(\bar{a} + c)(b + c) = (a + b)(\bar{a} + c) \qquad \text{Duality}$$

The theorems and postulates presented above can be used to simplify complex Boolean expressions into simple yet equivalent forms.

Example 2–1 ▰▰▰▰▰▰▰▰▰▰▰▰▰▰▰▰

Minimize the following Boolean function:

$$F = ac + (a + b)\bar{c}$$

Boolean algebra can be used to simplify this function. The postulates and theorems will not be given for the simplification. The reader should try to find the appropriate postulates and theorems used in simplifying.

$$F = ac + (a + b)\bar{c}$$
$$= ac + a\bar{c} + b\bar{c}$$
$$= a(c + \bar{c}) + b\bar{c}$$
$$= a + b\bar{c}$$

Example 2–2 ▰▰▰▰▰▰▰▰▰▰▰▰▰▰▰▰

Simplify the following Boolean expression:

$$F = [\overline{(\overline{a} + b)(ab + \overline{c})}]$$

The Boolean expression can be simplified as follows:

$$F = [\overline{(\overline{a} + b)(ab + \overline{c})}]$$
$$= \overline{(\overline{a} + b)} + \overline{(ab + \overline{c})}$$
$$= (\bar{a} + b) + (ab + \bar{c})$$
$$= \bar{a} + b + ab + \bar{c}$$
$$= \bar{a} + b + \bar{c}$$

Simplification of Boolean functions by the above postulates and theorems is not always an easy task, and one is never sure when the task is finished because the simplification process is heuristic and not systematic.

2.3

Binary Logic and Boolean Algebra

The goal in developing Boolean algebra is to be able to apply the results to the practical building of digital systems. The essence of Boolean algebra consists of Boolean constants, Boolean variables, Boolean operations, Boolean expressions, and Boolean functions. Since we are dealing only with Boolean algebra, most of the time we will drop the "Boolean" designation.

Boolean Variables and Operations

Binary logic takes on two values only, 0 and 1, which correspond to the same symbols in a switching algebra. These are the only two allowable Boolean constants. A *Boolean variable*, which is used to represent logic values, is a variable having one of the Boolean constants as its value. The conventions or symbols for representing Boolean variables are not standardized. Generally speaking, we use upper- and lowercase letters at the beginning and end of the alphabet. The outputs of digital circuits are most often denoted as f, F, g, and G, or other letters from the middle of the alphabet. However, in complex digital systems, mnemonics such as sequences of letters and numbers are often used to convey information that is meaningful to the engineers designing the system. For example, MEMWR may be used to denote the memory write signal.

A number of physical devices can be built to perform the Boolean "+" operation, the Boolean "·" operation, and the Boolean complement operation in binary logic. The three basic building blocks used to perform the binary logic are the following operations:

1. AND performs the "·" Boolean operator. For example, $a \cdot b = c$ or $ab = c$ is read "a AND b is equal to c."
2. OR performs the "+" Boolean operator. For example, $a + b = c$ is read "a OR b is equal to c."
3. NOT performs the complement or "⁻" Boolean operator. For example, $\bar{a} = b$ is read "NOT a is equal to b" or sometimes "b is the complement of a."

These three operations are not the only possible operations for operations dealing with two variables. For two variables, a total of 16 operations are possible. The definitions, names, and logic symbols for these 16 combinations are shown in Figure 2–4. Some of these operations are named, whereas others are not.

Output Combination	A	0	0	1	1	Operation Name	Operation Operator Symbol	Operation Gate Symbol
	B	0	1	0	1			
f_0		0	0	0	0	zero	0	
f_1		0	0	0	1	AND		
f_2		0	0	1	0			
f_3		0	0	1	1	A		
f_4		0	1	0	0			
f_5		0	1	0	1	B		
f_6		0	1	1	0	Exclusive OR	$\oplus$	
f_7		0	1	1	1	OR	$+$	
f_8		1	0	0	0	NOR	$\downarrow$	
f_9		1	0	0	1	Exclusive NOR	$\overline{\oplus}$	
f_{10}		1	0	1	0	$\overline{B}$	$^{-}$	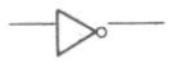
f_{11}		1	0	1	1			
f_{12}		1	1	0	0	$\overline{A}$	$^{-}$	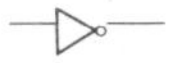
f_{13}		1	1	0	1			
f_{14}		1	1	1	0	NAND	$\uparrow$	
f_{15}		1	1	1	1	ONE	1	

Figure 2–4
Sixteen possible operations using two binary variables
(some combinations are unnamed)

The electronic digital circuits that perform the binary logic are called *logic circuits*, because with the proper inputs the logical control is performed. Any computing or control function can be accomplished by interconnecting logic circuits in different configurations to build complex digital systems. The symbols of the logic circuits that perform these logical operations of AND, OR, and NOT are given in Figure 2–5. These circuits are called *gates* and represent electronic hardware that produces a logic 1 or logic 0 at the output when the input represents logic-level 0's and 1's. The terminology is not that standard for these types of circuits, with digital circuits, switching circuits, logic circuits, and gates all being used. We will refer to the circuits as gates in most cases. The number of inputs to any OR or AND gate may vary, but practical limitations do exist based on the technology used. This will be discussed in the next chapter. The NOT gate is also called an *inverter*, because it inverts the signal.

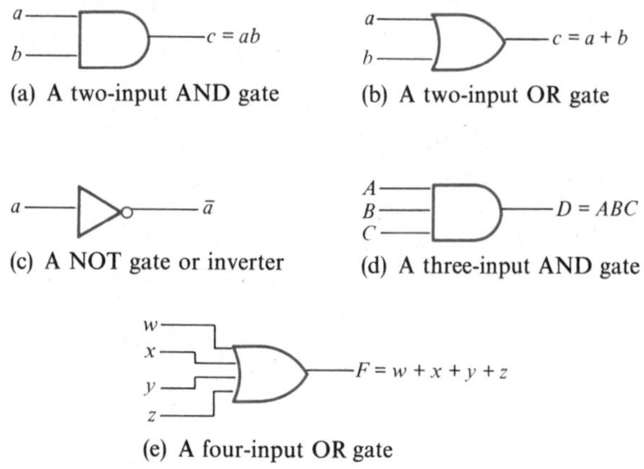

(a) A two-input AND gate $c = ab$

(b) A two-input OR gate $c = a + b$

(c) A NOT gate or inverter $\bar{a}$

(d) A three-input AND gate $D = ABC$

(e) A four-input OR gate $F = w + x + y + z$

Figure 2–5
Symbols for logical operations

The logic circuit composed of an AND gate followed by a NOT gate is called a NAND gate. A similar case holds for a NOT gate following an OR gate to implement a NOR gate. Figure 2–6 gives the symbols that are normally used for these gates. Many manufacturers of digital equipment use many of the equivalent symbols interchangeably. The reader should become familiar with all the different notations. While a circle on the output to a gate means that the output is inverted (complemented) before the other gates operate on it, a circle on the input to a gate similarly means that the input is inverted before it is operated upon.

Although much formal analysis has been given to the AND, OR, and NOT functions, two other useful functions, the Exclusive OR ($\oplus$) and Exclusive NOR

Figure 2–6
Symbols for (a) NAND and (b) NOR gates

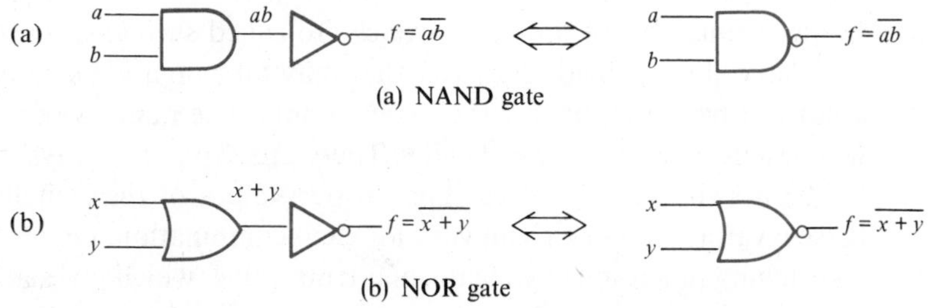

(a) ab $f = \overline{ab}$ $\Longleftrightarrow$ $f = \overline{ab}$

(a) NAND gate

(b) $x + y$ $f = \overline{x + y}$ $\Longleftrightarrow$ $f = \overline{x + y}$

(b) NOR gate

($\bar{\oplus}$) functions, are also frequently used in practice. The Exclusive OR function is

$$f = x\bar{y} + \bar{x}y = x \oplus y$$

where "$\oplus$" is called the Exclusive OR symbol. The Exclusive OR operation occurs many times in digital systems design. The properties of the Exclusive OR (XOR) and Exclusive NOR (XNOR) functions are listed below.

Commutative:

$$x \oplus y = y \oplus x$$

Associative:

$$x \oplus (y \oplus z) = (x \oplus y) \oplus z = x \oplus y \oplus z$$

Distributive:

$$x(y \oplus z) = (xy) \oplus (xz)$$

Other properties of the XOR and XNOR functions are summarized below:

EXCLUSIVE OR	EXCLUSIVE NOR
$0 \oplus 0 = 0$	$0 \bar{\oplus} 0 = 1$
$1 \oplus 1 = 0$	$1 \bar{\oplus} 1 = 1$
$1 \oplus 0 = 1$	$1 \bar{\oplus} 0 = 0$
$x \oplus 0 = x$	$x \bar{\oplus} 0 = \bar{x}$
$x \oplus 1 = \bar{x}$	$x \bar{\oplus} 1 = x$
$x \oplus x = 0$	$x \bar{\oplus} x = 1$
$x \oplus \bar{x} = 1$	$x \bar{\oplus} \bar{x} = 0$
$\bar{x} \oplus \bar{y} = x \oplus y$	$\bar{x} \bar{\oplus} \bar{y} = x \bar{\oplus} y$
$(x \oplus y) = x \bar{\oplus} y$	$\overline{(x \bar{\oplus} y)} = x \oplus y$

The Exclusive OR function is very useful and is discussed further in the next chapter.

Boolean Functions

Boolean functions of only two values are called *switching functions*. A *switching variable* will be defined as a letter that may take on a value of 0 or 1. These variables will be the inputs for the logic circuits. The number of variables or inputs will be a number that we will call n. There are 2^n possible ways of assigning values to the n switching variables. The output values of the switching function that these n variables take on can vary for each combination. One way of representing a switching function is by means of a truth table, which gives all combinations of

w	x	y	F
0	0	0	?
0	0	1	?
0	1	0	?
0	1	1	?
1	0	0	?
1	0	1	?
1	1	0	?
1	1	1	?

Figure 2–7

Truth table for a function with three switching variables

the input as shown in Figure 2–7. To define a specific function, the eight question marks could be specified as any combination of 0's or 1's. There are $2^8 = 2$ to the power of 2^3 ways of specifying the eight question marks. The value of F for any given input combination (row of the truth table) is the functional value for the corresponding combination of inputs. Two functions of the same switching variables that have the same truth table are equal and are called *equivalent functions*. Therefore, a truth table can be used to verify the equivalence or nonequivalence of two switching functions.

A switching function of n variables is defined as any one particular assignment of the functional values (0's or 1's) for all 2^n possible input combinations of the switching variables. Therefore, there are 2 to the power of 2^n distinct switching functions possible based on n input switching variables. Figure 2–8 gives the 2 to the power of $2^1 = 4$ different truth tables for a switching function of one switching variable.

Figure 2–8

The four possible switching functions with one switching variable

a	F
0	0
1	0

(a) $F = 0$

a	F
0	0
1	1

(b) $F = a$

a	F
0	1
1	0

(c) $F = \bar{a}$

a	F
0	1
1	1

(d) $F = 1$

Conversion of Switching Functions into Standard Sum-of-Products and Product-of-Sums Forms

The digital design engineer is normally faced with the job of implementing a switching function with the simplest or least expensive design. Most switching functions have many possible implementations, and there exists no known approach to arrive directly at the minimum or simplest form for an arbitrary function without an exhaustive search. Boolean algebra can be used to manipulate switching functions into numerous forms; however, the form to which most functions are minimized is the two-level realization. This involves AND gates driving a single OR gate or vice versa. This structure can be shown to have minimal delay, and well-known procedures exist to minimize the functions into this form.

A *literal* is defined as a switching variable or its complement (A, $\bar{A}$, x, $\bar{y}$, etc.). A *product term* is defined as a series of literals related by the AND operator ($A\bar{B}C$, $w\bar{x}\bar{y}$, $\bar{w}\bar{z}$, etc.). The product term is also called an *implicant*. A *sum term* is defined as a series of literals related by the OR operator ($w + x$, $\bar{x} + y + z$, $A + B$, etc.). A *normal term* is defined as a product or sum term in which no variable or its complement appears more than once within the term. A normal term implies that the "best" form is used, that is,

$$AA = A = A + A$$
$$x + \bar{x} = 1$$
$$y\bar{y} = 0$$

because a multiple occurrence of the same variable is either redundant or involves a trivial function. Now we will develop the form known as the *sum-of-products form*, which is literally a sum (through ORing) of product terms. There is also a *product-of-sums form*, which is obtained by ANDing a series of sum terms together. The development of these forms will be illustrated with several examples.

Example 2–3 ▬▬▬▬▬▬▬▬

Find the sum-of-products form for the following function:

$$
\begin{aligned}
F(w, x, y) &= (w + x\bar{y})(wx + y) \\
&= (w + x\bar{y})wx + (w + x\bar{y})y \\
&= wx + wx\bar{y} + wy \\
&= wx + wy
\end{aligned}
$$

More than one postulate or theorem may have been used at each step to arrive at a sum-of-products form. The correctness of the equivalence between the original and final expressions can be checked by the truth table technique.

Example 2–4 ▬▬▬▬▬▬▬▬▬▬▬▬▬▬▬▬▬▬▬▬▬▬▬▬▬

Convert the following function to a sum-of-products form:

$$
\begin{aligned}
F(a, b, c, d) &= \overline{(a + bc)}(a\bar{c} + (bd)) \\
&= (\bar{a}(\overline{bc}))(a\bar{c} + \bar{b} + \bar{d}) \\
&= \bar{a}(\bar{b} + \bar{c})(a\bar{c} + \bar{b} + \bar{d}) \\
&= (\bar{a}\bar{b} + \bar{a}\bar{c})(a\bar{c} + \bar{b} + \bar{d}) \\
&= \bar{a}\bar{b} + \bar{a}\bar{b}\bar{d} + \bar{a}\bar{b}\bar{c} + \bar{a}\bar{c}\bar{d} \\
&= \bar{a}\bar{b} + \bar{a}\bar{c}\bar{d}
\end{aligned}
$$

The sum-of-products form can also be expanded to ensure that each product term has exactly n literals and thus is a normal product term. The following example will show how this is accomplished.

Example 2–5 ▬▬▬▬▬▬▬▬▬▬▬▬▬▬▬▬▬▬▬▬▬▬▬▬▬

Convert the expressions in the two previous examples so that there are n literals in each product term. We will make repeated use of the relation

$$
a = a(b + \bar{b}) = ab + a\bar{b}
$$

to perform the expansion.

$$
\begin{aligned}
F(w, x, y) &= wx + wy \\
&= wx(y + \bar{y}) + wy(x + \bar{x}) \\
&= wxy + wx\bar{y} + wxy + w\bar{x}y \\
&= wxy + wx\bar{y} + w\bar{x}y \\
F(a, b, c, d) &= \bar{a}\bar{b} + \bar{a}\bar{c}\bar{d} \\
&= \bar{a}\bar{b}(c + \bar{c})(d + \bar{d}) + \bar{a}\bar{c}\bar{d}(b + \bar{b}) \\
&= \bar{a}\bar{b}cd + \bar{a}\bar{b}c\bar{d} + \bar{a}\bar{b}\bar{c}d + \bar{a}\bar{b}\bar{c}\bar{d} + \bar{a}\bar{b}\bar{c}\bar{d} + \bar{a}b\bar{c}\bar{d} \\
&= \bar{a}\bar{b}cd + \bar{a}\bar{b}c\bar{d} + \bar{a}\bar{b}\bar{c}d + \bar{a}\bar{b}\bar{c}\bar{d} + \bar{a}b\bar{c}\bar{d}
\end{aligned}
$$

By using the Boolean relations and expanding, it is possible to convert any switching function into a sum of normal products in which every product term contains as many literals as there are variables in the function. Such products are called *canonic products, standard products,* or *minterms.* The sum-of-products form, where each product is a standard product, is called a standard sum-of-products form, disjunctive normal form, or canonical sum-of-products form. We will use the term *standard sum-of-products form.*

It is also possible to expand any Boolean function into its dual form, i.e., the standard product-of-sums form. Example 2-6 shows the procedure.

Example 2-6 ▰▰▰▰▰▰▰▰▰▰▰▰▰▰▰▰▰▰▰▰▰▰

Express $F(A, B, C, D) = B + (A + C)(\bar{A} + D)$ in standard product-of-sums form.

$$F(A, B, C, D) = B + (A + C)(\bar{A} + D)$$
$$= (B + A + C)(B + \bar{A} + D)$$
$$= (A + B + C + D\bar{D})(\bar{A} + B + D + C\bar{C})$$
$$= (A + B + C + D)(A + B + C + \bar{D})$$
$$\cdot (\bar{A} + B + C + D)(\bar{A} + B + \bar{C} + D)$$

The sums containing the same number of literals as variables in the function are similarly called *canonical sums*, *standard sums*, or *maxterms*. The product-of-sums form, where each sum is a canonical sum, is called a standard product-of-sums form.

Generation of Minterms and Maxterms Directly

The expansion of switching functions into a standard sum-of-products form or a product-of-sums form by Boolean algebra is lengthy and prone to error. A shorthand notation can be developed directly from the truth table that completely

---

▰▰▰▰▰▰▰▰▰▰▰▰▰▰▰▰▰

Figure 2-9

A truth table showing the entries labeled with decimal equivalents

Decimal No.	w	x	y	f
0	0	0	0	0
1	0	0	1	1
2	0	1	0	1
3	0	1	1	0
4	1	0	0	1
5	1	0	1	1
6	1	1	0	0
7	1	1	1	0

$$f(w, x, y) = \sum(1, 2, 4, 5) \text{ or } \prod(0, 3, 6, 7)$$

specifies the function. The rows of the truth table can be numbered according to the decimal equivalents of the input combinations as shown in Figure 2–9. The first letter in the function variables will always be considered to be the most significant bit. Then the function can be specified in standard sum-of-products form for the rows when it has a value of 1,

$$f(w, x, y) = \sum(1, 2, 4, 5)$$

or in standard product-of-sums form for the rows when it has a value of 0,

$$f(w, x, y) = \prod(0, 3, 6, 7)$$

The symbols "$\sum$" and "$\prod$" are chosen to indicate a standard sum-of-products form and a standard product-of-sums form, respectively. The Boolean expressions for both cases are then

$$f(w, x, y) = \bar{w}\bar{x}y + \bar{w}x\bar{y} + w\bar{x}\bar{y} + w\bar{x}y$$
$$f(w, x, y) = (w + x + y)(w + \bar{x} + \bar{y})(\bar{w} + \bar{x} + y)(\bar{w} + \bar{x} + \bar{y}) \qquad (2–1)$$

Since each entry in the truth table is either a 1 or a 0, the decimal equivalent of each entry must be included and must be in either the sum series or the product series, but not in both or missing from both. In other words, if an entry is not in one series, it must be in the other series. Hence the two series are complementary.

The function will be 1 in a sum-of-products form whenever a product term is 1. In the standard sum-of-products form, each product term can be 1 for one and only one input combination. The first product term in (2–1) is 1 when $w = 0$, $x = 0$, and $y = 1$, which corresponds to the decimal number $1_{10} = 001_2$. This term is called minterm m_1. Other minterms and similarly defined maxterms are given in Figure 2–10. The maxterm is 0 for the corresponding row and 1 for all other rows.

Figure 2–10
Minterms and maxterms for a three-variable function,

Decimal No.	w	x	y	Minterm	Maxterm
0	0	0	0	$\bar{w}\bar{x}\bar{y} = m_0$	$w + x + y = M_0$
1	0	0	1	$\bar{w}\bar{x}y = m_1$	$w + x + \bar{y} = M_1$
2	0	1	0	$\bar{w}x\bar{y} = m_2$	$w + \bar{x} + y = M_2$
3	0	1	1	$\bar{w}xy = m_3$	$w + \bar{x} + \bar{y} = M_3$
4	1	0	0	$w\bar{x}\bar{y} = m_4$	$\bar{w} + x + y = M_4$
5	1	0	1	$w\bar{x}y = m_5$	$\bar{w} + x + \bar{y} = M_5$
6	1	1	0	$wx\bar{y} = m_6$	$\bar{w} + \bar{x} + y = M_6$
7	1	1	1	$wxy = m_7$	$\bar{w} + \bar{x} + \bar{y} = M_7$

The next example illustrates the use of decimal equivalents in Figure 2–10 to express a function in the various equivalent forms.

Example 2–7 ▬▬▬▬▬▬▬▬▬▬▬▬▬▬▬▬▬▬▬▬▬▬▬▬▬

Convert $f(w, x, y) = wx\bar{y} + \bar{w}\bar{x}y + w\bar{x}y + \bar{w}xy$ into minterm and maxterm forms.

The given function is already in the normal sum-of-products form, so the other minterm expressions are found as follows:

$$f(w, x, y) = wx\bar{y} + \bar{w}\bar{x}y + w\bar{x}y + \bar{w}xy$$
$$\begin{array}{cccc} 110 & 001 & 101 & 011 \\ 6 & 1 & 5 & 3 \end{array}$$
$$= \sum(1, 3, 5, 6)$$
$$= m_1 + m_3 + m_5 + m_6$$

Since the function must be 0 for the remaining input combinations, the maxterm expressions are

$$f(w, x, y) = \prod(0, 2, 4, 7)$$
$$= M_0 \cdot M_2 \cdot M_4 \cdot M_7$$
$$= (w + x + y)(w + \bar{x} + y)(\bar{w} + x + y)(\bar{w} + \bar{x} + \bar{y})$$

The calculation of minterms and maxterms for functions with more variables is an easy extension of Figure 2–10. The truth table provides the complete specification of any switching function. Thus, from the truth table we can express the given function in a standard sum-of-products form or a standard product-of-sums form, and hence the following theorem.

Theorem 2–13 _____

Any switching function of n variables $f(x_1, x_2, \ldots, x_n)$ may be expressed in a standard sum-of-products form or a standard product-of-sums form.

The expression of a switching function into a standard sum-of-products form in general makes the function more complex and costly to build. However, the minterm (or maxterm) list can be used as a starting point in the minimization process.

2.4

Minimization by Formulas

We can now use Boolean algebra to minimize or simplify Boolean functions. The following example is very simple, but it shows that the form of Boolean expression and a logic circuit interconnection is a one-to-one relationship.

Example 2–8

Minimize the logic circuit shown in Figure 2–11(a), which represents the Boolean function

$$f = ab + \bar{a}c + bc$$

In Theorem 2–12 we proved that

$$f = ab + \bar{a}c + bc = ab + \bar{a}c$$

Therefore, the circuit shown in Figure 2–11(b) is an equivalent circuit to the one shown in Figure 2–11(a). The two circuits are equivalent because for all input combinations, the outputs of the circuits are identical. As measured by the number of gates, the cost of implementing the second circuit is obviously less than the cost of the first.

Figure 2–11

Two equivalent circuits

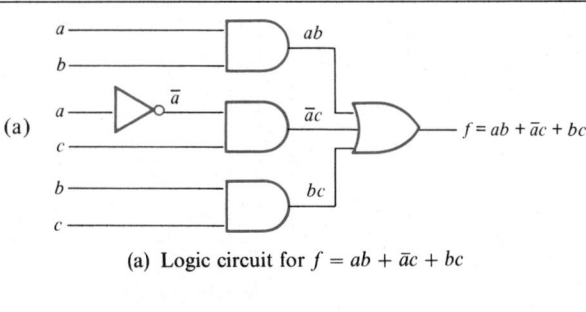

(a) Logic circuit for $f = ab + \bar{a}c + bc$

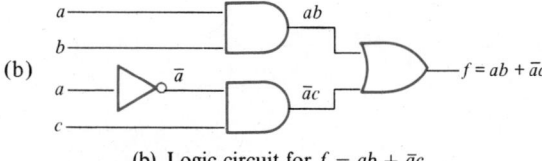

(b) Logic circuit for $f = ab + \bar{a}c$

Example 2-9 is more complex, and the initial approach is by trial and error.

Example 2–9 ▰▰▰▰▰▰▰▰▰▰▰▰▰▰▰▰▰▰▰▰

Minimize the logic circuit shown in Figure 2–12(a).

The circuit to be simplified is given in Figure 2–12(a). The expression for the circuit based on the diagram is

$$F = (ab + \bar{a}c) + (\bar{a}b + bc)$$

Boolean algebra can be used to simplify this circuit. The postulates and theorems will not be given for the simplification. The reader should try to identify the postulates and theorems used in simplifying.

$$\begin{aligned}
F &= (ab + \bar{a}c) + (\bar{a}b + bc) \\
&= ab + \bar{a}c + bc + \bar{a}b \\
&= ab + \bar{a}c + \bar{a}b \\
&= b(a + \bar{a}) + \bar{a}c \\
&= b + \bar{a}c
\end{aligned}$$

The simplified circuit is shown in Figure 2–12(b).

Figure 2–12
Circuit simplification

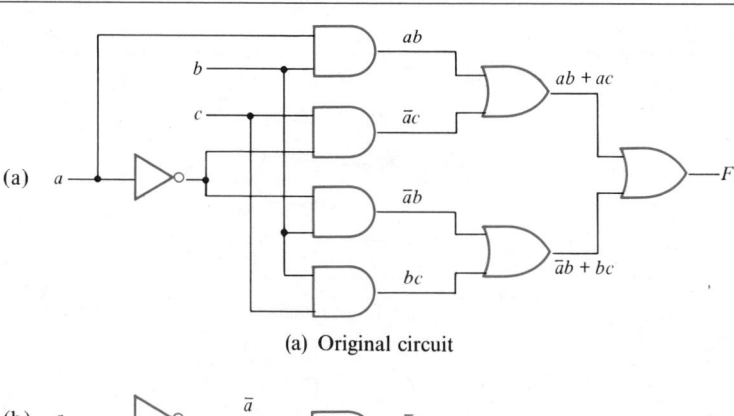

(a) Original circuit

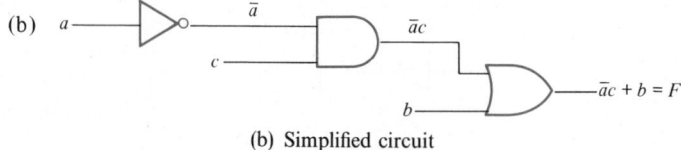

(b) Simplified circuit

Example 2–10

Simplify the following Boolean expression:

$$F = (\overline{a + b})(ab + \bar{a}c + bcd)$$

The Boolean expression can be simplified as follows:

$$
\begin{aligned}
F &= (\overline{a + b})(ab + \bar{a}c + bcd) \\
&= (\overline{a + b})(ab + \bar{a}c + bc + bcd) \\
&= (\overline{a + b})[ab + \bar{a}c + bc(1 + d)] \\
&= (\overline{a + b})[ab + \bar{a}c + bc] \\
&= (\overline{a + b})(ab + \bar{a}c) \\
&= \bar{a}\bar{b}(ab + \bar{a}c) \\
&= \bar{a}\bar{b}ab + \bar{a}\bar{b}\bar{a}c \\
&= \bar{a}a\bar{b}b + \bar{a}\bar{a}\bar{b}c \\
&= 0 + \bar{a}\bar{b}c \\
&= \bar{a}\bar{b}c
\end{aligned}
$$

2.5
Minimization by Karnaugh Map

The previous section showed how Boolean expressions can be simplified by using formulas and postulates. When the expression is complex, this procedure becomes very tedious and is prone to error. A graphic technique called the *Karnaugh map* technique is often used. The Karnaugh map is a very powerful tool for minimizing logic circuits in two-level sum-of-products or product-of-sums form. The effectiveness of the Karnaugh map lies in the ability of the human mind to visualize patterns in the map. The use of Karnaugh maps is not new to the solution of switching function circuits, but it is very helpful. The Karnaugh map may be thought of as a different form of the truth table. It is also similar to the Venn diagrams used in set theory.

Karnaugh Map Representation of Switching Functions

The Karnaugh map uses one square or cell for each minterm or maxterm, similar to the truth table which uses one row for each minterm. The two-, three-, and four-variable Karnaugh maps are given in Figure 2–13. The number in the upper right-hand corner of each cell represents the decimal equivalent of the binary

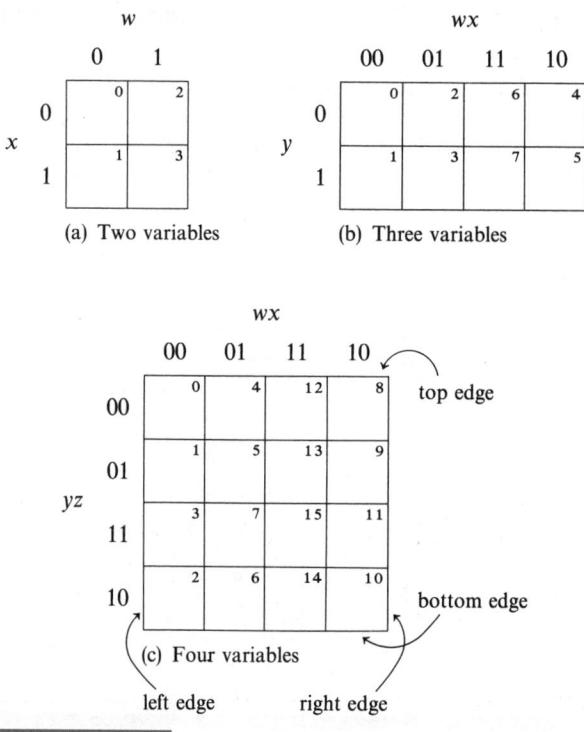

Figure 2–13
Karnaugh maps for two, three, and four variables

input. This is also the subscript attached to the corresponding minterm or max-term. The Karnaugh map is filled with a 1 for each cell when the corresponding minterm is present and a 0 when it is not present. Likewise, the Karnaugh map is filled with a 0 for each cell when the corresponding maxterm is present and a 1 when it is not present. Note the order and arrangement of the variables in the three- and four-variable maps. There is only one bit change between any cell and its adjacent cell. The reader should verify this fact for adjacent cells shown in Figure 2–13. For example, cell 15 in a four-variable map is adjacent to cell 7, cell 11, cell 13, and cell 14. This property holds even for those cells on the edge of the map. The left edge can be considered to be the same edge as the right edge. Likewise, the top edge can be considered to be the same edge as the bottom one. For example, cell 10 in a four-variable map is not only adjacent to cell 11 and cell 14, but also adjacent to cell 2 and cell 8. This is a very important property of the Karnaugh maps, and it very much facilitates the minimization process. This process will be illustrated by several examples.

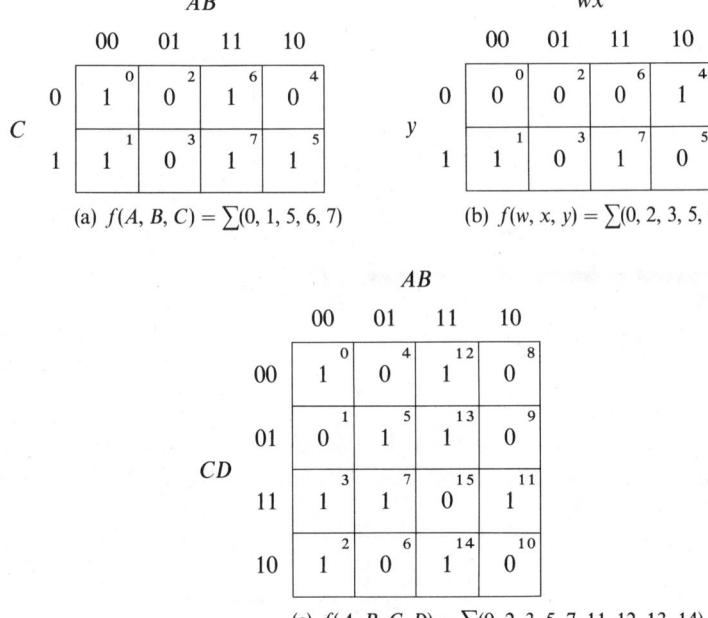

(a) $f(A, B, C) = \sum(0, 1, 5, 6, 7)$

(b) $f(w, x, y) = \sum(0, 2, 3, 5, 6)$

(c) $f(A, B, C, D) = \sum(0, 2, 3, 5, 7, 11, 12, 13, 14)$

Figure 2–14

Using Karnaugh maps to represent, two-variable, three-variable, and four-variable functions

Example 2–11

Construct Karnaugh maps for the following functions:

(a) $f(A, B, C) = \sum(0, 1, 5, 6, 7)$
(b) $f(w, x, y) = \prod(0, 2, 3, 5, 6)$
(c) $f(A, B, C, D) = \sum(0, 2, 3, 5, 7, 11, 12, 13, 14)$

The appropriately filled Karnaugh maps are given in Figure 2–14.

Example 2–12

Karnaugh maps can also be easily constructed directly from partially minimized sum-of-products or product-of-sums form. Find the Karnaugh map representation of the following function:

$$f(w, x, y) = w + \bar{x}y$$

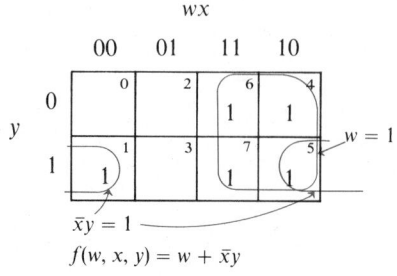

Figure 2–15
Karnaugh map for $f(w, x, y) = w + \bar{x}y$

The function is obviously 1 when $w = 1$, or when $x = 0$ and $y = 1$. Any cell that satisfies either condition is filled with a 1. The Karnaugh map is given in Figure 2–15.

Example 2–13

Construct the Karnaugh map for

$$f(w, x, y, z) = w(x + \bar{y} + z)(\bar{x} + \bar{z})$$

The Karnaugh map is given in Figure 2–16. Once the map is constructed, the maxterm and minterm lists can be obtained directly from the map.

Figure 2–16
Karnaugh map for $f(w, x, y, z) = w(x + \bar{y} + z)(\bar{x} + \bar{z})$

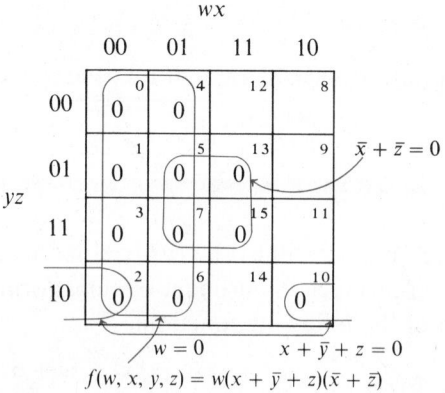

$$F(w, x, y, z) = \prod(0, 1, 2, 3, 4, 5, 6, 7, 10, 13, 15)$$
$$= \sum(8, 9, 11, 12, 14)$$

The variables used in the Karnaugh map may vary in different examples. The reader should be able to use any variables, but care must be exercised to keep all the variables in their proper sequence.

Minimization Using Karnaugh Maps

The cost of implementing a logic circuit is normally considered to be a function of the number of gates and the number of inputs to each gate. The goal of the minimization process is to try to reduce the number of gates and inputs. Previously, we discussed two-level circuits in the sum-of-products or product-of-sums form. A change in an input (variable) in these circuits must propagate through only two levels of gates to affect the output, thus greatly increasing the speed of operation and minimizing the delay. It is certainly possible to have less complex or expensive realizations with more levels; however, the design techniques are much harder to apply. We will define a sum-of-products form to be a minimal expression if there exists no other equivalent expression using fewer product terms or using the same number of product terms with fewer literals.

The implementation of a sum-of-products form requires one OR gate at the output and one AND gate for each product (or none if the product term is a single literal). Likewise, the implementation of a product-of-sums form requires one AND gate at the output and one OR gate for each product (or none if the sum term is a single literal).

Consider the following function:

$$f(a, b, c) = \sum(0, 2, 5, 7)$$
$$= \bar{a}\bar{b}\bar{c} + \bar{a}b\bar{c} + a\bar{b}c + abc$$

The corresponding Karnaugh map appears in Figure 2–17. By algebraic

---

Figure 2–17

Karnaugh map minimization for $f(a, b, c) = \sum(0, 2, 5, 7)$

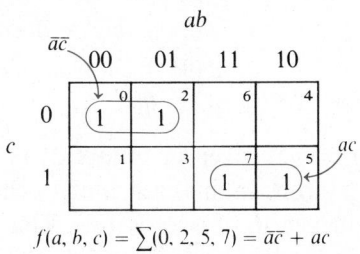

$$f(a, b, c) = \sum(0, 2, 5, 7) = \bar{a}\bar{c} + ac$$

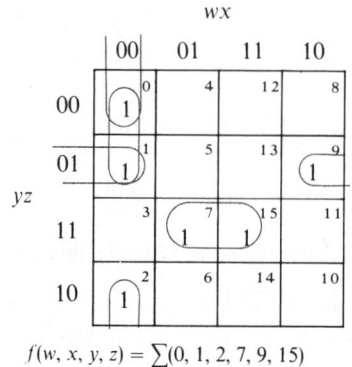

$$f(w, x, y, z) = \sum(0, 1, 2, 7, 9, 15)$$

Figure 2–18

Examples of adjacent 1-cells in a Karnaugh map for
$f(w, x, y, z) = \sum(0, 1, 2, 7, 9, 15)$

manipulation:

$$f(a, b, c) = \overline{a}\,\overline{c}(\overline{b} + b) + ac(\overline{b} + b)$$
$$= \overline{a}\,\overline{c} + ac$$

Note that in Figure 2–17 the circled groups of 1-cells correspond to the product terms obtained by algebraic manipulation. The sum-of-products form is obviously simpler than the standard sum-of-products form. Any pair of adjacent 1-cells in an n-variable Karnaugh map can be combined into a single product term of $n - 1$ literals. The Karnaugh map was constructed so that physically adjacent 1-cells would exhibit this property.

The minimization technique requires that the designer recognize all adjacent 1-cells. Figure 2–18 gives examples of adjacent 1-cells for a four-variable Karnaugh map. The function used here is $f(w, x, y, z) = \sum(0, 1, 2, 7, 9, 15)$. The adjacent 1-cells differ in one variable only. Note that the diagonal 1-cells 9 and 15 are not adjacent, because $x = 0$ and $y = 0$ for 9 and $x = 1$ and $y = 1$ for 15.

Let us now consider the following function:

$$f(a, b, c, d) = \sum(0, 1, 2, 3, 14, 15)$$
$$= \overline{a}\,\overline{b}\,\overline{c}\,\overline{d} + \overline{a}\,\overline{b}\,\overline{c}d + \overline{a}\,\overline{b}c\overline{d} + \overline{a}\,\overline{b}cd + abc\overline{d} + abcd$$
$$= \overline{a}\,\overline{b}(\overline{c}\,\overline{d} + \overline{c}d + c\overline{d} + cd) + abc(\overline{d} + d)$$
$$= \overline{a}\,\overline{b} + abc$$

The Karnaugh map is given in Figure 2–19. The two groupings correspond to the two product terms. In this case, four 1-cells that are adjacent in pairs combine to form a product term of two literals. This is a general process where any set of four adjacent 1-cells in which each 1-cell is adjacent to two other 1-cells in

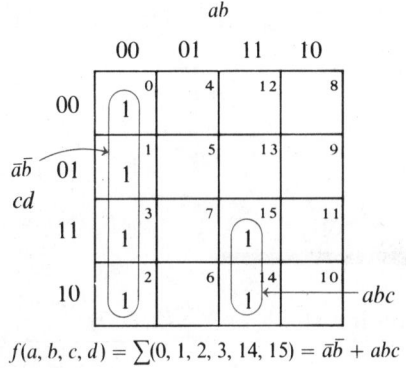

$$f(a, b, c, d) = \sum(0, 1, 2, 3, 14, 15) = \bar{a}\bar{b} + abc$$

Figure 2–19

Example of a set of four 1-cells in a four-variable
Karnaugh map

the set may be combined to form a product term of two literals less than the
number of input variables.

The process of obtaining larger groupings also exists. The general rule is that
a set of 2^i 1-cells may be grouped together if each of the cells is adjacent to i of
the other 1-cells in the set. Then a product term consisting of $n - i$ literals for
an n-input function can be formed that corresponds to this group. The product
term that corresponds to the largest grouping that is not a proper subgrouping
of another group is called a *prime implicant*.

The minimization process involves first finding the largest sets of 1-cells in
the Karnaugh map that are not proper subsets of another set and then selecting
a minimum set of these to cover all the 1-cells. Covering ensures that the circuit
output is 1 for all input combinations for which the function is 1. The general
rule for Karnaugh maps is first to select the largest grouping of 1-cells (or 0-cells)
that are covered by one and only one set; then select the next largest grouping
from the remaining 1's (or 0's) entries, using the covered entries only if needed
to complete the groupings. The process is repeated for smaller groupings until
no further groupings can be formed. It is not always evident which grouping will
yield the best minimal solution. Sometimes the final form may be obtained only
by trial and error. This process is now illustrated by Examples 2-14, 2-15, 2-16,
2-17, and 2-18.

Example 2–14

Simplify the following function:

$$f(a, b, c) = \sum(2, 6, 7)$$

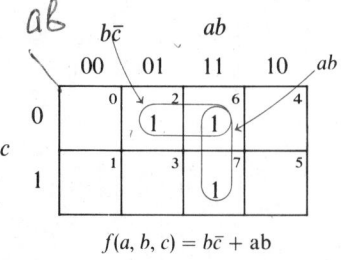

$$f(a, b, c) = b\bar{c} + ab$$

Figure 2–20

Three-input minimization for $f(a, b, c) = \sum(2, 6, 7)$

The solution is given in Figure 2–20. Note that both product terms make the function 1 for cell 6. However, this poses no problem, since $1 + 1 = 1$ in Boolean algebra.

Example 2–15

Simplify the following function:

$$f(w, x, y, z) = \sum(4, 5, 9, 11, 13, 15)$$

The Karnaugh map is given in Figure 2–21. All of the groupings are given. The minimal solution does not need the product term $x\bar{y}z$ because the other two product terms already cover the function for the 1-cells 5 and 13.

Figure 2–21

Four-input minimization with redundant prime implicants

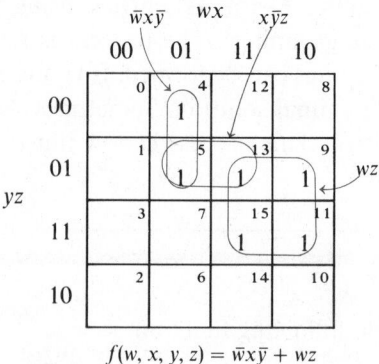

$$f(w, x, y, z) = \bar{w}x\bar{y} + wz$$

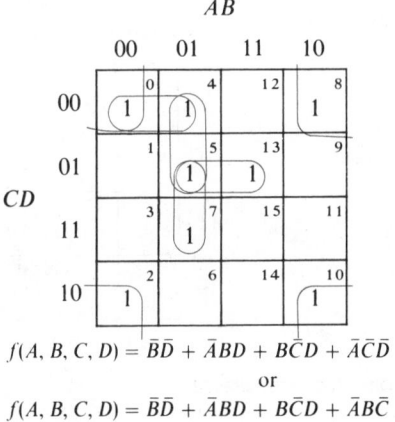

$$f(A, B, C, D) = \bar{B}\bar{D} + \bar{A}BD + B\bar{C}D + \bar{A}\bar{C}\bar{D}$$

or

$$f(A, B, C, D) = \bar{B}\bar{D} + \bar{A}BD + B\bar{C}D + \bar{A}B\bar{C}$$

Figure 2–22
Karnaugh map with two minimal solutions

Example 2–16

Simplify the following function:

$$f(A, B, C, D) = \sum(0, 2, 4, 5, 7, 8, 10, 13)$$

The Karnaugh map is given in Figure 2–22. Note the grouping of four 0's at the corners corresponding to the maxterm $\bar{B}\bar{D}$. The product terms $\bar{B}\bar{D}$, $\bar{A}BD$, and $B\bar{C}D$ must be used because each covers a 1-cell that is not covered by any other product term. These three product terms cover all 1-cells except the number 4 1-cell. This may be covered by $\bar{A}\bar{C}\bar{D}$ or $\bar{A}B\bar{C}$. Hence this example has two different minimal solutions.

Example 2–17

Simplify the following function:

$$f(w, x, y, z) = \sum(3, 4, 6, 7, 10, 13, 14, 15)$$

The Karnaugh map is given in Figure 2–23. Note in this example that the large set (xy) in the center is not used. One must not be hasty in using the large sets first and end up with a nonminimal solution.

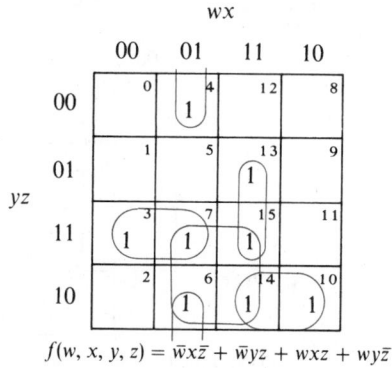

$$f(w, x, y, z) = \bar{w}x\bar{z} + \bar{w}yz + wxz + wy\bar{z}$$

Figure 2–23
Karnaugh map minimization that is missing the largest group

Example 2–18

Simplify the following function:

$$f(w, x, y) = \sum(1, 2, 3, 4, 5, 6)$$

The Karnaugh map of this function is shown in Figure 2–24. This example has no 1-cell, which is covered by one and only one product term. This is called a *cyclic* map. Normally, one proceeds by arbitrarily selecting a 1-cell, then choosing a product term that covers it, and then covering all remaining 1-cells in a heuristic manner. Next, one selects another product term that covers the initial 1-cell, and proceeds by covering all remaining 1-cells. When all is done, one selects the minimal solution. In this case, we select the 1-cell number 2. If $\bar{w}x$ is chosen to cover it, the remaining 1-cells can be covered by $w\bar{y}$ and $\bar{x}y$ to give

$$f(w, x, y) = \bar{w}x + w\bar{y} + \bar{x}y$$

Figure 2–24
A cyclic Karnaugh map

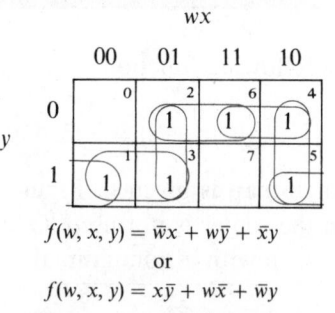

$$f(w, x, y) = \bar{w}x + w\bar{y} + \bar{x}y$$
$$\text{or}$$
$$f(w, x, y) = x\bar{y} + w\bar{x} + \bar{w}y$$

Now we choose $x\bar{y}$ to cover 1-cell number 2. Then $w\bar{x}$ and $\bar{w}y$ cover the remaining 1-cells to give

$$f(w, x, y) = w\bar{x} + \bar{w}y + x\bar{y}$$

For this example, both solutions are minimal.

Up to this point we have used Karnaugh maps to find a minimal sum-of-products form. A minimum product-of-sums form can be handled in a similar manner according to the principle of duality by grouping the 0-cells. In general, a minimal two-level solution requires that both the minimal sum-of-products form and the minimal product-of-sums form are found. The minimal realization is then chosen between the two.

Example 2–19 ▬▬▬▬▬▬▬▬▬▬▬▬▬▬▬▬▬▬▬▬

Simplify the following function:

$$f(w, x, y, z) = \sum(0, 3, 4, 7, 8, 11)$$

The Karnaugh maps with the minimal sum-of-products form and the minimal product-of-sums form are given in Figure 2–25. The minimal product-of-sums form needs only four gates, while the minimal sum-of-products form requires five gates with more inputs. Thus the minimal solution is the product-of-sums form.

Karnaugh maps can be constructed of five and six variables, but the simplification process is more difficult because it becomes harder to find all the groupings. A number of other texts, which can be found in the references, treat these cases.

Figure 2–25
Two-level minimal solution for $f(w, x, y, z) = \sum(0, 3, 4, 7, 8, 11)$

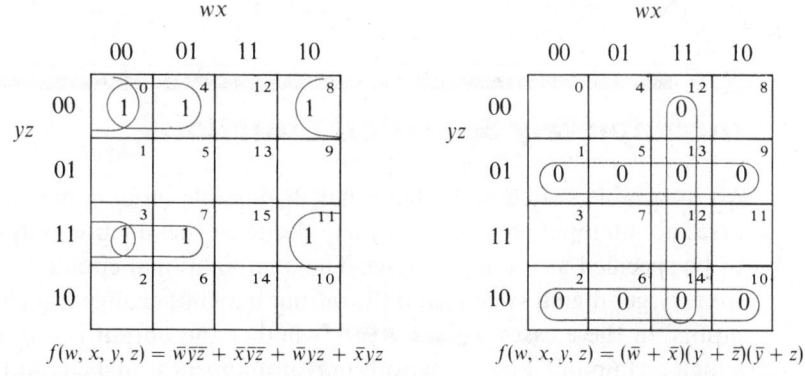

$$f(w, x, y, z) = \bar{w}\bar{y}\bar{z} + \bar{x}\bar{y}\bar{z} + \bar{w}yz + \bar{x}yz$$

$$f(w, x, y, z) = (\bar{w} + \bar{x})(y + \bar{z})(\bar{y} + z)$$

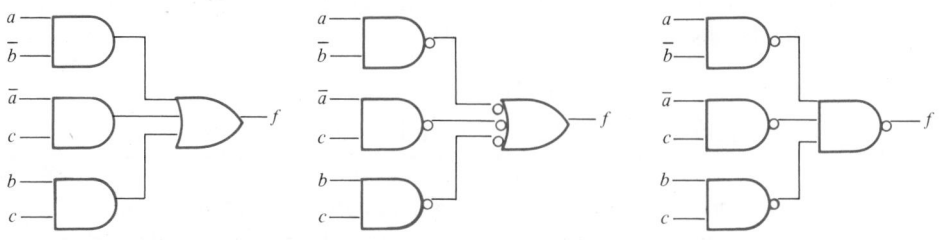

Figure 2–26
Equivalent circuits between AND-OR form and NAND-NAND form

We have discussed the sum-of-products and product-of-sums forms that use only AND and OR gates. Two levels of NAND gates can always replace a sum-of-products form as shown below.

$$f(a, b, c) = a\overline{b} + \overline{a}c + bc$$
$$= \overline{\overline{(a\overline{b} + \overline{a}c + bc)}}$$
$$= \overline{(\overline{a\overline{b}})(\overline{\overline{a}c})(\overline{bc})}$$
$$= (a \uparrow \overline{b}) \uparrow (\overline{a} \uparrow c) \uparrow (b \uparrow c)$$

This represents two levels of NAND gates as shown in Figure 2–26. (A bubble in Figure 2–26 indicates an inverter.) The final OR gate can be replaced by a NAND gate, since

$$\overline{x} + \overline{y} = \overline{\overline{\overline{x} + \overline{y}}}$$
$$= \overline{(\overline{\overline{x}})(\overline{\overline{y}})}$$
$$= x \uparrow y$$

Thus, all the two-level minimizations for the sum-of-products form can be implemented with NAND gates only. Likewise, NOR gates can always be used to replace OR gates followed by AND gates in the product-of-sums form. Note that the circuit topology or connection pattern remains the same while changing from the AND-OR form to the NAND-NAND form.

2.6

Incompletely Specified Functions

An n-variable switching function has 2^n possible input combinations. In certain cases not all input combinations may occur, or the function output may not need to be specified at a unique value. The combinational circuit is usually a part of some larger digital system, and the output may not change its behavior for certain inputs. In these cases we don't care whether the output is a 0 or a 1. A case in which certain input combinations may not appear is in a circuit that has only the

BCD inputs 0 to 9, but not 10 through 15. The output for "don't care" inputs will be represented by a d. Other symbols such as $\emptyset$, $-$, or $\times$ are also used.

When don't cares occur, we say that the output is unspecified, and the corresponding function is then said to be an incompletely specified function. The realization we pick will ensure only that the switching function is valid for the specified entries, but we will allow unspecified outputs to be selected to help minimize the cost. For the shorthand notation, we will make use of the symbol $\sum_d(\ )$ for those don't care entries in the minterm form and $\prod_d(\ )$ for those in the maxterm form.

Example 2–20 ▪▪▪▪▪▪▪▪▪▪▪▪▪▪▪▪▪▪▪▪▪▪▪▪▪

Find the decimal notation and a minimal solution for the function that accepts BCD inputs only and has a 1 output for all odd inputs.

The problem can be rephrased as minimizing the following function:

$$f(w, x, y, z) = \sum(1, 3, 5, 7, 9) + \sum_d(10, 11, 12, 13, 14, 15)$$

The minimization process is almost the same as before, except that no grouping made up exclusively of don't cares is used. Also, we do not need to use the don't care cells unless they help to reduce the number of groupings or increase the size of groupings. The solution by Karnaugh maps is given in Figure 2–27. The solution in this example shows that, without making use of the don't care entries, the minimal solution is not simple at all. Using the don't care entries, however, greatly simplifies the solution to $f(w, x, y, z) = z$, that is, the least significant bit.

Figure 2–27
Minimization with don't cares

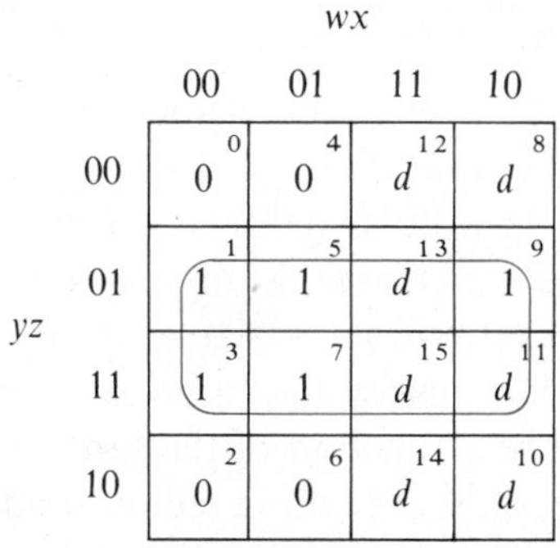

If d-cells are not used:
   $f(w, x, y, z) = \bar{w}z + \bar{x}\bar{y}z$
If d-cells are used:
   $f(w, x, y, z) = z$

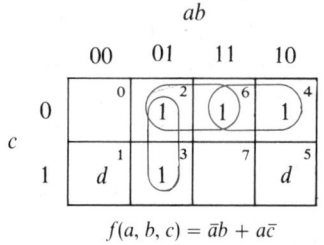

$$f(a, b, c) = \bar{a}b + a\bar{c}$$

Figure 2–28
Minimization with don't cares for $f(a, b, c) = \sum(2, 3, 4, 6) + \sum'_d(1, 5)$

Example 2–21

Obtain a minimal solution in the sum-of-products form for $f(a, b, c) = \sum(2, 3, 4, 6) + \sum_d(1, 5)$.

The solution is given in Figure 2–28. In this example, none of the don't care entries contributes to the final minimal solution.

Karnaugh maps of five and six variables are not discussed because of space constraints. They are also more difficult to use. For functions of more variables, it is helpful to have a technique that is much more systematic than the Karnaugh map. The most well known procedure is the Quine-McCluskey algorithm. This procedure starts from the minterm list and step by step finds the largest groupings to form the product terms or prime implicants. Once the prime implicants are determined, a prime implicant table is used, which finds the minimum set of prime implicants to cover the function. "Covering" just means that the function is 1 for all minterms in the original list. The Quine-McCluskey procedure is well adapted to computer minimization. Details on this procedure are given in the references.

There are other heuristic techniques that produce near-minimal results and take much less computer time and storage than the Quine-McCluskey algorithm. The minimization process becomes increasingly more time consuming when the number of inputs increases. Many cases exist that require minimization with 8 to 16, or even more, input variables. Furthermore, the Karnaugh map method and the Quine-McCluskey algorithm can be modified to minimize multiple-output functions that are a function of the same input variables. These techniques will not be discussed. They are more tedious and difficult than the single-output case. The multiple-output case occurs quite often in digital systems design such as code conversions and binary addition. In Chapter 6, we will give other techniques to implement combinational circuits using existing digital hardware. In many cases this will make some of the techniques in this chapter unnecessary.

2.7
Minimization by the Variable Entered Map Method

Another method that closely resembles the Karnaugh map method is the variable entered map approach. While the Karnaugh map method allows only 0's and 1's in the maps, the variable entered map method allows variables themselves to be entered directly onto the maps, thus resulting in a reduction of the size of the maps required. Based on the variable entered map method, an n-variable function can be described by an $(n-1)$-variable map using one of the variables in the map entries, or by an $(n-2)$-variable map using two of the variables in the map entries, etc. For many functions, their minimization processes using the variable entered map approach can be much simplified.

Representation of Functions in Variable Entered Maps

The basic idea in the variable entered map approach is to represent the values of a function in terms of its variables within the framework of the Karnaugh maps. Any variable and any number of variables can be chosen for map entry. A variable chosen for map entry is called a *map entered variable*. Once the variable is chosen, a Karnaugh map can be formed for the rest of the variables. The Karnaugh map formed will have less order than the function itself. Now for each entry of the map, a function related to the selected variable is entered. If there is only one map entered variable, then each map entry may represent the function value for one or two minterms. For example, consider the following function:

$$f(a, b, c) = \bar{a}\bar{b}\bar{c} + \bar{a}bc + a\bar{b}\bar{c} + a\bar{b}c$$

We will arbitrarily choose c as the map entered variable. The first term indicates that for the condition $a = 0$ and $b = 0$, the function $f = 1$ when $c = 0$. In this case there is only one minterm for this condition. For the condition $a = 0$ and $b = 1$, again there is only one minterm, and the function is 1 whenever $c = 1$. For the third condition, $a = 1$ and $b = 0$, there are two minterms. The function is 1 when $c = 0$ and when $c = 1$. Hence there are four possibilities altogether when one map entered variable is used. These four possibilities and their corresponding map entries are shown below:

	Map Entry
No minterms with this condition	0
$f = 1$ when $c = 0$	$\bar{c}$
$f = 1$ when $c = 1$	c
$f = 1$ when $c = 0$ or 1	$c + \bar{c}$

a / b	0	1
0	$f = 1$ when $c = 0$	$f = 1$ when $c = 0$ or when $c = 1$
1	$f = 1$ when $c = 1$	$f = 0$

(a) Intermediate form

a / b	0	1
0	$\bar{c}$	$c + \bar{c}$
1	c	0

(b) Final form

	$a = 0$	$a = 1$
$f = 1$	$\bar{b}\bar{c} + bc$	$\bar{b}\bar{c} + \bar{b}c$

(c) Using b and c as map entered variables

Figure 2–29
Representing $f(a, b, c) = \bar{a}\bar{b}\bar{c} + \bar{a}bc + a\bar{b}\bar{c} + a\bar{b}c$ using the variable entered map approach: (a) and (b) using c as the map entered variable; (c) using b and c as map entered variables

Using this rule, the above function can be represented as a two-variable Karnaugh map in an unsimplified form in Figure 2–29(a) and in a simplified form in Figure 2–29(b).

Example 2–22

Represent the following function using one map entered variable:

$$f(x, y, z) = \bar{x}\bar{y}z + xyz + x\bar{y}\bar{z} + \bar{x}y\bar{z} + \bar{x}yz$$

Any map entered variable can be used. Let us use x as the map entered variable. The resultant map is shown in Figure 2–30(a). Let us now use y as the map entered variable. The resultant map is shown in Figure 2–30(b). Finally, let us use z as the map entered variable. The resultant map is shown in Figure 2–30(c). All three maps in Figure 2–30 look very similar, further confirming the assertion stated earlier that any variable can be chosen to be used as the map entered variable.

There is really no reason why there should be only one map entered variable. The same function can be represented just as well using two map entered variables. If there are two map entered variables, then the order of the Karnaugh

y \ z	0	1
0	x	$\bar{x}$
1	$\bar{x}$	$x + \bar{x}$

(a) Using x as the map entered variable

x \ z	0	1
0	y	$\bar{y}$
1	$\bar{y} + y$	y

(b) Using y as the map entered variable

x \ y	0	1
0	z	$\bar{z}$
1	$\bar{z} + z$	z

(c) Using z as the map entered variable

Figure 2–30

Map for $f(x, y, z) = \bar{x}\bar{y}z + xyz + x\bar{y}\bar{z} + \bar{x}y\bar{z} + \bar{x}yz$ using the variable entered map method

maps can be reduced by 2. With two map entered variables, each map entry may represent up to four minterms. Using the same function as before, this time we will choose b and c as the map entered variables. We are now left with a one-variable Karnaugh map. We need to determine what goes into the map for the $a = 0$ condition and the $a = 1$ condition. For the $a = 0$ condition we have two minterms: $\bar{b}\bar{c}$ and bc. These two terms are then entered into the map. For the $a = 1$ condition we again have two minterms: $\bar{b}\bar{c}$ and $\bar{b}c$. These two terms are also entered into the map. The final form of the map with two map entered variables for this function is shown in Figure 2–29(c). Though this example is simple, it does illustrate the power of the variable entered map approach to reduce the final order of the maps in representing the given function.

The procedure for determining the map entries also applies to partially minimized functions. If a particular minterm is partially minimized, choosing the missing variables as map entered variables would be an excellent choice. Further, the missing variables can always be recovered by expanding the minterm to include the needed variables with an $(x + \bar{x})$ factor, where x is the desired variable. This is illustrated in Example 2-23.

Example 2–23 ▬▬▬▬▬▬▬▬▬▬▬▬▬▬▬▬▬▬▬▬▬▬

Represent the following function by the variable entered map approach:

$$f(x, y, z) = \bar{x}\bar{y} + x + x\bar{y}z + \bar{x}\bar{y}\bar{z}$$

Without expanding each term fully into its canonical representation, we observe immediately that either y or z could be used as map entered variables. We will choose z as the map entered variable in this case. Before the function can be used, we must make sure that each term contains specifications for x and y by supplying $(x + \bar{x})$ and $(y + \bar{y})$ to those terms with a missing x or y.

$$f(x, y, z) = \bar{x}\bar{y} + x(y + \bar{y}) + x\bar{y}z + \bar{x}\bar{y}\bar{z}$$

Now the function can be decomposed according to the map entered approach. When $x = 0$ and $y = 0$, $f = 1$ regardless (first term) and when $z = 0$ (last term). When $x = 0$ and $y = 1$, $f = 0$ regardless, because no minterms contain this particular condition. When $x = 1$ and $y = 0$, $f = 1$ regardless (second term) and when $z = 0$ (third term). Finally, when $x = 1$ and $y = 1$, $f = 0$ regardless (second term). The resultant map entries for this function are shown in Figure 2–31.

Minimization Procedures

Once the map entries are determined and the map is prepared, minimization of these maps in the map entered approach is quite similar to the Karnaugh map method. Instead of combining neighboring 1-cells or 0-cells as in the case of Karnaugh maps, we combine terms with the same map entries. If two adjacent cells have the same map entry, then these cells or parts of these cells can be combined to form larger groups. If there is more than one entry in a particular cell, the combination may include all entries or some of the entries. The same rules regarding the adjacency of corners and edges in the Karnaugh map also apply for the map entered approach.

Example 2–24

Minimize the following function using one map entered variable:

$$f(a, b, c) = \bar{a}\bar{b}\bar{c} + \bar{a}bc + a\bar{b}\bar{c} + a\bar{b}c$$

We have already analyzed this function and have constructed a two-variable map for this function. From Figure 2–29, we observe that the top two adjacent

Figure 2–31
Variable entered maps for a partially minimized function

x ╲ y	0	1
0	$1 + \bar{z}$	$1 + z$
1	0	1

$$f(x, y, z) = \bar{x}\bar{y} + x + x\bar{y}z + \bar{x}\bar{y}\bar{z}$$

cells have a common term, $\bar{c}$. Therefore they can be grouped together to form a simpler term. Since no other groupings are possible, we have the following as the final answer:

$$f(a, b, c) = \bar{b}\bar{c} + a\bar{b}c + \bar{a}bc$$

Note that for the $(a = 1, b = 0)$ cell, even though the larger grouping has taken care of one of the terms, the other product term in the cell must still be included in the final answer.

Example 2–25 ▰▰▰▰▰▰▰▰▰▰▰▰▰▰▰▰▰▰▰▰▰▰▰▰▰▰▰▰▰▰

Minimize the following function using one map entered variable:

$$f(a, b, c) = \bar{a}\bar{b}\bar{c} + \bar{a}bc + a\bar{b}\bar{c} + a\bar{b}c$$

For the sake of illustration and comparison, we will also show the minimization for the first function using the map constructed with two map entered variables. From Figure 2–29(c), we see that the common term is $\bar{b}\bar{c}$. These form a larger grouping. The final answer is the same as the one obtained above. The same common terms can be observed from the maps, resulting in the same solution regardless of how many and which map entered variables are chosen.

Example 2–26 ▰▰▰▰▰▰▰▰▰▰▰▰▰▰▰▰▰▰▰▰▰▰▰▰▰▰▰▰▰▰

Minimize the following function using one map entered variable:

$$f(x, y, z) = \bar{x}\bar{y}z + xyz + x\bar{y}\bar{z} + \bar{x}y\bar{z} + \bar{x}yz$$

The solution can be found from the maps obtained earlier. We will use z as the map entered variable as shown in Figure 2–30(c). The two bottom cells share a common term, z. The same common term is also shared by the two left cells. Therefore the minimal solution is

$$f(x, y, z) = \bar{x}z + yz + x\bar{y}\bar{z} + \bar{x}y\bar{z}$$

The reader should verify at this point that the same solution is obtained using any of the other variables as the map entered variable.

Minimization with Don't Cares

The map entered approach can also be applied effectively for incompletely specified functions. When don't cares are present, the same approach is taken in constructing the map entries. Those minterms with don't cares are also entered. In

w	x	y	z	f	Map Entry
0	0	0	0	0	
0	0	0	1	0	0
0	0	1	0	0	
0	0	1	1	1	z
0	1	0	0	1	
0	1	0	1	0	$\bar{z}$
0	1	1	0	1	
0	1	1	1	1	1
1	0	0	0	0	
1	0	0	1	x	zd
1	0	1	0	1	
1	0	1	1	x	$\bar{z} + zd$
1	1	0	0	x	
1	1	0	1	0	$\bar{z}d$
1	1	1	0	x	
1	1	1	1	x	$\bar{z}d + zd$

(a) Truth table representation

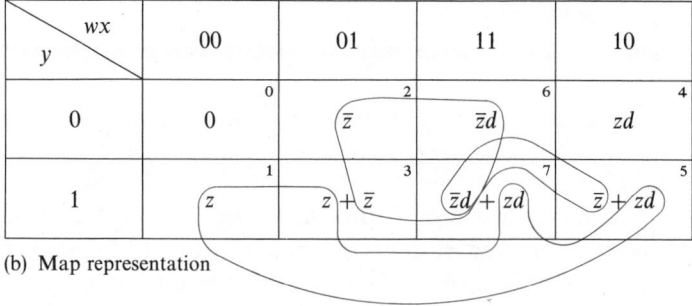

(b) Map representation

Figure 2–32
Map entered approach for incompletely specified functions

the minimization process, the entries with don't cares are used if needed and discarded if not used. Assume that we have the following function:

$$f(w, x, y, z) = \sum(3, 4, 6, 7, 10) + \sum_d(9, 11, 12, 14, 15)$$

The first step is to obtain the truth table for this function as shown in Figure 2–32(a). We arbitrarily choose z to be the map entered variable. The truth table is so arranged that the map entered variable is placed at the end as shown. The minterms are then paired and the map entries determined. A number of possibilities may arise. Let "d" be the don't care symbol. Since the function output may take on three values, 0, 1, and x (don't cares), there may be nine possibilities when pairing two minterms together:

$z = 0$	$z = 1$	Map Entry
0	0	0
0	1	z
0	x	zd
1	0	$\bar{z}$
1	1	1
1	x	$\bar{z} + zd$
x	0	$\bar{z}d$
x	1	$\bar{z}d + z$
x	x	$\bar{z}d + zd$

Once the map entries are determined, the map corresponding to the function can be constructed and is given in Figure 2–32(b). Minimization can now begin. In the grouping process, we need to pay attention only to the real minterms. Those with don't care symbols are used only when needed to complete a bigger grouping. From the figure, we see that cell 1 ($w, x, y = 0, 0, 1$) can be combined with cell 3 ($w, x, y = 0, 1, 1$), cell 5 ($w, x, y = 1, 0, 1$), and cell 7 ($w, x, y = 1, 1, 1$). In addition, cell 2 can be combined with cell 3, cell 6, and cell 7 to form another group. These two groups take care of all but one minterm in cell 5. Lastly, the remaining minterm in cell 5 can be combined with cell 7 to form another term. Hence the final minimized solution is

$$f(w, x, y, z) = yz + x\bar{z} + wy\bar{z}$$

The reader can verify that the same result is obtained if the problem is solved using the Karnaugh map method.

2.8

Application of Boolean Operations

In this section we will use the design techniques presented for Boolean operations on some useful applications. Specifically, we will examine the circuits for code conversions.

BCD to Excess-3 The conversion from BCD to the Excess-3 form can easily be done by defining the truth table and by minimizing the terms. Let a_3, a_2, a_1, a_0 represent the four BCD bits, and e_3, e_2, e_1, e_0 the Excess-3 bits. Assume that a_0

and e_0 are the least significant bits (LSBs). Then the following relations exist:

$$e_0 = \bar{a}_0$$
$$e_1 = a_0 a_1 + \bar{a}_0 \bar{a}_1$$
$$e_2 = a_0 \bar{a}_2 + a_1 \bar{a}_2 + \bar{a}_0 \bar{a}_1 a_2$$
$$e_3 = a_0 a_2 + a_1 a_2 + a_3$$

The above four formulas can be realized using standard SSI components. Their implementations will be left to the reader as an exercise.

Excess-3 to BCD   The conversion from Excess-3 back to BCD can be done in the same manner as the reverse conversion described above. Assume that a_0 and e_0 are the LSBs. Using the same notation, the following relations exist:

$$a_0 = \bar{e}_0$$
$$a_1 = e_0 \bar{e}_1 + \bar{e}_0 e_1$$
$$a_2 = \bar{e}_0 \bar{e}_2 + e_0 e_1 e_2 + \bar{e}_1 \bar{e}_2$$
$$a_3 = e_0 e_1 e_3 + e_2 e_3$$

These formulas can also be realized using standard SSI circuit components. Alternatively, the conversion from Excess-3 back to BCD can be accomplished by subtracting 3 from the BCD code.

BCD to Gray   The conversion from BCD to Gray can be done in similar fashion as before. Let the four binary bits of the Gray code be represented as g_3, g_2, g_1, and g_0, where g_0 is the least significant bit. Then the following relations exist:

$$g_0 = a_0 \oplus a_1$$
$$g_1 = a_1 \oplus a_2$$
$$g_2 = a_2 \oplus a_3$$
$$g_3 = a_3$$

The above solution uses the Exclusive OR operators. Canonical sum-of-products or product-of-sums form can also be used at the expense of a greater number of gates.

Gray to BCD   The conversion from Gray code to BCD can be accomplished as follows:

$$a_0 = g_0 \oplus g_1 \oplus g_2 \oplus g_3$$
$$a_1 = g_1 \oplus g_2 \oplus g_3$$
$$a_2 = g_2 \oplus g_3$$
$$a_3 = g_3$$

The above formulas for BCD-to-Gray and Gray-to-BCD conversions can be easily realized by a set of Exclusive OR gates. The above formulas can be further simplified if a_3 is used to generate a_2, and a_2 to generate a_1, etc., resulting in the following conversion formula:

$$a_i = g_i \oplus a_{i+1}$$

In contrast with the original formula, the last one uses fewer gates at the expense of increased timing due to delays through the $\oplus$ gates.

References

1. Ahmad, S. Imtiaz, and Kwok T. Fung. *Introduction to Computer Design and Implementation.* Computer Science Press, Rockville, Md., 1981.
2. Becher, William D. *Logical Design Using Integrated Circuits.* Hayden Book Company, Rochelle Park, N.J., 1977.
3. Birkhoff, G., and T. C. Bartee. *Modern Applied Algebra.* McGraw-Hill, New York, 1970.
4. Birkhoff, G., and S. Maclane, *A Survey of Modern Algebra,* 3d ed. Macmillan, New York, 1965.
5. Boole, G. *An Investigation of the Laws of Thought.* Dover, New York, 1954.
6. Booth, Taylor L. *Introduction to Computer Engineering Hardware and Software Design.* John Wiley & Sons, New York, 1984.
7. Carroll, Lewis. *The Complete Works.* Nonesuch Press, London, 1939.
8. Chu, Y. *Digital Computer Design Fundamentals.* McGraw-Hill, New York, 1962.
9. Dietmeyer, D. L. *Logic Design of Digital Systems,* 2d ed. Allyn & Bacon, Boston, 1978.
10. Fletcher, William I. *An Engineering Approach to Digital Design.* Prentice-Hall, Englewood Cliffs, N.J., 1980.
11. Greenfield, Joseph D. *Practical Digital Design Using ICs,* 2d ed. John Wiley & Sons, New York, 1983.
12. Hayes, John P. *Digital System Design and Microprocessors.* McGraw-Hill, New York, 1984.
13. Hill, Frederick J., and Gerald R. Peterson. *Introduction to Switching Theory and Logical Design,* 3d ed. John Wiley & Sons, New York, 1981.
14. Hill, F. J., and G. R. Peterson. *Digital Logic and Microprocessors.* John Wiley & Sons, New York, 1984.
15. Hohn, F. E. *Applied Boolean Algebra,* 2d ed. Macmillan, New York, 1966.
16. Humphrey, W. S., Jr. *Switching Circuits with Computer Applications.* McGraw-Hill, New York, 1958, chap. 4.
17. Huntington, E. V. Sets of Independent Postulates for the Algebra of Logic, *Trans. Am. Math. Soc.,* 5, 1904.
18. Karnaugh, M. A Map Method for Synthesis of Combinational Logic Circuits, *AIEE Trans. Commun. and Electron.,* 72, Part I, November 1953.
19. Kline, Raymond M. *Structured Digital Design Including MSI/LSI Components and Microprocessors.* Prentice-Hall, Englewood Cliffs, N.J., 1983.
20. Kohavi, Z. *Switching and Finite Automata Theory.* McGraw-Hill, New York, 1970.
21. Kostopoulos, G. K. *Digital Engineering.* John Wiley & Sons, New York, 1975.
22. Mano, M. Morris. *Digital Logic and Computer Design.* Prentice-Hall, Englewood Cliffs, N.J., 1979.
23. McCluskey, E. J., Jr. Minimization of Boolean Functions, *Bell System Tech. J.,* 35(6), November 1956.

24. McCluskey, E. J., Jr. *Introduction to the Theory of Switching Circuits.* McGraw-Hill, New York, 1965, chap. 4.
25. Paul, M. C., and S. H. Unger. Minimizing the Number of States in Incompletely Specified Sequential Switching Functions, *IRE Trans. Electron. Comput.*, EC-8(3): 356–357, September 1959.
26. Quine, W. V. The Problem of Simplifying Truth Functions, *Am. Math. Monthly*, 59(8), October 1952.
27. Rafiquzzaman, M. *Microcomputer Theory and Applications with the Intel-SDK-85.* John Wiley & Sons, New York, 1982.
28. Rhyne, V. T. *Fundamentals of Digital Systems Design.* Prentice-Hall, Englewood Cliffs, N.J., 1973.
29. Roth, Charles H., Jr. *Fundamentals of Logic Design*, 2d ed. West Publishing Co., St. Paul, Minn., 1979.
30. Shannon, C. E. A Symbolic Analysis of Relay and Switching Circuits, *Trans. AIEE*, 57, 1938.
31. Sheffer, H. M. A Set of Five Independent Postulates for Boolean Algebras, with Applications to Logical Constants, *Trans. Am. Math. Soc.*, 1913.
32. Stone, H. S. *Discrete Mathematical Structures and Applications.* Science Research Associates, Palo Alto, Calif., 1973.
33. Taub, Herbert. *Digital Circuits and Microprocessors.* McGraw-Hill, New York, 1982.
34. Tremblay, J. P., and R. Manochar. *Discrete Mathematical Structures and Applications to Computer Science.* McGraw-Hill, New York, 1975.
35. Whitesitt, J. E. *Boolean Algebra and Its Applications.* Addison-Wesley, Reading, Mass., 1961.
36. Wiatrowski, Claude A., and Charles H. House. *Logic Circuits and Microcomputer Systems.* McGraw-Hill, New York, 1980.

Problems

1. Examine the two tables below, which define the binary operators $\cdot$ and $+$ on a set of three elements $\{0, 1, a\}$.

$\cdot$	0	1	a		$+$	0	1	a
0	0	0	0		0	0	1	a
1	0	1	a		1	1	1	1
a	0	a	a		a	a	1	a

Show that this is not a Boolean algebra, and give all postulates that are not satisfied.

2. Simplify the following Boolean functions using the postulates and theorems given in the chapter.
   (a) $ab + \bar{a}b$
   (b) $\overline{xy + \bar{x}\bar{y}x}$
   (c) $(A + B)(\bar{A} + \bar{B})$
   (d) $wxy(\bar{x} + \bar{y} + w)$
   (e) $(a + b)(\bar{a} + c)(b + c)$

3. Show whether the following Boolean functions are equivalent or not by algebraic manipulation.

(a) $(w + \bar{x} + wx)(w + \bar{x})(\bar{w}x) = 0$

(b) $ab + ab + bcde = ab + ab$

(c) $wxyz + wxy\bar{z} + wx\bar{y}z + wx\bar{y}\bar{z} = wx$

4. Change the following Boolean functions into product-of-sums forms by algebraic manipulation.

(a) $\overline{(ab + \bar{a}\bar{c})}b$

(b) $\overline{(a + b + c)}\bar{a}c$

(c) $w\bar{x} + \bar{w}y + \bar{w}x$

5. Let the two functions F_1 and F_2 be specified in decimal notation as follows:

$$F_1(w, x, y, z) = \Sigma(0, 1, 2, 3, 7, 8, 11, 13)$$
$$F_2(w, x, y, z) = \Sigma(1, 4, 6, 7, 10, 13, 15)$$

Find $F_1 + F_2$, $F_1 \cdot F_2$, and $F_1 \oplus F_2$ in decimal notation.

6. Implement the following functions with gates in exactly the same structure as the expression.

(a) $\overline{(a\bar{b} + \bar{a}b)}bc$

(b) $\overline{wx + \bar{w}yz}$

(c) $\overline{\overline{abc}\overline{ab}\bar{c} + \bar{a}c}$

7. Find the Karnaugh maps for the following Boolean functions:

(a) $F(a, b, c) = \Sigma(0, 1, 2, 6, 7)$

(b) $F = wx + \bar{w}yz + \bar{y}z$

(c) $F(A, B, C, D) = \prod(1, 2, 3, 4, 5, 6, 9, 11, 13)$

(d) $F = (\overline{wx + y})(\overline{w\bar{y} + z})$

8. A complete gate set is usually defined as a set of gates that can implement any switching function. We have shown that any function can be implemented in a sum-of-products form. This implies that NOT, AND, and OR gates can implement any function and are a complete gate set. Which of the following are complete gate sets?

(a) NOT and AND

(c) NAND

(c) NOR and NAND

(d) AND and OR

(e) NOR

9. Show that

$$f(x_1, x_2, \ldots, x_n) = x_1 f(1, x_2, \ldots, x_n) + \bar{x}_1 f(0, x_2, \ldots, x_n)$$

for a switching function with two values, 0 and 1, for each x_1.

10. Find the minimum sum-of-products form using Karnaugh maps for the following functions:

(a) $F(a, b, c) = \Sigma(1, 2, 6, 7)$

(b) $f(w, x, y, z) = \Sigma(0, 2, 8, 10, 14, 15)$

(c) $F(A, B, C, D) = \Sigma(0, 1, 2, 3, 5, 7)$

(d) $f(x_1, x_2, x_3, x_4) = \prod(0, 1, 3, 7, 8, 11, 14, 15)$

11. Minimize the following functions with Karnaugh maps into a sum-of-products form:
    (a) $f(a, b, c) = \sum(1, 2, 3, 4) + \sum_d(6, 7)$
    (b) $F(w, x, y, z) = \sum(1, 3, 5, 7, 8, 9, 11) + \sum_d(12, 13, 15)$
    (c) $f(A, B, C, D) = \prod(0, 15)$

12. Use Karnaugh maps to find the minimal product-of-sums form.
    (a) $f(a, b, c, d) = \sum(0, 1, 2, 3, 5, 8, 10, 11, 13, 15)$
    (b) $F(w, x, y) = \sum(0, 7)$
    (c) $F(W, X, Y, Z) = \prod(1, 2, 3, 5, 7, 8, 10, 12, 15)$

13. Find the minimal two-level realization of each of the following functions:
    (a) $f(a, b, c, d) = \sum(5, 6, 9, 10) + \sum_d(0, 1, 2, 3)$
    (b) $f(w, x, y, z) = \sum(1, 2, 3, 5, 7, 12, 13) + \sum_d(4, 14, 15)$

14. Implement the minimal sum-of-products form found in Problem 10.

15. There are 2 to the power of $2^2 = 16$ two-input switching functions. List them in truth table form. Use a separate column for each function. Find a sum-of-products realization for each function.

16. Design a sum-of-products circuit that converts the BCD code to Excess-3 code. The code table is given in Chapter 1.

17. A full adder is a circuit that has three inputs, A_i, B_i, and C_{i-1}, and two outputs, S_i and C_i. S_i represents the sum of the three inputs and C_i represents the carry. S_i and C_i. S_i represents the sum of the three inputs and C_1 represents the carry. $S_i = 1$ if and only if an odd number of the inputs are 1. $C_i = 1$ if and only if two or three inputs are 1. Design a minimal sum-of-products form and a minimal product-of-sums form for the full adder.

18. A majority gate is a gate whose output is 1 if a majority of the inputs are 1. Design a two-, three-, and four-input majority gate in a sum-of-products form. Assume that a strict majority is needed.

19. Design a full adder from Exclusive OR and majority gates only.

20. Prove the commutative, associative, and distributive properties of the XOR and XNOR functions.

21. Prove the properties of the XOR/XNOR functions given in the text.

22. Given the two sets of formulas for converting between Gray and binary codes, calculate the timing delay in each case assuming that the timing delay through one $\oplus$ gate is d.

23. Repeat Problem 7 using the variable entered map approach.

24. Repeat Problem 10 using the variable entered map approach.

25. Repeat Problem 11 using the variable entered map approach.

26. Repeat Problem 12 using the variable entered map approach.

27. Repeat Problem 13 using the variable entered map approach.

Design Techniques with Combinational Circuits

3

In this chapter we first review the practical realization of Boolean operations. Discrepancies naturally occur in the transition from the theoretical realm to the real world. These differences are discussed. Then the different types of digital logic and their characteristics are presented showing the varieties available. Simple SSI logic chips are next introduced and described. Following this, the design techniques of basic combinational circuits using SSI components are discussed in detail. In many designs, mixed logic is often used. Their design techniques are next presented. The last section of this chapter briefly describes the operating characteristics of SSI components.

In the preceding chapters we have covered the fundamental principles of Boolean algebra and have shown how, given a problem, a Boolean function can be defined and represented by logical variables. However, this is only half of the job. The end product, of course, is an actual circuit that behaves exactly as specified by the given problem. In this chapter we will go further in our design process and look at the components that can be used to realize a Boolean function. Realizing a Boolean function means that digital hardware is connected in a systematic manner so that the output of the hardware is in a one-to-one correspondence with the desired function output. We will address the procedures for choosing the proper components so that one can realize or implement the solution of a digital problem. This is a practical digital design problem.

As we move from the ideal world of Boolean algebra to the real world with actual components, one immediate challenge is that things are not perfect in the real world. Therefore, intelligent choices and trade-offs have to be considered. Compromises are inevitable. A unique solution may not exist; rather, a family of solutions exists, each differing from the other according to the various economic and physical constraints imposed on the design. These limitations must be carefully considered and accounted for in order to have a successful design.

3.1

Representation of Boolean Parameters

Realization of Boolean functions involves finding a way to represent each of the Boolean parameters covered in previous chapters, including Boolean constants, variables, operations, and functions. One of the reasons that Boolean algebra is so popular and widely used is the fact that it can be easily realized. Since each Boolean constant or variable can have one of two logical states or levels, it can easily be represented by one of two corresponding physical states. For example, logical state one/zero can be realized by an on/off switch, an opened/closed door, a left/right position, a full/empty condition, a positive/negative value, a have/have not test, or a high/low level. In terms of electrical signals, typical implementations of logical state zero/one in positive logic are low/high voltages, presence/absence of current, or positive/negative direction of current flow. Boolean operations with these logical states can be realized by various circuit arrangements with semiconductor components such as diodes, transistors, and integrated circuit (IC) chips.

With the technology available today, Boolean parameters are normally realized electrically in positive logic as follows:

Constants   0 = low voltage level; 1 = high voltage level
Variables   Voltage level at given connection points

| Operations | Special circuit arrangements called gates |
| Functions | Voltage level at the output of some arrangements of gates |

The value of a logical constant is represented by either a low (ground) or high (supply, source) voltage level. The value of a logical variable is then given by the voltage level at any connection point that is the output of a particular circuit arrangement. Any such circuit arrangement that would respond to several inputs with different input voltage levels and produce a specified output voltage level in accordance with the elementary Boolean operation is called a *gate*. Specifically, a circuit that functions as a logical AND operation is called an AND gate, and one that functions as a logical OR operation is called an OR gate. A logical function can easily be represented by different arrangements of gates that form the basic relation or operation between various logical variables.

As mentioned before, the transition from the ideal world to the real world requires compromise. Many assumptions that were taken for granted or ignored must now be individually accounted for. Some of these considerations include varying voltage levels, existence of propagation delays, limited input and output capacities, and nonzero power consumption. Since voltage levels are used to represent the two logical states, a range of voltage values must be specified because voltage levels are never discrete. Noise margins must be established by separating the two voltage levels for noise immunity. Propagation delay refers to a finite amount of time required for the circuit's output to change states in response to a change in the input. Furthermore, the output waveform can easily be distorted when compared with the original input. Limited provisions for input and output capacities also impose a restriction: the output from a particular gate cannot be connected indiscriminantly to a large number of inputs without the risk of violating the specifications of other circuit parameters. In the following sections, these and many other practical restrictions that are examined in detail must be accounted for when using real logic circuits.

3.2
Integrated Circuit Characteristics

In the design and fabrication of IC chips, a transistor is the basic element. When these transistors are combined in such a way that the circuit behaves according to an elementary logical operation, the circuit is called a gate. Usually, for ease of handling and for convenience of packaging, several gates are combined together in a protective covering. The total package containing a number of gates is then called a *chip*. The chips are set in a board (called a printed circuit board or PC board) with interconnecting signal lines imprinted on it for communication among the various IC chips on the board. To form a system, these circuit boards are connected through a set of sockets to a *motherboard*, which serves as a communication point for the different boards. Sometimes the motherboard is called the *backplane*.

General Categories of Integrated Circuits

There are primarily two categories of ICs, analog and digital. There are also some special ICs that include both types of components, such as interface ICs and hybrid ICs. Each category of IC chip has its own characteristics and area of application.

Analog or *linear components* were among the first to be made into integrated circuits with the invention of the planar transistor, which is a transistor fabricated on a planar silicon substrate. Analog/linear ICs are characterized by the input and output of the IC being primarily analog signals. Analog/linear ICs have been designed for many applications. Some of the more popular ones include transistor and diode arrays, operational amplifiers, instrumentation amplifiers, radio-frequency amplifiers, video frequency amplifiers, voltage references, voltage regulators, analog comparators, analog multiplexers, motor drivers, and telecommunication circuits such as CODECS and tone converters. Detailed design procedures using these circuit elements are governed primarily by the specific application at hand and are outside the scope of this book.

In contrast to analog/linear ICs, *digital ICs* refer to those ICs with digital inputs and outputs. This group of circuits is most useful in dealing with binary-type computations. Within this group, there are chips to perform basic logic gate functions, arithmetic functions, data encoders, data decoders, digital multiplexers, digital demultiplexers, flip-flops, and latches. More complex chips include counters, shift registers, logic arrays, etc. The most sophisticated chips include microprocessors, memory chips, and peripheral controllers. The remainder of this book will be concerned largely with design techniques using chips in this category. One characteristic of digital ICs is the use of a clock signal to synchronize binary data operations, regardless of whether they are data input operations, data output operations, or internal data manipulations.

Interface ICs include those ICs that perform conversion or provide interaction between analog and digital components. It is not uncommon for both analog and digital signals to be present in the input and output set for this group of chips. Some typical examples of chips in this group are analog switches with digital control, analog-to-digital converters, digital-to-analog converters, display drivers, peripheral drivers, sense amplifiers, level translators, and level converters. Some of the circuits in this group will be studied in this book as they apply to designs that require an interface between the digital circuits and the real world.

Strictly speaking, *hybrid ICs* are not a different group of ICs, but a special subgroup of interface ICs. Hybrid ICs earn their name by having digital, analog, and possibly discrete components bonded and packaged together on one chip to meet the requirements of a particular design. Typical hybrid ICs are commonly found in analog-to-digital (A/D) converters and telecommunications. For example, a sophisticated A/D converter may have an analog sample-to-hold circuit, a converter circuit that is an interface circuit, and a digital microprocessor bus control logic all packaged together. The advantages of such a bundled package are reduction in chip count, noise interference immunity, and reduction in interconnection costs.

Custom ICs are special ICs designed for a particular customer to perform a specific function. In general, they are not available to the public. They are custom made to fit a specific requirement. For those who can afford them and justify their use because of the large quantity required, the advantages of using custom ICs are reduction in component count and protection of proprietary designs. Many custom ICs make use of logic arrays, to be discussed in Chapter 6.

Types of Integrated Circuits by Technology

In fabricating an IC chip, several technologies are available, each tailored to a specific working environment. The major considerations for using chips fabricated from different technologies are speed, power consumption, cost, and availability. Some of the major types are

TTL	Transistor-to-transistor logic
ECL	Emitter-coupled logic
MOS	Metal oxide semiconductor

These three are by no means the only technologies available, but they are the most popular and available. A comparison of the electrical characteristics of these three groups is given later.

Among these three types, TTL chips are the most popular because they were among the first ones available. Because of their widespread use, TTL chips are least expensive and most accessible compared with chips from the other technologies. TTL chips, which require a single 5-V power supply, are the most suitable for many applications.

Chips made by the ECL technology are used primarily in high-frequency applications because they have short propagation delays. Typical ECL chips have delays as short as 2 ns compared to 10 ns for TTL chips. It is not unusual to find ECL chips that operate at a frequency greater than 50 MHz. Because the ECL chips operate at considerably higher frequencies, in general these chips consume much more power than TTL chips.

Chips made by the MOS technology are particularly suited for low-power conditions and can operate with a variety of power supply voltages ranging from less than 5 V to 15 V. MOS chips are the slowest of the three technologies, with typical propagation delays of 20 ns. However, the advantage of MOS chips is their extremely low power consumption. There are various MOS technologies, such as CMOS for complementary MOS, NMOS, or HMOS. A typical CMOS chip consumes only one-tenth of a milliwatt when active and five thousandths of a milliwatt when on standby, compared to 10 mW for TTL chips and 20 mW for ECL chips. This extremely low power consumption makes CMOS chips very attractive for battery-operated applications.

Categories of Integrated Circuits by Complexity

In addition to differentiating IC chips by technology, complexity is another measure that can be used to group IC chips. Complexity is usually measured in terms of the number of equivalent gates involved in making a particular IC. The *number of equivalent gates* is that number of discrete elementary two-input logic gates required to achieve the same function as a particular IC. Though the number of gates or transistors is not always a good estimate of complexity, it is nevertheless a convenient and readily available measure. Basically, there are four categories:

1. SSI   Small-scale integration, with fewer than 12 gates on a chip
2. MSI   Medium-scale integration, with fewer than 100 gates on a chip
3. LSI   Large-scale integration, with fewer than 1,000 gates on a chip
4. VLSI   Very large scale integration, where the limit on the number of gates on a chip is still undefined

Some representative samples of these four categories and their usages are shown in Figure 3–1.

Figure 3–1

Characteristics of various IC categories according to circuit complexity

IC Category	Number of Equivalent Gates	Complexity	Examples
SSI	< 12	Basic gates	AND, NAND, NOR, etc.
		Two-level logic	AND-OR-INVERT gates
MSI	< 99	Basic gates	Exclusive OR/NOR
		Submodules	4-bit adders, comparators, encoders
LSI	< 999	Functional modules	Registers, FIFO stacks, scratchpad memories
VLSI	> 1,000	Major building blocks	Microprocessors, memory chips, peripheral controllers

Integrated circuit chips in the SSI category are usually a collection of basic gates performing elementary operations. The full range of IC chips in this category includes elementary Boolean operations such as the AND, OR, NOT, NAND, and NOR operations with different numbers of input variables. Gates that implement two-level logic such as AND-OR gates and AND-OR-INVERT gates are also available. SSI chips are useful mostly for implementing logic controls of signals, data, and even control lines themselves. For example, an AND gate is often used to "gate," i.e., to control a signal by passing or not passing its value. Control lines and data lines are used for time synchronization.

Chips in the MSI category are more complex and perform more popular and well-known functions such as counting, addition, and data decoding. These chips can be considered as basic submodules or building blocks of a digital system. For example, counters are often used in conjunction with the system clock to schedule and to synchronize events. Data decoders provide conversion between one form of data and another, e.g., from ten decimal lines to four 8421 BCD lines.

LSI chips are standard building blocks with rather complex and specific functions or utilities. Examples of LSI chips include memory chips, scratchpad memories, first-in, first-out (FIFO) arrays, and register arrays. In general, LSI chips form the basic modules of a digital system. Functionally, LSI chips differ from MSI chips in that there are many more equivalent gates in LSI chips.

VLSI chips are the major building blocks of a digital system and are often used as the centerpieces in the design. Chips in this category include microprocessors, peripheral processors, and peripheral controllers. VLSI chips seldom stand alone but are often supported by other chips. They are used in many applications for control operations that require a certain amount of intelligence and sophistication. Many hand-held toys are made with a single VLSI chip that contains the program logic coupled with only a small number of other interface chips.

Nomenclature

In order to identify the different types of chips, a common identification scheme consists of five parts—the manufacturer's code, the product line, the performance class, the logic family, and the package type. For TTL a typical code may appear as follows:

Manufacturer Code	Product Line	Performance Class	Logic Family	Package Type
SN		ALS	02	J
DM	54	L	04	N
MC	74	S	190	T
MM		LS	290	W

The first two or three alphabetic characters of the identification code specify the manufacturer and its product line. Some examples are

SN   Texas Instruments TTL products
DM   National Semiconductor
MC   Motorola
MM   Monolithic Memories

The second part of the identification code is a two-digit number that specifies the product line of the chip. Some manufacturers also use other alphanumeric codes to indicate various grades such as temperature range. For example, some typical codes are

54   Military grade of the TTL product line
74   Commercial grade of the TTL product line

The next part of the identification code specifies the performance class and is usually related to parameters such as speed and power:

(blank)   Regular
H   High-speed
L   Low-power
S   Schottky
LS   Low-power Schottky
AS   Advanced Schottky
ALS   Advanced low-power Schottky

Characteristics of the different performance classes are discussed in more detail later in this section.

The next part of the TTL identification is a numeric code that is unique for each kind of chip. This code identifies the type and number of gates in a particular chip. This code is often the same among all IC chip manufacturers to avoid confusion. For example:

02   Quad two-input NOR
04   Hex inverter
190   Synchronous up/down BCD counter
290   Decade counter

The last part of the TTL identification code specifies the package size and type. For example:

J   Ceramic dual in-line package (DIP)
N   Plastic DIP
T   Flat package
W   Ceramic flat package

This nomenclature is adopted by most chip manufacturers with minor variations. This code serves as a means of identifying the contents of a chip and provides all the necessary information for the user to find a suitable replacement for a particular chip when the need arises.

3.3

SSI Circuit Components

In the last section we saw how logical variables and operations can be represented by physical entities such as voltage levels and gates. In this section we take a brief look at the available circuit components in the SSI category. These SSI components are the basic building blocks for circuit realization. Because of manufacturing costs and other logistics problems, there are several restrictions regarding the availability of these SSI components. First, it is impractical to have an infinite variety of choices for SSI chips with all ranges of fan-in and fan-out. Second, the cost of the chips depends directly on the production volume; that is, the higher the production volume, the lower the cost per chip. Furthermore, conforming to standard sizes and standard numbers of pins often proves helpful in circuit board packaging and layout. Since every input and output terminal of a gate must be accessible outside the chip for connections, these input and output connections are made accessible by providing pins on the chip called *connection* or *terminal pins*. Therefore the number of gates that can be packed into a chip is related directly to and largely dependent on the number of available pins on a chip. These constraints often decide what gates are to be fabricated on a chip.

There are various families of SSI components. These can be grouped roughly into three categories:

1. Basic gates
2. Combination gates
3. Buffers and drivers

The first group consists of basic gates such as AND gates, OR gates, etc. The second group is composed of chips with two-level logic gates and chips with other simple combinations of gates. The last group consists of different kinds of buffers and drivers used for special purposes, such as bus drivers, peripheral drivers, and signal conversions. The summary table in Figure 3–2 shows the different types of chips and their corresponding identification number.

Basic Gates

The majority of SSI chips are basic gates such as NANDs, ANDs, NORs, ORs, and inverters. Each gate differs from the others in its function and fan-in provision. Figure 3–2 shows the most popular types of chips available according to gate function, fan-in provision, and output type. It is interesting to note that there is a wide selection of NAND gates and a somewhat narrower choice for

		Number of Inputs	Totem Pole	Open Collector	Tri-state
Basic gates	NAND	2 3 4 8 12 13	00/26/37 10 13/20/40 30 133	01/38 12 22	134
	AND	2 3 4	08 11 21	09 15	
	NOR	2 3 4 5	02/28 27 23, 25 260	03/33	
	OR	2	32		
	Inverters	1	04/14	05/06/16	240
Two-level combination gates	AOI	2-2 2-2-2-2 4-4 4-2-3-2	50/51 53/54 55 64	65	
	AO	2-2-2-3	52		
	Expanders	2-3-3-2 3 4	62 61 60		
Drivers	Drivers	1 2 4	 128 140	07/17	125/126/241
	Transceivers	1			242/243/245

Figure 3–2

A sampling of typical SSI circuit components

the others. This is because transistors used in making these IC chips inherently act as inverters. Hence, an AND gate is actually more complex in design than a NAND gate because the former has an extra inversion stage while the latter does not. Note also that in keeping the number of input/output connection pins the same, the number of gates in a chip decreases as the fan-in increases.

Example 3–1 ▉▉▉   ▉▉▉

Assume an application where the output equation has been obtained by any of the methods discussed in Chapter 2 as follows:

$$f(w, x, y) = \bar{x}y + \bar{w}x\bar{y} + w\bar{x}$$

Realize the function f using standard SSI components.

A straightforward realization of f is shown in Figure 3–3(a). The circuit realization is obtained by first determining the necessary kinds of gates and then by choosing chips from Figure 3–2 that have the proper fan-in as required by the function. For this particular example, the requirement is as follows:

Inverters		3
AND gates:	two-input	2
	three-input	1
OR gates:	three-input	1

We can readily see that we need one 74ALS04 for the three inverters, one 74ALS08 and one 74ALS11 for the AND gates, and one 74ALS32 for the OR gates. Only one 74ALS04 and one 74ALS08 are needed, because there are six inverters in the 74ALS04 chip and four two-input AND gates in the 74ALS08 chip. Once we have selected the chips, the circuit can be finalized by drawing the wiring diagram shown in Figure 3–3(a).

Since manufacturing costs are related directly to the number of chips used, a further step in the minimization problem when applied to circuit realization is to develop a design that realizes the required functions with not just the minimal number of gates, but more important, with the minimal number of chips. A systematic solution to this minimization problem is still not known, but there are several heuristic rules that can provide some shortcuts. Three of these rules will be introduced here. These are

1. Fan-in substitution
2. Gate substitution
3. Form substitution

When choosing the set of gates, it should be kept in mind that a gate with more inputs can always be used to realize the same operation as one requiring fewer inputs, provided that the unused inputs are terminated properly. This is called *fan-in substitution*. For AND gates, the unused inputs should be connected to a logical 1 (supply voltage). For OR gates, the unused inputs should be connected to a logical 0 (ground or low voltage). This is shown in Figure 3–4. Stray or unconnected inputs can cause strange results and should never be ignored in a proper design. With this principle in mind, the above example can be redone

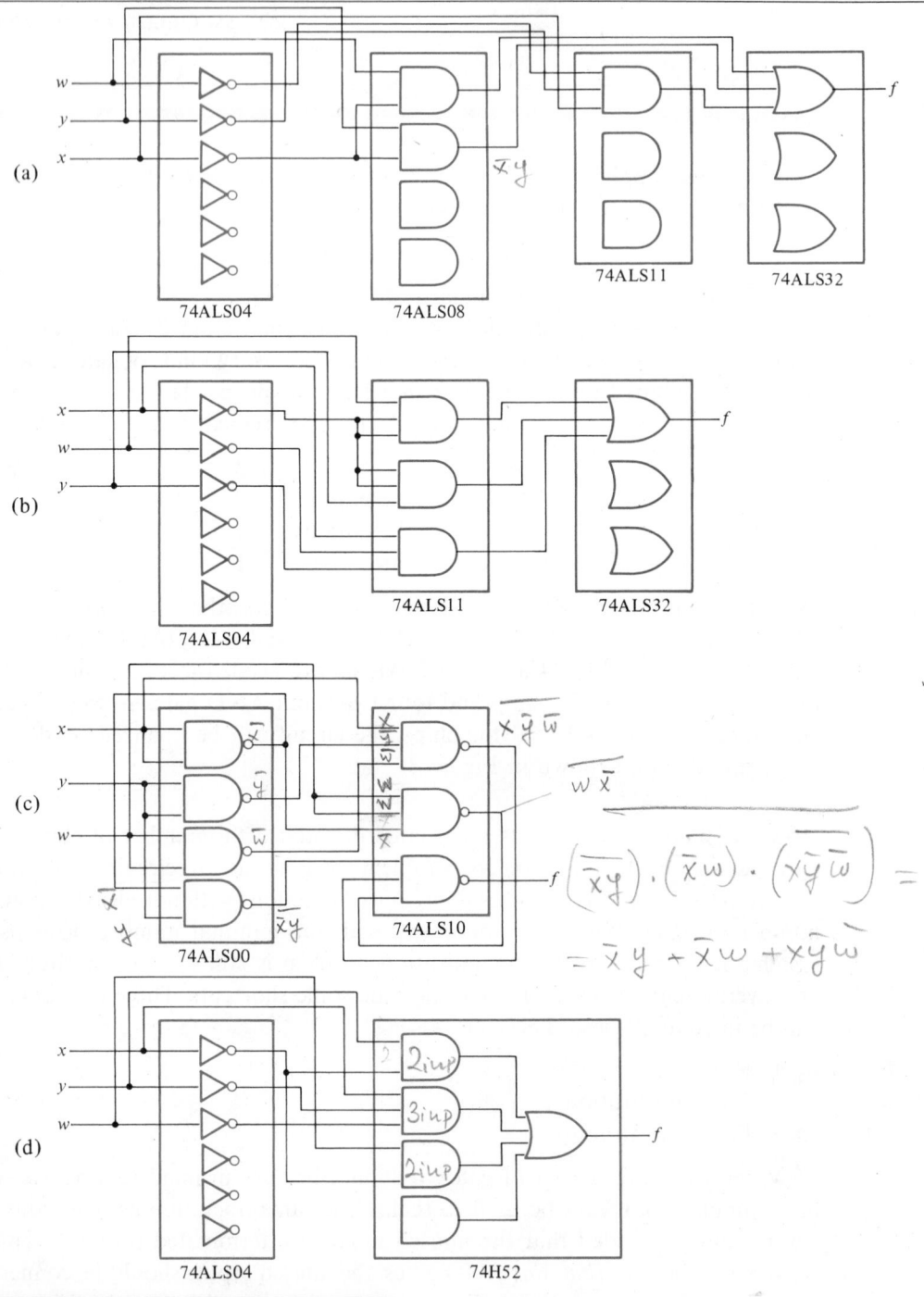

Figure 3–3

Four realizations of $f(w, x, y) = \bar{x}y + \bar{w}x\bar{y} + w\bar{x}$ (a) using no substitution,
(b) using fan-in substitution (c) using gate and form substitution, (d) using
a combination gate

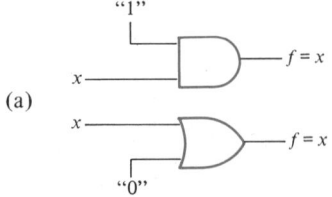

(a)

(a) Two-input gate to be used with one input

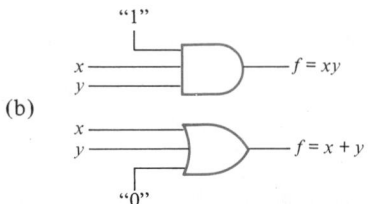

(b)

(b) Three-input gate to be used with two inputs

Figure 3–4
Connection of unused gates for fan-in substitution

by eliminating the two-input AND and performing those operations with the three-input AND gate. This is shown in Figure 3–3(b). It is readily seen that Figure 3–3(b) has one less chip and significantly fewer unused gates than the original design shown in Figure 3–3(a).

The second rule for function minimization is to use another type of gate to perform the same required functions. This is called *gate substitution*. For example, NAND and NOR gates can be used in such a way that they function as inverters when the inputs are properly connected. This is done either by tying all the inputs together or, alternatively, by tying unused inputs of the NAND (or NOR) gates to logical 1 (or 0) to reduce the fan-out requirement of the driving output gate as shown in Figure 3–5.

A more complex form of gate substitution involves changing the functional representation of the solution from one form to another. For example, changing AND-OR circuits to NAND-NAND form is quite frequently done so that only one type of chip, the NAND chip, is needed. With this in mind, the above example in AND-OR format is changed to the NAND-NAND format as follows:

$$f(w, x, y) = (\bar{x} \uparrow y) \uparrow (\bar{w} \uparrow x \uparrow \bar{y}) \uparrow (w \uparrow \bar{x})$$

and we observe immediately that even the inverters can be done away with by using the extra NAND gates as inverters, resulting in further reduction in the total number of chips required. This design is also shown in Figure 3–3(c).

It should be understood that applying the various methods for minimizing the number of chips discussed above will not always reduce the number of chips

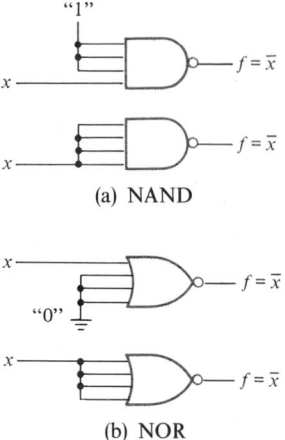

Figure 3–5

Gate substitutions using (a) NANDs and (b) NORs for inverters

required, i.e., the chip count. However, as demonstrated so vividly in this example, proper application of these principles many times can reduce the chip count in general to an acceptable level.

Combination Gates

Combination gates provide two levels of circuit logic. They are extremely useful because more than one kind of gate is fabricated on the same chip. For example, the 74H52 [shown in Figure 3–6(a)] is commonly called a 2-2-2-3 AND-OR (AO) chip because it has three two-input AND gates and one three-input AND gate with all the outputs of the four AND gates connected to an OR gate. Another example is the 74S64 shown in Figure 3–6(b), which is commonly known as the 4-2-3-2 AND-OR-INVERT (AOI) chip. This particular chip has a four-input, a two-input, a three-input, and a two-input AND gate all connected to a NOR gate. A sample list of these chips was given in Figure 3–2. Among the chips in this group, the AO and AOI gates are useful for realizing functions in the sum-of-products form. Using an AO chip with a set of inverters, the function f in Example 3–1 can now be readily realized using two chips as shown in Figure 3–3(d).

In order to retain flexibility, some of the AO and AOI gates are equipped with expandable inputs that can accept special outputs from chips called expanders. An *expandable gate* is equipped with a special input that is connected directly to its output section. An expander is a special gate whose input is TTL compatible but not its output. It is specially designed to mate with the special input on the expandable gate. Use of expanders provides a limited amount of

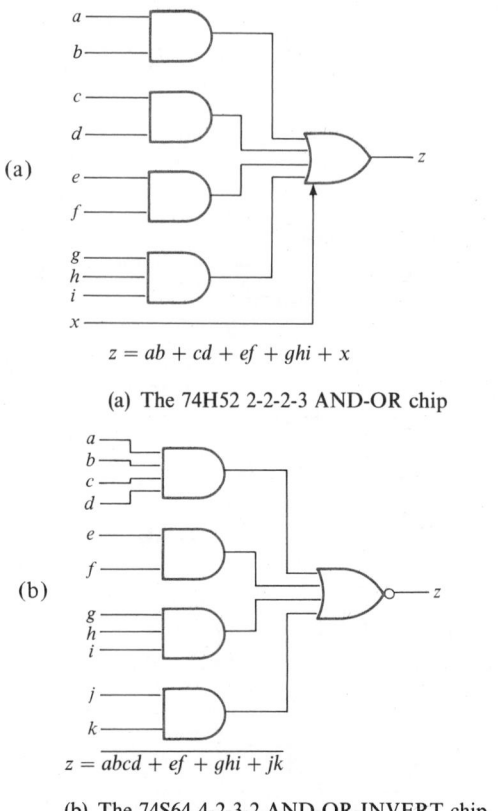

$$z = ab + cd + ef + ghi + x$$

(a) The 74H52 2-2-2-3 AND-OR chip

$$z = \overline{abcd + ef + ghi + jk}$$

(b) The 74S64 4-2-3-2 AND-OR-INVERT chip

Figure 3–6

Equivalent logic diagram of (a) the 74H52 2-2-2-3 AND-OR chip and (b) the 74S64 4-2-3-2 AND-OR-INVERT chip (Reprinted by permission of Texas Instruments)

customization so that existing AO and AOI gates can be applied to a wide variety of environments. Figure 3–7 shows two typical AND-OR functions achieved by combining the expander 74H62 with two different expandable gates—74H53 and 74LS55. It is evident from Figure 3–7 that a great variety of functions can be realized with these two combinations.

Example 3–2

Realize the following function $g(w, x, y, z)$:

$$g(w, x, y, z) = \overline{w}\overline{x}\,\overline{y}z + xz + wx\overline{y}z + \overline{x}y\overline{z}$$

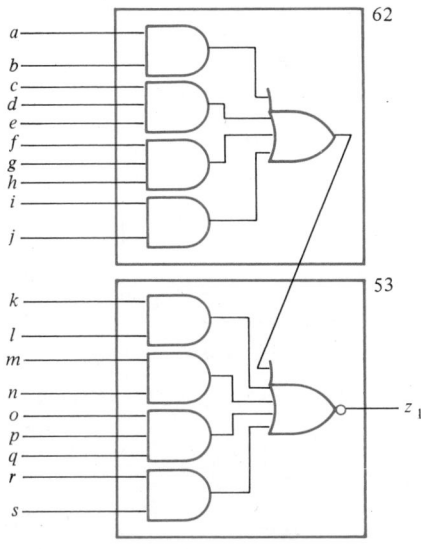

(a) 74H53 with 74H62

$$z_1 = \overline{ab + cde + fgh + ij + kl + mn + opq + rs}$$

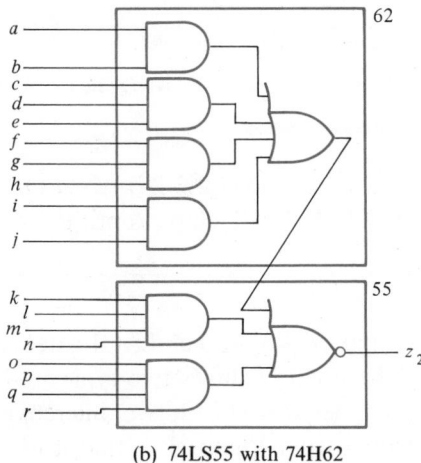

(b) 74LS55 with 74H62

$$z_2 = \overline{ab + cde + fgh + ij + klmn + opqr}$$

Figure 3–7
Use of expanders in expandable gates (a) 74H53 with 74H62; (b) 74LS55 with 74H62

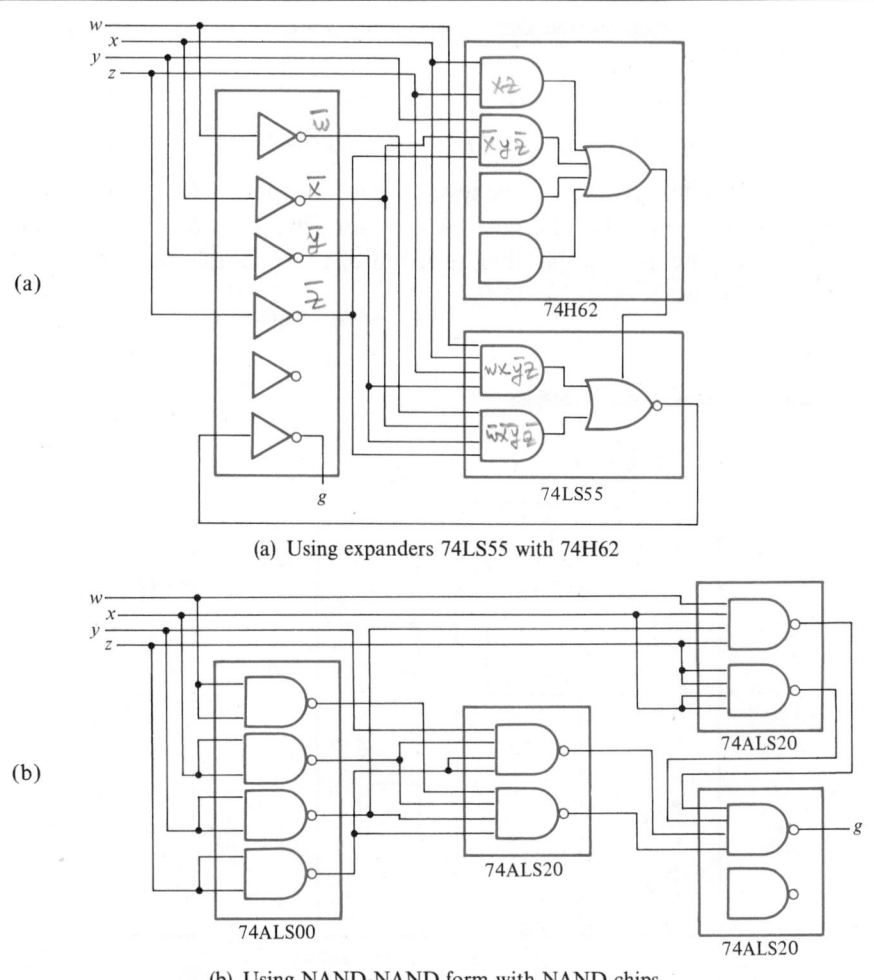

(a) Using expanders 74LS55 with 74H62

(b) Using NAND-NAND form with NAND chips

Figure 3–8

Realization of $g = \bar{w}\bar{x}\bar{y}z + xz + wx\bar{y}z + \bar{x}y\bar{z}$ (a) using expander 74LS55 with 74H62; (b) using NAND-NAND form

A quick look at the function $g(w, x, y, z)$ reveals that the first product term will not fit any one of the available gate configurations because a four-input AND gate is required. An expander is ideal for the job in this situation. Using an expander, the function can immediately be realized with three chips: an inverter 74ALS04, an expander 74H62, and an expandable gate 74LS55 as shown in Figure 3–8(a). If an expander is not available, then the NAND-NAND form gives the best realization, requiring four chips as shown in Figure 3–8(b).

Example 3–3

Realize the following function:

$$h(a, b, c, d) = \bar{a} + \bar{b}\bar{c}$$

This function gives a 1 whenever the 4-bit input (a, b, c, d) represents a true decimal digit and a 0 otherwise. A straightforward solution of this function is given in Figure 3–9(a). Another solution of this example using NANDs only is given in Figure 3–9(b).

Figure 3–9

Realization of $h(a, b, c, d) = \bar{a} + \bar{b}\bar{c}$

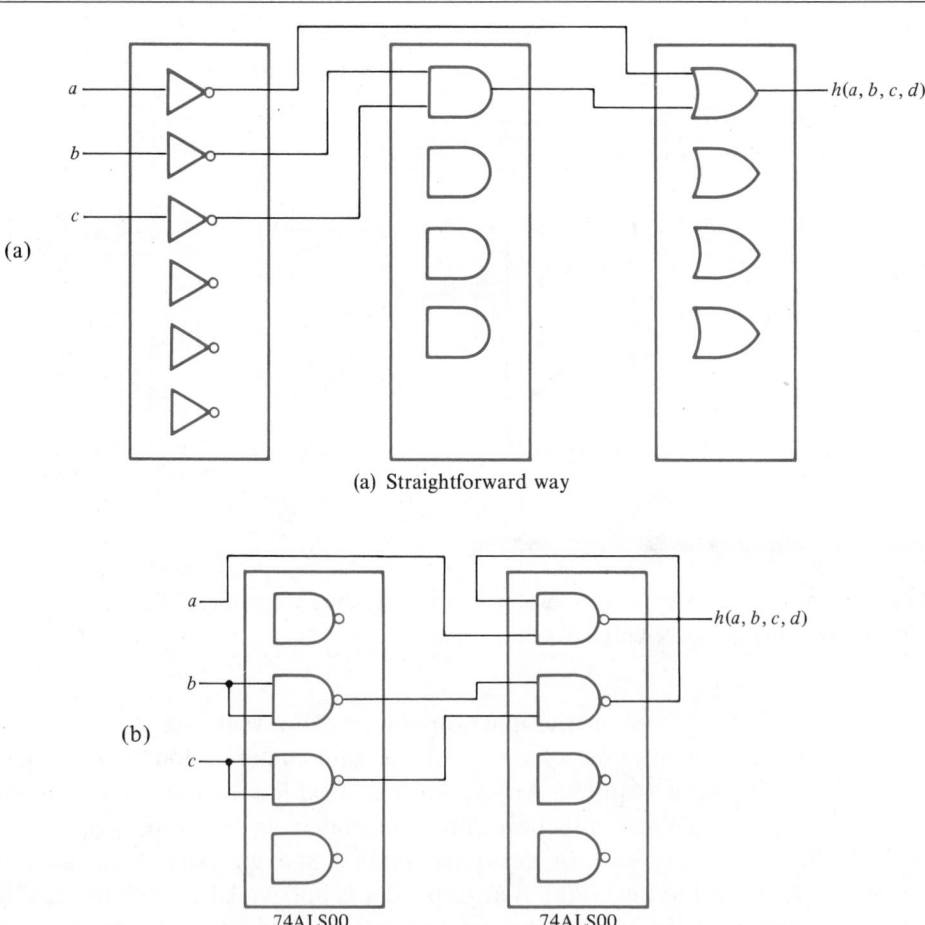

(a)

(a) Straightforward way

(b)

74ALS00   74ALS00

(b) NAND-NAND form

Example 3–4 ▰▰▰▰▰▰▰▰▰▰▰▰▰▰▰▰▰▰▰▰▰▰▰▰▰▰▰

Realize the following function:

$$F(x, y, z) = \bar{x}y\bar{z} + x\bar{y}z + \bar{x}\,\bar{y}\bar{z}$$

This function can be realized in a number of ways. Assume that we want to use only one type of gate, i.e., the NAND gate. Further assume that only three-input NAND gates (SN74ALS11) are to be used. A straightforward solution of this function is given in Figure 3–10(a). For comparison, a solution of this example using only three-input NANDs is shown in Figure 3–10(b).

Figure 3–10

Realization of $F(x, y, z) = \bar{x}y\bar{z} + x\bar{y}z + \bar{x}\,\bar{y}\bar{z}$

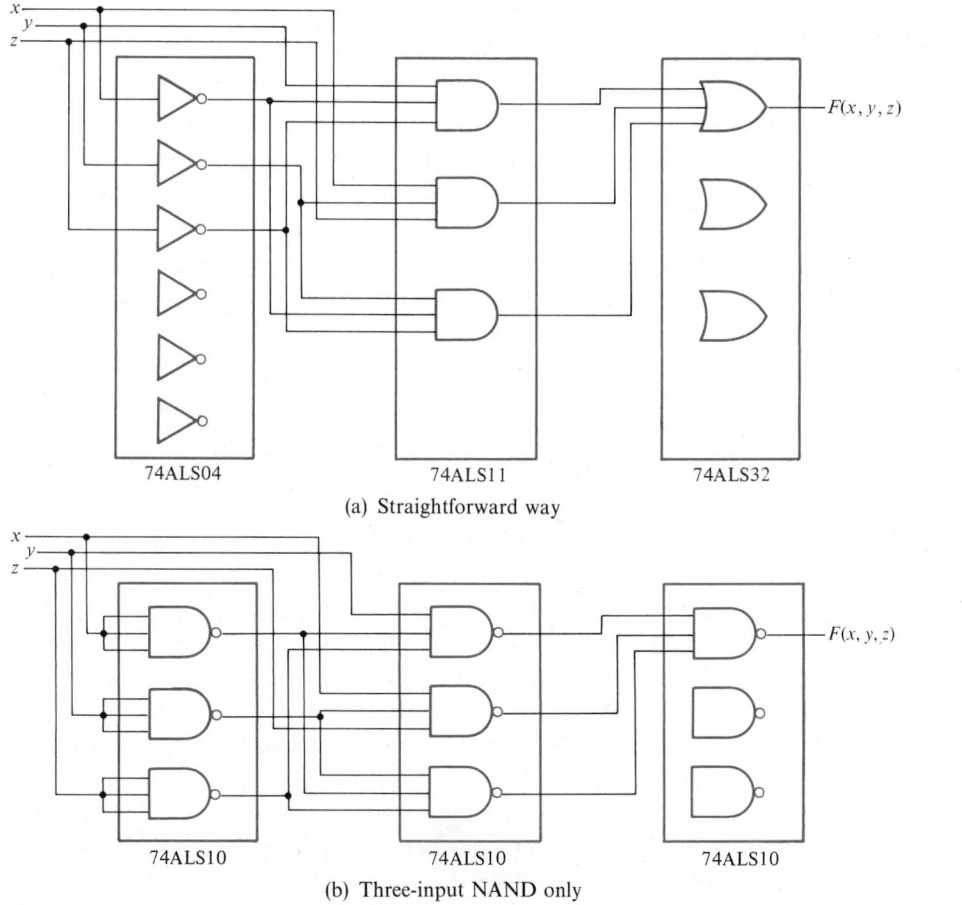

(a) Straightforward way

(b) Three-input NAND only

Buffers/Drivers

Buffers/drivers consist mainly of inverters and buffers with inverting or noninverting outputs. These chips have special output driving circuits that enable their output to be used in a variety of ways. Some of the more common ones include:

1. Buffers
2. Drivers
3. High-voltage drivers
4. Line drivers
5. Bus drivers
6. Line receivers
7. Transceivers

Buffers are gates designed with a special output section that can provide more driving current than ordinary TTL gates and are usually available in either inverting or noninverting form. Buffers are often used in situations requiring a large fan-out number that would exceed the capability of a normal gate output. For example, this situation is particularly common for clocked circuits because a large number of gates are operated and synchronized under a single clock signal. In this case, the clock signal must be buffered to provide enough driving current to all circuits and yet maintain synchronization by adhering to TTL specifications. Typical buffers are the 74ALS37 two-input NAND and the 7417 buffer. Other buffers are given in Figure 3–2.

A typical arrangement of buffers is shown in Figure 3–11(a). The primary signal is connected to a number of drivers, which are used in turn to drive the primary loads. Figure 3–11(b) shows another arrangement of buffers that can be used to drive the same number of loads. As long as timing is not critical, both arrangements will perform equally well. In high-frequency operations, however, the first arrangement is preferred over the second because all the signals from the first arrangement arrive at each of the loads more nearly at the same time. In the second arrangement, some signals have to pass through a number of additional gate delays, thus making it more difficult to synchronize the loads.

Drivers function in similar fashion as buffers with the exception that the driver outputs are usually adapted for higher current and voltage levels. These gates are particularly useful in voltage-level conversions when different chips with different technologies are used, such as TTL-to-CMOS or CMOS-to-TTL conversions. Digital interface to any analog product also requires a driver because the TTL output specification does not always match the analog input specification. Hence, a driver is used as an interface between the digital and analog domains for devices such as relays and analog switches. Because of their functions, the outputs of many of these chips are usually open collector outputs. Typical examples include the 7407 and 7417 drivers. Some applications require an interface between the normal digital system part and the other high-voltage devices such

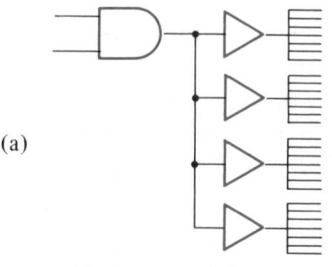

(a)

(a) Recommended arrangement with equal time delays

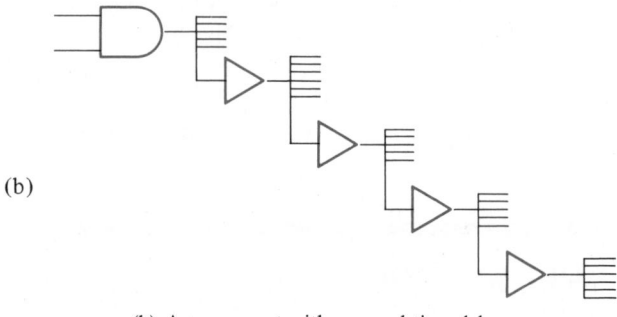

(b)

(b) Arrangement with unequal time delays

Figure 3–11
Arrangement of buffers: (a) recommended way with equal time delays;
(b) arrangement with unequal time delays

as fluorescent displays and appliance controls. Because of the exposure to high voltage with the possibility of high-voltage leakage, special protection circuitry is needed at the output to prevent saturation or harmful feedback to the gate circuitry. Typical examples of chips in this category are the 7407 and 7417 drivers.

Line drivers are special drivers that can provide (source) or accept (sink) large amounts of current when connected to a specific output load such as a 50-Ω or 75-Ω line. These chips can typically sink as much as 60 mA and source as much as 40 mA. This is particularly useful for driving communication lines or interface lines between different digital systems. Examples of these chips are the 74128 and 74140 chips.

Bus drivers are very similar to regular drivers with the exception that bus drivers usually have tri-state outputs. This means that for a regular driver the output is always asserted; i.e., the logic is always trying to drive a logical 0 or 1. In contrast, the output of a bus driver can be asserted when needed to drive the bus with a logical 0 and 1. In addition, it can be disabled into a high-impedance state so that other similar chips can be asserted to drive the bus. These chips are designed specifically for interfacing devices to the bus, and thus the name bus drivers. Typical examples of chips in this category are numerous, including the 74ALS125 and 74ALS240 chips.

Line receivers are actually drivers and are often used at the receiving end of a bus interface or connection point. Their primary use is to latch the communication signals transmitted over the bus and to amplify that signal for other chips to use. In this way, the loading to the bus appears as a single load, but the information coming out of the line receivers can be used by many chips. In many respects, line receivers are very similar to regular buffers.

Transceivers are actually a combination of line drivers and line receivers. They are used for interfacing a bidirectional data bus to a digital system, because the information must be read from the bus at certain times and transmitted to the bus at other times. Control is provided so that only the driver or only the receiver is active at any one time. Typical examples of transceivers are the 74ALS226, 74ALS242, and 74ALS245 chips.

3.4
SSI Design Procedures

We have discussed the various SSI components that can be used as building blocks to realize any logic functions. In this section we present the design procedure for solving logic problems from problem definition to circuit realization. The design procedure involves five steps:

1. Input definition
2. Output definition
3. Truth table construction
4. Output function minimization
5. Final circuit realization

These five steps are best illustrated and presented by an example. In this example, we want to design a voting machine that would give the majority votes from two persons.

Input Definition

When a logic problem is given, the first step is to define the problem by identifying clearly the inputs, the outputs, and the relation between the outputs and the inputs. When we say inputs and outputs here, we are referring to logical variables for inputs and outputs. Far too often the problem will be given in terms of physical entities and not in terms of logical variables. Therefore the inputs and outputs of the given problem must be identified from these physical entities and then mapped onto well-chosen logical variables in some meaningful way.

This example problem is rather vague in that not many specifics are given. In this case the problem definition stage must include intelligent choices to fill in the specific details. The task is to define the input and output in terms of logical variables. Since each person may cast either a "yes" or "no" vote, we can readily see that only one logical variable is needed for each person—a logical 1 to indicate a "yes" and a logical 0 to indicate a "no." Therefore there is a total of two logical variables as input to this machine and one logical variable as output to indicate the final outcome.

Suppose we now allow the condition that any one of the voting members may abstain from voting. Two logical variables for the input are now needed to identify the three choices from each member. Since there are altogether four distinct states to combine two logical variables and only three are needed to identify the input conditions, we may choose any three states to represent the input conditions. There is no easy way to decide how the three states should be chosen. Experience has shown that such state assignments are usually done heuristically, and many times, states are assigned based on the concept of functional modularity. In other words, states are assigned in such a way that each logical variable represents an independent function. This latter method usually gives acceptable results.

One possible assignment is shown in Figure 3–12(a). The choice there may appear to be somewhat arbitrary; however, assignments are really done according to their logic functions. The variable x represents whether a vote has been cast or not, and the variable y represents a "yes" or "no" vote. In other words, variable x serves as a qualifier for y and determines whether y should be used or not. For this particular assignment, the meaning of each variable can be clearly identified, although this is not always necessary. Any other assignment can be used just as well. Since one of the states is available but is not used, we can take advantage of this unused state for other purposes. For example, a combination of $x = 1$ and $y = 1$ received at the voting machine may be used to indicate that an error or invalid condition exists. This is shown in Figure 3–12(b). Another

Figure 3–12

Different ways to assign input logic states

x	y	Condition
0	0	"No" vote
0	1	"Yes" vote
1	0	Abstain
1	1	Not used

(a)

x	y	Condition
0	0	Abstain
0	1	"Yes" vote
1	0	"No" vote
1	1	Not used or Invalid

(b)

x	y	Condition
0	0	Not used or Error
0	1	Abstain
1	0	"No" vote
1	1	"Yes" vote

(c)

x	y	Condition
0	x	Abstain
1	0	"No" vote
1	1	"Yes" vote

(d)

possible assignment could result by making the requirement that $x = 0$ and $y = 0$ an invalid condition, thus ensuring that the signal must be coming through from the voting members. If the communication from the voting members is valid, neither x nor y should be zero, as indicated in Figure 3–12(c). Still another way to assign the choices is to annihilate the value of variable y—that is, a don't care where both 0 and 1 are okay, as long as variable x indicates that it is an abstain vote. This is shown in Figure 3-12(d). For the time being, we will choose the assignments given above in Figure 3–12(a), with the last condition $x = 1$ and $y = 1$ indicating reception of an invalid option.

Output Definition

For the output variables, we have a total of five possible outcomes—a majority "no" vote, a majority "yes" vote, an indeterminate case (a tie), an abstained vote if all voting members abstain, and an error condition when an invalid code is received. It is conceivable that the last three cases could be combined; the exact choice will depend on the application. Let us assume that a tie and an abstaining vote are the same in both cases since a re-vote is necessary, yet we still want to have an indication of an error condition so that erroneous results can be detected immediately. The output assignments of the choices can now be finalized as shown in Figure 3–13. In similar fashion, the reader should be able to arrive at other output definitions.

   To summarize, we will need two input logical variables for each voting member. They will be identified as x_1, y_1, x_2, and y_2. For the output variables, we have f and g to indicate the four possible outcomes. The assignment of input and output variables defines the specific details relating to the input and output choices and defines their relations with one another. Note that these relations are exhaustive and are more precise than the problem given.

Truth Table Construction

After the logical variables have been assigned, the next step is to construct a truth table showing the relationships for all combinations between the output

Figure 3–13
Assignment of output logic states

f	g	Condition
0	0	Composite "no"
0	1	Composite "yes"
1	0	"Tie" or abstain
1	1	Error

variables and the input variables. The truth table is a precise way of representing the problem in clear mathematical terms. There cannot be any ambiguities in the truth table, except for the don't care situation: All possible cases must be covered. The truth table for the voting machine is given in Figure 3–14. The construction of the truth table is fairly straightforward, and the assignment of values for the output variables directly follows the logic defined by the given problem.

Output Function Minimization

After the truth table has been constructed, the output functions can be minimized by any of the methods discussed in Chapter 2. If at all possible, both the sum-of-products form and the product-of-sums form should be obtained for comparison. The one that is most suitable for implementation is the one with the smallest number of available gates. This should be the form used for actual circuit realization. The solution with the smallest number of gates may not always provide the smallest number of chips, though it tends to do so in most cases. In applications where the number of chips is crucial to the design because of cost, power, or other reasons, the realization of both the sum-of-products and the product-of-sums forms should be worked out to determine which form produces the minimum

Figure 3–14
Truth table for the voting machine

x_1	y_1	x_2	y_2	f	g
0	0	0	0	0	0
0	0	0	1	1	0
0	0	1	0	0	0
0	0	1	1	1	1
0	1	0	0	1	0
0	1	0	1	0	1
0	1	1	0	0	1
0	1	1	1	1	1
1	0	0	0	0	0
1	0	0	1	0	1
1	0	1	0	1	0
1	0	1	1	1	1
1	1	0	0	1	1
1	1	0	1	1	1
1	1	1	0	1	1
1	1	1	1	1	1

number of chips. Typical minimization techniques give only a two-level representation. It is extremely difficult to find a simpler solution using multilevel logic, if, in fact, a simpler solution does exist. Intimate knowledge of the variables and their functional modularity many times will give a clue to how some variables can be grouped together meaningfully, thus reducing the final gate count.

For the voting machine example, the Karnaugh maps and the representations of the two output functions for both the sum-of-products and product-of-sums forms are given in Figure 3–15. It is seen that the sum-of-products forms for the two functions are simpler than the product-of-sums forms and are likely to require a simpler implementation. Furthermore, some of the terms for f and g in the sum-of-products form are the same, indicating that it is possible to share those terms in the final realization. So, without trying the product-of-sums forms, we will rule them out for implementation in this example.

Final Circuit Realization

After minimization of the output functions, the next step is to find the appropriate circuit components that will adequately realize the functions. This step is by no means simple, and the solution obtained is by no means unique. We have already seen that there are various implementation constraints that must be accounted for; notably, fan-in, fan-out, timing, etc. Furthermore, since the solution is not unique, additional constraints must be imposed as guidelines to help us determine which design is more suited for the current application. Different choices will have to be made if a design is to have the minimum cost, or if a design is to have the minimum number of chips, or if the design must only use a specific kind

Figure 3–15
Karnaugh maps for the voting machine

x_2y_2 \ x_1y_1 (f)	00	01	11	10
00	0	1	1	0
01	1	0	1	0
11	1	1	1	1
10	0	0	1	1

x_2y_2 \ x_1y_1 (g)	00	01	11	10
00	0	0	1	0
01	0	1	1	1
11	1	1	1	1
10	0	1	1	0

Sum-of-products form:
$$f = x_1y_1 + x_2y_2 + x_1x_2 + \bar{x}_1\bar{y}_1y_2 + \bar{x}_2y_1\bar{y}_2$$
$$g = x_1y_1 + x_2y_2 + x_1y_2 + y_1y_2 + x_2y_1$$

Product-of-sums form:
$$f = (\bar{x}_1 + y_1 + x_2)(x_1 + y_1 + y_2)(x_1 + \bar{x}_2 + y_2)(x_1 + \bar{y}_1 + x_2 + \bar{y}_2)$$
$$g = (y_1 + y_2)(x_1 + x_2 + y_2)(x_1 + y_1 + x_2)$$

of chip. These questions cannot be answered easily and the trade-offs must be made intelligently.

In addition to the three substitution rules discussed earlier, there are several other guidelines to follow in choosing the proper IC for realizing the functions.

1. Use chips fabricated from the same technology if possible, so that conversion is not necessary between chips from different technology types. Far too often, if chips from various technologies are mixed in the same design, more than one power supply voltage level is required. Furthermore, use of converters to match the specifications from different technology types will increase the chip count.

2. Use chips from the same family. This will help to provide uniformity in design and ease in matching input and output loads. For example, as much as possible use ALS chips throughout the design and convert only the output to other levels if needed, using converters for interfacing to analog components.

3. Design the circuit with the same set of chips if at all possible. This helps to keep the inventory down. In addition, spare gates from one part of the function can be used for another part of the function.

4. Always use an IC chip that can handle as much of the function to be realized as possible. This is important because manufacturing cost is usually related more to the number of connections required and the number of holes to be drilled on the circuit board than to the cost of using a more sophisticated type of IC chip.

These rules are generalizations of the various practical considerations. There may be additional or special requirements of the application that force one to choose a particular design over another.

For the example of the majority voting machine, three different designs are given in Figure 3–16. The circuit in Figure 3–16(a) is a straightforward design according to the requirement in the functions f and g. Note that we need inverters, various AND gates with different fan-in numbers, and various OR gates with different fan-in numbers. This design results in five chips and 34 external connections.

For the circuit shown in Figure 3–16(b), we impose a constraint that we will use only NAND gates but will allow chips with different fan-in numbers to be used. Since only NAND gates can be used, the function can be rewritten in NAND-NAND format as follows:

$$f(x_1, y_1, x_2, y_2) = (x_1 \uparrow y_1) \uparrow (x_2 \uparrow y_2) \uparrow (x_1 \uparrow x_2) \uparrow (\bar{x}_1 \uparrow \bar{y}_1 \uparrow y_2) \uparrow (y_1 \uparrow \bar{x}_2 \uparrow \bar{y}_2)$$
$$g(x_1, y_1, x_2, y_2) = (x_1 \uparrow y_1) \uparrow (x_2 \uparrow y_2) \uparrow (x_1 \uparrow y_2) \uparrow (y_1 \uparrow y_2) \uparrow (x_2 \uparrow y_1)$$

With this particular requirement, the design requires six chips and 40 interconnections. The chip count is higher because a dual five-input OR gates chip is not available.

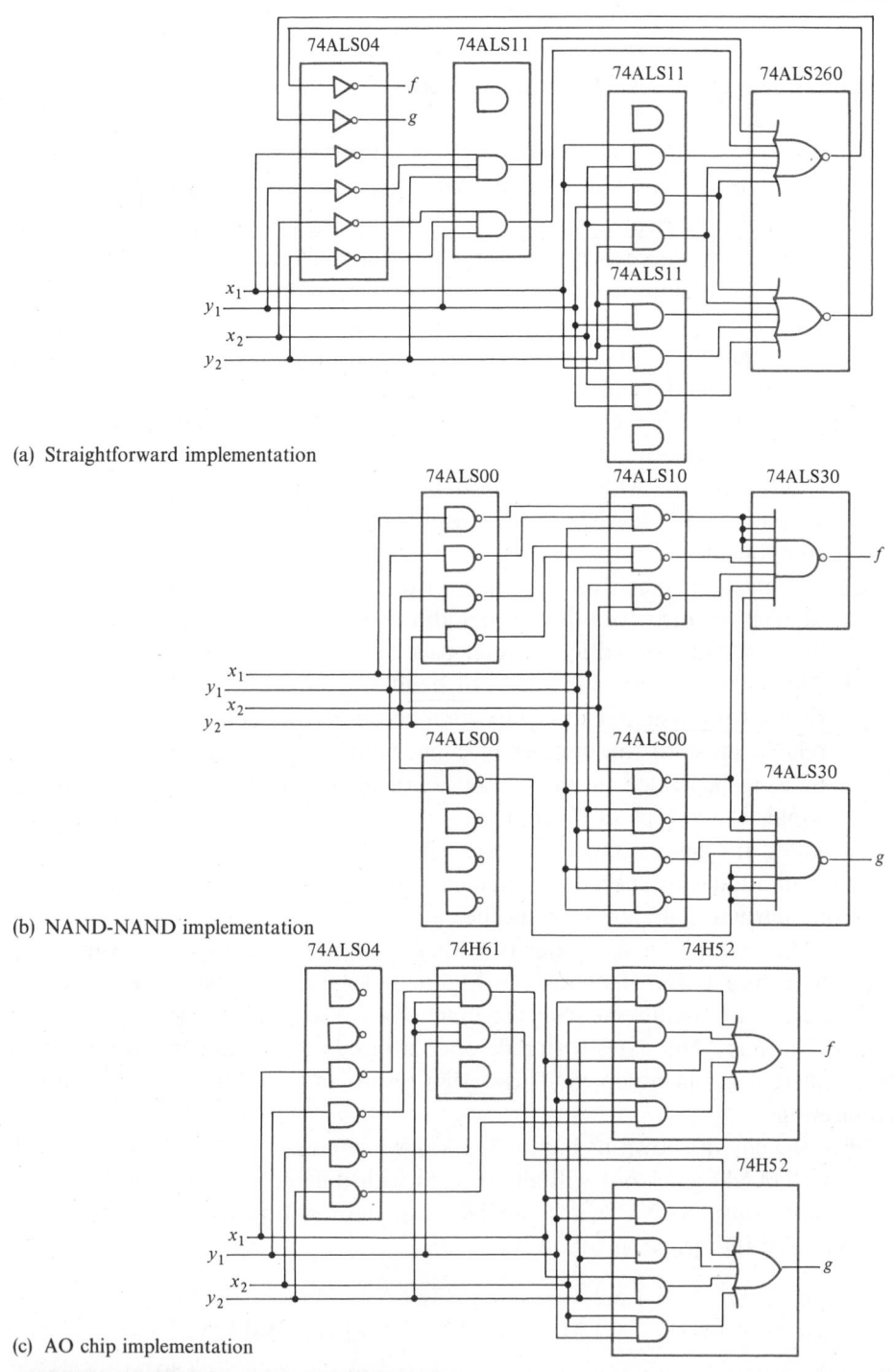

(a) Straightforward implementation

(b) NAND-NAND implementation

(c) AO chip implementation

Figure 3–16

Circuit realization for the voting machine using (a) straightforward
implementation, (b) NAND-NAND and (c) AO chip implementation

For the circuit shown in Figure 3–16(c), a chip is found that can take care of the bulk of the logic. By using the proper AO gates, we find the design simplified because most of the two-level logic is included in the AO gate. This design results in only four chips and 30 external connections.

From this example we can see that the rules given above generally give satisfactory results with the minimum number of chips and interconnections. We must stress, however, that there are no suitable techniques so far that would give the optimal conditions under any constraints.

Another Complete Design Example— A Mathematics Toy

In this section we go through the complete design process again to trace clearly all the steps required in starting a problem and carrying it through to the actual circuit realization.

Example 3–5 ▮▮▮

We would like to design a children's mathematics instructional toy (CMIT) that will provide a "yes" or "no" answer by lighting the proper "yes" or "no" light for two questions. The first question is to tell whether the particular decimal digit pressed is a prime number or not, and the second question is to tell whether the particular decimal digit pressed is even or not. A mode switch is provided to indicate which question is being solved.

With the problem given above, we will now go through all five steps and show how the problem can be carried out to the final circuit realization.

STEP 1
Input Definition
The input for this machine is composed of a decimal digit and a mode switch. The decimal digit will be represented by an 8421 BCD code with four logical variables—w, x, y, and z. The mode switch, which is used to indicate the question being solved, is represented by a single logical variable c. The first question is indicated by $c = 1$ and the second by $c = 0$. Hence there are five input variables altogether.

STEP 2
Output Definition
The output consists of only two signals: one to drive the "yes" light and the other to drive the "no" light. Since the two signals are always complementary, only one output signal is needed. We will call that f. Then, when $f = 1$, the "yes" light will go on; when $f = 0$, the "no" light will go on.

Figure 3–17

Truth table for a children's mathematics instructional toy

Mode	c	w	x	y	z	f
	0	0	0	0	0	1
	0	0	0	0	1	1
	0	0	0	1	0	1
	0	0	0	1	1	1
	0	0	1	0	0	0
	0	0	1	0	1	1
	0	0	1	1	0	0
	0	0	1	1	1	1
Prime numbers	0	1	0	0	0	0
	0	1	0	0	1	0
	0	1	0	1	0	x
	0	1	0	1	1	x
	0	1	1	0	0	x
	0	1	1	0	1	x
	0	1	1	1	0	x
	0	1	1	1	1	x
	1	0	0	0	0	1
	1	0	0	0	1	0
	1	0	0	1	0	1
	1	0	0	1	1	0
	1	0	1	0	0	1
	1	0	1	0	1	0
	1	0	1	1	0	1
	1	0	1	1	1	0
Divisible by 2	1	1	0	0	0	1
	1	1	0	0	1	0
	1	1	0	1	0	x
	1	1	0	1	1	x
	1	1	1	0	0	x
	1	1	1	0	1	x
	1	1	1	1	0	x
	1	1	1	1	1	x

STEP 3

Truth Table Construction

Construction of the truth table for the CMIT is straightforward and is given in Figure 3–17. Note that since only the valid decimal digits will be accepted and passed on to the input, the other invalid codes can be used to advantage as don't cares. It turns out that these don't care entries really help in simplifying the output function.

STEP 4

Output Function Minimization

From the truth table, the output function f can be minimized as shown in Figure 3–18. A careful comparison between the sum-of-products form and the product-of-sums form for the output function f reveals that the product-of-sums form is slightly simpler than the sum-of-products form. Based on that finding, we will use the product-of-sums form for circuit realization. The function is

$$f(c, w, x, z) = (c + \bar{x} + z)(c + \bar{w})(\bar{c} + \bar{z})$$

STEP 5

Circuit Realization

As before, we will present three different realizations of this function. In Figure 3–19(a) we have the straightforward realization of output function f using the AND-OR form. Since we cannot find a three-input OR gate, we will use a three-input NOR gate and use a spare inverter to take care of the extra inversion. This realization uses three chips.

Figure 3–18

Karnaugh map for the CMIT

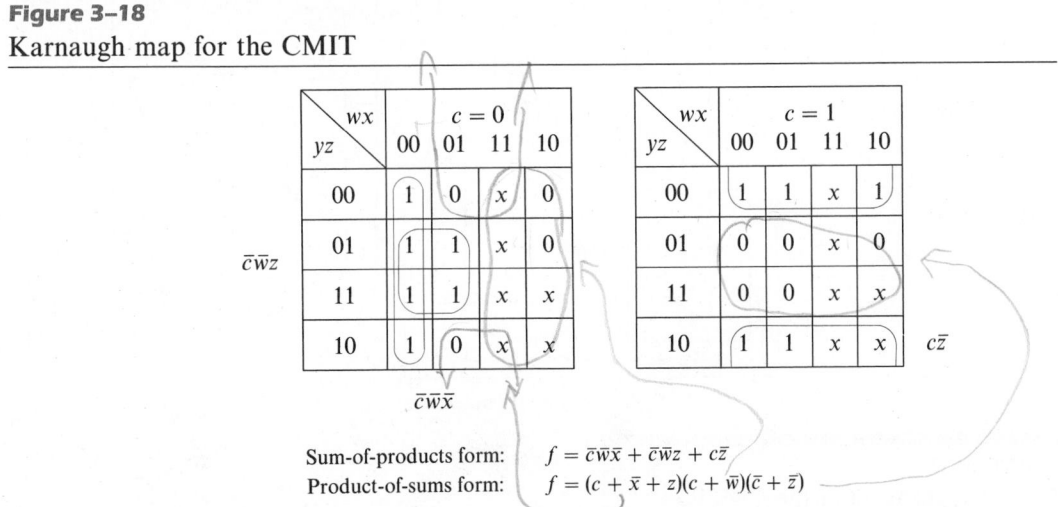

Sum-of-products form: $f = \bar{c}\bar{w}\bar{x} + \bar{c}\bar{w}z + c\bar{z}$

Product-of-sums form: $f = (c + \bar{x} + z)(c + \bar{w})(\bar{c} + \bar{z})$

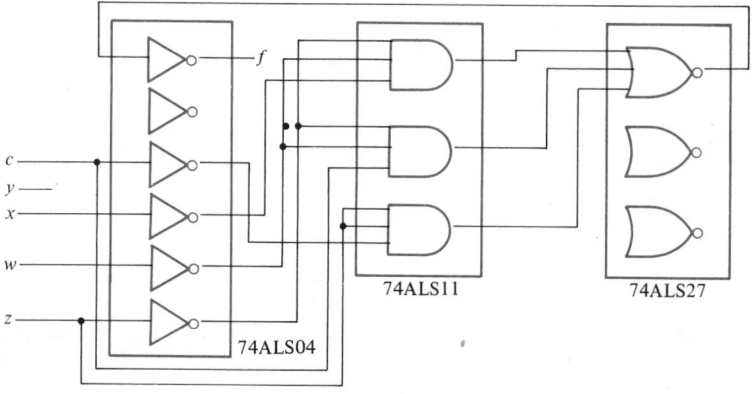

(a) Straightforward implementation

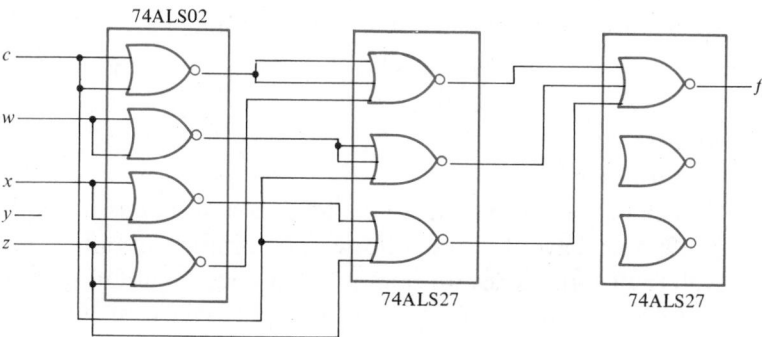

(b) NOR-NOR implementation

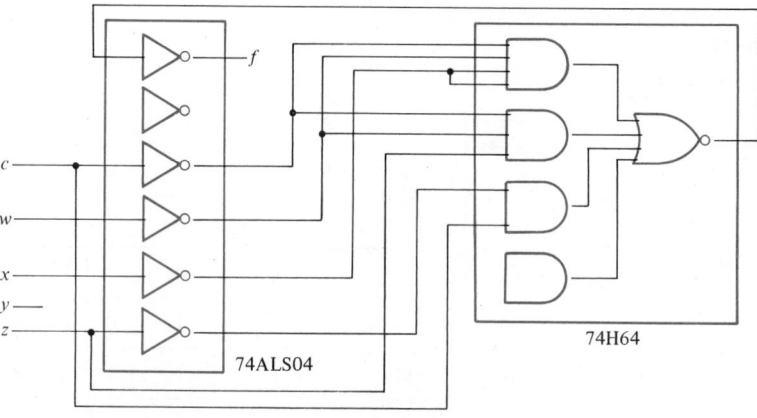

(c) AOI implementation

Figure 3-19

Circuit realization of the CMIT

In the second realization, given in Figure 3–19(b), we attempt to use all NOR gates by changing the OR-AND form to the NOR-NOR form as follows:

$$f(c, w, x, z) = (c \downarrow \bar{x} \downarrow z) \downarrow (c \downarrow \bar{w}) \downarrow (\bar{c} \downarrow \bar{z})$$

This, however, does not result in a reduction of the number of gates. The only benefit for this particular case is that this realization uses only two types of gates rather than three as in the first case. However, in the last realization, by using an inverter chip and an AO gate, we are able to obtain the required signal in a sum-of-products form as follows:

$$f(c, w, x, z) = \bar{c}\bar{w}\bar{x} + \bar{c}\bar{w}z + c\bar{z}$$

This realization is shown in Figure 3–19(c).

3.5

Logic Types and Voltage Symbolism

So far in this chapter, we have used only standard gates in positive logic. In other words, the logical states 0 and 1 are represented by low and high voltages. In this section we introduce other logic types and their symbolisms.

Logic Types

Throughout this section we have associated the two logical states 0 and 1 with low and high voltage levels, respectively. If we consider this more carefully, we can see that it is also possible to define logical states 0 and 1 as high and low voltage levels, respectively. In other words, the representation of a logic state as either a high or low voltage level is equally valid. We may interpret the voltage levels as follows:

Logic state 0   Low voltage
Logic state 1   High voltage

This is called *positive logic*. On the other hand, if we interpret the voltage levels in the opposite manner,

Logic state 0   High voltage
Logic state 1   Low voltage

this is called *negative logic*. The signals that use negative logic are sometimes referred to as *active low signals*. For clarity, the same logic is often used on both the input and output sides of a gate. It is possible, however, to have positive (or negative) logic on the input and negative (or positive) logic on the output. This is called *mixed logic*. Mixed logic is somewhat difficult to design, and should, in general, be avoided unless the particular application can be better solved in this manner.

The representation of logic states by voltage levels can be selected by the user. A given circuit will always function according to the way it is designed and will respond to input voltage levels only. It is up to the user to associate the high and low voltage levels to logical states 0 and 1 and label the gate according to its function under the selected interpretation rules of logic states and voltage levels. In other words, the interpretation of whether a particular circuit arrangement as a certain type of gate is totally dependent on the interpretation of the voltage levels.

Figure 3-20

Equivalence of gate definition for positive and negative logic

	Circuit Behavior			Positive Logic Interpretation			Negative Logic Interpretation		
	x_1	x_2	z	x_1	x_2	z	x_1	x_2	z
(a)	L	L	L	0	0	0	1	1	1
	L	H	L	0	1	0	1	0	1
	H	L	L	1	0	0	0	1	1
	H	H	H	1	1	1	0	0	0
				Positive AND			Negative OR		
	x_1	x_2	z	x_1	x_2	z	x_1	x_2	z
(b)	L	L	L	0	0	0	1	1	1
	L	H	H	0	1	1	1	0	0
	H	L	H	1	0	1	0	1	0
	H	H	H	1	1	1	0	0	0
				Positive OR			Negative AND		
	x_1	x_2	z	x_1	x_2	z	x_1	x_2	z
(c)	L	L	H	0	0	1	1	1	0
	L	H	H	0	1	1	1	0	0
	H	L	H	1	0	1	0	1	0
	H	H	L	1	1	0	0	0	1
				Positive NAND			Negative NOR		
	x_1	x_2	z	x_1	x_2	z	x_1	x_2	z
(d)	L	L	H	0	0	1	1	1	0
	L	H	L	0	1	0	1	0	1
	H	L	L	1	0	0	0	1	1
	H	H	L	1	1	0	0	0	1
				Positive NOR			Negative NAND		

As an example, we show in Figure 3–20(a) the behavior of a particular gate in terms of a truth table that depicts the way the output voltage level follows the changes in the input voltage level. If we apply the positive logic convention to interpret the voltage levels, we immediately obtain a truth table for an AND gate. However, if we apply the negative logic convention to interpret the voltage levels of this gate, we find that, instead of an AND gate, we have an OR gate. This shows clearly that the interpretation of the gate is dependent on the interpretation of the voltage levels. The interpretation of what a gate does must always be defined in terms of the particular type of logic used. To avoid confusion, most manufacturers always specify gates in terms of positive logic. Therefore, the gate in Figure 3–20(a) can be described as "positive AND" or "negative OR."

We have seen that the same gate can be interpreted to perform different operations according to the way the user represents the logic levels. To facilitate conversion between positive and negative logic systems, we show in Figure 3–20 the equivalent definitions for different gates in both positive and negative logic.

Voltage Symbolism

Inputs or outputs that are defined in terms of negative logic are related to their positive logic counterparts by an inverter that is denoted by a "bubble" on the signal line. In other words, if the input is in negative logic, then a bubble in the input signal line indicates that a positive logic signal is complemented. This is shown in Figure 3–21. Likewise, if the output is in negative logic, then a bubble

Figure 3–21
Use of "bubble" for negative logic

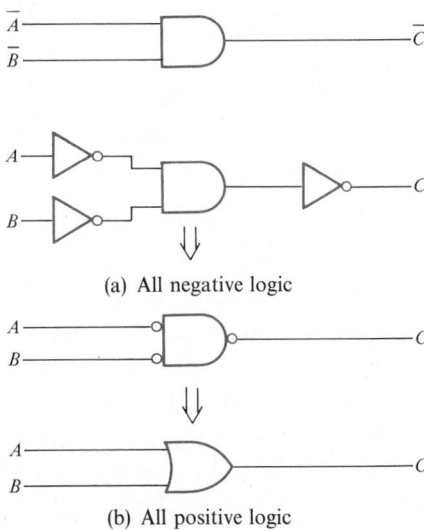

(a) All negative logic

(b) All positive logic

is put on the output signal line to indicate that the active high-output signal has been complemented.

Why is negative logic used? There are primarily two reasons. First, from the circuit standpoint, it is much easier to hold any unused output at a high voltage level than at a low voltage level. Recall that there must be current flowing into the output terminal in order for the output voltage level to become low. Hence, if there are many unused outputs, all would be held at a high voltage level, and only the selected ones that are to be asserted (or made active) will be held at a low voltage level. In essence, this is negative logic. This situation is particularly true and commonly found in priority encoders, data encoders and decoders, and multiplexers.

The second reason that negative logic is so often used is that it is easy to label signals in terms of their functions regardless of whether they are defined in either positive or negative logic. For example, if a particular output signal line represents the direction of data flow, it may be appropriately termed as $\text{IN}/\overline{\text{OUT}}$; that is, when the output is high, the flow of data is in, but when the signal is low, the flow of data is out. Therefore, in terms of the data flowing in with all its associated driving circuits, the signal line is in positive logic. However, in terms of the data going out with all its supporting circuits, the signal line is in negative logic.

3.6
Mixed-Logic Design

In many real applications, not all the variables are provided in positive logic form. Furthermore, a two-level sum-of-products or product-of-sums form may not always provide the best solution. Rather, the designer of digital systems is often faced with mixed-logic designs, that is, designs involving variables in both positive and negative logic format in multilevel NAND and NOR circuits. In this section we present mixed-logic designs, which are really not hard to analyze and synthesize.

NAND and NOR Transformations

Earlier we mentioned briefly the concept of form substitution for circuit minimization. Whereas a combination of AND, OR, and NOT gates can be used to represent any Boolean function, a combination of NAND gates alone or NOR gates alone can also be used to represent any Boolean function. The use of NAND and NOR gates is quite common. First, the fabrication and manufacturing of NAND and NOR gates are easier than the others. Second, it is often more appealing to stock only one type of chip rather than three different types of chips. Third, as will be seen later, both the NAND and the NOR gates can be used in place of AND, OR, and NOT gates.

Any arbitrary AND-OR or OR-AND circuit can be transformed to an equivalent NAND-NAND or NOR-NOR circuit. Recall that an AND gate can be replaced by a NOR gate by complementing all the inputs as shown in Figure 3–22(a):

$$ab = \overline{\overline{ab}}$$
$$= \overline{\overline{a} + \overline{b}}$$
$$= \overline{a} \downarrow \overline{b}$$

Likewise, an OR gate can be replaced by a NAND gate by complementing all the inputs as shown in Figure 3–22(b):

$$a + b = \overline{\overline{a + b}}$$
$$= \overline{\overline{ab}}$$
$$= \overline{a} \uparrow \overline{b}$$

Using the above two transformations, we can readily see that a two-level AND-OR circuit can be transformed into a two-level NAND-NAND circuit by inserting a pair of bubbles (i.e., inverters) between the two levels. The procedure is as follows:

1. Bracket all product terms—i.e., AND terms—together.
2. Complement any literals in the expression—i.e., a single term not ANDed to other terms.
3. Replace all AND and OR operators by the NAND operator.

A single literal must be complemented due to the bubble inserted between the two levels. The above procedure can be illustrated by Example 3-6.

Figure 3–22
(a) AND to NOR and (b) OR to NAND transformations

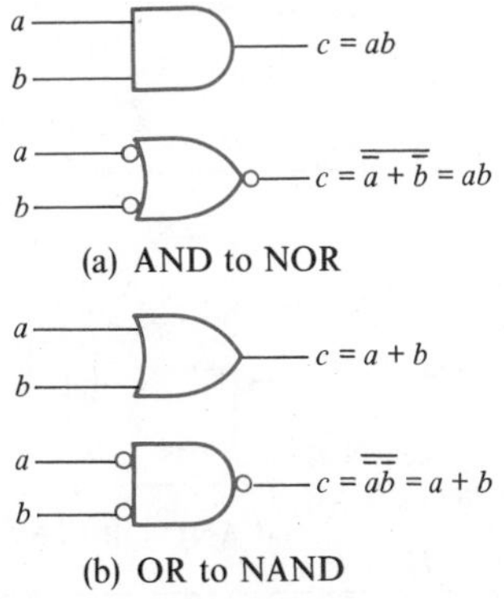

(a) AND to NOR

(b) OR to NAND

Example 3–6

Transform the following AND-OR circuit into the NAND-NAND form:

$$f(a, b, c, d) = a\bar{b} + \bar{a}c + d$$

$$\begin{aligned}
f(a, b, c, d) &= a\bar{b} + \bar{a}c + d \\
&= (a\bar{b}) + (\bar{a}c) + (d) \qquad \text{Step 1} \\
&= \overline{\overline{(a\bar{b}) + (\bar{a}c) + (d)}} \\
&= \overline{(\overline{a\bar{b}}) \cdot (\overline{\bar{a}c}) \cdot (\bar{d})} \qquad \text{Step 2} \\
&= \overline{(a\bar{b})} \uparrow \overline{(\bar{a}c)} \uparrow (\bar{d}) \qquad \text{Step 3} \\
&= (a \uparrow \bar{b}) \uparrow (\bar{a} \uparrow c) \uparrow \bar{d}
\end{aligned}$$

The transformation of the above AND-OR circuit into NAND-NAND form is shown in Figure 3–23(a). The parentheses in Step 1 serve to identify the product terms in the first level. Then a pair of bubbles is inserted in every connection

Figure 3–23

Transformation from AND-OR form (a) to NAND-NAND form and (b) to NOR-NOR form; from OR-AND form (c) to NAND-NAND form and (d) to NOR-NOR form

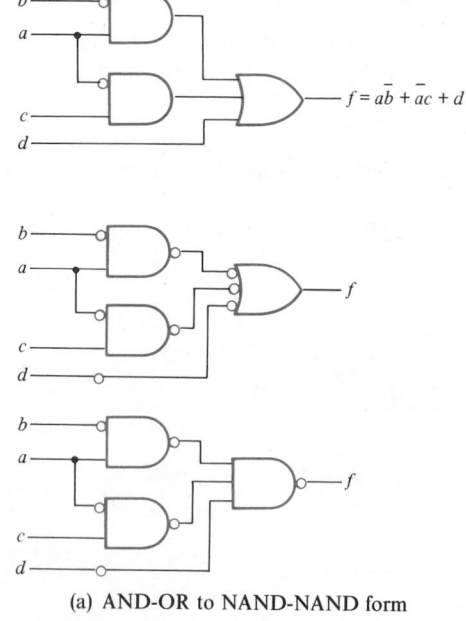

(a) AND-OR to NAND-NAND form

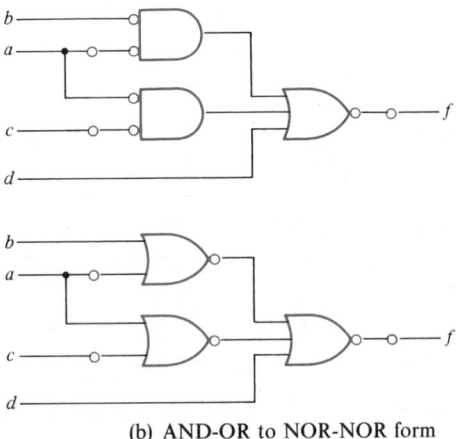

(b) AND-OR to NOR-NOR form

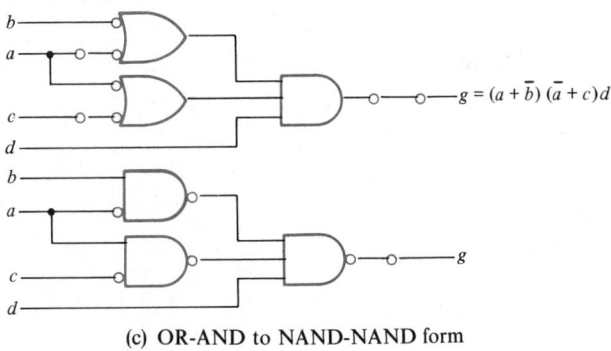

(c) OR-AND to NAND-NAND form

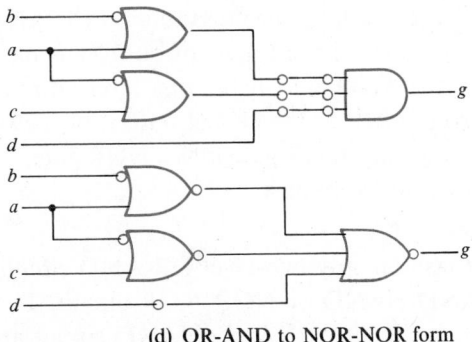

(d) OR-AND to NOR-NOR form

Figure 3–23 Continued

between the two levels. The first set of bubbles is used to turn the first-level AND gates into NAND gates. The second set of bubbles is used to transform the second-level OR gate into a NAND gate.

The same two-level AND-OR circuit can also be transformed into an equivalent two-level NOR-NOR form. The procedure is as follows:

1. Bracket all product terms together.
2. Complement all product terms except for any literals in the expression—i.e., a single term not ANDed to other terms.
3. Replace all AND and OR operators by the NOR operator.
4. Complement the final function output.

The above procedure for converting two-level AND-OR circuits to NOR-NOR circuits can be illustrated by the following example using the same function.

Example 3–7

Transform the following AND-OR circuit into the NOR-NOR form:

$$f(a, b, c, d) = a\bar{b} + \bar{a}c + d$$

$$
\begin{aligned}
f(a, b, c, d) &= a\bar{b} + \bar{a}c + d \\
&= (a\bar{b}) + (\bar{a}c) + (d) && \text{Step 1} \\
&= \overline{\overline{(a\bar{b})}} + \overline{\overline{(\bar{a}c)}} + (d) \\
&= \overline{\bar{a} + \bar{\bar{b}}} + \overline{\bar{\bar{a}} + \bar{c}} + d && \text{Step 2} \\
&= \overline{(\bar{a} \downarrow b)} + \overline{(a \downarrow \bar{c})} + d && \text{Step 3} \\
&= \overline{(\bar{a} \downarrow b) \downarrow (\bar{a} \downarrow c) \downarrow d}
\end{aligned}
$$

The transformation of the above AND-OR circuit into the NOR-NOR form is shown in Figure 3–23(b). The parentheses in Step 1 serve to identify the product terms in the first level. Then a pair of bubbles is inserted at each product term, thus changing the AND gate at the first level into a NOR gate. Another pair of bubbles is also inserted at the end of output of the OR gate at the second level, thus changing the output OR gate into a NOR gate. This completes the AND-OR to NOR-NOR transformation.

In similar fashion, any two-level OR-AND circuit can be transformed into a two-level NAND-NAND or NOR-NOR circuit. The steps for converting from an OR-AND circuit into a NAND-NAND circuit are as follows:

1. Bracket all sum terms—i.e., OR terms—together.
2. Complement all products in each sum term except for any single literals in the expression—i.e., a single term not ORed to other terms.

3. Replace all OR and AND operators by the NAND operator.
4. Complement the function output.

The steps for converting an OR-AND circuit into a NOR-NOR circuit are as follows:

1. Bracket all sum terms—i.e., OR terms—together.
2. Complement any single literals in the expression—i.e., a single term not ORed to other terms.
3. Replace all OR and AND operators by the NOR operator.

The logic of the last two conversions is similar to that for the AND-OR circuits and can be understood immediately by the principle of duality. The proof of these steps is left to the reader as an exercise.

Example 3–8

Transform the following OR-AND circuit into the NAND-NAND form:

$$g(a, b, c, d) = (a + \bar{b})(\bar{a} + c)d$$

$$\begin{aligned}
g(a, b, c, d) &= (a + \bar{b})(\bar{a} + c)d \\
&= (a + \bar{b}) \cdot (\bar{a} + c) \cdot d && \text{Step 1} \\
&= \overline{\overline{(a + \bar{b})}} \cdot \overline{\overline{(\bar{a} + c)}} \cdot d \\
&= \overline{(\overline{\bar{a}b})} \cdot \overline{(\overline{\bar{a}c})} \cdot d && \text{Step 2} \\
&= \overline{(\bar{a} \uparrow b)} \cdot \overline{(a \uparrow \bar{c})} \cdot d && \text{Step 3} \\
&= (\bar{a} \uparrow b) \uparrow (a \uparrow \bar{c}) \uparrow d && \text{Step 4}
\end{aligned}$$

Example 3–9

Transform the following OR-AND circuit into the NOR-NOR form:

$$g(a, b, c, d) = (a + \bar{b})(\bar{a} + c)d$$

$$\begin{aligned}
g(a, b, c, d) &= (a + \bar{b})(\bar{a} + c)d \\
&= (a + \bar{b}) \cdot (\bar{a} + c) \cdot d && \text{Step 1} \\
&= \overline{\overline{(a + \bar{b}) \cdot (\bar{a} + c) \cdot d}} \\
&= \overline{\overline{(a + \bar{b})} + \overline{(\bar{a} + c)} + \bar{d}} && \text{Step 2} \\
&= \overline{(a + \bar{b})} \downarrow \overline{(\bar{a} + c)} \downarrow \bar{d} && \text{Step 3} \\
&= (a \downarrow \bar{b}) \downarrow (\bar{a} \downarrow c) \downarrow \bar{d}
\end{aligned}$$

The transformation of the same OR-AND circuit into NAND-NAND and NOR-NOR forms is shown in Figures 3–23(c) and 3–23(d), respectively. The reader is urged to verify these step-by-step transformations graphically on the given figures.

If we compare the topology of an AND-OR circuit with its NAND-NAND and NOR-NOR equivalents, we immediately find that the topology of the AND-OR circuit is the same as the NAND-NAND circuit except for the complemented literal. In other words, if the inputs and outputs of the AND-OR circuit are of positive logic, the inputs and outputs of the corresponding NAND-NAND circuit are also of positive logic. The logic type of any single literal, however, must be inverted. This is not the case for the NOR-NOR circuit. The logic type of the inputs and outputs is inverted, while that of the literal remains the same. Hence an active high input in an AND-OR circuit will become active low in a NOR-NOR circuit. Likewise, comparing the topology of an OR-AND circuit with its NAND-NAND and NOR-NOR equivalents, the same type of logic inversion is also observed.

Analysis of Multilevel Circuits

In the previous section we showed that either the NAND gate or the NOR gate can function as an AND gate or an OR gate depending on whether the signals are active high or active low. The function of the NAND and NOR gates is then dependent on the logic type of the inputs, the logic type of the output, and the particular level at which the particular gate is situated. For example, when an AND-OR circuit is transformed into a NAND-NAND circuit, the NAND gates in the first level function as AND gates while the NAND gates in the second level function as OR gates.

Consider the multilevel circuit composed of NAND gates shown in Figure 3–24. We can number each level of NAND gates starting from the output gate as shown. This circuit can be broken down into two subcircuits. The first one includes the gates in levels 1 and 2, while the second includes the gates in levels 3 and 4. The output h can be found in terms of the input parameters as follows:

$$h = \overline{(z \uparrow f) \uparrow g}$$
$$= zf + \bar{g}$$
$$z = ab + cd + \bar{e}$$

Combining the above two expressions:

$$h = \underbrace{(a \cdot b}_{4} + \underbrace{c \cdot d}_{4} + \bar{e}) \cdot f + \bar{g}$$

with the brackets labeled 3, 2, and 1 grouping the expression progressively.

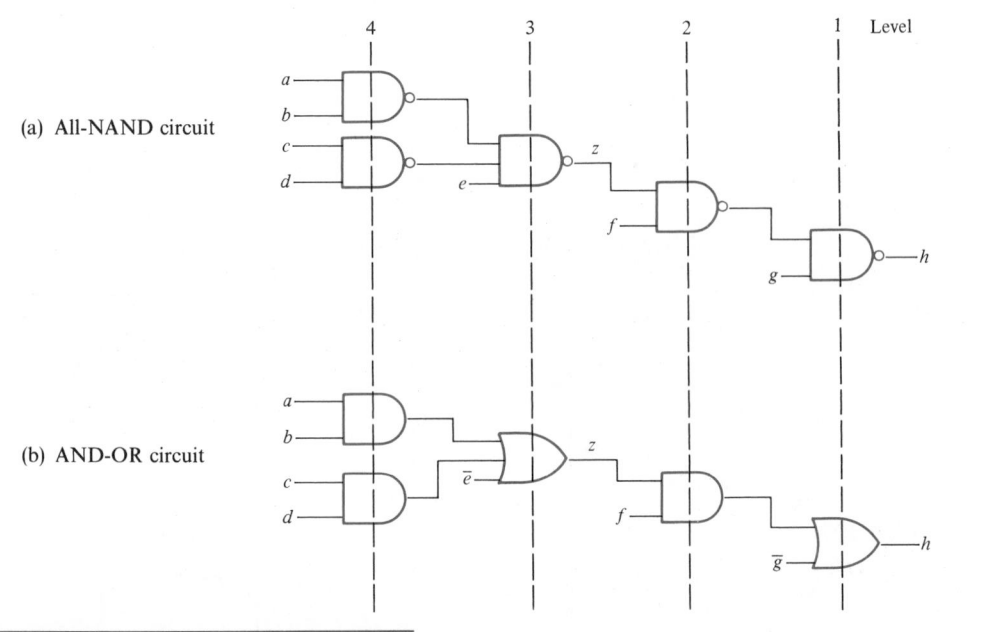

Figure 3–24
A multilevel circuit (All-NAND to AND-OR)

From the above diagram, note that odd levels of NAND gates perform the OR operation while even levels of NAND gates perform the AND operation. Also note that the literals are complemented when introduced in the odd levels and uncomplemented in the even levels. Applying these principles to the all-NAND circuit, we can redraw the circuit using AND and OR gates in mixed-logic notation. This is shown in Figure 3–24(b).

By the principle of duality, we can see that the same principle applies for the NOR-NOR circuits. Suppose that we have the NOR-NOR circuit shown in Figure 3–25(a). The output of the NOR-NOR circuit can be found as follows:

$$H = (y + D)(E + F)\bar{G}$$
$$y = (A + B)\bar{C}$$

or

$$H = ((A + B)\bar{C} + D)(E + F)\bar{G}$$

Hence we see that NOR gates in odd levels function as AND gates but in even levels as OR gates. In the same way, literals are introduced in the complemented form in the odd levels and in the uncomplemented form in the even levels. The

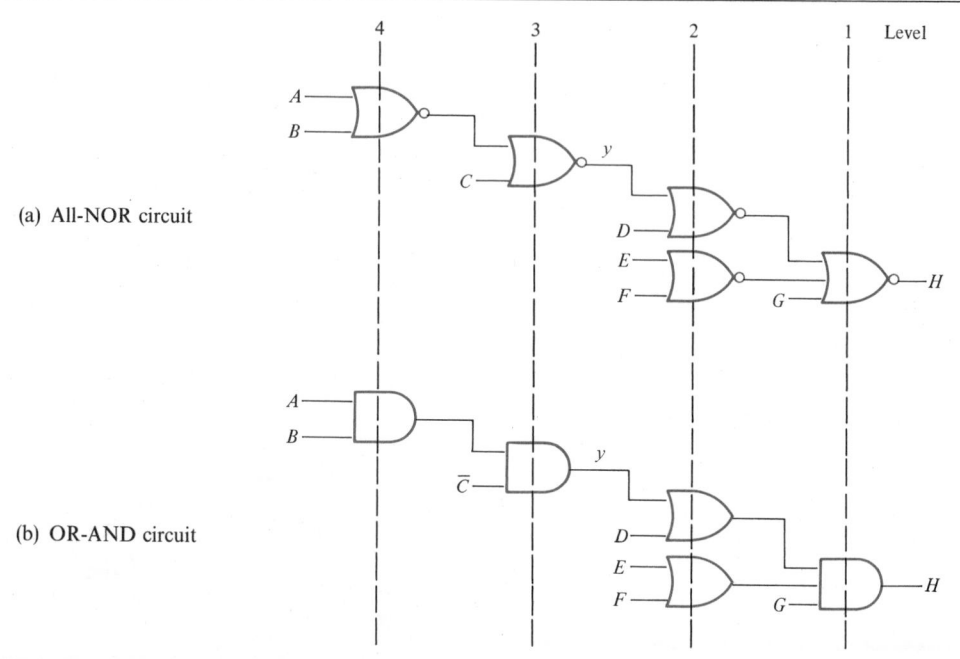

Figure 3–25

An All-NOR to OR-AND transformation

same circuit can now be redrawn using AND and OR gates in mixed-logic notation as shown in Figure 3–25(b).

Generalizing from Examples 3–8 and 3–9, we can see that NAND gates function as AND gates in even levels and as OR gates in odd levels. Likewise, the NOR gates function as OR gates in even levels and as AND gates in odd levels. This can be tabulated as shown in Figure 3–26. This is true when the logic

Figure 3–26

Gate functions based on levels

	Input Variable	NAND	NOR	Function Output
Odd levels	Complemented	OR	AND	Positive logic
Even levels	Uncomplemented	AND	OR	Negative logic

type of the output is the same as the logic type of the inputs. If the logic type of the output is opposite, then an extra inverter is needed at the output. This additional inverter becomes level 1, while all other levels are shifted by one.

For both examples given above, the type of gates used in the entire circuit is homogeneous throughout the circuit. The question now is whether the rule given in Figure 3–26 still holds when both NAND and NOR gates are present in multilevel circuits. This can be best shown by the following example.

Example 3-10

Analyze the multilevel circuit shown in Figure 3–27.

In this circuit, both NAND and NOR gates are present. We will again use temporary variables to represent the gate outputs and determine the behavior of this circuit:

$$h = \overline{\overline{(x + e)} + \overline{(f \cdot g)}}$$
$$= (x + e)fg$$

Figure 3–27

A multilevel circuit with both NAND and NOR gates

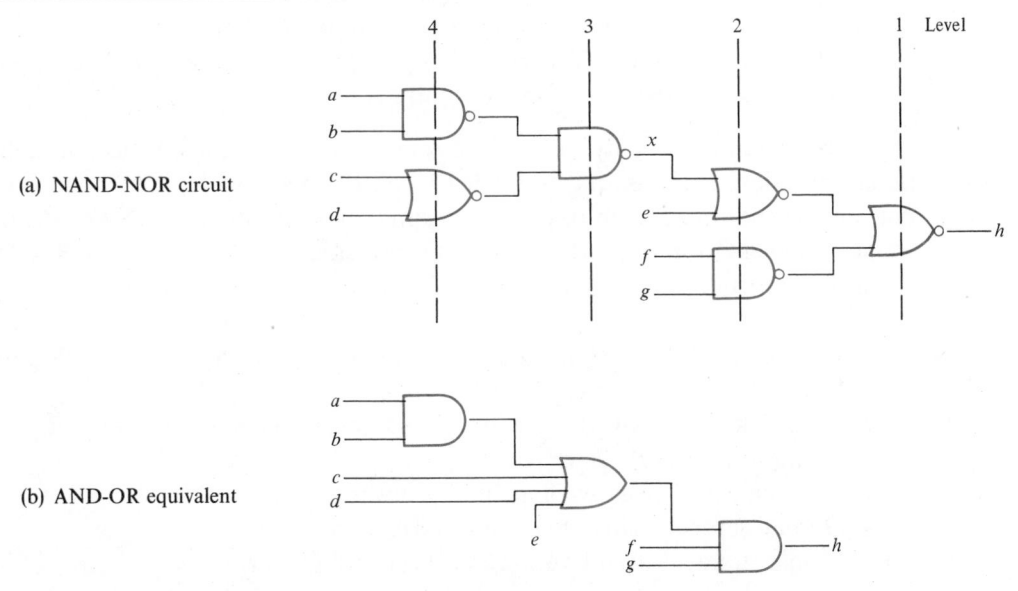

(a) NAND-NOR circuit

(b) AND-OR equivalent

$$x = \overline{(a \cdot b) \cdot (c + d)}$$
$$= ab + c + d$$
$$h = (ab + (c + d) + e)fg$$

The final expression for the function output, h, can be inferred directly from the circuit if one interprets the function of the NAND and NOR gates as either an AND gate or an OR gate according to the level they are in. The AND/OR representation is shown in Figure 3–27(b).

Synthesis of Multilevel Circuits

From the previous section, we have seen several properties of multilevel circuits using NAND and NOR gates with positive and negative logic inputs. Let us summarize these properties for positive logic function output:

1. On odd levels, a NAND gate functions as an OR gate while a NOR gate functions as an AND gate.
2. On even levels, a NAND gate functions as an AND gate while a NOR gate functions as an OR gate.
3. Inputs from odd levels become complemented in the final form.
4. Inputs from even levels are uncomplemented in the final form.
5. Gate outputs at odd levels are in positive logic form.
6. Gate outputs at even levels are in negative logic form.
7. An additional level is needed to invert the positive logic function output to obtain negative logic function outputs.

The above rules apply only for positive logic function outputs, hence an additional inversion at the output is needed to obtain negative logic function output. Implementation of sum-of-products expressions terminate in a NAND gate, while implementation of product-of-sums expressions terminate in a NOR gate. If negative function outputs are desired, the rules can be easily modified:

1. On even levels, a NAND gate functions as an OR gate while a NOR gate functions as an AND gate.
2. On odd levels, a NAND gate functions as an AND gate while a NOR gate functions as an OR gate.
3. On even levels, negative logic inputs are used.
4. On odd levels, positive logic inputs are used.
5. Gate outputs at even levels are in positive logic form.
6. Gate outputs at odd levels are in negative logic form.

7. An additional level is needed to invert the negative logic function output to obtain positive logic function outputs.

Another easy way to obtain negative logic function output is to start counting the function output level from level 2 instead of level 1.

Using the above rules, we can synthesize multilevel circuits. The synthesis procedure must account not only for the function to be implemented but also for the availability of complemented or uncomplemented inputs. Even though the same function is being implemented, the implementation may be different depending on whether complemented or uncomplemented inputs are available. The general synthesis procedure is as follows:

1. Minimize the expressions, if necessary.
2. Start at the output function. Determine the level of the output function and work backward toward the function inputs.
3. Use the complemented and uncomplemented inputs at the proper levels. For example, use uncomplemented inputs at even levels to obtain uncomplemented literals in the final form. Use complemented inputs at odd levels to obtain uncomplemented literals in the final form.

Example 3–11 ▰▰▰▰▰▰▰▰▰▰▰▰▰▰▰▰▰▰▰▰▰▰▰▰▰▰▰

Implement the following function in multilevel circuits with positive logic output assuming that only uncomplemented inputs are available:

$$f(a, b, c, d) = ab + \bar{a}cd + \bar{b}\bar{c}d$$

We want the function output to be in positive logic. Hence the output level is level 1. Since the expression is given in sum-of-products form, we will first construct the function using an AND-OR format as shown in Figure 3–28(a). This AND-OR circuit requires both complemented and uncomplemented forms of the input variables. Since the problem specifies that only uncomplemented forms are available, we must rearrange the circuit so that if a complemented literal occurs in the final expression, the same literal will be an input to gates at odd levels. Likewise, if the literal is uncomplemented in the final expression, then the literal will become an input to gates at even levels. From Figure 3–28(a), we see that the $\bar{a}$ input from gate 2 must be relocated to level 3. The $\bar{c}$ and $\bar{d}$ inputs in gate 3 must also be relocated to level 3. The final circuit is shown in Figure 3–28(b). The reader should verify at this point that the final circuit given in Figure 3–28(b) indeed produces the same expression as specified by the problem.

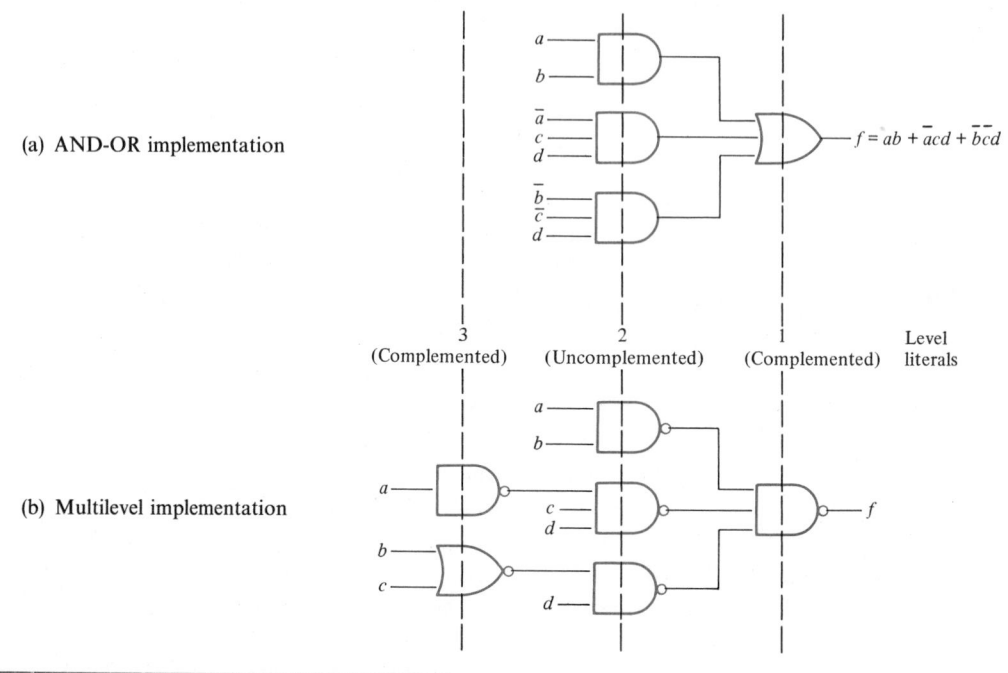

(a) AND-OR implementation

$f = ab + \bar{a}cd + \bar{b}\bar{c}d$

| 3 | 2 | 1 | Level |
| (Complemented) | (Uncomplemented) | (Complemented) | literals |

(b) Multilevel implementation

f

Figure 3–28
Synthesis of a multilevel circuit for $f(a, b, c, d) = ab + \bar{a}cd + \bar{b}\bar{c}d$

Example 3–12

Implement the same function as in the previous example in multilevel circuits but with negative logic function output assuming that only complemented inputs are available; that is,

$$f(a, b, c, d) = \overline{(ab + \bar{a}cd + \bar{b}\bar{c}d)}$$

Even though this is the same function as the previous example, the function output must be of negative logic as specified by the problem. This means that the function output level is level 2. The problem specifies that only negative logic variables are available. That means that if the variable is complemented in the final form, that variable must be used with a complemented input at even levels. If the variable is uncomplemented in the final form, that variable must be used with a complemented input at odd levels. The final circuit is shown in Figure 3–29.

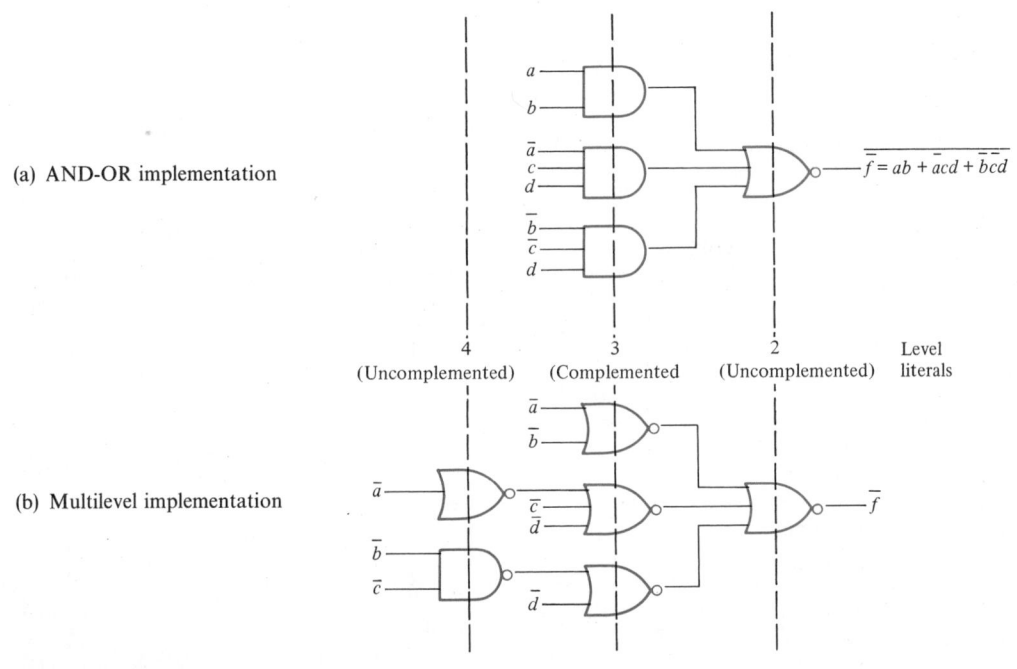

(a) AND-OR implementation

$$\overline{\overline{f}} = ab + \overline{a}cd + \overline{b}\overline{c}d$$

4 (Uncomplemented)   3 (Complemented   2 (Uncomplemented)   Level literals

(b) Multilevel implementation

Figure 3–29
Implementation of $f(a, b, c, d) = \overline{ab + \overline{a}cd + \overline{b}\overline{c}d}$ with only complemented inputs available

3.7

Characteristics of Transistor-Transistor Logic (TTL) Chips

Transistor-transistor logic (TTL) refers to a specified standard of input and output voltage and current conditions. All chips that conform to that standard are called *TTL chips*. Typical specifications for chips in the TTL, ECL, and CMOS family are shown in Figure 3–30.

Since the voltage level is not discrete, a range of voltage levels must be specified for each logical state of 0 and 1. For TTL, the input voltages for logic states 0 and 1 are 0.8 V or less and 2.4 V or more, respectively. The output voltage levels for logic states 0 and 1 are 0.4 V or less and 2.4 V or more, respectively. These two ranges of voltage levels provide a safety margin (called a *noise margin*) that ensures the output of one gate is interpreted correctly by the input of another

Characteristics			TTL		ECL		CMOS		Units
	Supply voltage		4.75–5.25		−7.0		5–15		V
	Maximum supply current		−100		−22		−0.02		mA
	Logic state		Input	Output	Input	Output	Input	Output	
	"0"	Voltage	0.8	0.5	−1.63	−1.48	1.5	0.05	V
		Current	−0.4	8	0.0005	22	−0.0001	0.5	mA
	"1"	Voltage	2.0	2.7	−1.11	−0.98	3.5	4.95	V
		Current	0.02	−0.4	0.5	−1.0	0.0001	−0.2	mA
Speed	Maximum operating frequency		33		160		10		MHz
	Maximum propagation delay		15		2.0		20		ns
Power consumption	Active		8		25		0.1		mW
	Standby		8		25		0.005		mW
Cost			Low		High		Medium		—
Availability			High		Low		Medium		—

Figure 3–30
Characteristics of various IC technology. The parameter values are taken from Texas Instruments' SN 74LS00 for TTL, Motorola's MC10100 for ECL, and National Semiconductor's MM 74C00 for CMOS

gate even in the presence of noise and electrical fluctuations. This is shown in Figure 3–31.

Characteristics of Different Performance Classes

Within the category of TTL gates, there are several different performance classes. Some of the popular ones are standard, high-speed (H), low-power (L), Schottky (S), low-power Schottky (LS), advanced Schottky (AS), and advanced low-power Schottky (ALS). The characteristics of each class are shown in Figure 3–32. The primary distinctions between performance classes are speed, power, and driving capabilities.

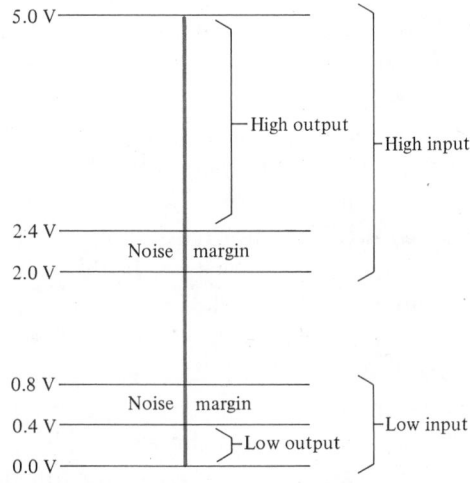

Figure 3–31
Voltage diagram showing the relationship between input and output
levels and the definition of noise margin for TTL chips

The standard chips were one of the first types on the market. The typical
propagation delay time through a gate is about 10 ns, and the usual power consumption is around 10 mW. Since their introduction, the standard chips have
most often been replaced by ALS-type chips, which are produced with a newer
technology. The high-speed (H) versions of TTL chips can operate at a higher

Figure 3–32
Characteristics of various performance classes of TTL gates

Performance Class:	Regular	H	L	S	LS	AS	ALS
Identification	7400	74H00	74L00	74S00	74LS00	74AS00	74ALS00
Propagation delay (ns)	10	6	33	3	9.5	1.5	4
Power dissipation (mW)	10	22	1	19	2	10	1
Output current logic "1" (μA)	−400	−500	−200	−1,000	−400	−2,000	−400
Output current logic "0" (mA)	16	20	36	20	8	20	8
Supply current (mA)	22	40	2	36	4.4	6	1
Speed–power product (pJ)	100	132	33	57	19	15	4
Operating frequency (MHz)	35	50	3	125	45	175	50

frequency at the expense of twice the power consumption. The increase in speed for the H chips is not as dramatic as for the ALS chips because the former are fabricated with an older technology. The H-type chips are almost always replaced by S-type chips today. The low-power (L or LP) version consumes the least amount of power among the TTL family at the sacrifice of having the slowest speed. These L chips require merely 1 mW. Still not as low as the CMOS chips, the L chips often suffice for certain low-power operations. This is particularly true for digital systems that do not require high-frequency operations. The Schottky (S) chips are among the fastest ones in the TTL family. Their speed rivals that of the ECL family. They provide TTL signals and at the same time consume less power than the ECL chips. Typically, S chips are used in critical areas of the design, where the frequency of the clock input is high and excessive propagation delays must be eliminated. The most popular type in the TTL family is the low-power Schottky (LS) class. It achieves speeds comparable to the standard chip but consumes almost as little power as the low-power version. With both speed and power in its favor, this class of chip is often the starting point in digital system designs. In terms of the speed–power product, the LS-type chips are the most efficient. The advanced Schottky (AS) chips are recently announced products that are faster and consume less power than the S-type chips. This is made possible by a new technology. The advanced low-power Schottky (ALS) is another family of fast chips made possible by newer technology. The ALS-type chips are made to replace the LS-type chips in places where timing is critical because the ALS-type chips are faster than the LS-type chips.

For comparing these performance classes, a speed–power curve is often used. A typical speed–power curve is shown in Figure 3–33. The goal is to discover a product that has the highest speed and consumes the least power. Graphically, such an ideal product will reside toward the lower left-hand corner of the figure. For real products, the speed and power are often inversely proportional to one another. It is evident from Figure 3–33 that an ALS-type chip possesses the characteristics most similar to an ideal chip when compared to the chips in the other performance classes.

Static Characteristics of TTL Gates

The primary elements in the performance of a gate are its input and output characteristics. Two main areas must be considered, namely the static characteristics and the dynamic characteristics. Static characteristics of a gate refer to the voltage and current levels at static or low-frequency conditions. Dynamic characteristics refer to the behavior of the gate when operated under high frequencies. Static characteristics include:

1. Fan-in
2. Resistive loading
3. Fan-out

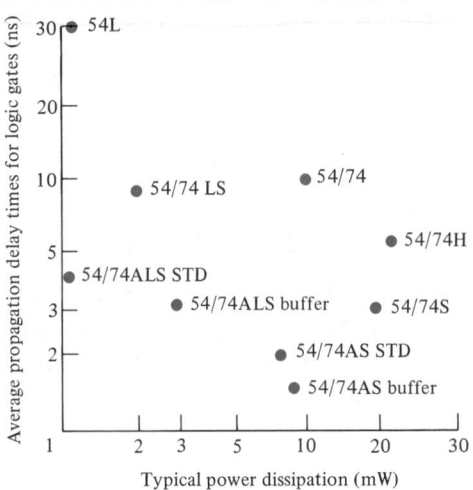

Figure 3–33
Speed–power relationship for TTL chips in various performance classes (Reprinted by permission of Texas Instruments)

The number of inputs for a particular gate corresponds directly to the gate's fabrication. This number is called *fan-in* and is usually determined at the time the chip is designed. In other words, we do not have a situation where there is an unlimited number of inputs. Hence, when using the chip, one must choose the right chip to perform the correct Boolean operations and also choose a chip that will provide the required fan-in.

Typically, the output of a gate in an IC chip would be connected to the inputs of other gates. Every connection of the output to an input of another gate is called an *output load*. Each additional connection constitutes an extra output load. Under static conditions, this loading is strictly resistive. In other words, current must be supplied by the output gate when driving a logical 1 to every output load. As more loads are connected, the output voltage will change. If the gate is driving a high at the gate output, the output voltage will gradually drop as more loads are connected. Similarly, when driving a low at the gate output, the output voltage will gradually rise as more loads are connected.

When the output is at a logical 1, the voltage drops as the number of loads is increased. Since the output voltage must be maintained above a certain level in order for it to be recognized as a logical 1, there is a definite limitation as to how many loads or output connections one particular output gate is capable of driving. Likewise, when driving a logical 0, there is a definite maximum upper bound on the output voltage level. This is shown in Figure 3–34. Typically, a standard TTL gate can drive 20 other similar gates. The number of other similar input gates that can be driven by a single output gate in such a way that all requirements of input and output specifications are satisfied is called *fan-out*. Figure 3–34 shows the fan-out for output gates from different performance classes when connected to other input gates from the same or different performance classes.

Dynamic Characteristics of TTL Gates

Whereas the static characteristics of a gate refer primarily to the voltage and current levels at static or low-frequency conditions, the dynamic characteristics refer to the behavior of the gate at high frequencies. The concerns here are related to timing, including:

1. Propagation delay
2. Maximum operating frequency
3. Capacitive loading

Parameters that can distort the waveform, such as capacitive loading and the rise/fall time, are also important.

In an ideal world, no delay time is assumed between any output changes corresponding to any input changes. In contrast, there does exist a finite delay for any changes in the input to propagate to the output in the practical world. This delay is called the *propagation delay* and is shown in Figure 3–35. When several gates are connected together, the simultaneous arrival of a particular signal at

Figure 3–34

Typical fan-out capability for TTL IC chips in various performance classes (Courtesy of Signetics Company, a division of North American Philips Corp.)

From	To				
	74LS	74	74H	74L	74S
74LS	20	5	4	40	4
74LS buffers	60	15	12	120	12
74	40	10	8	80	8
74 buffers	60	30	24	120	24
74H	50	12	10	100	10
74H buffers	75	37	30	150	30
74L	9	2	1	20	1
74S	50	12	10	100	10
74S buffers	150	37	30	150	30

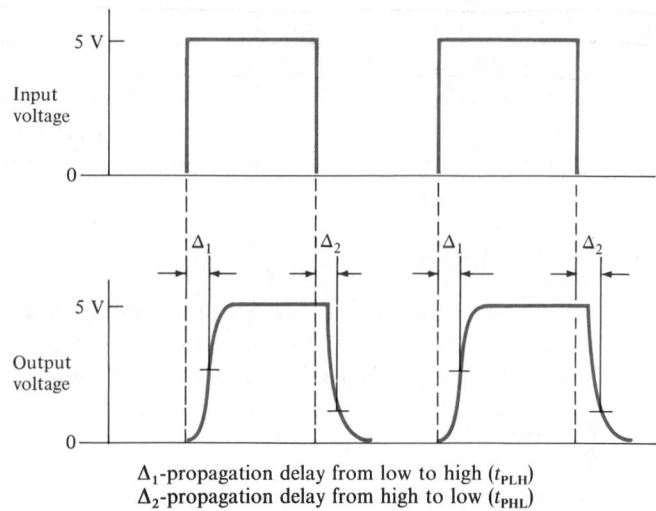

Δ_1-propagation delay from low to high (t_{PLH})
Δ_2-propagation delay from high to low (t_{PHL})

Figure 3–35
Propagation delay of a TTL gate

the gates' inputs can no longer be guaranteed; rather, it actually depends on the number of gates along the path that the signal must pass through and the length of the interconnecting wires. The total propagation delay is then the sum of each individual gate delay and the transmission time for the signal to traverse through all the interconnection paths. It is obvious that when inputs come from different places, their arrival times will seldom be simultaneous. Each takes a different path traversing through different gates. Time must be allowed for all signals to propagate through the network in order for the output to be meaningful.

There is another parasitic result related to the propagation delay. Because of the delay from the input to the output, a gate can no longer operate at unlimited frequency. Time must be allowed for the output to settle before the input can change again. This means that the gate can operate only within a specified frequency range. For a square-wave input, a perfectly squared input wave is not reproduced at the output; rather, a "low-passed" version of the input is reproduced. When the frequency of the input waveform increases beyond the allowable operating frequency, the output waveform can no longer rise and fall fast enough to the specified output levels. This cutoff condition defines the *maximum operating frequency.*

Similar to resistive loading, capacitance found externally at the input and output terminals and internally across the transistors cannot be ignored when operating at high frequencies. As the output is connected to more inputs, not only does the resistive load increase, the capacitance of the output load also increases. As a result, the waveform of the output voltage rises and falls exponentially, with a time constant corresponding to the number of output loads. On one hand, the resistive load lowers the final voltage level when all capacitors are charged; on the other hand, the capacitive load lowers the rate at which the output voltage can change, thus lowering the maximum operating frequency and increasing the rise/fall time.

Types of TTL Inputs

The specifications of a regular TTL input are well defined. To meet the requirements of other special applications, there are three other kinds of TTL input conditions that are worth mentioning. The four different types of TTL inputs are

1. Regular input
2. Strobed input
3. Schmitt-triggered input
4. Current-sensing input

Regular TTL input is the standard input configuration. For the most part, TTL input is voltage level sensitive. When the input voltage level is high, only a small amount of current is required to flow into the input node. When the input voltage is low, current flows out of the input node.

Synchronization of events is often desired. This is true when the input signals do not arrive at the same gate, or even at different gates of different parts of the circuit, at the same time. This discrepancy often arises because of propagation delay. If a particular signal is early, its arrival may cause unwanted changes in the output of a gate, causing glitches to occur and to propagate down the network of gates. Sometimes a circuit may have to wait an indefinite amount of time for a signal to arrive. In both cases it is desirable to have control over when the gate is to operate on the input signals and when it is to ignore the input signals. This control is accomplished by having a strobe that is ANDed with every input. If the strobe is low, no activities on any of the inputs will pass on to the output section of the gate. In other words, the input section is completely isolated from the rest of the gate. When the strobe is high, the gate inputs are read in as usual. The equivalent circuit of a strobed input gate is shown in Figure 3–36. Typical chips that have strobed inputs are the 74ALS23s and 74ALS25s.

In some applications where the input environment is particularly noisy or when the input is changing very slowly compared to the response time of the gate, it is possible to find the input voltage at a level that is between the accepted ranges for logic states 0 and 1. The output of a gate with regular TTL input in this case is unpredictable and may vary according to various environmental factors. This is the time when a gate with Schmitt-triggered input is needed. A graph of the output voltage versus the input voltage for a regular inverter is

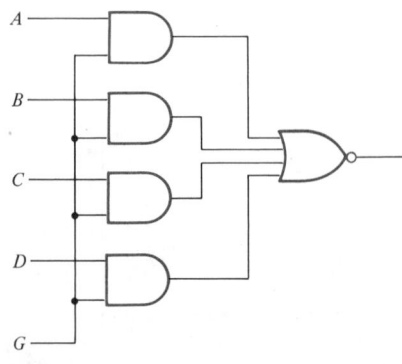

Figure 3–36
Equivalent circuit of a strobed-input gate strobe (74LS25)

shown in Figure 3–37(a). For Schmitt-triggered input, a hysteresis loop, as shown in Figure 3–37(b), is set at the input, so that, on the one hand, when the input is slowly rising, it must exceed 1.7 V before the output will change. On the other hand, once the output has changed, the input must drop below 0.8 V before the output will change again. This provides a guard band on the input so that the output will not fluctuate randomly due to noise while the input is in the transition region. The operation of an inverter with and without Schmitt-triggered input is shown in Figure 3–37(c) and 3–37(d), respectively.

Another kind of TTL input is the current-sensing input. These inputs are current sensitive rather than voltage sensitive and are ideal for applications requiring current-to-voltage conversion. Many sensors send out current rather than voltage in accordance with the input level. This sensor output current can be detected and converted to standard TTL form in a digital system by using chips with current-sensing input such as the 74ALS63 chip.

Types of TTL Outputs

In order to increase flexibility and to meet different circuit requirements, there are three kinds of TTL outputs:

1. Totem pole output
2. Open collector output
3. Tri-state output

The difference lies primarily in the way the output can be used.

The most common output type used is called the *totem pole output*. This is the typical output for TTL gates. This totem pole situation for TTL gates is ideal if the output is meant to be connected to other similar gate inputs.

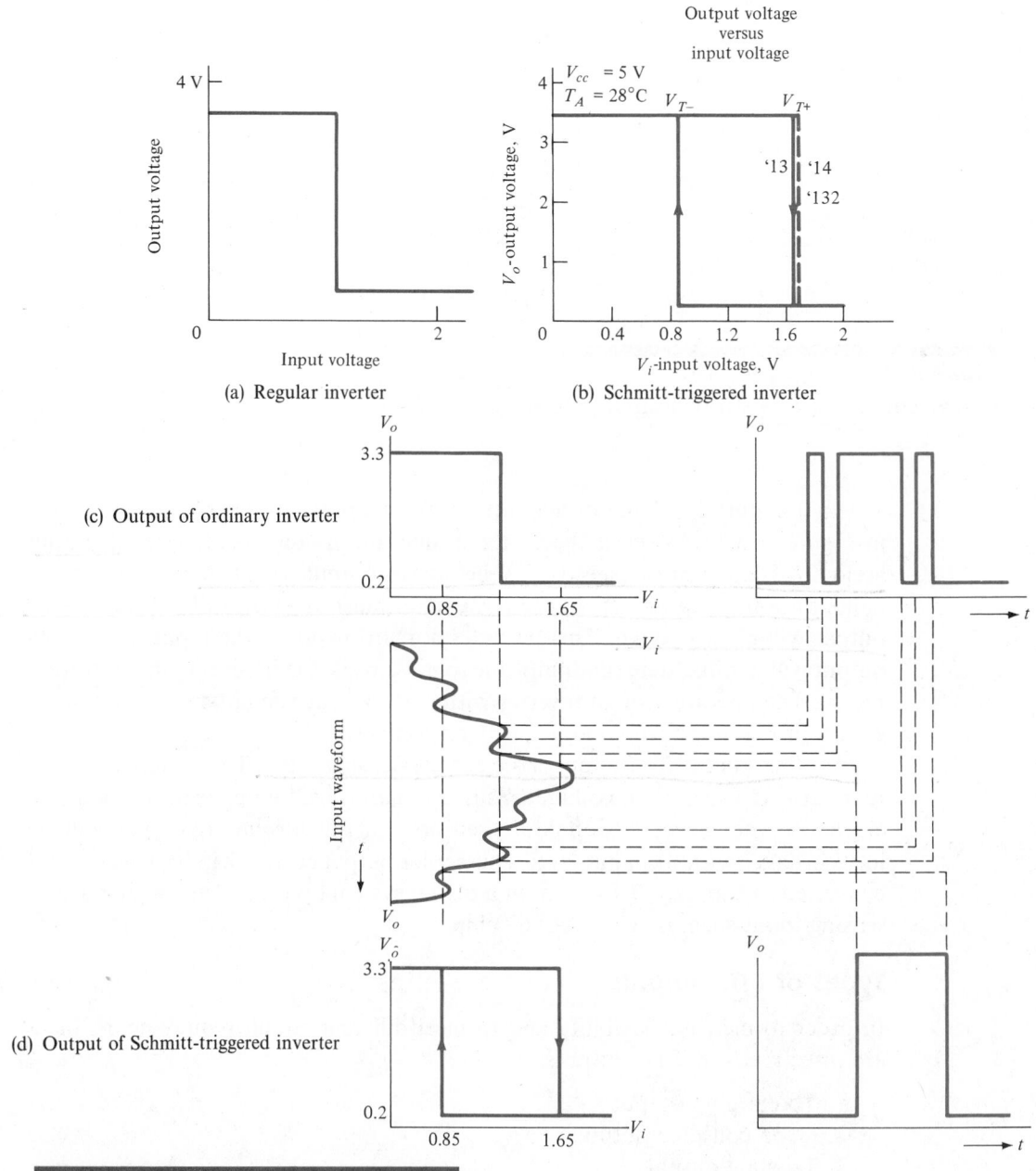

Figure 3–37

Plots showing the relationship between the input and output voltage levels (a) for a regular inverter, and (b) for a Schmitt-triggered inverter (Reprinted by permission of Texas Instruments). Typical output to a noisy input signal for (c) a regular inverter, and (d) a Schmitt-triggered inverter

The second kind of TTL output is called *open collector output* because part of the internal circuit is open and must be connected externally with a load resistor. This arrangement has three main advantages. First, if more output current is required, the load resistor can be tailored to supply the actual load. Second, the load resistor can be connected to a different supply voltage. In this way, the output can be used to drive a load with a different operating voltage. Third, open collector outputs can be connected through a load resistor to effect an OR operation among the outputs. This is called "wired-OR" logic.

The third kind of TTL output is called three-state output or *tri-state output*. As implied by the name, there are three output states instead of two. In addition to the two conventional logic states 0 and 1, the output can be effectively isolated from the connection by entering into a high-impedance state. In the high-impedance state, the output is practically disabled from the output connection. At this point, the output is not actively forcing its voltage level to be a logical 0 or 1; rather, the output "floats" so that other gates that are connected to the same junction point can drive the circuit. Tri-state outputs, therefore, are absolutely necessary whenever there is more than one driving gate connected to the same connection point. However, extreme caution is necessary when using tri-state logic so that not more than one gate is driving at the same time.

3.8
Manufacturers' Specifications

In this section we examine some typical manufacturers' specification sheets for SSI chips and become familiar with some of the industry jargon used. Description of a chip or gate usually involves specifying the input and output parameters of the gate. There are standard notations that describe these parameters. Some of these parameters are given and discussed below.

Input Parameters

For SSI chips, the input parameters are rather standardized. The primary input parameters are I_{ih}, I_{il}, V_{ih}, and V_{il}.

I_{ih} The amount of input current flowing into a gate when a high-level voltage is applied at the input.

I_{il} The amount of input current flowing into the gate when a low-level voltage is applied at the input. (Since the direction of current is always taken as positive when it is flowing into the gate at the input, I_{il} is usually negative, indicating that current is flowing out of the input gate when a low-level voltage is applied at the input.)

V_{ih} The input voltage level that must be applied to the input to produce a high voltage level; usually specified as the minimum voltage level.

V_{il} The low voltage level that must be applied at the input; usually specified as the maximum voltage level.

Output Parameters

The output section is often designed for different applications and may vary from application to application. We will look at the characteristics of each of the three types of output separately.

Totem Pole Output   When describing the output, the parameters may be separated into two categories: the static characteristics and the dynamic characteristics. The static characteristics include the following parameters:

I_{oh} The output current flowing into the gate when the output voltage level is high.

I_{ol} The output current flowing into the gate when the output voltage level is low.

V_{oh} The output voltage level that indicates a high output level. This parameter is usually given as the minimum voltage level.

V_{ol} The output voltage level that indicates a low output level. This parameter is usually given as the maximum voltage level.

t_h The hold time during which a signal must be retained at any other input terminal after an active transition occurs at a particular input terminal.

t_{su} The setup time, the time interval between the application of a signal maintained at a specified input terminal and before an active transition occurs at another input terminal.

t_{pd} The propagation delay time between the specified reference points on the input and output voltage waveforms during which the output voltage changes from one defined level to the other defined level.

When connecting chips together, it is imperative to check the fan-out requirement. This may be done by observing the current requirement of the input gates. The fan-out number can be calculated as follows:

$$N_{\text{fan-in}} = \min\left(\left[\frac{I_{ol}}{I_{il}}\right], \left[\frac{I_{oh}}{I_{ih}}\right]\right) \qquad (3-1)$$

The [ ] symbol indicates the largest integer that is smaller than the argument. The dynamic characteristics of an SSI gate include the following:

t_{plh} The propagation delay time between the specified reference points on the input and output voltage waveforms during which the output voltage changes from the defined low level to the defined high level.

t_{phl} The propagation delay time between the specified reference points on the input and output voltage waveforms during which the output voltage changes from the defined high level to the defined low level.

Open Collector Output The input and output characteristics for the open collector gates are very similar to the totem pole gates. When using open collector gates, the value for the external resistor R_L must be properly determined. This resistance value must be matched for both the high-level output condition and the low-level output condition. In Figure 3–38 we can see that when the output is high, the amount of current determined by the value of this external resistor R_L must be large enough to supply both the input current requirements for all the TTL loads and the output current requirements for all the gates connected to the same connection point. Furthermore, the value of R_L must be small enough to maintain the output voltage above the output high-level voltage requirement. Summing these requirements, the maximum value of R_L can be calculated as shown in the following equation:

$$R_{L(max)} = \frac{V_{cc} - V_{oh}}{(n \times I_{oh}) + (m \times I_{ih})} \tag{3-2}$$

where
$$V_{cc} = \text{supply voltage}$$
$$m = \text{number of TTL loads}$$
$$n = \text{number of open collector gates}$$

In the same way that the high-output voltage determines the maximum value of R_L, the low-output voltage also specifies the minimum value of R_L. When the

Figure 3–38

Circuit arrangement for calculating the maximum value of an external pull-up resistor (Reprinted by permission of Texas Instruments)

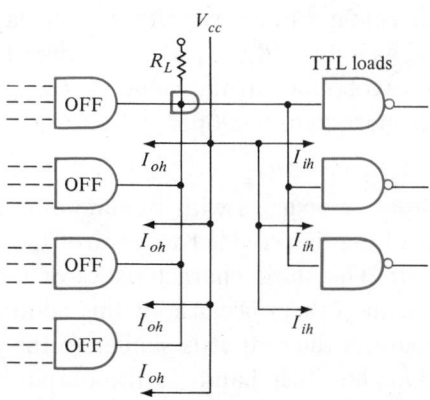

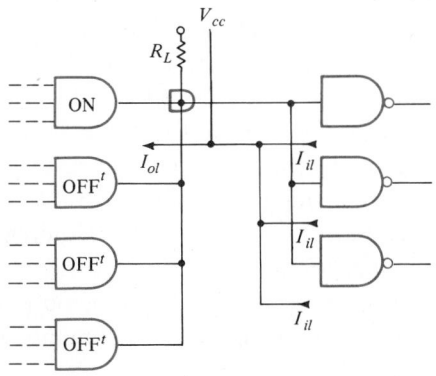

Figure 3–39

Circuit arrangement for calculating the minimum value of an external pull-up resistor (Reprinted by permission of Texas Instruments)

output is low, the output current from all the input gates and the current limited by R_L must flow into at least one of the open collector output gates as shown in Figure 3–39. Since this current flowing in must be less than a certain threshold in order to keep the low-level output voltage below the maximum value required, another constraint is put on the value of R_L as follows:

$$R_{L(\min)} = \frac{V_{cc} - V_{ol}}{I_{ol} - m * I_{il}} \tag{3-3}$$

With the above two equations, the value of R_L can be determined for any given circuit configuration. We simply calculate the maximum and minimum values of R_L. As long as $R_{L(\max)}$ is greater than $R_{L(\min)}$, we can safely pick a value between the two bounds. If the value of $R_{L(\max)}$ is less than $R_{L(\min)}$, then buffers or drivers or more than one open collector gate must be used.

Tri-state Output   For gates with tri-state output, an extra input line called the output control line is provided to control the driving capability of the output transistor pair. The static characteristics of the tri-state output are somewhat different than the others because of this control. On one hand, as the output control is enabled, the output is similar to the normal totem pole output characteristics. On the other hand, as the output control disables the output, the gate enters into a high-impedance stage.

References

1. Becher, William D. *Logical Design Using Integrated Circuits*. Hayden Book Co., Rochelle Park, N.J., 1977.
2. Booth, Taylor L. *Introduction to Computer Engineering Hardware and Software Design*. John Wiley & Sons, New York, 1984.
3. Dietmeyer, D. L. *Logic Design of Digital Systems*, 2d ed. Allyn & Bacon, Boston, 1978.
4. Fletcher, William I. *An Engineering Approach to Digital Design*. Prentice-Hall, Englewood Cliffs, N.J., 1980.
5. Greenfield, Joseph D. *Practical Digital Design Using ICs*, 2d ed. John Wiley & Sons, New York, 1983.
6. Hayes, John P. *Digital System Design and Microprocessors*. McGraw-Hill, New York, 1984.
7. Hill, Frederick J., and Gerald R. Peterson. *Introduction to Switching Theory and Logical Design*, 3d ed. John Wiley & Sons, New York, 1981.
8. Hill, F. J., and G. R. Peterson. *Digital Logic and Microprocessors*. John Wiley & Sons, New York, 1984.
9. Humphrey, W. S., Jr. *Switching Circuits with Computer Applications*. McGraw-Hill, New York, 1958, chap. 4.
10. Karnaugh, M. A Map Method for Synthesis of Combinational Logic Circuits, *Trans. AIEE, Commun. and Electronics*, 72, Part I, November 1953.
11. Kohavi, Z. *Switching and Finite Automata Theory*. McGraw-Hill, New York, 1970.
12. Kostopoulos, G. K. *Digital Engineering*. John Wiley & Sons, New York, 1975.
13. Lewin, Morton H. *Logic Design and Computer Organization*. Addison-Wesley, Reading, Mass., 1983.
14. Mano, M. Morris. *Digital Logic and Computer Design*. Prentice-Hall, Englewood Cliffs, N.J., 1979.
15. Mckay, Charles W. *Experimenting with MSI, LSI, IO, and Modular Memory Systems*. Prentice-Hall, Englewood Cliffs, N.J., 1981.
16. Nagle, H. T., Jr., B. D. Carrol, and J. D. Irwin. *An Introduction to Computer Logic*. Prentice-Hall, Englewood Cliffs, N.J., 1975.
17. National Semiconductor Corp. *CMOS Data Book*. National Semiconductor Corporation, 1981.
18. National Semiconductor Corp. *Logic Data Book*. National Semiconductor Corporation, 1981.
19. O'Connor, Patrick. *Digital and Microprocessor Technology*. Prentice-Hall, Englewood Cliffs, N.J., 1979.
20. Oleksy, Jerome E., and George B. Rutkowski. *Microprocessor and Digital Computer Technology*. Prentice-Hall, Englewood Cliffs, N.J., 1981.
21. Rhyne, V. T. *Fundamentals of Digital Systems Design*. Prentice-Hall, Englewood Cliffs, N.J., 1973.
22. Roth, Charles H., Jr. *Fundamentals of Logic Design*, 2d ed. West Publishing Co., St. Paul, Minn., 1979.
23. Rutkowski, George, and Jerome Oleksy. *Fundamentals of Digital Electronics: A Text Laboratory Manual*, 2d ed. Prentice-Hall, Englewood Cliffs, N.J., 1985.
24. Signetics Corp. *Signetics Logic—TTL Data Manual*. Signetics, 1978.
25. Taub, Herbert. *Digital Circuits and Microprocessors*. McGraw-Hill, New York, 1982.
26. Ware, W. H. *Digital Computer Technology and Design*, Vol. 1. John Wiley & Sons, New York, 1963.

27. Wiatrowski, Claude A., and Charles H. House. *Logic Circuits and Microcomputer Systems.* McGraw-Hill, New York, 1980.
28. Winkel, David, and Franklin Prosser. *The ART of Digital Design.* Prentice-Hall, Englewood Cliffs, N.J., 1980.

Experiment I

The Static Characteristics of TTL Output

OBJECTIVES

In this experiment we investigate the static characteristics of TTL output. Specifically, we want to measure the output voltages V_{ol} and V_{oh} as a function of different numbers of output loads and to deduce the effects of resistive loading.

EQUIPMENT

    1   power supply, 5 V
    1   multimeter, $3\frac{1}{2}$-digit accuracy

COMPONENTS

    5   74ALS04 hex inverters

PROCEDURE

1. Connect the input of one of the inverters to ground. Measure and record the output voltage. Then connect the same input to the power supply. Measure and record the output voltage again.
2. Connect the inputs of the other five inverters to the output of the first inverter. Measure and record the output voltage with the input at low and high levels as before.
3. Get another 74ALS04 inverter chip and connect the inputs of all six inverters to the output of the first inverter. Measure and record the output voltage with the input at low and high levels as before.
4. Repeat Step 3 for another three 74ALS04s, making a total of 29 inputs connected to the same output.

RESULTS

1. Tabulate and plot the output voltage as a function of the number of output loads when the input voltage level is low.
2. Tabulate and plot the output voltage as a function of the number of output loads when the input voltage level is high.
3. Deduce from your plots the fan-out number for the resistive loading of the gate. Be sure to check the voltage requirements on both plots.

4. Deduce from the behavior of these two curves the resistance value of the internal output resistor. Check the data book to see if your estimation is correct or not.

Experiment II

The Dynamic Characteristics of TTL Output

OBJECTIVES

In this experiment we investigate the dynamic characteristics of TTL output. Specifically, we want to measure the rise time and the final output voltage V_{oh} as a function of both operating frequency and the number of output loads. From the data, we want to deduce the effects of capacitive loading.

EQUIPMENT

1   oscilloscope, 200-MHz
1   square-wave generator, 200-MHz bandwidth (BW)
1   power supply, 5 V

COMPONENTS

5   74L04 hex inverters

PROCEDURE

1. Connect the input of one of the inverters to the signal generator. Set the frequency of the generator at 10 MHz. Measure from the oscilloscope and record the rise time (10–90%) and the final output voltage. Then set the frequency of the signal generator at 20, 30, 40, and 50 MHz; measure and record the rise time and the final output voltage for each of the frequencies.
2. Connect the inputs of the other five inverters to the output of the first inverter. Measure and record the rise time and the final output voltage for each of the five frequencies.
3. Get another 74ALS04 inverter chip and connect the inputs of all six inverters to the output of the first inverter. Measure and record the rise time and the final output voltage for all five frequencies.
4. Repeat Step 3 for another three 74ALS04s, making a total of 29 inputs connected to the same output.

RESULTS

1. Tabulate and plot the rise time for a particular input frequency as a function of the number of output loads. Use different plot lines for the various frequencies.

2. Tabulate and plot the rise time for the same number of output loads as a function of frequencies. Use different plot lines for the various numbers of output loads.

3. Deduce from your plots the fan-out number for the capacitive loading of the gate. Be sure to check the requirements on both plots.

4. Deduce from the behavior of these two curves the resistance value of the internal output resistor. Check the data book to see if your estimation is correct or not.

Experiment III

Design of a Majority Circuit

OBJECTIVES

In this experiment we take a practical problem and carry out the design from the input definition to the actual circuit realization. The problem is to implement the majority voting circuit discussed earlier in the text.

EQUIPMENT

  1   power supply, 5 V
  5   SPST switches
  5   LEDs
  5   220-Ω resistors

COMPONENTS

  2   74ALS00 quad two-input NANDs
  2   74ALS04 hex inverters
  2   74ALS08 quad two-input ANDs
  5   74ALS10 triple three-input NANDs

PROCEDURE

1. Define the inputs. Assign a logical variable to each member of the voting team. Determine the relationship between the voltage levels (high or low) and the outcome of each vote ("yes" or "no").

2. Define the output. Assign a logical variable to the output of the system to indicate the majority vote.

3. Construct the truth table.

4. Minimize the output function by Karnaugh maps or any other available methods. Obtain an expression for both the sum-of-products form and the product-of-sums form.

5. Choose the best set of SSI chips to realize the functions. Use all the substitu-

tion rules to change the output function to obtain a solution to realize the function with the least number of chips.

6. Wire the circuit to verify the design.
7. Calculate the cost and power consumption for your design.
8. Assume that you are restricted to using only one type of gate, e.g., NAND gates. How does this constraint change your design? Calculate the cost and power consumption of this design.
9. Assume that you are further restricted to using only 74ALS10 triple, three-input NANDs. How does this change your design? Calculate the cost and power consumption of this design.

RESULTS

1. Compare the costs of your three designs.
2. Compare the power consumption of your designs.
3. If your design must be portable, the power must now come from four AA batteries. Estimate the battery life of your designs.
4. If you change your design to use only L-type chips rather than ALS-type chips, how much increase in operational time do you gain?

Problems

1. With SSI chips, realize the following functions with an actual design using the sum-of-products form:
   (a) $f(a, b, c, d) = \sum(0, 1, 2, 4, 6, 13, 15)$
   (b) $f(a, b, c, d) = \sum(0, 2, 6, 8, 9, 10, 12, 14)$
   (c) $f(a, b, c, d) = \sum(1, 3, 4, 5, 6, 11, 12, 13, 14)$
   (d) $f(a, b, c, d) = \sum(4, 5, 7, 8, 15)$

2. With SSI chips, realize the following functions with an actual design using the product-of-sums form:
   (a) $f(a, b, c, d) = \prod(0, 1, 2, 4, 6, 13, 15)$
   (b) $f(a, b, c, d) = \prod(3, 4, 5, 6, 12, 13, 14, 15)$
   (c) $f(a, b, c, d) = \prod(0, 6, 7, 10, 11, 12)$
   (d) $f(a, b, c, d) = \prod(1, 3, 5, 7, 9, 11, 13, 15)$

3. Using SSI chips, realize the following functions with an actual design with the minimum number of chips:
   (a) $f(a, b, c, d) = \sum(2, 3, 5, 7, 9, 10)$
   (b) $f(a, b, c, d) = \prod(4, 7, 8, 9, 10, 11)$

4. Repeat Problem 3 using NAND gates only.

5. Repeat Problem 3 using NOR gates only.

6. Find an expression for a 2-bit binary half-adder and realize the switching function with standard SSI chips.

7. Realize a quad Exclusive-OR chip with standard SSI chips. (*Hint*: You need only three SSI chips to do this.)

8. Assume that one has four switches that indicate on or off conditions in TTL levels; i.e., on is high level and off is low level. Design a four-way control box so that action in any of the switches will change the current state of the light.

9. Design an SSI circuit to convert a 4-bit incoming code to another 4-bit outgoing code. The required code conversions are

	Incoming	to	Outgoing
(a)	BCD	to	Gray
(b)	Gray	to	BCD
(c)	BCD	to	1's complement BCD
(d)	1's complement	to	BCD

10. Design a three-person majority circuit. Each person is allowed a yes/no vote.

11. Design a three-person majority circuit. Each person is allowed the option to abstain or to vote with a yes/no vote.

12. The ignition can be turned on only if the driver's seatbelt is turned on. Furthermore, if a passenger is present, the passenger's seatbelt must also be on. Define clearly what are the input and output signals, then design the internal circuit.

13. Design a 4-bit parity generator.

14. Design a 5-bit parity checker. Four bits are allocated to the information bit and the additional bit is for the parity bit.

15. Design an elevator controller. Assume that there are four levels in the building. There is an up and a down button on each floor. Additionally, the current elevator direction is sensed by a binary variable as to whether the elevator is going up or down. The single output of the controller is two binary variables with the following indications:

z_1	z_2	Condition
0	0	Stop
0	1	No change
1	0	Go up
1	1	Go down

16. The seatbelt light must be on if the front passengers have not put on their belts. Note that if no passenger is present, the belt does not have to be buckled. Design the logic for this controller.

17. Use SSI circuit components to design a full adder.

18. Use SSI circuit components to design a 1's complementer and a 2's complementer.

19. Given the multilevel circuits shown in Figure P3–1, show that they are all equivalent circuits for a full adder. Compare the various implementations.

20. Given the multilevel circuits shown in Figure P3–2, find the logic expressions for the circuits and redesign these circuits in two-level implementations.

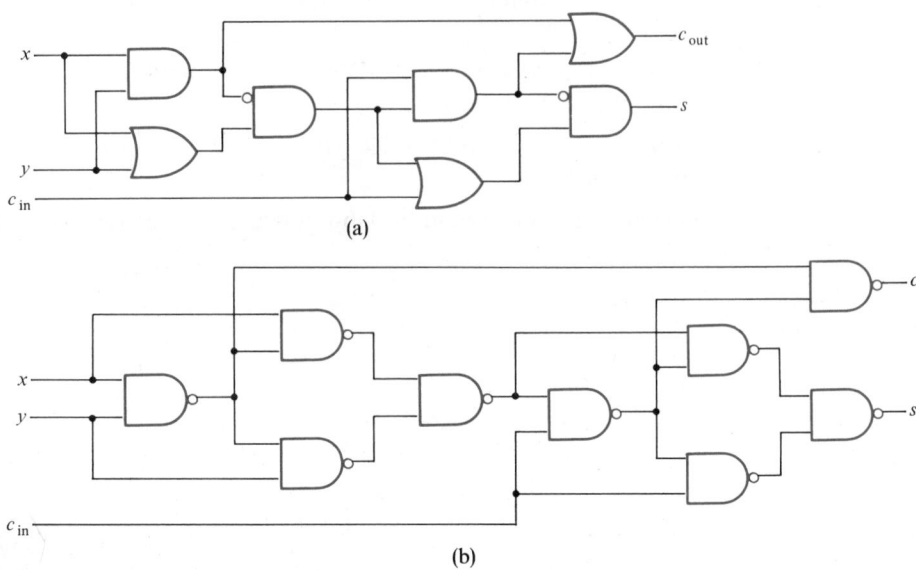

Figure P3-1
Equivalent circuits for a full adder

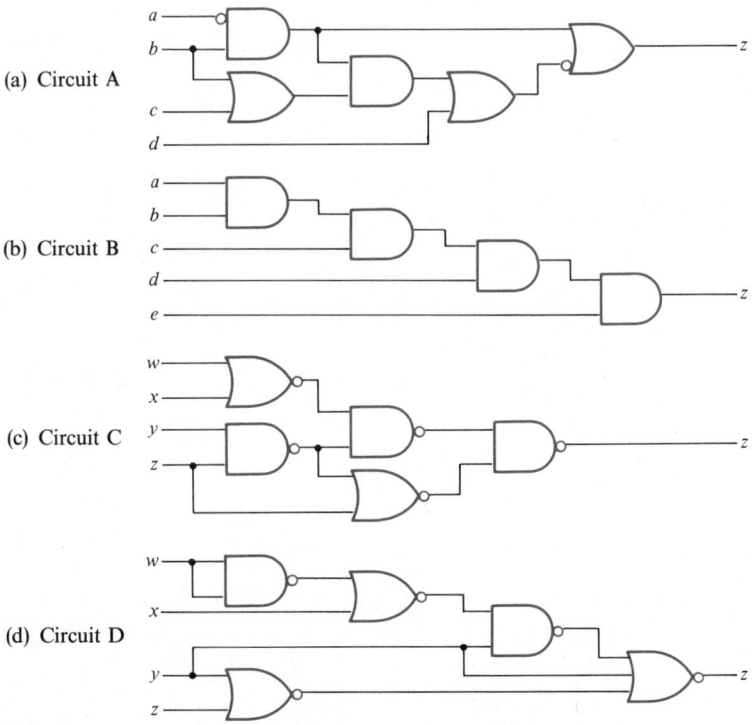

(a) Circuit A

(b) Circuit B

(c) Circuit C

(d) Circuit D

Figure P3-2
Multilevel circuits

21. Show how a two-input NAND gate or a two-input NOR gate can be used as an inverter.

22. Show how a set of two-input NAND gates can be used to replace a three-input and a five-input NAND gate.

23. Show how a set of two-input NOR gates can be used to replace a three-input and a five-input NOR gate.

24. Verify the information given in the fan-out figure between the different TTL logic types.

Advanced Design Techniques with Combinational Circuits

4

In this chapter we concentrate on medium-scale integration (MSI) components used in combinational circuits and show how they can be used in many digital system applications. First we examine closely the four major categories of MSI component functions for combinational circuits, including arithmetic operations, data selection operations, data formatting operations, and data communication operations. The components that perform these functions are the basic tools for MSI circuit design. Second, the design procedures for circuits using MSI components are presented and illustrated. The last part of the chapter is devoted to further examples of MSI circuits as applied to binary and decimal arithmetic circuits.

In the previous chapter, small-scale integration (SSI) components were introduced and design techniques for combinational circuits using these components were discussed in detail. In this chapter the primary emphasis is placed on medium-scale integration (MSI) components for combinational circuits and their design techniques. Components for sequential circuits will be treated in a later chapter. While SSI components are composed primarily of gates used to realize the basic set of Boolean functions, MSI components are composed of elementary functional units to do specific tasks such as data conversions, additions, etc. The MSI components are considerably more complicated than the basic gates and are designed to perform more specific and dedicated operations than SSI components.

There are many MSI chips with a variety of interconnected gates that implement commonly used combinational circuits. Minimizing the number of integrated circuits and external interconnections is a practical goal in digital system design because the cost of packaging is quite often determined in part by the number of integrated circuits. To achieve this goal, different design procedures must be mastered.

4.1

MSI Circuit Components

There are four main groups of MSI products applicable for combinational circuits. As shown in Figure 4–1, the functions of these four groups are

1. Arithmetic operations
2. Data format conversions
3. Data selections
4. Data communication

Arithmetic functions include operations such as addition, subtraction, multiplication, and magnitude comparison. Arithmetic chips can be used for additions and subtractions in various number representations, and along with other similar chips arranged in an array can be used for multiplications and divisions also. Data format conversion refers to the conversion of various data formats, such as binary to binary-coded decimal (BCD), BCD to display codes, etc. Data converters include encoders, decoders, and various kinds of display drivers. Being an integral part of any digital design, data selection is used frequently in governing and controlling the flow of information through a network. Data selectors include multiplexers, demultiplexers, and XOR gates. Data communication chips include mostly parity generators and parity checkers at the MSI level.

When describing MSI components in the following section, the discussion for each component is centered around the following four areas:

1. Equivalent circuit diagram
2. Truth table

Number of Operands	Category	Type	Group	Number
Dual operands	Arithmetic units	Adders	1-bit full adder 2-bit full adder 4-bit full adder	80, 183 82 83, 283
		Look-ahead carry generator	Look-ahead carry generator	182
		Comparators	Magnitude comparator	85
		ALUs	4-bit accumulator 4-bit ALU	281 181
		Multipliers	Multiplier Wallace tree	261 275
Single operands	Data format converters	Encoders	BCD priority encoder Octal priority encoder	147 148, 348
		Decoders	BCD-decimal decoder	45, 145
		Converters	BCD-7 segment	46, 246 47, 247 48, 248 49, 249
			Binary-BCD BCD-binary	185 184
	Data selectors	$\oplus/\overline{\oplus}$	Quad 2-input $\oplus$ Quad 2-input $\overline{\oplus}$ Quad 2-input $\oplus/\overline{\oplus}$	86, 136, 386 266 135
		T/C-∅/1	True/complement-∅/1	87
		Multiplexers	16-to-1 MUX 8-to-1 MUX 4-to-1 MUX 2-to-1 MUX	150 151, 251, 351 153, 253, 352, 353 98, 298, 398, 399 157, 257 158, 258
		Demultiplexers	4-to-16 DEMUX 4-to-10 BCD 4-to-10 Excess-3 3-to-8 2-to-4	154, 159 42 43, 44 138 139, 155, 156
	Data communication units	Parity generators/checkers	9-bit parity 8-bit parity generator/converter	280 180

Figure 4–1

Different categories of medium-scale integration (MSI) components

3. Literal description
4. Application

It should be understood that for MSI components, the literal description is by no means complete and exact. The only exact definition comes from the function table or truth table itself, which gives the precise output of the chip in response to any possible input combination.

4.2

MSI Components for Arithmetic Functions

From Figure 4–1 it is evident that there is a great variety of chips designed to perform many arithmetic functions. Only five of these most commonly used components are covered here in detail. These are listed below:

1. Full adders
2. Look-ahead carry generators
3. Magnitude comparators
4. Universal arithmetic and logic modules
5. Binary multipliers

Full Adders

A binary 1-bit full adder can be used to add two binary 1-bit numbers. The full adder is a combinational circuit with three 1-bit inputs [A, B, C(in)] and two 1-bit outputs [S, C(out)]. The truth table for a 1-bit binary adder is given in Figure 4–2(a). The Karnaugh map and sum-of-products realization are given in Figures 4–2(b) and 4–2(c), respectively. The full adder logic circuit has been extensively analyzed because of its wide use over the years. A full adder implemented with only two input NAND gates is given in Figure 4–2(d). The analysis of this circuit is left as a problem. This is by no means the only realization for a full adder. Many other designs have been proposed to perform the addition operation. Full adders are most frequently used for performing simple arithmetic operations in digital circuits where microprocessors are unnecessary. Full adders can also be used in many applications requiring the sum of different inputs.

74LS183—A Dual 1-Bit Full Adder An example of a full adder is the 74LS183 chip, which contains two 1-bit full adders. The two full adders are identical. Each has its own individual inputs and outputs. There are three input bits for each adder: the two 1-bit inputs, A and B, and the carry-in bit, C(in). Each adder also has two output bits: the sum output bit, S, and the corresponding carry-out bit, C(out).

A 4-bit full adder can be obtained directly by cascading four 1-bit full adders together. The carry-out from the first stage becomes the carry-in for the second

A	*B*	*C*(in)	*C*(out)	*S*
0	0	0	0	0
0	0	1	0	1
0	1	0	0	1
0	1	1	1	0
1	0	0	0	1
1	0	1	1	0
1	1	0	1	0
1	1	1	1	1

(a) Truth table

C(in) \ *AB*	00	01	11	10
0	0	1	0	1
1	1	0	1	0

$$S = \bar{A}\bar{B}C(\text{in}) + \bar{A}B\bar{C}(\text{in}) + A\bar{B}\bar{C}(\text{in}) + ABC(\text{in})$$

C(in) \ *AB*	00	01	11	10
0	0	0	1	0
1	0	1	1	1

$$C(\text{out}) = AB + BC(\text{in}) + AC(\text{in})$$

(b) Karnaugh maps

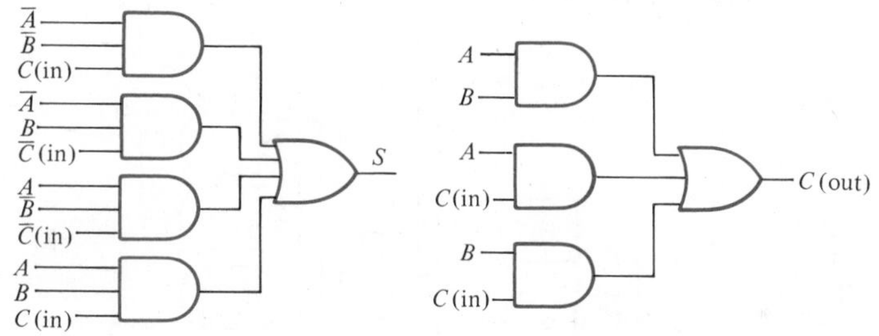

(c) Full adder implemented with sum of products

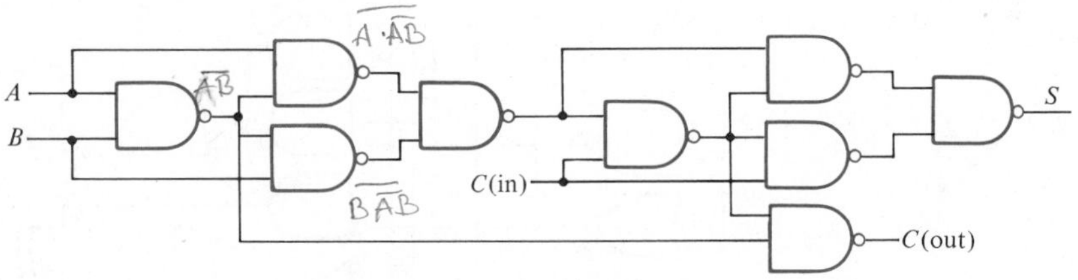

(d) Full adder with NAND gates

Figure 4–2
Implementation of a 1-bit full adder

stage, the carry-out from the second stage becomes the carry-in for the third stage, and so on. Adders connected in this way are called *ripple carry adders.* Alternatively, the carry-in for each stage and the final carry-out can be realized as a combinational circuit directly from the input signals. Adders connected in this way are called *parallel carry adders.* Ripple carry adders are in general slower

(a) Symbol

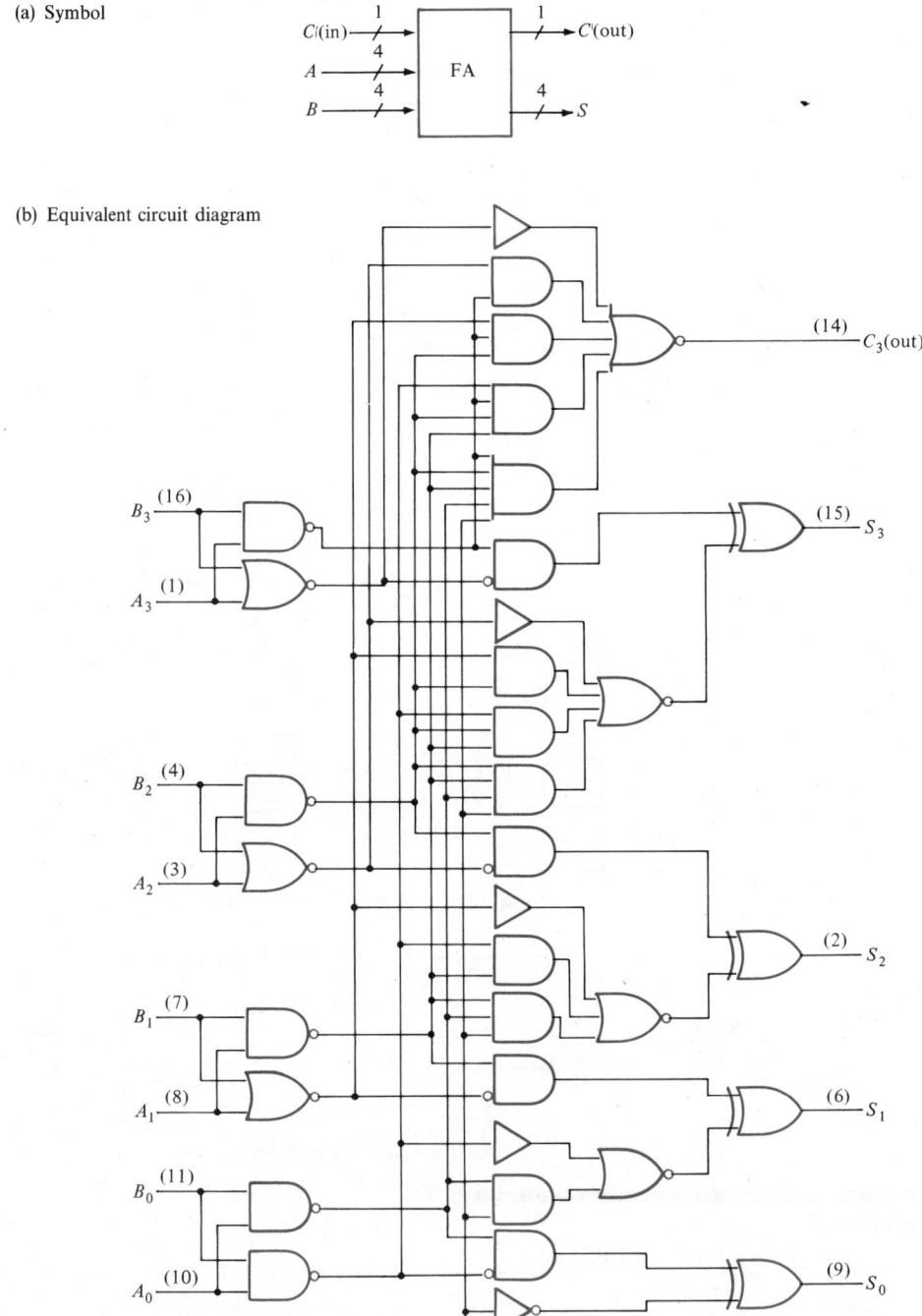

(b) Equivalent circuit diagram

Figure 4–3

A 4-bit full adder (74LS83) (Part b reprinted by permission of Texas Instruments)

than parallel carry adders because the carry must propagate from the least significant bit to the most significant bit. Hence addition is accomplished one bit at a time. Though ripple carry adders are slower, they are nevertheless much simpler in design than parallel carry adders.

74LS83—A 4-Bit Full Adder   In the 74LS83, in order to achieve faster operation, the carry from the most significant bit of this 4-bit adder is computed directly from the input operands and the input carry using combinational logic. The logic symbol and the equivalent circuit diagram of a 74LS83 full adder are shown in Figure 4–3. The input to the adder is the two 4-bit operands and the carry-in bit for the least significant bit. The output of the adder is another 4-bit sum and the carry-out bit from the most significant bit. The 4-bit cascaded full adder can be used for other combinational logic applications that may not be readily apparent.

When summing large numbers with a word length that is more that 4 bits long, several full adders must be used. The carry-out from the least significant adder then becomes the carry-in to the adder for the next significant bit. Figure 4–4 shows a typical connection of full adders for adding two 16-bit numbers using four 4-bit full adder chips in a ripple carry configuration.

Depending on the binary inputs, the main problem when cascading these full adders in the ripple carry configuration is that the carry-out bit for the most significant bit can be a function of the least significant inputs. This implies that the overall delay is equal to the delay of each full adder times the number of full

Figure 4–4

Typical connection of full adders for adding two 16-bit numbers in a ripple carry configuration

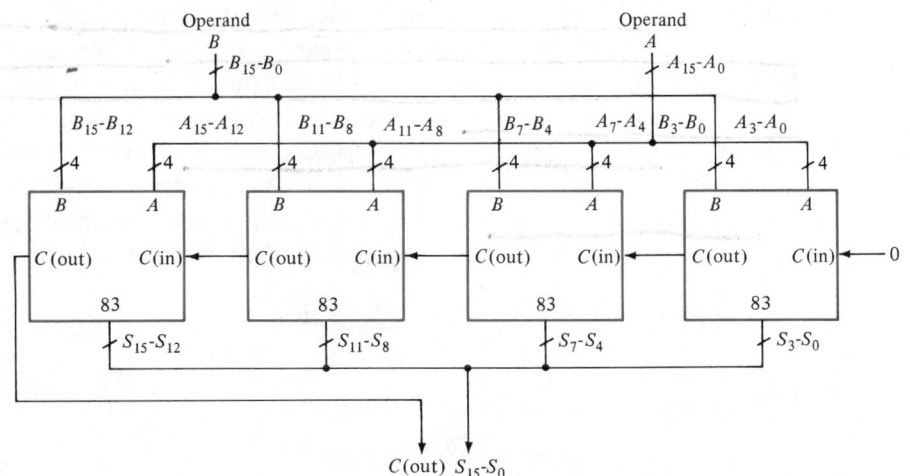

adders:

$$t_{total} = nt_{FA}$$ (4-1)

where n is the number of full adder chips and t_{FA} is the delay time of a full adder chip. Using typical time delays of 15 ns for each full adder, the maximum delay time for this ripple carry configuration for the whole chip with four full adders is 60 ns. If the number of adders is large because of a large number of bits, the complete addition process can be slowed down significantly.

Digital logic that makes repeated use of the same circuit is called *iterative logic*. The interconnection of the full adders is an example of iterative logic. The case where logic is expanded in one direction is called one-dimensional iterative logic. Two-dimensional iterative logic creates arrays. Iterative networks normally reduce the cost of the logic with an accompanying increase in delay. This is just one example of an engineering trade-off.

Example 4-1 ▬▬▬▬▬▬▬▬▬▬▬▬▬▬▬▬▬▬▬▬▬▬▬▬

Design a BCD-to-Excess-3 code converter.

The codes for the BCD and the Excess-3 have been given in Chapter 1. The Excess-3 code is obtained by adding 3 to the BCD code. Thus the conversion can be obtained by hard-wiring a 0011 in base 2 to one set of inputs in the 4-bit full adder and the BCD code to the other set of inputs.

Look-Ahead Carry Generators

For high-performance applications, a look-ahead carry generator is used to generate the carry-in bits beforehand without having each stage waiting for the result of the addition from the previous stage. Consider the truth table for a full adder with A, B, and $C(in)$ as inputs and with S and $C(out)$ as outputs as shown in Figure 4-5. When A and B are both zero, $C(out)$ will always be zero regardless of the value of $C(in)$. This is called the no-carry condition. When A and B are complements of one another, $C(out)$ depends on $C(in)$. If $C(in) = 0$, then $C(out) = 0$. If $C(in) = 1$, then $C(out) = 1$. In other words, the carry-in is propagated to the carry-out. This is called the *carry-propagate condition*. Lastly, if both A and B are 1, there is a carry-out regardless of the value of the carry-in. This is called the *carry-generate condition*. If we now designate P and G as the carry-propagate and carry-generate conditions, respectively, then P and G can be expressed as follows:

$$P = A \oplus B$$
$$G = AB$$

From P and G, one can define the carry-out as

$$C(out) = G + PC(in)$$

Condition	A	B	C(in)	C(out)	S	G	P
No-carry	0	0	0	0	0	0	0
	0	0	1	0	1	0	0
Carry-propagate	0	1	0	0	1	0	1
	0	1	1	1	0	0	1
	1	0	0	0	1	0	1
	1	0	1	1	0	0	1
Carry-generate	1	1	0	1	0	1	0
	1	1	1	1	1	1	0

carry-generate condition   carry-propagate condition

Figure 4–5

Truth table of a 1-bit binary addition showing the no-carry condition, the carry-propagate condition, and the carry-generate condition

Denoting the least significant bit of the operands as A_0 and B_0 and the original $C(in)$ as $C_0(in)$, the carry-out for the first bit is $C_0(out)$, which becomes the carry-in for the second bit $C_1(in)$. The general case for the ith bit is

$$C_i(out) = C_{i+1}(in)$$

The carry-in for the second bit can now be rewritten as

$$C_1(in) = C_0(out) = G_0 + P_0 C_0(in)$$

The carry-in for the third bit can also be written:

$$C_2(in) = C_1(out) = G_1 + P_1 C_1(in)$$

But $C_1(in)$ is exactly $C_0(out)$. Substituting the expression for $C_1(in)$ from the equation for the second bit, we obtain

$$C_2(in) = G_1 + P_1 G_0 + P_1 P_0 C_0(in)$$

The same method of substitution can be applied for the carry-in of the other higher-order bits:

$$C_3(in) = G_2 + P_2 G_1 + P_2 P_1 G_0 + P_2 P_1 P_0 C_0(in)$$
$$C_4(in) = G_3 + P_3 G_2 + P_3 P_2 G_1 + P_3 P_2 P_1 G_0 + P_3 P_2 P_1 P_0 C_0(in)$$

Note that the expressions for $C_1(in)$, $C_2(in)$, $C_3(in)$, and $C_4(in)$ are expressed as functions only of the input parameters, P's, G's, and the initial carry-in bit. This means that all the carry-out bits can be calculated from the input operands and

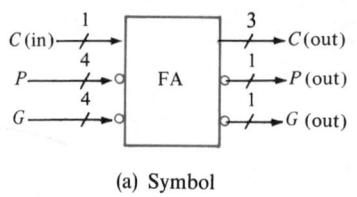

(a) Symbol

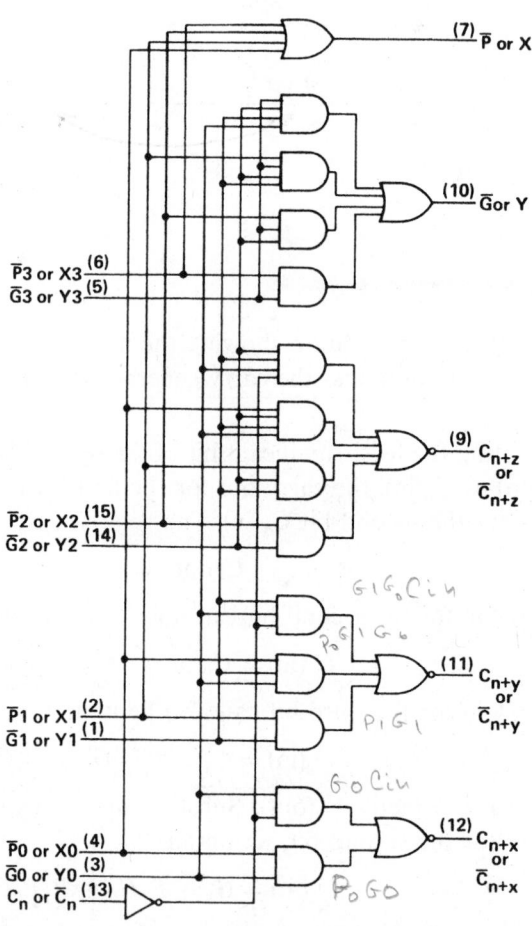

(b) Equivalent circuit

Figure 4-6

A 4-bit look-ahead carry generator (74LS182) (Part b reprinted by permission of Texas Instruments)

the initial carry-in at the same time using two-level combinational circuits with a maximum time of two gate delays. This is exactly what a look-ahead carry generator implements.

74ALS182—A 4-Bit Look-Ahead Carry Generator The use of a look-ahead carry generator approximates the behavior of a parallel carry configuration. The inputs to the look-ahead carry generator are the G's and P's for each adder. The G's and P's indicate the carry-generate and carry-propagate conditions. The outputs of the look-ahead carry generator are the individual carry-in bits. The logic symbol and the equivalent circuit diagram of a look-ahead carry generator are shown in Figure 4–6. The inputs to this chip are the P and G functions of the four input bits. The initial carry-in for the least significant bit is also required. The outputs for this chip are the three carry-out lines for the first three bits beginning from the least significant bits. Additional outputs consist of the P and G signals that can be used for higher levels of look-ahead carry operations when connecting the look-ahead carry generator in a tree structure.

Example 4–2

Use a look-ahead carry generator in conjunction with full adders to add two 16-bit numbers.

With the use of the look-ahead carry generator, the addition of two 16-bit numbers can be done as shown in Figure 4–7. Since the numbers to be added are 16 bits long, four 4-bit full adder (FA) chips are required. A single look-ahead carry generator (LACG) is needed for these four adder chips. The timing of this design can be calculated as

$$t_{\text{total}} = t_{\text{XOR}} + \log_4 n \times t_{\text{LACG}} + t_{\text{FA}} \qquad (4–2)$$

where n is the number of full adder chips. For this configuration, the total delay time is

$$t_{\text{total}} = (10 \text{ ns}) + (1 \times 21 \text{ ns}) + (15 \text{ ns})$$
$$= 46 \text{ ns}$$

For comparison, note that the delay time for four full adders connected in the ripple carry fashion is 60 ns as calculated before. The advantage of using the look-ahead carry generator is obvious. This increase in speed is achieved largely because all four adders are adding their respective pieces of the number simultaneously. Comparing (4–1) with (4–2), one can see that the use of look-ahead carry generators results in a logarithmic increase in time rather than a linear increase in time as in the straight connection case. For numbers with larger word sizes, the look-ahead carry generators can themselves be cascaded, as shown in Figure 4–8.

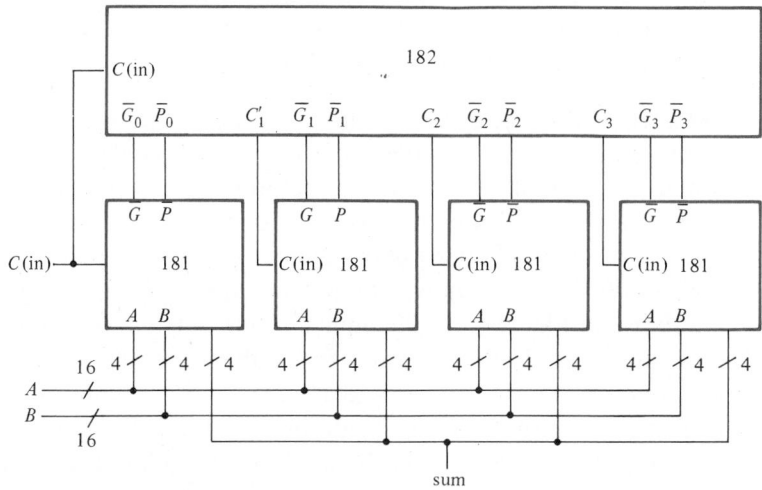

Figure 4-7
Connecting four ALUs together through a look-ahead carry generator
for fast addition

Magnitude Comparators

A magnitude comparator is used primarily to compare numbers for decision
making. In other words, its main function is to compare the magnitude of two
numbers in unsigned representation, and to determine if one is larger, smaller,
or equal to the other. Let the two input numbers supplied to the magnitude comparator be A and B. Then the three binary outputs are $(A > B)_{out}$, $(A = B)_{out}$,
and $(A < B)_{out}$. A comparator circuit for comparing two n-bit numbers has 2 to

Figure 4-8
Cascading layers of look-ahead carry generators for maximum
efficiency (Reprinted by permission of Texas Instruments)

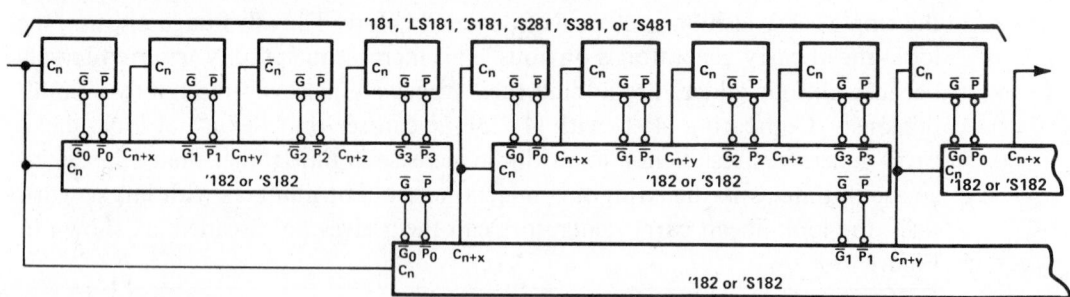

the power of 2^n entries in the truth table. The sum-of-products procedure becomes hard to handle even when $n = 3$. A possible solution for large n is to construct an iterative circuit.

The procedure for comparing two binary numbers begins by examining the most significant bits of the numbers. Let these bits be A_{n-1} and B_{n-1}. If they are different, then we already know if $A > B$ or $A < B$. If they are the same, then we must examine the next most significant bits, bits A_{n-2} and B_{n-2}, etc. Let us design an iterative cell that has inputs A_i and B_i, $(A < B)_{\text{in}}$, $(A > B)_{\text{in}}$, and $(A = B)_{\text{in}}$, and outputs $(A < B)_{\text{out}}$, $(A = B)_{\text{out}}$, and $(A > B)_{\text{out}}$. The output definitions are such that $(A < B)_{\text{in}} = 1$ implies that $A < B$, and $(A > B)_{\text{in}} = 1$ implies that $A > B$, and $(A = B)_{\text{in}} = 1$ implies that $A = B$ up to that point for all the less significant bits. Then the following equations are true:

$$(A < B)_{\text{out}} = (A < B)_{\text{in}}(A_i \oplus B_i) + \overline{A}_i B_i$$
$$(A > B)_{\text{out}} = (A > B)_{\text{in}}(A_i \oplus B_i) + A_i \overline{B}_i$$
$$(A = B)_{\text{out}} = (A = B)_{\text{in}}(A_i \oplus B_i)$$

If there are no cascading inputs from less significant bits, then the cascading inputs must be terminated properly. The $(A < B)_{\text{in}}$ and the $(A > B)_{\text{in}}$ lines must be connected to a logical 0, while the $(A = B)_{\text{in}}$ line must be connected to a logical 1. Note that only one of the inputs $(A < B)_{\text{in}}$, $(A > B)_{\text{in}}$, and $(A = B)_{\text{in}}$ can be 1 at any time.

74LS85—A 4-Bit Magnitude Comparator   The logic symbol and the function table of the 74LS85 4-bit magnitude comparator are shown in Figure 4–9. Inputs to this chip consist of two 4-bit operands to be compared. In addition, three cascading inputs are labeled $(A < B)_{\text{in}}$, $(A = B)_{\text{in}}$, and $(A > B)_{\text{in}}$. These inputs are used to connect these comparators together, and their function will be explained later. There are three outputs: $(A < B)_{\text{out}}$, $(A = B)_{\text{out}}$, and $(A > B)_{\text{out}}$ lines. The $(A < B)_{\text{out}}$ line is an active high line and is asserted whenever the magnitude of A is less than but not equal to B. The $(A = B)_{\text{out}}$ line is an active high line and is asserted whenever the magnitude of A is equal to B provided that the cascading inputs are terminated properly. Likewise, the $(A > B)_{\text{out}}$ line is an active high line and is asserted whenever the magnitude of A is greater than B. Of these three output lines, only one can be true at any time. The one that indicates the true condition for the two input operands is driven high.

In normal applications, the cascading inputs can be connected to the least significant bits of the inputs with the $(A = B)_{\text{in}}$ line tied high; or they can be terminated properly by connecting the $(A = B)_{\text{in}}$ line to high, the $(A > B)_{\text{in}}$ line low, and the $(A < B)_{\text{in}}$ line low. For the purpose of comparing numbers of larger word size, three cascading inputs are provided. These are the $(A = B)_{\text{in}}$ line, the $(A < B)_{\text{in}}$ line, and the $(A > B)_{\text{in}}$ line. These cascading inputs of the comparators for more significant bits can be connected directly to the outputs of the comparators from less significant bits. When the magnitudes of the two input

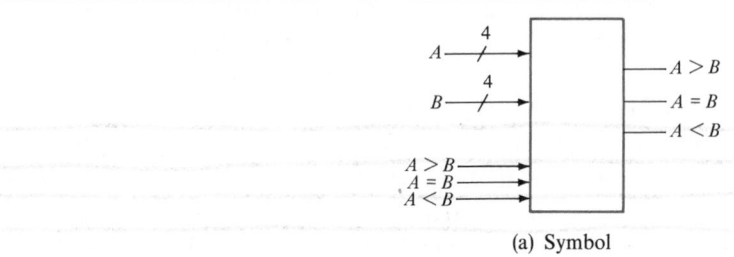

(a) Symbol

Comparing Inputs				Cascading Inputs			Outputs		
A_3, B_3	A_2, B_2	A_1, B_1	A_0, B_0	$A > B$	$A < B$	$A = B$	$A > B$	$A < B$	$A = B$
$A_3 > B_3$	X	X	X	X	X	X	H	L	L
$A_3 < B_3$	X	X	X	X	X	X	L	H	L
$A_3 = B_3$	$A_2 > B_2$	X	X	X	X	X	H	L	L
$A_3 = B_3$	$A_2 < B_2$	X	X	X	X	X	L	H	L
$A_3 = B_2$	$A_2 = B_2$	$A_1 > B_1$	X	X	X	X	H	L	L
$A_3 = B_3$	$A_2 = B_2$	$A_1 < B_1$	X	X	X	X	L	H	L
$A_3 = B_3$	$A_2 = B_2$	$A_1 = B_1$	$A_0 > B_0$	X	X	X	H	L	L
$A_3 = B_3$	$A_2 = B_2$	$A_1 = B_1$	$A_0 < B_0$	X	X	X	L	H	L
$A_3 = B_3$	$A_2 = B_2$	$A_1 = B_1$	$A_0 = B_0$	H	L	L	H	L	L
$A_3 = B_3$	$A_2 = B_2$	$A_1 = B_1$	$A_0 = B_0$	L	H	L	L	H	L
$A_3 = B_3$	$A_2 = B_2$	$A_1 = B_1$	$A_0 = B_0$	L	L	H	L	L	H
$A_3 = B_3$	$A_2 = B_2$	$A_1 = B_1$	$A_0 = B_0$	X	X	H	L	L	H
$A_3 = B_3$	$A_2 = B_2$	$A_1 = B_1$	$A_0 = B_0$	H	H	L	L	L	L
$A_3 = B_3$	$A_2 = B_2$	$A_1 = B_1$	$A_0 = B_0$	L	L	L	H	H	L

(b) Function table

Figure 4–9

A 4-bit magnitude comparator (74LS85) (Part b reprinted by permission of Texas Instruments)

operands are unequal, these cascading inputs are ignored, and the output lines reflect the result of the current comparison. However, if the magnitudes of the two operands are equal, then the output lines reflect what is on the input cascading lines, which normally comes from the result of the comparison of the less significant bits or the termination configuration.

Example 4–3

Design a comparator for comparing two 24-bit numbers.

There are a number of ways to connect the comparators for comparing two 24-bit numbers. One possible configuration for comparing two 24-bit numbers in a linear fashion is shown in Figure 4–10(a). The timing in this case is

$$t_{\text{total}} = mt_c$$

where m is the number of comparators and t_c is the delay time through each. The number of comparators required is obtained by dividing the word length by 4. This linear cascading configuration of comparators yields a comparison time of 6×27 ns $= 162$ ns. For numbers with larger word size, the comparators themselves can be cascaded for faster response as shown in Figure 4–10(b). The timing delay in this case is

$$t_{\text{total}} = (\log_4 m + 1) \times t_c$$

The delay time for the parallel configuration is 2×27 ns $= 54$ ns. This represents a tremendous savings in time.

Universal Arithmetic and Logic Modules

Universal logic modules (ULMs) are chips that perform a variety of functions as determined by a set of control lines. Usually both arithmetic and logical functions are available. One advantage of using ULMs is that the chip can be used in many applications or in many areas of the same design, thus reducing the need for a large inventory of specialized chips. Furthermore, if there should be an engineering field change of a product already in service, the use of ULMs may sometimes minimize the extent of changes necessary.

74AS181—A 4-Bit Universal Arithmetic Logic Unit A 4-bit universal arithmetic logic unit (ALU) and function generator can perform both arithmetic and logical operations. The reason for such a chip is to reduce the necessary inventory of chips for various applications. The exact operation desired is governed by the four select lines, a mode control, and the current value of the carry-in bit. The logic symbol and the function table of the 74AS181 are shown in Figure 4–11. The normal data input parameters are the two 4-bit operands and the carry-in bit. A number of output lines are available, however. There is the normal 4-bit function output and the carry-out bit from the most significant bit. But in addition, the cascading $\bar{G}$ and $\bar{P}$ lines are also given to facilitate the use of a multilevel, look-ahead carry generator. One of the magnitude comparator output lines, the $(A = B)_{\text{out}}$ line, is also provided. Since there are a number of functions performed by the universal logic module, the chosen function is selected by five lines: a mode line and four function select lines. The four select lines, S_0–S_3, control the function to be performed. The mode bit, M, directs the unit to perform logical or arithmetic operations. When $M = 1$, the ALU performs logical operations. When $M = 0$, the ALU performs arithmetic operations. In conjunction with the mode control, the four select lines are used to choose one of 16 arithmetic or logical operations. Some typical arithmetic operations are addition, subtraction, and magnitude comparison of the input operands taking into account the current

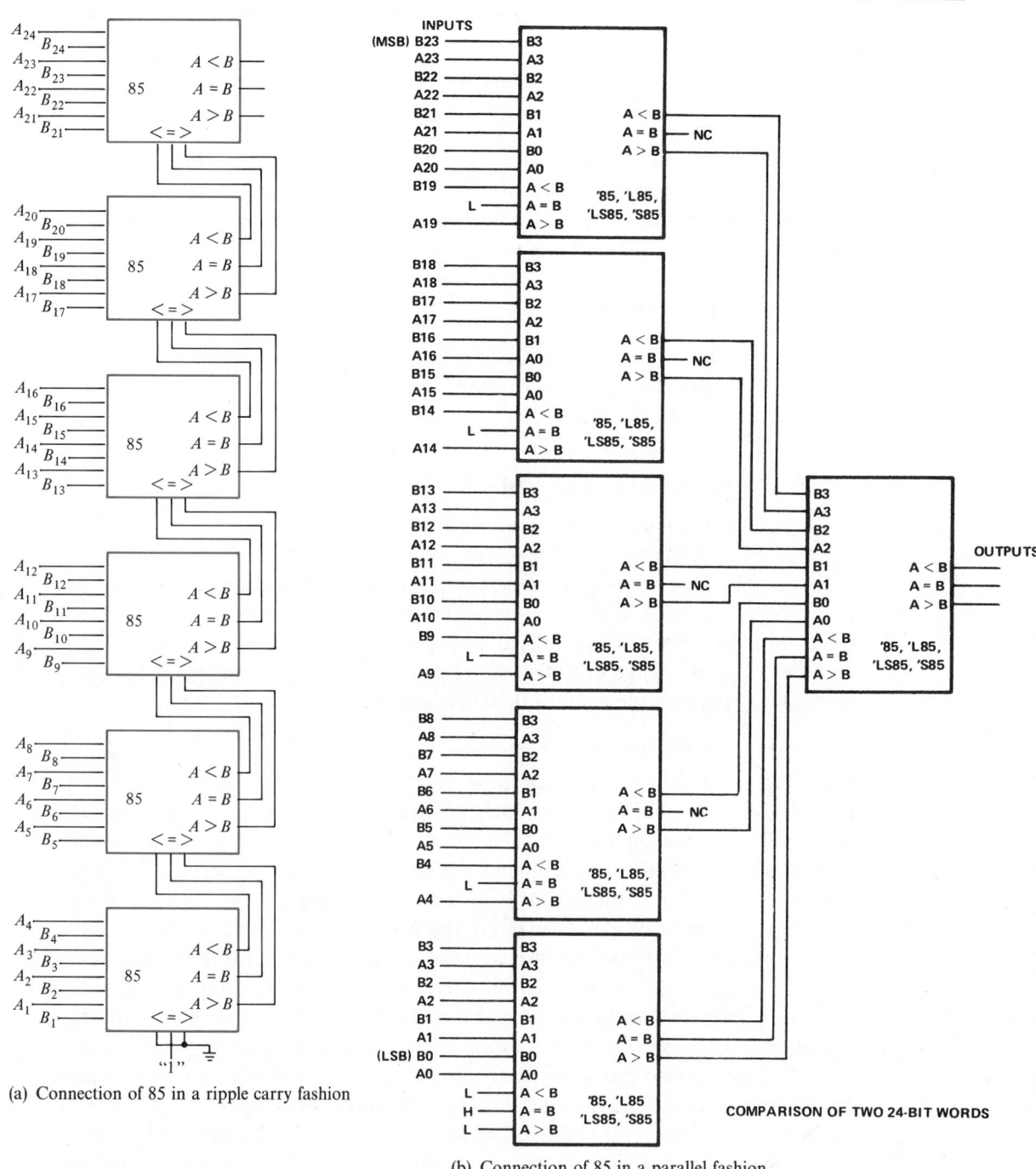

(a) Connection of 85 in a ripple carry fashion

INPUTS

(b) Connection of 85 in a parallel fashion

COMPARISON OF TWO 24-BIT WORDS

Figure 4–10

A typical connection of magnitude comparators for comparing two 24-bit numbers (Part b reprinted by permission of Texas Instruments)

value of the carry-in bit. Typical logical operations include the basic Boolean operations, exclusive OR and exclusive NOR operations. The carry-in bit is ignored for logical operations.

Example 4–4 ▰▰▰▰▰▰▰▰▰▰▰▰▰▰▰▰▰▰▰▰

Use ALUs and look-ahead carry generators to add two 24-bit numbers. Compare the timing with a similar circuit made of full adders.

The configuration of the ALUs and the look-ahead carry generators is similar to that of the full adders. However, more connections must be made in this case to configure the ALU for the addition operations. The delay time for four ALUs connected in parallel fashion is $15 + 41 = 56$ ns, while that in the ripple carry fashion is $4 \times 41 = 164$ ns.

Binary Multipliers

Multiplication of binary numbers can be implemented in combinational circuits by means of multiplier chips. If the number of bits for the input operands is fixed, the multiplier chip can be implemented simply by defining the truth table and expressing the outputs as a function of the inputs. This is not a practical solution, however, first because the circuit connection would not be the same for different word length, and second because the circuit becomes extremely complex even with small word length. The use of multiplier chips is attractive because the same type of chips can be used in an array fashion for any word length. Expansion of these multiplier chips for any number of bits, however, is not simple and requires

Figure 4–11

A 4-bit universal arithmetic logic unit (74AS181) (Part b reprinted by permission of Texas Instruments)

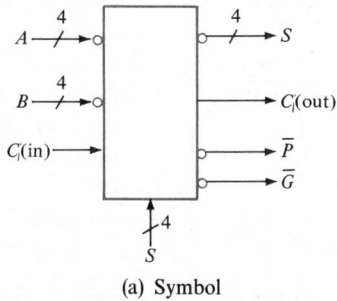

(a) Symbol

(continued)

SELECTION				ACTIVE-LOW DATA		
				M = H	M = L; ARITHMETIC OPERATIONS	
S3	S2	S1	S0	LOGIC FUNCTIONS	Cn = L (no carry)	Cn = H (with carry)
L	L	L	L	$F = \overline{A}$	F = A MINUS 1	F = A
L	L	L	H	$F = \overline{AB}$	F = AB MINUS 1	F = AB
L	L	H	L	$F = \overline{A} + B$	$F = A\overline{B}$ MINUS 1	$F = A\overline{B}$
L	L	H	H	F = 1	F = MINUS 1 (2's COMP)	F = ZERO
L	H	L	L	$F = \overline{A + B}$	$F = A \text{ PLUS } (A + \overline{B})$	$F = A \text{ PLUS } (A + \overline{B}) \text{ PLUS } 1$
L	H	L	H	$F = \overline{B}$	$F = AB \text{ PLUS } (A + \overline{B})$	$F = AB \text{ PLUS } (A + \overline{B}) \text{ PLUS } 1$
L	H	H	L	$F = A \oplus B$	F = A MINUS B MINUS 1	F = A MINUS B
L	H	H	H	$F = A + \overline{B}$	$F = A + \overline{B}$	$F = (A + \overline{B}) \text{ PLUS } 1$
H	L	L	L	$F = \overline{A}B$	F = A PLUS (A + B)	F = A PLUS (A + B) PLUS 1
H	L	L	H	$F = A \oplus B$	F = A PLUS B	F = A PLUS B PLUS 1
H	L	H	L	F = B	$F = A\overline{B} \text{ PLUS } (A + B)$	$F = A\overline{B} \text{ PLUS } (A + B) \text{ PLUS } 1$
H	L	H	H	F = A + B	F = (A + B)	F = (A + B) PLUS 1
H	H	L	L	F = 0	F = A	F = A PLUS A PLUS 1
H	H	L	H	$F = A\overline{B}$	F = AB PLUS A	F = AB PLUS A PLUS 1
H	H	H	L	F = AB	$F = A\overline{B} \text{ PLUS } A$	$F = A\overline{B} \text{ PLUS } A \text{ PLUS } 1$
H	H	H	H	F = A	F = A	F = A PLUS

SELECTION				ACTIVE-HIGH DATA		
				M = H	M = L; ARITHMETIC OPERATIONS	
S3	S2	S1	S0	LOGIC FUNCTIONS	$\overline{C}_n = H$ (no carry)	$\overline{C}_n = L$ (with carry)
L	L	L	L	$F = \overline{A}$	F = A	F = A PLUS 1
L	L	L	H	$F = \overline{A + B}$	F = A + B	F = (A + B) PLUS 1
L	L	H	L	$F = \overline{A}B$	$F = A + \overline{B}$	$F = (A + \overline{B}) \text{ PLUS } 1$
L	L	H	H	F = 0	F = MINUS 1 (2's COMPL)	F = ZERO
L	H	L	L	$F = \overline{AB}$	$F = A \text{ PLUS } A\overline{B}$	$F = A \text{ PLUS } A\overline{B} \text{ PLUS } 1$
L	H	L	H	$F = \overline{B}$	$F = (A + B) \text{ PLUS } A\overline{B}$	$F = (A + B) \text{ PLUS } A\overline{B} \text{ PLUS } 1$
L	H	H	L	$F = A \oplus B$	F = A MINUS B MINUS 1	F = A MINUS B
L	H	H	H	$F = A\overline{B}$	$F = A\overline{B}$ MINUS 1	$F = A\overline{B}$
H	L	L	L	$F = \overline{A} + B$	F = A PLUS AB	F = A PLUS AB PLUS 1
H	L	L	H	$F = A \oplus B$	F = A PLUS B	F = A PLUS B PLUS 1
H	L	H	L	F = B	$F = (A + \overline{B}) \text{ PLUS } AB$	$F = (A + \overline{B}) \text{ PLUS } AB \text{ PLUS } 1$
H	L	H	H	F = AB	F = AB MINUS 1	F = AB
H	H	L	L	F = 1	F = A	F = A PLUS A PLUS 1
H	H	L	H	$F = A + \overline{B}$	F = (A + B) PLUS A	F = (A + B) PLUS A PLUS 1
H	H	H	L	F = A + B	$F = (A + \overline{B}) \text{ PLUS } A$	$F = (A + \overline{B}) \text{ PLUS } A \text{ PLUS } 1$
H	H	H	H	F = A	F = A MINUS 1	F = A

Figure 4–11 Continued

the use of at least one normal carry-propagating full adder to sum up the partial sums.

Assume that we have a 4 × 4-bit multiplier chip that produces an 8-bit product. Now consider the multiplication of two 8-bit numbers, A and B. If the two 8-bit numbers are broken into 4-bit parts, namely A_h, A_l, B_h, and B_l, four 4-bit multiplications are needed as shown in Figure 4–12.

$$A = A_h \times 2^4 + A_l$$
$$B = B_h \times 2^4 + B_l$$
$$P = A \times B = A_h \times B_h \times 2^8 + (A_h \times B_l + A_l \times B_h) \times 2^4 + (A_l \times B_l)$$

Figure 4–12

Multiplication of two 8-bit numbers (Reprinted by permission of Texas Instruments)

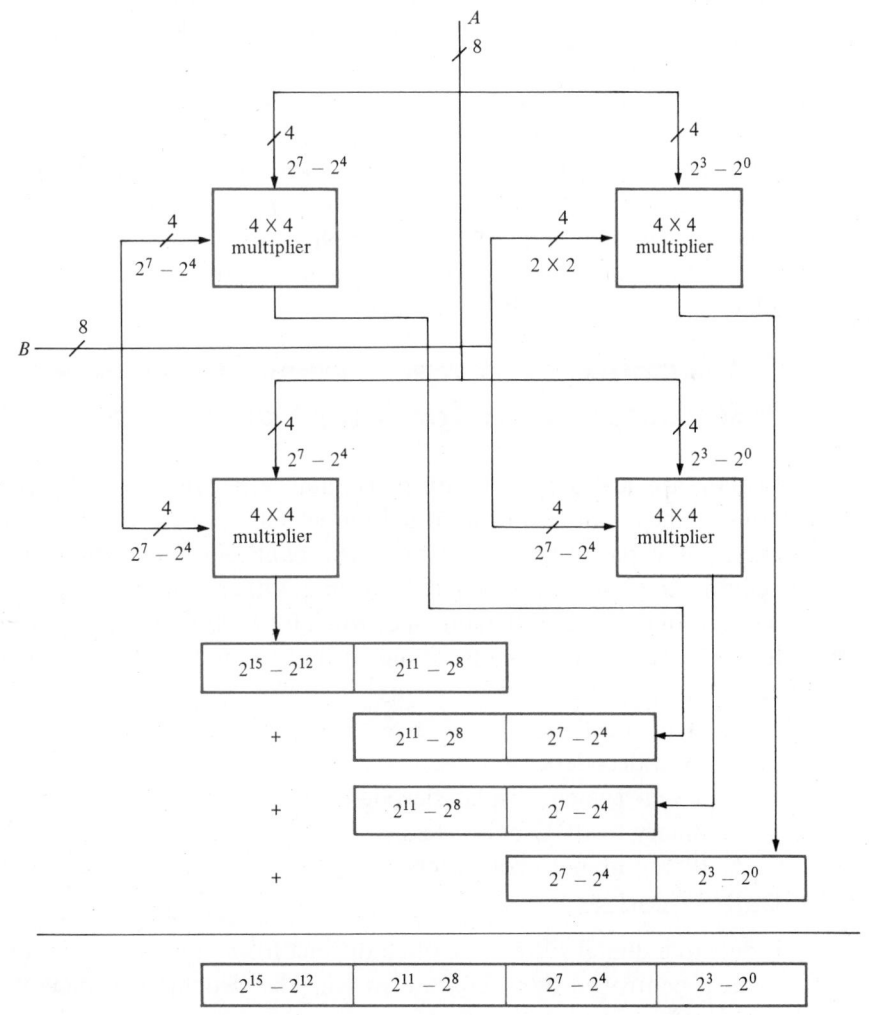

The four resulting partial products must now be summed after being properly shifted in positions to form the final product. Since both A and B are 8-bit numbers, the product will be 16 bits. Denoting P_{15} as the most significant bit and P_0 as the least significant bit of the partial product, $A_h \times B_h$ is an 8-bit product to be added to bits P_8 to P_{15}. The two middle partial sums are to be added to bits P_4 to P_{11}. The last term is to be added to bits P_0 to P_8. Note that the terms do overlap. The summation requires a total of five full adder chips. Another way to perform the addition is by means of carry-save adders set in a Wallace tree format. Interested readers are directed to the references at the end of the chapter for further details.

74LS274—A 4 × 4-Bit Binary Multiplier   High-speed binary unsigned multiplication can be performed using 4 × 4-bit binary multipliers such as the 74LS274. Though a lot of operations are performed by this chip, the input and output interface is extremely simple. The inputs consist of two 4-bit operands to be multiplied together. The output consists of an 8-bit product. For proper control of the output and the operation of the chip, two enable lines are provided. These select lines must be asserted high before the multiplication process begins.

If the number of bits in the original multiplier and multiplicand is larger than 4, they must be divided into smaller (4-bit) units. The partial product of each pair of 4-bit units between the multiplier and multiplicand is generated by a binary multiplier. These partial products are then summed by an array of full adders to produce the final product.

4.3

MSI Components for Data Format Conversion

In many applications, the inputs to a digital system may be in different representations as they come from sensors, from switches, or from other devices. For example, a keypad generally operates as a bank of switches that may or may not be in binary-coded decimal (BCD) format. Other numeric keypads are usually a set of distinct switches, having one switch for each numeral. It is often necessary, therefore, to convert data from one format to another. Some popular data conversion chips are

1. Data encoders
2. Data decoders
3. Seven-segment display decoders/drivers
4. Binary-to-BCD converters
5. BCD-to-binary converters

Data Encoders

Inputs to a digital system are often distinct from one another in a way that each input is controlled by a separate switch. The switches can be collectively taken as a binary number with only one switch being active at any time. When the number of inputs becomes large, it is quite common to convert this binary format to the BCD format. The conversion from many lines in a binary format to a lesser number of lines in the BCD format is called *data encoding*.

74LS147—A BCD Priority Encoder A typical encoder is the BCD priority encoder chip, the 74LS147. The logic symbol and the function table are shown in Figure 4–13. Inputs to the priority encoder consist of nine lines because this is a decimal encoder. For a binary encoder, there will be 15 input lines. These input lines are normally active low. Priority of these input lines is preset. The highest priority is the I_9 line, and the lowest priority is the I_1 line. Only nine input lines are needed for a decimal encoder because I_0, the line with the lowest priority, is the default and is assumed active always until superceded by other, higher-priority lines. There are four output lines. These are the four BCD outputs that represent the BCD code of the highest active line.

Figure 4–13

A priority encoder (74LS147) (Part b reprinted by permission of Texas Instruments)

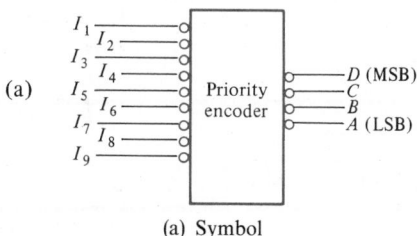

(a) Symbol

Inputs									Outputs			
1	2	3	4	5	6	7	8	9	D	C	B	A
H	H	H	H	H	H	H	H	H	H	H	H	H
X	X	X	X	X	X	X	X	L	L	H	H	L
X	X	X	X	X	X	X	L	H	L	H	H	H
X	X	X	X	X	X	L	H	H	H	L	L	L
X	X	X	X	X	L	H	H	H	H	L	L	H
X	X	X	X	L	H	H	H	H	H	L	H	L
X	X	X	L	H	H	H	H	H	H	L	H	H
X	X	L	H	H	H	H	H	H	H	H	L	L
X	L	H	H	H	H	H	H	H	H	H	L	H
L	H	H	H	H	H	H	H	H	H	H	H	L

(b) Function table

From the truth table, it is evident that the BCD output always indicates which input line is asserted in an active low environment. An ambiguous situation may arise if more than one input line is asserted. This ambiguity is resolved by forcing the BCD output to report only that signal line with the highest priority. This is why the encoder is also called a priority encoder. Because of this built-in priority, this type of encoder is frequently used in priority interrupt systems to indicate the highest active interrupt level and to maintain priority by blocking any lower-level interrupt requests and passing only the interrupt requests from the highest level.

Figure 4–14
BCD-to-decimal decoder (74LS145) (Part b reprinted by permission of Texas Instruments)

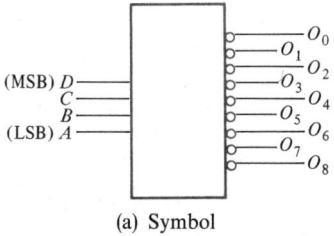

(a) Symbol

No.	Inputs				Outputs									
	D	C	B	A	0	1	2	3	4	5	6	7	8	9
0	L	L	L	L	L	H	H	H	H	H	H	H	H	H
1	L	L	L	H	H	L	H	H	H	H	H	H	H	H
2	L	L	H	L	H	H	L	H	H	H	H	H	H	H
3	L	L	H	H	H	H	H	L	H	H	H	H	H	H
4	L	H	L	L	H	H	H	H	L	H	H	H	H	H
5	L	H	L	H	H	H	H	H	H	L	H	H	H	H
6	L	H	H	L	H	H	H	H	H	H	L	H	H	H
7	L	H	H	H	H	H	H	H	H	H	H	L	H	H
8	H	L	L	L	H	H	H	H	H	H	H	H	L	H
9	H	L	L	H	H	H	H	H	H	H	H	H	H	L
Invalid	H	L	H	L	H	H	H	H	H	H	H	H	H	H
	H	L	H	H	H	H	H	H	H	H	H	H	H	H
	H	H	L	L	H	H	H	H	H	H	H	H	H	H
	H	H	L	H	H	H	H	H	H	H	H	H	H	H
	H	H	H	L	H	H	H	H	H	H	H	H	H	H
	H	H	H	H	H	H	H	H	H	H	H	H	H	H

(b) Function table

Data Decoders

A number of occasions exist when unique binary inputs must be decoded to initiate some circuit action. In general, an n-bit binary decoder can decode up to 2^n distinct output lines. The decoders will have n inputs and m outputs. The value of m is less than or equal to 2^n. The outputs are labeled $\bar{Y}_0, \bar{Y}_1, \ldots, \bar{Y}_{m-1}$, where $\bar{Y}_i$ is asserted if, and only if, the binary input is equal to i. Most decoders are made with the selected output equal to 0 and all other outputs equal to 1. This is in fact the negative logic notation. Many decoders also have an enable input that can force all outputs of a decoder into the nondecoded state. The enable inputs allow the decoders to be configured in an expandable manner.

74ALS145—A BCD-to-Decimal Decoder Functionally, the decoder operates in the opposite direction compared to the priority encoder. The latter encodes from 10 lines to 4 lines, while the former decodes from 4 lines back to 10 lines. The logic symbol and the function table of the BCD-to-decimal decoder are shown in Figure 4–14. The inputs to the decoder are only the four select lines. For a decimal decoder there are 10 output lines. For a binary decoder there will be 16 output lines. These output lines are active low and are asserted low when selected by the input select lines.

Example 4–5

Use two 3-to-8 decoders to form a 4-to-16 decoder.

To form a 4-to-16 decoder, three of the control bits are routed to each of the two 3-to-8 decoders. The remaining control bit is used to decode which of the decoders should be active. This is shown in Figure 4–15.

Figure 4–15

Using two 3-to-8 decoders to form a single 4-to-16 decoder

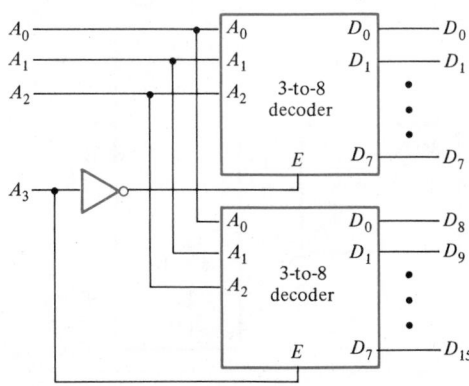

Example 4–6

Design a bedside home monitor using the BCD-to-decimal decoder. By entering the appropriate BCD code of a lamp or appliance, the home monitor system will automatically activate the corresponding device.

This method of device code entry using the BCD code is particularly attractive when there are more than 10 devices. The complete circuit can be implemented with the addition of a few sequential circuit components. Regardless of how many devices there are, only one keypad will suffice. In this case, the decoder is used to interpret the 4-bit BCD code and activate the proper line as shown in Figure 4–16.

Seven-Segment Display Decoders/Drivers

In a typical decimal data entry system, the input may come from a series of switches that are to be converted to the BCD format for manipulation. It is also desirable to display the data as they are being entered. Many common displays are seven-segment light-emitting diodes (LEDs) or liquid crystal displays (LCDs). To drive a seven-segment display, seven signal lines are required, one for each segment element of the display. The seven-segment decoder accepts a BCD digit as input and activates the seven output lines accordingly.

74LS46–74LS49 and 74LS246–74LS249 Series—BCD-to-Seven-Segment Display Decoders/Drivers   The BCD-to-seven-segment display decoders/drivers are also decoders, but they are special in that the output lines are decoded in such a way that they form patterns of decimal digits. Unlike normal decoders, more than one output line can be activated at the same time when needed to indicate the

Figure 4–16
Block diagram of a simple bedside home monitor using a
BCD-to-decimal decoder

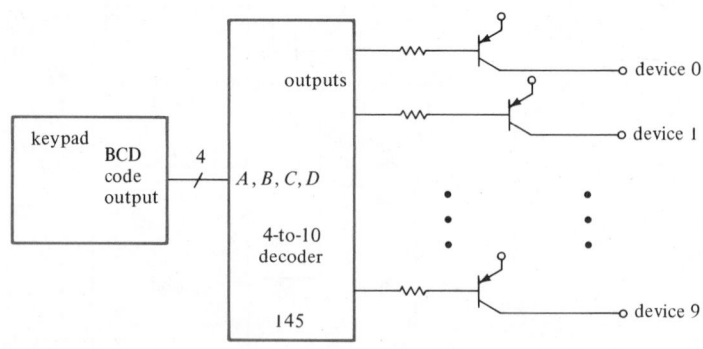

pattern of a decimal digit. The logic symbol and the function table of a BCD-to-seven-segment display decoder/driver are shown in Figure 4–17. A typical seven-segment display of the numerals with selected special characters is also shown in Figure 4–17(c).

The primary differences among the 74LS46, the 74LS47, the 74LS48, and the 74LS49 chips are the output logic levels (active low or active high) and the output configuration (open collector or with internal pull-up resistors). Other differences include the maximum amount of sink current and the maximum driving output voltage levels. The chips in the 24X series are functionally equivalent to the basic series.

There are four input lines to the decoder, the four select lines that determine which digit pattern is to be placed on the output. There are of course seven output lines, each corresponding to a particular segment of the display. Some versions have an additional output line for the decimal point. In order to provide further support for the display, there are three additional control lines provided by the chip. The first is the ripple blanking input, RBI. When asserted low, the RBI input indicates that the leading digit is zero and that the output display should be suppressed. A second control line is the blanking input or ripple blanking output (BI/RBO) line. This line should be used as a blanking input if the decoder represents the most significant digit or as a ripple blanking output for connection to the RBI of the less significant digit. For normal operation, the BI must be left open or held high for digit display. Leading (or trailing) zero suppression occurs by connecting the RBI of the first digit to zero, thus suppressing the display, and by cascading the RBO from the most (or least) significant digit to the RBI of the next most (or least) significant digit. If the first digit in the cascading chain is not zero, then RBO from the first digit is high. This signal propagates down the chain enabling all the rest of the displays, even if they are zero. If the first digit in the chain is zero, then the first digit is blanked. The signal RBO from the first digit is low, causing the next digit to blank if its display is also a zero.

The third control signal is the lamp test, $\overline{LT}$. When this line is asserted low, all segments of the digits are activated so that unlit segments of the display can be checked. In order to check if all segments of the display are working or not, a lamp test is provided. When the BI line is open or held high, the $\overline{LT}$ line can be pulsed to turn on all segments of the display for a visual check to ensure that all display segments are working.

Example 4–7 ▬▬▬▬▬▬▬▬▬▬▬▬▬▬▬▬▬▬▬▬▬▬▬▬

Design a two-digit display with leading zero blanking for displaying two decimal digits ranging between 00 and 99.

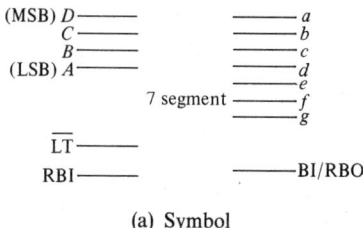

(a) Symbol

Decimal or Function	Inputs							Outputs						
	LT	RBI	D	C	B	A	BI/RBO	a	b	c	d	e	f	g
0	H	H	L	L	L	L	H	ON	ON	ON	ON	ON	ON	OFF
1	H	X	L	L	L	H	H	OFF	ON	ON	OFF	OFF	OFF	OFF
2	H	X	L	L	H	L	H	ON	ON	OFF	ON	ON	OFF	ON
3	H	X	L	L	H	H	H	ON	ON	ON	ON	OFF	OFF	ON
4	H	X	L	H	L	L	H	OFF	ON	ON	OFF	OFF	ON	ON
5	H	X	L	H	L	H	H	ON	OFF	ON	ON	OFF	ON	ON
6	H	X	L	H	H	L	H	ON	OFF	ON	ON	ON	ON	ON
7	H	X	L	H	H	H	H	ON	ON	ON	OFF	OFF	OFF	OFF
8	H	X	H	L	L	L	H	ON	ON	ON	ON	ON	ON	ON
9	H	X	H	L	L	H	H	ON	ON	ON	ON	OFF	ON	ON
10	H	X	H	L	H	L	H	OFF	OFF	OFF	ON	ON	OFF	ON
11	H	X	H	L	H	H	H	OFF	OFF	ON	ON	OFF	OFF	ON
12	H	X	H	H	L	L	H	OFF	ON	OFF	OFF	OFF	ON	ON
13	H	X	H	H	L	H	H	ON	OFF	OFF	ON	OFF	ON	ON
14	H	X	H	H	H	L	H	OFF	OFF	OFF	ON	ON	ON	ON
15	H	X	H	H	H	H	H	OFF	OFF	OFF	OFF	OFF	OFF	OFF
BI	X	X	X	X	X	X	L	OFF	OFF	OFF	OFF	OFF	OFF	OFF
RBI	H	L	L	L	L	L	L	OFF	OFF	OFF	OFF	OFF	OFF	OFF
LT	L	X	X	X	X	X	H	ON	ON	ON	ON	ON	ON	ON

(b) Function table

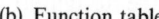
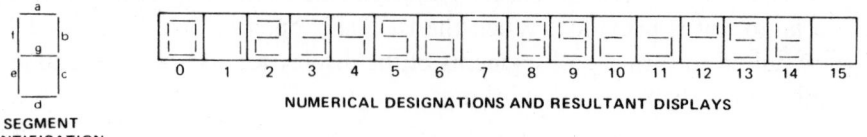

SEGMENT
IDENTIFICATION

NUMERICAL DESIGNATIONS AND RESULTANT DISPLAYS

(c) Display format

Figure 4-17

A BCD-to-seven-segment display decoder/driver (74LS247) (Parts b and c reprinted by permission of Texas Instruments)

A typical arrangement of the display decoder chips is shown in Figure 4–18. A switch is also added for the lamp test. Note that it is not necessary to actually detect the zero condition for the leading digit with external SSI chips.

Example 4–8 ▰▰▰▰▰▰▰▰▰▰▰▰▰▰▰▰▰▰▰▰▰▰▰▰▰

Design a home security system using the priority encoder. Display the activated sensor by an LED.

An example of a priority encoder is a home security system. Sensors for movement, smoke, water, or intrusion may be located throughout the house. Each sensor, acting as a switch, is connected to the priority encoder. As any sensor is tripped, the encoder will report the decimal number of the active switch on a display. A simple block diagram of such a system is shown in Figure 4–19.

Binary-to-BCD Converters

Most numeric data entries and displays are in decimal format, while most arithmetic operations are in binary format. This discrepancy requires the use of a

Figure 4–18

Typical arrangement of BCD-to-seven-segment display decoders to provide for lamp test and leading zero blanking

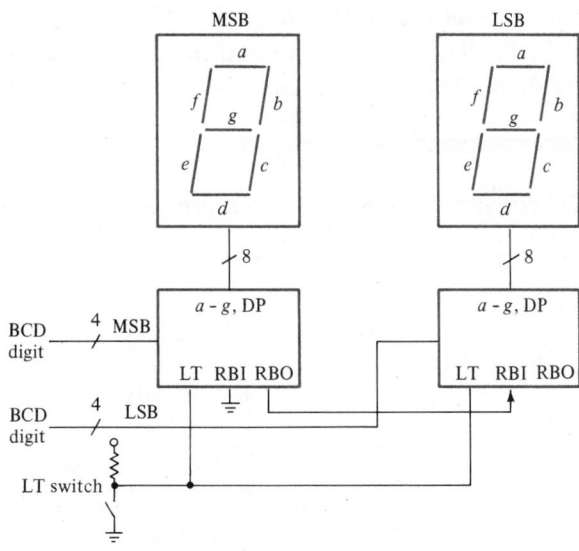

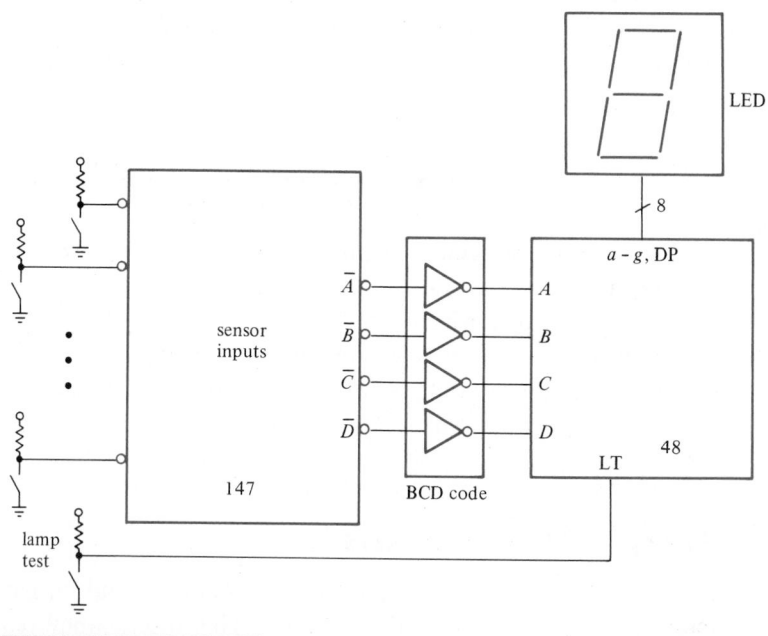

Figure 4–19

Block diagram of a simple home security system using a priority encoder

BCD-to-binary converter at the input of the computation section and a binary-to-BCD converter at the output.

74185—A Binary-to-BCD Converter   An example of a binary-to-BCD converter is the 74185. The logic symbol and the function table of the 74185 are given in Figure 4–20. The 74185 binary-to-BCD converter is designed for 6-bit binary

Figure 4–20

A binary-to-BCD converter (74185) (Part b reprinted by permission of Texas Instruments)

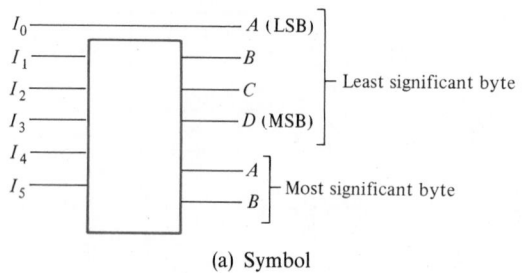

(a) Symbol

Binary Words	Inputs						Outputs							
	Binary Select					Enable								
	E	D	C	B	A	G	Y_8	Y_7	Y_6	Y_5	Y_4	Y_3	Y_2	Y_1
0–1	L	L	L	L	L	L	H	H	L	L	L	L	L	L
2–3	L	L	L	L	H	L	H	H	L	L	L	L	L	H
4–5	L	L	L	H	L	L	H	H	L	L	L	L	H	L
6–7	L	L	L	H	H	L	H	H	L	L	L	L	H	H
8–9	L	L	H	L	L	L	H	H	L	L	L	H	L	L
10–11	L	L	H	L	H	L	H	H	L	L	H	L	L	L
12–13	L	L	H	H	L	L	H	H	L	L	H	L	L	H
14–15	L	L	H	H	H	L	H	H	L	L	H	L	H	L
16–17	L	H	L	L	L	L	H	H	L	L	H	L	H	H
18–19	L	H	L	L	H	L	H	H	L	L	H	H	L	L
20–21	L	H	L	H	L	L	H	H	L	H	L	L	L	L
22–23	L	H	L	H	H	L	H	H	L	H	L	L	L	H
24–25	L	H	H	L	L	L	H	H	L	H	L	L	H	L
26–27	L	H	H	L	H	L	H	H	L	H	L	L	H	H
28–29	L	H	H	H	L	L	H	H	L	H	L	H	L	L
30–31	L	H	H	H	H	L	H	H	L	H	H	L	L	L
32–33	H	L	L	L	L	L	H	H	L	H	H	L	L	H
34–35	H	L	L	L	H	L	H	H	L	H	H	L	H	L
36–37	H	L	L	H	L	L	H	H	L	H	H	L	H	H
38–39	H	L	L	H	H	L	H	H	L	H	H	H	L	L
40–41	H	L	H	L	L	L	H	H	H	L	L	L	L	L
42–43	H	L	H	L	H	L	H	H	H	L	L	L	L	H
44–45	H	L	H	H	L	L	H	H	H	L	L	L	H	L
46–47	H	L	H	H	H	L	H	H	H	L	L	L	H	H
48–49	H	H	L	L	L	L	H	H	H	L	L	H	L	L
50–51	H	H	L	L	H	L	H	H	H	L	H	L	L	L
52–53	H	H	L	H	L	L	H	H	H	L	H	L	L	H
54–55	H	H	L	H	H	L	H	H	H	L	H	L	H	L
56–57	H	H	H	L	L	L	H	H	H	L	H	L	H	H
58–59	H	H	H	L	H	L	H	H	H	L	H	H	L	L
60–61	H	H	H	H	L	L	H	H	H	H	L	L	L	L
62–63	H	H	H	H	H	L	H	H	H	H	L	L	L	H
All	X	X	X	X	X	H	H	H	H	H	H	H	H	H

(b) Function table

Figure 4–20 Continued

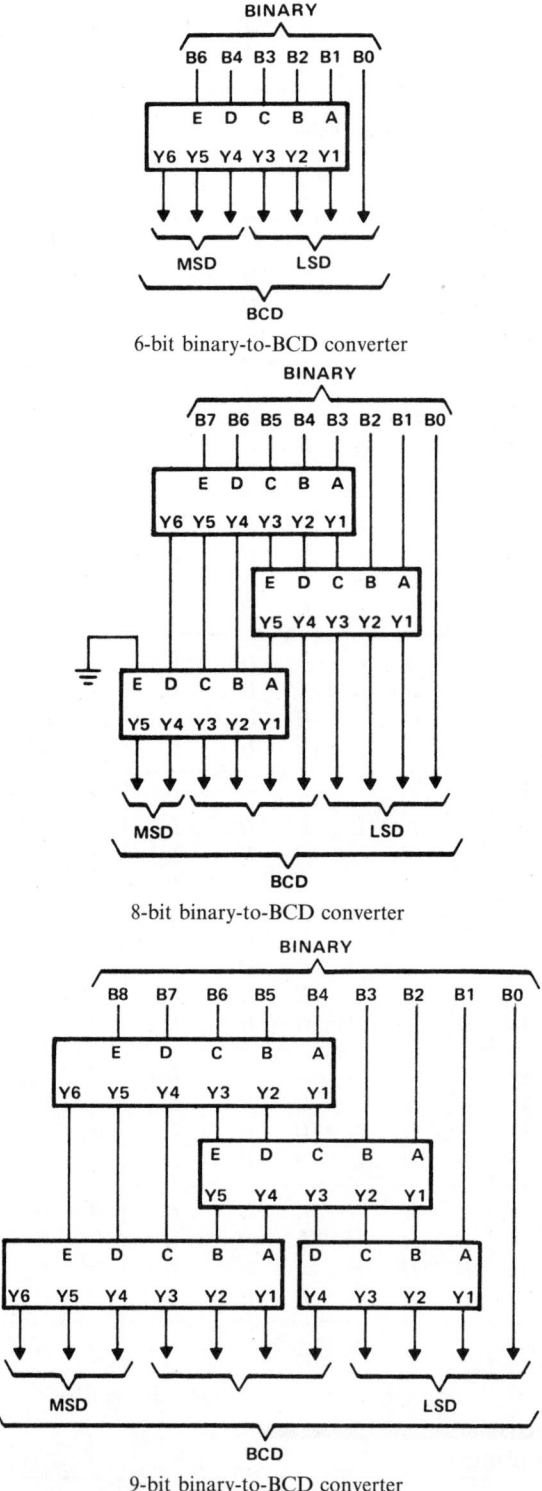

6-bit binary-to-BCD converter

8-bit binary-to-BCD converter

9-bit binary-to-BCD converter

Figure 4-21

Array connection of binary-to-BCD converters for larger word size
(Reprinted by permission of Texas Instruments)

numbers. However, only the five most significant bits are needed. The least significant bit of a binary number is always the same as the least significant bit of the least significant digit. Hence there are only five binary inputs. The largest decimal number to be represented by a 6-bit binary number is 63, which requires two decimal digits. But since the most significant decimal digit is 6, which is 0110, the leading bit is always 0. This means that there should be 6 bits in the output. Three bits belong to the least significant decimal digit as shown in Figure 4–21 and three others are for the second decimal digit with a leading zero supplied. For control of the chip, an enable line is provided. When the enable $\bar{E}$ is asserted low, the conversion will be performed. When the enable signal $\bar{E}$ is high, no signals will be provided at the output. To convert binary numbers that are

Figure 4–22

A BCD-to-binary convector (74184) function table (Reprinted by permission of Texas Instruments)

BCD Words	Inputs						Outputs				
	E	D	C	B	A	G	Y_5	Y_4	Y_3	Y_2	Y_1
0–1	L	L	L	L	L	L	L	L	L	L	L
2–3	L	L	L	L	H	L	L	L	L	L	H
4–5	L	L	L	H	L	L	L	L	L	H	L
6–7	L	L	L	H	H	L	L	L	L	H	H
8–9	L	L	H	L	L	L	L	L	H	L	L
10–11	L	H	L	L	L	L	L	L	H	L	H
12–13	L	H	L	L	H	L	L	L	H	H	L
14–15	L	H	L	H	L	L	L	L	H	H	H
16–17	L	H	L	H	H	L	L	H	L	L	L
18–19	L	H	H	L	L	L	L	H	L	L	H
20–21	H	L	L	L	L	L	L	H	L	H	L
22–23	H	L	L	L	H	L	L	H	L	H	H
24–25	H	L	L	H	L	L	L	H	H	L	L
26–27	H	L	L	H	H	L	L	H	H	L	H
28–29	H	L	H	L	L	L	L	H	H	H	L
30–31	H	H	L	L	L	L	L	H	H	H	H
32–33	H	H	L	L	H	L	H	L	L	L	L
34–35	H	H	L	H	L	L	H	L	L	L	H
36–37	H	H	L	H	H	L	H	L	L	H	L
38–39	H	H	H	L	L	L	H	L	L	H	H
ANY	X	X	X	X	X	H	H	H	H	H	H

larger than 6 bits, these converters are connected in an array format as shown in Figure 4–21. The array is highly regular and hierarchical and can be adapted to any word length.

BCD-to-Binary Converters

A BCD-to-binary converter performs the opposite conversion. The input to the converter is a BCD number and the output is a binary number. One such converter is the 74184.

74184—A BCD-to-Binary Converter   The logic symbol and the function table of the 74184, a BCD-to-binary converter, is given in Figure 4–22. There are only five input lines for the BCD number. As is the case with the binary-to-BCD converter, the least significant bit of the least significant decimal digit is not used. Hence

Figure 4–23

Array connection of BCD-to-binary converters for large word size
(Reprinted by permission of Texas Instruments)

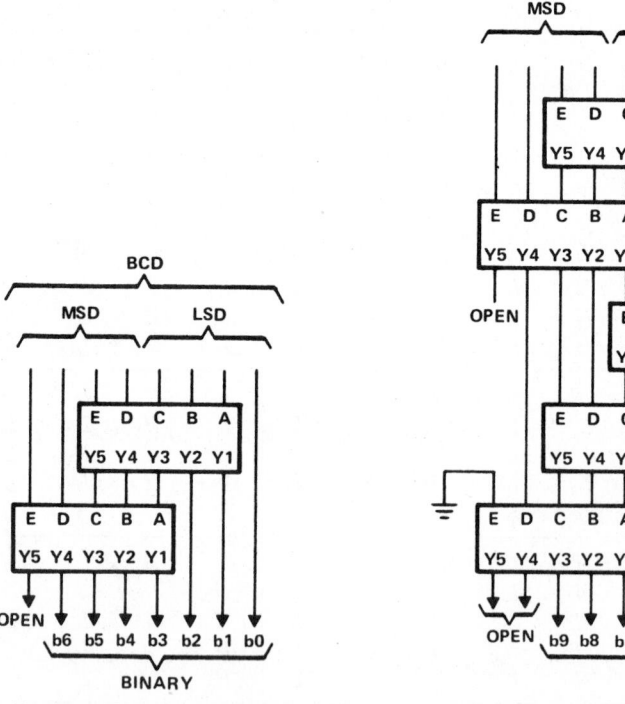

BCD-to-binary converter for two BCD decades   BCD-to-binary converter for three BCD decades

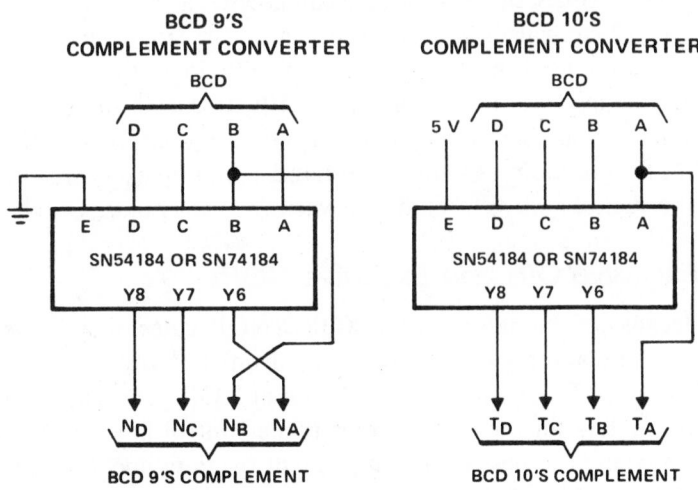

Figure 4–24

Connection pattern for a BCD-to-binary converter (SN74LS184) to
provide BCD 9's and 10's complement conversions (Reprinted by
permission of Texas Instruments)

the three least significant bits belong to the least significant BCD digit. The next
two belong to the most significant BCD digit, with the other two leading bits
assumed to be zero. Hence the largest input number in BCD format is 39_{10}, which
requires a total of six binary bits for the output. BCD-to-binary converters can
also be cascaded to convert larger numbers as shown in Figure 4–23.

An additional feature incorporated into the BCD-to-binary converter is three
extra output bits programmed to provide the BCD 9's or 10's complement conversions, which are discussed later in this section. The proper connection of these
three bits for BCD complement operations is shown in Figure 4–24.

4.4

MSI Components for Data Selection and Control

Data selection and control is another major category of chips in MSI components.
There are four main groups of chips in this category:

1. XORs and XNORs
2. True/complement-zero/one devices
3. Data multiplexers
4. Data demultiplexers

The XORs and XNORs are categorized as MSI chips because there are more than 12 equivalent gates on each chip. The true/complement-zero/one device is frequently used in conjunction with full adders to perform signed addition and subtraction. Multiplexers are frequently used as the input of a digital system in the same way as their analog counterparts are used for selecting a particular signal line for processing. On the other hand, demultiplexers are often used as the output of a digital system to control a variety of solenoids, switches, or actuators.

Exclusive ORs and Exclusive NORs

Because of the nature of the XOR/XNOR gates, they are widely used for complements and for comparisons based on the property of the XOR and XNOR functions. For example, in a two-input XOR gate, one input may be connected to the data line, while the other is the control line. In this case the control line can be used to pass or to complement the data line. For comparisons, both data lines are connected to the XOR gate. If both inputs are the same, the XOR gate reports a low level. If both inputs are not the same, the output of the XOR gate is high, indicating a discrepancy. In the case of the XOR/XNOR gate, an additional control line is provided to determine whether an XOR or an XNOR is to be performed.

74LS135—A Quad XOR/XNOR Chip The circuit diagram and function table of the XOR/XNOR chip are shown in Figure 4–25. Since each gate may function as either an XOR or an XNOR, in addition to the two 1-bit input operands, there is an additional gate control that determines whether the function to be performed is an XOR or an XNOR operation.

Example 4–9 ▪▪▪▪▪▪▪▪▪▪▪▪▪▪▪▪▪▪▪▪▪▪▪▪▪

Use quad XOR/XNOR chips to build a device that would take the 1's complement of a signed 4-bit binary number.

Such a device is often used to turn binary subtraction into addition. The device is inserted in-line (in series) with the subtrahend so that it is complemented before being added. The circuit configuration for this device is shown in Figure 4–26(a).

Example 4–10 ▪▪▪▪▪▪▪▪▪▪▪▪▪▪▪▪▪▪▪▪▪▪▪▪

Use quad XOR/XNOR chips to build a device that would take the 1's complement of a signed 4-bit binary number if the mode bit is 1 (indicating subtraction) and would leave the number unchanged if the mode bit is 0 (indicating addition).

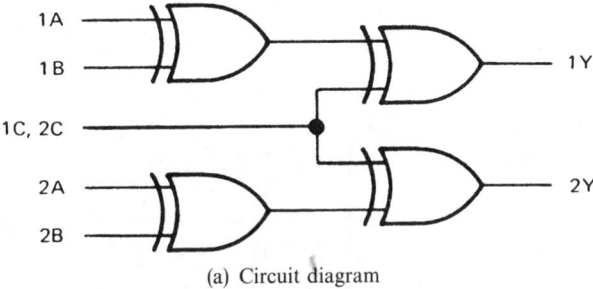

(a) Circuit diagram

Inputs			Output
A	B	C	Y
L	L	L	L
L	H	L	H
H	L	L	H
H	H	L	L
L	L	H	H
L	H	H	L
H	L	H	L
H	H	H	H

(b) Function table

Figure 4–25
A quad Exclusive OR/NOR chip (74LS135): (a) circuit diagram;
(b) function table (Reprinted by permission of Texas Instruments)

In reality, this device makes the task of addition and subtraction automatic. If an addition is called for as indicated by the mode bit, the inserted device will not complement the number. If a subtraction is called for, the inserted device will take the complement of the subtrahend before the addition is performed. The circuit is shown in Figure 4–26(b) with the C input connected to the mode bit.

True/Complement-Zero/One Devices

In signed arithmetic it is often necessary to complement one of the operands. Besides the XOR/XNOR gates, a true/complement-zero/one device can also be used. The true/complement portion of this device is basically performed by XOR gates, and the zero/one operation is an additional feature that is often useful in

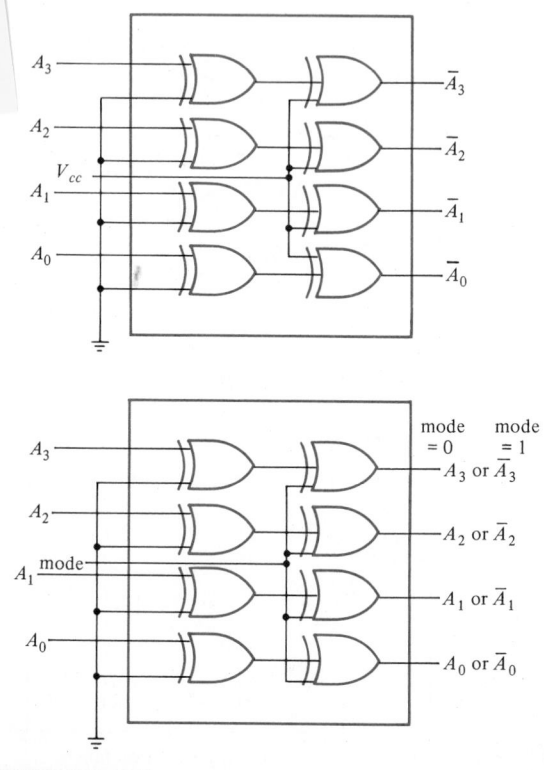

Figure 4–26

Using Exclusive OR/NOR chips to take the 1's complement of a 4-bit number

binary and decimal arithmetic. A null operation can be obtained by adding zero to preserve the original operand. An increment operation adds zero with an initial carry. A decrement operation is performed by adding -1 or all 1's in the 2's complement representation.

74LS87—A True/Complement-Zero/One Device The equivalent circuit diagram and function table of the 74LS87 true/complement-zero/one device are shown in Figure 4–27. The chip is made to operate on 4-bit operands. There are four input lines for the input operand and four output lines. There are two control lines, B and C, that govern what comes out of the output lines. This device is extremely useful because the output can be selected from one of four sources: the input operand itself, the input operand complemented, all 0's, and all 1's.

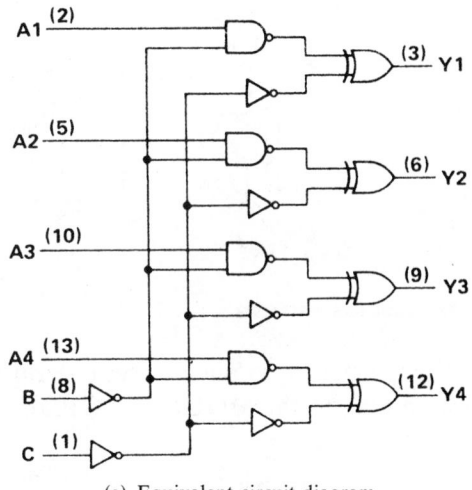

(a) Equivalent circuit diagram

Control Inputs		Outputs			
B	C	Y_1	Y_2	Y_3	Y_4
L	L	$\overline{A_1}$	$\overline{A_2}$	$\overline{A_3}$	$\overline{A_4}$
L	H	A_1	A_2	A_3	A_4
H	L	H	H	H	H
H	H	L	L	L	L

(b) Function table

Figure 4–27

A true/complement-zero/one device (74LS87) (Reprinted by permission of Texas Instruments)

Example 4–11

Use the true/complement-zero/one device to perform a generalized addition for numbers in the 1's complement format. When MODE is 0, an addition is performed. When MODE is 1, subtraction is performed. A second OP signal is also provided, so that addition or subtraction is performed only when OP is 1. When OP is 0, a null operation is performed.

The circuit configuration is shown in Figure 4–28. The circuit is similar to the previous one except that the MODE signal is connected through additional logic to the C input of the true/complement-zero/one device. The OP signal is

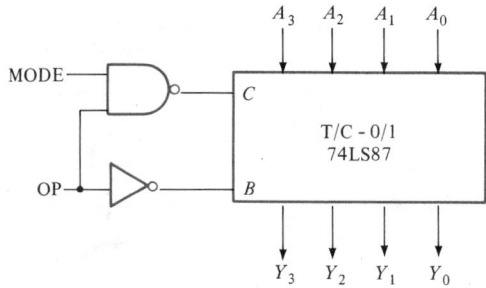

Figure 4–28

Connection of a true/complement-zero/one device to perform a generalized addition according to the MODE and OP input bits

connected to the B input of the device. Additional operations such as increment and decrement operations can be performed by the same circuit with minor modifications; see the problems at the end of this chapter.

Data Multiplexers

Multiplexing is commonly used to share a single communication line with more than one signal on the common line. The sharing can be either in the frequency domain or in the time domain. A digital multiplexer works only in the time domain. A digital multiplexer is a combinational circuit that selects binary information among 2^n input lines and directs it to the output controlled by the n selection lines. For some applications, a multiplexer can also function as a data selector.

74LS150—A 16-to-1 Multiplexer A 16-to-1 multiplexer provides selection of one out of 16 input data lines. Additionally, it can also operate as a Boolean function generator. In the data selection mode, the output level reflects any one of the 16 possible input levels according to what is specified by the four select lines (A, B, C, D). The enable input also facilitates the ability to expand the multiplexer for more input lines. The input I_i is directed to the output when $(DCBA)_2 = i_{10}$ through the selected AND gate and then through the output OR gate. Hence the inputs can be selectively strobed in and processed as necessary.

In a time-multiplexing mode, the select lines are connected to a counter that provides a continuous count from 0 to 15, thus creating 16 time slots with the output level sequentially taking on a value that is the same as the input level in each time slot. The symbol and the function table of a 16-to-1 multiplexer are shown in Figure 4–29. Since this is a binary multiplexer, there are 16 input lines and 1 output data line. To select the source of the output line, there are four select

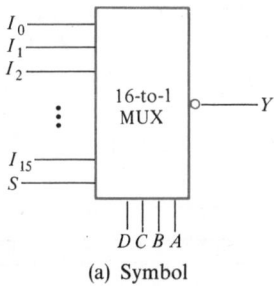

(a) Symbol

Inputs					Output
Select				Strobe	
D	C	B	A	S	Y
X	X	X	X	H	H
L	L	L	L	L	$\bar{I}_0$
L	L	L	H	L	$\bar{I}_1$
L	L	H	L	L	$\bar{I}_2$
L	L	H	H	L	$\bar{I}_3$
L	H	L	L	L	$\bar{I}_4$
L	H	L	H	L	$\bar{I}_5$
L	H	H	L	L	$\bar{I}_6$
L	H	H	H	L	$\bar{I}_7$
H	L	L	L	L	$\bar{I}_8$
H	L	L	H	L	$\bar{I}_9$
H	L	H	L	L	$\bar{I}_{10}$
H	L	H	H	L	$\bar{I}_{11}$
H	H	L	L	L	$\bar{I}_{12}$
H	H	L	H	L	$\bar{I}_{13}$
H	H	H	L	L	$\bar{I}_{14}$
H	H	H	H	L	$\bar{I}_{15}$

(b) Function table

Figure 4–29

A 16-to-1 multiplexer (74LS150) (Part b reprinted by permission of Texas Instruments)

lines. The input status of that input line with the same binary equivalent as the select lines is reported on the output data line. There is also an input strobe signal, $\overline{\text{STROBE}}$. Output data are valid only when the strobe is low.

A multiplexer can be used for many applications. Besides multiplexing signals, a multiplexer can be used as a gate to control signal flow or as a function generator for combinational circuits. A multiplexer can be used to effectively implement a combinational circuit. For a switching function of $n + 1$ variables, n of these are connected to the selection lines and the remaining input can be used as an input

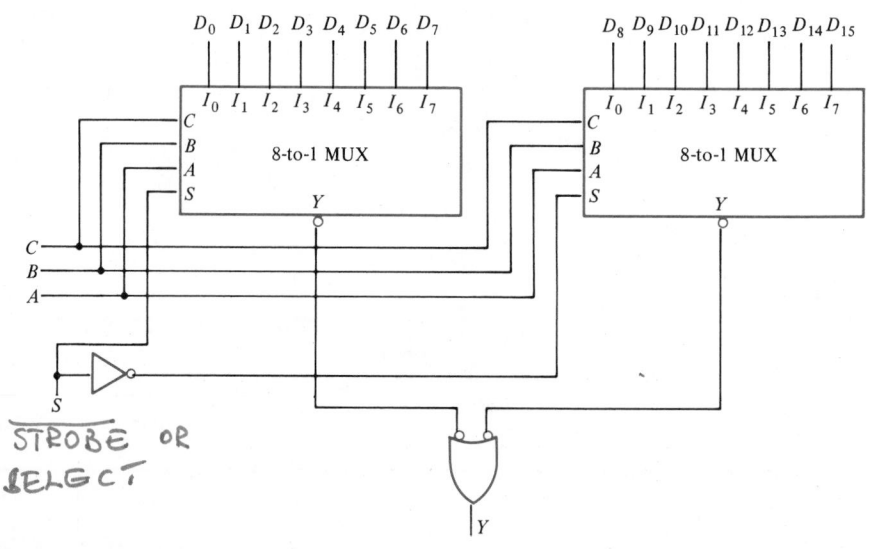

Figure 4-30

Using two 8-to-1 multiplexers to form a single 16-to-1 multiplexer

to the multiplexer. Then, by proper use of the data inputs, any switching function of $n + 1$ variables can be implemented with a 2^n to 1 multiplexer. The input connections are determined in a way similar to the variable entered map approach described earlier.

Example 4-12

Construct a 16-to-1 multiplexer from two 8-to-1 multiplexers.

A 16-to-1 multiplexer can be constructed using the enable signal from two 8-to-1 multiplexers. The circuit is given in Figure 4-30.

Example 4-13

Assume that we have a four-variable function as given by the truth table in Figure 4-31(a). Realize the function with a 16-to-1 multiplexer.

The four input variables are connected to the four select lines, which in turn determine which of the 16 inputs the output should be. The 16 inputs are connected to a logic low or high state according to the function desired as specified

w	x	y	z	$\overline{f}$
0	0	0	0	0
0	0	0	1	1
0	0	1	0	1
0	0	1	1	0
0	1	0	0	1
0	1	0	1	0
0	1	1	0	0
0	1	1	1	1
1	0	0	0	1
1	0	0	1	0
1	0	1	0	0
1	0	1	1	1
1	1	0	0	0
1	1	0	1	1
1	1	1	0	1
1	1	1	1	0

(a) Truth table for function $f(w, x, y, z)$

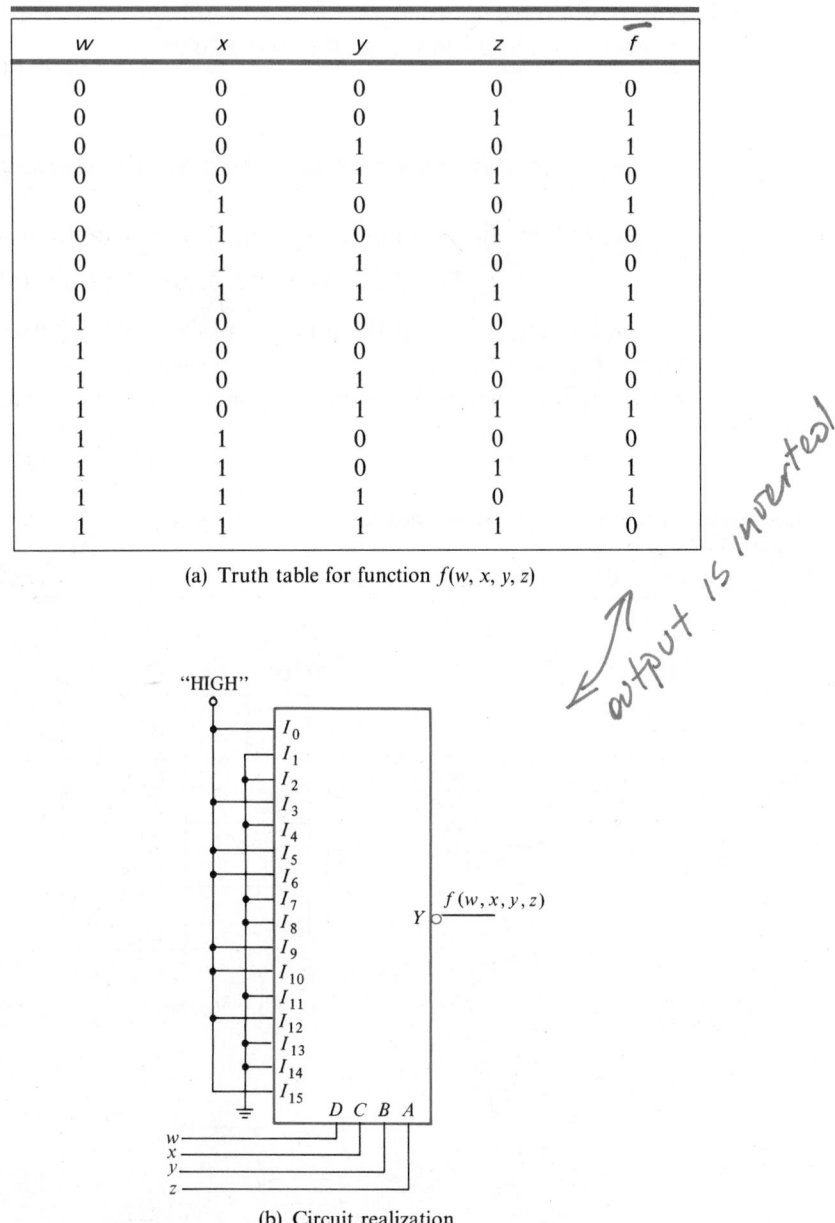

output is inverted

(b) Circuit realization

Figure 4–31
The 16-to-1 multiplexer used as a four-variable function generator

by the truth table in order. This is shown in Figure 4–31(b). Other configurations are also possible, as shown in the next two examples.

Example 4–14

Realize the following function using an 8-to-1 multiplexer:

$$F(D, C, B, A) = \sum(2, 3, 5, 6, 9, 12, 14, 15)$$

The Karnaugh map of the function is shown in Figure 4–32(a). Connect the inputs D, C, and B to the select lines as shown in Figure 4–32(b). When $DCB = 000$, the I_0 input will be fed to the output. The Karnaugh map cells 0 and 1, which

Figure 4–32

Using a multiplexer to implement a four-variable function

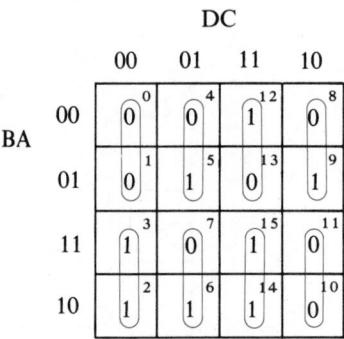

$$F(A, B, C, D) = \sum(2, 3, 5, 6, 9, 12, 14, 15)$$

(a) Karnaugh map

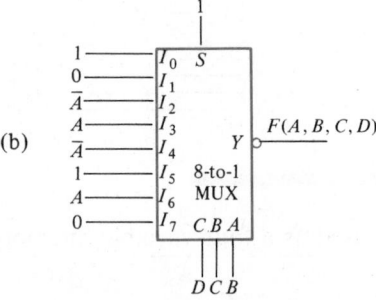

(b) Multiplexer implementation

are circled together and labeled with a 0, represent the value of $F = 0$ for both cells. Thus, we apply a 1 to input I_0. Similarly, when $DCB = 001$, the output must represent the A value for cells 2 and 3. Thus $I_1 = 0$. The case when $DCB = 010$ is represented by cells 4 and 5, which has a value of 0 when $A = 0$ and 1 when $A = 1$. Thus $I_2 = \overline{A}$. Similarly, $I_3 = A$ by examining cells 6 and 7. The final circuit is shown in Figure 4–32(b).

The advantage of this method is that only one integrated circuit (IC) chip and an additional inverter are needed to implement any four-input function with an 8-to-1 multiplexer regardless of the function. The actual inputs used for the selection inputs is left up to the designer. It is possible in some applications that the inverter may be eliminated entirely. This certainly should be looked at when implementing the function. The technique shown is very easy to modify for three-, five-, and six-input functions. We will consider another example.

Example 4–15

Given a five-variable function with the truth table shown in Figure 4–33(a), realize this function with a single inverter and a single 16-to-1 multiplexer.

Any four of the variables can be connected to the four select lines as before. The fifth variable is used to determine the connection of the 16 inputs to the multiplexer. The two states in the truth table having the same combination of the first four variables but differing by the fifth variable may be depicted by any one of the following four combinations:

	$l = 0$	$l = 1$	Use as Inputs
Function outputs:	0	0	1
	0	1	$\overline{l}$
	1	0	l
	1	1	0

Then the input to the multiplexer is also shown as above. Only one additional inverter is needed to obtain the complement of variable l as shown in Figure 4–33(b).

Data Demultiplexers

A 4-to-16 demultiplexer is functionally opposite to the 16-to-1 multiplexer. In operation, a demultiplexer may be used as a decoder, a demultiplexer, and a Boolean function generator.

m	n	o	p	f ($l=0$)	f ($l=1$)	MUX Input
0	0	0	0	1	1	1
0	0	0	1	0	0	0
0	0	1	0	1	0	$\bar{l}$
0	0	1	1	0	1	l
0	1	0	0	1	0	$\bar{l}$
0	1	0	1	0	0	0
0	1	1	0	1	1	1
0	1	1	1	1	1	1
1	0	0	0	0	1	l
1	0	0	1	0	1	l
1	0	1	0	0	0	0
1	0	1	1	1	0	$\bar{l}$
1	1	0	0	0	0	0
1	1	0	1	0	1	l
1	1	1	0	1	0	$\bar{l}$
1	1	1	1	1	1	1

(a) Truth table

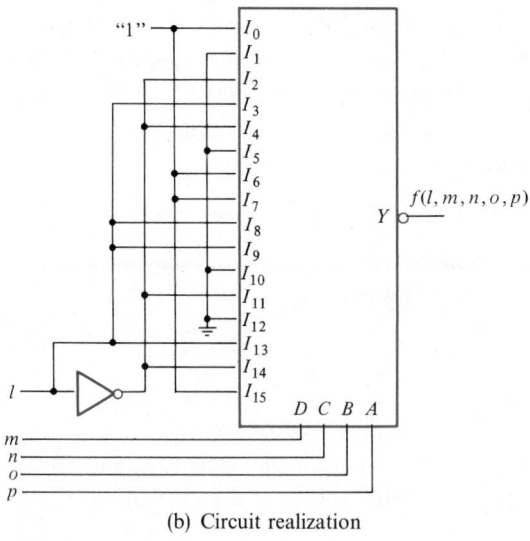

(b) Circuit realization

Figure 4–33
The 16-to-1 multiplexer used as a five-variable function generator

74LS154—A 4-to-16 Demultiplexer   The logic symbol and the function table of a 74LS154, 4-to-16 demultiplexer are shown in Figure 4–34. A demultiplexer has only four select lines as input. For a binary 4-to-16 demultiplexer, there are 16 output lines. In general, they are asserted low when selected by the input select lines. There are two enable lines. Both lines are active low, and data are valid on the output only when $\bar{G}_1\bar{G}_2 = 0$. In many cases, one of the enable lines is used as

Figure 4–34

A 4-to-16 demultiplexer (74LS154) (Part b reprinted by permission of Texas Instruments)

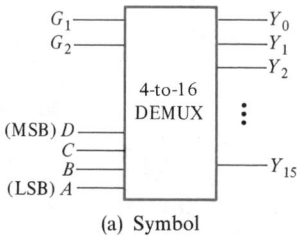

(a) Symbol

Inputs						Outputs															
G_1	G_2	D	C	B	A	0	1	2	3	4	5	6	7	8	9	10	11	12	13	14	15
L	L	L	L	L	L	L	H	H	H	H	H	H	H	H	H	H	H	H	H	H	H
L	L	L	L	L	H	H	L	H	H	H	H	H	H	H	H	H	H	H	H	H	H
L	L	L	L	H	L	H	H	L	H	H	H	H	H	H	H	H	H	H	H	H	H
L	L	L	L	H	H	H	H	H	L	H	H	H	H	H	H	H	H	H	H	H	H
L	L	L	H	L	L	H	H	H	H	L	H	H	H	H	H	H	H	H	H	H	H
L	L	L	H	L	H	H	H	H	H	H	L	H	H	H	H	H	H	H	H	H	H
L	L	L	H	H	L	H	H	H	H	H	H	L	H	H	H	H	H	H	H	H	H
L	L	L	H	H	H	H	H	H	H	H	H	H	L	H	H	H	H	H	H	H	H
L	L	H	L	L	L	H	H	H	H	H	H	H	H	L	H	H	H	H	H	H	H
L	L	H	L	L	H	H	H	H	H	H	H	H	H	H	L	H	H	H	H	H	H
L	L	H	L	H	L	H	H	H	H	H	H	H	H	H	H	L	H	H	H	H	H
L	L	H	L	H	H	H	H	H	H	H	H	H	H	H	H	H	L	H	H	H	H
L	L	H	H	L	L	H	H	H	H	H	H	H	H	H	H	H	H	L	H	H	H
L	L	H	H	L	H	H	H	H	H	H	H	H	H	H	H	H	H	H	L	H	H
L	L	H	H	H	L	H	H	H	H	H	H	H	H	H	H	H	H	H	H	L	H
L	L	H	H	H	H	H	H	H	H	H	H	H	H	H	H	H	H	H	H	H	L
L	H	X	X	X	X	H	H	H	H	H	H	H	H	H	H	H	H	H	H	H	H
H	L	X	X	X	X	H	H	H	H	H	H	H	H	H	H	H	H	H	H	H	H
H	H	X	X	X	X	H	H	H	H	H	H	H	H	H	H	H	H	H	H	H	H

(b) Function table

the data input line. When the data line is high corresponding to the chip in the deselected state, all output lines are high. When the data line is low, corresponding to the chip in the selected state, the selected output line is low.

There are many applications for this demultiplexer chip. As a decoder, the chip decodes the four BCD inputs into 16 mutually exclusive outputs. As a demultiplexer, the same count sequence driving the multiplexer is fed into the demultiplexer, and the output of the multiplexer is fed into one of the strobe inputs. As a four-variable Boolean function generator, the four variables are connected to the four select lines. The proper output lines are then gathered together into an OR gate to form the desired Boolean function output.

Example 4–16

Realize the following four-variable Boolean function:

$$f(w, x, y, z) = \sum(1, 2, 4, 7, 8, 11, 13, 14)$$

The realization of the demultiplexer as a four-variable function generator is done by summing outputs through a NAND gate as shown in Figure 4–35. If a particular truth table entry in sum-of-products form is 1, then that corresponding output is summed. The summing is done through a NAND gate because the out-

---

Figure 4–35

The 4-to-16 demultiplexer used as a four-variable function generator

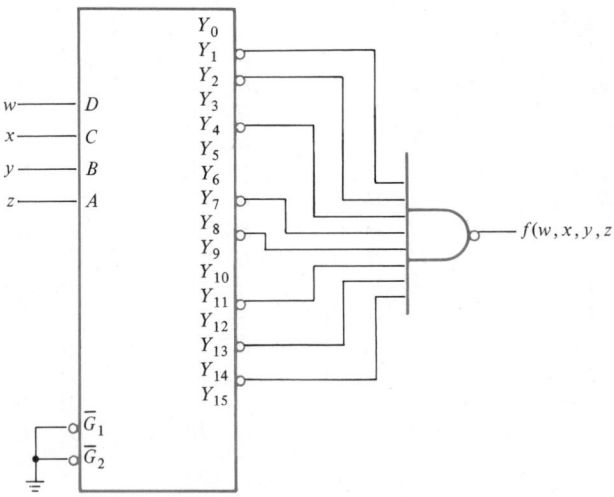

puts of the demultiplexer are active low. The same technique can also be applied to functions in the product-of-sums form. In the latter case, those maxterm entries that are 0 are gathered together through an AND gate to realize the function.

Example 4–17 ▬▬▬▬▬▬▬▬▬▬▬▬▬▬▬▬▬▬

Design a simple 16-line data multiplexer and demultiplexer system.

A 16-line multiplexer-demultiplexer system requires four select lines. These select lines can be generated by a clock or a counter. For connection between the two ends—i.e., between the multiplexer and the demultiplexer—only a single data line is required. However, the select lines must also be sent with the data line. It is possible to send only the clock signal and not the select lines or even further multiplex the clock into the data line. A simple solution is presented in Figure 4–36(a). A timing analysis of this system is given in Figure 4–36(b).

Figure 4–36
A digital multiplexing-demultiplexing communication system

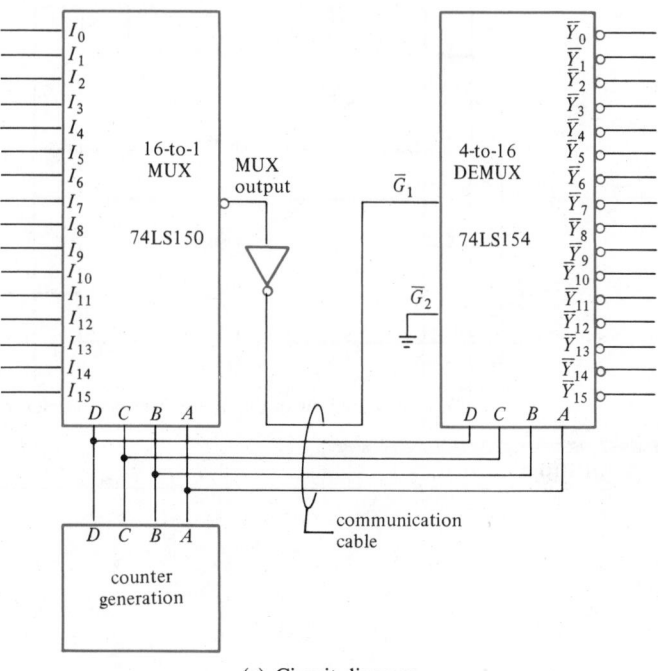

(a) Circuit diagram

(continued)

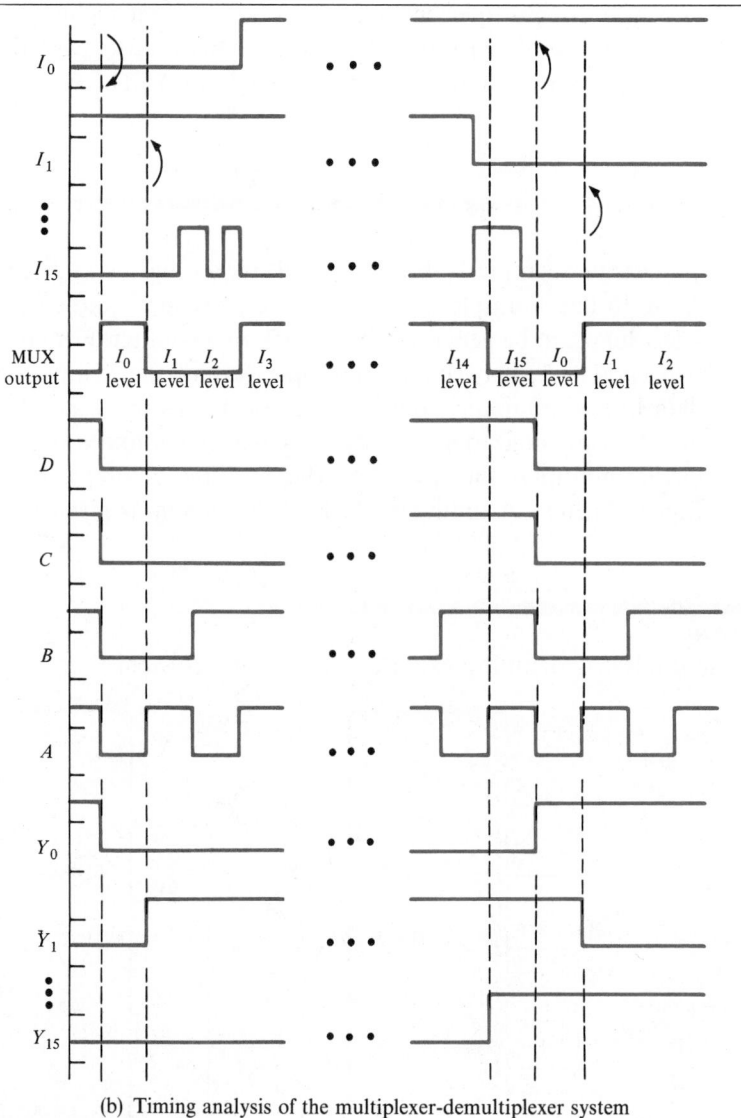

(b) Timing analysis of the multiplexer-demultiplexer system

Figure 4-36 Continued

4.5

MSI Components for Data Communication

For data communication, parity is an important parameter. The *parity bit* is an additional bit appended to the data bits so that the total number of 1's in the combined string is either even or odd. If the total number of 1's including the parity bit is even, this is called even parity; if the total number is odd, then it is called odd parity. An example of a parity generator/checker is the 74ALS280. This chip is designed for both generating the parity bit and checking for correct parity. The parity function is generated by performing the XOR function on all the data bits. For even and odd parity, we have

$$P_e = a_7 \oplus a_6 \oplus a_5 \oplus a_4 \oplus a_3 \oplus a_2 \oplus a_1 \oplus a_0$$
$$P_o = a_7 \oplus a_6 \oplus a_5 \oplus a_4 \oplus a_3 \oplus a_2 \oplus a_1 \oplus a_0 \oplus 1$$

The XOR function essentially sums the number of 1 bits in a modulo 2 fashion. If two data bits are 1, the result of the XOR function is 0. If only one data bit is 1, then the result of the XOR function is 1.

74AS280—A 9-Bit Odd/Even Parity Generator/Checker

An example of a 9-bit odd/even parity generator/checker is the 74AS280. There are altogether nine data input lines. Most data transmission takes place in the form of byte-sized units of information. Hence the nine input bits account for the eight information bits and the additional parity bit. There are two output lines, labeled $\sum_{even}$ and $\sum_{odd}$. One is always the complement of the other. To generate even parity for 8-bit data operands, let the extra input bit be 1. Then the $\sum_{even}$ and $\sum_{odd}$ outputs will represent the correct parity for even and odd parity, respectively. To check parity, all 9 bits are used. The output lines $\sum_{even}$ and $\sum_{odd}$ are then used to determine if the parity is correct or not.

Example 4–18

Design the circuit connection for a 9-bit parity generator/checker.

The circuit connection is shown in Figure 4–37. Note that the two output lines show merely whether the sum of 1's is even or odd. The interpretation and use are totally dependent on the application. For example, for parity generation with even parity in positive logic, the $\sum_{even}$ line is used. For parity checking with odd parity in positive logic, the $\sum_{odd}$ line is used.

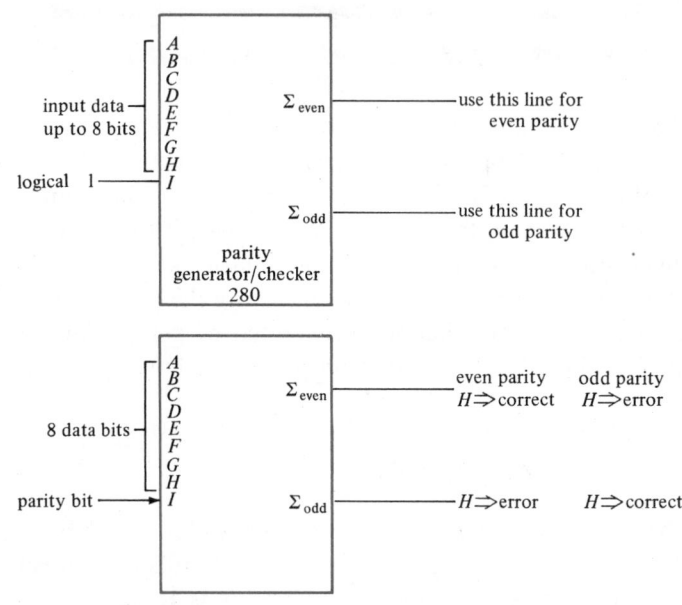

Figure 4–37

Connection of 9-bit parity generator and checker for both odd and even parity

With word sizes larger than 8 bits, parity checking and generation can be performed by cascading the 74LS280 together either in a parallel fashion or in a ripple fashion. For maximum efficiency, the parallel fashion saves more time. Both configurations are shown in Figure 4–38.

4.6

MSI Design Procedures

In the last section we provided a brief discussion of the MSI components and their use in some typical applications. In this section we concentrate on design techniques using MSI components and demonstrate how these products can be combined to make meaningful and useful systems.

Functional Block Diagram

When given a problem to solve or a system to design, the best way to start is to break down the problem into smaller functional blocks. Usually, there are at least three main sections: the input section, the processing section, and the output

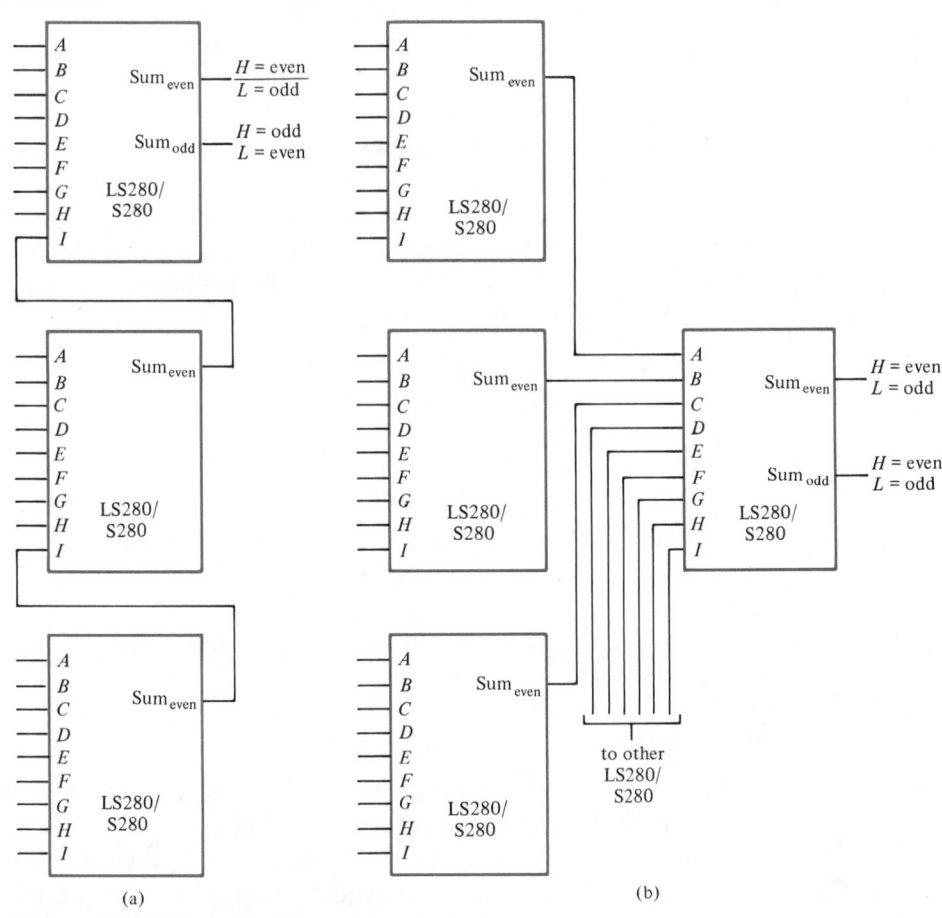

Figure 4–38

The 9-bit odd/even parity generators/checkers for larger word sizes (a) in ripple fashion and (b) in parallel fashion (Part b reprinted by permission of Texas Instruments)

section. Each of these sections may be further broken down into smaller subsections. Depending on the complexity of the problem, the processing section may be further divided into various stages. Dividing the tasks into smaller pieces allows the designer to concentrate on a smaller part of the total problem. This is the traditional "divide and conquer" technique, but done so in a top-down, hierarchical manner.

In dividing the task to be done, the general rule is to cluster the design around the major components of the system with various supporting chips surrounding them. The initial design should center around these major components, which are chosen carefully to perform the required operations. The detailed design of each block then follows with the design of the necessary supporting chips to supply the needed signals to these major components. Typically, supporting chips are SSI components and are sometimes called "glue logic," because they serve to link the major components together.

For each block, clearly identify the input and output signals, their properties, and the relationships between the input and the output signals. This is particularly important for the input and output sections. In identifying the signals, the following parameters should be considered:

- Voltage levels—TTL, CMOS, or others
- Active levels—active low or active high
- Fan-in—for input signals
- Fan-out—for output signals
- Current limits
- Setup time
- Hold time
- Delay tolerance

Once these signals are properly defined, the detail design can proceed within the specified limits.

System Timing

When the design becomes complicated, i.e., when there are various paths between input and output, it is necessary to include a timing analysis to ensure that the signals are synchronized at the end. Unlike synchronous circuits (discussed in the next chapter), where the signals are synchronized by a master clock, combinational circuits themselves do not necessarily require a clock. Therefore, a timing analysis is necessary to determine the setup and hold time required for each input signal, the worst-case signal propagation delay time from input to output, and the hold time for the output signals.

Control of a combinational circuit is usually indicated by a change in input conditions. Based on the newly supplied information, the circuit performs the specified operations and reports the result after the necessary propagation time delays. It is advisable that no further changes in the input be allowed until the circuit finishes its present processing based on the current set of information and arrives at some stable point.

Complete Design Example—A Coin Exchanger

In order to demonstrate the procedures of MSI design techniques, a complete design example is provided here. The problem at hand is to design a coin exchanger for a slot machine. The coin exchanger will accept nickels, dimes, and quarters.

The coin exchanger will then calculate the total amount of money inserted at one time, and deduct the price of the product. A light on the front panel indicates an insufficient amount of money received. If the amount of money inserted is sufficient, the light is turned off, the product door is unlatched, and any extra change is returned. The price of the product may be 35 cents, 40 cents, or 45 cents and can be varied by service personnel with a switch on the control panel.

In order to keep the problem simple, a few assumptions are made. First, the inserted coins are automatically identified and stacked in the proper holders. The holders are lined with photodiodes so that as the coins are successively stacked, the diodes, acting as switches, are also successively blocked and turned off. This simple mechanism provides a way to count the number of nickels, dimes, and quarters. Second, both the insufficient-amount light and the latch on the door are activated by solenoids designed to accept regular TTL output signals. Third, only nickels and dimes are returned as change. The return of nickels and dimes is taken care of by a special mechanism that accepts binary input signals indicating the number of dimes and/or nickels to return and dispenses the coins accordingly.

Functional Block Diagram With the above assumptions, the coin exchanger problem can now be broken up into smaller pieces in the form of a functional block diagram. Basically, there are three main sections: the input section, the computation section, and the output section. The input section is responsible for determining the number of nickels, dimes, and quarters that have been inserted. The computation section is designed to calculate the amount of money inserted and the amount of change to be given back. The output section controls the insufficient-amount light and the latch for the product door. The functional block diagram is shown in Figure 4–39.

Figure 4–39

Functional block diagram for a coin exchanger

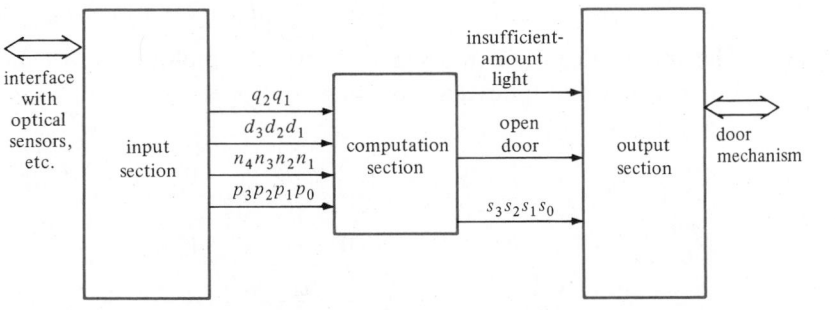

The Input Section   For the input section we have three banks of optical switches, with one bank for each stack of coins. Since the price of the products is less than 45 cents, no more than two quarters are needed or accepted. Likewise, no more than five dimes or nine nickels are needed to cover the full price. Therefore, only two binary lines are needed for the quarters, five for the dimes, and nine for the nickels. These individual lines can be converted into BCD format by three BCD priority encoders. These encoders are shown in Figure 4–40. Note that only two binary bits (q_1, q_2) are required to indicate the number of quarters inserted, three binary bits (d_1, d_2, d_3) for the number of dimes, and four binary bits (n_1, n_2, n_3, n_4) for the number of nickels.

For the price of the product, a simple switch arrangement showing the number of nickels will suffice. Since the price must be subtracted from the total, they are thus represented as negative numbers in 2's complement format:

Price	Switch Settings				Operation
35 cents	1	0	0	1	subtract 7 nickels
40 cents	1	0	0	0	subtract 8 nickels
45 cents	0	1	1	1	subtract 9 nickels

A three-pole, four-throw switch with the proper connection is sufficient to generate the correct values for the four binary variables (p_3, p_2, p_1, p_0) needed to indicate the current price of the products. Thus a total of 13 BCD lines make up the output of the first section. These outputs become the inputs to the computation section.

The Computation Section   In the computation section, the first step is to sum the coins with respect to their weights to find the total amount deposited. Realizing that all the weights are multiples of 5 cents, we can then work in terms of 5-cent tokens. The weight of a quarter is five tokens; of a dime, two; and of a nickel, one. The total amount is then:

$$T = q_1 q_0 \times 5 + d_2 d_1 d_0 \times 2 + n_3 n_2 n_1 n_0$$

The above calculation appears to require two multiplications and three additions. For the first multiplication, the longhand operation gives

$$
\begin{array}{rcccccc}
 & & & 0 & 0 & q_1 & q_0 \\
\times & & & & 1 & 0 & 1 \\
\hline
 & & & 0 & 0 & q_1 & q_0 \\
 & & 0 & 0 & 0 & 0 & \\
 & 0 & 0 & q_1 & q_0 & & \\
\hline
 & 0 & 0 & q_1 & q_0 & q_1 & q_0 \\
\end{array}
$$

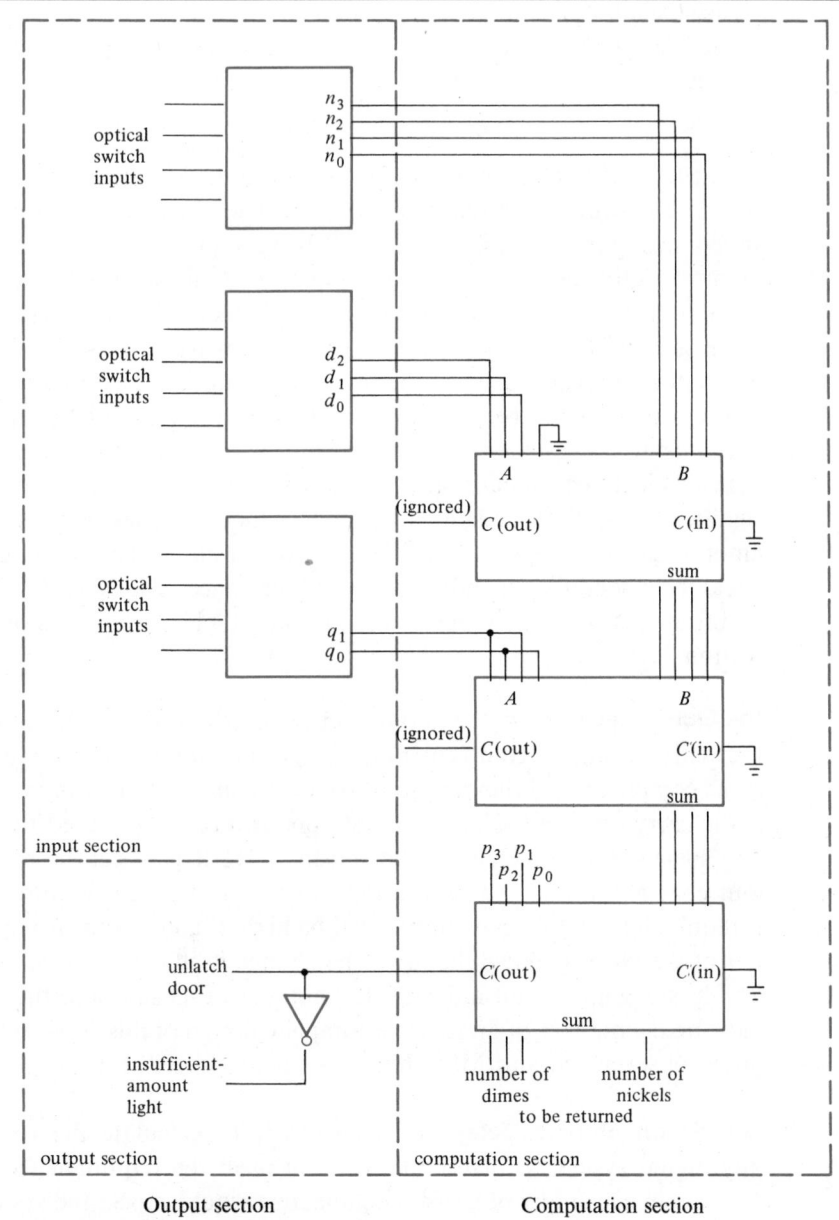

Figure 4–40

Complete circuit design of the coin exchanger

Hence the multiplication of a 2-bit number (q_1, q_0) by 5 is simply a rearrangement of the 2 bits as shown (q_1, q_0, q_1, q_0). For the second multiplication, multiplication by 2 is simply a shift to the left by one binary bit:

$$0 \; d_2 \; d_1 \; d_0 \times 0 \; 0 \; 1 \; 0 = d_2 \; d_1 \; d_0 \; 0$$

So the total can be computed with just two additions using two 4-bit full adders. No multiplications are required after all. The circuit to perform these tasks is shown in Figure 4–40, with the result being represented by 4 bits—i.e., a maximum of 15 tokens or 75 cents. An overflow condition should not occur, and all carry-out bits are ignored. We assume that the machine is fast enough (or the buyer is careful enough) so that no carry-out will occur.

In order to determine whether a sufficient amount has been entered, the price of the product must be subtracted from the total amount by another 4-bit full adder using 2's complement. If the result is negative, then the amount inserted is insufficient. This condition is indicated by lack of a carry-out from the most significant bit. If the result is positive, which includes zero, then a sufficient amount has been deposited, and the positive result is also the change to be given back. This condition is indicated by the presence of a carry-out from the most significant bit. This carry-out and the resultant 4 bits are passed on to the output section.

The Output Section   In the output section, note that the insufficient-amount light is exactly the logical complement of the product-door unlatch signal. The former is the complement of the carry-out from the computation section, while the latter is the carry-out itself. Therefore, only one inverter is required to generate these two signals. For the change mechanism, a NOR gate can be connected to the four change bits (s_3, s_2, s_1, s_0) to indicate if any change is required. If no change is required, the NOR gate output will be high. If any change is required, the output of the NOR gate will be low. The change itself can be in nickels and dimes. The least significant bit indicates the number of nickels, and the next three bits indicate the number of dimes. The complete design of this coin exchanger is composed of seven SSI and MSI chips. This is illustrated in Figure 4–40.

Calculation of Time Delay   It is necessary to calculate the response time of any digital system to determine if it will meet the requirements of the system. For a coin exchanger of a slot machine, one would expect the system to respond fast enough so that externally, the insufficient-amount light would go off soon enough to prevent the customer from dropping in any more coins, and internally, the system would close off the drop chute and not accept any more coins.

The time delay is always calculated for the worst case. In the design shown in Figure 4–40, the longest delay route is through the decoders (DECs), the three full adders (FAs), and the inverter (INV). This is shown below:

$$t_{\text{total}} = t_{\text{DEC}} + 3 \times t_{\text{FA}} + t_{\text{INV}}$$

Using typical propagation delay times, the total delay is 65 ns. This time delay should be very low for this example.

4.7
MSI Circuit Applications

In this section we illustrate the use of MSI components through several applications. Specifically, we examine circuits for arithmetic operations. The arithmetic using the binary numbering system was presented in Chapter 1. Here we want to demonstrate how these algorithms can be realized using standard MSI components.

Addition/Subtraction in 2's Complement Representation

Addition and subtraction in 2's complement representation can be accomplished by cascading 4-bit full adders (74LS283) or arithmetic logic units (74ALS181). Subtraction is done merely by changing the sign of the second operand. Hence, additional logic is required to complement the magnitude before the addition/subtraction. Complementing the second operand is performed by a series of XORs (74ALS86) or by a true/complement-zero/one device (74ALS87). The circuit for addition and subtraction in 2's complement representation is shown in Figure 4–41. Note that the carry-in for the least significant bit is taken into account during the complement operation, thus eliminating the need for an end-around carry.

Figure 4–41
MSI circuit for binary addition/subtraction in 2's complement form

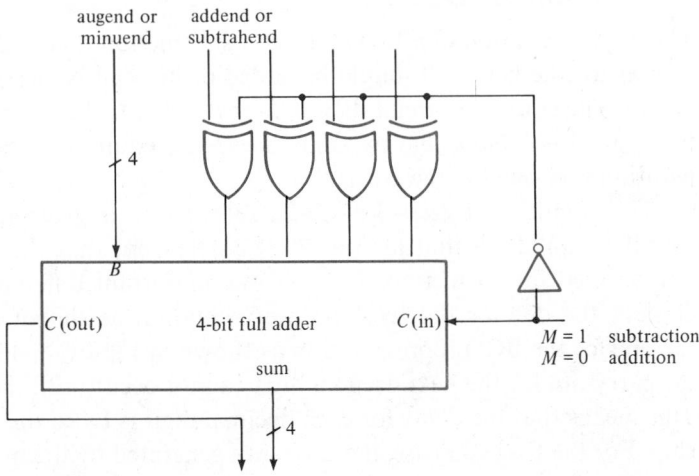

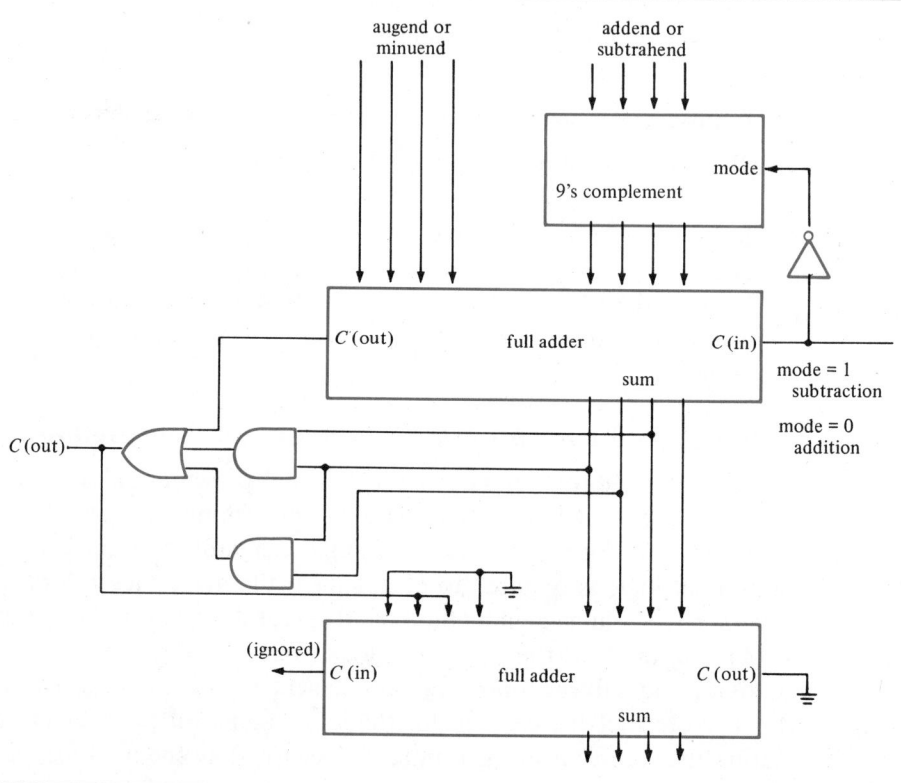

Figure 4–42
Circuit diagram to perform BCD addition/subtraction

BCD Arithmetic Operations

The implementation of a BCD addition/subtraction is straightforward. The decision as to whether 0110 should be added or not can be decoded by a simple SSI circuit. The complete circuit diagram for the BCD addition/subtraction is shown in Figure 4–42. Note that two full adders are required for each digit, one for the addition and one for the correction.

The circuit for Excess-3 addition/subtraction is given in Figure 4–43. This circuit is similar to that for the BCD case except that the second full adder is always used for correction. There is one important difference between the two circuits, the one for the Excess-3 representation as shown in Figure 4–43 and the one for the BCD representation as shown in Figure 4–42. For the BCD case, the carry-out for the next decimal digit cannot occur until the correction is done. This means that the delay for each decimal digit is twice the delay of a full adder chip. For the Excess-3 case, the carry-out generated by the first addition is always

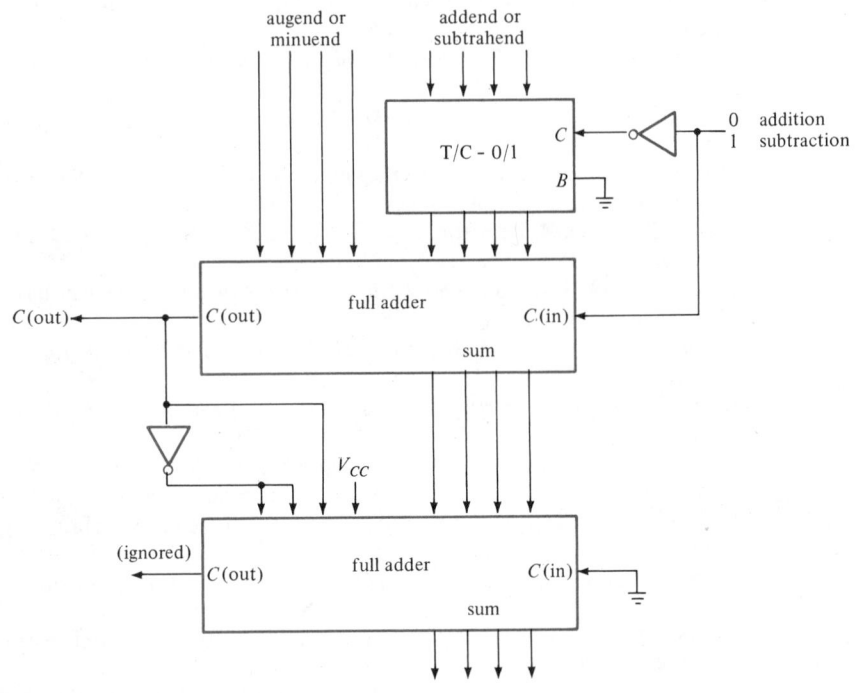

Figure 4–43
Circuit diagram for an Excess-3 addition/subtraction

the true carry-out bit for the present decimal digit and can be used immediately as the carry-in bit for the next digit. This means that the delay for each decimal digit is just the delay of one full adder chip.

References

1. Becher, William D. *Logical Design Using Integrated Circuits.* Hayden Book Co., Rochelle Park, N.J., 1977.
2. Boyce, Jefferson C. *Digital Logic Operation and Analysis.* Prentice-Hall, Englewood Cliffs, N.J., 1982.
3. Chu, Y. *Digital Computer Design Fundamentals.* McGraw-Hill, New York, 1962.
4. Comer, David J. *Digital Logic and State Machine Design.* Holt, Rinehart and Winston, New York, 1984.
5. Dietmeyer, D. L. *Logic Design of Digital Systems,* 2d ed. Allyn & Bacon, Boston, 1978.
6. Fletcher, William I. *An Engineering Approach to Digital Design.* Prentice-Hall, Englewood Cliffs, N.J., 1980.
7. Flores, I. *The Logic of Computer Arithmetic.* Prentice-Hall, Englewood Cliffs, N.J., 1962.

8. Friedman, Arthur D. *Fundamentals of Logic Design and Switching Theory*. Computer Science Press, Rockville, Md., 1986.

9. Greenfield, Joseph D. *Practical Digital Design Using ICs*, 2d ed. John Wiley & Sons, New York, 1983.

10. Hayes, John P. *Digital System Design and Microprocessors*. McGraw-Hill, New York, 1984.

11. Hill, Frederick J., and Gerald R. Peterson. *Introduction to Switching Theory and Logical Design*, 3d ed. John Wiley & Sons, New York, 1981.

12. Hill, F. J., and G. R. Peterson. *Digital Logic and Microprocessors*. John Wiley & Sons, New York, 1984.

13. Humphrey, W. S., Jr. *Switching Circuits with Computer Applications*. McGraw-Hill, New York, 1958, chap. 4.

14. Johnson, E. L., and M. A. Karim. *Digital Design, A Pragmatic Approach*. PWS Publishers, Boston, 1987.

15. Kline, Raymond M. *Structured Digital Design Including MSI/LSI Components and Microprocessors*. Prentice-Hall, Englewood Cliffs, N.J., 1983.

16. Kohavi, Z. *Switching and Finite Automata Theory*. McGraw-Hill, New York, 1970.

17. Kostopoulos, G. K. *Digital Engineering*. John Wiley & Sons, New York, 1975.

18. Lee, Samuel C. *Digital Circuits and Logic Design*. Prentice-Hall, Englewood Cliffs, N.J., 1976.

19. Lewin, Morton H. *Logic Design and Computer Organization*. Addison-Wesley, Reading, Mass., 1983.

20. Mano, M. Morris. *Digital Logic and Computer Design*. Prentice-Hall, Englewood Cliffs, N.J., 1979.

21. Mano, M. Morris. *Digital Design*. Prentice-Hall, Englewood Cliffs, N.J., 1984.

22. McCluskey, E. J., Jr. *Introduction to the Theory of Switching Circuits*. McGraw-Hill, New York, 1965, chap. 4.

23. McCluskey, E. J. *Logic Design Principles with Emphasis on Testable Semicustom Circuits*. Prentice-Hall, Englewood Cliffs, N.J., 1986.

24. Mckay, Charles W. *Experimenting with MSI, LSI, IO, and Modular Memory Systems*. Prentice-Hall, Englewood Cliffs, N.J., 1981.

25. Nagle, H. T., Jr., B. D. Carrol, and J. D. Irwin. *An Introduction to Computer Logic*. Prentice-Hall, Englewood Cliffs, N.J., 1975.

26. National Semiconductor Corp. *CMOS Data Book*. National Semiconductor, Santa Clara, Calif., 1981.

27. National Semiconductor Corp. *Logic Data Book*. National Semiconductor, Santa Clara, Calif., 1981.

28. O'Connor, Patrick. *Digital and Microprocessor Technology*. Prentice-Hall, Englewood Cliffs, N.J., 1979.

29. Parr, E. A. *The Logic Designer's Guidebook*. McGraw-Hill, New York, 1984.

30. Peatman, John B. *Digital Hardware Design*. McGraw-Hill, New York, 1980.

31. Rhyne, V. T. *Fundamentals of Digital Systems Design*. Prentice-Hall, Englewood Cliffs, N.J., 1973.

32. Roth, Charles H., Jr. *Fundamentals of Logic Design*, 3d ed. West Publishing Co., St. Paul, Minn., 1985.

33. Rutkowski, George, and Jerome Oleksy. *Fundamentals of Digital Electronics: A Text Laboratory Manual*, 2d ed. Prentice-Hall, Englewood Cliffs, N.J., 1985.

34. Signetics Corp. *Signetics Logic—TTL Data Manual*. Signetics, Sunnyvale, Calif., 1978.

35. Taub, Herbert. *Digital Circuits and Microprocessors*. McGraw-Hill, New York, 1982.

36. Texas Instruments. *The TTL Data Book for Design Engineers*, 2d ed. Texas Instruments, Dallas, Tex., 1976.
37. Wiatrowski, Claude A., and Charles H. House. *Logic Circuits and Microcomputer Systems*. McGraw-Hill, New York, 1980.
38. Wilkinson, Barry. *Digital System Design*. Prentice-Hall, Englewood Cliffs, N.J., 1987.
39. Williams, Gerald E. *Digital Technology*. Science Research Associates, Chicago, 1977.
40. Winkel, David, and Franklin Prosser. *The ART of Digital Design*. Prentice-Hall, Englewood Cliffs, N.J., 1980.

Experiment I

Design of a Home Security Controller

OBJECTIVES

Design a home security controller that will control 10 sensors around the house and report the code number of the activated sensor on an LED display.

EQUIPMENT

10 SPST switches
1   74LS147 BCD priority encoder
1   74LS48 display decoder
1   74ALS04 hex inverter
1   digital LED
10 220-Ω pull-up resistors
1   5-V power supply

PROCEDURE

1. Connect the switches with pull-up resistors to the priority encoder as shown in Figure E4–1.
2. Connect the output of the priority encoder to the display decoder.
3. Connect the output of the display decoder to the digital LED.
4. Put in a lamp test.

RESULTS

1. Verify that the lamp test is working.
2. Verify that as any of the switches are closed, the proper switch code is shown on the digital LED.
3. Verify that if two switches are pressed, the switch code of the one with the highest priority is shown on the digital LED.
4. Suppose that there are more than 10 switches. How can the circuit be modified to handle this situation? How easily can the circuit be expanded?

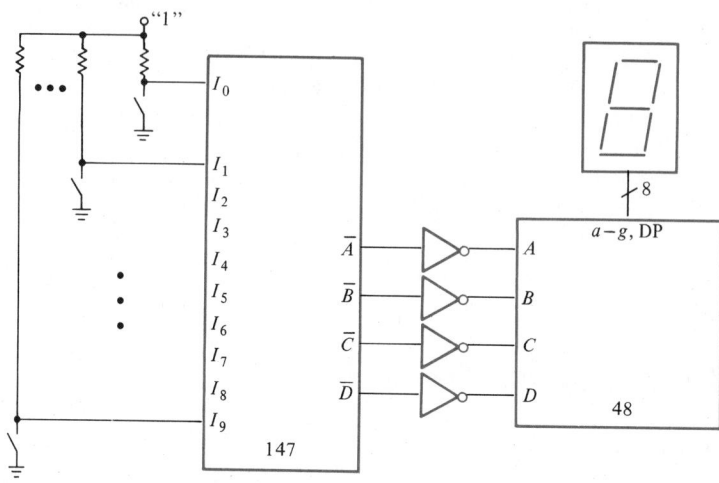

Figure E4–1
Circuit diagram for Experiment I, the design of a house security controller

Experiment II

Design of a Bedside Home Monitor System

OBJECTIVES

Design a bedside home monitor system that will activate an intercom at 10 different locations around the house. When the user wants to communicate with a certain room, he or she simply presses the code number of that room. The code number is displayed on a digital LED display, and an LED for the corresponding room lights up, indicating that communication is established.

EQUIPMENT

1	keypad or 10 switches
1	74LS147 BCD priority encoder
1	74LS154 BCD-to-binary decoder
1	74LS48 display decoder
1	74ALS04 hex inverter
10	LEDs
10	2N2907 PNP transistors
1	digital LED
20	220-Ω pull-up resistors
1	5-V power supply

PROCEDURE

1. Connect the keypad or switches with pull-up resistors to the priority encoder as shown in Figure E4–2.
2. Connect the priority encoder to the display decoder.
3. Connect the display decoder to the digital LED.
4. Connect the output of the priority encoder to the BCD-to-binary decoder.
5. Connect the output of the binary decoder to the 10 LEDs through pull-up resistors.
6. Connect the lamp test.

Figure E4–2

Circuit diagram for Experiment II, the design of a bedside home monitor system

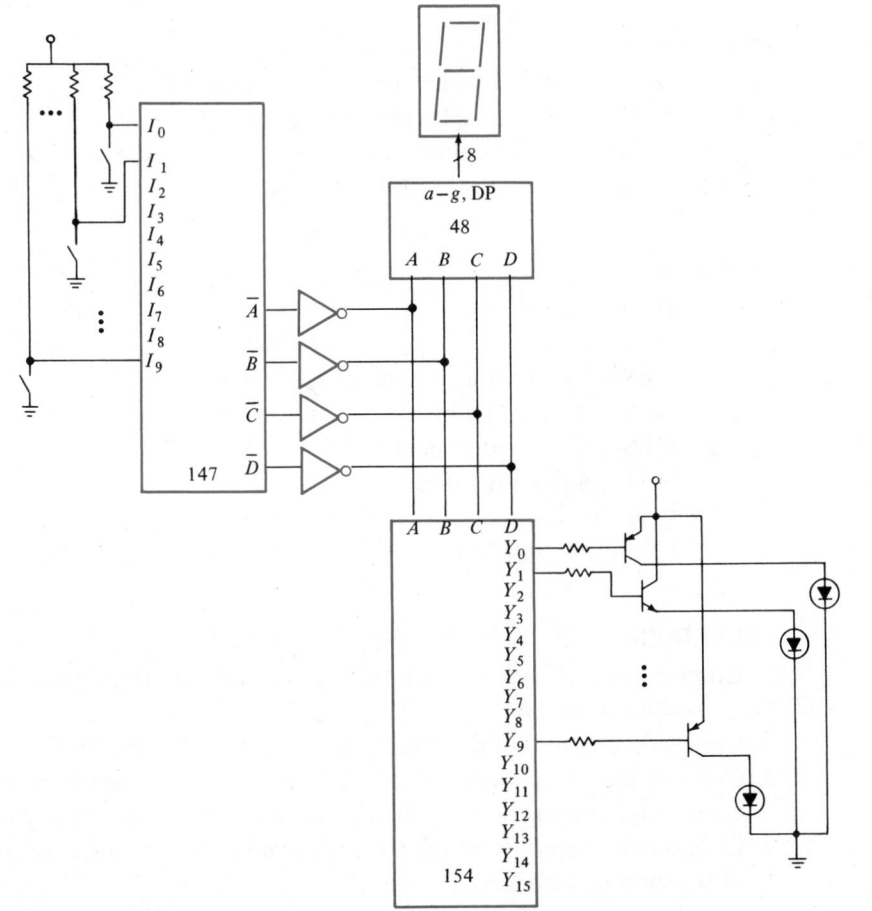

RESULTS

1. Verify that the lamp test is working.
2. Verify that as any switch on the keypad is activated, the proper switch code is shown on the digital LED, and that the proper LED is also lighted.
3. Verify that if two switches are pressed, the switch code of the one with the highest priority is shown on the digital LED, and only one LED, that of the highest priority, is lighted.
4. Suppose that there are more than 10 rooms. How can the circuit be modified to handle this situation? How easily can the circuit be expanded?

Experiment III

Design of a Super Coin Exchanger

OBJECTIVES

Design a super coin exchanger that will handle products with different prices as selected by the customer. The prices of the products are 35 cents, 40 cents, or 45 cents. Report the current amount of money inserted, the price of the product chosen, and the change in multiples of nickels.

EQUIPMENT

  12 SPST switches
   1 4P3T switch
   3 74LS147 BCD priority encoders
   3 74LS83 4-bit full adders
   3 74LS48 display decoders
   2 74ALS04 hex inverters
   3 digital LEDs
   1 5-V power supply

PROCEDURE

1. Connect the switches with pull-up resistors to the three priority encoders as shown in Figure E4–3.
2. Connect the output of the three priority encoders to the three 4-bit full adders.
3. Connect the 4P3T switch to indicate the proper price of the products, and connect the output of the switch directly to one of the display decoders.
4. Connect the output of the full adder that indicates the amount deposited to one of the display decoders.

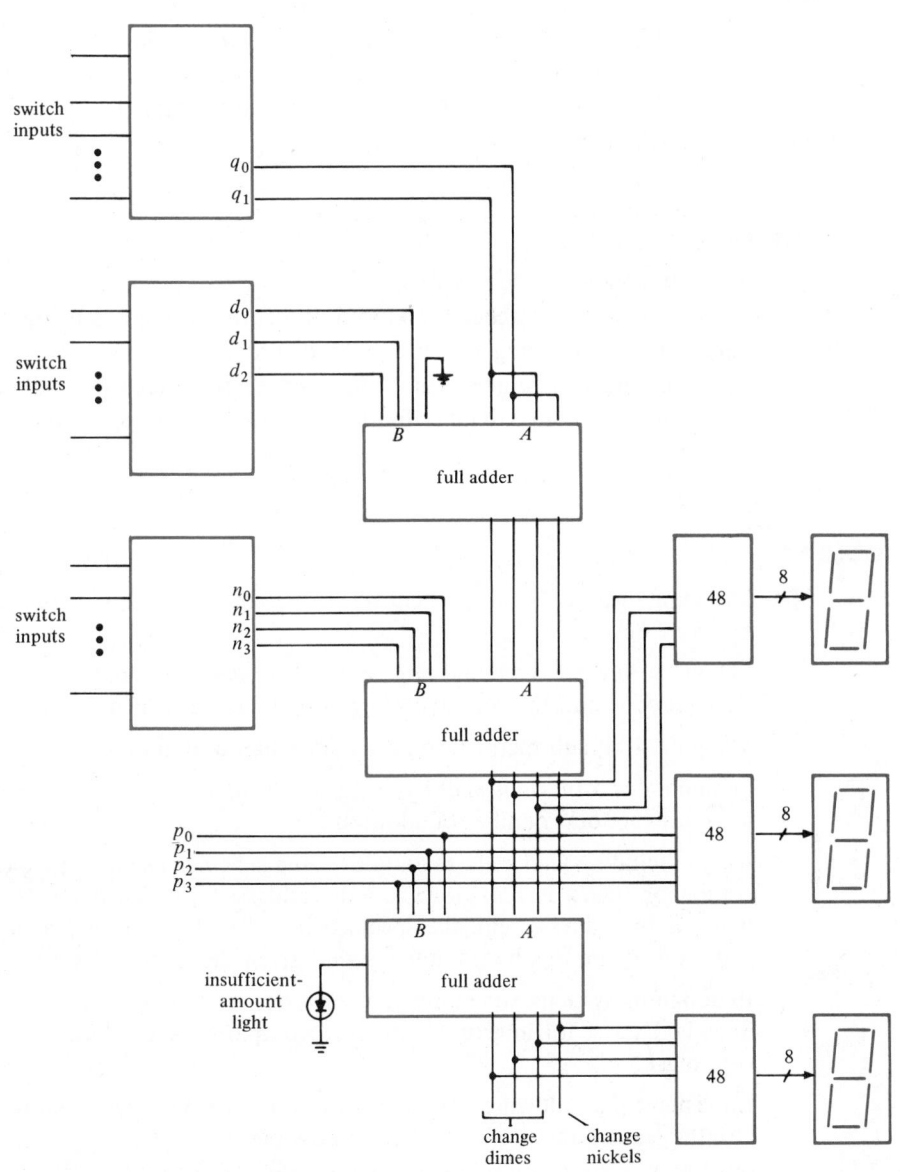

Figure E4–3
Circuit diagram for Experiment III, the design of a super coin exchanger

5. Connect the output of the full adder that indicates the change to another display decoder. The display should be blanked if insufficient money is inserted. This can be achieved by means of a BI input in the display decoder.
6. Connect all three display decoders to the three digital LEDs.
7. Connect the lamp test.

RESULTS

1. Verify that the lamp test works.
2. Verify that as the switches are closed, indicating the coins being inserted, the proper amount is being shown on the display.
3. Verify that, when a sufficient amount has been inserted, the product door latch opens and the change display lights, showing the correct change.

Problems

1. Design a 1's and 2's complement device by using adders and true/complement-zero/one devices.

2. Design a key-activated lighting system. A user need only enter a one-digit decimal code and the corresponding device will be turned on/off.

3. Design a 4-bit subtractor using a binary adder and additional logic.

4. Design a 4×4-bit multiplier by using adders to accumulate the partial sum and to count the total number of additions.

5. Many digital systems require an addressing scheme. The address of a particular station or board is sent down on the address bus. When the board address matches the address sent, the specified board will spring into action. Devise a way to generate this board active signal given the address lines.

6. In designing systems, sometimes a mask is desired to mask out unwanted signals. Such is the case for interrupts, for word comparisons, etc. Design such an interrupt mask.

7. Implement the following functions with only one multiplexer and one inverter. Use the least number of multiplexers possible.
   (a) $f(w, x, y, z) = \sum(0, 1, 5, 7, 8, 9, 10, 12, 14)$
   (b) $f(A, B, C, D, E) = \sum(4, 5, 7, 11, 12, 13, 15, 20, 21, 22, 26, 29, 31)$
   (c) $f(x_1, x_2, x_3) = \sum(1, 2, 4, 7)$
   (d) $f(a, b, c, d) = \prod(0, 2, 6, 9, 10, 11, 12, 15)$

8. Binary multiplication can be done with the use of a two-dimensional iterative network composed of full adders and AND gates. Examine the structure of the following example and sketch the iterative circuit.

```
                              1  0  1  1
              ×               1  1 ·0  1
                              1  0  1  1
                           0  0  0  0
                        1  0  1  1
                     1  0  1  1
                  1  0  0  0  1  1  1  1
```

Note how each row is obtained and then how the lines are summed.

9. Design a BCD-to-Excess-3 converter with the sum-of-products or product-of-sums realization.

10. Design an Excess-3-to-BCD converter with the sum-of-products or product-of-sums realization.

11. Design an Excess-3-to-BCD converter using full adders.

12. Modify the circuit that performs 1's complement with the true/complement-zero/one device to perform 2's complement addition and subtraction instead.

13. Modify the circuit that performs the 1's complement addition with the true/complement-zero/one device to include all of the following operations:
    (a) No operations
    (b) Adding 0s
    (c) Subtracting 0s
    (d) Incrementing by 1
    (e) Decrementing by 1

14. Implement the following combinational circuits with a multiplexer of the correct size:
    (a) $f(w, x, y, z) = \sum(0, 1, 2, 3, 7, 9, 11, 14)$
    (b) $f(A, B, C, D) = \prod(1, 3, 5, 7, 9, 11, 13, 15)$
    (c) $f(a, b, c, d) = \sum(4, 5, 6, 7, 8, 9, 10) + \sum_d(13, 15)$

15. Modify the coin exchanger problem in such a way that the prices of the products may be 35 cents, 40 cents, or 45 cents depending on the product chosen. (This is the paper design part of Experiment III.)

16. Design a circuit to convert a decimal number into its 9's complement form using:
    (a) SSI components
    (b) MSI components

17. You are asked to design an adder for a new computer system. The word length of this new computer system is 80 bits long. Since the adder is a part of the whole, it is crucial to make your circuit operate as fast as possible. Try to come up with at least three designs in block diagram form and calculate the delay time for each scheme.

18. Calculate the delay for multiplying two n-bit operands for n varying from 4 bits to 32 bits.

Sequential Circuit Design

5

In this chapter we first present the general concept of the Mealy and Moore types of sequential circuits. Flip-flops are then introduced as storage elements. We next examine the procedures for analyzing and synthesizing sequential circuits. The second half of this chapter is devoted to the introduction of basic MSI sequential components. In particular, details on flip-flops, counters, latches, and shift registers and their usage are presented.

The digital circuits examined in the previous chapters are mostly combinational circuits, in that the output at any point in time is only a function of its present input at the same point in time. If the set of inputs is presented later at another point in time, the identical output will be there because the output is only a function of the present inputs. These are called *memoryless circuits* and are time-invariant. Many digital systems have memory elements together with combinational circuits to produce digital circuits whose outputs are functions of both present and past inputs. Digital circuits with memory are called *sequential logic circuits, sequential circuits*, or *sequential machines*. Almost all digital systems make extensive use of sequential circuits to reduce the cost.

5.1
Sequential Circuits

In general, a combinational circuit can be represented simply by the relationship

$$Z = f(X)$$

where Z is a vector of output variables, X is a vector of input variables, and f is the functional relationship between the input and output parameters, i.e., the combinational circuit that relates X to Z. A sequential circuit consists of combinational circuits and memory elements connected together to form a feedback path. The block diagram of a sequential circuit is shown in Figure 5–1. The feedback path and the memory devices give the circuit its sequential nature. The memory elements can store binary information and retain the values indefinitely until they are changed. The binary information in the memory elements determines the state of a sequential circuit.

A sequential circuit typically receives inputs from another digital system or from the external world. The present outputs of a sequential circuit, as shown in Figure 5–1, are a function of both the inputs and the memory elements. In addition, the next state of the memory element inputs is also a function of the

Figure 5–1

Block diagram of a general sequential circuit

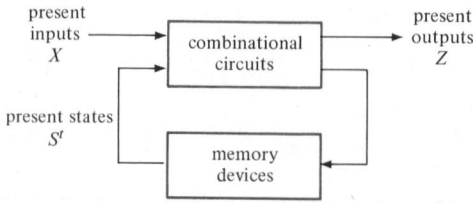

present inputs and their current state. This shows that a sequential circuit is a machine that responds to a time sequence of inputs, outputs, and the state of the memory elements. In other words, a sequential circuit can be represented as

$$Z = f(X, S^t)$$
$$S^{t+1} = g(X, S^t)$$

where Z represents the present outputs, X the present inputs, S^t the present states, and S^{t+1} the next states.

Types of Sequential Circuits

There are two main types of sequential circuits, and their classification depends on the structure of the memory elements. If the memory elements are controlled by an external signal called a clock such that the memory elements can be changed only at specified times, the sequential circuit is called a *synchronous* sequential circuit. For obvious reasons, a synchronous sequential circuit is also called a *clocked* sequential circuit.

Sequential circuits that do not have a clock are called *asynchronous* sequential circuits. The memory elements in an asynchronous sequential circuit change state whenever one or more of the inputs changes. Therefore, a change in the input to the sequential circuit can cause the inputs to a memory element to change, and thus the state of the memory elements may also change. This change is applied to the combination circuit, which in turn may change the memory elements again. Thus, if unchecked, a sequential circuit input change can cause a sequence of state changes. The relative delays in the circuit elements may cause the final stable states to vary, or the state may continue to oscillate. The clocks in the synchronous sequential circuits and memory elements are so designed that each memory element may have at most one change in state for each clock signal. This eliminates the instability problems and uncertain operation of the asynchronous sequential circuits that were just described.

Asynchronous sequential circuits are inherently faster than the synchronous type, and therefore attract the interest of researchers. The implementation difficulties encountered with variable delays, etc., prevent their widespread acceptance in most digital systems. Synchronous sequential circuits are easy to design because the clock controls the state changes, and because of their widespread use we will concentrate on circuits of this type.

Representations of Sequential Circuits

We have seen that combinational circuits are completely described by truth tables; similarly, sequential circuits are completely specified by state transition diagrams and state transition tables. For example, a sequential circuit may be designed to accept three inputs, generate two outputs, and be found in one of three states; i.e.,

$$X = [x_1, x_2, x_3]$$
$$Z = [z_1, z_2]$$
$$S = [s_1, s_2, s_3]$$

A typical state transition diagram and its corresponding state transition table for this sequential circuit are shown in Figures 5–2(a) and 5–2(b), respectively.

A state transition diagram is a graph with the nodes representing the allowable states that the machine can take on. Joining the nodes are arcs that represent the transitions or paths from one node to another based on the inputs, the x's. The outputs, z's, are also given for each transition. For example, in Figure 5–2(a), the machine can be found in three states, s_1, s_2, and s_3, respectively. At state s_1, the next state of the machine can be either s_1, s_2, or s_3 depending on whether the input is x_3, x_1, or x_2, respectively. From Figure 5–2(a), we can see that from

Figure 5–2

Typical (a) state transition diagram and (b) state transition table for a sequential circuit

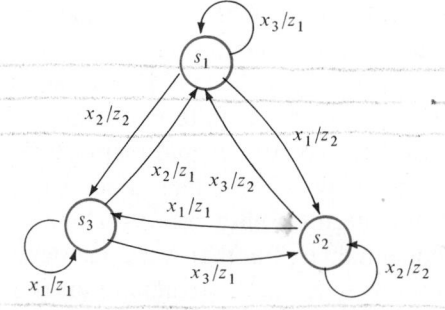

(a) State transition diagram

Present State	Present Input		
	x_1	x_2	x_3
s_1	s_2/z_2	s_3/z_2	s_1/z_1
s_2	s_3/z_1	s_2/z_2	s_1/z_2
s_3	s_3/z_1	s_1/z_1	s_2/z_1

(b) State transition table

state s_1, the machine will, upon the x_1 input, go to state s_2 with a z_2 output, or upon the x_2 input, go to state s_3 with a z_2 output, or, upon the x_3 input, go to state s_1 with a z_1 output. The transitions from nodes s_2 and s_3 can be represented in like manner. Since there are three possible inputs, there must be three arcs representing transitions from each state. Hence each arc is tagged with the identifying input that causes that transition and the corresponding output for that transition. Note also that it is perfectly legitimate for the arcs to loop back to the originating node.

In a similar fashion, the information contained in the state transition diagram can be represented in a tabular form called the state transition table. In the state transition table, all possible states of the machine are listed on the left, and all possible inputs are listed on the top. For each individual state, the next state and the output are specified for every possible input. The reader should verify that the state transition diagram in Figure 5–2(a) conveys the same information as the state transition table in Figure 5–2(b).

Example 5–1

Construct the state transition diagram and the state transition table for a decimal up counter.

Since the counter to be represented is a decimal counter, it is obvious that there are 10 states:

$$S = [s_0, s_2, \ldots, s_9]$$

There are no real inputs except the clock for this case. If the clock is present, the counter will count. If the clock stops, counting will also stop. The state of the counter is the count itself. The only counter output is the carry-out. Hence the state transition diagram can be drawn simply as shown in Figure 5–3(a). We will assume that there is no specific starting point for the counter, but the counter will start counting at any state up to state 9, go back to zero, and start counting up again. The state transition table for this counter can be constructed and is shown in Figure 5–3(b).

Example 5–2

Construct the state transition diagram and state transition table for a binary sequence detector that outputs a 1 if and only if the input sequence is 110_2. The sequence must begin with a 1, be followed by another 1, and then terminate with a 0.

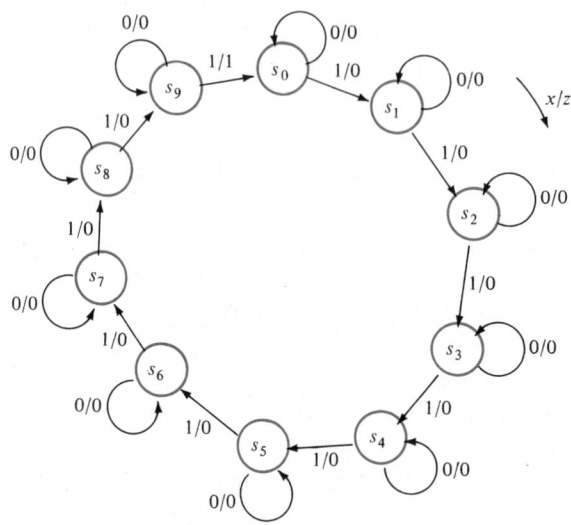

(a) State transition diagram

	Present Input	
Present State	0	1
s_0	$s_0/0$	$s_1/0$
s_1	$s_1/0$	$s_2/0$
s_2	$s_2/0$	$s_3/0$
s_3	$s_3/0$	$s_4/0$
s_4	$s_4/0$	$s_5/0$
s_5	$s_5/0$	$s_6/0$
s_6	$s_6/0$	$s_7/0$
s_7	$s_7/0$	$s_8/0$
s_8	$s_8/0$	$s_9/0$
s_9	$s_9/0$	$s_0/1$

(b) State transition table

Figure 5–3

(a) State transition diagram and (b) state transition table for a decimal
up counter

The input is clearly a binary sequence, which can be either 0 or 1. The output
is also a binary variable that is always 0 and is 1 only when the proper sequence
has been detected. It is not clear initially how many states are needed before the
state transition diagram is constructed. This is not important, because the number of states needed will become evident as we go along. The input and output

vectors can be stated as

$$X = [0, 1]$$
$$Z = [0, 1]$$

The machine must begin at an idle state. Call this state s_1. At this state, the machine is looking for the first incoming binary digit to be a 1. If the first digit is a 0, there is no possibility for it to be the beginning of the sequence and the machine remains in state s_1 to look for the next binary digit. If the first digit is a 1, then it is possible for that digit to be the beginning of the sequence. In other words, when a 1 is detected, the machine changes state to s_2 to reflect the new status. At state s_2, the machine is again looking for another 1. Upon finding the second incoming digit to be a 0, the sequence is off and the machine returns to the original state. If, however, the second digit is a 1, then the machine must change state to s_3 to reflect that 11_2 has been detected. At this point, if the third digit coming in is a 0, the sequence has occurred and the output is changed to 1, and the machine returns to s_1 for the next sequence. If the third digit is a 1, then the sequence may still be on if the next digit is a 0. Hence the machine must stay at state s_3 without generating any output. From this analysis, we see that we have three states:

$$S = [s_1, s_2, s_3]$$

The state transition diagram for the above logic is shown in Figure 5–4(a). The corresponding state transition table is given in Figure 5–4(b).

Mealy and Moore Representations

There are two basic representations of sequential machines, the Mealy and Moore types. They differ in the way the present output is determined. The Mealy-type machine can be described by the following set of equations:

$$Z = f(X, S^t)$$
$$S^{t+1} = g(X, S^t)$$

The discussion of sequential circuits so far in this chapter has been predominantly Mealy-type machines. The important thing to note here is that the outputs and the state changes are dependent on both the present state, S^t, and the present input, X. Many times, the output function is rather complex because of this dependency. The Moore-type machine can be described by the following set of equations:

$$Z = f(S^t)$$
$$S^{t+1} = g(X, S^t)$$

It is obvious that the output is dependent only on the state of the sequential machine and is not directly dependent on the present input or the transition

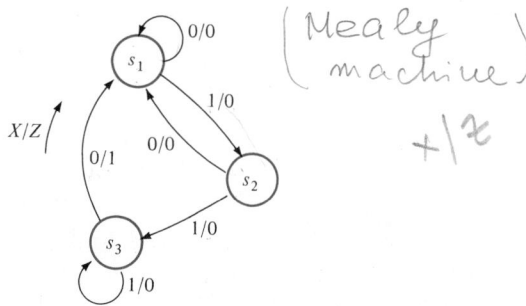

(a) State transition diagram

	Present Input	
Present Output	0	1
s_1	$s_1/0$	$s_2/0$
s_2	$s_1/0$	$s_3/0$
s_3	$s_1/1$	$s_3/0$

(b) State transition table

Figure 5–4

(a) State transition diagram and (b) state transition table for a binary sequence detector that recognizes 110_2

itself. In other words, there is a specific output associated with each state regardless of how the machine gets there. A typical state transition diagram and its corresponding state transition table for a Moore machine are given in Figure 5–5. Note that the output function is not specified in the transition arcs but is specified in association with the state only. The reader should verify that the information in the state transition diagram is the same as that in the state transition table.

Example 5–3

Express the binary sequence detector in Example 5–2 in a Moore sequential form.

The binary sequence to be detected is 110_2. As in Example 5–2, one seldom knows a priori how many states there will be. We will again start with the idle state s_1, where there is no output—i.e., a 0 output. When a binary digit is received, if it is a 0, then the machine remains in the idle state with a 0 output. If the

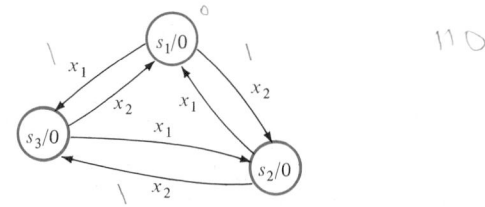

(a) State transition diagram (Moore machine)

Present State	Present Input		Present Output
	x_1	x_2	z
	Next state		
s_1	s_3	s_2	0
s_2	s_1	s_3	1
s_3	s_2	s_1	1
	Next state		

(b) State transition table (Moore machine)

Figure 5–5

(a) State transition diagram and (b) state transition table for a Moore-type sequential machine

received digit is a 1, then the machine changes its state to s_2 with a 0 output associated with s_2. Upon detecting the next 1, the machine will be in state s_3 with a 0 output. For the last digit, if it is a 1, then the sequence is off and the machine goes back to the idle state with a 0 output. If the last digit detected is a 0, then the sequence is completed. Instead of returning to the idle state with an output as in the Mealy-type machine, the machine must go to another state s_4 with an associated 1 output. In all practicality, the state s_4 is the same as s_1 except for the output. The transition diagram and the transition table for the Moore machine are shown in Figure 5–6.

5.2

Flip-flops

The basic memory element that can maintain (i.e., store) a binary value indefinitely is called a *flip-flop*. There are many different types of flip-flops, but we will consider only the main types that are presently available in integrated circuits.

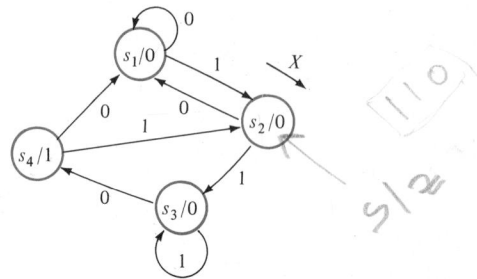

(a) State transition diagram

Present State	Present Input		Present Output
	$x = 0$	$x = 1$	z
	Next state		
s_1	s_1	s_2	0
s_2	s_1	s_3	0
s_3	s_4	s_3	0
s_4	s_1	s_2	1

(b) State transition table

Figure 5–6

(a) State transition diagram and (b) state transition table for the binary sequence detector in Moore form

Basic Design of a Flip-flop

There are many equivalent designs for a flip-flop. We will cover a number of these designs.

Cross-coupled NOR Flip-flop   The basic flip-flop is constructed of two cross-coupled NOR gates. This flip-flop is called an *asynchronous* flip-flop because there is no clock input to govern its operation. The flip-flop has only two inputs, set (S) and reset (R); and two outputs, called Q and $\bar{Q}$. The logic circuit of a basic flip-flop is given in Figure 5–7(a). A timing diagram is a time sequence of input changes that cause corresponding changes in the output. Timing diagrams are very helpful in the analysis of sequential circuits. The timing diagram for the basic cross-coupled NOR gate flip-flop is given in Figure 5–7(b). The two outputs Q and $\bar{Q}$ of the flip-flop are always complements of each other, except when both inputs are 1.

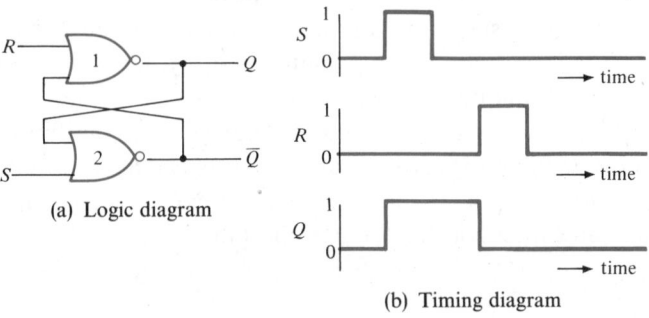

(a) Logic diagram

(b) Timing diagram

S	R	Q^{t+1}	$\bar{Q}^{t+1}$
0	0	Q^t	$\bar{Q}^t$
0	1	0	1
1	0	1	0
1	1	(Not allowed)	

(c) Truth table

Figure 5–7
Basic *RS* flip-flop with cross-coupled NOR gate

Let us analyze the flip-flop operation. Assume initially that the outputs are $Q = 0$, $\bar{Q} = 1$, and the inputs are $S = 0$ and $R = 0$. Let S change from 0 to 1. Then the output NOR gate 2 will go to zero, that is, $\bar{Q} = 0$. This forces the output of NOR gate 1 to a 1. Therefore, both inputs into NOR gate 2 are 1 and both inputs into NOR gate 1 are 0, and no further changes will take place. Now if S changes from 1 to 0, Q will remain at 1 and $\bar{Q}$ will be 0, which is a stable state. Hence the state of Q will remain at 1 indefinitely. Similarly, a $0 \rightarrow 1 \rightarrow 0$ transition (pulse) on R will change Q to 0 and $\bar{Q}$ to 1. Then the flip-flop will remain in that state indefinitely. This action is illustrated by the timing diagram in Figure 5–7(b).

In addition to timing diagrams, the operation of a flip-flop can be represented by a truth table, which is given in Figure 5–7(c). The superscript $t + 1$ represents the future time and t represents the present time. The expression $Q^{t+1} = Q^t$ implies that the output at the next time instant is the same as that at the last instant. In other words, the circuit is stable and will not change anymore. The first row in the truth table means that in the absence of any inputs, if $Q = 1$, it will remain that way; or if $Q = 0$, it will remain that way. The flip-flop is said to be in a *memory state* in this condition, because the output value Q is maintained. When S is set to 1 and R is made 0, the output Q will be set to 1 also. This is the *set condition.* When $S = 0$ and $R = 1$, the output Q is reset. This is the *reset condition.* The circuit designer usually does not make S and R equal to 1 simultaneously, because if both are made 1 simultaneously, no stable state can be found. Hence this input condition is not allowed.

It is obvious that the next output Q^{t+1} of the flip-flop is a function of the inputs S, R, and Q^t. A Karnaugh map of the next output in terms of these input parameters can be made as shown in Figure 5–8. The map is then minimized to yield an expression for the next output Q^{t+1}:

$$Q^{t+1} = S + \bar{R}Q^t \tag{5-1}$$

Equation (5–1) is called the flip-flop characteristic equation for the set-reset flip-flop (SRFF) because it shows how the next output of the SRFF can be obtained from the present output Q and the present input conditions.

The flip-flop is sometimes said to be in the set state when $Q = 1$ and in the reset state when $Q = 0$. Therefore, the S (set) input will set the flip-flop and the R (reset) input will reset the flip-flop. This explains the input designations. The nomenclature for these inputs is not unique: The set input is sometimes called the preset input, and the reset input is also called the clear input.

Cross-coupled NAND Flip-flop   Cross-coupled NAND gates can be used for the basic flip-flop as shown in Figure 5–9(a). The timing diagram, which is given in Figure 5–9(b), in this case shows that the second time $S = 0$, the output Q does not change. Similarly, when $Q = 1$ and then $S = 1$ in a cross-coupled NOR gate flip-flop, no further changes in Q will result. Sometimes when cross-coupled NAND gates are used, inputs are designated $\bar{S}$ and $\bar{R}$, because 0s, not 1s, on these inputs may change the state of the flip-flop. The truth table is given in Figure 5–9(c).

Figure 5–8

Karnaugh map of the output of an *SR* flip-flop with respect to its inputs

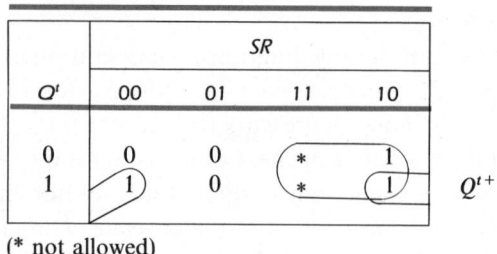

(* not allowed)

$$Q^{t+1} = S + \bar{R}Q^t$$

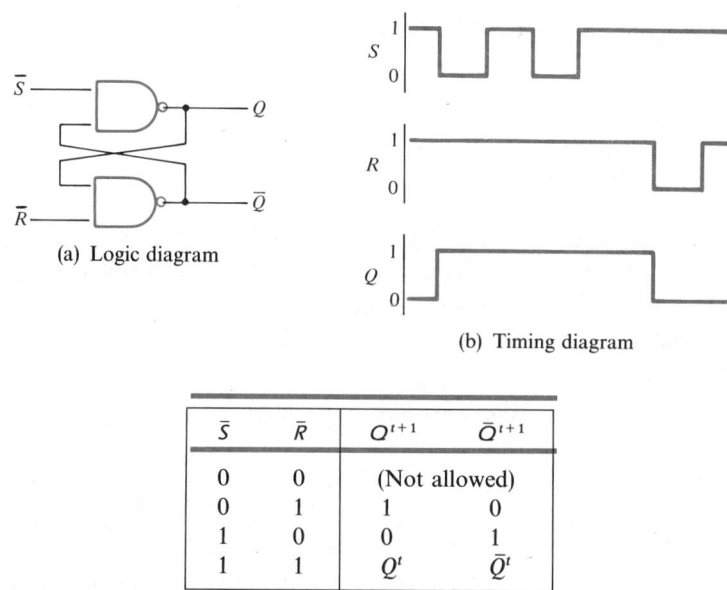

(a) Logic diagram

(b) Timing diagram

$\bar{S}$	$\bar{R}$	Q^{t+1}	$\bar{Q}^{t+1}$
0	0	(Not allowed)	
0	1	1	0
1	0	0	1
1	1	Q^t	$\bar{Q}^t$

(c) Truth table

Figure 5–9

Basic *RS* flip-flop with cross-coupled NAND gates

Clocked Flip-flop A clocked flip-flop can be made with the addition of two AND gates to the basic flip-flop as shown in Figure 5–10. The normal operation of this flip-flop is to fix the *R* and *S* inputs, then give a $0 \rightarrow 1 \rightarrow 0$ pulse on the *CP* (clock pulse) input. This flip-flop cannot change state unless $CP = 1$. For proper operation of this flip-flop the *CP* input must be at 1 long enough to allow the state of the flip-flop to change. The change of any flip-flop in a sequential circuit may cause a further change in one of the *R* or *S* inputs. If the *CP* input is still 1 when

Figure 5–10

A clocked *RS* flip-flop

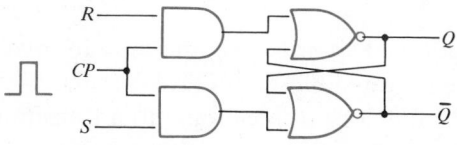

these changes occur on R and S, further changes could result. To prevent this, in a well-designed circuit, the width of the CP pulse must be carefully controlled to allow at most one change in any flip-flop for each clock pulse. The control of the CP width is difficult, and this type of clocked flip-flop is seldom used. This type of clock control is commonly called a level mode-triggered flip-flop because the flip-flop is triggered whenever the clock stays in the high level.

Master–Slave Flip-flop   A master–slave flip-flop is another way of constructing a flip-flop in which the pulse width of the clock is not crucial. When master–slave flip-flops are used in sequential circuits, the combinational logic is connected to the Q and $\bar{Q}$ outputs. As shown in Figure 5–11(a), the master–slave flip-flop contains two basic flip-flops. The first is called the master flip-flop (gates 3 and 4) and the second, the slave flip-flop (gates 7 and 8). The clock CP is initially kept at 0 when the S and R inputs are being changed by external circuitry. When CP goes to 1, the information on S and R is transferred into the master flip-flop through gates 1 and 2. Though the master flip-flop changes state to reflect the present input, the outputs of the master flip-flop are blocked by gates 5 and 6 from changing Q or $\bar{Q}$. Therefore, the clock may remain high indefinitely without any further state changes. When CP goes low (i.e., $CP = 0$), the state of the master flip-flop is transferred through gates 5 and 6 to the slave flip-flop, which then changes the state of the slave flip-flop. When $CP = 0$, however, any further changes on S and R are blocked from a further change on either the master or slave flip-flop. The symbol (a triangle) on the clock CP input in Figure 5–11(b) indicates that the outputs Q and $\bar{Q}$ will change in response to a positive going or a $0 \rightarrow 1$ transition on the clock input.

A master–slave flip-flop is often used when timing is critical. When flip-flops are connected together, ideally all flip-flops should receive the clock signal at the same time and change the outputs at the same time. For a level mode-clocked flip-flop, the output may change as long as the clock level is high. When the output of the first flip-flop is connected to a second flip-flop, it is possible that the change in the output of the first flip-flop will affect the state of the second flip-flop. The master–slave flip-flop provides a solution to this racing problem by isolating the changes in the inputs from simultaneously changing the outputs. This is shown in the timing diagrams in Figure 5–11. The use of the master–slave flip-flop ensures that all flip-flops connected to the same clock signal sample the inputs at the same time and initiate any changes at the outputs at the same time.

Edge-Triggered Flip-flop   Another way to solve the racing problem is by means of an edge-triggered flip-flop. This is another way to design clocked flip-flops by making the flip-flops respond only to a transition ($0 \rightarrow 1$ or $1 \rightarrow 0$) on the clock input and not allow the outputs (Q and $\bar{Q}$) to change again until the next similar

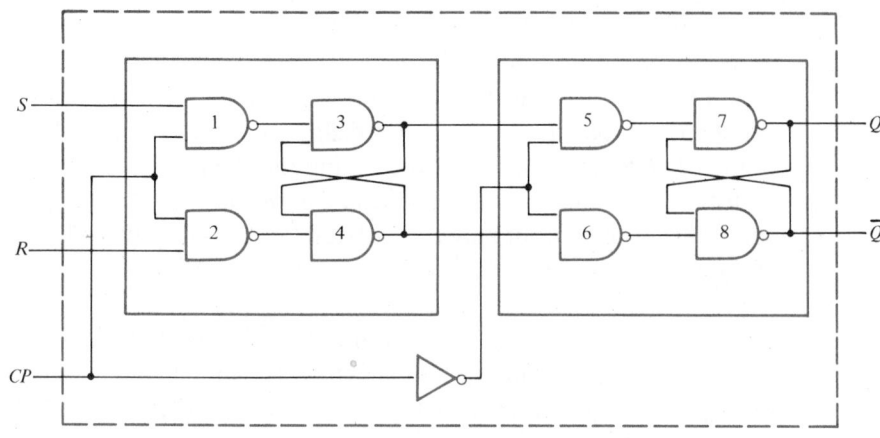

(a) Clocked master–slave SRFF

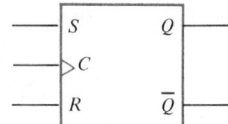

(b) Logic symbol for a master–slave SRFF

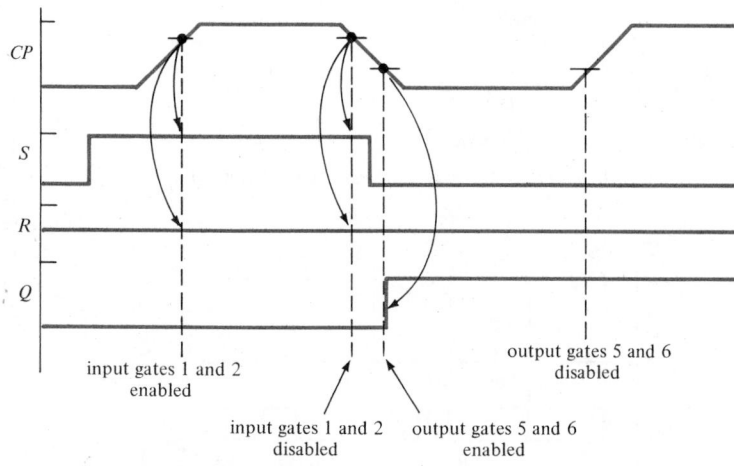

input gates 1 and 2
enabled

input gates 1 and 2
disabled

output gates 5 and 6
enabled

output gates 5 and 6
disabled

(c) Timing diagram showing the operation of a master–slave SRFF

Figure 5–11
Master–slave flip-flop

transition. In other words, the flip-flop responds not to the level of the clock input but to a transition in the clock input. This could be on the positive ($0 \rightarrow 1$) transition for the positive edge-triggered flip-flop or on the negative ($1 \rightarrow 0$) transition for the negative edge-triggered flip-flop. Figure 5–12(a) defines the transitions and shapes. A positive edge-triggered flip-flop is readily available in integrated circuits. The logic symbol and the timing diagram for this flip-flop are given in Figures 5–12(b) and 5–12(c). The triangle on the *CP* (clock) input logic symbol indicates that the output may change state on a positive ($0 \rightarrow 1$) transition on the *CP* input.

Complete Flip-flop   A complete flip-flop is a clocked flip-flop equipped not only with two synchronous inputs that are either level mode or edge-triggered but also two asynchronous inputs called PRESET and CLEAR, which take precedence over the synchronous inputs. These inputs are asynchronous in nature and are completely independent of the clock in operation; they are useful for initializing the flip-flop or clearing the flip-flop asynchronously. Figures 5–13(a) and 5–13(b) show a typical schematic and the logic symbol for a complete *SR* flip-flop. Note that the PRESET and CLEAR inputs, which are normally in negative logic, are

Figure 5–12
An edge-triggered *SR* flip-flop

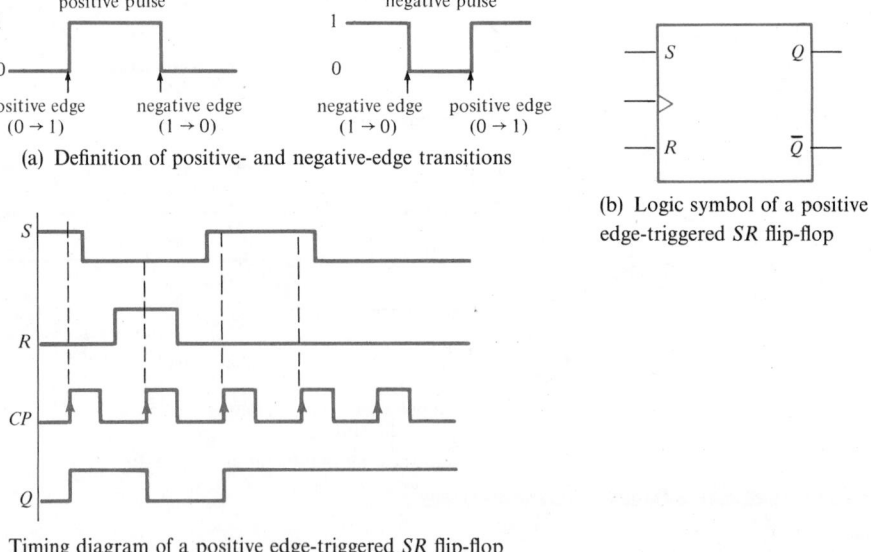

(a) Definition of positive- and negative-edge transitions

(b) Logic symbol of a positive edge-triggered *SR* flip-flop

(c) Timing diagram of a positive edge-triggered *SR* flip-flop

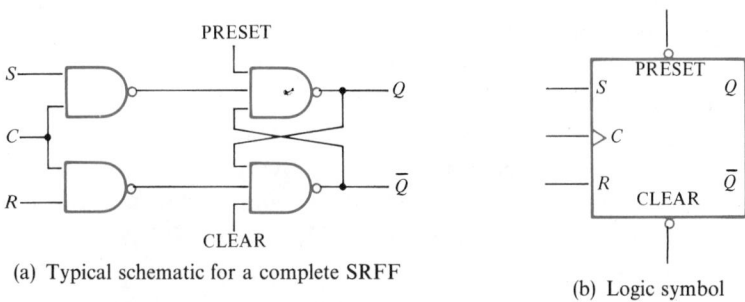

(a) Typical schematic for a complete SRFF

(b) Logic symbol

Inputs						Output
PRESET	CLEAR	CLOCK	S	R	Q^t	Q^{t+1}
0	1	d	d	d	d	1
1	0	d	d	d	d	0
1	1	$\uparrow$	0	0	0	0
1	1	$\uparrow$	0	0	1	1
1	1	$\uparrow$	0	1	d	0
1	1	$\uparrow$	1	0	d	1

(c) Truth table

Figure 5–13
A complete *SR* flip-flop (not master–slave or edge-triggered)

connected directly to the output gates. When the PRESET input is asserted, the output Q is high. Likewise, when the CLEAR input is asserted, the output Q is cleared. The truth table for an SRFF including the PRESET and CLEAR inputs is shown in Figure 5–13(c).

Types of Flip-flops

There are several types of flip-flops besides the basic *SR* flip-flop. These flip-flops vary in the number of inputs and their function. The most common types of flip-flops are:

SR flip-flop
D flip-flop
T flip-flop
JK flip-flop

We have already studied the SRFF. In the following sections we will examine the other three types of flip-flops.

D Flip-flop   The schematic and logic symbol of a delay (D) flip-flop (DFF) are given in Figures 5–14(a) and 5–14(b). The characteristic table (sometimes called a truth table) for the DFF is given in Figure 5–14(c), and the characteristic equation in Figure 5–14(d). This table gives the next value of the Q output when D and Q are fixed, following a $0 \to 1$ transition on the clock. The next value of Q is just a delayed version of D. This is why it is called a D flip-flop.

The operation of the DFF can be analyzed as follows. The value of S and R will determine the output Q uniquely. When $CP = 0$, then $S = R = 1$ and Q will be in the memory mode (i.e., $Q^{t+1} = Q^t$). Now assume that $CP = 0$ and $D = 1$; then $A = 1$ and $B = 0$, and let CP go from $0 \to 1$. This causes S to go to 0, with all other values staying the same ($A = 1$, $B = 0$, $R = 1$), which makes $Q = 1$. We have just shown that a $0 \to 1$ transition on CP with $D = 1$ makes $Q = 1$. Similarly, a $0 \to 1$ on CP with $D = 0$ makes $Q = 0$. Let us continue to examine the situation with $CP = 1$ and D now changing from $1 \to 0$. This makes $B = 1$ but will not change A, S, or R, which implies that Q still equals 1. This is true regardless of

Figure 5–14
An edge-triggered D flip-flop

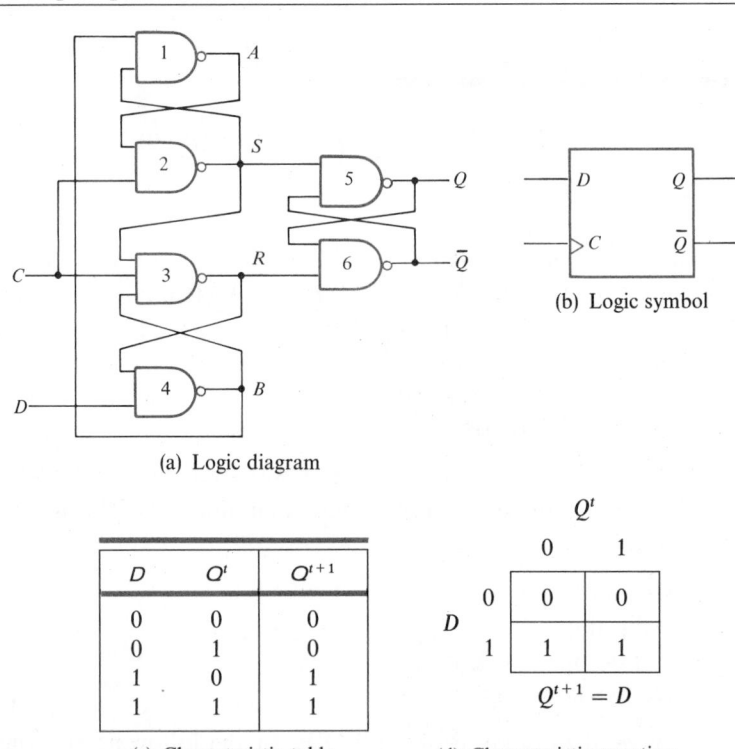

(a) Logic diagram

(b) Logic symbol

D	Q^t	Q^{t+1}
0	0	0
0	1	0
1	0	1
1	1	1

(c) Characteristic table

$$Q^{t+1} = D$$

(d) Characteristic equation

the number of changes on D when $CP = 1$. Thus, in this case the output Q will change only in response to the $0 \rightarrow 1$ transition on C, but it never changes when CP stays at 1. A similar situation occurs when CP goes from $0 \rightarrow 1$ when $D = 0$. Then, when CP goes back to 0, this forces $S = R = 1$, so the outputs will not change until another $0 \rightarrow 1$ transition on C. By minimizing the Karnaugh map of the DFF characteristic table, shown in Figure 5–14(d), we get the following characteristic equation for the DFF:

$$Q^{t+1} = D$$

The logic symbol for a complete DFF with PRESET and CLEAR is shown in Figure 5–15.

T **Flip-flop** The T flip-flop (TFF) has only one input, namely the T input. The schematic and notation for a TFF are shown in Figures 5–16(a) and 5–16(b). When the T input is 1, the output of the flip-flop will change state at every clock cycle. When the T input is 0, the output remains the same for all clock cycles.

Figure 5–15
A D flip-flop with asynchronous PRESET and CLEAR

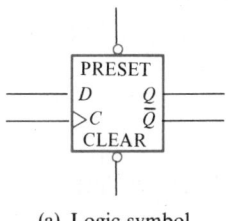

(a) Logic symbol

Inputs					Output
CLEAR	PRESET	CLOCK	D	Q^t	Q^{t+1}
0	1	d	d	d	0
1	0	d	d	d	1
1	1	↑	0	0	0
1	1	↑	0	1	0
1	1	↑	1	0	1
1	1	↑	1	1	1

(b) Function table

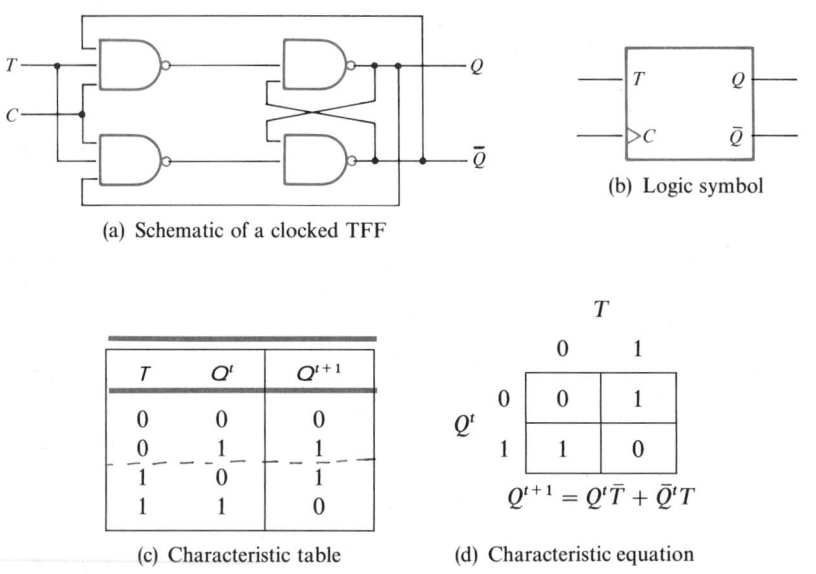

(a) Schematic of a clocked TFF

(b) Logic symbol

(c) Characteristic table

(d) Characteristic equation

Figure 5–16
A T flip-flop

The TFF characteristic table and the TFF characteristic equation are shown in Figures 5–16(c) and 5–16(d), respectively. By minimizing the Karnaugh map of the characteristic table, the TFF characteristic equation can be obtained:

$$Q^{t+1} = T \cdot \bar{Q}^t + \bar{T} \cdot Q^t$$

JK **Flip-flop** The *JK* flip-flop (JKFF) is the most versatile type of flip-flops available. The JKFF has two inputs, *J* and *K*, which are similar to the *S* and *R* inputs of the SRFF in that the *J* input can be used to set the output while the *K* can be used to reset the output. The main difference is that while both the *S* and *R* inputs cannot be asserted high at the same time, the *J* and *K* inputs can be asserted high simultaneously. For this condition, the output will change state. The schematic, the logic symbol, the characteristic table, and the characteristic equation for the JKFF are given in Figures 5–17(a), 5–17(b), 5–17(c), and 5–17(d) respectively. The JKFF characteristic equation can be obtained by minimizing the JKFF characteristic table:

$$Q^{t+1} = J \cdot \bar{Q}^t + \bar{K} \cdot Q^t$$

Careful examination of the JKFF characteristic table shows that the characteristic tables for the SRFF, the TFF, and the DFF are all contained in the JKFF. In fact, it is possible to fashion the JKFF to behave like one of the other three flip-flops. This is shown in Figure 5–18. In Figure 5–18(a), we see that the

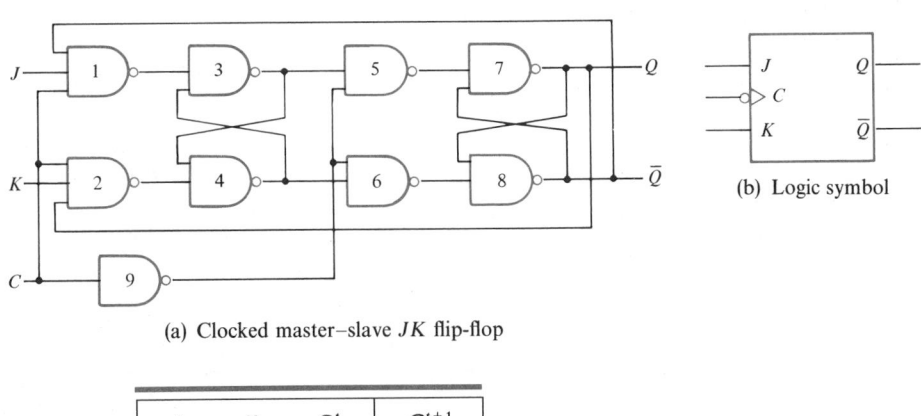

(a) Clocked master–slave JK flip-flop

(b) Logic symbol

J	K	Q^t	Q^{t+1}
0	0	0	0
0	0	1	1
0	1	0	0
0	1	1	0
1	0	0	1
1	0	1	1
1	1	0	1
1	1	1	0

(c) Characteristic table

(d) Characteristic equation

$$Q^{t+1} = J\bar{Q}^t + \bar{K}Q^t$$

Figure 5–17
A JK master–slave flip-flop

JKFF functions exactly like an SRFF. Since both the S and R inputs are not allowed to be asserted high simultaneously, there are no changes needed on the JKFF to make it behave like an SRFF. In other words, a JKFF is an SRFF and more. The portion of the JKFF characteristic table that simulates the SRFF is shown in Figure 5–18(a). In Figure 5–18(b), we see that by connecting J and K together, we obtain a TFF. This is evident from the solid portion of the JKFF characteristic table. By connecting a single input to J and its complement to K, we obtain a DFF. This is shown in Figure 5–18(c).

Some Typical Flip-flops

Some representative flip-flops are listed in Figure 5–19. Most commonly available flip-flops are SR latches, JKFFs, or DFFs. TFFs are seldom found because a JKFF is a TFF with the inputs J and K tied together, as shown earlier. If only the Q output of a flip-flop is available, it is said to have a *single rail output*. If both the Q and $\bar{Q}$ outputs are available, then the flip-flop is said to have a *double rail output*.

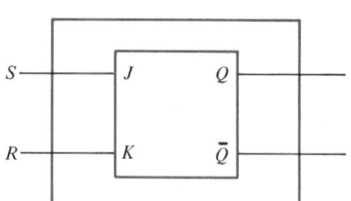

J	K	Q^{t+1}
0	0	Q^t
0	1	0
1	0	1
1	1	$\bar{Q}^t$

(a) Using a JKFF as an SRFF

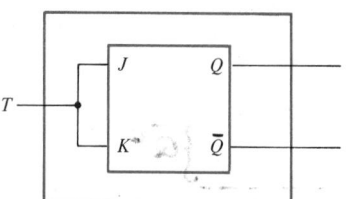

J	K	Q^{t+1}
0	1	0
1	0	1
1	1	$\bar{Q}^t$
0	0	Q^t

(b) Using a JKFF as a TFF

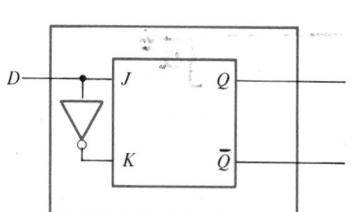

J	K	Q^{t+1}
0	0	Q^t
0	1	0
1	0	1
1	1	$\bar{Q}^t$

(c) Using a JKFF as a DFF

Figure 5–18 Using a JKFF as an SRFF, a TFF, and a DFF

Figure 5–19 Some representative flip-flops

	Flip-Flops	Single	Dual	Quad	Hex	Octal
JK	Edge-triggered	70, 101, 102	73, 76, 78, 103, 106, 107, 108, 109, 112, 113, 114	276, 376		
	Pulse-triggered	71, 72, 104, 105	73, 76, 78, 107			
D	Non-inverting		74	171, 175, 379	174, 378	374, 574, 273
	Inverting					534, 564, 576, 577, 876, 879
SR				279		

5.3

Analysis of Synchronous Sequential Circuits

The present output of a sequential circuit is a function of its present and past time sequence of inputs. Sequential circuits are constructed using combinational components and flip-flops. The sequential circuits are normally analyzed with the use of tables that determine the next state and present outputs as functions of present inputs and present states. The analysis of sequential circuits can be summarized in the following eight steps:

1. Identify the inputs (x's), outputs (z's), and coded states (y's) of the sequential circuit.
2. Obtain the output equations as a function of the present inputs and present states from the given sequential circuit.
3. Obtain the flip-flop excitation equations as a function of the present inputs and the present states from the circuit for each flip-flop.
4. Construct the flip-flop excitation table from the flip-flop excitation equations for all possible input states.
5. Construct the next-state table from the flip-flop excitation table.
6. Merge the output functions to the next-state table to form the coded state transition table.
7. Construct the state transition table from the coded state transition table.
8. Construct the state transition diagram from the state transition table.

The above procedures are best illustrated step by step with an example.

Example 5–4

Analyze the sequential circuit in Figure 5–20.

The circuit in Figure 5–20 is an example of a synchronous sequential circuit using JK master–slave flip-flops. The external inputs will always be labeled with x's, the state with y's, and the output with z's. This is a fairly standard nomenclature. The circuit diagram does not show all the connections, because this tends to clutter the circuit diagram. For instance, the y_2 input on gate 1 implies that a direct connection is made from the Q_2 output of the second flip-flop to this gate.

STEP 1

Input, Output, and State Variable Identification

In this example there is one input x, two state variables y_1 and y_2, and one output z. Hence:

$$X = [x = 0, x = 1]$$
$$Z = [z = 0, z = 1]$$
$$Y = [y_1 y_2 = 00, 01, 10, 11]$$

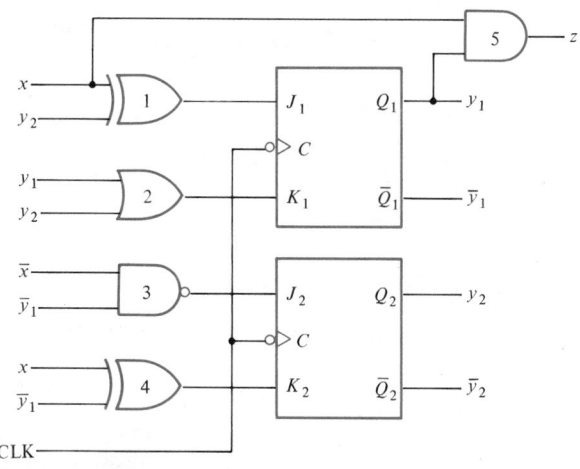

Figure 5–20

Example of a synchronous sequential circuit

Since there are two state variables, there are a maximum of four states. It is not possible to know beforehand whether all the states are used or not. At this point, all the states must be analyzed.

STEP 2

Output Equations

Since there is only one output variable z, there is only one corresponding output equation that can be obtained from the circuit connection:

$$z = xy_1$$

The present output is a function of the present state given by the output of the flip-flops and the present input.

STEP 3

Flip-flop Excitation Equations

The flip-flop excitation equations are obtained by writing down the circuit connections for every flip-flop input. This is done for all flip-flops. The flip-flop excitation equations show the inputs to the flip-flops as a function of the present state and the present input. In other words, the flip-flop excitation equations show how the flip-flops are being excited. In this example there are two flip-flops, each having two inputs. Hence there are four flip-flop excitation equations:

$$J_1 = x \oplus y_2$$
$$K_1 = y_1 + y_2$$

$$J_2 = \bar{x} \uparrow \bar{y}_1$$
$$K_2 = x \oplus \bar{y}_1$$

STEP 4

Flip-flop Excitation Table

From the flip-flop excitation equations, the flip-flop excitation table is constructed. It is the first table to be constructed in the analysis and is shown in Figure 5–21(a). This table represents a Karnaugh map showing the excitation of the flip-flops under all possible entries of the present states and present inputs. Hence J_1, K_1, J_2, and K_2 are evaluated under all possible input conditions.

STEP 5

Next-State Table

The next-state table shown in Figure 5–21(b) is obtained by applying the JKFF characteristic table to the flip-flop excitation table. The notation Y_i^{t+1} (with uppercase Y) indicates the next state of y_i^t (with lowercase y). The next-state table gives, for any current state $y_1^t y_2^t$, the next state $Y_1^{t+1}\ Y_2^{t+1}$ after the clock input. For the sake of clarity, the superscripts (t or $t + 1$) will be dropped on y or Y so that lowercase y's will always indicate the present-state variables while uppercase Y's will always indicate next-state variables.

STEP 6

Coded State Transition Table

The coded state transition table is obtained by merging the present output into the next-state table. The present output is obtained from the output equations previously obtained in Step 2. For every possible present state and present input, the present output is evaluated and is entered into the table. The coded state transition table thus obtained is shown in Figure 5–21(c). With both the next state and the present output entered into the table, we have essentially obtained the coded state transition table.

STEP 7

State Transition Table

The next step is to construct the real state transition table. The state transition table is obtained from an arbitrary assignment of distinct symbols to each distinct state, i.e., to each distinct binary combination of $y_1 y_2$. The state transition table for our example is given in Figure 5–21(d). From the sequential circuit, the coded state transition table can be obtained uniquely by the first six steps. These six steps are straightforward. The assignment of distinct states to distinct symbols, however, cannot be done from the knowledge of the sequential circuit alone.

x

y_1y_2	0	1
00	00, 01, 0	10, 10, 0
01	11, 01, 0	01, 10, 0
11	11, 10, 0	01, 11, 1
10	01, 10, 0	11, 11, 1

J_1K_1, J_2K_2, z

(a) Flip-flop excitation table

x

y_1y_2	0	1
00	00	11
01	10	01
11	01	00
10	01	01

Y_1Y_2

(b) Next-state table

x

y_1y_2	0	1
00	00/0	11/0
01	10/0	01/0
11	01/0	00/1
10	01/0	01/1

Y_1Y_2/z

(c) Coded state transition table

x

y_1y_2	S	0	1
00 — a		a, 0	c, 0
01 — b		d, 0	b, 0
11 — c		b, 0	a, 1
10 — d		b, 0	b, 1

S, z

(d) State transition table

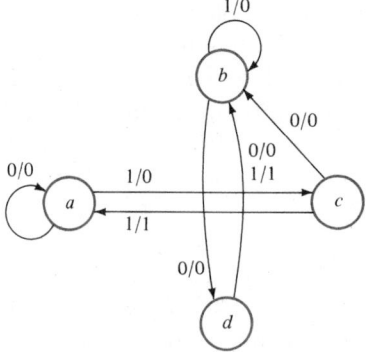

(e) State transition diagram

Figure 5–21

Analysis of synchronous sequential circuit of Figure 5–20

STEP 8

State Transition Diagram

The final step is to construct the state transition diagram. The state transition diagram, shown in Figure 5–21(e), is obtained from the state transition table by using a node (circle) for each state. The transitions are represented by an arc, a line drawn with an arrow, from the present state to the next state labeled with the input and present output. The state transition diagram and table present the same information but in a different form. Depending on the example, the state transition diagram may give a better visual indication of the circuit action.

The state transition table or diagram can now be used repeatedly to determine the state and output sequence for a given input sequence. For instance, assume that the input sequence $x = 0, 1, 1, 1, 0, 0, 1$ is applied when the sequential circuit is in state a initially. Then

Input sequence x: 0 1 1 1 0 0 1

State sequence s: $a\ a\ c\ a\ c\ b\ d\ b$

Output sequence z: 0 0 1 0 0 0 1

The analysis of any sequential circuit is carried out in a similar manner with 2^n columns for n inputs, 2^m rows for m state variables. The excitation table and transition table are identical for D flip-flops because D equals the next state of the flip-flop. To illustrate the procedures more clearly, we will analyze another sequential circuit.

Example 5–5 ▬▬▬▬▬▬▬▬▬▬▬▬▬▬▬▬▬▬▬▬

Analyze the sequential circuit in Figure 5–22.

The analysis of this sequential circuit can be done in the same manner as the one shown above. The inputs, outputs, and states are first identified:

$$X = [x = 0, x = 1]$$
$$Z = [z = 0, z = 1]$$
$$Y = [y_1 y_2 = 00, 01, 10, 11]$$

Then the output equations and the flip-flop excitation equations are found:

$$z = y_1 + \bar{y}_2$$
$$J_1 = x$$
$$K_1 = y_2$$
$$J_2 = \bar{y}_1$$
$$K_2 = x$$

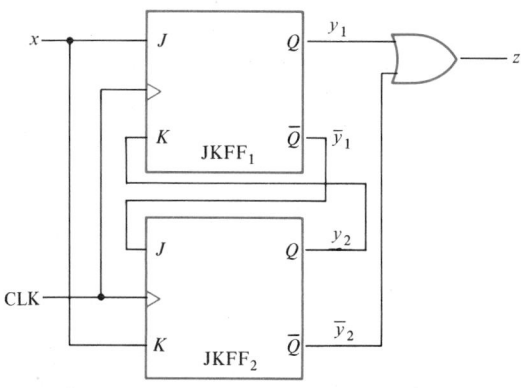

Figure 5–22
A typical sequential circuit diagram

By knowing the number of state variables, the flip-flop excitation table can be constructed readily as shown in Figure 5–23(a). Applying the flip-flop characteristic equations, the next-state table shown in Figure 5–23(b) and the present output can be obtained resulting in the coded state transition table shown in Figure 5–23(c). The extent of the mechanical procedures stops here. The interpretation of the coded state transition table and its conversion into the state transition table or the state transition diagram requires further knowledge of the problem. With only the sequential circuit given, one can merely and arbitrarily assign states and draw the state transition table and state transition diagram as shown in Figures 5–23(d) and 5–23(e), respectively.

5.4
Synthesis of Sequential Machines

The synthesis of synchronous sequential machines is essentially the reverse process of the analysis procedure; in general, however, the synthesis process is much more difficult. There are a total of 10 steps involved.

1. Construct the state transition diagram that accurately represents the word problem.
2. Construct the state transition table.
3. Reduce the number of states.
4. Assign the binary equivalents for each state.
5. Construct the coded state transition table.
6. Determine the type of flip-flop to be used.

x

y_1y_2	0	1
00	00, 10	10, 11
01	01, 10	11, 11
10	00, 00	10, 01
11	01, 00	11, 01

J_1K_1, J_2K_2

(a) Flip-flop excitation table

x

y_1y_2	0	1
00	01	11
01	01	10
10	10	10
11	01	00

Y_1Y_2

(b) Next-state table

x

y_1y_2	0	1
00	01/1	11/1
01	01/0	10/0
10	10/1	10/1
11	01/1	00/1

Y_1Y_2/z

(c) Coded state transition table

x

S	0	1
s_0	$s_1/1$	$s_3/1$
s_1	$s_1/0$	$s_2/0$
s_2	$s_2/1$	$s_2/1$
s_3	$s_1/1$	$s_0/1$

(d) State transition table

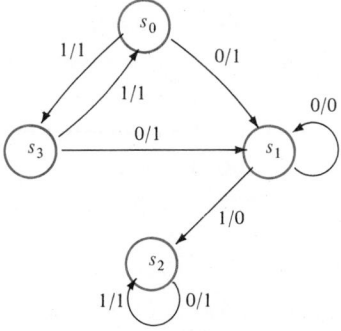

(e) State transition diagram

Figure 5–23
Analysis of the circuit in Figure 5–21

7. Construct the flip-flop excitation table.
8. Minimize the flip-flop excitation table to obtain the flip-flop excitation equations.
9. Minimize the output table to obtain the output equations.
10. Draw the circuit diagram.

To illustrate the design process, an example is given.

Example 5–6 ▬▬▬▬▬▬▬▬▬▬▬▬▬▬▬▬▬▬▬▬▬▬▬

Construct a state transition diagram that will produce a 1 on the output z when four or more consecutive 1 inputs on x after at least one 0 input has occurred. Assume that the inputs on x are synchronized with the clock pulses.

STEP 1

State Transition Diagram

The first part in the synthesis procedure involves converting a word statement into a state transition diagram. This process usually requires a reasonable amount of ingenuity and skill to make this translation correctly. The word statement may be ambiguous, and the designer may have to make some assumptions. To construct the state transition diagram, one usually starts with an initial state as in Figure 5–24(a). Then states are added as the problem is represented and translated into the state transition diagram. For this problem, additional states are added to detect the sequence 0, 1, 1, 1, 1, which will make the output $z = 1$.

The partial state transition diagram shown in Figure 5–24(b) will perform correctly with this sequence and with continuous 1's after this sequence. The state diagram must now be completed with the possible addition of states when 0 inputs follow 1 inputs. Figure 5–24(c) shows the final state transition diagram. This circuit, with the output as a function of the present state and the present input, is a Mealy-type sequential circuit representation. One might notice that state 6 might have been eliminated by directing state 5 back to itself with a 1 input. Sometimes it is not obvious that additional states are needed. When in doubt, define a new state so as not to introduce an error.

STEP 2

State Transition Table

When the state transition diagram accurately represents the word problem, the state transition table can be constructed next as shown in Figure 5–24(d). It is possible, of course, to construct the state transition table directly from the word problem.

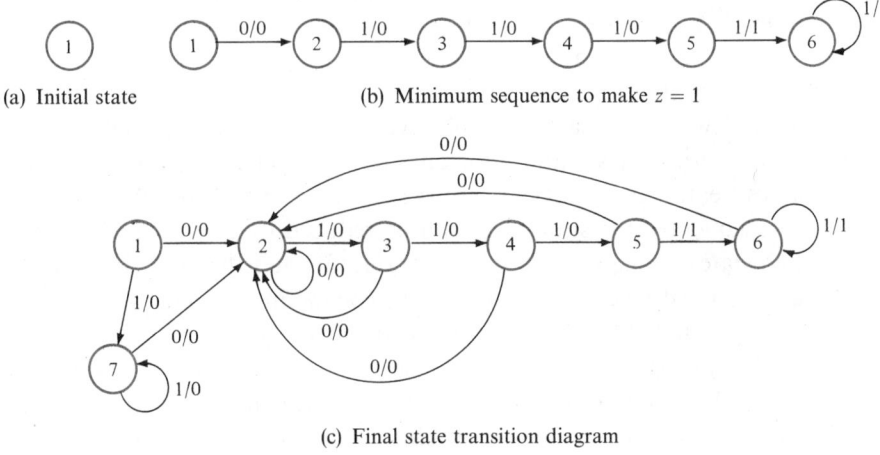

(a) Initial state

(b) Minimum sequence to make $z = 1$

(c) Final state transition diagram

	x	
s	0	1
1	2, 0	7, 0
2	2, 0	3, 0
3	2, 0	4, 0
4	2, 0	5, 0
5	2, 0	6, 1
6	2, 0	6, 1
7	2, 0	7, 0

S, z

(d) Final state transition table

Figure 5–24
State diagram design for Example 5–6

STEP 3

State Reduction

Once the state transition table has been obtained, the next step in the synthesis procedure is to reduce the number of states in the state transition table. The reduction of states may result in a reduction in the number of flip-flops and possibly in the complexity of the combinational circuit itself. The number of flip-flops needed is equal to the smallest integer greater than or equal to $\log_2$ of the number of states. For example, with three or four states, two flip-flops are required; with five to eight states, three flip-flops are required; etc.

The minimization procedure for fully specified sequential circuits is relatively easy when compared to the case of incompletely specified circuits. A fully specified circuit has all of the next states and outputs defined for every state and input.

Incompletely specified circuits will have one or more entries with don't care (unspecified) next states or outputs.

For any given input sequence, it is the output sequence and not the sequence of internal states that is the important relationship for a sequential circuit. Therefore, if two states, say a and b, have the same output sequence when the circuit is started in states a and b for any input sequence, then we say that a and b are equivalent. If two states are equivalent, then one of the states can be eliminated. Consequently, the state table can be reduced by one state. It can be shown that two states are equivalent when the outputs and the next states from both states are identical when presented with the same inputs. This process is much easier to apply than checking all input sequences.

Let us examine the state transition table in Figure 5–24(d). States 5 and 6 have the same outputs for $x = 0$ and $x = 1$, being $z = 0$ and $z = 1$, respectively. Also, since the next states for each input are identical, states 5 and 6 are equivalent. Furthermore, states 1 and 7 are equivalent. Thus, we can eliminate states 6 and 7, replacing an entry of 6 with 5 and 7 with 1 in the reduced table. The reduced state table is given in Figure 5–25(a).

Reducing the number of states may not result in reducing the number of flip-flops or the amount of combinational logic, but in most cases the circuit complexity is reduced.

STEP 4

State Assignment

The assignment of binary combinations to states is called *state assignment*. The conversion from a state transition table to a coded state transition table is ac-

Figure 5–25

Synthesis of sequential circuit for Example 5–6

s	x 0	1
1	2, 0	1, 0
2	2, 0	3, 0
3	2, 0	4, 0
4	2, 0	5, 0
5	2, 0	5, 1

S, z

(a) Reduced state transition table

s	$y_1 y_2 y_3$	x 0	1
1 —	000	001, 0	000, 0
2 —	001	001, 0	011, 0
3 —	011	001, 0	010, 0
4 —	010	001, 0	110, 0
5 —	110	001, 0	110, 1
	111	ddd, d	ddd, d
	101	ddd, d	ddd, d
	100	ddd, d	ddd, d

$Y_1 Y_2 Y_3, z$

(b) Coded state transition table

complished by representing each state by a binary number. The assignment of binary variables to the states will, in general, affect the cost of the final combinational circuit.

There are two problems associated with state assignment. The first problem is that no known general procedure will guarantee minimal cost. If the synthesis process is carried out correctly, the circuit will always function correctly for any state assignment where a unique binary combination is assigned to each state. The second problem in state assignment is that unused binary values must be handled properly. One method is to let all unused binary values with no associated states be represented by don't care entries. This will help to reduce the combinational circuit costs. The rationale is that these unused states will never occur. The problem is that, upon power up, the current state of the machine can be anything. It is possible that the machine will be trapped in one of these unused states and unable to get out. A safer method is to let all unused states be the same as the idle state. For this example, we will assume that the unused states produce don't care entries to simplify the circuit. Let the state assignments be chosen as follows:

S	y_1	y_2	y_3
1	0	0	0
2	0	0	1
3	0	1	1
4	0	1	0
5	1	1	0

STEP 5
Coded State Transition Table

When a satisfactory state assignment is obtained, the coded state transition table can be constructed next. This is shown in Figure 5–25(b).

STEP 6
Selection of Flip-flop Type

The selection of the flip-flop type is up to the designer. We have covered the SRFF, the TFF, the JKFF, and the DFF. Any type of flip-flop can be used to implement the sequential circuit. The number of flip-flops remains the same, regardless of which type of flip-flop is used. The difference in using different types of flip-flops is in the final circuit complexity. It is not possible to predict which type of flip-flop will give the simplest circuit without actually working out the complete problem to the end. From experience, it has been seen that the JKFF, because of its versatility, often gives simpler results than the other types of flip-flops. When we say simpler, we mean fewer supporting SSI chips to realize the flip-flop excitation functions. The disadvantage of using the JKFF is that there are two inputs to each *JK* flip-flop rather than the one required for the TFF or the DFF.

STEP 7

Flip-flop Excitation Table

The excitation table is obtained from the coded state transition table once the type of flip-flop has been selected. The flip-flop excitation table gives the necessary inputs of the flip-flops for all possible states and present inputs. What is given, however, in the coded state transition table are the outputs of the flip-flops for every present state in response to a particular present input. What is needed are the flip-flop excitation tables or equations to translate between the two.

The flip-flop excitation tables for each type of flip-flop are very helpful in the synthesis process. The flip-flop excitation tables are given in Figure 5–26. The *JK* flip-flop has four don't care conditions. These don't cares occur because the next state is not a function of this input. The flip-flop excitation tables can be minimized to obtain expressions for the inputs based on the present state and the next state. This is done and is given below:

$$
\begin{aligned}
\text{SRFF} \quad & \text{If } Q^t = 0, \text{ then } S = Q^{t+1}, S \cdot R = 0 \\
& \qquad\qquad\qquad R = \text{don't care if } S = 0 \\
& \text{If } Q^t = 1, \text{ then } R = \bar{Q}^{t+1}, S \cdot R = 0 \\
& \qquad\qquad\qquad S = \text{don't care if } R = 0 \\
\text{TFF} \quad & T = Q^t \oplus Q^{t+1} \\
\text{DFF} \quad & D = Q^{t+1} \\
\text{JKFF} \quad & \text{If } Q^t = 0, \text{ then } J = Q^{t+1}, K = \text{don't care} \\
& \text{If } Q^t = 1, \text{ then } K = \bar{Q}^{t+1}, J = \text{don't care}
\end{aligned}
$$

Equations for the SRFF and JKFF are somewhat complex because the inputs *S* and *R* or *J* and *K* may take on three values: 0, 1, or don't cares. The reader is urged to derive and verify the above equations.

Figure 5–26

Flip-flop excitation tables

Q^t	Q^{t+1}	D		Q^t	Q^{t+1}	J	K		Q^t	Q^{t+1}	T		Q^t	Q^{t+1}	J	K
0	0	0		0	0	0	d		0	0	0		0	0	0	d
0	1	1		0	1	1	d		0	1	1		0	1	1	0
1	0	0		1	0	d	1		1	0	1		1	0	0	1
1	1	1		1	1	d	0		1	1	0		1	1	d	0

(a) *D* flip-flop (b) *JK* flip-flop (c) *T* flip-flop (d) *SR* flip-flop

Let us use the D flip-flop to implement the sequential circuit for this example. The flip-flop output Q is generally associated with the state variable y. Since $D = Q^{t+1}$ for a D flip-flop and $Q^{t+1} = Y$, the flip-flop excitation table as shown in Figure 5–27(a) is almost the same as the coded state transition table.

STEP 8
Flip-flop Excitation Equation

After obtaining the flip-flop excitation table, the flip-flop excitation equation for each input of every flip-flop is obtained by minimizing the corresponding

Figure 5–27
Excitation table and output map for Example 5–6

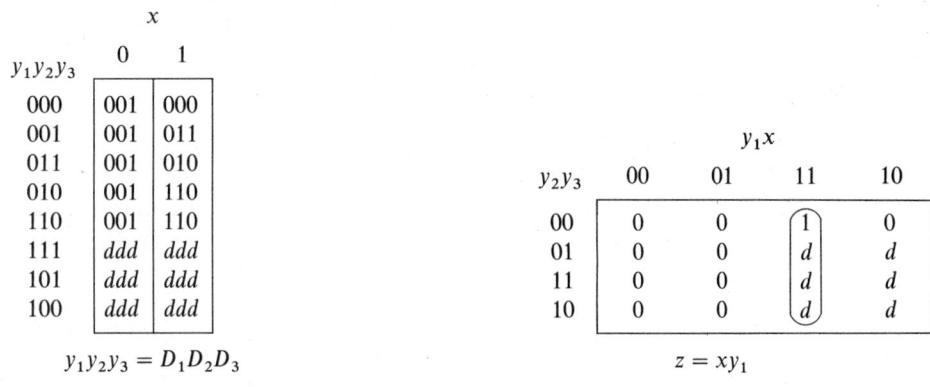

$$y_1y_2y_3 = D_1D_2D_3$$

(a) Flip-flop excitation table

$$z = xy_1$$

(b) Karnaugh map for the output function

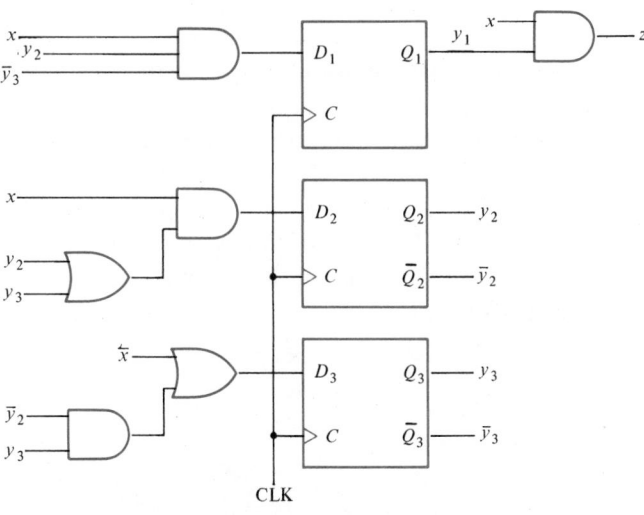

CLK

(c) Final circuit for the transition table in Figure 5–18

Karnaugh map. This can be done directly on the flip-flop excitation table, or, alternatively, the reader might want to convert these to four separate Karnaugh maps before simplifying. For our example using D flip-flops, the flip-flop excitation equations are

$$D_1 = xy_2\bar{y}_3$$
$$D_2 = x(y_2 + y_3)$$
$$D_3 = \bar{x} + \bar{y}_2y_3$$

STEP 9

Flip-flop Output Equation

The next step is to minimize the outputs in a Karnaugh mapping. This can be done either as a separate map or directly on the coded state transition table. For the sake of clarity, the minimization is done on the output table Karnaugh map shown in Figure 5–27(b). The resultant output equation is

$$Z = xy_1$$

STEP 10

Sequential Circuit Diagram

The last step is to draw the sequential circuit corresponding to the flip-flop excitation equations and the output equations. This is shown in Figure 5–27(c).

We have passed through the steps showing the implementation of the problem in a sequential circuit using DFFs. We have mentioned that the use of JKFFs often gives a simpler solution. In the next example, we will use a JKFF for comparison.

Example 5–7

Implement the sequential circuit in Example 5–6 using JKFFs instead of DFFs.

The first five steps are identical to the previous example. In Step 6, we have decided to use a JKFF. The flip-flop excitation table is shown in Figure 5–28(a). The Karnaugh maps for the flip-flop inputs are not shown. The flip-flop equations after minimization are

$$J_1 = xy_2\bar{y}_3$$
$$K_1 = \bar{x}$$
$$J_2 = xy_3$$
$$K_2 = \bar{x}$$
$$J_3 = \bar{x}$$
$$K_3 = xy_2$$

$$x$$

$y_1y_2y_3$	0			1		
000	0d	0d	1d	0d	0d	0d
001	0d	0d	d0	0d	1d	d0
011	0d	d1	d0	0d	d0	d1
010	0d	d1	1d	1d	d0	0d
110	d1	d1	1d	d0	d0	0d
111	dd	dd	dd	dd	dd	dd
101	dd	dd	dd	dd	dd	dd
100	dd	dd	dd	dd	dd	dd

$$J_1K_1, J_2K_2, J_3K_3$$

(a) Flip-flop excitation table using a JK flip-flop

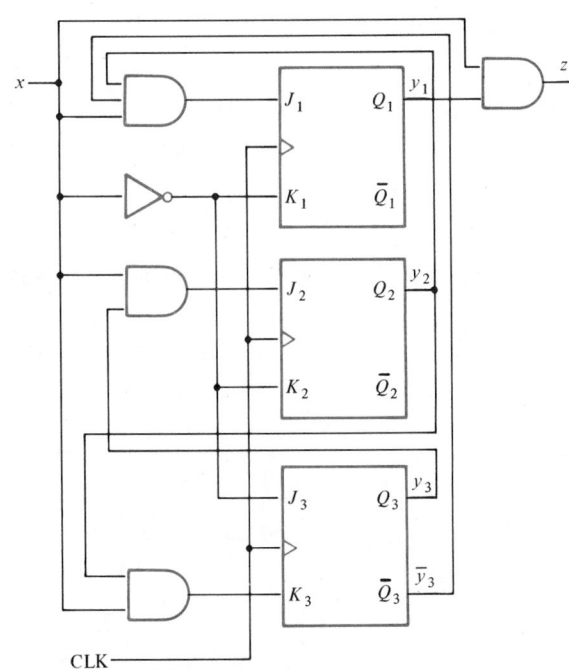

(b) Circuit implementation of Example 5–6 using a JK flip-flop

Figure 5–28
Solution of Example 5–6 using a JK flip-flop

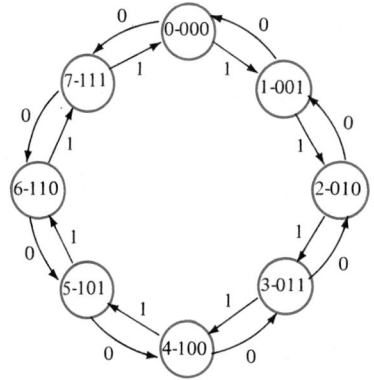

(a) State transition diagram for up/down counter

s	x		$z_1 z_2 z_3$
	0	1	
0	7	1	000
1	0	2	001
2	1	3	010
3	2	4	011
4	3	5	100
5	4	6	101
6	5	7	.110
7	6	0	111

S

(b) State transition table for up/down counter

Present State			x						x						
			0			1			0			1		
y_1	y_2	y_3	Y_1	Y_2	Y_3	Y_1	Y_2	Y_3	T_1	T_2	T_3	T_1	T_2	T_3
0	0	0	1	1	1	0	0	1	1	1	1	0	0	1
0	0	1	0	0	0	0	1	0	0	0	1	0	1	1
0	1	0	0	0	1	0	1	1	0	1	1	0	0	1
0	1	1	0	1	0	1	0	0	0	0	1	1	1	1
1	0	0	0	1	1	1	0	1	1	1	1	0	0	1
1	0	1	1	0	0	1	1	0	0	0	1	0	1	1
1	1	0	1	0	1	1	1	1	0	1	1	0	0	1
1	1	1	1	1	0	0	0	0	0	0	1	1	1	1

(c) Flip-flop excitation table using a TFF

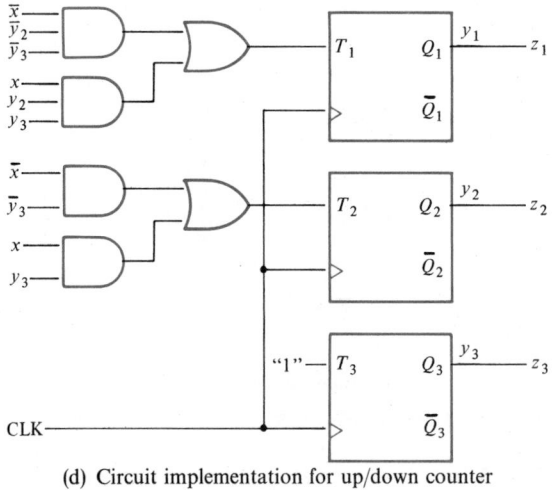

(d) Circuit implementation for up/down counter

Figure 5–29

(a) State transition diagram, (b) state transition table, (c) flip-flop excitation table, and (d) circuit implementation for up/down counter in Example 5–8

The circuit diagram is shown in Figure 5–28(b). It is clear from Figure 5–27(c) and Figure 5–28(b) that the implementation using JKFFs is much simpler than that using the DFFs.

Example 5–8

Construct a state diagram that has three outputs, z_1, z_2, and z_3, which increase in binary value when input x is 1 and decrease in binary value when $x = 0$.

This is called an *up/down counter*. The outputs will go from 0 to 7 then back to 0, etc., when x is 1. This is often called a *mod 8 arithmetic device*. Assume that the machine is a Moore-type machine; i.e., the output is a function of the state only. The state transition diagram and state transition table in Figures 5–29(a) and 5–29(b) will execute correctly according to the word problem. The flip-flop excitation table using TFFs is given in Figure 5–29(c). The final circuit diagram is shown in Figure 5–29(d).

Example 5–9

Find a state diagram for a synchronous sequential circuit with two inputs, x_1 and x_2, and one output, z. The output $z = 1$ only when $x_1 = 1$ and $x_2 = 1$ after the input sequence $x_1 x_2 = 00, 01, 10, 11$.

The state transition table and state transition diagram are given in Figures 5–30(a) and 5–30(b), respectively. Note that in this example, state a is reached

Figure 5–30
(a) State transition diagram and (b) state transition table for Example 5–9

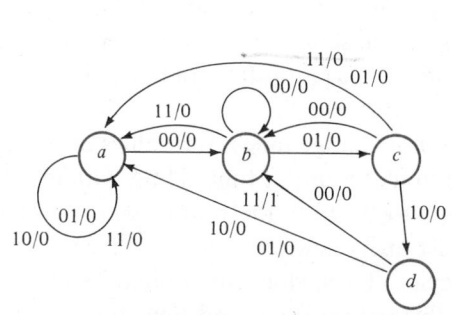

(a) State transition diagram

	$x_1 x_2$			
s	00	01	11	10
a	$b, 0$	$a, 0$	$a, 0$	$a, 0$
b	$b, 0$	$c, 0$	$a, 0$	$a, 0$
c	$b, 0$	$a, 0$	$a, 0$	$d, 0$
d	$b, 0$	$a, 0$	$a, 1$	$a, 0$

S, z

(b) State transition table

by any inputs other than $x_1 = 0$, $x_2 = 0$ that are not in the turn-on sequence, and state b is reached only by input $x_1 = x_2 = 0$.

5.5

Design Considerations with Sequential Circuits

The main purpose of this chapter is to familiarize the digital designer with the concept of sequential circuits. We have covered only fully specified synchronous sequential machines, because a full treatment of incompletely specified machines would be extremely lengthy. Several of the references listed at the end of the chapter include excellent treatments. An example of a word statement of an incompletely specified sequential circuit would be one in which the output is specified every third input only.

Synchronous sequential circuits are much easier to build because the next state of the flip-flop is completely determined by the flip-flop inputs whenever the clock pulse or transition occurs. However, whenever an input is changed in the asynchronous sequential circuit, the individual flip-flop may change immediately and in different order depending on the relative delays. Unless extreme care is taken, the behavior of the circuit is adversely affected by unequal delays, which may cause the sequential circuit to reach the wrong final state. The conversion of word statements into state transition tables is relatively easy in asynchronous sequential circuits. The state assignments and gate realizations, however, become much more difficult.

Setup and Hold Times

A synchronous sequential circuit uses a clock pulse to control flip-flop changes when all inputs are stable. A flip-flop will have deterministic circuit operation when the flip-flop inputs are stable before and after the transition on the clock. The final state of a flip-flop is impossible to determine when the clock and D or JK inputs change simultaneously. Figure 5–31 defines the time that D must be stable before the clock transition as the *setup time* and the time after as the *hold time*. A similar definition will hold for a JK flip-flop and also for any other type of memory device.

Digital logic gate outputs will not change unless the values of the inputs have been changed for a certain minimum amount of time. Thus, if an inverter's input is pulsed with a very short $0 \rightarrow 1 \rightarrow 0$ pulse, the output will not change state. The input to all logic must persist at its new value for a sufficient time to overcome the inertial delay of the device. Thus, clocks in a digital system must have a minimum value for clock width for both high and low states. The setup and hold times together with gate and flip-flop delays will determine maximum clock frequency.

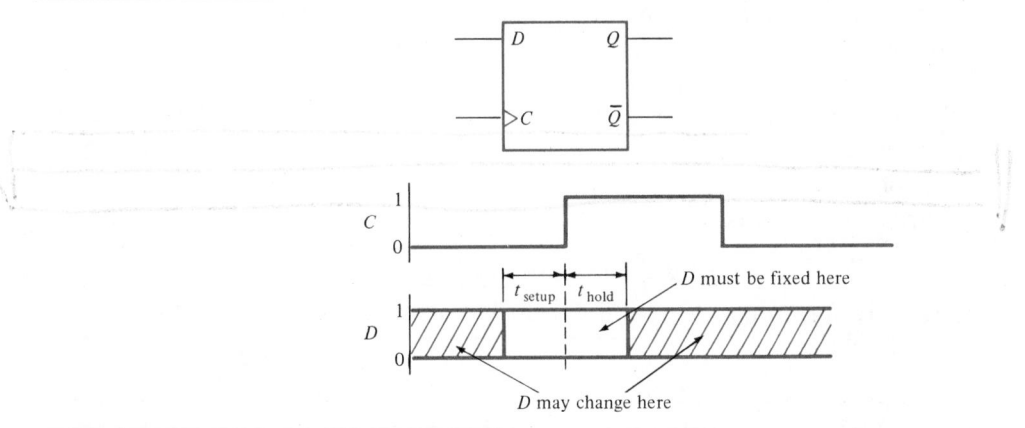

Figure 5–31
Definition of setup and hold times for a D flip-flop

Example 5–10

Find the maximum clock frequency for the circuit in Figure 5–32(a).
The clock period t_{clock} must satisfy two constraints. The first is that

$$t_{clock} \geq t_{high} + t_{low}$$

Figure 5–32
Clock frequency considerations for a sequential circuit

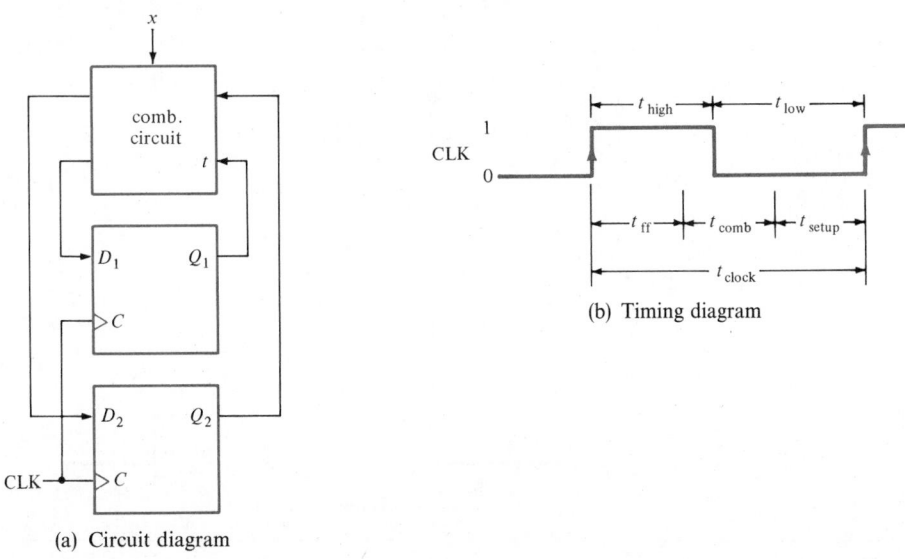

(a) Circuit diagram

where t_{high} and t_{low} are the minimum time durations that the clock must be high and low, respectively. Second, the circuit must stabilize from the previous positive clock transition before the next positive clock transition can occur to ensure reliable operation. Thus the time from one active transition on the clock to the next must be longer than the maximum flip-flop delay $t_{ff(max)}$, plus the maximum delay of the combinational circuit $t_{comb(max)}$ plus the setup time t_{setup} as shown in Figure 5–32(b). Thus,

$$t_{clock} \geq t_{ff(max)} + t_{comb(max)} + t_{setup}$$

Obviously, for proper operation the clock must satisfy both constraints.

Example 5–11

Find the minimum clock time for the sequential circuit shown in Figure 5–33(a) with the following specifications. Assume minimum high and low clock times of 20 nanoseconds (ns) each, and maximum flip-flop and combinational circuit delays of 30 ns and 20 ns, respectively. Assume that the setup time is 10 ns.

Figure 5–33

An example of clock skew

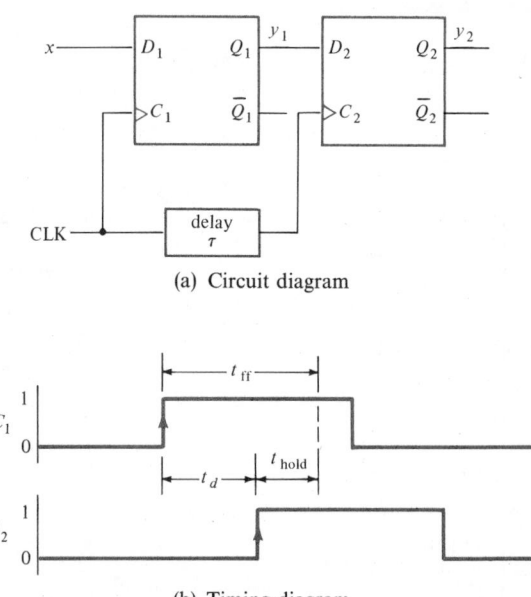

(a) Circuit diagram

(b) Timing diagram

The period can be calculated simply by applying the above formula. Then

$$t_{clock} \geq t_{high} + t_{low} = 20 \text{ ns} + 20 \text{ ns} = 40 \text{ ns}$$
$$t_{clock} \geq t_{ff(max)} + t_{comb(max)} + t_{setup}$$
$$= 30 \text{ ns} + 20 \text{ ns} + 10 \text{ ns}$$
$$= 60 \text{ ns}$$

The minimum clock time is, therefore, 60 ns or a maximum frequency of 16.67 megahertz (MHz).

The clocks of synchronous sequential circuits are typically driven in parallel. However, unequal wire lengths or different gate delays when clocks must be driven from different gate sources because of fan-out limitations may cause one clock transition to be delayed from the other. This situation is referred to as *clock skew*, as depicted in Figure 5–33(b). Assume that $x = 0$, $y_1 = 1$, and $y_2 = 0$ when a clock transition occurs. This will cause the first flip-flop to change state to $y_1 = 0$. We would like the effects of this change not to be felt at D_2 until the hold time has passed after the $0 \rightarrow 1$ transition on C_2 occurs. Thus

$$t_d + t_{hold} \leq t_{ff(min)}$$

where $t_{ff(min)}$ is the minimum flip-flop delay. This circuit represents the worst case for clock skew, and hence

$$t_d \leq t_{ff(min)} - t_{hold}$$

5.6
Counters

Counters are used to regulate and control circuits and events by generating the proper timing signals. The counters repeat the state after a prescribed number of clock pulses, say N, which is sometimes said to divide by N or by mod N. Mod N is the remainder after dividing by N so, if the counter is in state n after m clock pulses, it will also be in state n after $m + N$, $m + 2N$, etc., clock pulses. One of the most common types of sequential circuits is a counter. The counter goes through a prescribed sequence of states upon the application of clock pulses. This sequence may be specified as the binary values from 0 to $2^n - 1$ and then repeated. Other counters may be designed so that the exact state sequence is not important, but the number of states is crucial. Many techniques have been developed to design various types of counters. In general, these techniques have been rendered obsolete with the widespread availability of MSI counters, which are very flexible and easy to use.

Categories of Counters

The states of the counters are usually represented by the decimal equivalents of the binary bits of the flip-flop outputs. The bits corresponding to each flip-flop must be specified from the least significant bit to the most significant bit. A counter that goes through the sequence $0, 1, 2, \ldots, 2^n - 1$, and back to 0, is called a *binary up counter*. If the sequence is reversed, it is a *binary down counter*. Similarly, a counter that sequences $0, 1, 2, \ldots, 10^n - 1, 0, 1$, etc., is called a *decade up counter*.

The general specification of a counter can be shown in Figuure 5–34. A counter may be specified as either asynchronous or synchronous. Of the synchronous types, a counter may be of ripple carry or of parallel carry type. The ripple counters use the output of one flip-flop to generate the clock transition on the next flip-flop. Some people call the ripple counter an asynchronous counter, but this is different from asynchronous loads and circuits. We will stick with the ripple designation. The synchronous counter implies that all clocks on every flip-flop are connected together. In terms of the count sequence, a counter may be binary, decade, octal, or from any starting number to any ending number. Finally, the direction of a counter may be up, down, or either up or down depending on the instantaneous value of a control input.

Asynchronous (Ripple) Counters

An example of a binary ripple up counter is given in Figure 5–35(a). The $1 \to 0$ transition on the clock input driven by the Q output of the previous flip-flop causes the flip-flop to change state. The flip-flop takes t_t seconds to stabilize after the $1 \to 0$ clock transition. The timing diagram is given in Figure 5–35(b). Assume that the counter is in state $A_2 A_1 A_0 = 011$ when the $1 \to 0$ transition occurs. This causes A_0 to produce a $1 \to 0$ transition, which causes A_1 to produce a $1 \to 0$ transition and, in turn, A_2 produces a $0 \to 1$ transition. The effect ripples from one flip-flop to the next, and it takes $3 \times t_t$ seconds for the counter to stabilize. In general, for an n-bit ripple counter, it will take $n \times t$ seconds for the counter to stabilize. This is the major disadvantage of a ripple counter. This counter is obviously a binary up counter, because the states follow the sequence $0, 1, 2, 3, 4, 5, \ldots, 14, 15, 0, 1, 2, \ldots$, and so on.

Figure 5–34
Categories of counters

$$
\left\{
\begin{array}{l}
\text{asynchronous (ripple)} \\
\text{synchronous} \left\{ \begin{array}{l} \text{ripple carry} \\ \text{parallel carry} \end{array} \right\}
\end{array}
\right\}
\left\{
\begin{array}{l}
\text{binary} \\
\text{octal} \\
\text{decimal} \\
\text{special} \\
\text{etc.}
\end{array}
\right\}
\left\{
\begin{array}{l}
\text{up} \\
\text{down} \\
\text{up/down}
\end{array}
\right\}
\text{counter}
$$

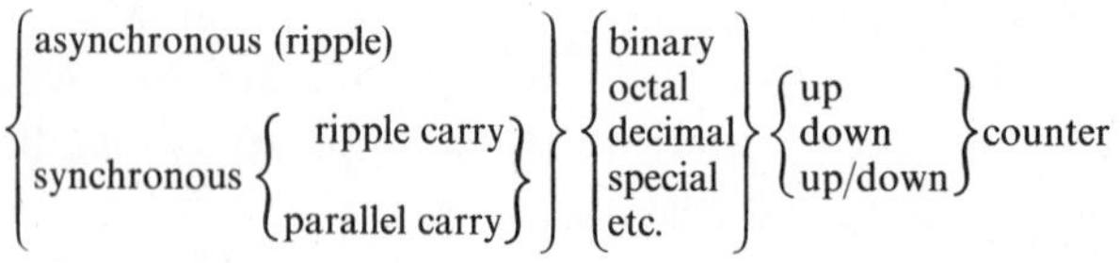

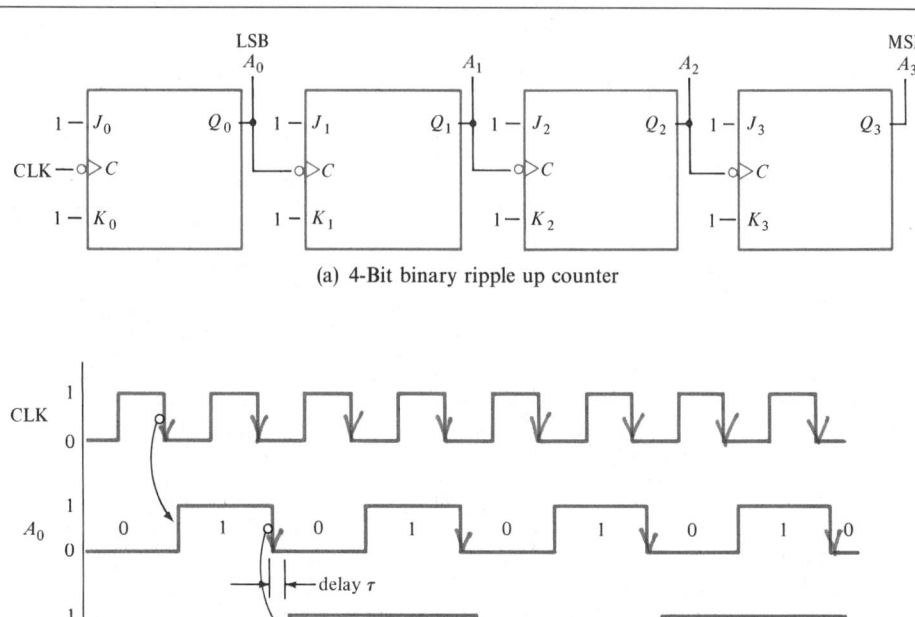

(a) 4-Bit binary ripple up counter

(b) Timing diagram

Figure 5–35
A binary ripple up counter

The counter shown in Figure 5–35 is a binary counter because the count sequence is the binary sequence itself. The same counter can be made to count in decimal simply by changing the count sequence so that when the count reaches 9, the next count will be 0. This can be done simply by detecting when Q_3 and Q_1 go to 1. At that point, immediately reset all flip-flops. The schematic and timing diagrams of this technique are shown in Figures 5–36(a) and 5–36(b).

The same technique can be extended so that the counter is made to count starting from any number and stopping at any number. This technique, however, is not often used in asynchronous counters because of the possibility of races, which cause unpredictable results.

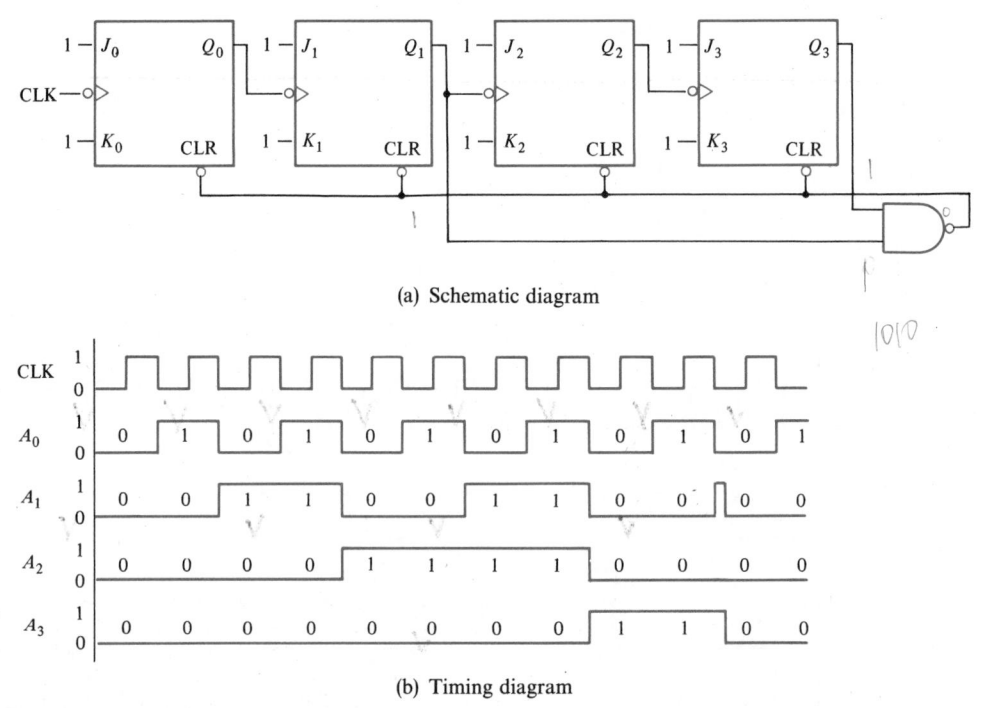

(a) Schematic diagram

(b) Timing diagram

Figure 5–36

A decimal ripple up counter

Up/Down Control   The direction of up/down counting can be changed easily by taking the output from the $\bar{Q}$ side, or connecting the clock signal to the $\bar{Q}$ side, or by using negative-edge flip-flops rather than positive-edge flip-flops. The count sequence of a binary up counter at the Q and $\bar{Q}$ outputs is given by

Q_3	Q_2	Q_1	Q_0	$\bar{Q}_3$	$\bar{Q}_2$	$\bar{Q}_1$	$\bar{Q}_0$
0	0	0	0	1	1	1	1
0	0	0	1	1	1	1	0
0	0	1	0	1	1	0	1
0	0	1	1	1	1	0	0
0	1	0	0	1	0	1	1
⋮				⋮			
1	1	0	0	0	0	1	1
1	1	0	1	0	0	1	0
1	1	1	0	0	0	0	1
1	1	1	1	0	0	0	0

Note that as the outputs on the Q side are counting up, the outputs on the $\bar{Q}$ side are counting down.

If the ripple clock is taken from the $\bar{Q}$ side rather than the Q side, the counter will count down instead of up. This is evident from the timing diagram in Figure 5–37(a). Similarly, if positive-edge flip-flops are used instead of negative-edge flip-flops with the clock signal taken from the Q side, the count sequence will also change direction. This is shown in Figure 5–37(b).

Figure 5–37
Effect of clock connection on the count sequence

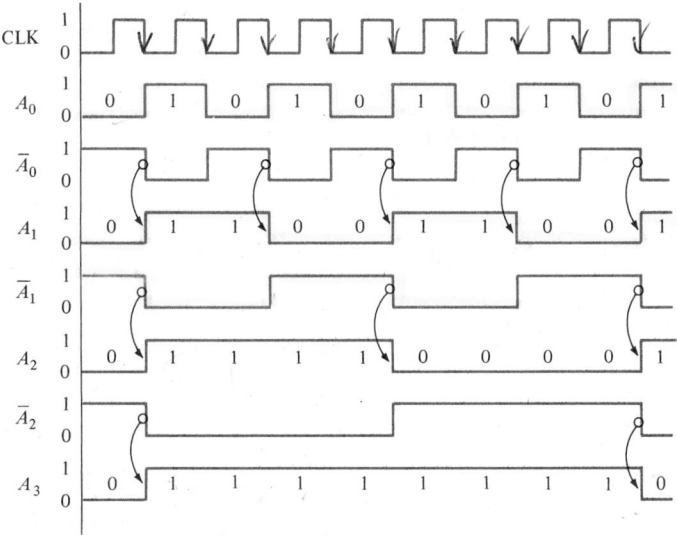

(a) Timing diagram of a ripple counter with the clock taken from the $\bar{Q}$ side

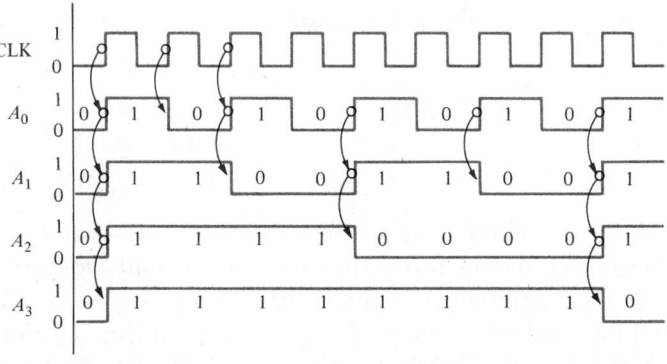

(b) Timing diagram of a ripple counter with positive-edge flip-flops

Synchronous Counters

Synchronous counters are connected with all clocks in parallel to ensure faster responses. The techniques presented at the beginning of this chapter for synthesizing synchronous circuits can be used to design all types of synchronous counters. However, since a number of MSI counters are general-purpose in nature, we will concentrate on them. A synchronous counter is made by observing that whenever all the least significant bits are 1, the next digit must change. This is shown below:

Q_3	Q_2	Q_1	Q_0	
0	0	0	0	
0	0	0	1	Q_0 is 1, hence Q_1 must change.
0	0	1	0	
0	0	1	1	Q_0 is 1, hence Q_1 must change.
				Q_0 and Q_1 are 1, hence Q_2 must change.
0	1	0	0	
0	1	0	1	Q_0 is 1, hence Q_1 must change.
0	1	1	0	
0	1	1	1	Q_0 is 1, hence Q_1 must change.
				Q_0 and Q_1 are 1, hence Q_2 must change.
				Q_0, Q_1, and Q_2 are 1, hence Q_3 must change.
1	0	0	0	
1	0	0	1	Q_0 is 1, hence Q_1 must change.
1	0	1	0	
1	0	1	1	Q_0 is 1, hence Q_1 must change.
				Q_0 and Q_1 are 1, hence Q_2 must change.
1	1	0	0	
1	1	0	1	Q_0 is 1, hence Q_1 must change.
1	1	1	0	
1	1	1	1	Q_0 is 1, hence Q_1 must change.
				Q_0 and Q_1 are 1, hence Q_2 must change.
				Q_0, Q_1, and Q_2 are 1, hence Q_3 must change.

If the counter is made of TFFs, then we can write down immediately the conditions for change for each flip-flop:

$$T_0 = \text{CLK}$$
$$T_1 = T_0 \cdot Q_0 = \text{CLK} \cdot Q_0$$
$$T_2 = T_1 \cdot Q_1 = \text{CLK} \cdot Q_0 \cdot Q_1$$
$$T_3 = T_2 \cdot Q_2 = \text{CLK} \cdot Q_0 \cdot Q_1 \cdot Q_2$$

The general block diagram for a synchronous counter is given in Figure 5–38. The reader should prove that the same circuit would be obtained if one follows the traditional sequential circuit design with the state transition diagram and the state transition table discussed in previous sections.

The operation of the counter is as follows. The clock input works on the positive $0 \rightarrow 1$ transition. The signals Q_3, Q_2, Q_1, and Q_0 are the outputs of the flip-flops with Q_0 the least significant bit and Q_3 the most significant bit. The CLEAR

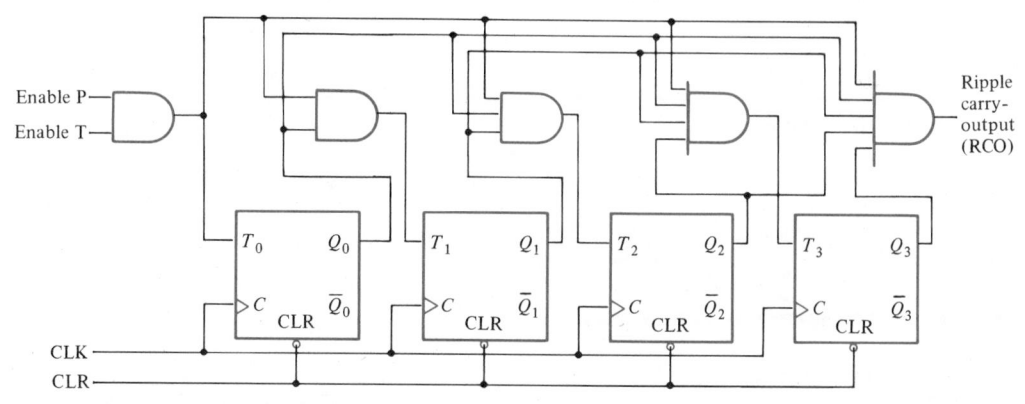

Figure 5–38
Synchronous counter with parallel carry

input will clear the counter when low, which is synchronous or asynchronous for the different types. The counter will count only when Enable P and Enable T are 1. The ripple carry-output (RCO) is 1 only when the terminal count is reached and when Enable T is 1. The load input (not shown) works in conjunction with the data inputs to parallel-load or preset the counters. The inputs A, B, C, and D will appear at the Q_0, Q_1, Q_2, and Q_3 outputs, respectively, after a positive transition on the clock when the load is 0. We also assume that CLEAR is high during this time.

The basic circuit of a synchronous counter is the carry generation. In Figure 5–38, the carry is generated in parallel. It is also possible to generate the carry in a simpler form by rippling the output of the first AND gate to the input of the second as shown in Figure 5–39. This is called a ripple carry synchronous

Figure 5–39
Synchronous counter with ripple carry

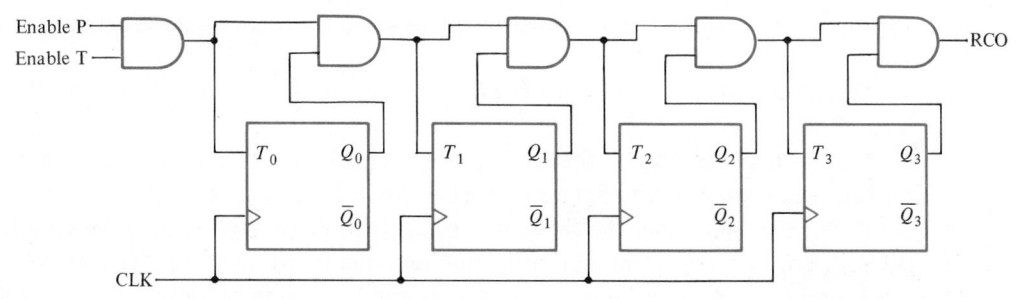

Counters	Decade		Binary	
	Up	Up/Down	Up	Up/Down
Synchronous	160, 162 560, 668 690, 692	168, 190 192, 508 696, 698	161, 163, 561, 669, 691, 693 169, 569, 697, 699	867, 869
Asynchronous	90, 68, 176, 196, 290		93, 69 177, 197, 293	

Figure 5–40
Summary of some representative counters

counter. The penalty for the ripple carry case is that sufficient time between the clock pulse must be allowed for the ripple to propagate from the least significant bit to the most significant bit.

Examples of Asynchronous and Synchronous Counters

In this section we examine some sample products for both asynchronous and synchronous counters. A summary of some representative counters is given in Figure 5–40.

Asynchronous Counters   An example of an asynchronous counter is the SN74177. The counter is asynchronous because not all flip-flops are connected to the same clock input. The SN74176 is also an asynchronous counter but is connected specially for the normal decimal format or for a biquinary format. Both formats yield a mod 10 counter. The normal decimal format gives the BCD representations of the 10 decimal digits. The biquinary counting does not give the normal BCD format. Instead, it gives a balance square-wave output on the most significant bit, with five counts at 0 and five counts at 1. The equivalent circuit diagram and function tables for the SN74176 are given in Figure 5–41.

Example 5–12

Design an asynchronous up counter using the SN74176 that counts from 3 to 12.

For counts from 3 to 12, the counter counts normally. At count 13, the counter must quickly reset itself to 3. Hence, we can use a NAND gate to detect when the count is 1101_2 and reset the flip-flops to 0011_2 immediately so that the counter stays in count 13 only momentarily but remains in count 3 for the rest of the clock period. The time the counter is in count 13 is called a *glitch*. We are able

SN54176, SN74176

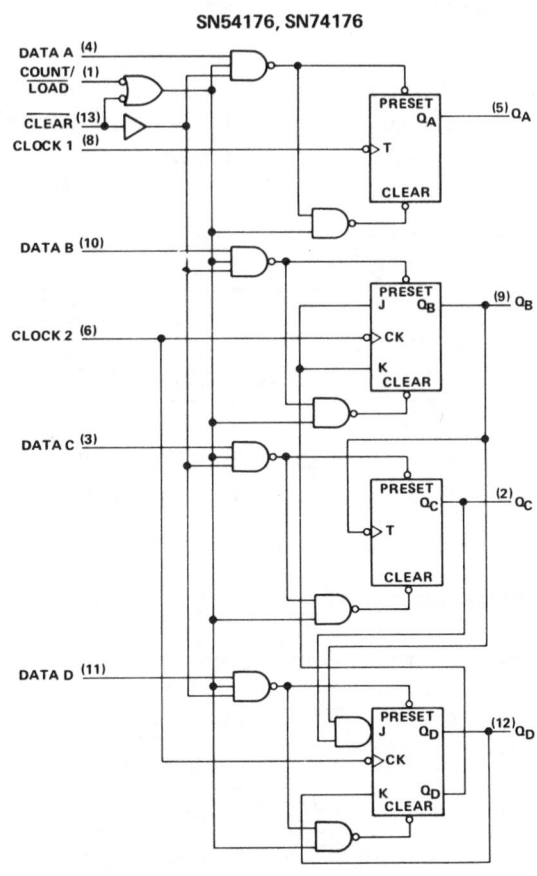

DATA A (4)
COUNT/ (1)
LOAD
$\overline{\text{CLEAR}}$ (13)
CLOCK 1 (8)

PRESET
Q_A (5) Q_A
T
CLEAR

DATA B (10)

CLOCK 2 (6)

PRESET
J Q_B (9) Q_B
CK
K
CLEAR

DATA C (3)

PRESET
Q_C (2) Q_C
T
CLEAR

DATA D (11)

PRESET
J Q_D (12) Q_D
CK
K $\overline{Q_D}$
CLEAR

(a) Equivalent circuit diagram

DECADE (BCD)
(See Note A)

COUNT	OUTPUT			
	Q_D	Q_C	Q_B	Q_A
0	L	L	L	L
1	L	L	L	H
2	L	L	H	L
3	L	L	H	H
4	L	H	L	L
5	L	H	L	H
6	L	H	H	L
7	L	H	H	H
8	H	L	L	L
9	H	L	L	H

BI-QUINARY (5-2)
(See Note B)

COUNT	OUTPUT			
	Q_A	Q_D	Q_C	Q_B
0	L	L	L	L
1	L	L	L	H
2	L	L	H	L
3	L	L	H	H
4	L	H	L	L
5	H	L	L	L
6	H	L	L	H
7	H	L	H	L
8	H	L	H	H
9	H	H	L	L

H = high level, L = low level

NOTES: A. Output Q_A connected to clock-2 input.
B. Output Q_D connected to clock-1 input.

(b) Function table

Figure 5–41

Equivalent circuit diagram and truth tables for the SN74176, an
asynchronous decade up counter (Reprinted by permission of Texas
Instruments)

to reset the counter immediately because of the PRESET and CLEAR inputs, which are asynchronous inputs. Note that a NAND gate is used because the PRESET and CLEAR inputs are active low. The action to be performed on each flip-flop is shown below:

Q_D	Q_C	Q_B	Q_A
1	1	0	1
0	0	1	1
CLEAR	CLEAR	PRESET	No change

Since the output of the first flip-flop, that is, Q_D, was 1 and must be changed to 0, this flip-flop must be cleared. The same action is done to the second flip-flop.

Figure 5–42

An asynchronous up counter that counts from 3 to 12

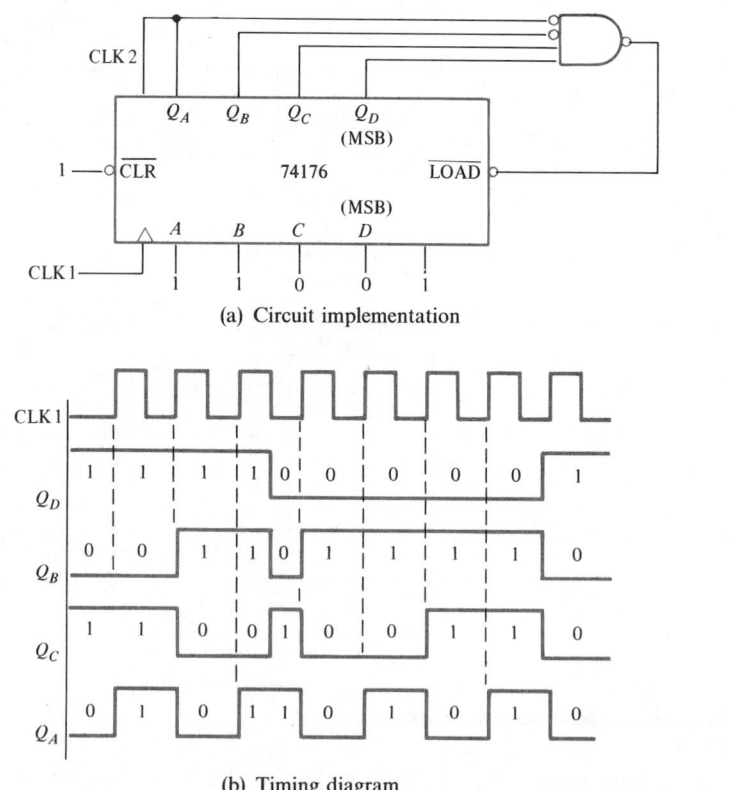

(a) Circuit implementation

(b) Timing diagram

For flip-flop B, the original state is 0 and must be changed to 1 by activating the PRESET line. The last flip-flop requires no change. The circuit connection and timing diagram for this counter are shown in Figure 5–42.

Synchronous Counters   Synchronous counters in general operate at a higher clock frequency than asynchronous counters because all flip-flops change state at the same time. For the synchronous counters, we will examine three popular series: the 74ALS160–74ALS163 series, the 74ALS168–74ALS169 series, and the 74ALS190–74ALS193 series.

The 74ALS160–74ALS163 series   For the 74ALS160–74ALS163 series, the differences in the four counters are as follows:

Counter	Clear	Modulus	Sequence
74ALS160	Asynchronous	10	$0, 1, \ldots, 8, 9, 0, 1, \ldots$
74ALS161	Asynchronous	16	$0, 1, \ldots, 14, 15, 0, 1, \ldots$
74ALS162	Synchronous	10	$0, 1, \ldots, 8, 9, 0, 1, \ldots$
74ALS163	Synchronous	16	$0, 1, \ldots, 14, 15, 0, 1, \ldots$

The logic symbol and the associated timing diagram of the 74ALS163 are shown in Figures 5–43(a) and 5–43(b). The ripple carry-out is 1 whenever the state is 9 for the 74ALS163 and 74ALS162 and Enable T is 1. For the 74ALS161 and 74ALS163, RCO = 1 for state 15 and Enable T = 1.

The counters with no external modifications will count in their normal sequence. The 74ALS163 can be modified to change the modulus of the counter with the addition of some external hardware. A count sequence of $0, 1, 2, \ldots,$ $N - 1, 0, 1, 2$, etc., can be obtained by driving the CLEAR input to zero when state $N - 1$ is reached, $N \leq 16$. This produces a counter with mod N capabilities. There are two ways to modify the normal count sequence. First, the normal count sequence can be forced to change asynchronously through the PRESET and CLEAR inputs. Second, the sequence can be changed synchronously through the LOAD input or through the flip-flop inputs. The first way is the same as that for the asynchronous counters. We will discuss the synchronous way through the following examples.

Example 5–13

Design a counter using the 74ALS163 that is mod 13 with the count sequence $0, 1, \ldots, 10, 11, 12, 0, 1$, etc. The logic that forces the CLEAR to 0 for state 12 must be 1 for the other states $0, 1, \ldots, 11$. The inputs 13, 14, and 15 will not occur;

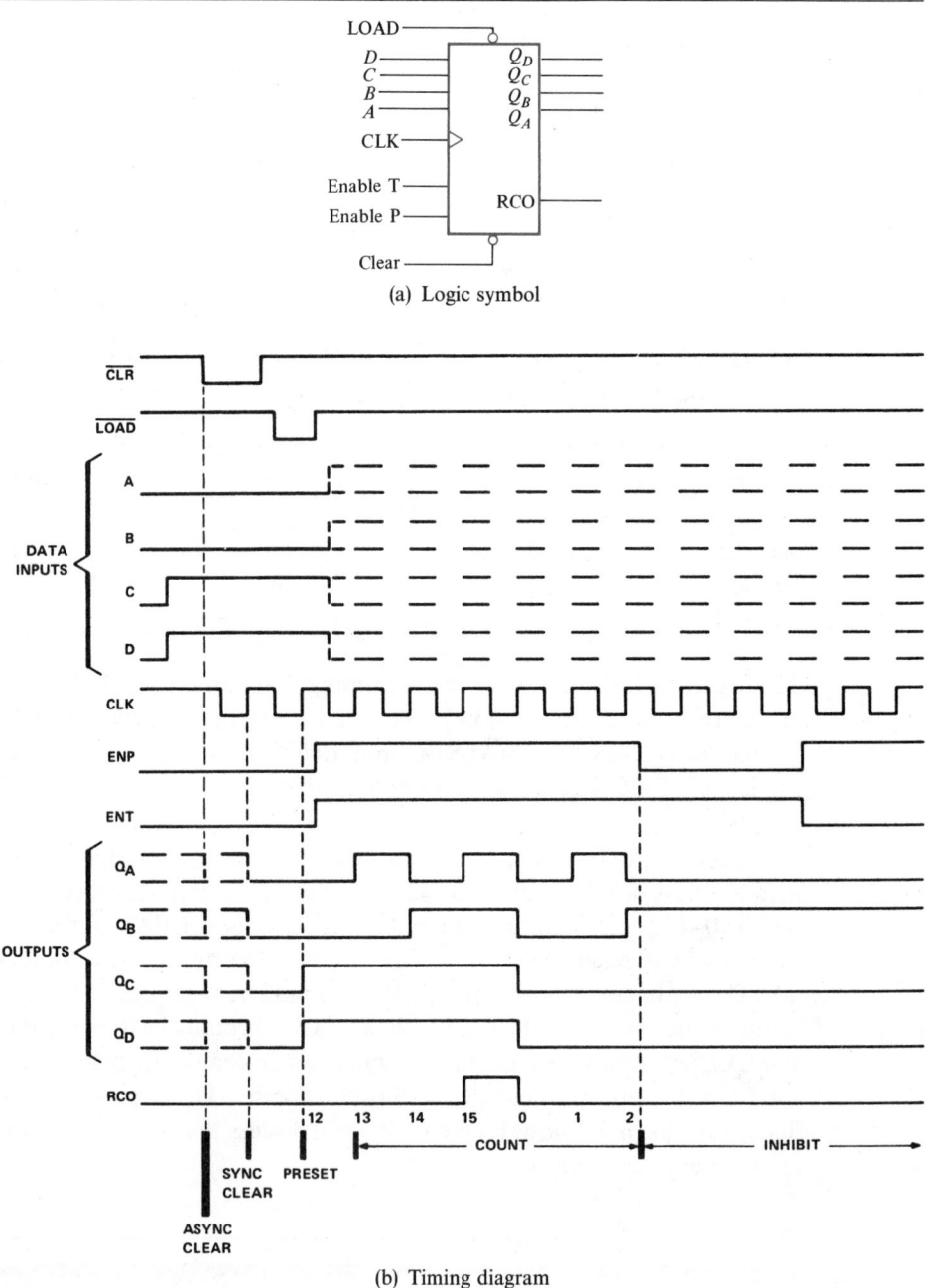

(a) Logic symbol

(b) Timing diagram

Figure 5–43

The 74ALS163 synchronous 4-bit binary up counter (Part b reprinted by permission of Texas Instruments)

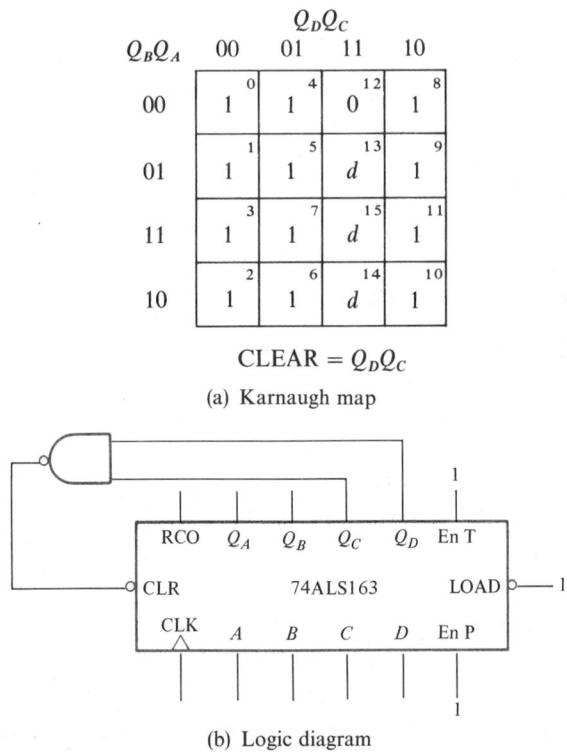

(a) Karnaugh map

$$CLEAR = Q_D Q_C$$

(b) Logic diagram

Figure 5–44

A modulus-13 counter

thus, they are don't cares. The Karnaugh map is given in Figure 5–44(a) and the logic diagram in Figure 5–44(b).

The maximum amount of external hardware for this technique with a modulus less than 16 is one NAND with a maximum of three inputs. The 74ALS162 could be used for this technique if the modulus was less than 10. The method used in this example would produce a different modulus if the CLEAR were asynchronous. This is left as a problem at the end of the chapter. The actual sequence of states is not crucial in some applications when only the modulus is given. The 74ALS163 counter can use the RCO to load a given number M to produce the sequence $M, M + 1, \ldots, 14, 15, M$, etc. The modulus for this counter is $16 - M$.

Example 5–14

Design a counter using the 74ALS163 with the sequence of states $7, 8, 9, \ldots,$ $14, 15, 7, 8$, etc.

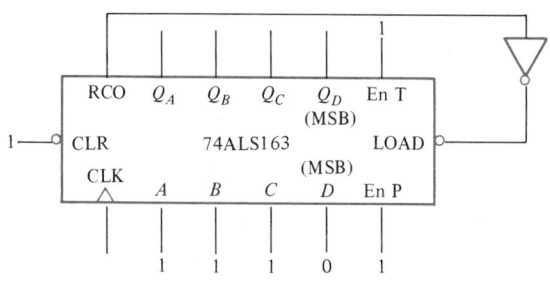

Figure 5–45

A Mod 9 counter with states 7 to 15

The modulus of the counter in this case is 9. An inverter can be used from the RCO to the load input. The inputs D, C, B, and A are connected to 0111, which is 7. The circuit is given in Figure 5–45. The 74ALS161 could be used equally as well with this type of counter because all these counters have synchronous load.

Example 5–15

Cascade the 74ALS161–74ALS163 counters together to form a 16-bit counter.

The Enable T and P together with the RCO output allow expansion of this counter to more bits. The general expansion is shown in Figure 5–46, which works with any type of counter in the 74160 line. What is shown in Figure 5–46 actually represents a 16-bit counter with a natural modulus of $2^{16} = 65,536$ when the 74ALS163 counters are used. The structure is still synchronous with all clocks connected in parallel.

The operation of this 16-bit counter is as follows. The least significant counter (first counter) will count when the Enable P and T are 1. The next most significant counter (second counter) will count when RCO = 1 from the first counter. This occurs when Enable T = 1 and the count is 15 (or 9) in the least significant counter. The next most significant counter (third counter) will count when the RCO from both the first and second counters is 1 simultaneously. This implies that the first and second counters are both 15 (or 9) and Enable T is 1 into the first counter. Thus, any counter in the chain will count when the less significant counters ahead of it are all at their terminal count, which is either 15 for mod 16 or 9 for mod 10. The technique for changing the modulus for cascaded counters is the same as for individual counters. Some problems are given at the end of the chapter.

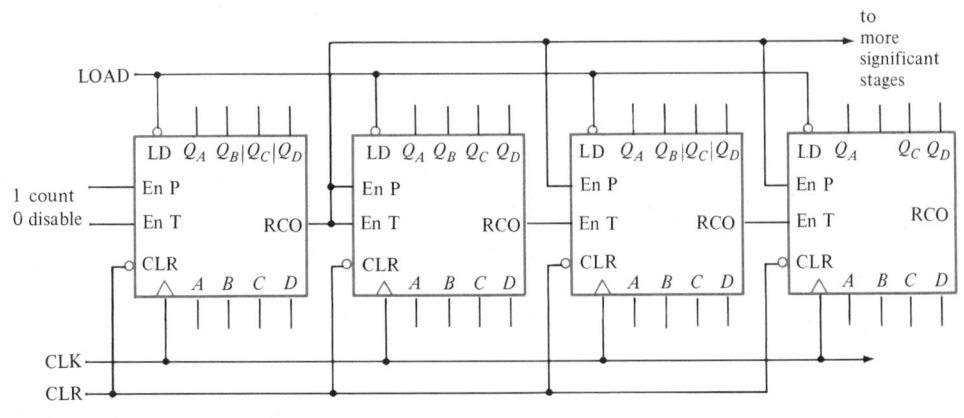

Figure 5–46

Cascading of MSI 74261–74163 counters (Reprinted by permission of Texas Instruments)

The 74ALS168–74ALS169 series   The 74ALS168 and 74ALS169 are synchronous up/down counters that have numerous uses. The 74ALS168 and 74ALS169 counters are similar to the 74ALS160–74ALS163 series of counters except for the following. The CLEAR input is replaced with an UP/DOWN input. This input is needed to make the counter count up when it is 1 and down when it is 0. The RCO is low instead of high at the terminal count. Note that RCO = 0 when UP/DOWN = 1, Enable T = 0, and the state is 15 (or 9 for mod 10). RCO = 0 also when UP/DOWN = 0, Enable T = 0, and the state is 0, which represents a borrow condition. The counter also counts when Enable P and T are both 0 instead of 1. The equivalent circuit and the timing diagram are given in Figure 5–47. The 74ALS168 is a decade counter and the 74ALS169 is a mod 16 binary counter. The cascading of these counters is done in a manner very similar to the 74ALS160–74ALS163 counters.

The 74ALS190–74ALS193 series   The 74ALS190 and 74ALS191 are up/down decade and mod 16 binary counters, respectively. These are also popular counters and feature asynchronous loading with up/down mode control. These counters, when cascaded, do not operate in a synchronous manner anymore. Some up/down counters use two clocks, one for up and the other for down counts. The 74ALS192 and 74ALS193 are decade and mod 16 binary counters that are good examples of this type. For some cases, it is advantageous to use this type of up/down counter. These two counters feature asynchronous load and are not synchronous when they are cascaded. The logic symbol and timing diagram are given in Figure 5–48. There are other counters in the 74 logic line and also in CMOS circuits. We have highlighted only the most common types.

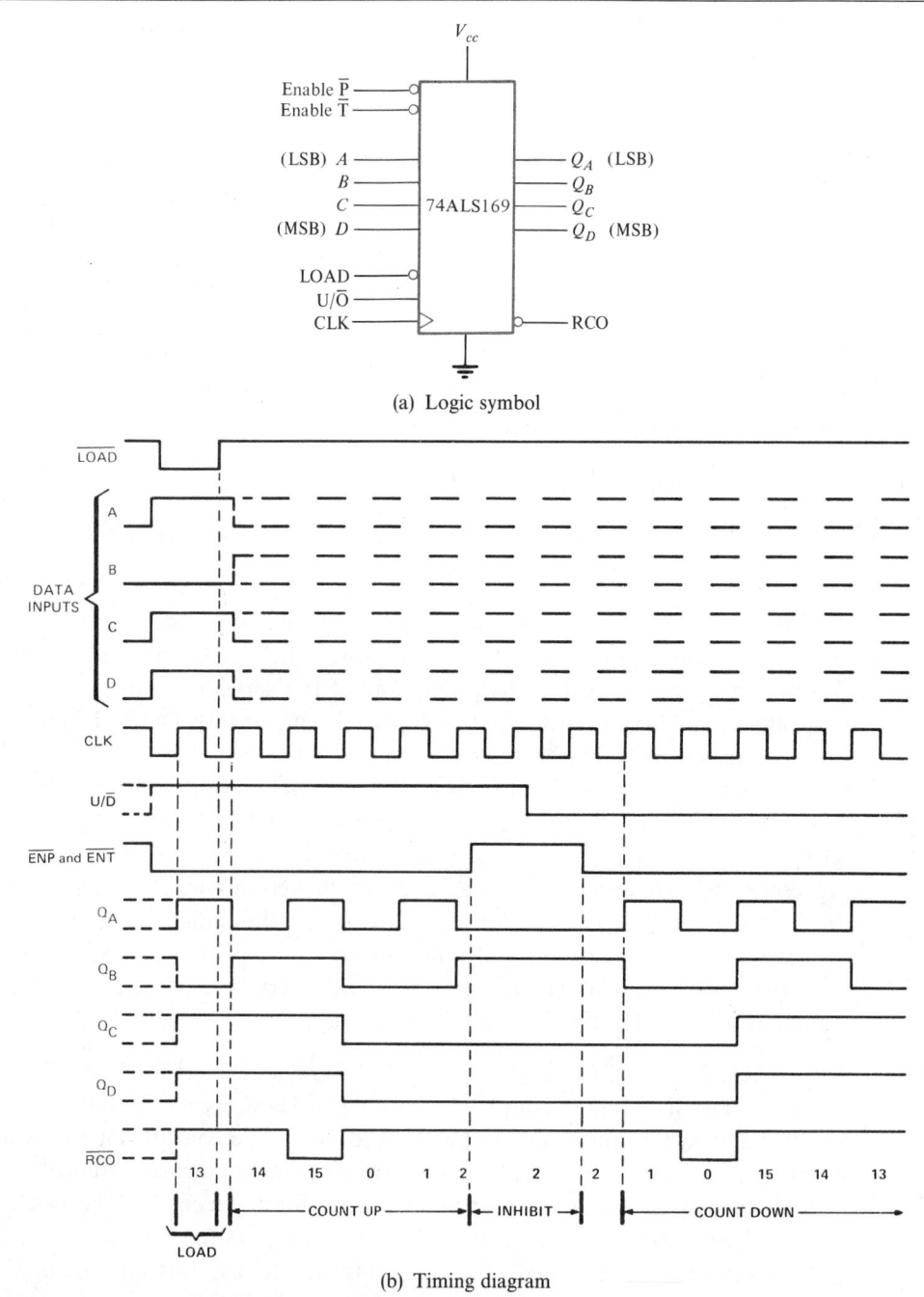

(a) Logic symbol

(b) Timing diagram

Figure 5–47

(a) Logic symbol and (b) timing diagram for the 74ALS169 counter
(Part b reprinted by permission of Texas Instruments)

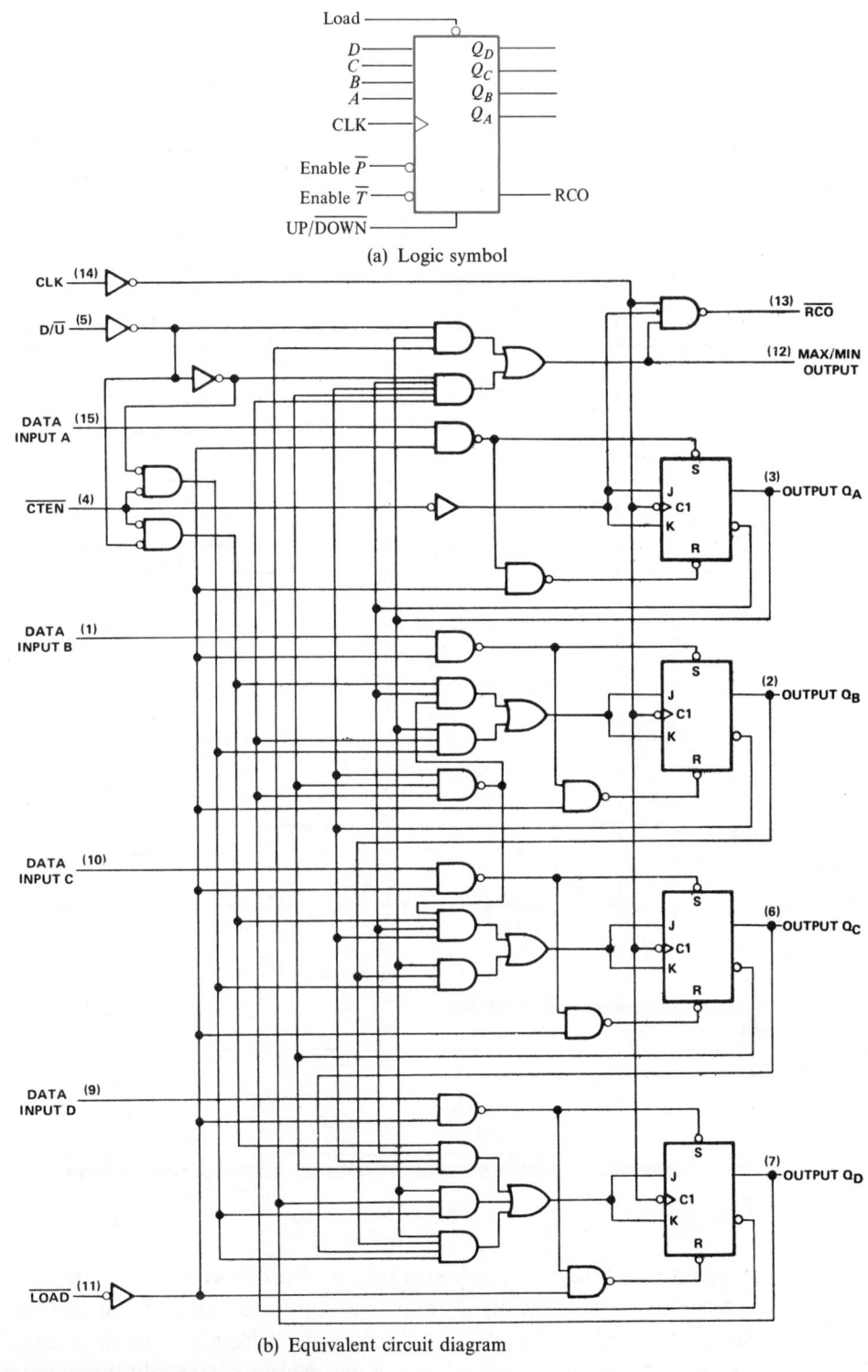

(a) Logic symbol

(b) Equivalent circuit diagram

Figure 5-48 (a) Logic symbol, (b) equivalent circuit diagram, and (c) timing diagram for the SN74ALS190 synchronous decade counter (Parts b and c reprinted by permission of Texas Instruments) (*continued*)

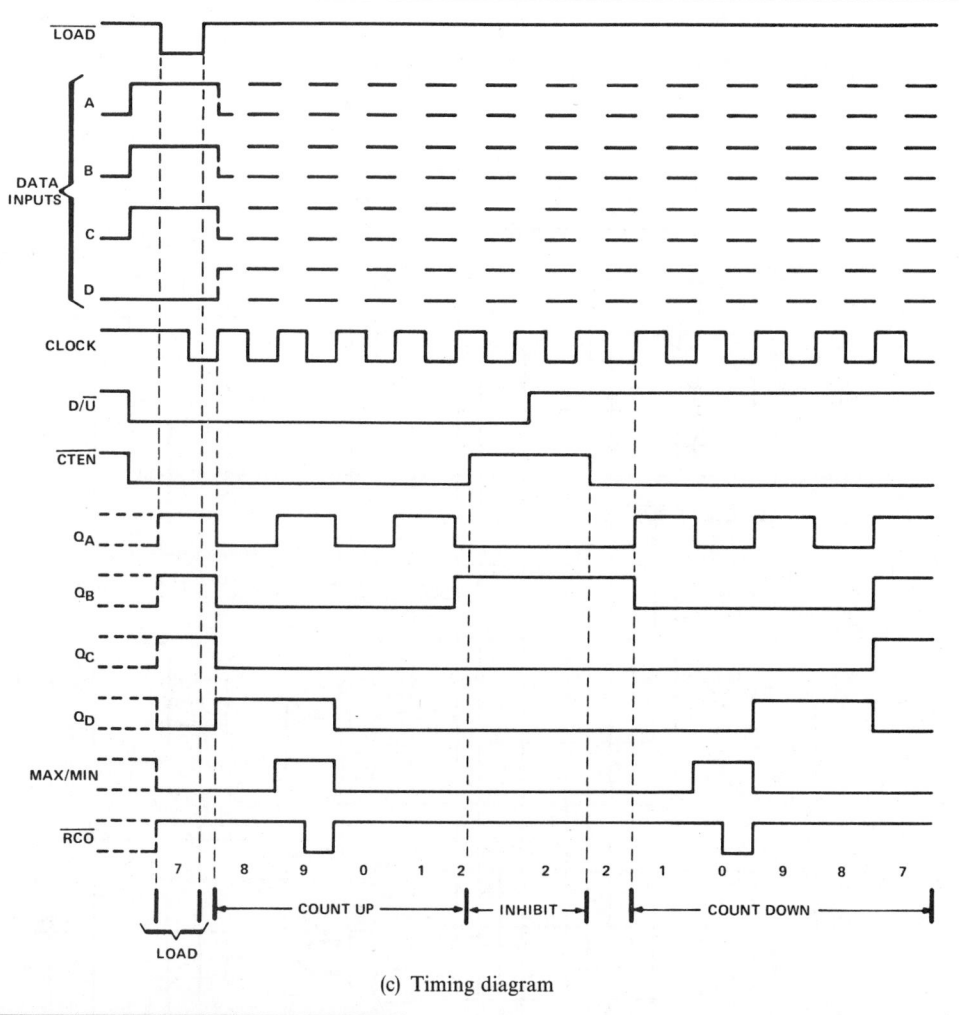

(c) Timing diagram

Figure 5-48   Continued

5.7

Registers

A *register* is a set of binary storage units for storing binary information. The storage units are typically a number of flip-flops contained in one MSI component. The number of flip-flops typically ranges from four to eight, corresponding to the same number of stored bits. Some registers have additional gates that can affect the circuit operation. We will study two main types of registers:

1. Latches
2. Shift registers

Latches

Latches are typically registers used for asynchronously loading information, storing the data, and outputting the data upon demand. An example of a 4-bit register is given in Figure 5–49. This register is constructed from four D flip-flops. The clock input is sometimes labeled with a G to indicate a gating pulse for latches. Latches should never be used in a sequential circuit where feedback occurs unless proper control signals are provided.

There are two basic ways to load information into a synchronous flip-flop that has available the asynchronous SET and CLEAR inputs. The flip-flop in Figure 5–50(a) is an example of a synchronous load operation. The initial information X is loaded into the D flip-flop, that is, $Q = X$ when LOAD = 1, after the positive transition on the CLOCK occurs. This is called synchronous load, because the CLOCK input is used. The flip-flop in Figure 5–50(b) is an example of an asynchronous load, because the CLOCK input does not affect the

Figure 5–49
A 4-bit register

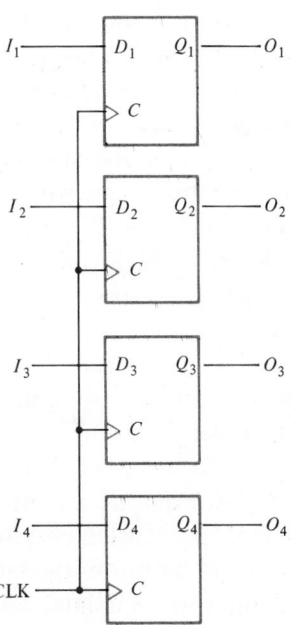

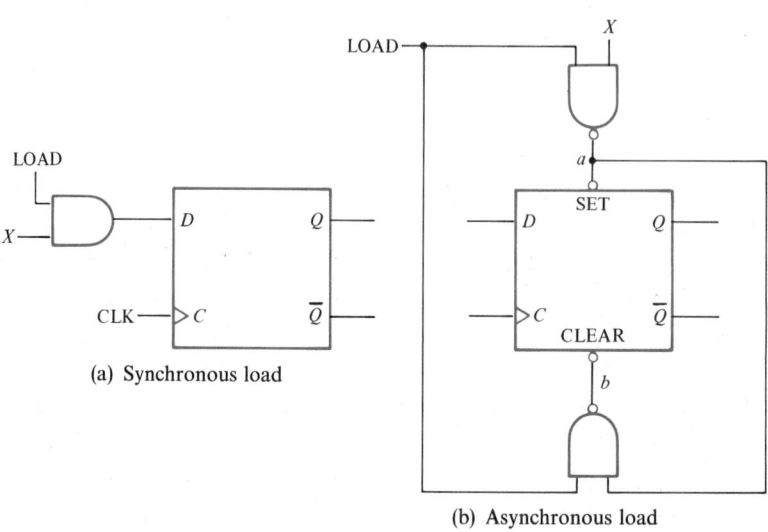

(a) Synchronous load

(b) Asynchronous load

Figure 5–50

Two different loading methods for a flip-flop

direct SET and RESET (CLEAR) inputs. Note that when LOAD = 1, $a = \bar{X}$ and $b = X$; otherwise $a = b = 1$. The reader is urged to write down the logic conditions for the PRESET and CLEAR inputs and verify that the proper information is loaded into the flip-flops. This method of loading is similar to the loading of a latch with the LOAD input acting as a gating pulse.

The D inputs usually have more combinational circuitry than is shown in Figure 5–50(a), but any time the information affects D, we will say that it is a synchronous load operation. This will also hold true with JK flip-flops. Some registers also have CLEAR or PRESET inputs that will make all flip-flops 0 or 1, respectively. These CLEARs and PRESETs can also be of the synchronous or asynchronous type. The types will become apparent with some of the next registers discussed.

Examples of Latches Some representative examples of latches are given in Figure 5–51. They differ from one another according to the word length, the type of outputs, and their functions. The word length may vary from 1 to 16 bits in a single chip. The latch output may be of totem pole (dual-state) type or more commonly of tri-state type. Most latches allow transparent operation. This means that when the input and output of the latch are activated at the same time, the output reflects the input with minimal delay, thus giving the impression that the latch is not there.

Latches	Quad	Octal	9-Bit	10-Bit
Non-inverting	75, 77, 375, 279	268, 373, 573, 100, 116, 873, 845	843	841
Inverting		533, 563, 580, 846, 880	844	842

Figure 5–51
Some representative registers

The 74ALS373–74ALS374 series The 74ALS373–74ALS374 series contains octal *D*-type latches. Whereas the 74ALS373 is level mode-triggered, the 74ALS374 is positive edge-triggered. Both have separate output control, which can also be tri-stated for connection to the bus. The equivalent circuit and the truth table for the output control of the 74ALS374 are shown in Figure 5–52.

The 74LS604–74LS607 series The 74LS604–74LS607 series contains octal two-input multiplexed latches. The equivalent circuit and function table are shown in Figure 5–53. There are two sets of octal latches in the chip. Either set of latches can be selected for output.

Shift Registers

A *shift register* is a set of flip-flops that can transfer information serially from one flip-flop to the next. There are five major types of shift registers:

1. Serial-in, serial-out
2. Serial-in, parallel-out
3. Parallel-in, serial-out
4. Parallel-in, parallel-out
5. Bidirectional parallel-in, parallel-out

Each of these five kinds of shift registers is now examined in detail. Some representative examples of shift registers are listed in Figure 5–54.

A typical example of a serial-in, serial-out shift register is the 74ALS94. The circuit schematic and register function table are shown in Figure 5–55. Data are serially shifted into the register from the leftmost flip-flop and serially shifted out from the rightmost flip-flop. A typical timing diagram for this register, given in Figure 5–55(c), shows how input data travel through the register from one end to the other.

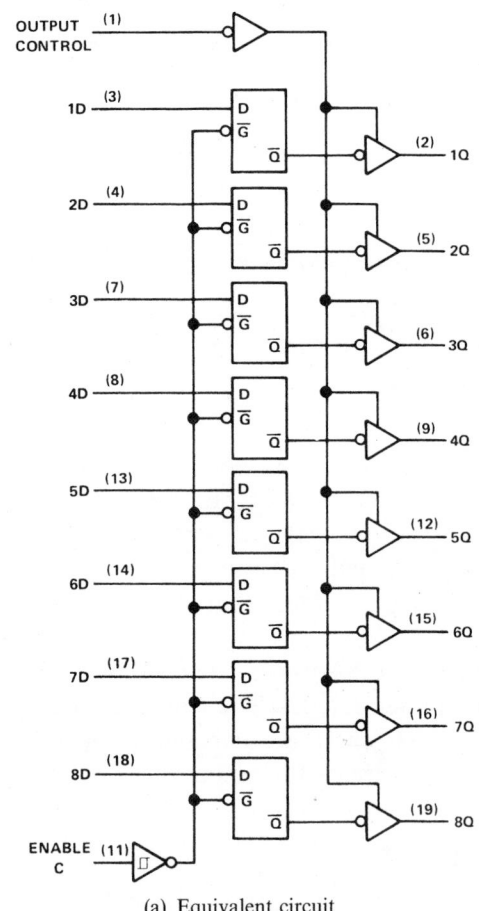

(a) Equivalent circuit

Output Control	Enable G	D	Output
L	H	H	H
L	H	L	L
L	L	X	Q_0
H	X	X	Z

(b) Function table

Figure 5–52

(a) Equivalent circuit and (b) function table of the 74ALS374 latches
(Reprinted by permission of Texas Instruments)

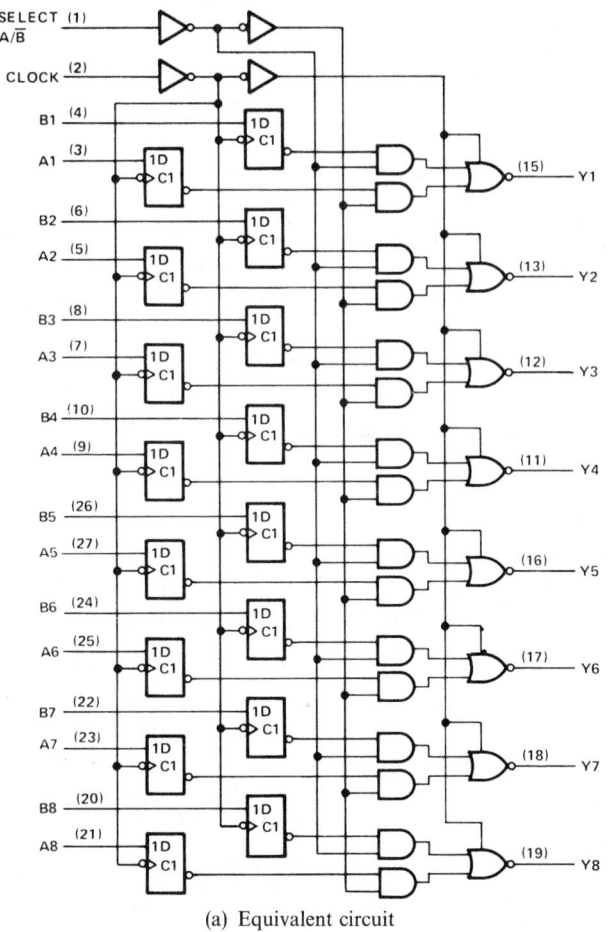

(a) Equivalent circuit

Inputs				Outputs
A_1–A_8	B_1–B_8	Select $A/\bar{B}$	CLOCK	Y_1–Y_8
A data	B data	L	↑	B data
A data	B data	H	↑	A data
X	X	X	L	Z or OFF
X	X	L	H	B register stored data
X	X	H	H	A register stored data

(b) Function table

Figure 5–53

(a) Equivalent circuit and (b) function table for the 74LS604 latches
(Reprinted by permission of Texas Instruments)

Shift Registers	Quad	Octal	16-Bit
Serial-in, Serial-out	94	91	
Serial-in, Parallel-out		164	673
Parallel-in, Serial-out		165, 166	674
Parallel-in, Parallel-out	95, 99, 295	199	
	671, 672		
	178, 179, 195		
	395		
Parallel-in, Parallel-out (bidirectional)	194	198, 299, 323	

Figure 5–54

Some representative shift registers

Figure 5–55

(a) Equivalent circuit, (b) function tables, and (c) timing diagram for the 74AS94 serial-in, serial-out shift register (Reprinted by permission of Texas Instruments)

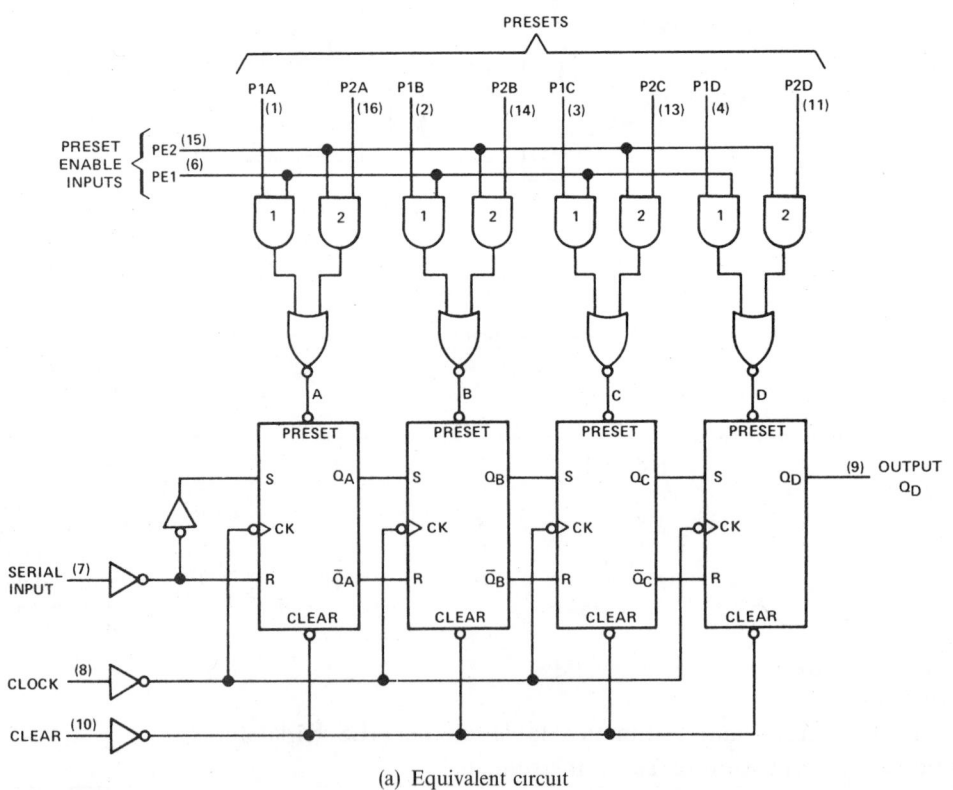

(a) Equivalent circuit

Preset Function Table (Bit A, Typical of All)				
Preset Inputs				Internal
PE_1	P_1A	PE_2	P_2A	Preset A
L	X	L	X	H (inactive)
L	X	X	L	H (inactive)
X	L	L	X	H (inactive)
X	L	X	L	H (inactive)
H	H	X	X	L (active)
X	X	H	H	L (active)

Register Function Table										
Internal Presets				Inputs			Internal Outputs			Output
A	B	C	D	Clear	Clock	Serial	Q_A	Q_B	Q_C	Q_D
H	H	H	H	H	X	X	L	L	L	L
L	L	L	L	L	X	X	H	H	H	H
H	H	H	H	L	L	X	Q_{A0}	Q_{B0}	Q_{C0}	Q_{D0}
L	H	L	H	L	L	X	H	Q_{B0}	H	Q_{D0}
H	H	H	H	L	$\uparrow$	H	H	Q_{An}	Q_{Bn}	Q_{Cn}
H	H	H	H	L	$\uparrow$	L	L	Q_{An}	Q_{Bn}	Q_{Cn}

(b) Function tables

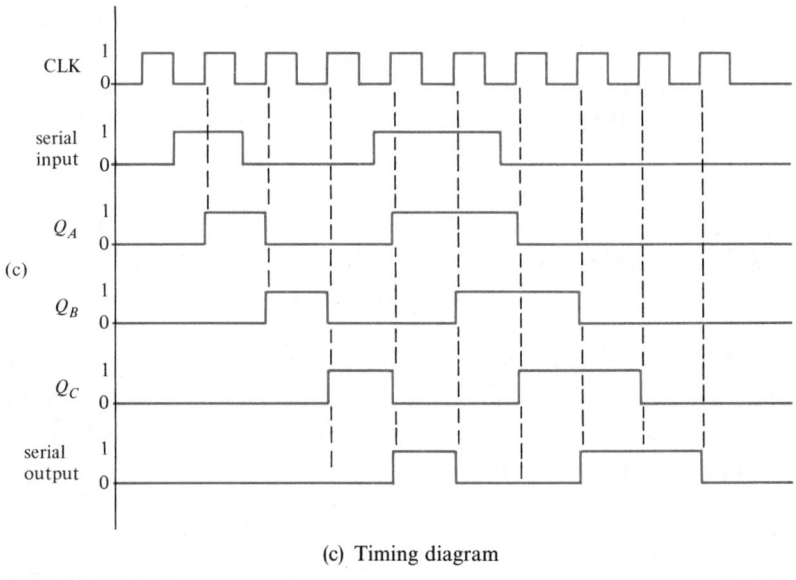

(c)

(c) Timing diagram

Figure 5–55 Continued

The 4-bit shift register in Figure 5–55 has synchronous parallel load, asynchronous clear, and serial out. The internal multiplexer is a quad 2-to-1 multiplexer where the outputs $Y_i = B_i$ when $PE_2 = 1$ and $Y_i = A_i$ when $PE_1 = 1$ for $i = 1, 2, 3, 4$.

A positive clock transition will then shift the data or information to the right.

An example of a serial-in, parallel-out shift register is the 74ALS164. The 74ALS164 is a 4-bit shift register with an asynchronous clear. Whenever the clear equals zero (active low condition), the outputs $Q_0 = Q_1 = Q_2 = Q_3 = 0$. The output $Q_0 =$ serial-in, $Q_1 = Q_0$, $Q_2 = Q_1$, and $Q_3 = Q_2$ after the positive transition occurs on the clock input. The serial-in, parallel-out shift register is excellent for converting serial data from a transmission line to parallel form, which is more useful for a computer system.

An example of a parallel-in, serial-out shift register is the 74ALS165. Data can be loaded into the register by asynchronously presetting or clearing each flip-flop. When the control line SHIFT/LOAD is activated to be a 1, then $Q_1 = I_1$, $Q_2 = I_2$, $Q_3 = I_3$, and $Q_4 = I_4$ without the need of a clock pulse. Hence, this is an asynchronous parallel load. The 4-bit shift register in Figure 5–55 will shift to the right when a positive transition occurs on the clock and the control line SHIFT/LOAD is zero. The serial-in can be connected to the serial-out of another stage to allow expansion to 8 bits, when the clock and the SHIFT/LOAD lines are connected together in parallel. This type of shift register works well for converting parallel information to serial, which typically appears on transmission lines. The asynchronous parallel load is very useful when the information is not synchronized with the clock. The synchronous parallel load is used in those cases where the information and clock changes are synchronized.

The parallel-in, parallel-out shift register is useful for performing parallel-to-serial data and serial-to-parallel data. This shift register only shifts data from the serial-in side to the serial-out side. In function, it is very similar to the bidirectional parallel-in, parallel-out ones discussed next.

The bidirectional parallel-in, parallel-out shift register is identical to the parallel-in, parallel-out kind except that the latter shifts in only one direction while the former can shift data in both directions. An example is the 74198. The equivalent circuit and function table are shown in Figure 5–56. The timing diagram is also given in Figure 5–56.

5.8

Applications of Sequential Circuits

One possible application of sequential circuits is code conversions. In many applications, a machine must interact with human operators. If the machine must operate in the binary system, frequent conversions between the decimal system and the binary system are required. Such conversions can be cumbersome and wasteful of time. It is desirable to perform the arithmetic in decimal directly. In this section, we will first present the different representations of decimal digits

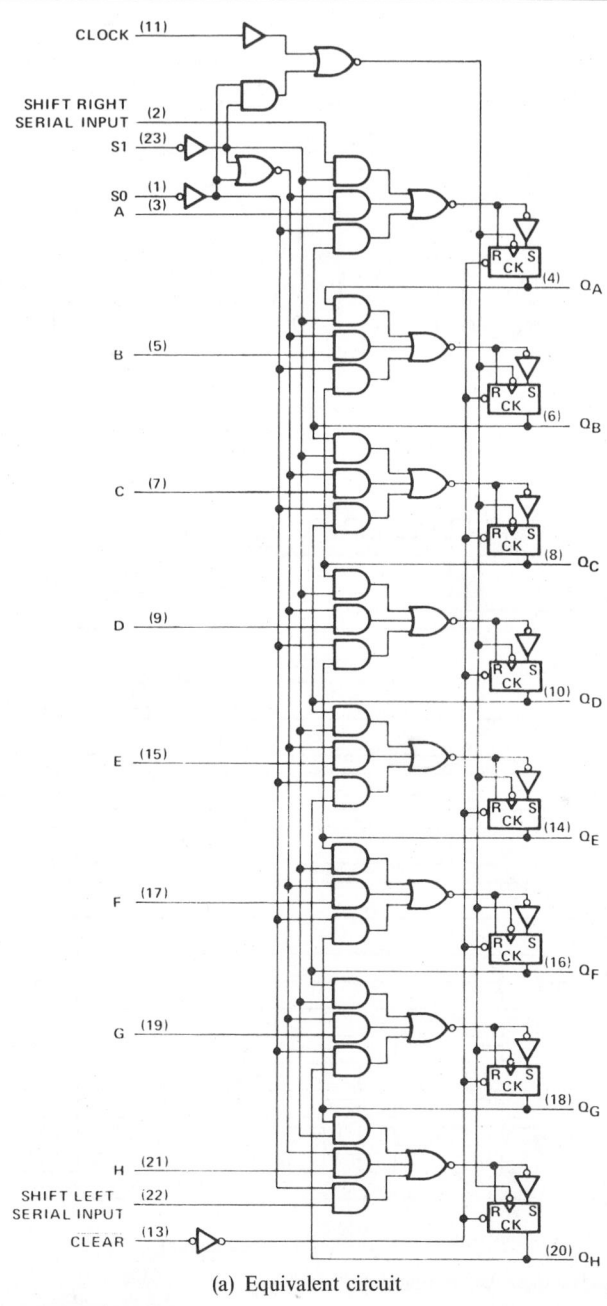

(a) Equivalent circuit

Figure 5–56

A bidirectional parallel-in, parallel-out shift register, 198 (Reprinted by permission of Texas Instruments)

(*continued*)

Inputs							Outputs			
	Mode			Serial		Parallel				
Clear	S_1	S_0	Clock	Left	Right	$A \cdots H$	Q_A	Q_B	$\cdots$ Q_G	Q_H
L	X	X	X	X	X	X	L	L	L	L
H	X	X	L	X	X	X	Q_{A0}	Q_{B0}	Q_{G0}	Q_{H0}
H	H	H	↑	X	X	$a \cdots h$	a	b	g	h
H	L	H	↑	X	H	X	H	Q_{An}	Q_{Fn}	Q_{Gn}
H	L	H	↑	X	L	X	L	Q_{An}	Q_{Fn}	Q_{Gn}
H	H	L	↑	H	X	X	Q_{Bn}	Q_{Cn}	Q_{Hn}	H
H	H	L	↑	L	X	X	Q_{Bn}	Q_{Cn}	Q_{Hn}	L
H	L	L	X	X	X	X	Q_{A0}	Q_{B0}	Q_{G0}	Q_{H0}

(b) Function table

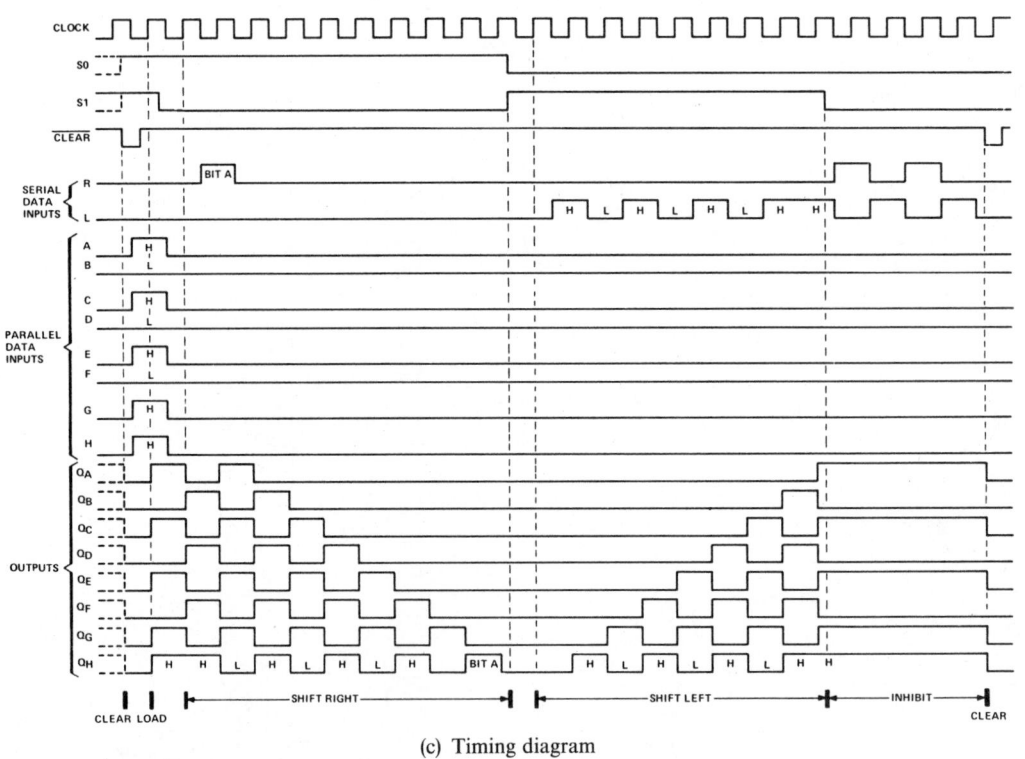

(c) Timing diagram

Figure 5–56 Continued

and then the conversions between them. Following that, we will deal with algorithms for performing the arithmetic directly in decimals.

Binary to BCD

The conversion from binary to BCD is straightforward and parallels the conversion from binary to decimal discussed in Chapter 1. The algorithm is as follows:

1. Shift the most significant bit of the binary BCD number pair left by 1 bit so that the MSB of the binary number becomes the LSB of the BCD number.

2. Examine each BCD decade. If the sum is greater than 4, add 3 to the BCD decade.

3. Repeat Steps 1 and 2 until the least significant binary bit has been shifted into the least significant BCD location.

Example 5–16

Convert 00101110_2 to a BCD number.

The steps of this binary-to-BCD conversion using the algorithm described above are shown in Figure 5–57.

The above algorithm is suited for software implementation. Because of readily available binary-to-BCD (74ALS185) converters discussed earlier, the conversion using MSI components is rather simple and has been shown already in Chapter 3.

Figure 5–57

An example of a binary-to-BCD conversion

BCD	Binary	Description
	00101110	Original number
0	0101110	Shift, no adjust
00	101110	Shift, no adjust
001	01110	Shift, no adjust
0010	1110	Shift, no adjust
0/0101	110	Shift
0/1000	110	Adjust
01/0001	10	Shift, no adjust
010/0011	0	Shift, no adjust
0100/0110		End

BCD to Binary

The conversion from BCD to binary parallels that from decimal to binary. The algorithm is given below:

1. Shift the least significant bit of the BCD binary number pair right by 1 bit so that the LSB of the BCD number becomes the MSB of the binary number.
2. Examine each BCD decade. If any BCD decade is greater than 7, subtract 3 from that 4-bit decade.
3. Repeat Steps 1 and 2 until all the bits in the BCD number have been shifted out.

Example 5–17

Convert the BCD number 0111/0101 to binary representation.

The steps of this BCD-to-binary conversion using the above algorithm are shown in Figure 5–58.

This conversion is also made simple with ready-made BCD-to-binary (74ALS184) converters discussed earlier.

References

1. Boyce, Jefferson C. *Digital Logic Operation and Analysis.* Prentice-Hall, Englewood Cliffs, N.J., 1982.
2. Comer, David J. *Digital Logic and State Machine Design.* Holt, Rinehart and Winston, New York, 1984.

Figure 5–58
An example of a BCD-to-binary conversion

BCD	Binary	Explanation
0111/0101		Original number
011/1010	1	Shift
011/0111	1	Adjust
01/1011	11	Shift
01/1000	11	Adjust
0/1100	011	Shift
0/1001	011	Adjust
0100	1011	Shift, no adjust
010	01011	Shift, no adjust
01	001011	Shift, no adjust
0	1001011	End

3. Dietmeyer, D. L. *Logic Design of Digital Systems*, 2d ed. Allyn & Bacon, Boston, 1978.

4. Greenfield, Joseph D. *Practical Digital Design Using ICs*, 2d ed. John Wiley & Sons, New York, 1983.

5. Hayes, John P. *Digital System Design and Microprocessors.* McGraw-Hill, New York, 1984.

6. Hill, Frederick J., and Gerald R. Peterson. *Introduction to Switching Theory and Logical Design*, 3d ed. John Wiley & Sons, New York, 1981.

7. Hill, F. J., and G. R. Peterson. *Digital Logic and Microprocessors.* John Wiley & Sons, New York, 1984.

8. Huffman, D. A. The Synthesis of Sequential Switching Circuits, *J. Franklin Inst.*, 257(3):161–190; 257(4):275–303, 1954.

9. Humphrey, W. S., Jr. *Switching Circuits with Computer Applications.* McGraw-Hill, New York, 1958, chap. 4.

10. Johnson, E. L., and M. A. Karim. *Digital Design, A Pragmatic Approach.* PWS Publishers, Boston, 1987.

11. Kline, Raymond M. *Structured Digital Design Including MSI/LSI Components and Microprocessors.* Prentice-Hall, Englewood Cliffs, N.J., 1983.

12. Kohavi, Z. *Switching and Finite Automata Theory.* McGraw-Hill, New York, 1970.

13. Kostopoulos, G. K. *Digital Engineering.* John Wiley & Sons, New York, 1975.

14. Lee, Samuel C., *Digital Circuits and Logic Design.* Prentice-Hall, Englewood Cliffs, N.J., 1976.

15. Lewin, Morton H. *Logic Design and Computer Organization.* Addison-Wesley, Reading, Mass., 1983.

16. Mano, M. Morris. *Digital Logic and Computer Design.* Prentice-Hall, Englewood Cliffs, N.J., 1979.

17. Mano, M. Morris. *Digital Design.* Prentice-Hall, Englewood Cliffs, N.J., 1984.

18. McCluskey, E. J., Jr. *Introduction to the Theory of Switching Circuits.* McGraw-Hill, New York, 1965, chap. 4.

19. McCluskey, E. J. *Logic Design Principles with Emphasis on Testable Semicustom Circuits.* Prentice-Hall, Englewood Cliffs, N.J., 1986.

20. Mckay, Charles W. *Experimenting with MSI, LSI, IO, and Modular Memory Systems.* Prentice-Hall, Englewood Cliffs, N.J., 1981.

21. Mealy, G. H. A Method for Synthesizing Sequential Circuits, *Bell System Tech. J.*, 34(5):1045–1080, 1955.

22. Moore, E. F. Gedanken Experiments on Sequential Machines, in C. E. Shannon and J. McCarthy (Eds.), *Automata Studies.* Princeton University Press, Princeton, N.J., 1956.

23. Nagle, H. T., Jr., B. D. Carrol, and J. D. Irwin. *An Introduction to Computer Logic.* Prentice-Hall, Englewood Cliffs, N.J., 1975.

24. National Semiconductor Corp. *CMOS Data Book.* National Semiconductor, Santa Clara, Calif., 1981.

25. National Semiconductor Corp. *Logic Data Book.* National Semiconductor, Santa Clara, Calif., 1981.

26. National Semiconductor Corp. *PAL Data Book.* National Semiconductor, Santa Clara, Calif., 1982.

27. Parr, E. A. *The Logic Designer's Guidebook.* McGraw-Hill, New York, 1984.

28. Paul, M. C., and S. H. Unger. Minimizing the Number of States in Incompletely Specified Sequential Switching Functions, *IRE Trans. Electronic Computers*, EC-8(3):356–357, September 1959.

29. Rhyne, V. T. *Fundamentals of Digital Systems Design.* Prentice-Hall, Englewood Cliffs, N.J., 1973.
30. Roth, Charles H., Jr. *Fundamentals of Logic Design*, 3d ed. West Publishing Co., St. Paul, Minn., 1985.
31. Rutkowski, George, and Jerome Oleksy. *Fundamentals of Digital Electronics: A Text Laboratory Manual*, 2d ed. Prentice-Hall, Englewood Cliffs, N.J., 1985.
32. Shannon, C. E. A Symbolic Analysis of Relay and Switching Circuits, *Trans. AIEE*, 57, 1938.
33. Signetics Corp. *Signetics Logic—TTL Data Manual.* Signetics, Sunnyvale, Calif., 1978.
34. Stearns, R. E., and J. Hartmanis. On the State Assignment Problem for Sequential Machines, II, *IRE Trans. Electronic Computers*, EC-10(4):593–603, December 1961.
35. Taub, Herbert. *Digital Circuits and Microprocessors.* McGraw-Hill, New York, 1982.
36. Texas Instruments, Inc. *The TTL Data Book for Design Engineers*, 2d ed. Texas Instruments, Dallas, Tex. 1976.
37. Wiatrowski, Claude A., and Charles H. House. *Logic Circuits and Microcomputer Systems.* McGraw-Hill, New York, 1980.
38. Wilkinson, Barry. *Digital System Design.* Prentice-Hall, Englewood Cliffs, N.J., 1987.
39. Williams, Gerald E. *Digital Technology.* Science Research Associates, Chicago, 1977.
40. Winkel, David, and Franklin Prosser. *The ART of Digital Design.* Prentice-Hall, Englewood Cliffs, N.J., 1980.

Experiment I

Design of a Sequence Detector

OBJECTIVES

In this experiment, we want to design a sequence detector that can also be used as a digital combination lock.

EQUIPMENT

1   power supply, 5 V

COMPONENTS

3   74ALS177 4-bit counters
2   74ALS04 hex inverters
5   74ALS10 triple three-input NANDs

PROCEDURE

1. Choose a 3-bit sequence.
2. Design the state transition diagram and state transition table for the chosen sequence.
3. Derive the coded state transition table and the flip-flop excitation table.
4. Minimize the output equations and the flip-flop excitation equations.

5. Connect the circuit.
6. Verify the operation of the circuit.

RESULTS

1. Can the sequence be easily changed?
2. Are there any other ways to approach this problem in light of the requirement that the sequence can be easily changed?
3. Compare the chip count and number of interconnections between the described way and any other designs.

Experiment II

Design of a Super Coin Exchanger

OBJECTIVES

In this experiment, we want to design a super coin exchanger by means of sequential circuit components to help count the number of coins received and to dispense the necessary change.

EQUIPMENT

   1   power supply, 5 V

COMPONENTS

   3   74ALS177 4-bit counters
   2   74ALS04 hex inverters
   5   74ALS10 triple three-input NANDs

PROCEDURE

1. To count the number of coins in each category, we will use counters. Every time a coin is dropped, the proper counter is activated and incremented by one.
2. Verify the operation of the counters by depressing momentary switches for each counter to simulate the dropping of coins.
3. Connect the counters to the coin exchanger given in Chapter 4.
4. Verify the operation of the coin exchanger.

RESULTS

1. How should the counters be cleared after the product and the change are dispensed?
2. How easy is it to do away with the full adders and use counters and SSI glue for this problem?

Experiment III

Design of a Variable Frequency Counter

OBJECTIVES

In this experiment, we want to design a highly accurate counter with adjustable frequency. The frequency should be accurate to 1 Hz for an operating range between 1 Hz and 1 MHz.

EQUIPMENT

   1   power supply, 5 V

COMPONENTS

   3   74ALS177 4-bit counters
   2   74ALS04 hex inverters
   5   74ALS10 triple three-input NANDs

PROCEDURE

1. Connect the four counters together in a cascading format. Verify the operation of the counter by checking the outputs of the counter. Each successive output pin should have a frequency that doubles the previous pin.
2. Increase the primary clock frequency to ensure that the counter will operate at 1 MHz.
3. Choose a frequency. Determine what number to detect on the counter output and at what number to preset the counter. Make the connections.
4. Verify the operation of the counter.

RESULTS

1. What is the maximum clock rate at which the counter can work?
2. Compare the accuracy of your counter at 1 Hz and at 1 MHz. Is the accuracy the same? Why?

Problems

1. The JK master–slave flip-flop assumes that the JK inputs are fixed when the clock is high. What type of circuit action results when J and/or K changes when the clock is high? Specify Q^{t+1} in terms of Q^t, J, and K.
2. Analyze the synchronous sequential circuit in Figure P5–1. Find the excitation, transition, and state tables, and also give the state diagram.

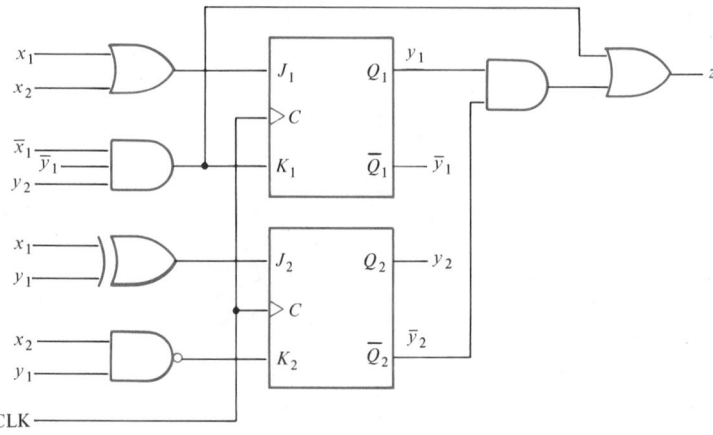

Figure P5–1

A synchronous sequential circuit

3. Derive a state diagram and state table for a synchronous sequential circuit that has one input x and one output z, where $z = 1$ after three or more successive 1 inputs in a row or 0 inputs in a row; otherwise $z = 0$.

4. Derive a state diagram and state table for a synchronous sequential circuit that has one input x and one output z. Assume that x^t and x^{t-1} are the present and immediate past inputs on x. The output $z = 1$ if $x^{t-1}x^t = 01$ or 10; otherwise $z = 0$.

5. Find the state diagram and state table for a one-input x and one-output z synchronous sequential circuit. The output $z^t = x^{t-2}$ (i.e., the output equals the input from two clock periods earlier).

6. Design a state table and diagram for a two-input, x_1 and x_2, single-output z synchronous sequential circuit. Assume that the input x_1 is the most significant bit and x_2 is the least significant bit. The output $z = 1$ if and only if $x_1^t x_2^t \geq x_1^{t-1}x_2^{t-1}$. Thus, the output is 1 when the present input meets or exceeds the previous input.

7. Let x^t, x^{t-1}, and x^{t-2} be the present and previous two inputs. The output z satisfies the relationship

$$z^t = \bar{x}^t x^{t-1}\bar{x}^{t-2} + x^t\bar{x}^{t-1}x^{t-2}$$

Find the state table and diagram.

8. Minimize the state table in Figure P5–2.

9. Minimize the state table found in Problem 3.

10. Minimize the state table found in Problem 5.

11. Minimize the state table found in Problem 6.

12. Find a circuit realization of the minimized state table in Figure P5–3. Implement using JK flip-flops.

$$x$$

s	0	1
a	$a, 0$	$a, 0$
b	$a, 1$	$f, 1$
c	$d, 0$	$e, 0$
d	$a, 1$	$g, 0$
e	$b, 0$	$c, 0$
f	$d, 0$	$e, 0$
g	$b, 0$	$c, 0$

$$S, z$$

Figure P5-2
State transition table for Problem 8

$$x_1 x_2$$

s	00	01	11	10
1	$1, 0$	$2, 0$	$3, 0$	$4, 1$
2	$2, 0$	$1, 0$	$2, 1$	$1, 0$
3	$4, 1$	$1, 0$	$2, 0$	$1, 0$
4	$3, 0$	$1, 1$	$2, 0$	$1, 0$

$$S, z$$

Figure P5-3
State transition table for Problem 12

13. Repeat Problem 12 using D flip-flops.

14. Find the circuit for the counter-type synchronous sequential circuit transition table shown in Figure P5-4 with no input other than the clock. The counter should have three flip-flops whose Q outputs should count in the binary sequence

$y_1 y_2 y_3$		$z_1 z_2 z_3$
000	001	000
001	010	001
011	100	011
010	011	010
110	111	110
111	000	111
101	110	101
100	101	100

$$Y_1 Y_2 Y_3$$

Figure P5-4
State transition table with clock as only input for Problem 14

000, 001, 010, 011, 100, 101, 110, 111, 000, 001, etc. Therefore the sequence is fixed. Use *JK* flip-flops.

15. Repeat Problem 14 using *D* flip-flops.

16. The circuit in Figure P5–5 is used for debouncing a switch. Assume that the switch is a single-pole, double-throw pushbutton switch which in the rest position makes contact with the normally closed (NC) terminal. When the pushbutton is depressed, it will make contact with the normally open (NO) terminal. Switches typically have contact bounce, in that a switch, after making initial contact with a terminal, will bounce off, make contact and bounce off, make contact and bounce off again a number of times before remaining solidly on the terminal. Assume that this is a break-before-make switch, which implies that the contact with the NC terminal is broken before the contact is made on the NO terminal when the pushbutton is depressed. Also, assume that no further contact is made with the NC terminal when the pushbutton is depressed until the pushbutton is released. The break-before-make action works the same way when the switch is released. Show by timing diagrams that for each depression of the pushbutton switch, regardless of how many times the switch contacts bounce, the Q output has only one $0 \rightarrow 1$ transition. Similarly, show the case when the button is released.

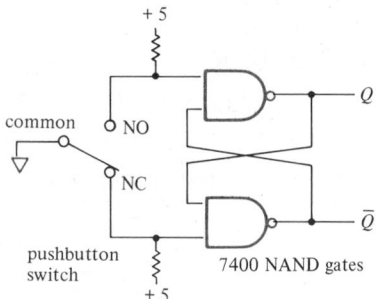

Figure P5–5
A switch debouncer

17. Design counters that use the minimum number of 74163 counters and external logic to count the following sequences:
    (a) 0, 1, 2, . . . , 126, 127, 128, 0, 1, etc.
    (b) 100, 101, . . . , 253, 254, 255, 100, 101, etc.
    (c) 5, 6, 7, . . . , 11, 12, 5, 6, 7, etc.

18. Assume that you have modified a 74163 counter with an additional gate that counts as

$$0, 1, 2, 3, \ldots, 10 , 11, 0, 1, 2, \text{etc.}$$

What happens due to a noise spike or when power is initially applied, if state 13 is reached? Will the counter go back to the correct sequence?

19. Assume that you are using a 74161 counter with asynchronous clear instead of a 74163 with synchronous clear, which drives the clear low when $Q_3 = Q_4 = 1$. The normal count sequence for the 74163 would be

$$0, 1, 2, \ldots, 10, 11, 12, 0, 1, 2, \ldots$$

which is a mod 13 counter. What count sequence results when the 74161 counter is used? What is the modulus?

20. Design a 4-bit shift register that has parallel synchronous load, shift-right, shift-left, and that can also shift two positions to the left. Make sure that it is expandable and has parallel-out and serial-in. Carefully show all inputs and outputs that are needed. How many pin packages will be needed?

21. Assume that you have a synchronous sequential circuit composed of 74LS74 D flip-flops with two levels of combinational logic of 74LS00 gates only. Calculate the maximum clock frequency.

22. The 74168 is an up/down synchronous decade counter with an up/down control. Modify this counter to count in the following sequence:

$$0, 1, 2, 3, 4, 5, 6, 7, 8, 9, 8, 7, 6, 5, 4, 3, 2, 1, 0, 1, 2, \text{ etc.}$$

You will need additional logic, including gates and a flip-flop.

23. Design a counter using the 74163 counter where the modulus of the counter is controlled by the bits stored in a 4-bit register. (*Hint*: The *A*, *B*, *C*, and *D* inputs can be loaded at the appropriate time from this register.)

24. Draw the timing diagrams for the 74163 including the LOAD, CLEAR, and Enable P and T inputs.

25. Show that the three circuits in Figure P5–6 are equivalent circuits of an RSFF.

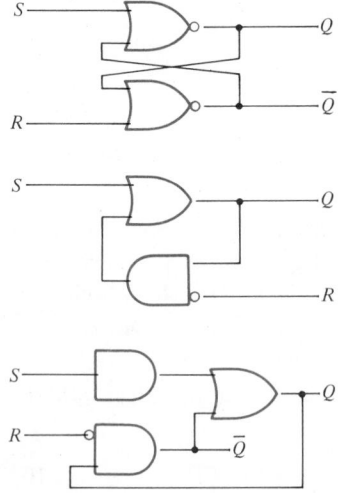

Figure P5–6
Equivalent circuits of an *RS* flip-flop

Design Techniques with Logic Array Components

6

In this chapter we examine digital components in the large-scale integration category. More specifically, several types of logic arrays and their design procedures are presented. The use of logic arrays can greatly simplify the design of random combinational and sequential logic circuits. The simplification often results in a reduction of the total chip count for the design.

In previous chapters we discussed the design of both combinational and sequential circuits using circuit components from small-scale integration (SSI) and medium-scale integration (MSI). In this chapter we discuss components from large-scale integration (LSI) and their design techniques. LSI components are typically composed of chips with more than 100 but fewer than 1,000 gates in each component. Some components in this category are logic arrays. While some of the early memory chips are indeed LSI components, the latest memory components should be more accurately categorized as very-large-scale integration (VLSI) components because of the large number of equivalent gates in each chip. For the sake of continuity and completeness, memory components are discussed later.

6.1

Logic Arrays

A popular kind of LSI chip is the logic array. They are very useful for generating complex multi-input and multi-output Boolean functions and for realizing sequential circuits. Inasmuch as the multiplexers or the demultiplexers can be used to realize a single output function, a single logic array can be used to realize one or more functions at the same time.

Definition of a Logic Array

A logic array (LA) is composed of two parts—an AND matrix and an OR matrix. Hence they are sometimes called an AND/OR/read-only memory (AND-OR-ROM). They are read-only because the connections, once made, cannot be undone. A simplified block diagram is shown in Figure 6–1(a). The input signals are usually routed immediately to an inverter to provide internal circuits with both the complemented and the uncomplemented signals. Hence, for every input, one of the two signal lines, the complemented or the uncomplemented line, will be asserted. Intersecting these input lines are a number of vertical crosslines. A number of these vertical crosslines forming a group terminate in an AND gate through fusable links. There are several groups of these vertical crosslines, which is why this part is called the AND matrix. The effect of the output of these AND gates is to implement the minterms, with each AND gate generating a single minterm.

The outputs of these AND gates become the input to another matrix called the OR matrix because the AND gate outputs are intersected by groups of horizontal crosslines that terminate in OR gates through fusable links as shown in Figure 6–1(a). The effect of the output of these horizontal crosslines is to generate sums of minterms. To realize a function, the OR gate selectively sums the chosen product lines. The output of the OR gate is then the sum of various

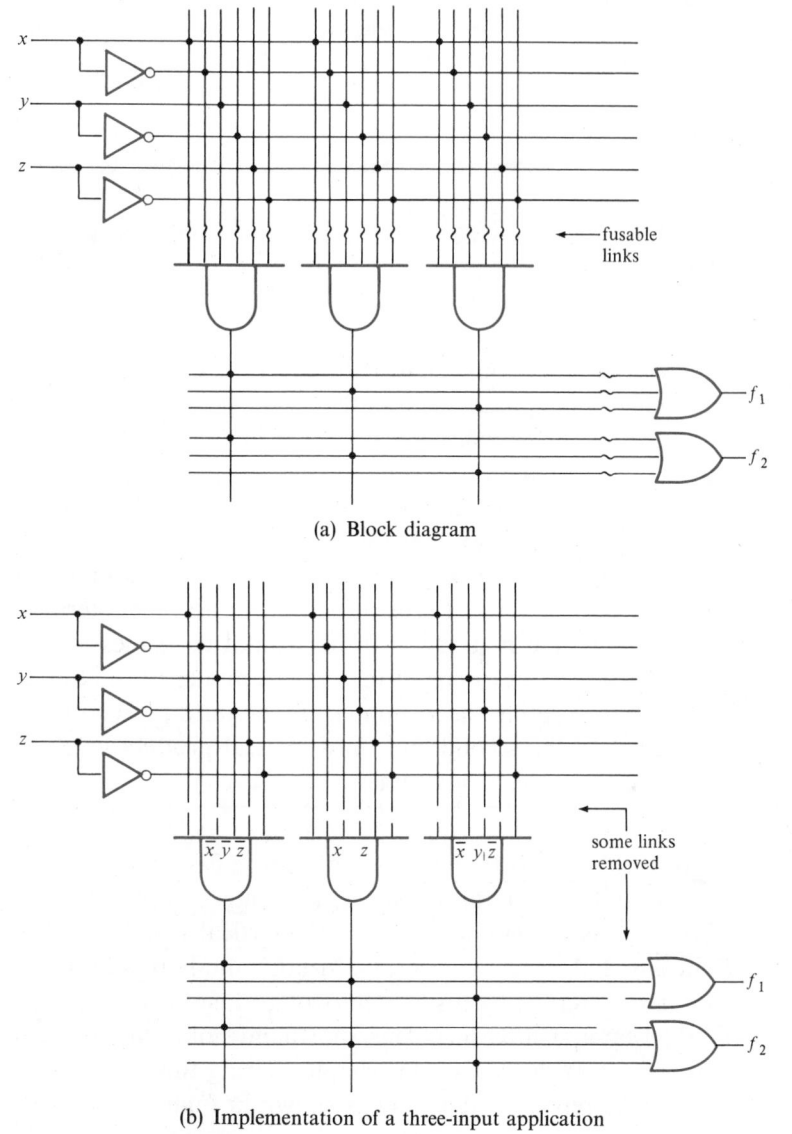

(a) Block diagram

(b) Implementation of a three-input application

Figure 6–1
Logic arrays: (a) block diagram; (b) implementation of a three-input application

product lines and is thus the realization of the desired function. Hence each output of the OR gate is a realization of a single-output combinational circuit. Usually a number of OR gates are provided for multiple outputs.

Example 6–1

Realize the following two functions by a logic array:

$$f_1 = \overline{x}\overline{y}\overline{z} + xz$$
$$f_2 = \overline{x}y\overline{z} + xz$$

These two functions can be realized by the logic array shown in Figure 6–1(b). Examination of f_1 and f_2 shows that there is one common minterm. Hence there are three distinct minterms. These two functions can be realized with a three-input (x, y, z), three-product line $(\overline{x}\overline{y}\overline{z}, xz, \overline{x}y\overline{z})$ and two-output (f_1, f_2) logic array. Figure 6–1(b) shows that the first minterm $\overline{x}\overline{y}\overline{z}$ is realized using the first group of vertical crosslines by blowing out the fuses on the x, y, and z lines, leaving the fuses intact on the $\overline{x}$, $\overline{y}$, and $\overline{z}$ lines. The second minterm is realized by the second group of vertical crosslines and the third minterm by the third group. To realize f_1, the first group of horizontal crosslines is connected to the first vertical product line (representing the $\overline{x}\overline{y}\overline{z}$ minterm) and the second vertical product line (representing the xz minterm). The third product line is left unconnected. To realize f_2, the second and third vertical product lines are connected to the second group of horizontal lines.

In reality, a shorthand notation is more frequently used instead of groups of vertical and horizontal crosslines. A group of vertical lines terminating in an AND gate is represented by a single vertical line for each group as shown in Figure 6–2(a). Likewise, a group of horizontal crosslines terminating in an OR gate is represented by a single horizontal line for that group. An intersection is marked by a slanted line. This shorthand notation will be used for describing logic arrays in the rest of this chapter. Using this shorthand notation, Example 6–1 can be represented simply as shown in Figure 6–2(b).

Example 6–2

Implement the following three functions in a logic array:

$$F_1(w, x, y) = \sum(1, 2, 3, 6)$$
$$F_2(w, x, y) = \sum(2, 4, 6)$$
$$F_3(w, x, y) = \sum(0, 2, 5, 7)$$

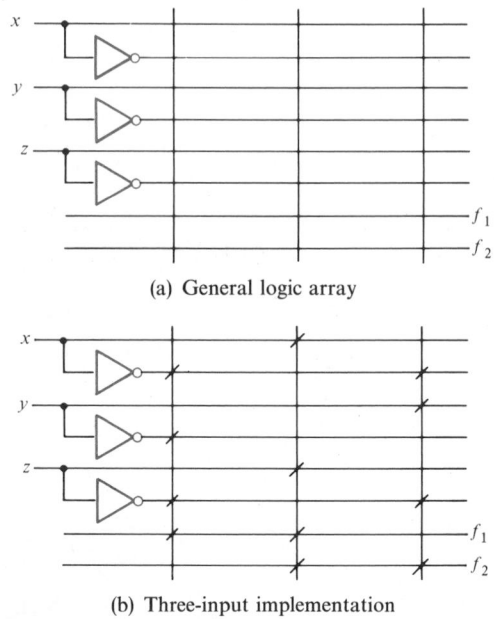

(a) General logic array

(b) Three-input implementation

Figure 6–2
Shorthand notation for (a) a general logic array; (b) the three-input implementation

The Karnaugh maps of these three functions are given in Figure 6–3(a). The sum-of-products form of these three functions are

$$F_1(w, x, y) = \bar{w}y + x\bar{y}$$
$$F_2(w, x, y) = x\bar{y} + w\bar{y}$$
$$F_3(w, x, y) = \bar{w}\bar{y} + wy$$
$$\bar{F}_3(w, x, y) = \bar{w}y + w\bar{y}$$

$F_3 = \bar{w}\,\bar{y} + y\,w$

$F_3 = (w + y) \cdot (w + \bar{y})$

The Karnaugh map for $\bar{F}_3$ is also given. Note that the total number of product terms needed to implement F_1, F_2, and $\bar{F}_3$ is three, though the implementation of F_1, F_2, and F_3 would require five distinct products. The problem of minimizing a multiple-output function where some functions are complemented and others are not is a difficult and yet unsolved problem. The solution for this example is done by trial and error. The logic array implementation for these equations is given in Figure 6–3(b). The function $\bar{F}_3$ is implemented as a sum of products, and the output inverter is used to obtain F_3.

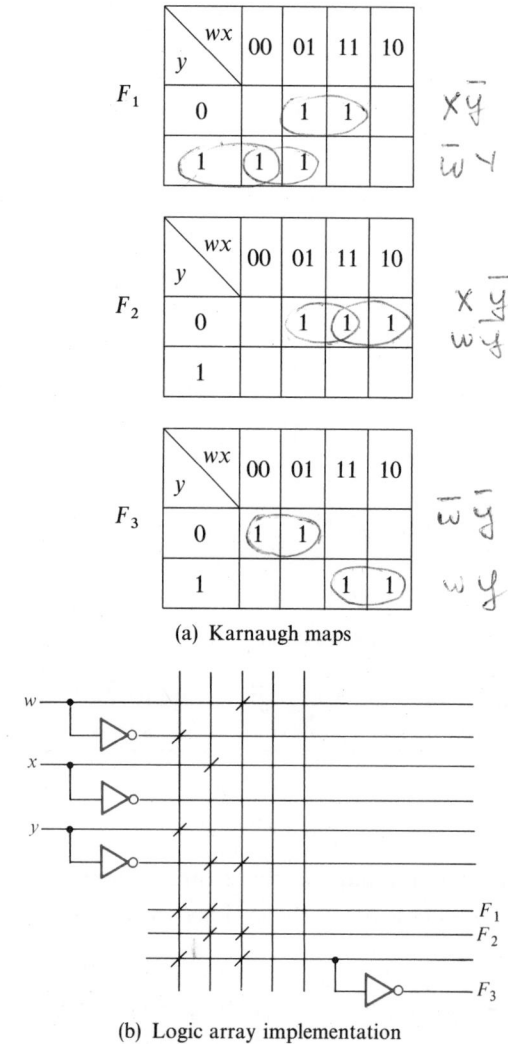

(a) Karnaugh maps

(b) Logic array implementation

Figure 6–3
Karnaugh maps and logic array implementation for the
multiple-output-function example

Types of Logic Arrays

There are many different kinds of logic arrays. Among them, three are most
popular:

1. Programmable array of logic (PAL)
2. Programmable logic array (PLA)
3. Programmable logic sequencer (PLS)

In addition to these three types of logic arrays, the programmable read-only memories (PROMs) are quite similar in structure. Figure 6–4 shows the differences among these four types.

A PROM, as shown in Figure 6–4(a), has a fixed arrangement in the AND array. The fixed arrangement is used for address decoding. The OR array contains the information to be stored and is left to be programmed by the user.

Figure 6–4

The differences between a programmable read-only memory and the three types of logic arrays, showing (a) a programmable read-only memory (PROM), (b) a programmable array of logic (PAL), (c) a programmable logic array (PLA), and (d) a programmable logic sequencer (PLS) (Parts a, b, c, copyright © Advanced Micro Devices, Inc., 1983. Reprinted with permission of copyright owner. All rights reserved.)

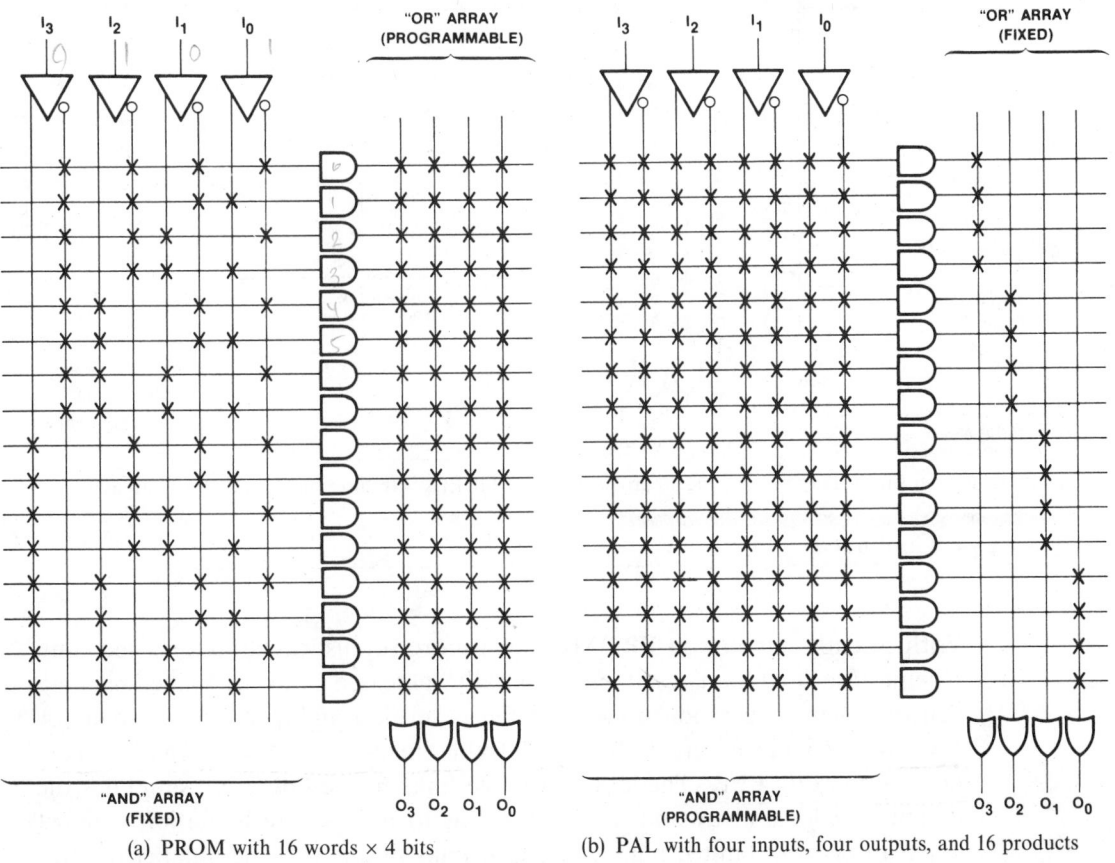

(a) PROM with 16 words × 4 bits

(b) PAL with four inputs, four outputs, and 16 products

(continued)

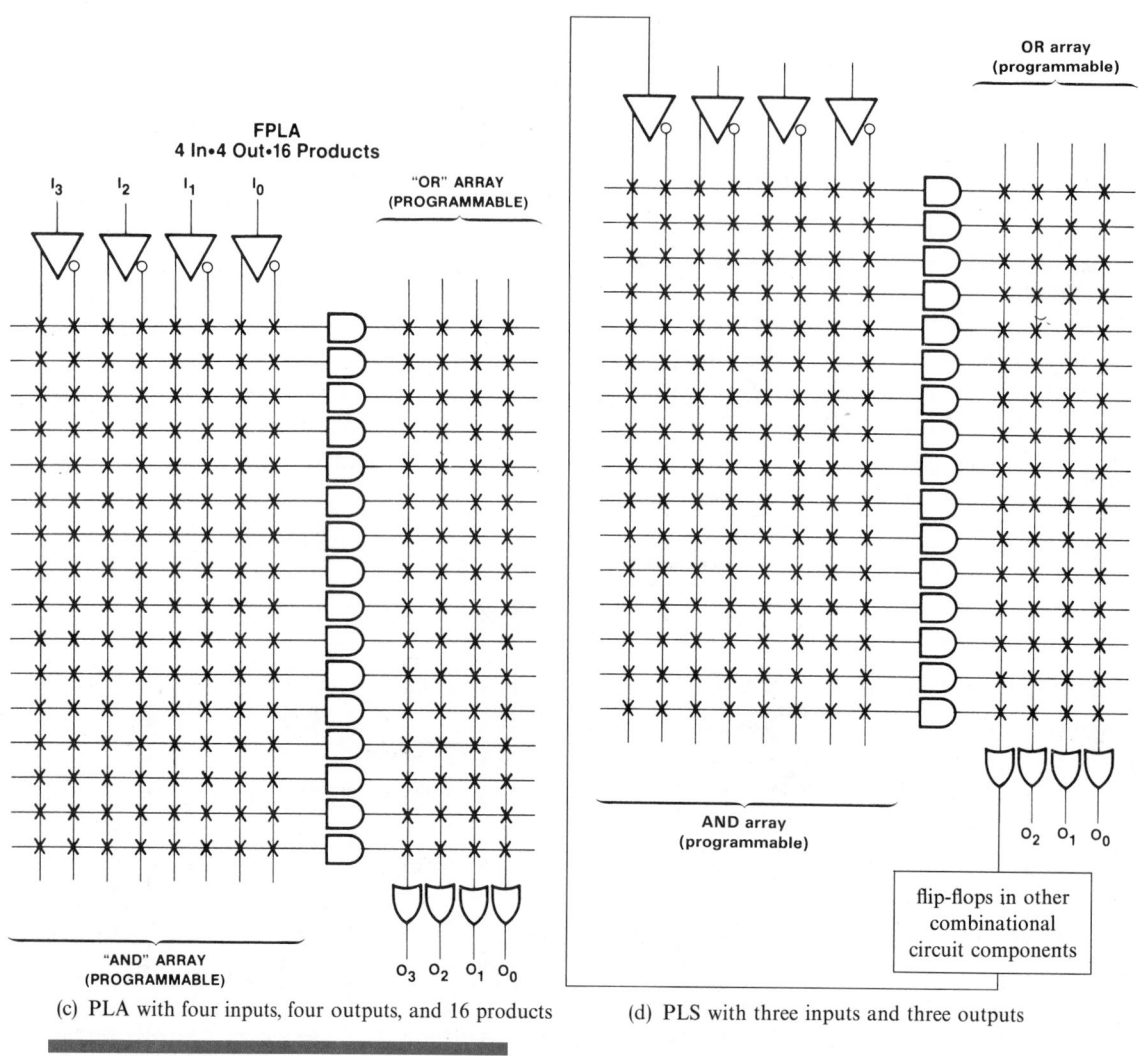

(c) PLA with four inputs, four outputs, and 16 products

(d) PLS with three inputs and three outputs

Figure 6–4 Continued

With recently announced PROMs, much more sophisticated address decoding schemes are used as a result of the greatly increased need for higher packing density. Memory components are treated more thoroughly in later chapters. A PAL, as shown in Figure 6–4(b), has fusable links on the AND array but fixed links in the OR array. The lack of fusable links in the OR array simplifies the design of a logic array and has proved not to be a severe limitation in most applications. The limited number of fixed links means that the number of minterms for each output function is limited. Furthermore, the same minterm for one function cannot be used for another output function. A PLA, as shown in Figure 6–4(c), has fusable links in both the AND array and the OR array to be programmed by the user as desired. Since there are no restrictions on the connections in the AND and OR arrays, a minterm can be produced with as many input variables as are available on the chip, and an output function can be connected to as many product lines as are provided on the chip. Lastly, a PLS, as shown in Figure 6–4(d), not only has fusable links in both the AND and OR arrays, it also has flip-flops and other combinational circuit components connected to the output of the OR gates for realizing sequential functions.

Output Configurations

The input to a logic array is usually the variables themselves. There are, however, many different types of outputs available in logic arrays. These include:

1. Normal output
2. Programmable bidirectional input/output (I/O)
3. Registered output
4. Registered XOR output
5. Registered arithmetic output

The simplified logic diagrams for the outputs of these five versions of logic arrays are shown in Figure 6–5. Most programmable arrays of logic and programmable logic arrays have normal outputs or programmable bidirectional input/output pins. Programmable logic sequencers have registered outputs.

The normal output of a PAL is composed of an OR or NOR output gate as shown in Figure 6–5(a). The OR gate provides output for positive logic functions, while the NOR gate provides output for negative logic functions. For some logic arrays, both uncomplemented and complemented outputs are made available.

In Figure 6–5(b), the block diagram of a logic array with programmable I/O is shown. One of the product terms in the AND array is used to control the tri-state driver of the PAL outputs. This means that the bidirectional pin can be used as an input or output pin depending on the input variables. This feature is useful to provide more flexibility to the number of I/O pins or to provide user- and/or program-definable bidirectional input/output on the same pin. The latter is needed for interfacing with other logic components equipped with bidirectional pins. This option provides maximum flexibility in PAL usage. The output of the OR gate when enabled can be used in two ways:

1. As an output to realize a function
2. As an input in a feedback condition to affect other minterms

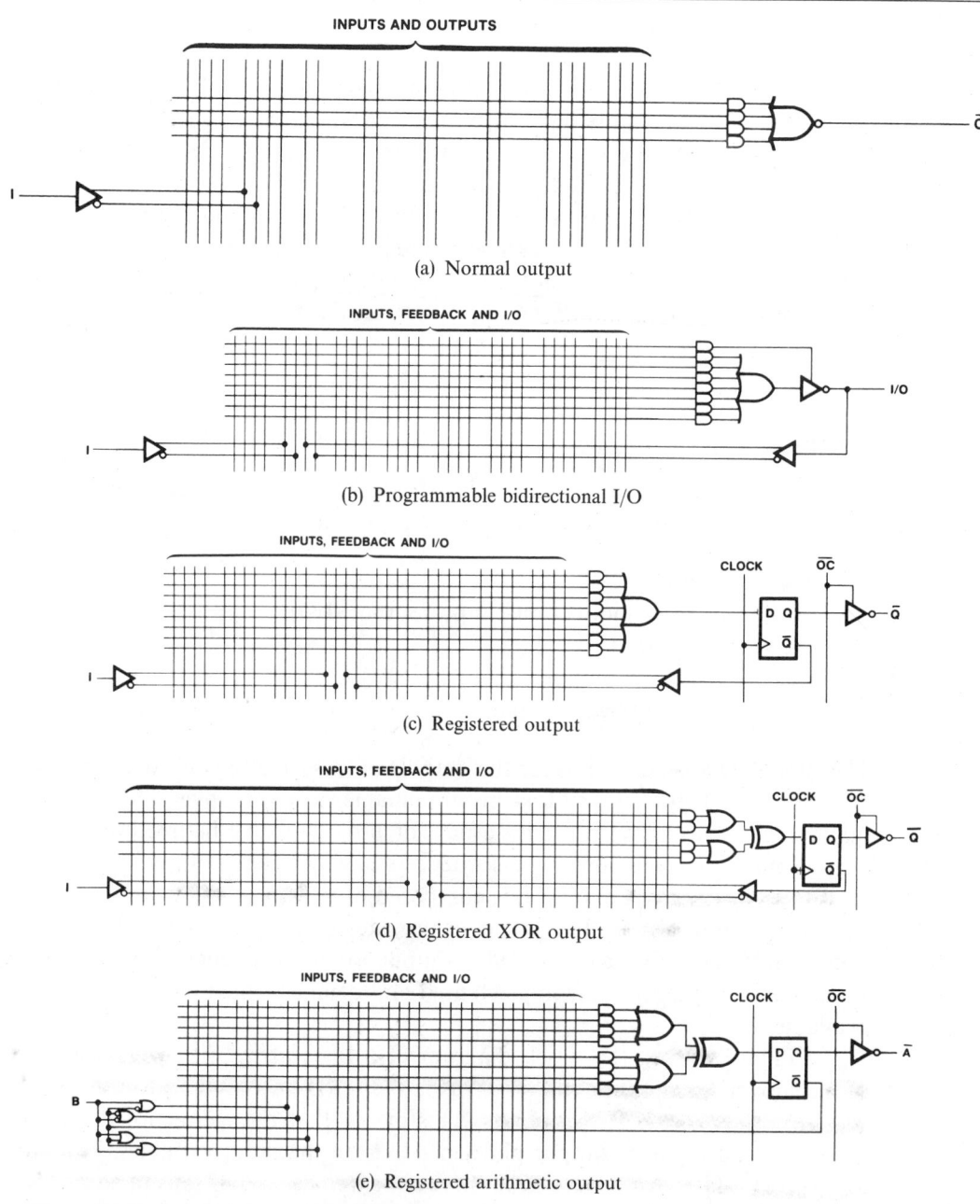

(a) Normal output

(b) Programmable bidirectional I/O

(c) Registered output

(d) Registered XOR output

(e) Registered arithmetic output

Figure 6–5

Simplified block diagrams of different types of logic array outputs:
(a) normal output, (b) programmable bidirectional I/O, (c) registered
output, (d) registered XOR output, and (e) registered arithmetic output

Alternatively, the output of the OR gate can be disabled and the pin can be used just like any other ordinary input variable.

In Figure 6–5(c), the block diagram of a logic array with registered output is shown. In other words, the output is stored in a D flip-flop. New information is registered at the control of the input clock. In principle, this type of logic array is similar in operation to the PLS discussed later. The D flip-flop provides a buffer to hold and store the output of the OR gate after being synchronized by an external clock. The flip-flop output can then be enabled as a logic array output or fed back as an input. The provision of the flip-flop means that the PAL can also be used to realize synchronous sequential circuits.

Figure 6–6

Arithmetic functions performed with the output feedback in logic arrays (Copyright © Advanced Micro Devices, Inc., 1983. Reprinted with permission of copyright owner. All rights reserved.)

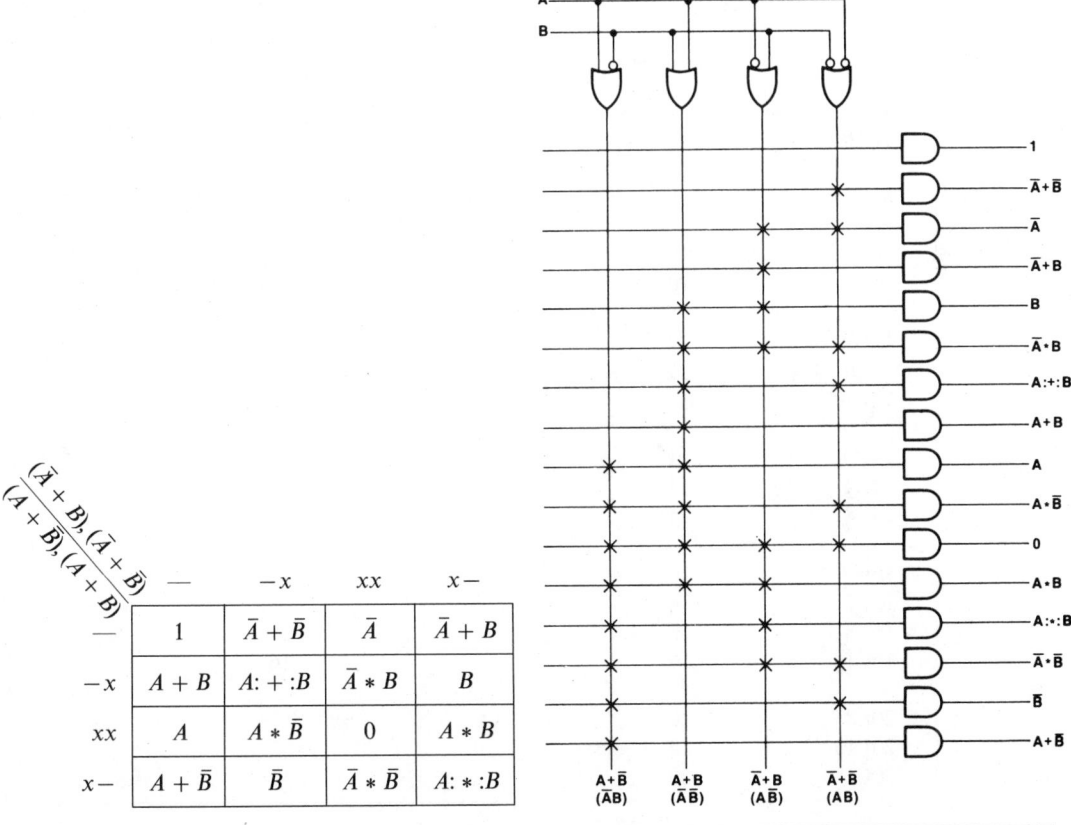

	—	$-x$	xx	$x-$
—	1	$\bar{A} + \bar{B}$	$\bar{A}$	$\bar{A} + B$
$-x$	$A + \bar{B}$	$A: + :B$	$\bar{A} * B$	B
xx	A	$A * \bar{B}$	0	$A * B$
$x-$	$A + \bar{B}$	$\bar{B}$	$\bar{A} * \bar{B}$	$A: * :B$

Another type of PAL output is the registered XOR output as shown in Figure 6–5(d). Instead of a normal OR gate terminating the OR array, the output is taken as the XOR of two functions. In Figure 6–5(d), the registered output $\bar{Q}$ of a logic array is the XOR of two functions, each consisting of two minterms. The availability of an XOR at the output simplifies the needed operations for many applications, particularly those dealing with counters.

When dealing with arithmetic operands, additional arithmetic operations are highly desirable. These arithmetic operations can be effectively provided by including additional logic using the input variables and the output variables as shown in Figure 6–5(e). A typical logic arrangement and associated arithmetic operations at the input to the AND array are shown in Figure 6–6. This additional logic is useful when performing fast arithmetic operations.

Figure 6–7
Examples of programmable arrays of logic (PALs)

Device Number	Number of Inputs	Number of Outputs	Number of Product Terms per Output	Output Polarity
6L16	6	16	1	−
8L14	8	14	1	−
10H8	10	8	2	+
10L8	10	8	2	−
12H6	12	6	2,4	+
12L6	12	6	2,4	−
12L10	12	10	2	−
14H4	14	4	4	+
14L4	14	4	4	−
14L8	14	8	2,4	−
16H2	16	2	8	+
16L2	16	2	8	−
16L8	16	8	7	−
16C1	16	2	16	+, −
20L8	20	8	7	−
20L10	20	10	3	−
20C1	20	2	16	+, −

6.2

Programmable Array of Logic (PAL)

Programmable arrays of logic are very popular because they are easy to use. In PALs, the AND array contains fusable links that the user may use to specify the product terms for the particular application in mind. Though the OR array is fixed, a wide range of PALs with different combinations of AND and OR arrays is available to the designer. Hence specifying the OR terms becomes one of simply choosing the proper PAL.

Types of PALs

PALs are characterized primarily by the number of input variables and the number of output functions. Some PALs, however, also have a number of programmable input or output signals and internal flip-flops. Some representative examples of PALs are shown in Figure 6–7.

The PAL12L6 PAL

An example of a PAL is the PAL12L6. It has 12 input variables and six output terms. Among the six output terms, two are four-input ORs and four are two-input ORs. A simplified block diagram is shown in Figure 6–8. There are 12 uncomplemented input lines. They will be buffered and complemented internally. In the PAL12L6 there are provisions for 16 minterms, indicated in Figure 6–8 by horizontal product lines that terminate in AND gates. The output of the AND gates of the first four horizontal product lines, labeled 8, 9, 10, and 11, forms a group and goes into a four-input OR gate. The output of this OR gate thus becomes one of the function outputs. This means that this output function can be formed for up to four minterms with each minterm consisting of zero to 12 input variables.

PAL Design Procedures

To realize an output function, choose an OR gate with an adequate number of inputs coming from the AND gates. The minterms are realized by leaving the fuse connected between the input variables and the product lines. All other fuses on the same product line should be made disconnected. If only three minterms are needed in a four-input OR gate, the fourth input can be nullified by blowing all the fuses along the product line. We will now illustrate the use of a PAL with an example. The design technique to implement a set of functions using a PAL is shown below in a step-by-step manner.

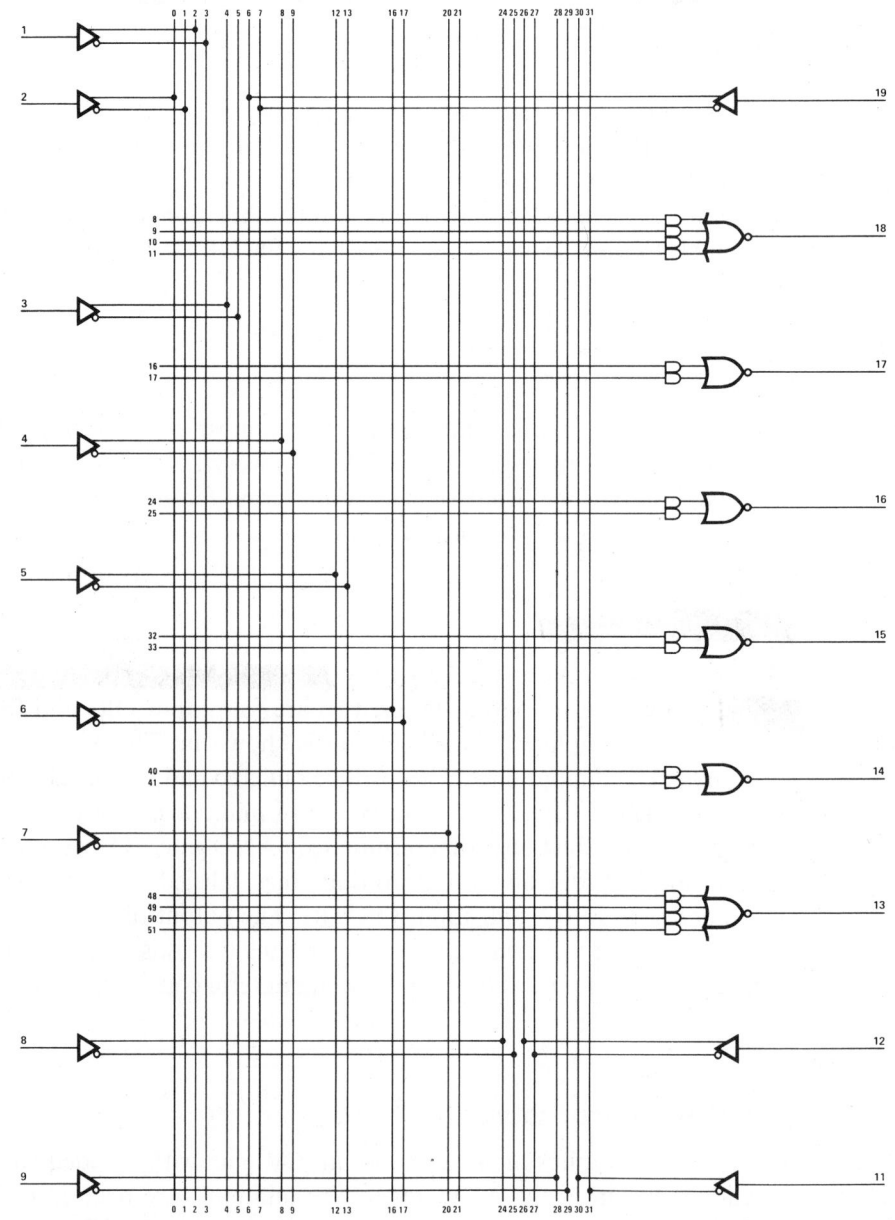

Figure 6-8

A simplified block diagram of the PAL12L6 programmable array of logic (Copyright © Advanced Micro Devices, Inc., 1983. Reprinted with permission of copyright owner. All rights reserved.)

Example 6-3 ▬▬▬▬▬▬▬▬▬▬▬▬▬▬▬▬▬▬▬▬▬▬

Implement the combinational circuit shown in Figure 6-9 with a PAL.

The following is a set of six output functions with nine input variables according to the combinational circuit shown. The six output equations are

$$O_1 = \overline{I}_1$$
$$O_2 = I_1 + I_2 + I_3$$
$$O_3 = I_1 I_2 I_3$$
$$O_4 = I_1 \oplus I_2$$
$$O_5 = I_1 \uparrow I_2 \uparrow I_3 \longrightarrow NAND$$
$$O_6 = I_1 \downarrow I_2 \downarrow I_3 \longrightarrow NOR$$

Three steps are required when implementing combinational circuits in PAL:

1. Choose the correct PAL for the job.
2. Rewrite the output equations in a sum-of-products or a product-of-sums form.
3. Design the connections for the AND and OR arrays.

These three steps are now illustrated.

Figure 6-9

An example of a combinational circuit to be implemented by a programmable array of logic (PAL)

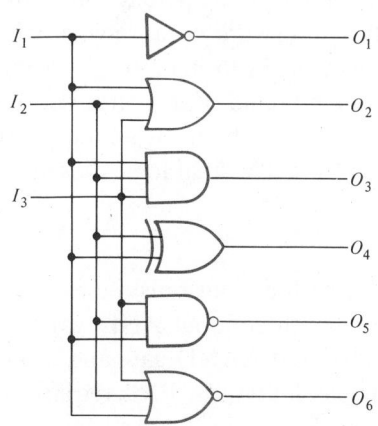

STEP 1

Choose the Correct PAL for the Job

The first step is to choose the correct PAL for the job. This is crucial because the OR array is fixed for PALs. Examination of the six output functions shows that we need a three-input OR gate. Hence the minimum requirement for a proper PAL here is 10 inputs, six outputs with one output from a three-input gate. Several PALs satisfy this requirement. The smallest PAL that can do the job is the PAL12L6.

STEP 2

Rewrite the Output Equations in a Sum-of-Products
or a Product-of-Sums Form

After choosing the proper PAL, we notice that the output is active low. In other words, the product terms are summed through a NOR gate rather than an OR gate. So we must complement the outputs and rewrite them in standard sum-of-products form as follows:

$$\overline{O}_1 = I_1$$
$$\overline{O}_2 = \overline{I}_1\overline{I}_2\overline{I}_3$$
$$\overline{O}_3 = \overline{I}_1 + \overline{I}_2 + \overline{I}_3$$
$$\overline{O}_4 = I_1I_2 + \overline{I}_1\overline{I}_2$$
$$\overline{O}_5 = I_1I_2I_3$$
$$\overline{O}_6 = I_1 + I_2 + I_3$$

STEP 3

Design the Connections for the AND and
OR Arrays

The last step is to design the actual connections to realize the output equations. This step is straightforward because whether the connections are made or not is determined largely by the form of the given equations. For a small set of functions, this may be done by hand. For a large set of functions, standard computer programs have been written by most manufacturers that automatically generate the connection diagrams. The final logic diagram for the problem is shown in Figure 6–10.

If the six output functions must be realized using SSI components, one would need an inverter, a three-input AND gate, a three-input OR gate, an exclusive OR gate, a three-input NAND gate and a three-input NOR gate. This totals six chips. The implementation in PAL requires only a single chip to realize all six functions. Larger PALs can be used for more complicated functions, generally resulting in a drastic reduction in the number of chips required.

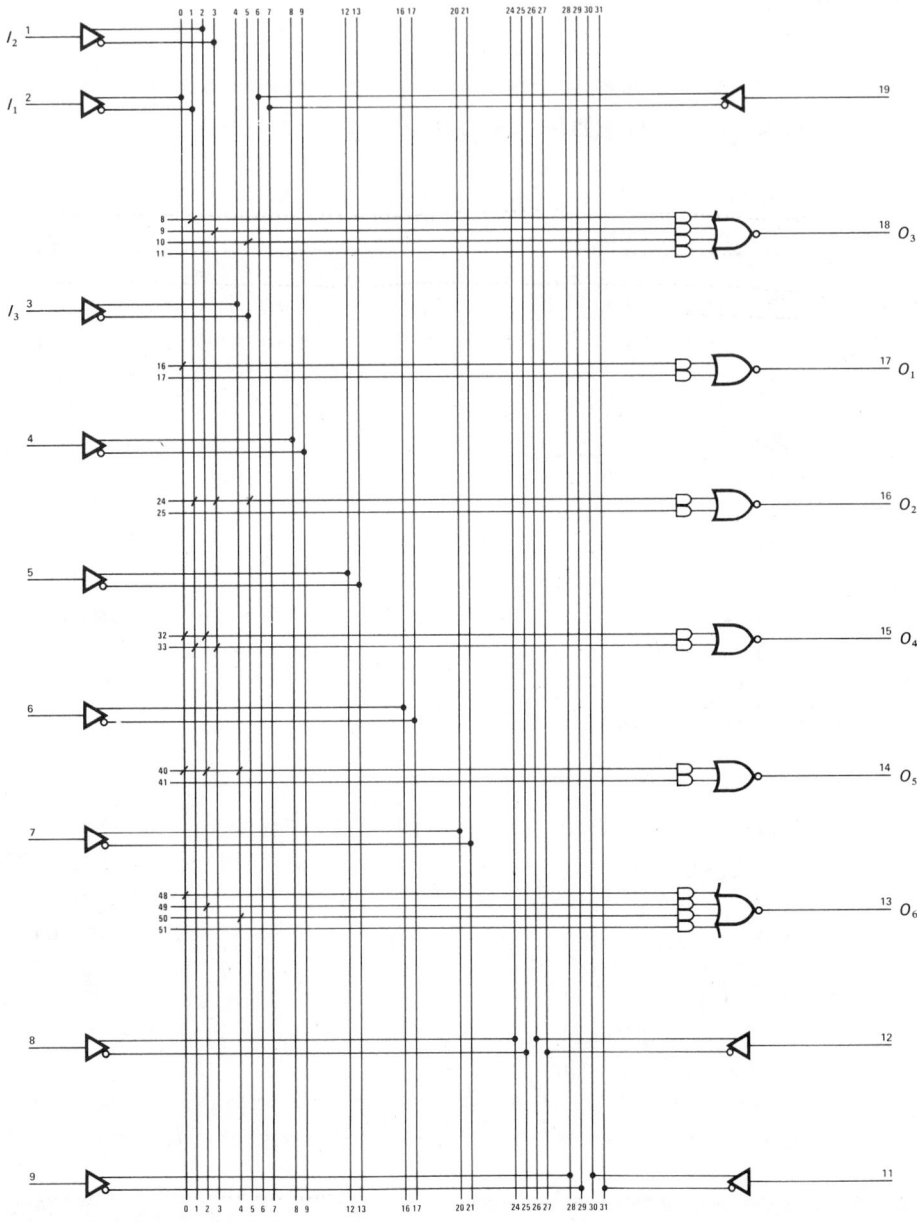

Figure 6–10

Final AND-OR connection for the programmable array of logic that implements the combinational circuit of Example 6–2 (Modified from *Programmable Array Logic Handbook* © Advanced Micro Devices, Inc., 1983. Reprinted with permission of copyright owner. All rights reserved.)

6.3

Programmable Logic Array (PLA)

A programmable logic array (PLA) is an LSI component and differs from the PAL in that both the AND array and the OR array can be programmed to realize a multiple-output function in a sum-of-products form or in a product-of-sums form. A typical PLA has many input variables, a large number of product terms, and many outputs. It should be apparent that programming a PLA often requires the sum-of-products minimization (or near-minimum results) of multiple-output functions with a large number of inputs and outputs. Obviously, Karnaugh map techniques will not be sufficient for this minimization task. Therefore the designer will need to rely on computer-aided minimization techniques.

There are various programmable logic arrays available. Figure 6–11 shows some representative products. Programmable logic arrays are also characterized by the number of input variables, the number of AND gates (minterm capacity), and the number of OR gates (summing capacity or output capacity). The number of AND gates determines the total number of distinct minterms that can be realized. The number of OR gates determines the number of functions that can be realized.

The TIFPLA839 PLA

An example of a PLA is the TIFPLA839, with 14 inputs, 32 product terms and six outputs. The logic symbol and an equivalent block diagram for the TIFPLA programmable logic array are shown in Figure 6–12. This particular chip may handle up to 14 input variables. In addition, the six sum-of-products output lines can be programmed either in active low or active high condition. For proper control of the chip, there are two chip enable lines. The $\overline{CE_1}/\overline{FE}$ line is normally used as a chip enable line. It is also used to program the PLA. The $\overline{CE_2}$ is a second chip enable line.

The TIFPLA839 PLA has 14 inputs and six outputs. Each input is buffered and complemented internally and can be connected to each of the 32 product

Figure 6–11

Examples of programmable logic arrays (PLAs)

Device Number	Number of Inputs	Number of Product Terms	Number of Outputs	Output Polarity
839	14	32	6	+, −
840	14	32	6	+, −

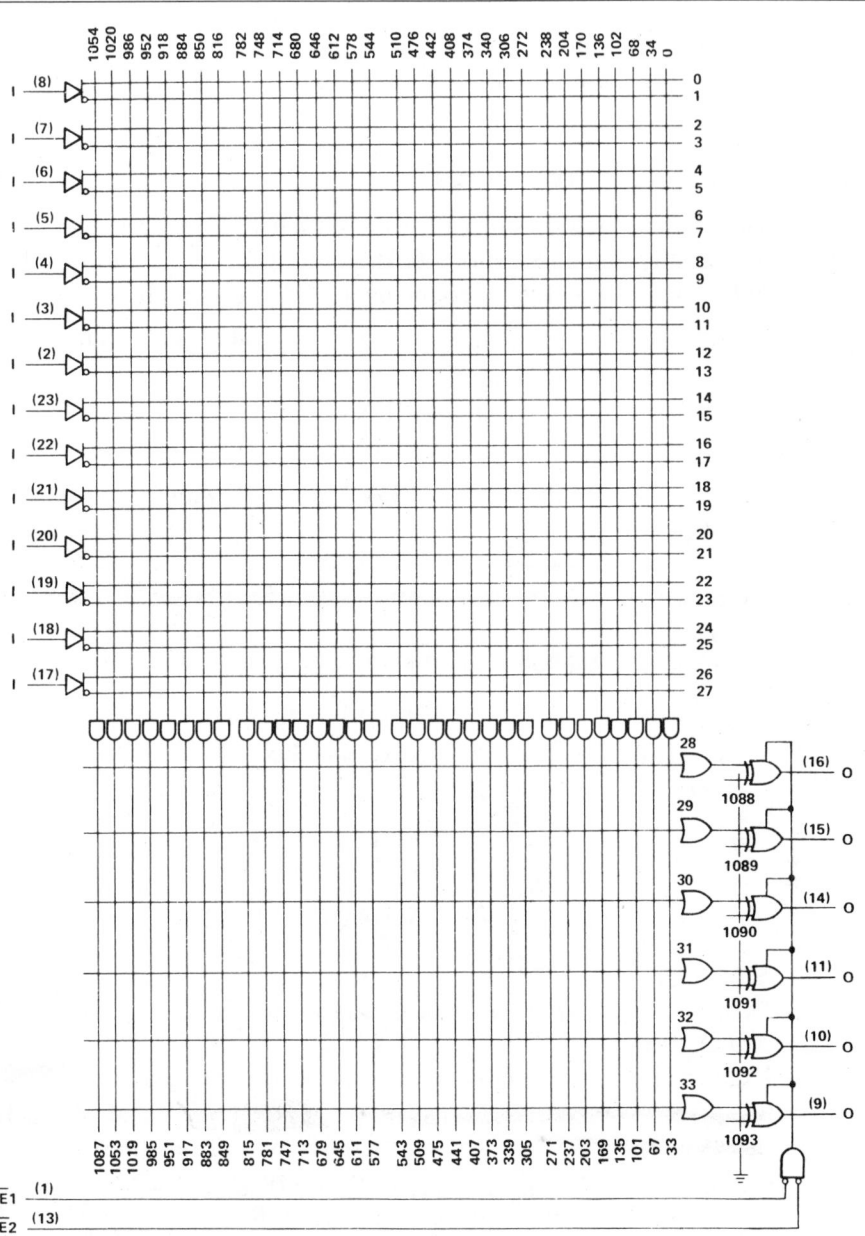

Figure 6–12

Equivalent block diagram of the TIFPLA839 programmable logic array
(Reprinted by permission of Texas Instruments)

terms. Each product term can in turn be connected to each of the six outputs. Each of the output functions can be individually programmed to be inverting or noninverting. In addition, all six output functions are controlled by two chip enable lines.

PLA Design Procedures

The design procedures of a PLA are similar to that of the PAL. Once minimized or partially minimized forms of the output functions are obtained, the steps to follow in realizing the functions by a PLA are as follows:

1. Determine the number of input variables, the number of minterms, and the number of outputs.
2. Choose the appropriate PLA.
3. Determine the proper connection to the AND and OR array.

We will illustrate the use of a PLA by the following example.

Example 6–4

Implement the following functions using a PLA:

$$F_1 = x_1 x_2 x_3 + x_4 x_5 x_6$$
$$F_2 = x_3 \overline{x}_6 x_7 + \overline{x}_1 x_2 + x_5$$
$$F_3 = x_2 + x_4 + x_6 + x_8 + x_{10}$$
$$F_4 = x_5 x_6 x_7 x_8 x_9 x_{10}$$
$$F_5 = x_1 x_3 x_5 x_7 + x_2 x_4 + x_6 x_8$$
$$F_6 = x_1 x_2 x_3 + x_6 + x_2 x_4$$

STEP 1
Determine the Number of Input Variables, Minterms, and Outputs

Examination of the above equations shows that there are 10 input variables $(x_1, x_2, x_3, x_4, x_5, x_6, x_7, x_8, x_9, x_{10})$, 18 minterms, and six output variables $(F_1, F_2, F_3, F_4, F_5, F_6)$. Among these six output functions, the required output configurations are one one-input output OR gate, one two-input OR gate, three three-input OR gates, and one five-input OR gate.

STEP 2
Choose the Appropriate PLA

After identifying the proper requirements to implement the function, the next step is to choose the proper PLA to do the job. From the list of PLAs in Figure 6–11, we can see that the TIFPLA839 will fit the above requirements.

STEP 3

Determine the Proper Connection to the AND and OR Array

Once the proper PLA has been chosen, the connection to the PLA can be specified according to the given equations. This is shown in Figure 6–13. Note that no new product terms for F_6 are required. Unused inputs can be left unconnected. To prevent spurious information from these unconnected inputs, all internal connections should be left open.

Note that if the six functions in Example 6-4 were implemented by SSI components, many chips would be required. To realize the minterms alone, we would need two inverters, three two-input AND gates, three three-input AND gates, one four-input AND gate, and one six-input AND gate. For the summation we would need one two-input OR gate, three three-input OR gates, and one five-input OR gate. Using standard SSI chips, this implementation would require a total of seven chips.

6.4

Programmable Logic Sequencer (PLS)

Programmable logic sequencers (PLSs) differ from programmable logic arrays in that flip-flops are provided at the output of some of the OR gates. While programmable logic arrays are useful for implementing combinational circuits, programmable logic sequencers are ideal for sequential circuits. Both the complemented and uncomplemented outputs of the flip-flops are available for the AND matrix in the same way as the inputs are. Note that these output flip-flops can be used to indicate the state of a sequential machine. Therefore, a programmable logic sequencer can be used to fully realize a complete sequential machine. There are various PLSs available. Some of these are shown in Figure 6–14. The main differences among the various PLSs are the number of input variables, the number of product terms, the number of flip-flops, and the number of output functions.

The TIB82S105 PLS

An example of a PLS is the TIB82S105, with 12 input variables, 32 product terms, six output functions, and four *JK* flip-flops, each with complete output feedback connections. The logic diagram is shown in Figure 6–15. An equivalent block diagram for the TIB82S105 programmable logic sequencer is shown in Figure 6–16. There are 16 input lines for input variables. The eight ouptut lines can be enabled or disabled as a group under external control. For control, the PRE/OE

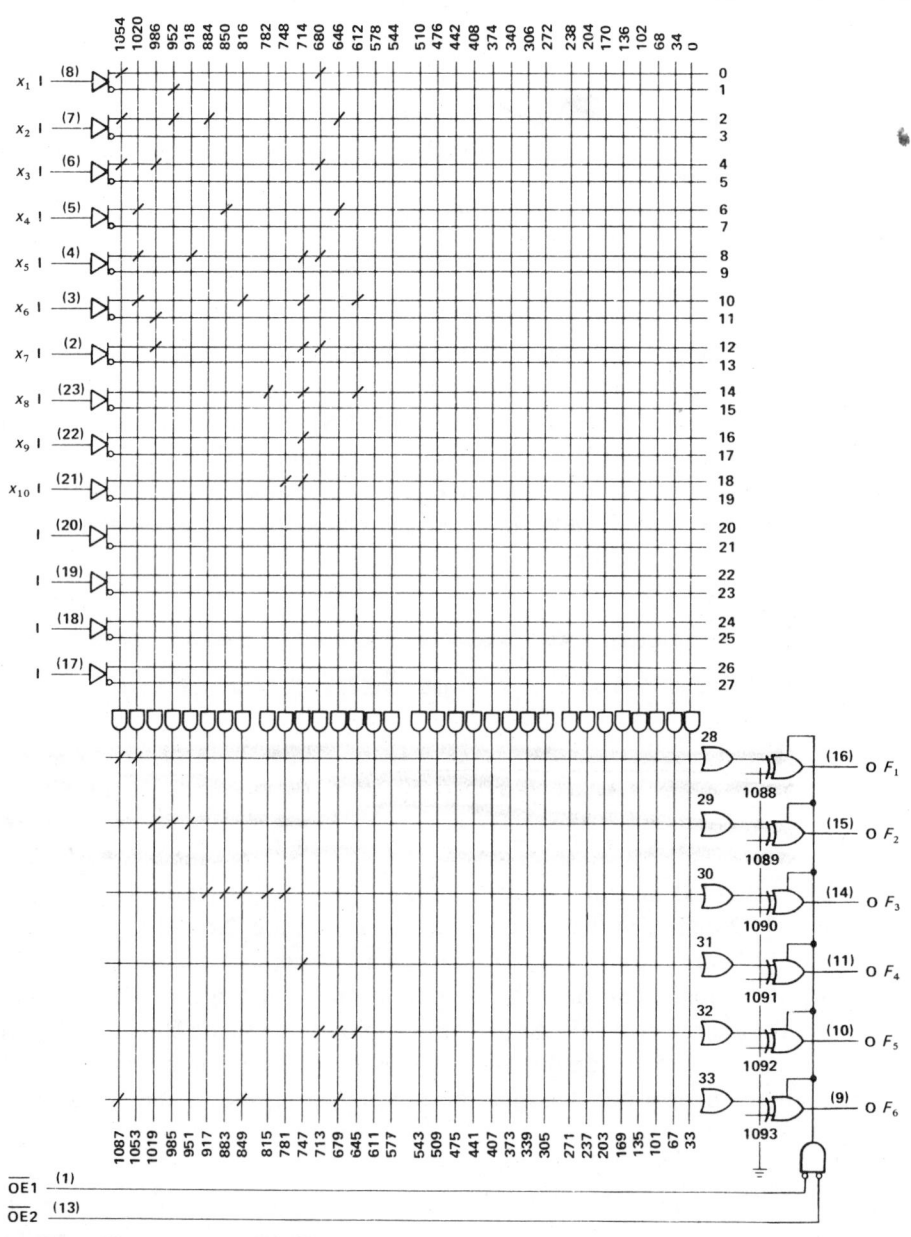

Figure 6–13

PLA implementation of the six-output function in Example 6–4
(Modified from *Programmable Array Logic Handbook* © Advanced
Micro Devices, Inc., 1983. Reprinted with permission of copyright
owner. All rights reserved.)

Device Number	Number of Inputs	Number of Outputs	Number of Flip-flops	Number of Product Terms per Output	Output Polarity
16R8	16	8	8	8	–
16R6	16	8	6	8	–
16R4	16	8	4	8	–
20R8	20	8	8	8	–
20R6	20	8	6	8	–
20R4	20	8	4	8	–
20RS10	20	10	10	0–16	–
22V10	22	10	0–10	8–16	–
32R16	32	16	16	0–16	–

Figure 6–14
Examples of programmable logic sequencers (PLSs)

Figure 6–15
Logic diagram of the TIB82S105 programmable logic sequencer
(Reprinted by permission of Texas Instruments)

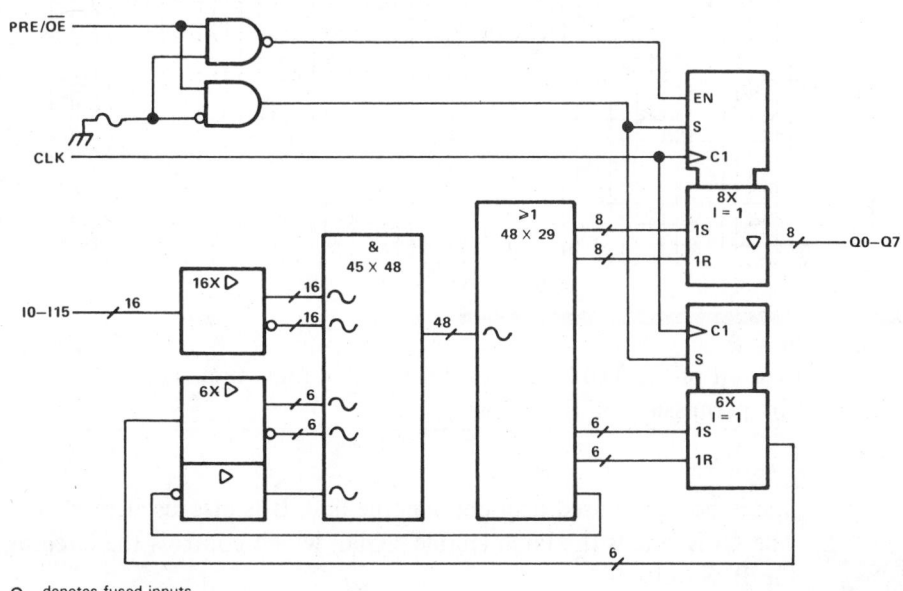

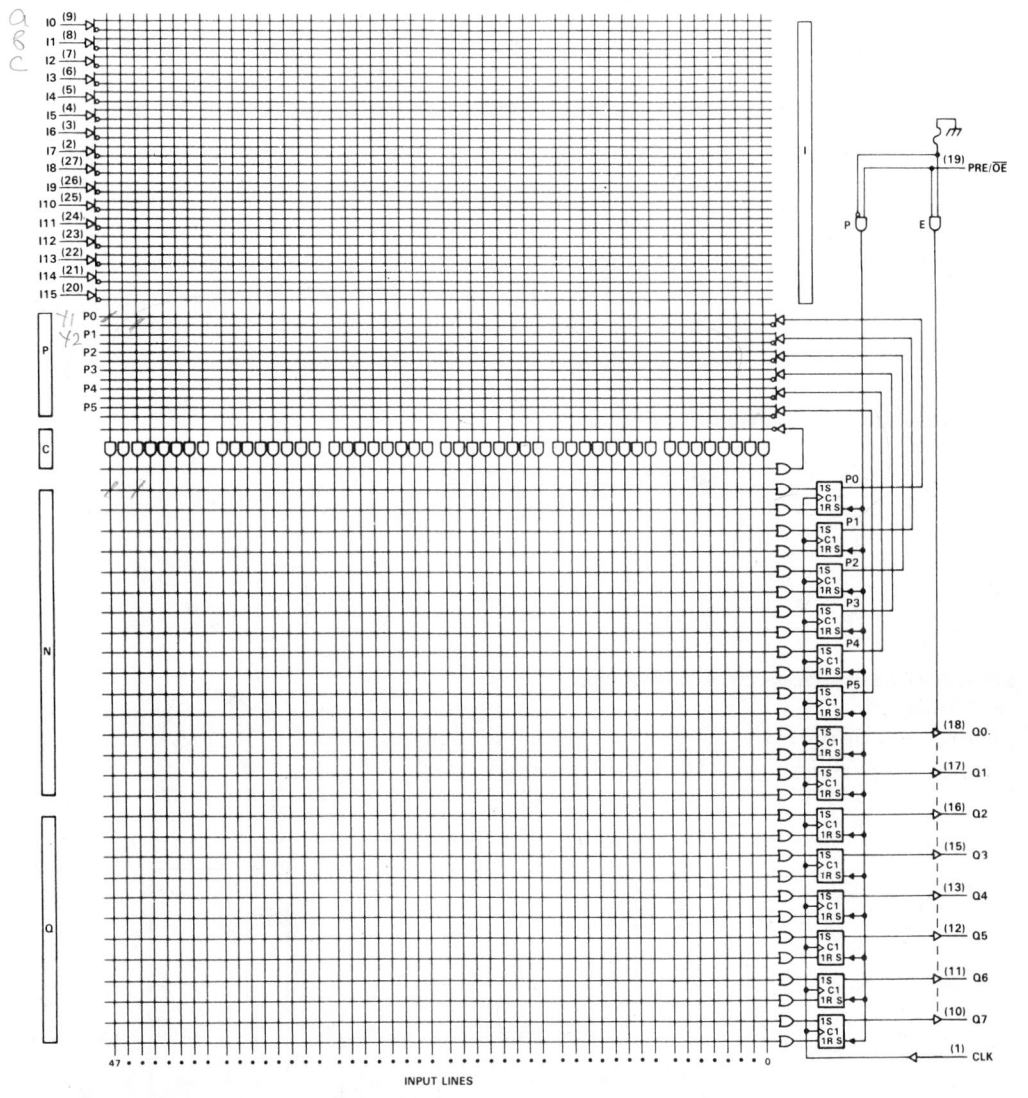

Figure 6–16
Equivalent circuit of the TIB82S105 programmable logic sequencer
(Reprinted by permission of Texas Instruments)

line is normally used as a chip enable line. It is also used to program the PLA. The CLK line is the latch enable signal, which controls the latching function of the PLS outputs.

The TIB82S105 is designed with 16 possible input variables. Each input signal is buffered and complemented internally. There are 48 product terms. Connected to these product terms are 8 output fractions and 6 sets of input parameters for the SR flip-flops. The S and R inputs of each SR flip-flop can be individually programmed. The outputs of the SR flip-flops are fed back to the AND array as additional input variables.

PLS Design Procedures

The design procedures for implementing sequential circuits on the PLS are similar in principle to those for implementing combinational circuits on the PLAs. The steps are as follows:

1. Determine the number of input variables, the number of state variables, the number of minterms including the state equations, and the number of output functions.
2. Choose the proper PLS.
3. Implement the sequential circuit on the chosen PLS.

We will now illustrate the design procedures by some examples.

Example 6–5

Implement the following sequential circuit.

Input parameters: a, b, c

State equations: $S_1 = ay_1 + by_2$
$$R_1 = y_1$$
$$S_2 = by_1 + cy_2$$
$$R_2 = y_2$$

Output equations: $z_1 = abc$
$$z_2 = ay_1y_2$$

STEP 1
Determine the Number of Input Variables, State
Variables, Minterms, and Output Functions
There are three input variables (a, b, c), two state variables (y_1, y_2), two output functions (z_1, z_2), and four state update equations (S_1, R_1, S_2, R_2). A total of eight minterms is required to realize the state equations and the output functions.

STEP 2
Choose the Proper PLS
From Figure 6–14 we can see that several different PLSs will fit the requirement of this problem. One possible PLS is the TIB82S105 PLS.

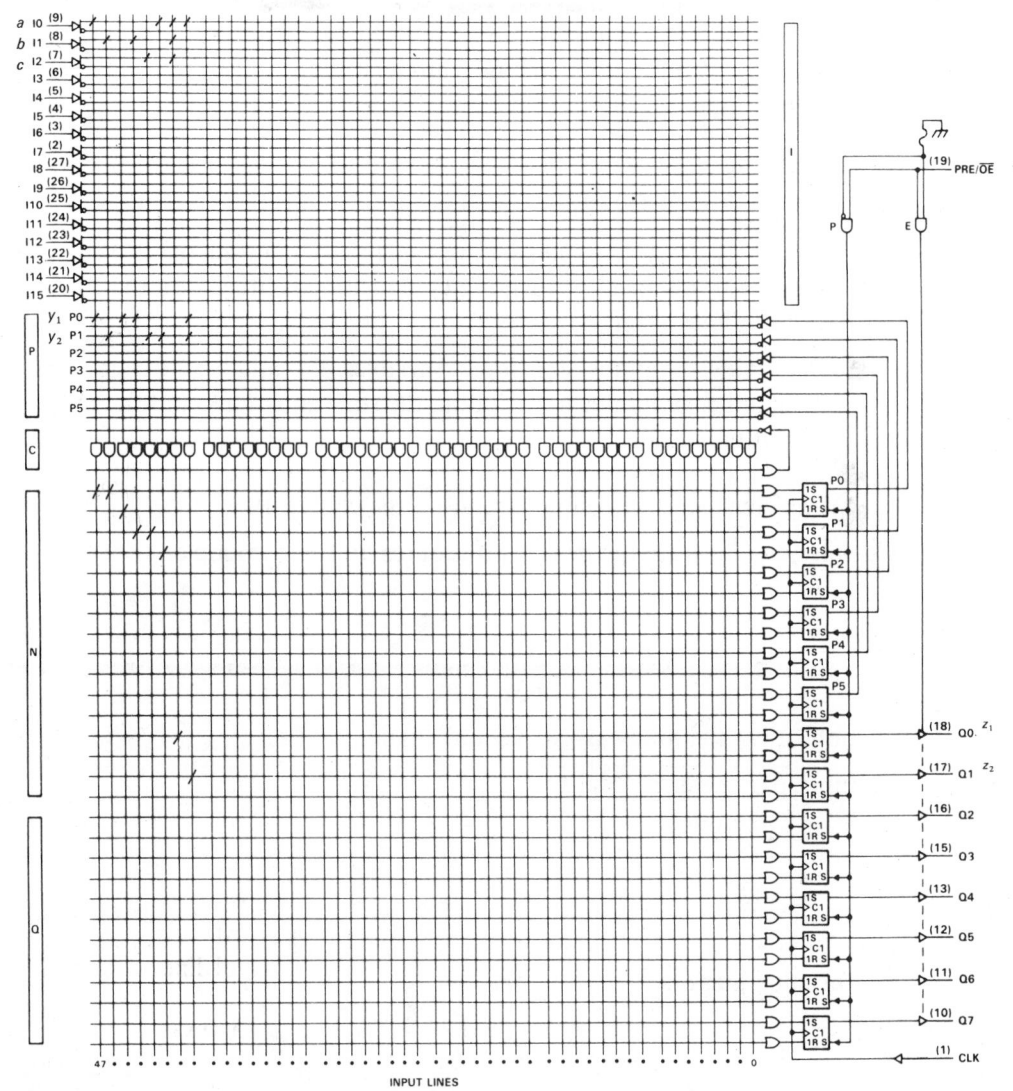

Figure 6–17

PLS implementation of the sequential circuit in Example 6–5 (Modified from *Programmable Array Logic Handbook* © Advanced Micro Devices, Inc., 1983. Reprinted with permission of copyright owner. All rights reserved.)

STEP 3

Implement the Sequential Circuit on the Chosen PLS

The implementation of the PLS is basically the same as that for the PLA. The state equations can be treated as output equations but must be realized in output OR gates that terminate into the flip-flops. The outputs of the flip-flops are considered to be the same as input variables. The complete implementation is shown in Figure 6–17. Note that if this circuit is implemented by conventional SSI chips, a minimum of four chips is required.

Example 6–6 ▬▬▬▬▬▬▬▬▬▬▬▬▬▬▬▬▬▬▬▬▬▬▬▬▬

Implement an octal up/down counter with parallel load.

The purpose of this example is to illustrate the use of registered XOR outputs. The counter to be implemented may count up, count down, be loaded with new inputs, or be put on hold. For high-speed operation, we will choose the synchronous design. A block diagram of the counter is shown in Figure 6–18(a). In addition to 8 bits of parallel input lines and 8 bits of parallel output lines containing the present count values, additional data and control lines are used for expansion and control. For the purpose of expansion, a carry-out bit is provided for cascading to counters of more significant bits. Likewise, a carry-in bit is needed for carry propagation from less significant bits. A clock is needed to activate the flip-flops and to provide the overall timing of the counter. Two additional control lines are used to provide the needed functions. A $\overline{\text{LD}}$ line is used to load the input data when asserted low. When the $\overline{\text{LD}}$ is asserted low, the input data are loaded into the counter. When the $\overline{\text{LD}}$ line is asserted high, the counter may hold the present value if CBI is low or may count if CBI is high. If the counter is to count, a UD line is used to indicate the direction of counting. An $\overline{\text{OC}}$ line is used to control the output drivers. The functions of these control lines are shown in Figure 6–18(b).

To realize the counter, the output flip-flops are used to hold the present count value. We need only to determine the necessary inputs to these flip-flops according to the control signals. We have already seen how the inputs should change for synchronous circuits. For the least significant bit, the input of the D flip-flop should be

$$\bar{D}_0 = (\overline{\text{LD}}\,\bar{Q}_0 + \text{LD}\,\bar{D}_0) \oplus (\overline{\text{LD}}\,\text{UD}\,\text{CBI} + \overline{\text{LD}}\,\overline{\text{UD}}\,\text{CBI})$$

Since the output is always inverted, we will always load the complement of the input. The first pair of parentheses identifies the hold and load conditions. When $\overline{\text{LD}}$ is asserted low, the present output of the flip-flop is chosen. When $\overline{\text{LD}}$ is asserted high, the input data are chosen. The above rule is true except for the condition given in the second pair of parentheses when the counting function is

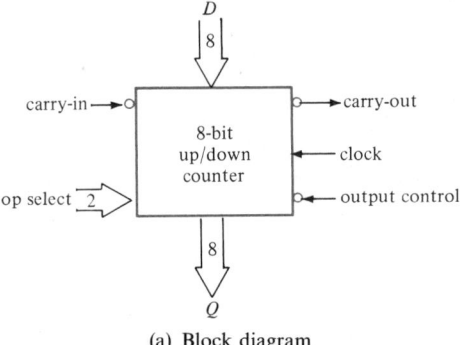

(a) Block diagram

$\overline{OC}$	CLK	$\overline{LD}$	$\overline{UD}$	$\overline{CBI}$	D_7-D_0	Q_7-Q_0	Operation
H	X	X	X	X	X	Z	Hi-Z
L	C	L	X	X	D	L	Load
L	C	H	L	H	X	Q	Hold
L	C	H	L	L	X	$Q+1$	Increment
L	C	H	H	H	X	Q	Hold
L	C	H	H	L	X	$Q-1$	Decrement

(b) Functions of control signals

Figure 6–18
An up/down counter: (a) block diagram; (b) functions of control signals
(Copyright © Advanced Micro Devices, Inc., 1983. Reprinted with
permission of copyright owner. All rights reserved.)

desired. When UD is 1, the count value goes up. When UD is 0, the count value
goes down. Since this is the least significant bit, counting up and counting down are
the same. The incrementing or decrementing also depends on the carry-input CBI
and $\overline{LD} = 0$ condition. For the next two counter bits, the expressions for the flip-
flop inputs are

$$\overline{D}_1 = (\overline{LD}\,\overline{Q}_1 + LD\,\overline{D}_1) \oplus (\overline{LD}\,UD\,CBI\,Q_0 + \overline{LD}\,\overline{UD}\,CBI\,\overline{Q}_0)$$
$$D_2 = (\overline{LD}\,\overline{Q}_2 + LD\,\overline{D}_2) \oplus (\overline{LD}\,UD\,CBI\,Q_0\,Q_1 + \overline{LD}\,\overline{UD}\,CBI\,Q_0\,Q_1)$$

The other bits can be obtained in similar fashion. Having found all the flip-flop
input equations, we can easily implement the required operations on a programmable logic array. The PAL20X8 is chosen, and the logic implementation is
shown in Figure 6–19. (This example is taken from the *PAL Handbook*, 3rd
ed., Monolithic Memories Inc., 1983.)

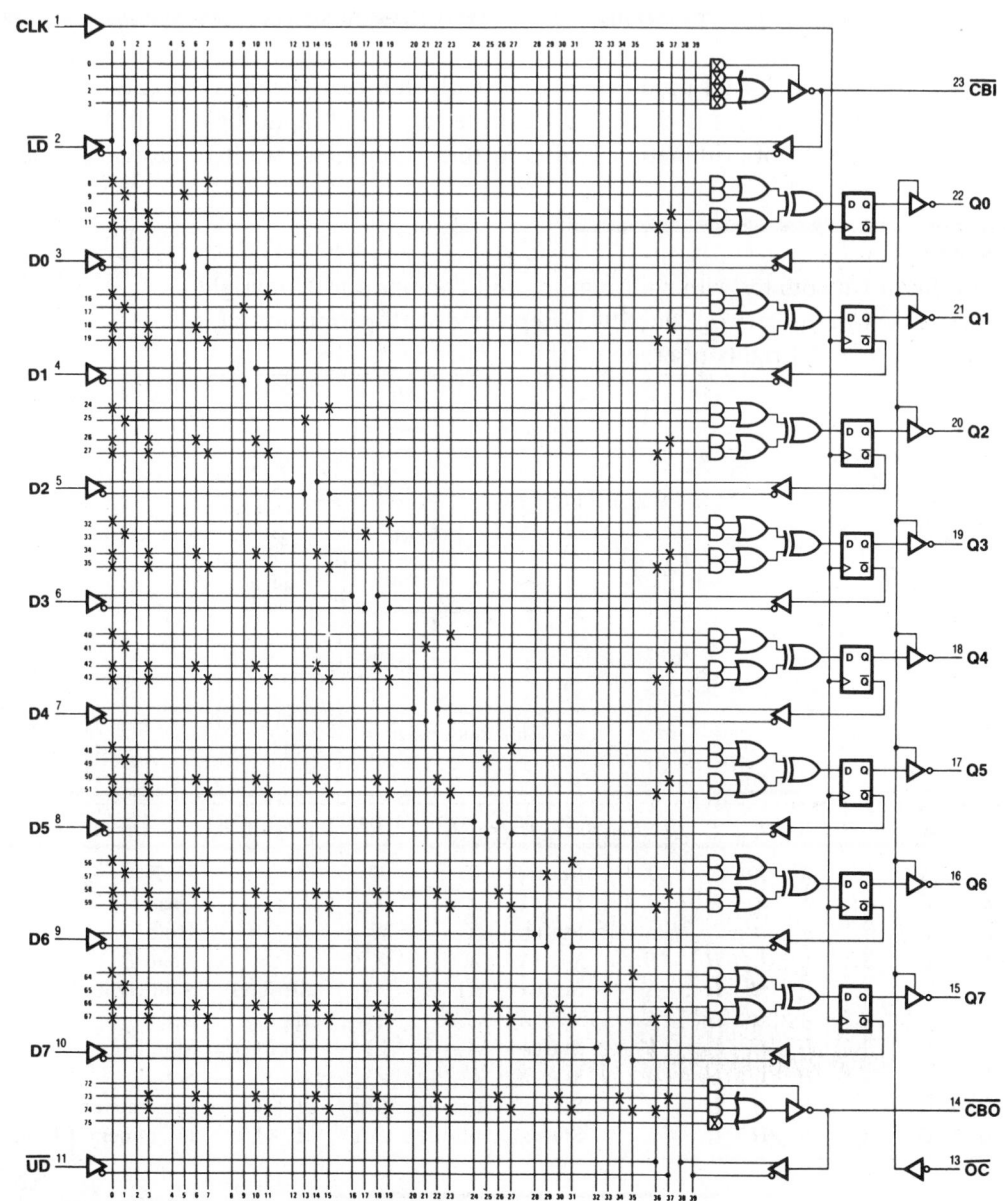

Figure 6-19

Implementation of an octal up/down counter on PAL20X8 (Copyright
© Advanced Micro Devices, Inc., 1983. Reprinted with permission of
copyright owner. All rights reserved.)

Example 6–7 ▬▬▬▬▬▬▬▬▬▬▬▬▬▬▬▬▬▬▬▬▬▬▬▬▬▬▬▬

Design a 4-bit up/down counter with shift register and comparator.

This example is chosen to illustrate the use of input logic in PLAs to aid in the design of arithmetic circuits. In this example we want to design a general-

Figure 6–20

A 4-bit up/down counter with shift register and comparator (Copyright © Advanced Micro Devices, Inc., 1983. Reprinted with permission of copyright owner. All rights reserved.)

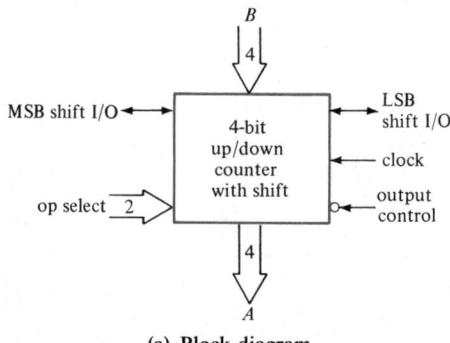

(a) Block diagram

$\overline{OC}$	CLK	CLR	I_2	I_1	I_0	B_3-B_0	$\overline{GE}$	$\overline{LE}$	$\overline{LIO}$	$\overline{RIO}$	A_3-A_0	Operation
H	X	X	X	X	X	X	Status	X	X	Z	Hi-Z
L	C	H	X	X	X	X	X	X	X	X	L	Clear
L	C	L	L	L	L	X	Status	X	X	A	Hold
L	C	L	L	L	H	B	X	X	X	X	B	Load B
L	C	L	L	H	L	X	Status	RI	A_0	SR(RIO)	Shift right
L	C	L	L	H	H	X	X	X	Z	Z	H	Set
L	C	L	H	L	L	X	Status	A_3	LI	SL(LIO)	Shift left
L	C	L	H	L	H	X	X	X	H	Z	H	Set
L	C	L	H	H	L	X	Status	C(out)	C(in)	$A+1$	Increment if C(in)
L	C	L	H	H	H	X	Status	B(out)	B(in)	$A-1$	Decrement if B(in)

Comparison	$\overline{GE}$	$\overline{LE}$
B is greater than A	L	H
B is equal to A	L	L
B is less than A	H	L

(b) Functions of control signals

purpose register that can be used as an up/down counter, as a shift register, and as a comparator between the stored operand and the input operand. A block diagram of the device and a table of control functions are given in Figure 6–20. The input to this device is a 4-bit data bus B, and the output is a 4-bit output bus A. The output is normally stored in the internal flip-flops. There are two shift bits for shifting data bits into and out of both ends of the device. They are the $\overline{\text{LIO}}$ and RIO, for left-side input/output and right-side input/output. For arithmetic operations, the LIO line is also used to indicate any carry-out or borrow-out conditions. In addition, there are two output lines that report the result of comparisons, the $\overline{\text{GE}}$ and the $\overline{\text{LE}}$ lines. The $\overline{\text{GE}}$ line is asserted low when the input operand B is greater than or equal to the internal operand A. Likewise, the $\overline{\text{LE}}$ line is asserted low when the input operand B is less than or equal to the stored operand A. For control of the device, three control signals are used. The functions performed are shown in Figure 6–20(b).

Having defined the above operations, the next step is to write down the necessary input to each flip-flop based on all the input conditions and the desired instruction. The input conditions to the least significant flip-flop can be expressed as follows:

$$
\begin{aligned}
\overline{A}_0 = (&\overline{I}_2\overline{I}_1\overline{I}_0\overline{A}_0\overline{\text{CLR}} && \text{Hold } A_0 \\
+ &\overline{I}_2\overline{I}_1I_0\overline{B}_0\overline{\text{CLR}} && \text{Load } B_0 \\
+ &\overline{I}_2I_1\overline{I}_0\overline{A}_1\overline{\text{CLR}} && \text{Shift right} \\
+ &I_2\overline{I}_1\overline{A}_0\overline{\text{CLR}}) && \text{Hold} \\
\oplus & \\
(&I_2\overline{I}_1\overline{I}_0\overline{\text{RIO}} && \text{Shift left} \\
+ &I_2I_1\overline{I}_0\text{RIO} && \text{Increment if carry-in} \\
+ &I_2I_1I_0\text{RIO} && \text{Decrement if borrow-in} \\
+ &\text{CLR}) && \text{Clear}
\end{aligned}
$$

The expression and the corresponding explanation for bit A_0 are given above. The same logic can be applied to the other three bits. These are given below:

$$
\begin{aligned}
\overline{A}_1 = (&\overline{I}_2\overline{I}_1\overline{I}_0\overline{A}_1\overline{\text{CLR}} && \text{Hold } A_1 \\
+ &\overline{I}_2\overline{I}_1I_0\overline{B}_1\overline{\text{CLR}} && \text{Load } B_1 \\
+ &\overline{I}_2I_1\overline{I}_0\overline{A}_2\overline{\text{CLR}} && \text{Shift right} \\
+ &I_2I_1\overline{A}_1\overline{\text{CLR}}) && \text{Hold} \\
\oplus & \\
(&I_2\overline{I}_1\overline{I}_0\overline{A}_0 && \text{Shift left} \\
+ &I_2I_1\overline{I}_0A_0\text{RIO} && \text{Increment if carry-in} \\
+ &I_2I_1I_0\overline{A}_0\text{RIO} && \text{Decrement if borrow-in} \\
+ &\text{CLR}) && \text{Clear}
\end{aligned}
$$

$$\bar{A}_2 = (\bar{I}_2\bar{I}_1\bar{I}_0\bar{A}_2\overline{CLR} \qquad \text{Hold } A_2$$
$$+ \bar{I}_2\bar{I}_1 I_0\bar{B}_2\overline{CLR} \qquad \text{Load } B_2$$
$$+ \bar{I}_2 I_1\bar{I}_0\bar{A}_3\overline{CLR} \qquad \text{Shift right}$$
$$+ I_2 I_1\bar{A}_2\overline{CLR}) \qquad \text{Hold}$$
$$\oplus$$
$$(I_2\bar{I}_1\bar{I}_0\bar{A}_1 \qquad \text{Shift left}$$
$$+ I_2 I_1\bar{I}_0 A_1 A_0 RIO \qquad \text{Increment if carry-in}$$
$$+ I_2 I_1 I_0\ \bar{A}_1\bar{A}_0 RIO \qquad \text{Decrement if borrow-in}$$
$$+ CLR) \qquad \text{Clear}$$

$$\bar{A}_3 = (\bar{I}_2\bar{I}_1\bar{I}_0\bar{A}_3\overline{CLR} \qquad \text{Hold } A_3$$
$$+ \bar{I}_2\bar{I}_1 I_0\bar{B}_3\overline{CLR} \qquad \text{Load } B_3$$
$$+ \bar{I}_2 I_1\bar{I}_0\overline{LIO}\,\overline{CLR} \qquad \text{Shift right}$$
$$+ I_2 I_1\bar{A}_3\overline{CLR}) \qquad \text{Hold}$$
$$\oplus$$
$$(I_2\bar{I}_1\bar{I}_0\bar{A}_2 \qquad \text{Shift left}$$
$$+ I_2 I_1\bar{I}_0 A_2 A_1 A_0 RIO \qquad \text{Increment if carry-in}$$
$$+ I_2 I_1 I_0\bar{A}_2\bar{A}_1\bar{A}_0 RIO \qquad \text{Decrement if borrow-in}$$
$$+ CLR) \qquad \text{Clear}$$

For the two shift input/output lines, the expressions are much simpler:

$$\text{IF}(\bar{I}_2 I_1\bar{I}_0)\,RIO = A_0$$
$$\text{IF}(I_2)\,LIO = I_2\bar{I}_1\bar{I}_0 A_3 \qquad \text{Shift left out}$$
$$+ I_2 I_1\bar{I}_0 A_3 A_2 A_1 A_0 RIO \qquad \text{Carry-out}$$
$$+ I_2 I_1 I_0\bar{A}_3\bar{A}_2\bar{A}_1\bar{A}_0 RIO \qquad \text{Borrow-out}$$

The two status lines can also be obtained in the following expressions:

$$\text{LE} = (A_3\bar{B}_3) \qquad\qquad\qquad B_3 < A_3$$
$$+ (A_3 \bar{\oplus} B_3)(A_2\bar{B}_2) \qquad\qquad B_2 < A_2$$
$$+ (A_3 \bar{\oplus} B_3)(A_2 \bar{\oplus} B_2)(A_1\bar{B}_1) \qquad B_1 < A_1$$
$$+ (A_3 \bar{\oplus} B_3)(A_2 \bar{\oplus} B_2)(A_1 \bar{\oplus} B_1)(A_0\bar{B}_0) \qquad B_0 < A_0$$
$$+ (A_3 \bar{\oplus} B_3)(A_2 \bar{\oplus} B_2)(A_1 \bar{\oplus} B_1)(A_0 \bar{\oplus} B_0) \qquad B = A$$

$$\text{GE} = (\bar{A}_3 B_3) \qquad\qquad\qquad B_3 > A_3$$
$$+ (A_3 \bar{\oplus} B_3)(\bar{A}_2 B_2) \qquad\qquad B_2 > A_2$$
$$+ (A_3 \bar{\oplus} B_3)(A_2 \bar{\oplus} B_2)(\bar{A}_1 B_1) \qquad B_1 > A_1$$
$$+ (A_3 \bar{\oplus} B_3)(A_2 \bar{\oplus} B_2)(A_1 \bar{\oplus} B_1)(\bar{A}_0 B_0) \qquad B_0 > A_0$$
$$+ (A_3 \bar{\oplus} B_3)(A_2 \bar{\oplus} B_2)(A_1 \bar{\oplus} B_1)(A_0 \bar{\oplus} B_0) \qquad B = A$$

A number of PLAs can be used to realize these functions. Note that many of the PLAs with registered arithmetic logic outputs have input logic that seems to fit

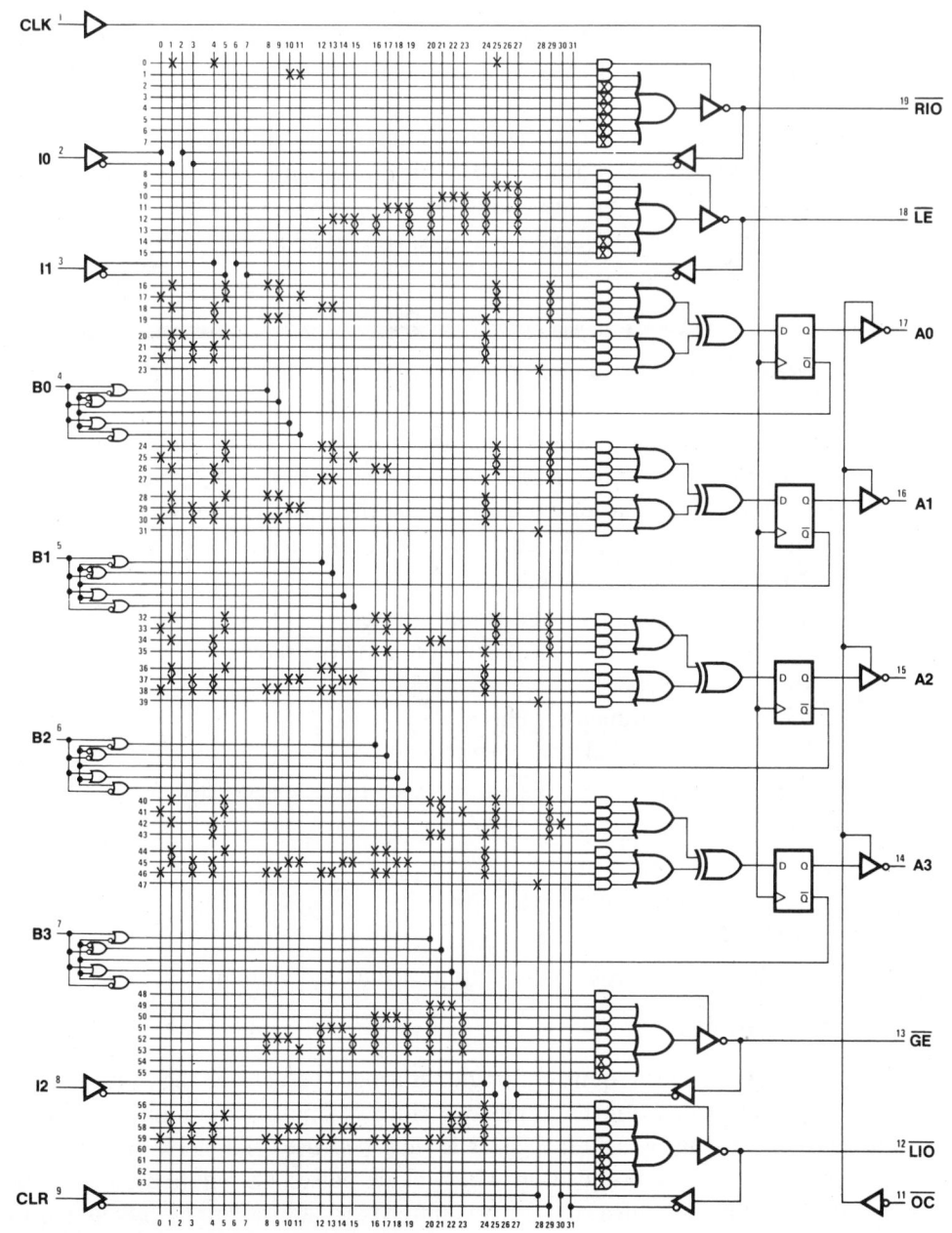

Figure 6–21

Implementation of a 4-bit up/down counter with shift register and comparator on PAL16X4 (Copyright © Advanced Micro Devices, Inc., 1983. Reprinted with permission of copyright owner. All rights reserved.)

what is needed here. In particular, the XOR functions and the complement functions of the input and stored variables can be conveniently obtained here. Hence, we will use the PAL16X4 PLA with registered arithmetic logic outputs to realize the required functions. This is shown in Figure 6–21. If the above functions were implemented using SSI circuit components, a large number of chips would be required. However, when using a PLA for implementation, only a single chip is needed to realize all the functions. (This example is taken from the *PAL Handbook*, 3rd ed., Monolithic Memories Inc., 1983.)

6.5

Applications of Logic Arrays

Logic arrays have been found to be useful in many applications. Because they have multiple-input and multiple-output lines, a frequent application of logic arrays is to combine and/or extend many of the MSI functions discussed in previous chapters. Some popular combinations and extensions include:

1. 4-bit shift register/comparator
2. 4-bit counter with two-input multiplexer
3. BCD-to-decimal decoder/bar graph display driver
4. Interface controller
5. 32-bit cyclic redundancy check
6. 8-bit error detection and correction

As examples of logic arrays, we will present three specific applications in more detail:

1. A barrel shifter
2. A binary-to-BCD converter
3. A traffic signal controller

We will examine each one of these applications individually.

A Barrel Shifter

Many digital systems require shifting operations. To achieve the shifting, usually shift registers are used if the number of shifts is small. If the number of shifts is large, then shifting by a shift register is slow because each clock pulse shifts the operand by only one bit position. Many clock pulses are required to accomplish a multiple shift. The idea of a barrel shifter is to shift as many bits as required within a single clock pulse, as shown in Figure 6–22(a). To achieve any number of shifts, all possible shifts are prewired in the shifter. Since the number of shifts is usually known beforehand and is provided as an input to a barrel shifter, the shifter merely chooses the data bit with the proper shift length, as shown in Figure 6–22(b).

Example 6–8

Design an octal registered barrel shifter.

The problem calls for an 8-bit shifter. This means that the maximun distance is 8 bits. In other words, the shift length can be coded as three control bits. The

Figure 6–22

A barrel shifter

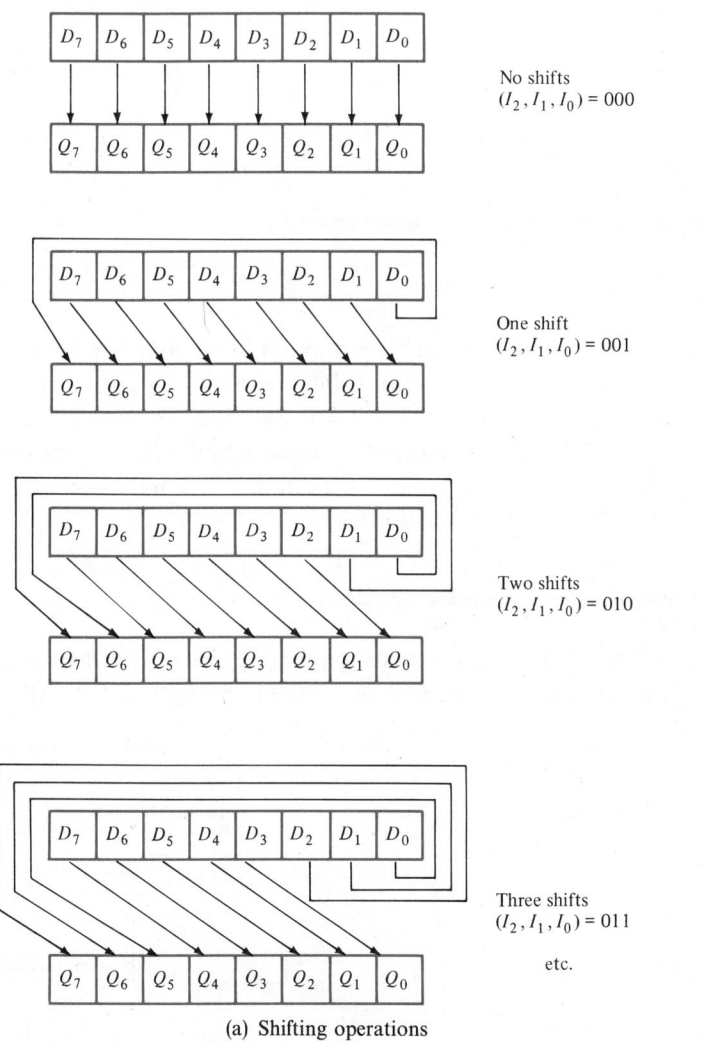

(a) Shifting operations

(*continued*)

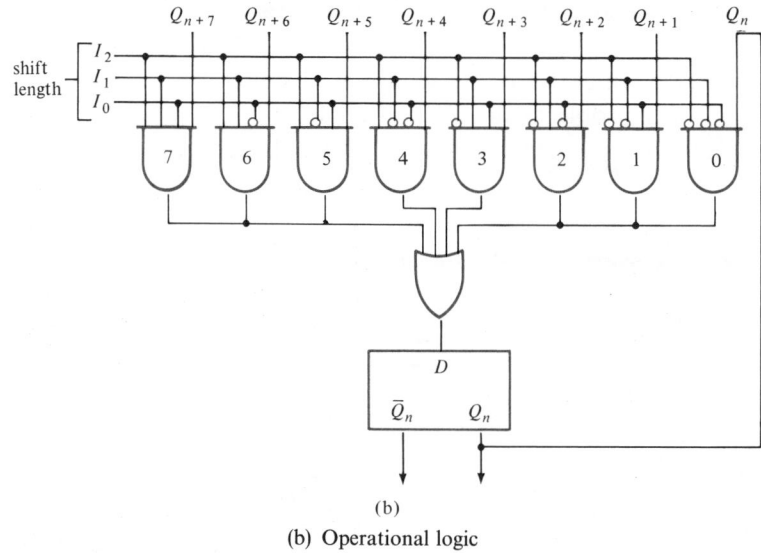

(b)

(b) Operational logic

Figure 6-22 Continued

barrel shifter can be represented in block diagram format as in Figure 6–23. The input to the shifter is an 8-bit operand that is strobed into the internal register whenever the enable signal $\bar{E}$ is asserted. The output of the shifter is an 8-bit bus controlled by the output control signal, $\overline{OC}$. There are three operation select lines and a clock input to the shifter. The operation select lines choose the number of shifts to be performed.

Figure 6-23

Block diagram of an octal shifter (Copyright © Advanced Micro Devices, Inc., 1983. Reprinted with permission of the copyright owner. All rights reserved.)

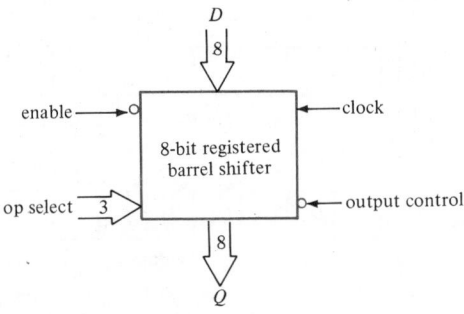

The inputs to each internal flip-flop are rather simple. In principle, any one of the decoded select lines ANDed with the proper data bit becomes one of the inputs. The inputs for $\bar{Q}_0$, the least significant bit, and $\bar{Q}_1$ are

$$\begin{aligned}
\bar{Q}_0 = \ & \bar{E}\bar{I}_2\bar{I}_1\bar{I}_0\bar{D}_0 && \text{Shift 0 places} \\
+ \ & \bar{E}\bar{I}_2\bar{I}_1 I_0\bar{D}_1 && \text{Shift 1 place} \\
+ \ & \bar{E}\bar{I}_2 I_1\bar{I}_0\bar{D}_2 && \text{Shift 2 places} \\
+ \ & \bar{E}\bar{I}_2 I_1 I_0\bar{D}_3 && \text{Shift 3 places} \\
+ \ & \bar{E} I_2\bar{I}_1\bar{I}_0\bar{D}_4 && \text{Shift 4 places} \\
+ \ & \bar{E} I_2\bar{I}_1 I_0\bar{D}_5 && \text{Shift 5 places} \\
+ \ & \bar{E} I_2 I_1\bar{I}_0\bar{D}_6 && \text{Shift 6 places} \\
+ \ & \bar{E} I_2 I_1 I_0\bar{D}_7 && \text{Shift 7 places} \\[6pt]
\bar{Q}_1 = \ & \bar{E}\bar{I}_2\bar{I}_1\bar{I}_0\bar{D}_1 && \text{Shift 0 places} \\
+ \ & \bar{E}\bar{I}_2\bar{I}_1 I_0\bar{D}_2 && \text{Shift 1 place} \\
+ \ & \bar{E}\bar{I}_2 I_1\bar{I}_0\bar{D}_3 && \text{Shift 2 places} \\
+ \ & \bar{E}\bar{I}_2 I_1 I_0\bar{D}_4 && \text{Shift 3 places} \\
+ \ & \bar{E} I_2\bar{I}_1\bar{I}_0\bar{D}_5 && \text{Shift 4 places} \\
+ \ & \bar{E} I_2\bar{I}_1 I_0\bar{D}_6 && \text{Shift 5 places} \\
+ \ & \bar{E} I_2 I_1\bar{I}_0\bar{D}_7 && \text{Shift 6 places} \\
+ \ & \bar{E} I_2 I_1 I_0\bar{D}_0 && \text{Shift 7 places}
\end{aligned}$$

etc.

The expressions for the rest of the bits follow the same pattern and can be written down in the same pattern.

From the above expressions, we observe a total of 12 inputs: 8 data bits, 3 shift select bits, and an enable signal. There are eight minterms for each bit, giving a total of 64 minterms. Of course, there are eight output variables. A PAL20R8 is ideal for this application. The final implementation of the barrel shifter is shown in Figure 6–24. (This example is taken from the *PAL Handbook*, 3rd ed., Monolithic Memories Inc., 1983.)

A Binary-to-BCD Converter

In addition to arithmetic operations, binary-to-BCD and BCD-to-binary conversions are also important operations that are frequently performed. We have already presented the conversions from binary to BCD and from BCD to binary using MSI components. Though the operation is fast and efficient, a large number of MSI chips is used to perform the conversion when the number of binary bits is large. In fact, the number of MSI components required grows on the order of n^2, where n is the number of bits. This is often the case for parallel implementation. Using the concept of PLAs, a serial version of the conversion can be effectively and efficiently implemented.

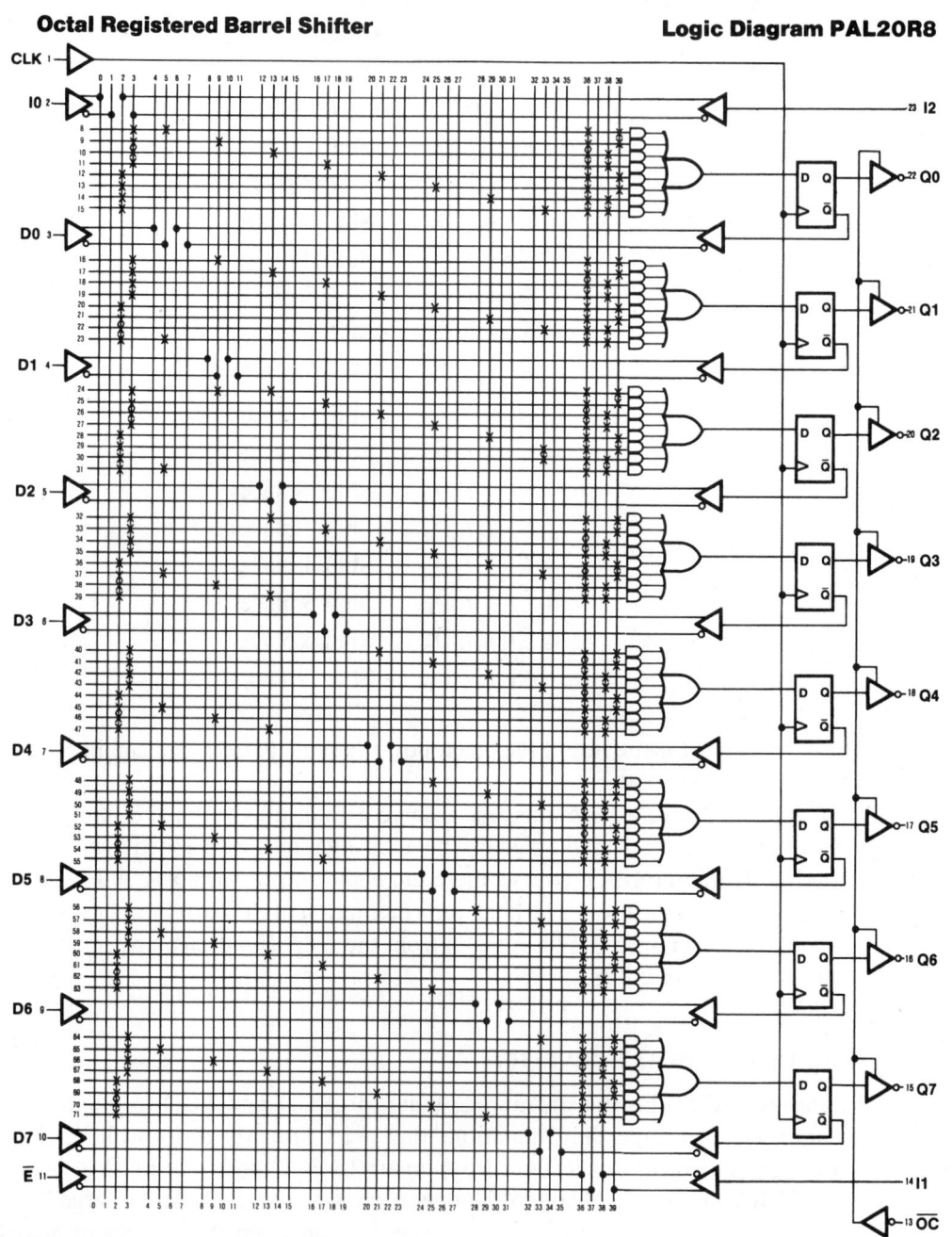

Figure 6–24

Implementation of an 8-bit barrel shifter on PAL20R8 (Copyright ©
Advanced Micro Devices, Inc., 1983. Reprinted with permission of
copyright owner. All rights reserved.)

In a serial conversion, the binary bits are shifted to the left one at a time. As the bits are shifted out of the binary number, they are shifted into a BCD number as shown in Figure 6–25. The shifting is equivalent to multiplying the BCD number by 2. Initially, the BCD number is 0. First, the most significant bit of the binary is shifted into the BCD digit. This bit is then stored in the least significant digit. Then the least significant digit is multiplied by 2 and added to the new bit shifted in from the right, the binary number. As long as the number in each BCD digit is less than or equal to 4, multiplication by 2 with the possible carry-in bit will result in a valid decimal digit. However, if the BCD digit is greater than or equal to 5, then multiplication by 2 must result in a carry-out bit in the decimal system. However, shifting the BCD digit to the left by 1 bit may or may not result in a carry-out. If no carry-out bit is generated, then an invalid code is obtained. If a carry-out bit is generated, then the residual digit is still invalid because a carry-out in binary takes away 16 while a carry-out in the decimal system takes away only 10. This is a discrepancy of 6. Hence, whenever the BCD digit is greater than or equal to 5, a 3 must be added before the left shift to compensate for the discrepancy. This modification must be performed whenever further shifts are performed. When all shifts are performed, i.e., when all bits in the binary number have been shifted into the BCD number, then no further modifications are needed in this last step, even if the BCD digits are greater than or equal to 5.

The block diagram of a binary-to-BCD converter is shown in Figure 6–26(a). The input to the converter is the previous BCD digit and the carry-in bit. The output of the converter is two BCD digits. An additional clear signal, CLR, is used to clear all output pins before the conversion begins. The function of the

Figure 6–25
Binary-to-BCD conversion

		1011011	Original number
	0001	011011	Shift left
	0010	11011	Shift left
	(0101)	1011	Shift left
0000	(1011)	011	Shift left
0001	0001	011	Correction
0010	0010	11	Shift left
0100	(0101)	1	Shift left
1000	(1011)		Shift left
1001	0001		Correction

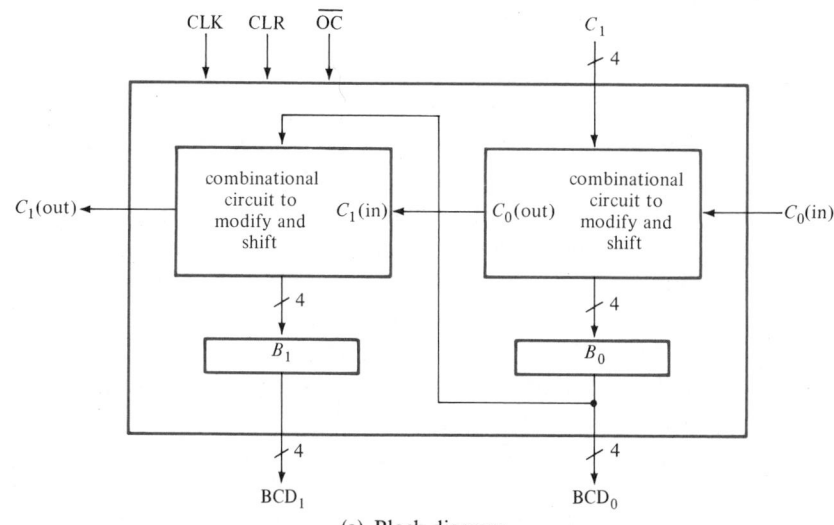

(a) Block diagram

| Present State | | | Next State |
B_3-B_0	$C(in)$	$C(out)$	B_3-B_0
0000	0	0	0000
0000	1	0	0001
0001	0	0	0010
0001	1	0	0011
0010	0	0	0100
0010	1	0	0101
0011	0	0	0110
0011	1	0	0111
0100	0	0	1000
0100	1	0	1001
0101	0	1	0000
0101	1	1	0001
0110	0	1	0010
0110	1	1	0011
0111	0	1	0100
0111	1	1	0101
1000	0	1	0110
1000	1	1	0111
1001	0	1	1000
1001	1	1	1001

(b) Table showing carry generation and correction of invalid decimal codes

Figure 6-26 A binary-to-BCD converter

converter is mostly used to modify the BCD digits and to shift the digits. The modification and the shifting operation can be realized in the form of a truth table as shown in Figure 6–26(b). The truth table has five input variables and five output variables. The input variables are the four BCD bits before the shift and modification and the incoming carry-in bit. The output variables are the four modified and shifted bits and the associated carry-out bit.

From the truth table, we can realize the output pins in the following expressions:

$$B_{00} = \overline{CLR}C(\text{in}) + \overline{CLR}C_{13} + \overline{CLR}C_{12}C_{10} + \overline{CLR}C_{12}C_{11}$$
$$B_{01} = \overline{CLR}\,\overline{B}_{03}\overline{B}_{02}B_{00} + \overline{CLR}B_{03}\,\overline{B}_{00} + \overline{CLR}B_{02}B_{01}\overline{B}_{00}$$
$$B_{02} = \overline{CLR}B_{01}B_{00} + \overline{CLR}\,\overline{B}_{02}B_{01} + \overline{CLR}B_{03}\,\overline{B}_{00}$$
$$B_{03} = \overline{CLR}B_{02}\,\overline{B}_{01}\overline{B}_{00} + \overline{CLR}B_{03}B_{00}$$

The expressions for the second BCD digit are similar to those for the first BCD digit given above. To realize these eight functions, we observe that we need a PAL with six input variables, 24 minterms, and eight registered output functions. The PAL16R8 fits this description. Using the PAL16R8, these eight functions are implemented as shown in Figure 6–27.

One advantage of this PAL implementation of the converter is that the converter is expandable to accommodate any input operand size. Since each converter produces two BCD digits, additional digits are obtained simply by cascading the PALs together in a linear fashion as shown in Figure 6–28. The original binary number is shifted only into the first converter chip, which accounts for the two least significant digits. The second BCD digit and the carry-out bit from the first PAL are then connected to the second PAL, which produces the next two BCD digits. These outputs are further connected to other PALs for more digits if needed. Note that the number of PALs needed grows linearly as a function of the word size. This is a significant improvement over the MSI technique because the number of chips in the latter scheme grows as the square of the word size. (This example is taken from the *PAL Handbook*, 3rd ed., Monolithic Memories Inc., 1983.)

A Traffic Signal Controller

In this example we will use the design of a traffic signal controller to illustrate the design techniques of sequential circuits using PALs. The scenario for a simple traffic light is shown in Figure 6–29(a). There are two sensors for each of the two one-way streets. These are labeled $\overline{SENA}$ and $\overline{SENB}$. There are also two traffic lights: $\overline{REDA}$, $\overline{YELA}$, and $\overline{GRNA}$ associated with $\overline{SENA}$ for one of the streets, and $\overline{REDB}$, $\overline{YELB}$, and $\overline{GRNB}$ associated with $\overline{SENB}$ for the other street. The object of the controller is to stop the traffic on the other street by the red light and to provide the green light to the street when its associated sensor is tripped. We can easily identify the input and output parameters of this traffic

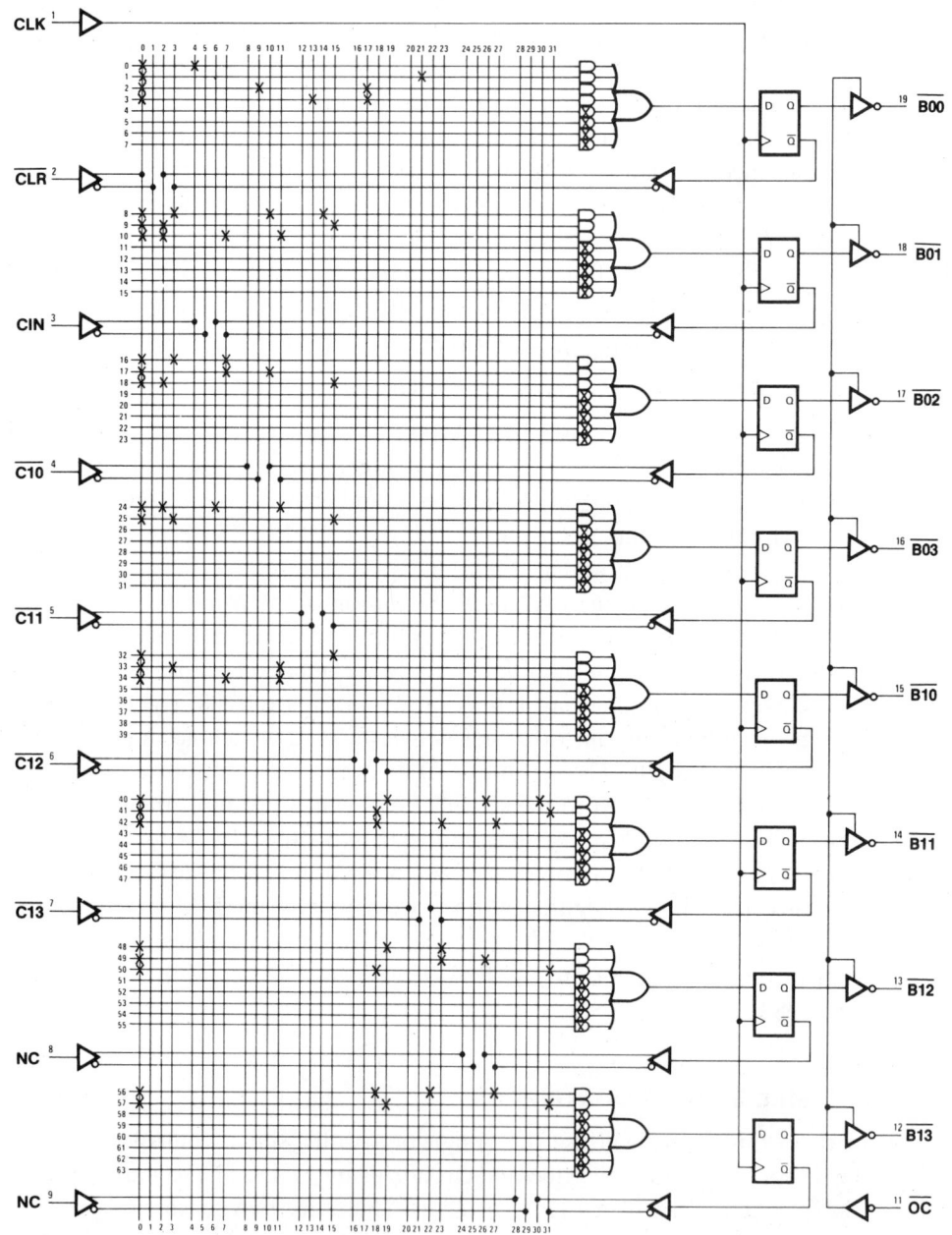

Figure 6–27
Binary-to-BCD conversion using PAL16R8 (Copyright © Advanced
Micro Devices, Inc., 1983. Reprinted with permission of copyright
owner. All rights reserved.)

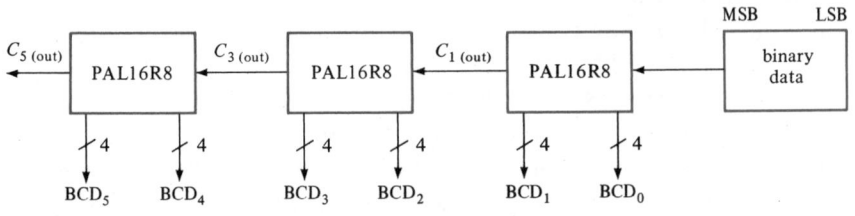

Figure 6–28
Configuration for larger word size in binary-to-BCD conversion

Figure 6–29
Scenario and block diagram for the traffic signal controller (Copyright ©
Advanced Micro Devices, Inc., 1983. Reprinted with permission of
copyright owner. All rights reserved.)

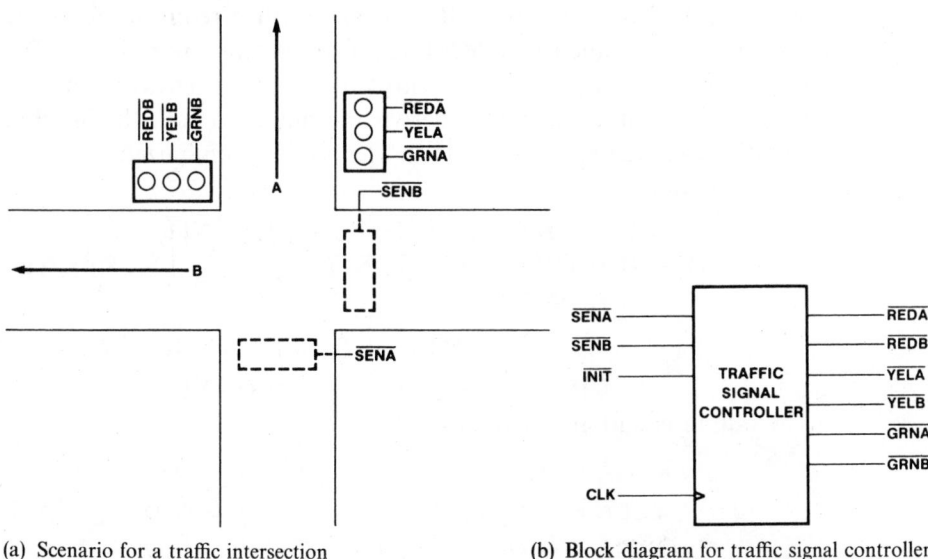

(a) Scenario for a traffic intersection   (b) Block diagram for traffic signal controller

controller. A general block diagram is shown in Figure 6–29(b). The inputs to the controller are the two sensors and an initialization signal to make sure that the lights are synchronized in the proper sequence to start with. The outputs of the controller are simply the six signals for the two traffic lights.

The relationship between the input sensory data and the output signals can be represented using the state transition diagram. This must be so because a certain progression must be followed when switching the lights. For example, one must be careful not to switch from green to red immediately when the sensor is tripped. The state transition diagram for the traffic controller is shown in Figure 6–30(a), and the corresponding state transition table is shown in Figure 6–30(b). We will assume that when the initialization signal comes, the traffic controller will be reset to the initial state s_0. This is the default condition, with the green light on street A and the red light on street B. Since there are two sensors, four possibilities may result. If sensor A is continually tripped and sensor B is not, the light should remain green on street A. If sensor B is tripped but not sensor A, then the controller should honor the request for street B quickly and enter a short green cycle for street A. If no one trips any sensors or if both sensors are tripped, then the controller will enter a long cycle for street A. Once the controller enters a cycle, either long or short, all sensory inputs are ignored until the cycle is finished, which happens when the other street has the green light. The same light-switching logic applies for the other street.

If we now define a long green cycle to be four clock cycles and a short green cycle to have only one clock cycle, then a total of 11 states are needed. These 11 states are coded in the usual BCD manner, thus requiring four flip-flops. The state assignments and the coded state transition table are shown in Figure 6–30(c). Choosing D flip-flops for our controller, we can progress in the usual manner for sequential circuit design to derive and minimize both the flip-flop excitation equations and the output equations. The partially minimized next-state equations are as follows:

$$Q_3 = Q_3\bar{Q}_0\overline{INIT} + Q_3\bar{Q}_2\overline{INIT} + \bar{Q}_3Q_2Q_0\overline{INIT}$$
$$Q_2 = Q_1Q_0\overline{INIT} + Q_2\bar{Q}_1\bar{Q}_0\overline{INIT} + \bar{Q}_3\bar{Q}_1\bar{Q}_0\text{SENA SENB}\overline{INIT}$$
$$Q_1 = Q_1\bar{Q}_0\overline{INIT} + \bar{Q}_2\bar{Q}_1Q_0\overline{INIT}$$
$$Q_0 = \bar{Q}_0\overline{\text{SENA SENB}}\,\overline{INIT} + \bar{Q}_0\text{SENA SENB}\overline{INIT}$$
$$+ Q_1\bar{Q}_0\overline{INIT} + Q_2\bar{Q}_0\overline{INIT} + Q_3\bar{Q}_0\overline{INIT}$$

The output equations for the lights are

$$\text{REDA} = Q_3 \qquad\qquad \text{REDB} = \bar{Q}_3$$
$$\text{YELA} = \bar{Q}_3Q_2Q_0 \qquad\qquad \text{YELB} = Q_3Q_2Q_0$$
$$\text{GRNA} = \bar{Q}_3\bar{Q}_2 + \bar{Q}_3Q_1\bar{Q}_0 \qquad \text{GRNB} = Q_3\bar{Q}_2 + Q_3\bar{Q}_1\bar{Q}_0$$

The equations above indicate that there are only three input variables, 21 minterms, and six output variables. The PAL16R4 appears to be sufficient for

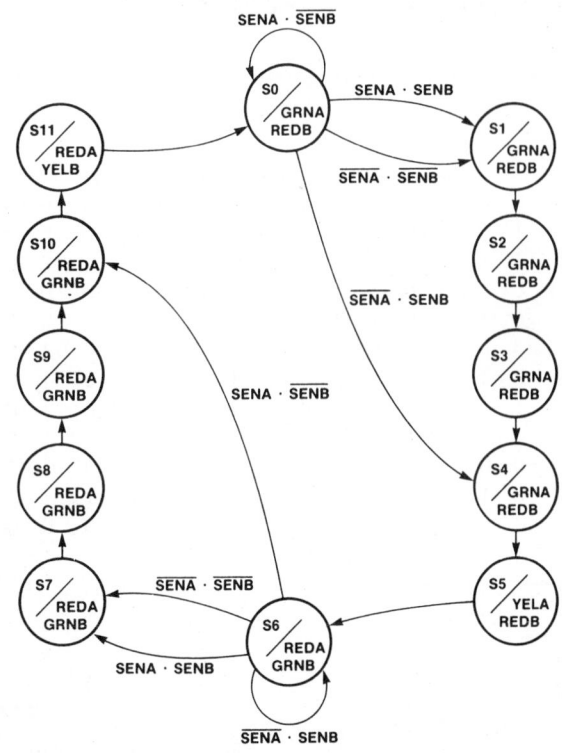

(a) State transition diagram

	Next State	Input				Output					
Current State		SENA, SENB				REDA	YELA	GRNA	REDB	YELB	GRNB
		00	01	10	11						
S_0		S_1	S_4	S_0	S_1	0	0	1	1	0	0
S_1		S_2	S_2	S_2	S_2	0	0	1	1	0	0
S_2		S_3	S_3	S_3	S_3	0	0	1	1	0	0
S_3		S_4	S_4	S_4	S_4	0	0	1	1	0	0
S_4		S_5	S_5	S_5	S_5	0	0	1	1	0	0
S_5		S_6	S_6	S_6	S_6	0	1	0	1	0	0
S_6		S_7	S_6	S_{10}	S_7	1	0	0	0	0	1
S_7		S_8	S_8	S_8	S_8	1	0	0	0	0	1
S_8		S_9	S_9	S_9	S_9	1	0	0	0	0	1
S_9		S_{10}	S_{10}	S_{10}	S_{10}	1	0	0	0	0	1
S_{10}		S_{11}	S_{11}	S_{11}	S_{11}	1	0	0	0	0	1
S_{11}		S_0	S_0	S_0	S_0	1	0	0	0	1	0

(b) State transition table

Figure 6–30
State transition diagram and table for the traffic signal controller
(Copyright © Advanced Micro Devices, Inc., 1983. Reprinted with
permission of copyright owner. All rights reserved.)

(*continued*)

Current State				Input/Next State								Outputs					
				00				01				10			11			
0 0 0 0	0 0 0 1	0 1 0 0	0 0 0	0 0 0 0 1	0 0 1 1 0 0
0 0 0 1	0 0 1 0	0 0 1 0	0 0 1	0 0 0 1 0	0 0 1 1 0 0
0 0 1 0	0 0 1 1	0 0 1 1	0 0 1	1 0 0 1 1	0 0 1 1 0 0
0 0 1 1	0 1 0 0	0 1 0 0	0 1 0	0 0 1 0 0	0 0 1 1 0 0
0 1 0 0	0 1 0 1	0 1 0 1	0 1 0	1 0 1 0 1	0 0 1 1 0 0
0 1 0 1	1 0 0 0	1 0 0 0	1 0 0	0 1 0 0 0	0 1 0 1 0 0
1 0 0 0	1 0 0 1	1 0 0 0	1 1 0	0 1 0 0 1	1 0 0 0 0 1
1 0 0 1	1 0 1 0	1 0 1 0	1 0 1	0 1 0 1 0	1 0 0 0 0 1
1 0 1 0	1 0 1 1	1 0 1 1	1 0 1	1 1 0 1 1	1 0 0 0 0 1
1 0 1 1	1 1 0 0	1 1 0 0	1 1 0	0 1 1 0 0	1 0 0 0 0 1
1 1 0 0	1 1 0 1	1 1 0 1	1 1 0	1 1 1 0 1	1 0 0 0 0 1
1 1 0 1	0 0 0 0	0 0 0 0	0 0 0	0 0 0 0 0	1 0 0 0 1 0

State	Q_3	Q_2	Q_1	Q_0
S_0	0	0	0	0
S_1	0	0	0	1
S_2	0	0	1	0
S_3	0	0	1	1
S_4	0	1	0	0
S_5	0	1	0	1
S_6	1	0	0	0
S_7	1	0	0	1
S_8	1	0	1	0
S_9	1	0	1	1
S_{10}	1	1	0	0
S_{11}	1	1	0	1

Input = SENA, SENB
Current/Next State = $Q_3, Q_2, Q_1, Q_0/Q_{3+}, Q_{2+}, Q_{1+}, Q_{0+}$
Output = REDA, YELA, GRNA, REDB, YELB, GRNB

(c) State assignment no. 1/transition table

Figure 6–30 Continued

this purpose. These equations are implemented on the PAL16R4 and shown in Figure 6–31.

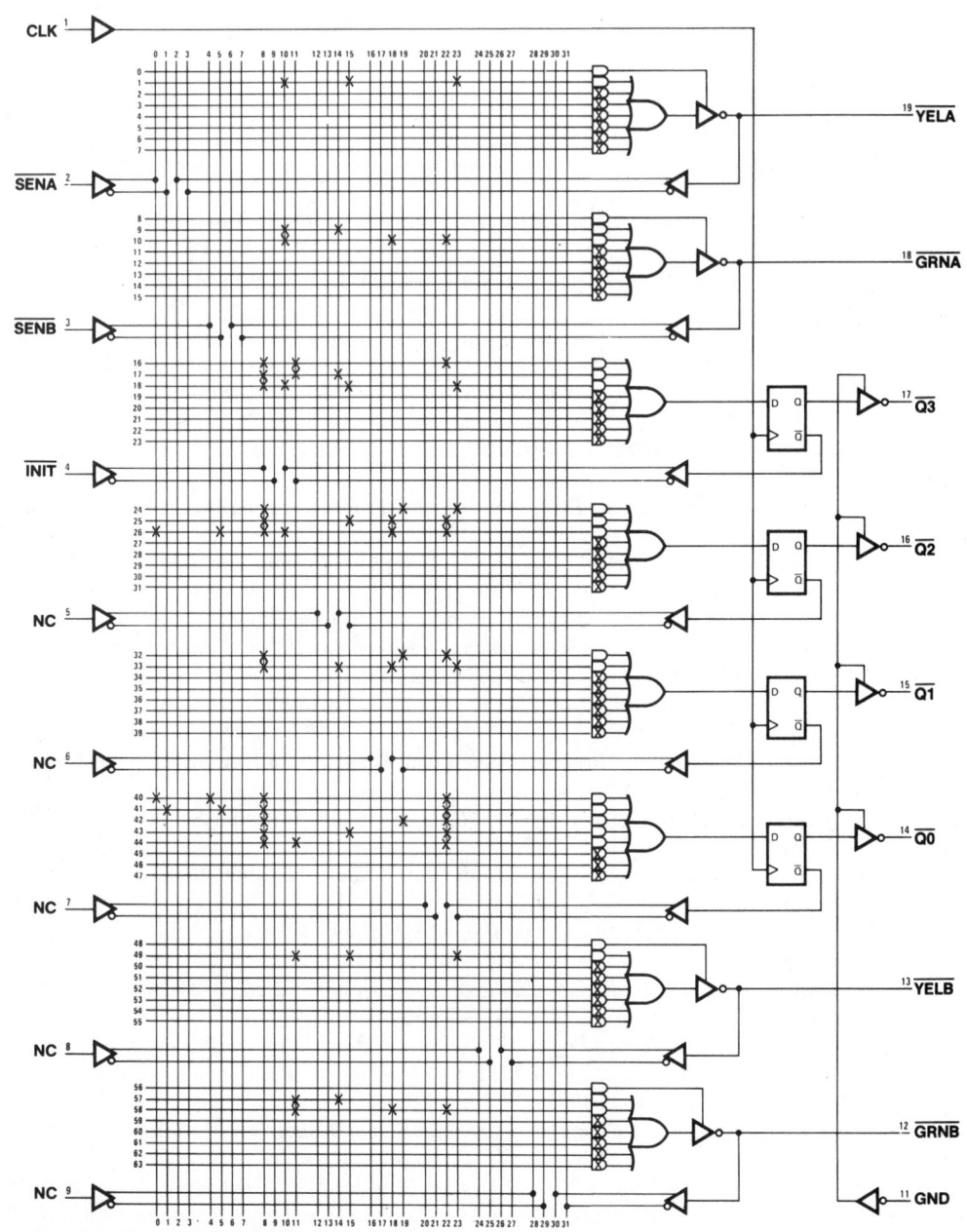

Figure 6–31
Implementation of the traffic signal controller using PAL16R4 (Copyright ©
Advanced Micro Devices, Inc., 1983. Reprinted with permission of copyright
owner. All rights reserved.)

References

1. Advanced Micro Devices. *Programmable Array Logic Handbook.* Advanced Micro Devices, Sunnyvale, Calif., 1984.
2. Becher, William D. *Logical Design Using Integrated Circuits.* Hayden Book Co. Rochelle Park, N.J., 1977.
3. Booth, Taylor L. *Introduction to Computer Engineering Hardware and Software Design.* John Wiley & Sons, New York, 1984.
4. Boyce, Jefferson C. *Digital Logic Operation and Analysis.* Prentice-Hall, Englewood Cliffs, N.J., 1982.
5. Comer, David J. *Digital Logic and State Machine Design.* Holt, Rinehart and Winston, New York, 1984.
6. Greenfield, Joseph D. *Practical Digital Design Using ICs*, 2d ed. John Wiley & Sons, New York, 1983.
7. Hayes, John P. *Digital System Design and Microprocessors.* McGraw-Hill, New York, 1984.
8. Johnson, E. L., and M. A. Karim. *Digital Design, A Pragmatic Approach.* PWS Publishers, Boston, 1987.
9. Lewin, Morton H. *Logic Design and Computer Organization.* Addison-Wesley, Reading, Mass., 1983.
10. Mano, M. Morris. *Digital Logic and Computer Design.* Prentice-Hall, Englewood Cliffs, N.J., 1979.
11. Mano, M. Morris. *Digital Design.* Prentice-Hall, Englewood Cliffs, N.J., 1984.
12. McCluskey, E. J. *Logic Design Principles with Emphasis on Testable Semicustom Circuits.* Prentice-Hall, Englewood Cliffs, N.J., 1986.
13. Monolithic Memories. *PAL Handbook*, 3d ed. Monolithi Memories, Santa Clara, Calif., 1983.
14. Monolithic Memories. *Programmable Logic Handbook*, 4th ed. Monolithic Memories, Santa Clara, Calif., 1985.
15. National Semiconductor Corp. *CMOS Data Book.* National Semiconductor, Santa Clara, Calif., 1981.
16. National Semiconductor Corp. *Logic Data Book.* National Semiconductor, Santa Clara, Calif., 1981.
17. National Semiconductor Corp. *PAL Data Book.* National Semiconductor, Santa Clara, Calif., 1982.
18. Roth, Charles H., Jr. *Fundamentals of Logic Design*, 3d ed. West Publishing Co., St. Paul, Minn., 1985.
19. Signetics Corp. *Signetics Logic—TTL Data Manual.* Signetics, Sunnyvale, Calif., 1978.
20. Texas Instruments, Inc. *The TTL Data Book for Design Engineers*, 2d ed. Texas Instruments, Dallas, Tex., 1976.
21. Ware, W. H. *Digital Computer Technology and Design*, Vol. 1. John Wiley & Sons, New York, 1963.
22. Wilkinson, Barry. *Digital System Design.* Prentice-Hall, Englewood Cliffs, N.J., 1987.
23. Williams, Gerald E. *Digital Technology.* Science Research Associates, Chicago, 1977.
24. Winkel, David, and Franklin Prosser. *The ART of Digital Design.* Prentice-Hall, Englewood Cliffs, N.J., 1980.

Problems

1. Design a data converter using a PAL for the following functions:
   (a) ASCII-to-EBCDIC converter
   (b) BCD-to-Gray converter
   (c) An 8-bit binary-to-decimal converter
   (d) A BCD-to-binary converter

2. Repeat Problem 1 but design with a PLA.

3. Design a universal BCD code converter using either a PLA or a PAL. There are eight inputs. Four of the inputs correspond to the four BCD bits. Another two inputs give the input format, and the last two inputs give the output format. The input or output format is as follows:

0 0	8-4-2-1 BCD
0 1	Excess-3
1 0	Gray
1 1	Modified Gray

4. Redo Example 6–2 using a PLA and using SSI chips. Compare the results.

5. Redo Example 6–4 using a PAL and using SSI chips. Compare the results.

6. Design a square-root converter with 10 input bits and 5 output bits. Use another output to indicate any error conditions.

7. Using PLA or PAL, implement the following random logic:
   (a) $x = abc + \bar{a}\bar{b}e$
       $y = ab\bar{d}\bar{e} + \bar{a} + \bar{d}$
       $z = \bar{b}\bar{d}\bar{e} + bd$
   (b) $F_1 = x\bar{y}z + w\bar{y}\bar{z} + wz$
       $F_2 = wxyz + wx$
       $F_3 = wxyz + wxy + xyz$
       $F_4 = wz$
       $F_5 = xyz + wz$

8. Design a 6-bit counter using a PLS.

9. Redo Example 6–5 using SSI chips and compare the results.

10. Using a PLS, design a combinational lock that responds to a three-octal-digit sequence such as 5, 2, 7.

11. If the octal digits in the combinational lock can be preset and stored in a latch, design the circuitry using a PLS.

12. Design a 4-bit full adder using a PLA.

13. Draw a flowchart of a computer program that would automatically generate the PLA design when the truth tables of the output functions are given in terms of the input variables.

Fundamentals of Microprocessors

7

In this chapter we present a brief overview of a generic microprocessor system, including what the principal components in the processor are and how these components are connected to form a working system. The first part of the chapter deals with the register transfer language, which we will use to describe operations within the processor. The specifications of operations, instructions, and even the processor itself can be greatly simplified using the register transfer language. We next examine the central processing unit (processor) and discuss its operation and relationship with memory and the outside world. We will describe the memory and the input/output devices briefly since they will be presented with greater detail in later chapters.

In previous chapters we discussed design techniques for small-scale integration (SSI) components, medium-scale integration (MSI) components, and large-scale integration (LSI) components. As solid-state technology advanced, LSI components evolved into very-large-scale integration (VLSI) components, packing more and more equivalent gates into the same chip size. We have seen that LSI components are specific functional units that are designed for particular purposes. SSI and MSI components function like logic that holds (or connects) the LSI components together. As circuits become more and more complex, the number of LSI chips increases accordingly. It seems to be an attractive solution to combine the LSI components and their associated control circuits in one package. This is in essence the underlying motivation for the development of microprocessors. Beginning in the early 1970s, microprocessors have been designed, fabricated, and used as principal components in many digital systems. General concepts of microprocessor systems are discussed in this chapter.

7.1

Microprocessor Evolution

Since the announcement of the Intel 4004, which can be considered the beginning of the microprocessor industry, there have been three distinct generations of microprocessor development. The 8-bit microprocessor is often referred to as a *first-generation microprocessor*. The 16-bit microprocessor can be considered a *second-generation microprocessor*. The 32-bit microprocessor, naturally, represents the *third generation* of microprocessor development. Third-generation machines are outside the scope of this book. Some of the differences among the three microprocessor generations are shown in Figure 7–1.

The *first-generation microprocessors* can be characterized as 8-bit machines capable of supporting an array of peripherals through an operating system. These 8-bit systems, in general, have 8-bit data registers and 16-bit address registers both internally and externally. Examples of first-generation microprocessors are the Intel I8085, Zilog Z80, and Motorola MC6800.

The *second-generation microprocessors* refer primarily to 16-bit machines with an internal data path that is 16 bits wide or more. Externally, the data path can be either 8 bits wide, as in the Intel 8088 or Motorola MC68008, or 16 bits wide, as in the Intel 8086, Zilog Z8000, or Motorola MC68000. In addition to doubling the width of the data path, the second-generation microprocessors are designed to support more complex operating system requirements.

The *third-generation microprocessors* refer to machines with 32-bit data paths both internally and externally. These chips are sometimes called *micromainframes* because the complexity of their designs can be compared to the complexity of mainframes. The performance levels of the third-generation microprocessors are

	Microprocessor Generation		
	First	Second	Third
Word length (bits)	8	16	32
Data operands (bits)	8/16	8/16/32	32
Data bus width (bits)	8	8/16	32
Address bus width (bits)	16	24	32
Arithmetic operations	+, −	+, −, *, /	+, −, *, /
String operations	No	No	Yes
Virtual memory support	No	No	Yes
Examples	8080/8085	8086/80286	80386
	6800	68000	68030
	Z80		

Figure 7–1

Characteristics of three generations of microprocessor technology

comparable to current high-performance minicomputers, yet their costs are substantially lower. Examples of these chips include the Intel 80386, Motorola MC68030, Zilog Z80000, National Semiconductor NS32032, and many others.

7.2

Register Transfer Language (RTL)

To more precisely describe the operations of a computer system, a *register transfer language* (RTL) can be used. The underlying principle of a digital system and a microprocessor system is to transfer data from one place to another. As the data are moved, they are being manipulated and transformed. In fact, most processors operate by moving data around. The data may be instructions or operands. As the operands are moved around, the necessary data manipulation operations are performed. As instructions are moved around, the program is executed. Therefore, the register transfer language is used to describe the movement of data around the processor.

The basic format of the RTL is rather simple. There are three major parts in an RTL program:

Identification
Declarations
Operations

The *identification* part is mainly used to collectively denominate the sequence of operations to be performed. The *declarations* specify the type of hardware used in the program segment. The *operations* are the actual coding of the data movement involved in the sequence.

Identification

In the identification, we give a name to a group of operations that accomplishes a certain task. For example, we may have a sequence of operations to multiply two numbers together or we may have another sequence to shift an operand by a number of bits. Each of these sequences is given a sequence name in the format

```
SEQUENCE name
```

where "SEQUENCE" is the keyword and "name" is the given name.

Declarations

The next part of the program following the identification is the declaration. In a way, declarations are used to identify the characteristics of the hardware. In most cases, the design is on the register level and specifications follow on the MSI level. Typical hardware components and their associated specifications for registers, counters, and arithmetic logic units (ALUs) are as follows:

1. Registers
   (a) Types:
       - Latching register
       - Shifting register
       - Counter register
   (b) Input:
       - Serial
       - Parallel
   (c) Output:
       - Single rail
       - Double rail
   (d) Length:
       - n bits
2. Counters
   (a) Types:
       - Binary or decimal
       - Synchronous or asynchronous
   (b) Input:
       - Synchronous or asynchronous load
   (c) Output:
       - Single rail
       - Double rail
       - Zero detect

    (d) Length:
       • n bits
3. Adders
    (a) Types:
       • Ripple carry or parallel carry
       • Binary or decimal
    (b) Input:
       • Input carry
    (c) Output:
       • Output carry
       • Overflow detection
    (d) Length:
       • n bits

The declaration is of utmost importance, because it actually specifies the hardware to be used to execute the sequence of operations. The results of many operations may be different depending on the hardware-associated parameters. For example, the length of a register determines the maximum size of the numbers allowed, and care must be exercised to ensure that an overflow condition does not occur during operations.

Operations

The last part of the sequence is the list of operations. These operations are elementary and typically involve data movements or manipulations. The general format for an operation can be written simply as

```
IF condition true then z ← f(X = {x₁, x₂, x₃, ..., xₙ})
```

where z is the output operand, f is the operation, and X is the input operand vector composed of many input operands, x_1, x_2, etc. The back arrow denotes replacement. Hence the above statement conveys the idea that when the stated condition is met, perform operation f on the input operands X and put the results in z, replacing what was in z before.

    Since each operation performs only a small part of the needed task, sometimes a register-level operation is referred to as a *microoperation*. There are a number of microoperations within a processor, including:

    Data transfer operations
    Data manipulation operations
    Conditional operations

Data transfer operations deal with data movement from one part of the processor to another. *Data manipulation operations* transform data from one form to another. *Conditional operations* are used for testing the data so that subsequent program execution may be altered based on the test results.

7.3

Data Transfer Operations in RTL

Data transfer operations are the most basic operations in a digital system. There are many types of data transfer operations, depending on what is required. Some of the more common data transfer operations are:

Interregister transfers
Bus transfers
Memory transfers

Interregister transfers are used when transferring data between registers. *Bus transfers* are used when many registers are interconnected. *Memory transfers* are used when dealing with memory.

Interregister Transfers

In an interregister data transfer operation, data from one register are transferred to another register when the clock is activated. The symbol for an interregister transfer is

$$(A) \leftarrow (B)$$

In this case, B is the source register and A is the destination register. Note that the previous contents of A are assumed lost, replaced by the contents of B. The previous contents of B remain, as they have not been replaced.

An interregister transfer can be realized by a direct connection between each bit in B to each bit in A. The connection should pass through a tri-state buffer controlled by a gating signal as shown in Figure 7–2(a). When a transfer is desired, the gating signal is pulsed so that the tri-state buffers are activated to transfer the bitwise information from B to A. Such a connection is unidirectional. If a similar data transfer is also desired from A to B, then another similar set of connections must be made from A to B. Figure 7–2(b) shows how the connection would be made if data in either register B or C can be transferred into A.

The arrangement for connection in the interregister transfer is effective but impractical when there are many registers to be interconnected. If there are *m* registers each *n* bits wide, there would have to be $m(m - 1)/2$ connections each *n* bits wide in order to allow data transfer between any two registers. The number of connections grows with the square of the number of registers. Many processors have a large number of registers. For example, the Intel 8085 has seven 8-bit data registers. The Motorola 68000 has sixteen 32-bit data and address registers. It is easy to see that direct interregister connections are not practical for these processors. One solution is to transfer all data to a common temporary register, then transfer the data from the temporary register to a destination register. In

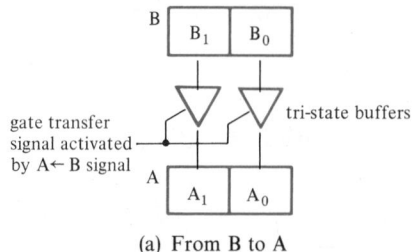

(a) From B to A

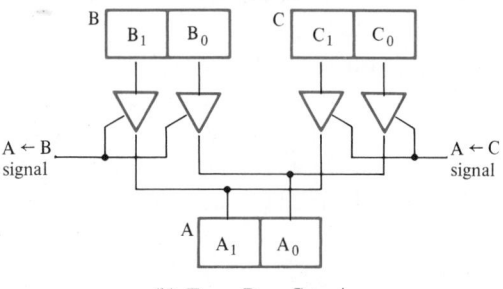

(b) From B or C to A

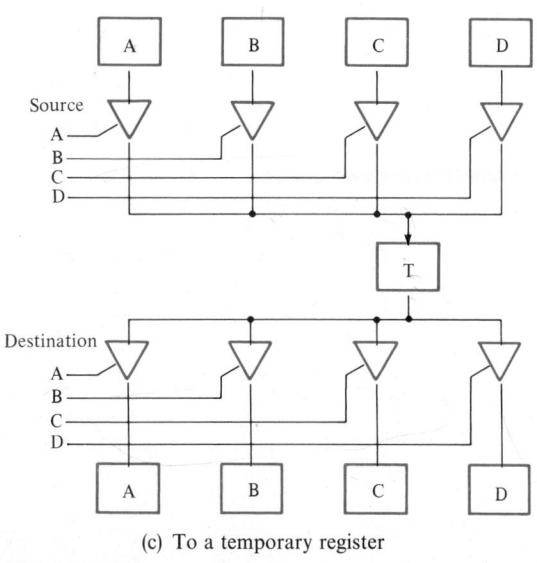

(c) To a temporary register

Figure 7-2

Interregister transfer: (a) from B to A; (b) from B or C to A; (c) to a
temporary register

other words:

$$(T) \leftarrow (B)$$
$$(A) \leftarrow (T)$$

This is shown in Figure 7–2(c). A better solution, however, is to use bus transfers.

Bus Transfers

A bus transfer makes use of a *bus*, which is a common connection point for all registers. In other words, the output of all registers is connected to a common point, called the *bus*, through tri-state buffers. At the same time, the bus is also connected to the gated input of all the registers on the bus. This is shown in Figure 7–3. When a transfer is to be made, the tri-state buffer of the source register is activated so that information contained in that source register alone is allowed to flow onto the bus. At the same time, the gates on the input of the destination register or registers are activated so that the information currently present on the bus can be read in. The sequence of events in a bus transfer can be represented as

For $(A) \leftarrow (B)$:

$$\text{bus} \leftarrow (B)$$
$$(A) \leftarrow \text{bus}$$

or simply

$$(A) \leftarrow \text{bus} \leftarrow (B)$$

Since the transfer to the bus is really transparent and simultaneous, and since

Figure 7–3

Bus transfers

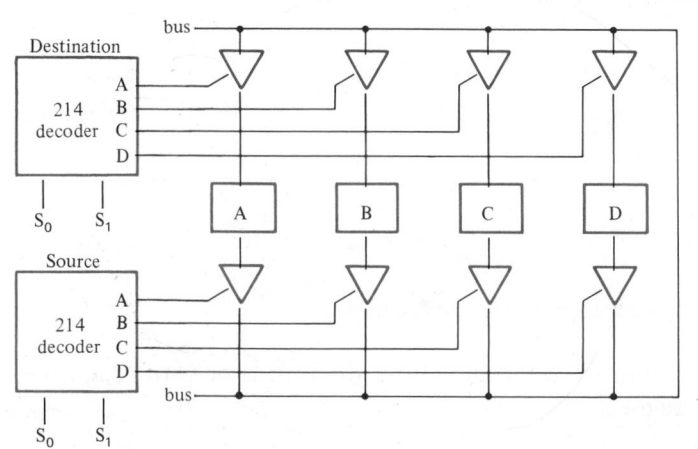

many times it is not necessary to know whether the transfer is an interregister transfer or a bus transfer, we will not use a separate symbol for a bus transfer.

When many registers are connected to the bus, care must be taken to ensure that only one register is allowed to transfer its contents onto the bus. If there are two or more registers actively doing so, the information will be jumbled. Decoders can be used effectively in this case. Timing is also important: the destination register must be reading from the bus at the same time that the source register is dumping its contents onto the bus; otherwise there will be nothing on the bus to read. It is not necessary to limit the number of registers that can read from the bus at the same time. In fact, all the registers can be activated to read from the bus if desired. In most cases, however, only one specified destination register should read the bus.

Memory Transfers

Most interregister transfers and bus transfers are internal and take place within the processor. There is also a tight connection between the processor and the memory. The instructions and the data operands are all stored in memory, because the data storage capability within the processor is usually limited to only a few temporary operands. At this point, we only state that to retrieve an operand from the memory, the processor must supply the address of that operand. Likewise, to store an operand in the memory, the processor must supply both the operand to be stored and its address.

In RTL, we will represent a memory fetch as

$$(A) \leftarrow M(address)$$

and a memory store as

$$M(address) \leftarrow (A)$$

In most cases the memory address is stored in a specific register called a *memory address register* (MAR), and the data to be retrieved from or to be sent to the memory are also typically stored in a special register called the *memory data register* (MDR). Hence a memory fetch can also be written as

$$(MDR) \leftarrow M(MAR)$$

and a memory store as

$$M(MAR) \leftarrow (MDR)$$

7.4
Data Manipulation Operations in RTL

Another major category of operations is data manipulation operations. These are operations that transform, convert, or manipulate input source data operands before producing destination operands. There are at least three general types of data manipulation operations:

    Arithmetic operations
    Shifting operations
    Logical operations

Arithmetic operations include basic addition, subtraction, and other numeric operations. *Shifting operations* include the shift and rotate functions. *Logical operations* include all the logical relations.

Arithmetic Operations

Of all the data manipulation operations, the arithmetic operation group is the largest. Arithmetic operations always have operands. Most take on two operands, whereas some take on only one operand. Typical arithmetic operations include the following:

1. Dual-operand operations:
   (a) Addition
   (b) Subtraction
   (c) Multiplication
   (d) Division
2. Single-operand operations:
   (a) Unary minus
   (b) Trigonometric functions
   (c) Transcendental functions
   (d) Logarithmic functions
   (e) Exponential functions

Dual-operand operations are most common. The four functions—addition, subtraction, multiplication, and division—are used in everyday life. *Single-operand operations* are used less often. The unary minus is used whenever the negative of a number is desired. Other operations such as trigonometric and transcendental functions are found only in special-purpose machines.

The RTL format for arithmetic operations consists of the operator and its associated operands. It is not possible to create a new symbol for all kinds of functions. If the symbol of the operation is known, the RTL format is

$$(A) \leftarrow (B) + (C)$$

where A is the destination operand, B and C are the source operands, and the symbol + indicates addition. The usual symbols are used for subtraction, multiplication, and division. If a symbol is not used, the RTL format is

$$(A) \leftarrow \text{Function}(B, C, D, \ldots)$$

where "function" is the name of the operation to be performed on the source operands within the parentheses. One advantage of this form is that the function can be any operation (as long as it can be realized by hardware). For example, a bit-reversal operation is best described this way. Another advantage of this form is that it permits as many source operands as needed. Any number of operands can be enclosed by the parentheses. Some examples of arithmetic operations are:

$(A) \leftarrow -(B)$

$(A) \leftarrow (B) + (C)$

$(A) \leftarrow Add(B, C, D, E, F)$ that is, $(A) = (B) + (C) + (D) + (E) + (F)$

$(A) \leftarrow Min(B, C, D, E, F, G)$ that is, $(A) = $ minimum value of $(B), (C), (D), (E),$
                                         $(F), $ or (G)

In most processors, the basic arithmetic operations for addition and subtraction are always implemented. Many processors do not have provisions for multiplication or division. Some processors are designed to perform multiplications only. More recent processors are designed to perform both multiplications and divisions. Since subtraction can be considered a generalized form of addition, the implementation of a generalized adder is shown in Figure 7–4. Control lines are assumed to be supplied to the adder to determine whether the current operation is an addition or a subtraction.

The arrowhead shape is taken to be a dual-operand operator for addition and subtraction. In general cases, it is also taken loosely to represent an arithmetic logic unit. For specialized functions, a rectangle will serve as well, as long as the box is clearly labeled with the function it is supposed to perform.

Shifting Operations

An important data manipulation function is the shifting operation. There are many uses for shifting. Shifting can be used to approximate simple arithmetic

Figure 7–4

A generalized adder for addition and subtraction

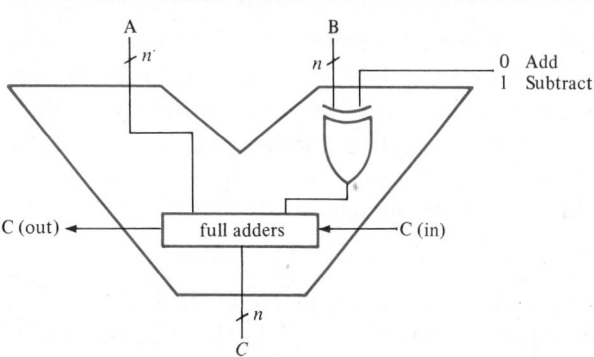

operations. For example, addition to the same operand is equivalent to shifting the operand left by 1 bit. Multiplication by 2 is equivalent to shifting the multiplicand left by 1 bit. Multiplication by 4 is equivalent to shifting the multiplicand left by 2 bits. Multiplication by 2^m is equivalent to shifting the multiplicand left by m bits. Likewise, division by 2 is equivalent to shifting the dividend right by 1 bit. Division by 4 is equivalent to shifting the dividend right by 2 bits. Division by 2^m is equivalent to shifting the dividend right by m bits. Hence shifting is often used for scaling both before and after arithmetic operations. Shifting is also used for alignment when bits at different positions must be aligned before arithmetic operations can be performed. This would be the case for floating-point numbers.

There are two basic kinds of shifting operations: *shifts* and *rotates*. *Shifts* are further divided into arithmetic shifts and logical shifts. The simplest kind of shifts are *logical shifts*. Assume that the operand is in register A, which is n bits wide with A_{n-1} being the most significant bit and A_0 the least significant bit. If we wish to shift register A to the right by 1 bit in the logical mode, the operation may be expressed as

$$(A_{n-1} \leftarrow 0, A_{n-2} \leftarrow A_{n-1}, \ldots, A_1 \leftarrow A_2, A_0 \leftarrow A_1)$$

The parentheses indicate that all operations are performed at the same time as a group. A logical shift with 2 bits can be expressed in similar fashion. When shifting right in the logical mode, zeroes are shifted in from the left and whatever is shifted out to the right is discarded. Likewise, a logical shift to the left by 1 bit can be expressed as

$$(A_{n-1} \leftarrow A_{n-2}, \ldots, A_2 \leftarrow A_1, A_1 \leftarrow A_0, A_0 \leftarrow 0)$$

When shifting left in the logical mode, zeroes are shifted in from the right and whatever is shifted out on the left is discarded. This is shown in Figure 7–5.

If the operand is signed, then the most significant bit is the sign bit. Often it is desirable to preserve the sign bit while shifting. This is called *shifting in the arithmetic mode*. This is the case when dividing or multiplying a signed number by powers of 2. An arithmetic right shift operation may be expressed as

$$(A_{n-1} \leftarrow A_{n-1}, A_{n-2} \leftarrow A_{n-1}, A_{n-3} \leftarrow A_{n-2}, \ldots, A_1 \leftarrow A_2, A_0 \leftarrow A_1)$$

Note that the sign bit is preserved even though the whole operand has been shifted to the right by 1 bit. This is why the operation is called an *arithmetic right shift*. The reader should verify that if an arithmetic right shift by 1 bit operation is performed on a negative number, the result is still a negative number and is equivalent to dividing the same negative number by 2. Likewise, an *arithmetic left shift* is represented as

$$(A_{n-1} \leftarrow A_{n-2}, \ldots, A_2 \leftarrow A_1, A_1 \leftarrow A_0, A_0 \leftarrow 0)$$

Both operations are shown in Figure 7–6.

We have seen that both logical and arithmetic shifts are destructive operations, because whatever is shifted out of the register is discarded and lost. In many

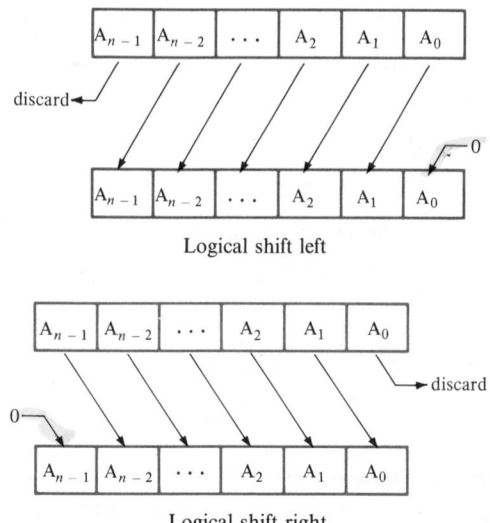

Logical shift left

Logical shift right

Figure 7–5
Logical left/right shift

Figure 7–6
Arithmetic left/right shift

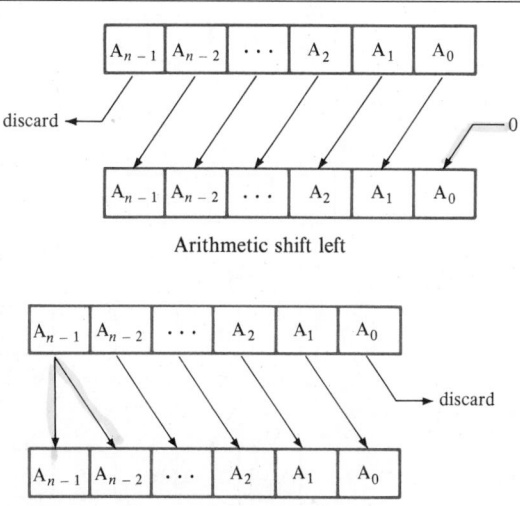

Arithmetic shift left

Arithmetic shift right

applications, particularly those dealing with BCD numbers, the other parts of the register may still contain needed information. A rotate instruction is useful in these circumstances. In a *rotate operation*, whatever is shifted out of one end of the register is fed back to the other end of the register. Hence no data bits are lost. A rotate to the right by 1 bit can be expressed as

$$(A_{n-1} \leftarrow A_0, A_{n-2} \leftarrow A_{n-1}, \ldots, A_1 \leftarrow A_2, A_0 \leftarrow A_1)$$

and a rotate to the left by 1 bit is

$$(A_{n-1} \leftarrow A_{n-2}, A_{n-2} \leftarrow A_{n-3}, \ldots, A_1 \leftarrow A_0, A_0 \leftarrow A_{n-1})$$

The rotate operations are shown in Figure 7–7. Rotate operations for multiple bits are performed in a similar fashion.

There are a number of parameters that must be specified for a shift or rotate operation:

1. The type: shifting or rotating
2. The mode: arithmetic or logical (for shifting only)
3. The direction: left or right
4. The amount: number of bits

The first specification is to determine whether a shifting or rotating operation is desired. If a shifting operation is chosen, then either arithmetic or logical shifts must be further specified. Then the direction of the shift or rotate operation must

Figure 7–7
Rotate left/right operations

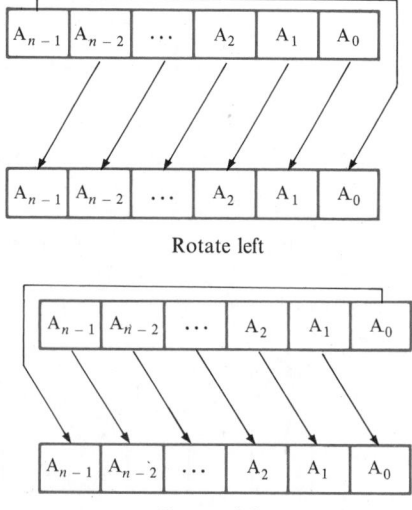

Rotate left

Rotate right

be stated, and finally the number of bits to be shifted must be specified. In general, mnemonics are used to specify the shifts. The RTL format we have presented is accurate but cumbersome. Mnemonics are simple to remember and to write down. The mnemonics we will use are:

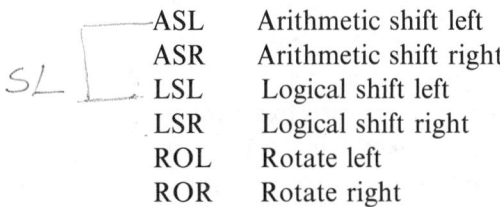

ASL   Arithmetic shift left
ASR   Arithmetic shift right
LSL   Logical shift left
LSR   Logical shift right
ROL   Rotate left
ROR   Rotate right

A combination of these mnemonics and the number of shifts will specify the operations to be performed. For example:

ASL(A, 3)   Arithmetic shift A to the left by 3 bits
ROR(B, 2)   Rotate register B to the right by 2 bits

Logical Operations

Logical operations are also important for many data manipulation operations. Though logical operations are usually defined in terms of a single Boolean variable, we will always assume bitwise operation when registers are referenced in logical operations. In other words, the AND operation between registers B and C is the bitwise AND between corresponding bits in registers B and C. This is shown in Figure 7–8. With this in mind, we may freely use logical operations on registers provided that the operands are all of the same length; that is, if they all have the same number of bits.

Figure 7–8
Logical AND operation

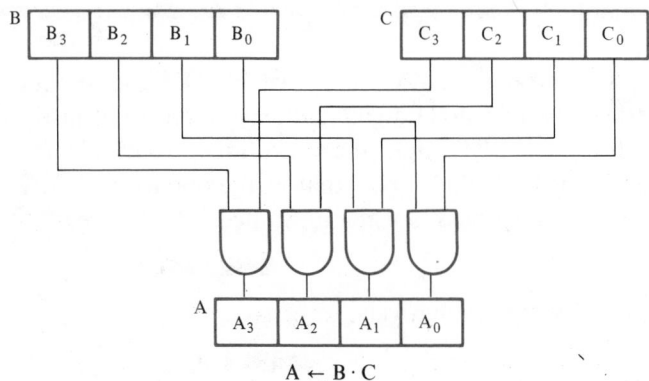

$$A \leftarrow B \cdot C$$

Common logical operations include the single-operand operation NOT and the dual-operand operations AND, OR, and XOR. Because of the confusion between the logical complement and arithmetic negation, the programmer must be careful about the mnemonics used to indicate the logical complement (i.e., the 1's complement) and the arithmetic complement (i.e., the 2's complement) operations because they are different. We will use the mnemonics "NOT" to indicate the logical complement and "NEG" for the arithmetic complement. The NOT operation requires only a single operand and is used to complement every bit of the designated register. This operation is useful when a test must be performed on the reverse state of the given condition.

The AND operation is also called a *masking operation,* because the AND operation can be used to mask out unwanted bits in a bit string by zeroing them out and leaving the desired bits unchanged for later testing. This technique is commonly used when the status register of a particular device is read. The status register contains many bits that indicate the current status of different parts of the device. If we are interested only in testing a single bit in the bit string, one way is to shift the desired bit into the most significant bit and test if the number is negative or not. If the number is negative, then the bit must be set. If the number is positive, then the bit must be reset. Testing a single bit this way is time-consuming and clumsy. A better solution is to use the AND operation to mask out the unwanted bits and then test the result for zero. For example, suppose that we want to test bit 5 in an 8-bit word. We use the following mask operation:

$$00100000 \,\&\, abcdefgh = 00c00000$$

For clarity, we will use the "&" symbol for the AND operation. The mask zeroes all unwanted bits, leaving a single bit to be tested. If the result is zero, then c must be 0. If the result is nonzero, then c must be 1.

Example 7-1 ▰▰▰▰▰▰▰▰▰▰▰▰▰▰▰▰▰▰▰▰▰▰▰▰▰▰▰▰▰▰

An 8-bit register A contains two BCD digits. What operations are needed to retrieve the two BCD digits and leave them right-justified?

For each BCD digit, the operations are different. Let us look at the most significant BCD digit first—that is, the one to the left. We need to shift the digit into place and mask out the unwanted bits. Hence:

$$LSR(A, 4)$$

will suffice, or alternatively, to be sure,

$$LSR(A, 4)$$
$$(A) \leftarrow (A) \,\&\, 0FH$$

The mask is 0FH, where the trailing H indicates hexadecimal notation. No shifting is necessary to retrieve the least significant BCD digit, but a masking operation is needed:

$$(A) \leftarrow (A) \ \& \ 0FH$$

The OR operation is commonly used to set any desired bit in a register. If a particular bit must be set, a mask with the corresponding bit set is ORed with the register. The result is that the desired bit in the register is set. For example, we do not know whether bit 3 is set in register A or not, but we would like to have it set. We could choose 00001000 as a mask. Since bit 3 of the mask is set, the same bit in the resultant bit must be set. The rest of the bits will remain the same; that is,

$$00001000 \vee abcdefgh = abcd1fgh$$

where $\vee$ is the symbol for the OR operation performed on the whole register.

Like the AND operation, the XOR operation has many uses. The first and most obvious is to perform the bitwise Exclusive OR operation between two given operands. Second, the XOR operation also indicates sameness, whereas the Exclusive NOR operation denotes difference. In other words, if the corresponding bits between the two operands are the same, the result of the XOR operation is 0. However, if the corresponding bits are different, the result of the XOR operation will be 1. Using this property, one can use the XOR function to check if a particular operand has the same bit pattern as the mask. If the bit pattern is the same, the result will be 0 in every bit location. If some bits differ, then a 1 will result at each location where the bits differ. Third, as a special application of the above property, a register performing an XOR operation on itself will always result in 0's, which is one of the many ways to clear a register.

Example 7–2 ▬▬▬▬▬▬▬▬▬▬▬▬

Show the result of the XOR function for different register contents if the mask used is 00001111.

Register Contents	Result
00000000	00001111
00000011	00001100
00001111	00000000
00111111	00110000
11111111	11110000

Example 7–3

Mask out all bits except bit 2 and bit 6 of an 8-bit word.

To mask out the unwanted bits, one merely needs to set them to zero. This can be accomplished by a bit-by-bit logical AND operation between the data and the mask. Choosing the mask to be 01000100_2, the resultant logical AND operation is

$$01000100_2 \, \& \, b_7b_6b_5b_4b_3b_2b_1b_0 = 0b_6000b_200$$

Once the unwanted bits are masked out, the value of the flag can be tested by checking for a zero value in the resultant word.

Example 7–4

Set bit 7 of a word.

To set a bit, simply logically OR the word with the mask. For this example, we will choose the mask to be 10000000_2.

$$10000000 \vee b_7b_6b_5b_4b_3b_2b_1b_0 = 1b_6b_5b_4b_3b_2b_1b_0$$

7.5

Conditional Operations in RTL

For a program segment to be meaningful, there must be provisions for testing operands and making decisions based on the results of the test. This is facilitated in the RTL format using the IF-THEN-ELSE construct. The format is

```
IF (condition is true)
THEN BEGIN
       .
       .

       .
       END
ELSE BEGIN
       .
       .

       .
       END
```

The above construct says that the given logical expression, which is the stated condition for the following operations, is tested. If the condition is found to be true, then the set of operations between the first BEGIN ... END block belonging to the THEN clause is executed. If the condition is found to be false, then the set of operations between the second BEGIN ... END block belonging to the ELSE clause is executed. In practice, the words "is true" are usually omitted and are assumed.

Example 7–5

Three numbers are stored in registers A, B, and C. Assume that the three numbers are the three coefficients of a quadratic equation. Write a sequence of operations to find the two square roots, placing them in registers D and E. Should the answer be complex, stop the sequence of operations and send an error message.

We are given the three coefficients of a quadratic equation. The solution is well known and is simply

$$ax^2 + bx + c = 0$$
$$x_1 = (-b + \sqrt{(b^2 - 4*a*c)})/(2*a)$$
$$x_2 = (-b - \sqrt{(b^2 - 4*a*c)})/(2*a)$$

We will need some hardware in order to perform the necessary operations. Assume that there is an arithmetic logic unit available for the four arithmetic functions: addition, subtraction, multiplication, and division. Also assume that there is a "black box" that will give the square root of its input operand. We will denote the function of the black box as SQRT. We can also draw in the five required registers and fill in the interconnections between them. A possible block diagram for this problem is given in Figure 7–9. Assume that the parameters a, b, and c

---

Figure 7–9

Block diagram for quadratic roots

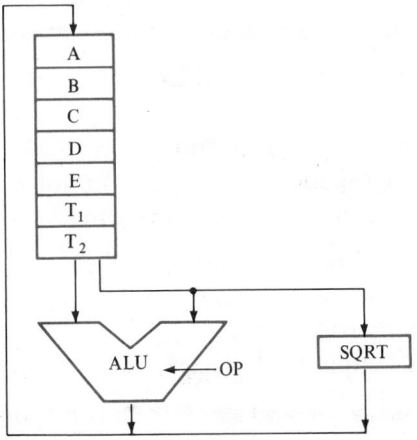

are stored in registers A, B, and C, respectively, and the resultant roots x_1 and x_2 in registers D and E, respectively. With the hardware identified, we are now ready to write the operations in RTL format:

```
SEQUENCE square_root_operation
Registers (n bit) A, B, C, D, E, T₁, T₂
ALU (n bit)
SQRT (n bit)
T₁ ← B * B
T₂ ← A
LSL(T₁, T₂)
T₂ ← T₂ * C
T₁ ← T₁ − T₂
IF T₁ > 0
THEN BEGIN
        T₁ ← SQRT(T₁)
        T₂ ← − B
        T₂ ← T₂ + T₁
        ASR(T₂, 1)
        D ← T₂ / A
        T₂ ← − B
        T₂ ← T₂ − T₁
        ASR(T₂, 1)
        E ← T₂ / A
        END
ELSE error
END_SEQUENCE
```

The above sequence of operations is easy to follow. Note that it is necessary to create two temporary registers, T_1 and T_2, for storing intermediate values, though one may be eliminated if registers D and E are used in the process.

7.6

Computer Organization

With the above brief introduction to the RTL, we will now begin to present the architecture of a simple generic processor for a computer system. A general computer system has four main parts as shown in Figure 7–10:

The central processing unit (CPU)
The processor system bus
The memory (M)
The input/output (I/O) devices

The *central processing unit* (CPU) is responsible for processing information and controlling every part of the computer system. It is usually divided into two

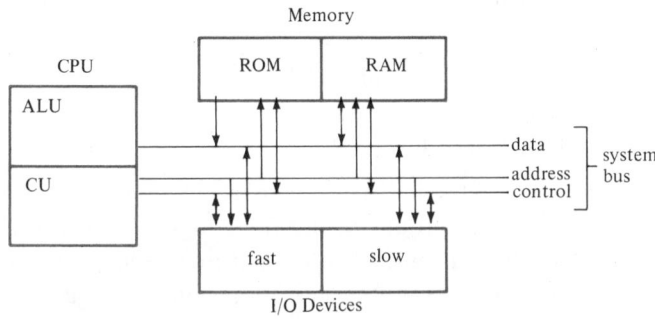

Figure 7-10
Organization of a generic computer system

parts: the *arithmetic logic unit* (ALU) and the *control unit* (CU). The ALU is responsible for handling the computational and data transfer aspects of the CPU. The CU is responsible for decoding and controlling the execution of the program instructions by sending out control signals to activate the appropriate parts of the computer at the proper time.

Memory is an indispensible part of a computer in that programs are read and computational results are stored there for later recall. There are two kinds of memory: *read-only memories* (ROMs) and *read/write memories* (RAMs). ROMs are primarily for program storage, whereas RAMs are for temporary data storage and hence are often called *scratchpad memory*.

The third component of a computer system is the *input and output* (I/O) *devices*. There are two kinds of I/O devices. The first kind can be described as fast devices and includes disks, tapes, and high-speed communication devices. The second kind of I/O devices are slow devices. The latter include terminals for user interactions, printers, and other interactive peripheral devices. These I/O devices present the computational result of the computer system to the outside environment.

Though the architecture of different microprocessors may appear to be vastly different, there are features that are common to all microprocessors. In this section we describe a general processor that is hypothetical yet functionally similar to a majority of the microprocessors available today. We will now develop a hypothetical machine from scratch, starting with the necessary registers in the CPU.

The Central Processing Unit

The CPU is the most important part of the whole processor. The main function of the CPU is to manipulate data given to it. The CPU is composed of four major components:

Registers
The arithmetic logic unit (ALU)
The control unit (CU)
The internal data bus (IDB)

Registers   One of the indispensible parts of a processor is a set of registers in the CPU. *Registers* are groups of flip-flops logically grouped together to represent some unit of information such as a word, an integer, or a character. Many registers are needed in a processor. Registers more commonly associated with the ALU are:

The accumulator (AC)
General-purpose registers (GPRs)
Special-purpose registers (SPRs)
Index registers (XRs)
The status register (SR)
Stack pointers (SPs)

The different registers and their interconnections and interrelations for a hypothetical processor are shown in Figure 7–11.

An *accumulator* (AC) is a register that can hold a data item. It is customary to hold the data for one of the two operands required for arithmetic, logical, or any dual-operand operation in the AC. The results of these operations can also be conveniently stored at a designated place such as the accumulator. The length of the accumulator defines the machine word size. It is helpful to store temporary information and other computational operands in temporary registers to be used again in later computations. This practice tends to decrease the number of memory fetches when the same constant or operand is used repeatedly.

The number of *general-purpose registers* (GPRs) varies for different processors. Each general-purpose register has a name or a designated register number for program reference. In some machines, a set of GPRs is provided instead of the AC, so that data manipulation can be performed with any one of the GPRs. This affords greater flexibility from the standpoint of the user.

Special-purpose registers (SPRs) are data or address registers that are used for specific purposes. For example, some registers can be used only as address pointers, or address extension registers, or data extension registers, and so on. The use of special-purpose registers varies greatly from one processor design to another.

Index registers (IRs) are useful for storing the index or count values in loops, tables, and subscript operations. They aid in providing offsets for address calculations and in program counting that is often used in loops. The main characteristic of index registers is the fact that they can be incremented or decremented explicitly by an instruction or sometimes implicitly whenever they are being referenced by an instruction.

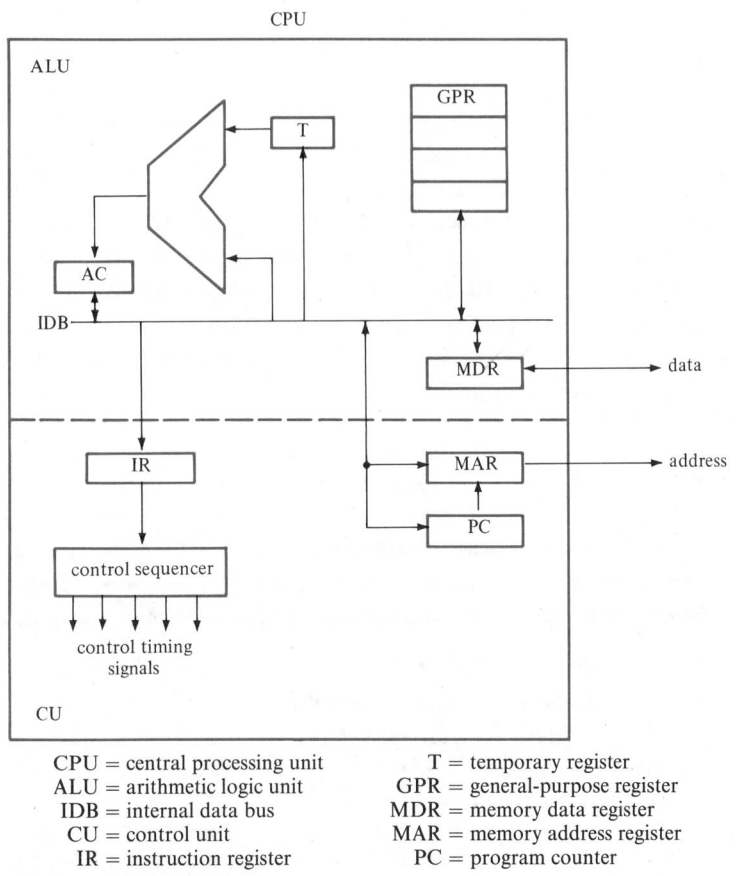

CPU

ALU

T

GPR

AC

IDB

MDR — data

IR

MAR — address

PC

control sequencer

control timing
signals

CU

CPU = central processing unit   T = temporary register
ALU = arithmetic logic unit   GPR = general-purpose register
IDB = internal data bus   MDR = memory data register
CU = control unit   MAR = memory address register
IR = instruction register   PC = program counter

Figure 7–11
Interconnections of internal registers in the CPU

The *status register* (SR) contains flags that report the condition of the machine and the results of the current computation. The current status of a machine includes flags such as the zero flag, overflow flag, carry flag, sign flag, and so on. The contents of the flags can be used later by conditional instructions to determine the proper direction of program flow.

Stack pointers (SPs) are registers that contain addresses of a stack, which is a last-in, first-out data structure. Stacks are used mainly for subroutine linkages and interrupt processing. In addition, the stack can also be used for temporary data storage. As items are put onto the top or bottom of the stack, the stack pointer is incremented or decremented accordingly in such a way that the stack pointer is always pointing to the last available item. As the most recently stacked information is retrieved, the stack pointer is also changed to point to the next available item on the stack.

The Arithmetic Logic Unit   The ALU constitutes the main computational part of the processor. It is designed to perform arithmetic operations such as additions, subtractions, multiplications, and divisions for binary and BCD numbers. Note that a particular processor may not have all the arithmetic functions built into its hardware. The ALU is also designed to perform logical operations such as AND operations, OR operations, and compare operations. Other operations such as shifts and rotates may also be included in the ALU or may be designed as auxiliary modules in support of the main ALU. It is convenient to refer to all these support modules collectively as the ALU.

The Control Unit   The registers in the ALU are usually quite accessible to the programmer and are used primarily for data manipulation and computation. The following set of registers, however, is associated with the control unit and is related to the internal operation of the machine. Hence these registers are generally not readily accessible to the programmer. Registers in the CU include:

Program counter (PC)
Memory address register (MAR)
Memory data register (MDR)
Instruction register (IR)
Temporary registers (Ts)

The *program counter* (PC) is an address register that points to the next instruction that is to be executed. Upon initial power-up, manual reset, or interrupt acknowledge, the program counter will be set at a specific starting location. Program execution will begin at the specified location. Under normal conditions the program counter is incremented after every instruction, so that instructions are executed sequentially until the sequence is explicitly changed by a conditional branching instruction. The PC is a key element in the CU, because the PC governs and sequences the program flow by pointing to the next instruction to be executed. Any abrupt change in program direction requires a new address to be loaded into the PC.

The *memory address register* (MAR) is the register that holds the memory address for the current memory operation. The MAR is connected to the address bus, which supplies the address to the memory system. In most cases the address bus is unidirectional, with the processor supplying the address to the memory system. The memory address may come from various registers. If the address in the MAR comes from the PC, then the memory read operation is called an *instruction fetch operation*. If the address in the MAR originates from another register, then the operation may be a memory read operation to get an operand from memory or a memory write operation to store an operand back to memory.

The *memory data register* (MDR) contains the data to be written to or read from the memory. It is connected to the external data bus for receiving external data and also to the internal data bus. The MDR is used to route the received data from memory to the appropriate registers in the case of a memory read operation or to route the data from the appropriate register to memory in the case of a memory write operation.

When the memory read operation is an instruction fetch operation, the data received and buffered in the MDR is the code for the current instruction. This code is called the *operation code* (OPCODE). For instruction decoding, the OPCODE is transferred from the MDR to the *instruction register* (IR) and saved in the IR for the duration of the current instruction cycle from the proper execution of the entire instruction.

In addition to these registers, many other *temporary registers* (Ts) are used in the internal architecture of a processor. These Ts are used to buffer the data and to synchronize data arriving from different parts of the processor. Most Ts are transparent to the programmer and cannot be referenced nor used directly by an assembly programmer.

Besides the listed internal registers, a major part of the CU is the sequence control unit, which is connected to the IR. The *sequence control unit* (SCU) decodes the OPCODE stored in the IR and generates the proper control signals in a predetermined sequence to activate the different parts of the processor to complete the necessary computation as required by the instruction.

The Internal Data Bus   It is helpful to connect these registers together. This may be done either through a single *internal data bus* (IDB) or through several IDBs. Information is passed from one register to another through the IDB. A single IDB that connects to every register component in the processor provides design flexibility and, ultimately and indirectly, versatility and ease in assembly programming. A typical connection of these registers has already been shown in Figure 7–11. Where there are several GPRs, a decoder is needed to arbitrate and activate the proper one for reading or storing.

The Processor System Bus

The operation of a processor depends not only on the internal registers and their interconnections but, more importantly, on the interface from the processor to the outside world. The latter is accomplished by a set of signals collectively called the *system bus*, as shown in Figure 7–12. Although the IDB is an internal channel for data transfers within the processor, the processor system bus is an external channel for data transfers from the processor to the outside environment. In most

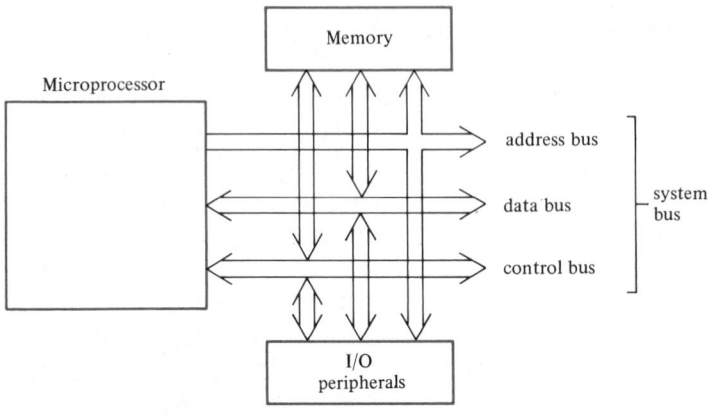

Figure 7–12
Microprocessor system bus

hardware designs, the system bus contains three parts:

The address bus
The data bus
The control bus

The Address Bus The address bus is always unidirectional, coming from the processor to the memory subsystem and the I/O subsystem. The address bus contains the memory address for memory operations or the I/O address for activating the proper I/O peripheral. The amount of memory that can be directly addressable depends on the width of the address bus. Most popular 8-bit microprocessors have a 16-bit address bus that can address up to a maximum of $2^{16} =$ 65,536 memory locations. Other recent microprocessors have an address bus that is 20, 24, or even 32 bits wide. Figure 7–13 shows the relationship between the width of the address bus and the corresponding maximum amount of directly addressable memory.

The Data Bus The data bus is always bidirectional, so that the processor can read information from or write data to the memory and the I/O peripherals. The direction of data flow in the data bus is governed by additional signal lines included as part of the control bus. The width of the data bus may be 8 bits, 16 bits, or 32 bits, depending on the machine type. The bus is usually as wide as the internal registers, but some microprocessors have a data bus that is only half as wide as the internal registers. These microprocessors require more than one memory cycle to fetch an operand.

Address Bus Width (bits)	Maximum Address (words)
8	256
10	1K = 1024
12	4K
14	16K
16	64K
18	256K
20	1M = 1024K
22	4M
24	16M
26	64M
28	256M
30	1G = 1024M
32	4G

Figure 7–13
Address bus width and its relation to maximum memory size

The Control Bus The control bus contains control signals that provide an orderly transfer of information between the processor and the memory or I/O peripherals. Data transfers may be synchronous with respect to a system clock, or asynchronous (relying solely on other control signals for timing reference). Control lines can be input, output, or input/output lines. Output control lines from the processor are used to inform peripherals about the processor's impending actions. Typical output control signals include the memory read line, the memory write line, the I/O read line, the I/O write line, and the processor status lines. There are also input control signals that report the status of peripherals and pending data transfers. These input control signals are handshaking signals that notify the processor concerning the arrival of data on the data bus. Both input and output control signals are needed in order to complete a data transfer.

The Memory

One of the main events of the bus is accessing the memory for information: instructions and data for the instructions. Without the memory, the processor is not able to process a large amount of information simply because there is no place in the CPU to store all the information. Hence there is a close relationship between the processor and the memory. There are two types of memory operations:

Memory read
Memory write

In memory read operations, the memory is interrogated for information. In memory write operations, the memory is supplied with information for safe-keeping.

The Memory Read Cycle Most accesses to memory and I/O are read and write operations. A *memory read cycle* is a series of events leading to an interrogation of the memory and a transfer of stored information from the memory to the processor. In terms of the RTL format, a memory read operation can be summarized as

$$(MAR) \leftarrow IR \text{ (or other sources)}$$
$$(MDR) \leftarrow M(MAR)$$

A typical flowchart for a memory read cycle is shown in Figure 7–14. The processor first obtains the address and stores it in the MAR. Then the address bus is asserted with the memory address. At the same time, a control signal is asserted to indicate a valid address on the address lines. In addition, other control lines are asserted to indicate whether this is a read operation or a write operation, and whether this is an I/O or a memory operation. At this point, another input signal line is pulsed to see if data from memory are ready. If so, the information on the data bus is read and latched, thus completing the memory read operation.

Figure 7–14
Flowchart of a memory read operation

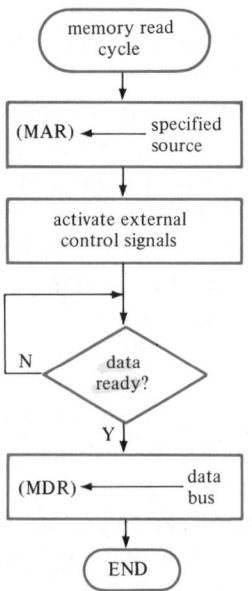

The I/O read operation is very similar to the memory read operation except that a different set of signal pins is involved in the control bus.

The Memory Write Cycle In a *memory write cycle*, the processor supplies both the address and the data that are to be latched by the memory and written into memory. In the RTL format, the memory write operation can be expressed as

$$(MAR) \leftarrow (IR) \text{ (or other sources)}$$
$$(MDR) \leftarrow (GPR) \text{ (or other sources)}$$
$$M(MAR) \leftarrow (MDR)$$

Except for the direction of data transfer in the data bus, most of the signals involved in the memory write cycle and I/O write cycle are the same as the memory read cycle. A flowchart of the memory write operation is shown in Figure 7–15.

The Input/Output (I/O) Devices

The processor system bus also connects the processor to the input/output devices, which are the primary means for the computer system to communicate with the outside world. The input/output devices include input devices such as terminals,

Figure 7–15
Flowchart of a memory write operation

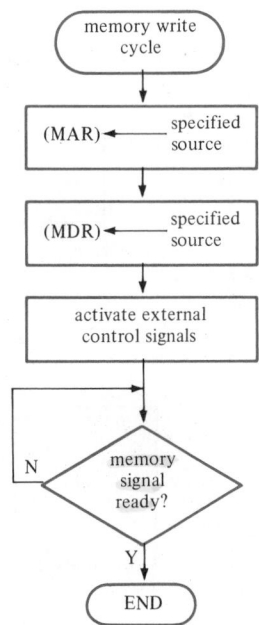

card readers, tape drives, disk drives, and so on. Output devices include printers, plotters, tape drives, and disk drives. There are also communication devices and instrument interfaces. The I/O devices are also called *peripherals* because they are on the periphery of the computer system, surrounding the processor itself.

The input/output devices are in many ways similar to the memory subsystem. Each peripheral device appears to the processor as a series of I/O ports. An I/O port is like a memory location with a distinct I/O address. An I/O port usually has the same word length as the processor itself. Communication between the processor and the I/O devices is accomplished through data transfer between the ports. This is shown in Figure 7–16. When the processor wants to read from an I/O port, the process issues an I/O address and the appropriate I/O control lines. In return, the device with the matching I/O address asserts the data bus with the contents of the requested port. This is called an I/O read cycle and can be written in RTL as

$$(\text{MAR}) \leftarrow \text{I/O address from IR (or other sources)}$$
$$(\text{MDR}) \leftarrow \text{I/O(MAR)}$$

Likewise, in an I/O write cycle, the processor supplies both the I/O address and the data to be written to the system bus. The designated peripheral device with the matching I/O address then latches the information from the data bus and

Figure 7–16

I/O ports

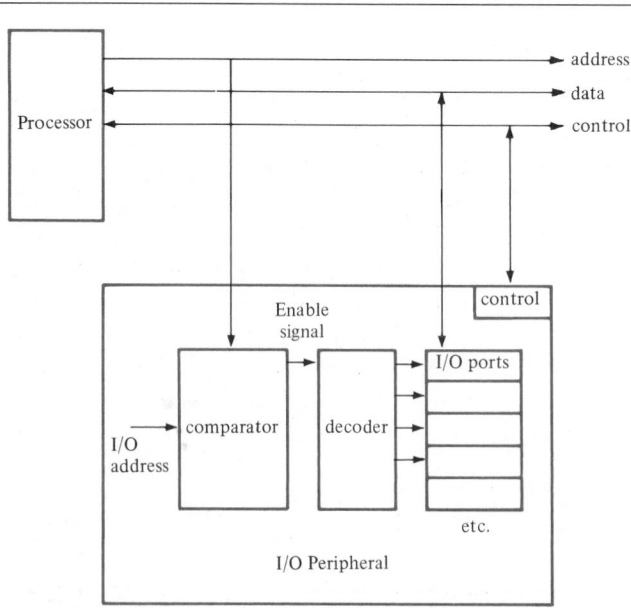

saves it for later use:

$$(MAR) \leftarrow I/O \text{ address from IR (or other sources)}$$
$$(MDR) \leftarrow (GPR) \text{ (or other sources)}$$
$$I/O(MAR) \leftarrow (MDR)$$

In general, there are two types of information passed onto I/O devices: control and data. *Control information* refers to information that directs the I/O device to perform a certain operation. Sometimes control information is called *commands*. *Data* refers simply to information that the device must act upon. In a printer, for example, *control information* refers to the baud rate, the number of data bits, and so on. *Data* are the actual characters to be printed. Likewise, there are two types of information received from the I/O devices: status and data. *Status* refers to information related to the state of the device, and *data* refers simply to data received by the device to be used by the processor. In the printer scenario, *status information* includes the paper out flag, the printer ready flag, and so on; but there will be no data received from the printer.

7.7
Processor Operation

The operation of the processor is composed entirely of operations dealing with data stored in registers. When data are moved from one register to another, the data are being operated upon. We will look at the operation of a simple processor, processor cycles, and instructions.

Operation of a Simple Processor

We can examine the operation of a simple processor by tracing the steps required to add two numbers in memory and to put the result back into memory. Assume that the two operands are stored in memory locations X and Y, and the result is to be stored in Z. We also assume that the OPCODE for this instruction has been fetched and decoded. To complete the instruction, several steps are required to fetch the operands, perform the operation, and store the resultant operand. Before the numbers can be added together, they must be transferred from memory into the processor. Two memory transfers are required. The first is used to obtain the operand stored at location X, and the second to obtain the operand at location Y. The destination registers for these two operands are the AC and the T registers. The steps to perform the above are listed below:

1. Transfer the first operand into AC.
2. Transfer the second operand into the processor and temporarily store it at T.
3. Activate the adder in the ALU to perform the addition.
4. Store the result of the addition in the AC back into memory.

STEP 1

Transfer the first operand into AC.

The first step is a memory read operation. From the IR, the address of A is obtained and loaded into the MAR. The memory is interrogated and the data received at the MDR are channeled through the IDB directly to the AC register to be stored there. In the RTL format, the needed operations are:

$$(MAR) \leftarrow (X)$$
$$(MDR) \leftarrow M(MAR)$$
$$(AC) \leftarrow (MDR)$$

The result of the first memory read cycle is shown in Figure 7–17(a).

STEP 2

Transfer the second operand into T.

The next step is to transfer the second operand into the processor through the MDR. This second operand is held at the T register temporarily. In RTL format:

$$(MAR) \leftarrow (Y)$$
$$(MDR) \leftarrow M(MAR)$$
$$(T) \leftarrow (MDR)$$

The result of the second memory read cycle is shown in Figure 7–17(b).

STEP 3

Add the numbers.

Once the operands have been fetched from memory and placed in the AC and T registers, the ALU can be activated to perform the required operation. In RTL format, this is shown as

$$(AC) \leftarrow (AC) + (T)$$

The state of the machine after this step is shown in Figure 7–17(c).

STEP 4

Store the result back into memory.

After the addition, the result from the adder is stored in the AC. This result must now be channeled back to the MDR to be written back into memory at memory location Z. This is a memory write operation. The address of location Z is obtained from the IR as part of the instruction. In RTL format:

$$(MAR) \leftarrow (Z)$$
$$(MDR) \leftarrow (AC)$$
$$M(MAR) \leftarrow (MDR)$$

The state of the machine after this step is shown in Figure 7–17(d).

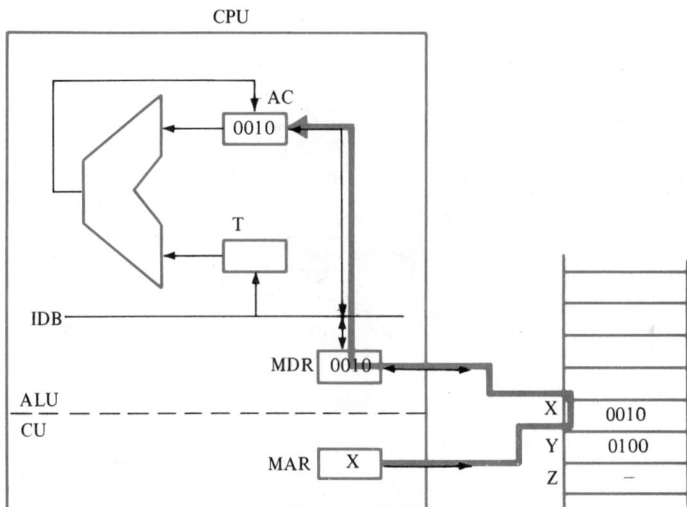

(a) Fetching memory operand X into AC

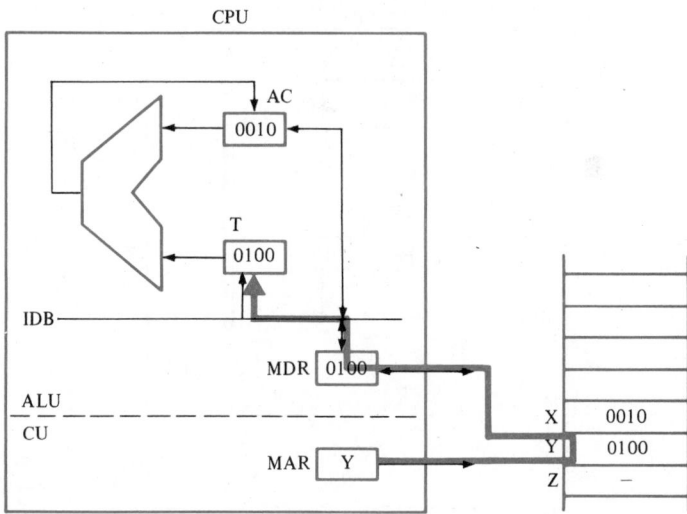

(b) Fetching memory operand Y into T

Figure 7-17

Operations required to add two memory operands

(*continued*)

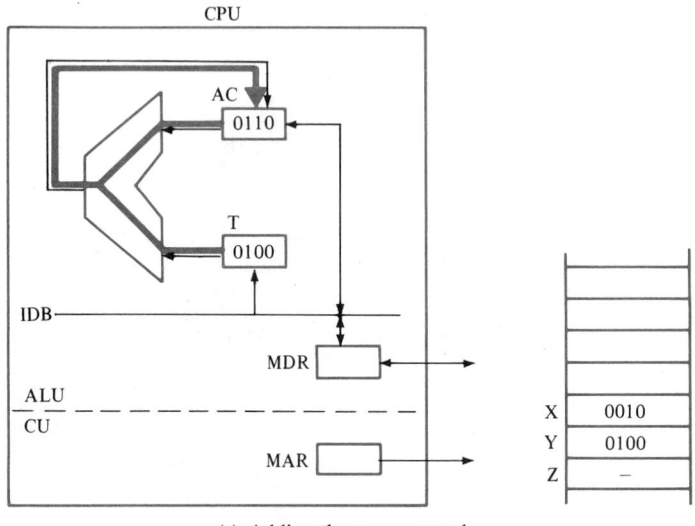

(c) Adding the two operands

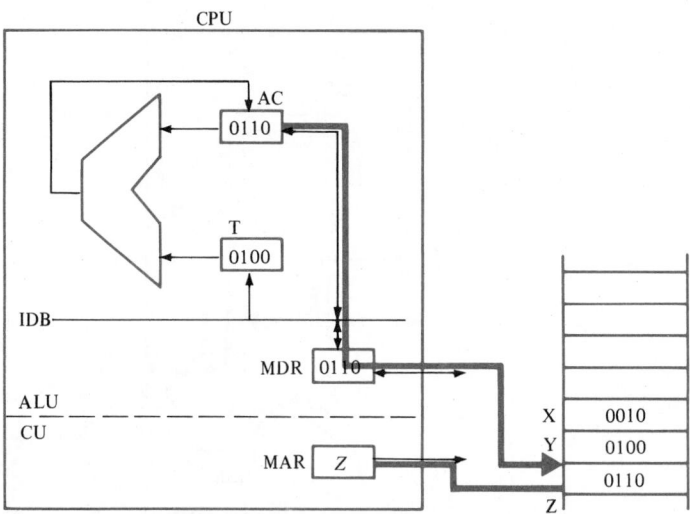

(d) Storing result to memory location Z

Figure 7–17 Continued

Processor Cycles

In many ways, the normal operation of a processor is similar to the addition operation presented above. In general, the operation of a processor can be considered as repetitions of four primary cycles:

1. The OPCODE fetch cycle
2. The operand fetch cycle
3. The instruction execute cycle
4. The operand store cycle

The OPCODE Fetch Cycle Every instruction begins with the OPCODE fetch cycle. During the OPCODE fetch cycle, the contents of the PC are transferred to the MAR. When the proper control signals are activated, the memory responds by reading the stored information and placing it onto the external bus. This information is received by the MDR and then transferred to the IR. This is shown in Figure 7–18. During the memory cycle, the PC is incremented to point to the

Figure 7–18
Action paths in an OPCODE fetch cycle

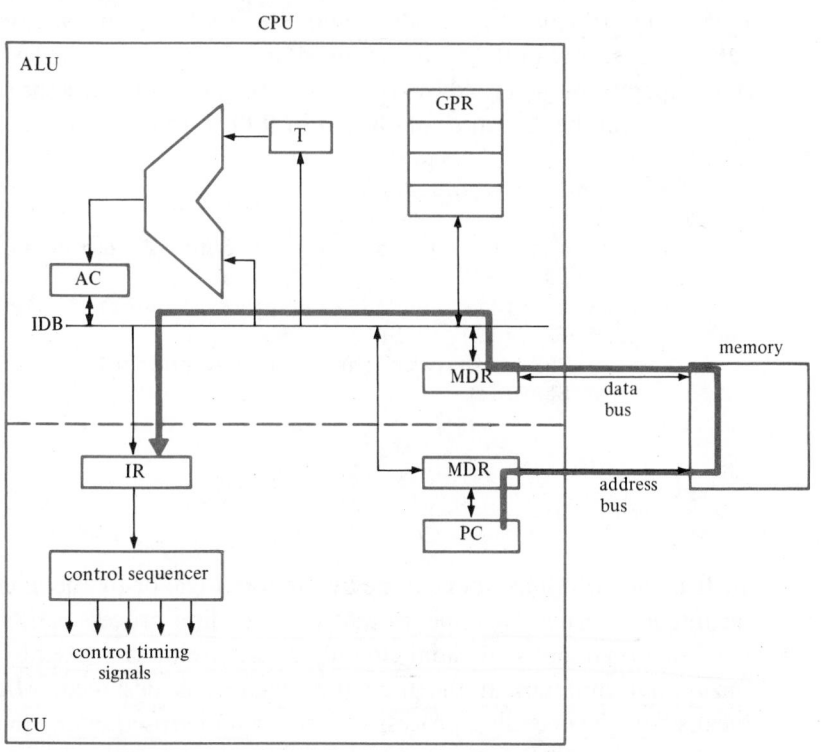

next instruction. This is the instruction or OPCODE fetch cycle. In RTL format, this set of operations can be described as

$$(MAR) \leftarrow (PC)$$
$$(MDR) \leftarrow M(MAR)$$
$$(IR) \leftarrow (MDR)$$
$$(PC) \leftarrow (PC) + 1$$

The Operand Fetch Cycle   After the OPCODE fetch cycle, a partial decoding of the OPCODE follows in order to determine the type of addressing mode used for the operands in this instruction. The first thing to find out is whether any operand or operands are required at all. If no operand is required, the rest of the operand fetch cycle is skipped. If operands are required, then there will be an operand fetch cycle performed for each required operand. During the operand fetch cycle, the specification of the addressing mode is decoded and the operand is fetched from one of the registers or from memory as specified by the addressing mode. A flowchart of the operand decoding process is shown in Figure 7–19. Detailed information on the addressing modes is presented later in the next chapter.

The Instruction Execute Cycle   The instruction execute cycle is composed of two operations: instruction decoding and execution. In instruction decoding, the OPCODE stored in the IR is examined and the proper sequence of register-level microoperations governed by control signals is generated by the sequence control unit. This can be shown symbolically in RTL format as follows:

```
SEQUENCE instruction_decode
IF ((IR)='ADD')
THEN generate the ADD sequence of control signals
ELSE IF ((IR)='SUB')
     THEN generate the SUB sequence of control signals
     ELSE IF ((IR)='COMP')
          THEN generate the COMP sequence of control signals
          ELSE IF .
                         .
                         .

          END
     END
END
```

Instruction decoding can be done by random logic or by the microprogramming technique. *Random logic* means that the decoding process is done by hardwired combinational and sequential circuits. These circuits are fixed by standard minimization techniques at the time the machine is designed. *Microprogramming* means that the decoding process is done by a microsequencer that executes a sequence of microinstructions, which specifies the internal operations of the ma-

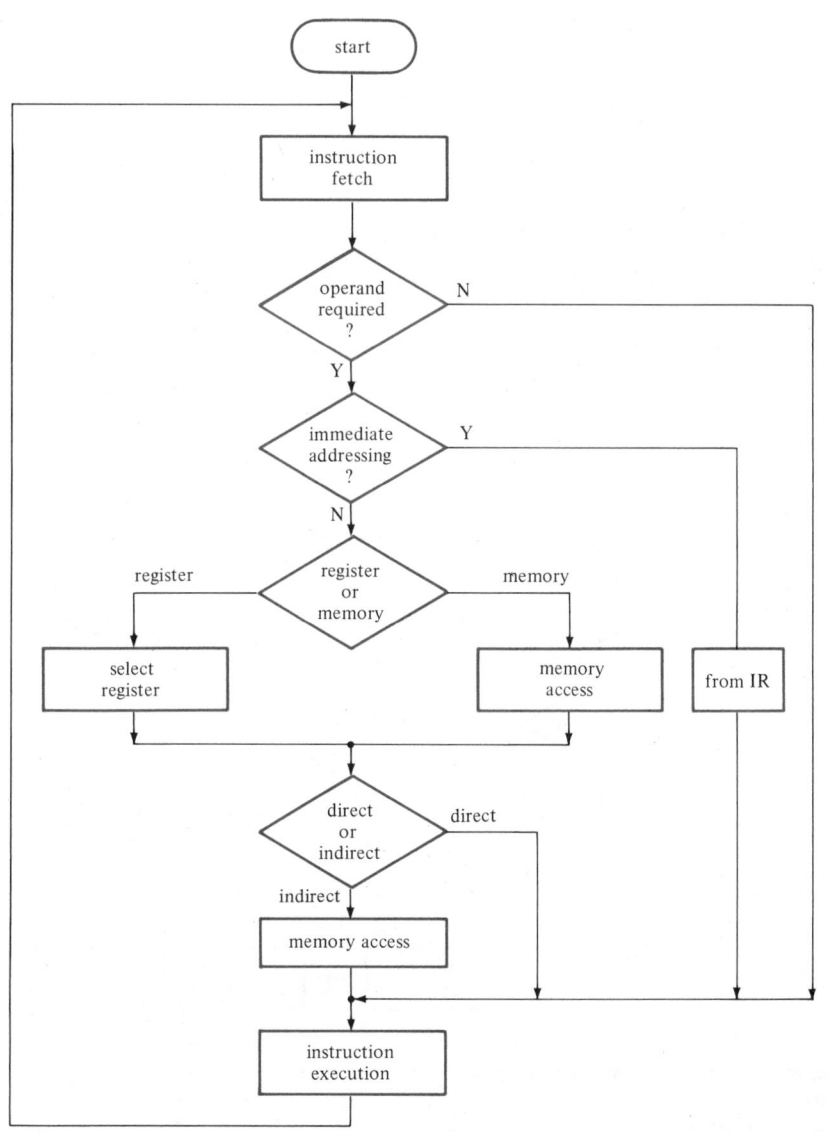

Figure 7–19

Flowchart for the operand decoding process

chine. In the instruction execute cycle, the processor begins by examining the instruction contained in the IR. A sequence of symbolic decision points is traversed to narrow down the exact operations specified by the current instruction. A flowchart of the instruction decoding and instruction execute cycle is shown in

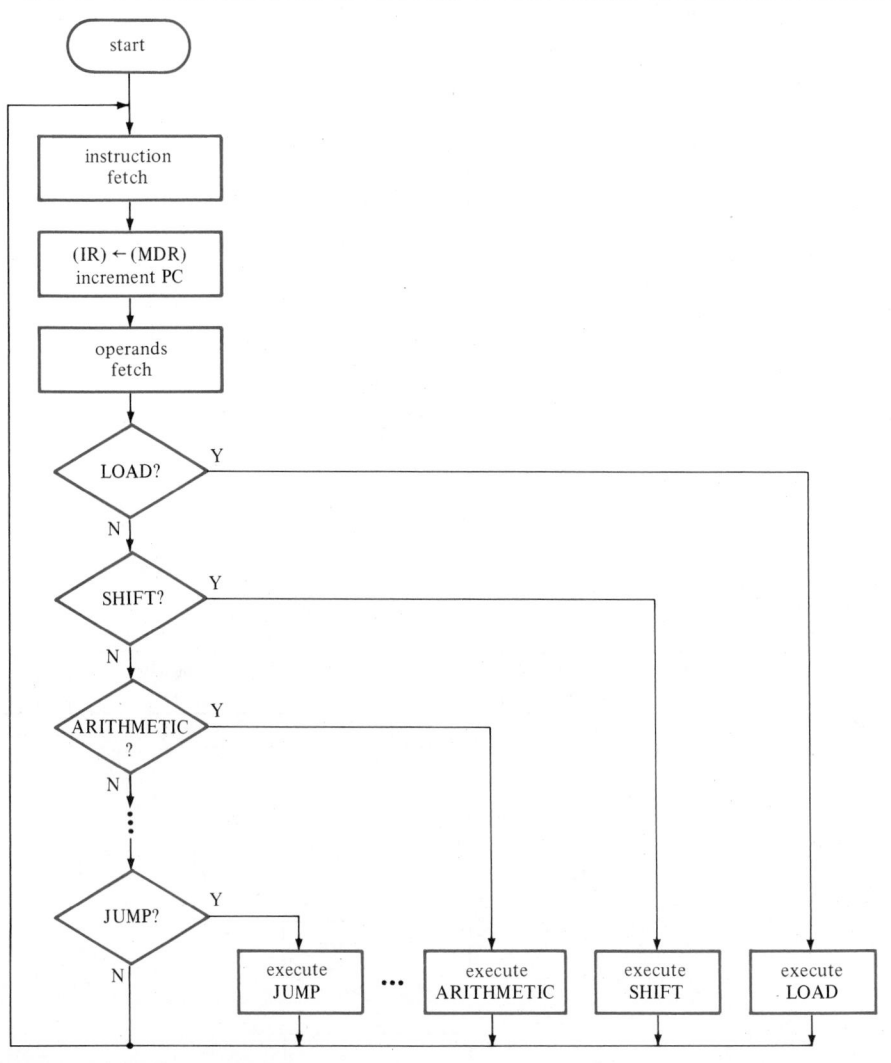

Figure 7-20
Flowchart for the instruction decoding process

Figure 7-20. After decoding the instruction, additional time and memory cycles may be needed in order to execute the instruction completely.

The Operand Store Cycle   After the specified operations have been performed, the resultant operand must be stored back into memory for later use. This is the operand store cycle and is the same as the operand fetch cycle except that we are storing operands into memory rather than retrieving operands from memory.

Instructions

As the processor progresses through the four cycles, instructions are first fetched and placed onto the IR. As the instruction is being decoded, operands are fetched at the same time. When all the operands have been fetched, instruction execution begins. Finally, when the specified operation is done, the operands are written back into their destination. We can readily see that two types of information come from the memory system: first, instructions that tell the processor what to perform next; second, data operands that are to be operated upon. We will examine the instruction types in this section and the data types in the following section.

The strength of a processor depends on the instructions, because it is the instructions that govern the sequence of operations to be performed on the data. During the instruction fetch cycle, the complete specification for one instruction is fetched. In practice, if the length of the instruction is long, several memory cycles may be necessary to retrieve the complete instruction. The format of an instruction can be

```
OPCODE     Operands
```

An *instruction* is composed of a single operation code (or OPCODE) and as many operands as necessary for the operator to work on. Each instruction has only one operator. Some operators require no operands, some one, whereas many others require two or more. An example of an operator with no operand is the NOP instruction, an instruction that does nothing. The EI instruction for enabling the interrupt system (discussed later) is another example. An example of a single-operand operator is the complement (COM) instruction, used for taking the complement of a register. Shifting operations are also examples of single-operand instructions. Most of the arithmetic and logical operations require two operands. In practice, there are actually three operand specifications: two for the source operands and one for the destination operands.

The collection of all possible instructions that can be performed by the processor is called an *instruction set*. Within an instruction set, individual instructions can be categorized into several groups:

1. Data movement instructions
2. Arithmetic instructions
3. Logical instructions
4. Shift/rotate instructions
5. Bit manipulation instructions
6. Bit string/field instructions
7. Program control instructions
8. System control instructions

The *data movement instructions* provide the basic instructions for retrieving data from memory to one of the processor registers, moving the data among the processor registers, and storing the data back to memory from one of the processor registers. All processors typically have a rich set of data movement instructions. Other special functions such as sign extension and register swaps may or may not be implemented, depending on the design philosophy of the particular processor in question.

Arithmetic and *logical instructions* are common to all processors. This group includes arithmetic operations for binary unsigned numbers, binary signed numbers, and decimal (BCD) numbers. Most of the earlier processors provided hardware only for binary addition and subtraction. Most recent high-performance processors are equipped with hardware multipliers and dividers for signed and unsigned numbers. Logical operations are also found in most processors. Most common logical instructions include AND, OR, and complement operations. Some processors also have the Exclusive OR function.

Shift and *rotate operations* are also commonly implemented in most processors. Earlier processors provided only single bit shifts. More advanced processors allow multiple bit shifts. *Bit manipulation instructions* are useful when dealing with a single bit within a word. *Bit string instructions* allow operations to be performed on a string of binary bits of variable length.

The most important of all instructions are program control instructions. *Program control instructions* allow the sequence of program flow to change depending on the data being operated upon, thus making the program more flexible. When a decision point is encountered, the data are examined and the condition of the data is reported in the flags or the condition code registers. Typical condition codes include the *zero flag*, the *overflow/underflow flag*, the *carry flag*, and the *sign flag*. These flags are set by comparisons that are performed by subtractions. Based on the result of the comparisons reported as status in the flags, the program may be directed to continue its normal sequence or have an abnormal termination by taking on a new sequence of instructions.

The last group of instructions are *system control instructions*. This set of instructions provides the ability for maintenance and housekeeping of the processor itself and the system as a whole. This set of instructions is rather specific to each processor and varies greatly from one processor to another.

7.8

Data Types

Most processors use the binary format to represent information. Therefore, a binary bit is the basic unit. Groups of bits together can collectively represent numbers, characters, and so on. Other data structures can be constructed from groups of numbers or characters to form arrays, strings, queues, trees, and so on. In this section we discuss the basic data structures most often used in processors.

Several binary quantities are used in microprocessors, including

1. Bits
2. Nibbles
3. Bytes
4. Words

A *bit* represents one of two states, such as on or off, yes or no, true or false, and so on. In microprocessor systems, a bit is most commonly used as a flag to indicate a certain condition or as a logical variable to indicate a true or false condition. A *nibble* is a 4-bit quantity. It is most commonly used to represent a BCD digit. Very few microprocessors have internal registers that are 4 bits wide. Hence a nibble representing a BCD digit is usually collected together with other BCD digits. A *byte* is an 8-bit quantity. Since the first-generation microprocessors are 8-bit processors, and since a byte can represent an alphanumeric character, the use of a byte is most common. A byte can be used to represent an unsigned integer from 0 to 255, or a signed integer from -128 to 127 in 2's complement format, or an alphanumeric character. A *word* is the basic width of the internal registers of a microprocessor. It can be as little as 1 byte wide or as much as 80 or more bits long. With the possible exception of the status registers, all internal data registers are usually of uniform width—that is, as wide as a word.

Information in the microprocessor can be roughly categorized into the following two types: numeric and non-numeric. Numeric data can be fixed-point arithmetic or floating-point arithmetic. Non-numeric data can be logical, alphanumeric, or address.

Numeric Data Types

Arithmetic data represent information related to arithmetic quantities and are most common in microprocessor systems. Figure 7–21 shows a breakdown of the different types of arithmetic operations. Arithmetic data can be divided into binary representation and binary-coded decimal (BCD) representation, as discussed in Chapter 1. For binary data, the 2's complement signed representation is most commonly used. For BCD data, the 10's complement signed representation is commonly adopted.

Representation of numeric or arithmetic data can be further broken down into integer (or fixed-point) and fractional (or floating-point) formats. In the *fixed-point representation,* all quantities are considered to be integers, with the fractional point either at the left or at the right of the representation as shown in Figure 7–22(a). In the *floating-point representation,* two parts are required to specify a particular arithmetic quantity completely: the mantissa and the exponent. While the mantissa specifies the magnitude, the fractional point can be

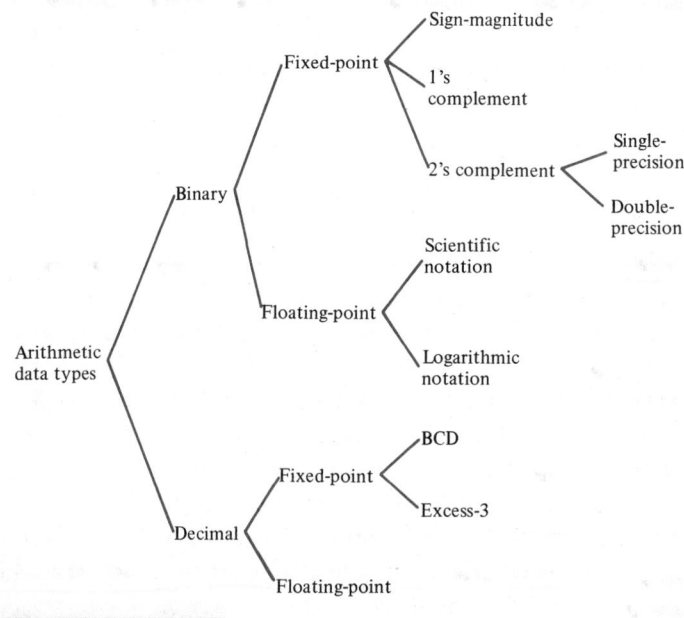

Figure 7–21
Arithmetic data types

Figure 7–22
Fixed-point and floating-point numbers

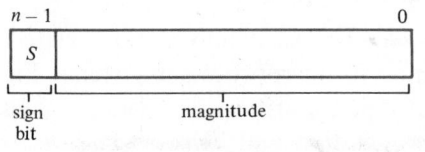

(a) Fixed-point representation

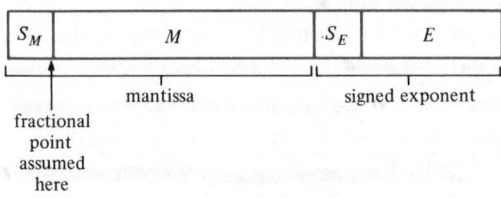

(b) Floating-point representation

placed anywhere in the mantissa with the exact location specified by the exponent as shown in Figure 7–22(b).

Fixed-Point Arithmetic   In fixed-point arithmetic, integers are usually represented by multiples of bytes. Though signed numbers can be represented in sign-magnitude form, 1's complement form, and 2's complement form, most microprocessors and computer systems use the 2's complement form. It is customary to use the most significant bit as the sign bit.

Depending on the word length of the computer, the number of bytes used to represent single-precision numbers is usually limited to the word size of the computer. This may be 1 byte, 2 bytes, 4 bytes, or even as much as 16 bytes long. If the numbers are longer than the word size, they are called multiple-precision numbers. The exact number of bytes varies for different computer systems.

Floating-Point Arithmetic   In floating-point arithmetic, fractions are represented by the mantissa and the exponent. Both the mantissa and the exponent may be signed, although only the sign of the mantissa truly represents the sign of a number. Arithmetic operations involving fractional numbers represented in floating-point format are not as simple as those involving integers. For addition and subtraction of floating-point numbers, the following steps are required:

1. Compare the two exponents.
2. Align by shifting one of the two mantissas in such a way that the two exponents are the same.
3. Perform the addition or subtraction.
4. Align the resultant mantissa into standard form by eliminating leading zeroes and adjusting the exponent accordingly.

The above process is illustrated by the following example.

Example 7–6 ▪▬▬▬▬▬▬▬▬▬▬▬▬▬▬▬▬▬▬▬▬▬▬▬▬

Add 656.789_{10} to 99555.12_{10}.

Assume that there are seven significant digits in the mantissa representation. The two decimal numbers are first represented in BCD floating-point format as

$$99555.12 = 0.9955512 \times 10^5$$
$$656.789 = 0.6567890 \times 10^3$$

STEP 1

Compare the two exponents.

The two exponents are obviously different. Therefore the subtraction cannot be carried out until both exponents are the same. The difference in the exponents is $5 - 3 = 2$.

STEP 2

Align the mantissas.

Either the augend or the addend can be shifted. In this case we arbitrarily choose to shift the augend.

$$0.9955512 \times 10^5 = 99.55512 \times 10^3$$

Since the difference between the exponents of the augend and the addend is positive, the augend is larger than the addend. Hence the augend is shifted to the left to reduce the exponent of the augend. If the difference of the exponents is negative, then the augend must be shifted to the right. Note that any shifting often means a loss of accuracy on one of the operands.

STEP 3

Perform the addition.

After the exponents are aligned, the required operation can be carried out.

$$(99.55512 + 0.6567890) \times 10^3 = 100.2119 \times 10^3$$

The addition should have resulted in 100.2119090×10^3, but because of the limitation of seven significant digits, the actual result is rounded off to 100.2119×10^3. Round-off and truncation errors can be very significant in floating-point arithmetic. In extreme conditions, subtraction of two floating-point numbers that are very close to one another in magnitude may result in grossly inaccurate results.

STEP 4

Realign the exponent.

The result in step 3 is not in standard form because the fractional point is not to the left of the mantissa. Hence the fractional point must be shifted to the left by three positions and the exponent incremented by three. The final form of the result is

$$100.2119^3 = 0.1002119 \times 10^6$$

The procedures for multiplication and division of floating-point numbers are somewhat simpler than those for addition and subtraction. For multiplication, the steps are

1. Multiply the two mantissas to form the mantissa of the product.
2. Add the two exponents to form the exponent of the product.
3. Realign the product into standard form.

The steps for a floating-point division are exactly the same as those for multiplication with the exception that the mantissa of the dividend is divided by that of the divisor and the exponent of the divisor is subtracted from that of the dividend.

Non-Numeric Data Types

Information is not always represented in numeric format. Non-numeric data include logical data, alphanumeric data, and addresses. *Logical data* are used to represent test conditions and test results. *Alphanumeric data* are textual data representing characters, graphics, and so on. *Addresses* are pointers that indicate the location of the operands in memory.

Logical Data Types   Logical data are usually represented by the bits themselves. For example:

0   False
1   True

Sometimes, a word containing all 0's or all 1's is also used to represent the true/false condition.

In many applications, groups of logical bits or flags are collected and placed together in a word. This necessitates the different logical operations. Operations involving logical data types are primarily Boolean operators such as the logical AND, the logical OR, the Exclusive OR, and so on. These logical operations have already been discussed in detail in Chapter 2.

Alphanumeric Data Types   Alphanumeric data are mostly characters or textual information. It is commonly accepted that a character is internally represented by a byte, though externally a character may be represented by a 6-bit display code (Control Data Corp. format), 7-bit ASCII, or 8-bit EBCDIC. Some of these codes have been presented in Chapter 1. Alphanumeric data types can also be used to represent color graphics. A pixel in normal black-and-white graphics in raster format can be represented by a single bit. For color, additional bits are needed to indicate the contributions from red, green, and blue. More bits to indicate the intensity of each basic color are sometimes used to obtain different shades and hues of color. Hence a collection of these bits into a word may be used to represent a single pixel in a picture.

Address Data Types   Addresses are used primarily to identify memory operands. The width of the address determines the maximum number of distinct memory locations directly addressable by the processor. The number of address bits used by a processor may or may not be the same as the word length. For example, the 8085 microprocessor uses 8-bit registers to hold data values but 16-bit registers to handle addresses. On the other hand, the 68010 has 32-bit data and address registers. Addresses are used and manipulated when dealing with loops, arrays, tables, and other data structures presented in the next section.

7.9

Data Structures

Data items can be collected together and identified as an entity with an associated name for reference. There are many data structures, some of which include:

Array/Vector/Table/List
String
Queue
Dequeue
Stack
Linked list/Tree
File

Array/Vector/Table/List

An *array* is a finite and ordered collection of data items of the same type. The contents of an array can be numbers, addresses, names, or other parameters. The principal characteristics of an array are that the items are homogeneous in nature and are positioned one next to another in an ordered, possibly linear, fashion. The size of an array is usually known at the time of its creation and is not expected to change dynamically. Examples of arrays are shown in Figure 7–23. A particular item in the array is identified by the name of the array and the subscript of the array. In implementation, the name of the array is the starting address and the subscript is the offset between the starting address of the array and the address of the desired element.

A one-dimensional array is linear in its dimension and can be simply implemented as an ordered sequence. A two-dimensional array can be linearized into a one-dimensional array by subscript calculation. For example, a two-dimensional array with m rows and n columns can be represented by a one-dimensional array with $(m \times n)$ elements. Any element in the original two-dimensional array with subscripts j and k is equivalent to subscript i in the new array, where

$$i = (k - 1) * m + j$$

Other multidimensional arrays can be linearized in the same manner.

String

A *string* is a collection of text or alphanumeric characters. A string is identified by a name and its associated length. A particular character is identified by the offset from the starting address. Sometimes a small group of consecutive characters inside a string is collectively referred to as a *substring*, as shown in Figure 7–24. The format of a substring is expressed as

```
SUBSTRING = STRING(a:b)
```

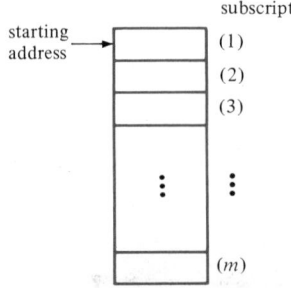

(a) One-dimensional array

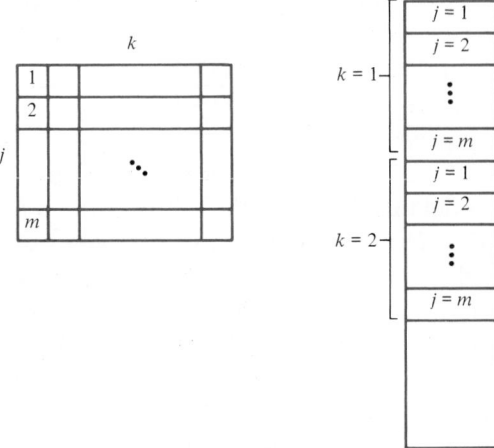

(b) Two-dimensional array

Figure 7–23
Array/table

Figure 7–24
Strings, substrings, and their relation

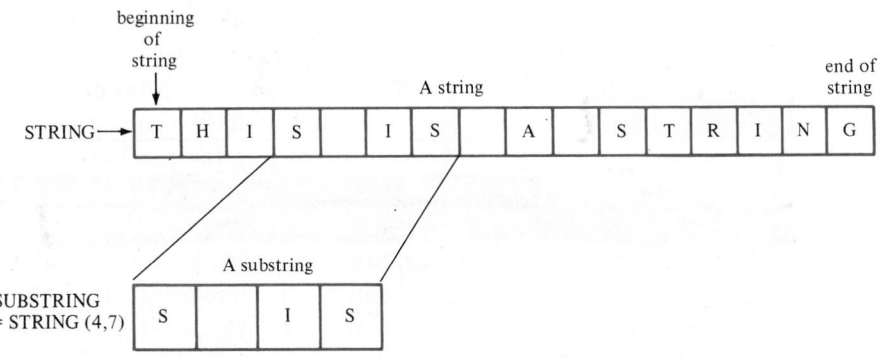

where STRING is the name of the string, SUBSTRING is the name of the substring, and the variables *a* and *b* represent the lower and upper limits, respectively, of the substring within the parent string.

Queue

A *queue* is similar to an array but with dynamic length. A queue is identified by two pointers, a last-data-in pointer and a next-data-out pointer, as shown in Figure 7–25(a). A queue grows in one direction, either toward upper memory or toward lower memory. Assume that the queue grows toward the high end of the memory space. When the queue is first created, the last-data-in pointer is smaller than the next-data-out pointer. This indicates the null condition, which occurs when there are no data items stored in the queue. When there is only one data item in the queue, both pointers are identical. When another data item arrives, the last-data-in pointer will be numerically larger than the next-data-out pointer.

When a data item arrives, the last-data-in pointer is first updated to the next available slot. The incoming data item is then stored into the queue at the location pointed to by the last-data-in pointer as shown in Figure 7–25(b). Similarly, when the next data item is recalled, the data pointed to by the next-data-out pointer is first retrieved and the next-data-out pointer is then subsequently updated as shown in Figure 7–25(c). Because data items enter from one end of the queue and exit from the other, the order of the items is preserved. The queue merely acts as a buffer to store the incoming data temporarily until it is called for later in the process. Hence it is often referred to as a first-in, first-out (FIFO) buffer. In other words, the data items are recalled in exactly the same order as they come in.

The memory space between the next-data-out pointer and the last-data-in pointer identifies the filled space where data items are stored in the queue. Hence the number of items in the queue is given by N, where

$$N = \text{last-data-in} - \text{next-data-out} + 1$$

A queue tends to creep over the available memory space as items are stored and retrieved. The solution is to limit the usable space of the queue and reuse the space where items have been retrieved. This is the concept of a circular queue. A *circular queue* is one that wraps around by limiting the queue to stay within a certain section of memory. This is done by imposing an upper limit and a lower limit on the pointers. As any one of the pointers reaches the lower limit, the same pointer is immediately replaced by the upper limit and vice versa. This is shown in Figure 7–26(a). The implementation of a circular queue is shown in Figure 7–26(b). After adding two more items to the circular queue as shown in Figure 7–26(b), the lower limit is reached and the last-data-in pointer is immediately replaced by the upper limit before further items are added to the queue. The result of adding three more items to the queue is shown in Figure 7–26(c). The flow-

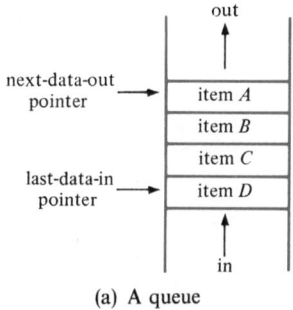

(a) A queue

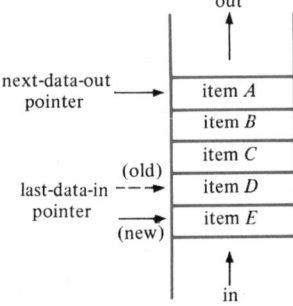

(b) Putting an item into a queue

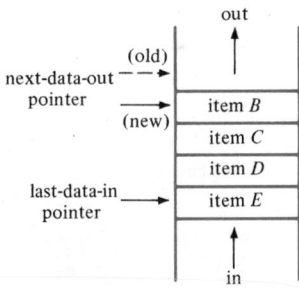

(c) Retrieving an item from the queue

Figure 7–25
(a) A queue; (b) a queue after addition of an item; (c) a queue after removal
of an item

charts for adding items to and deleting items from a circular queue are shown
in Figures 7–27(a) and 7–27(b).

Dequeue

A dequeue is similar to a queue. While a queue is single-ended, a *dequeue* is
double-ended. For a queue, the data can only enter from one end and exit from

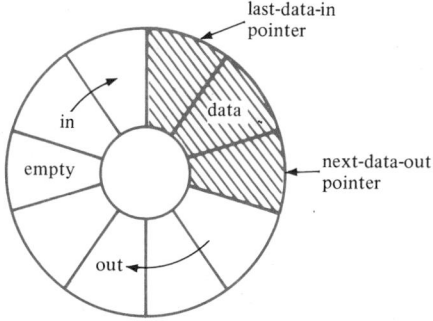

(a) A circular queue with 10 slots

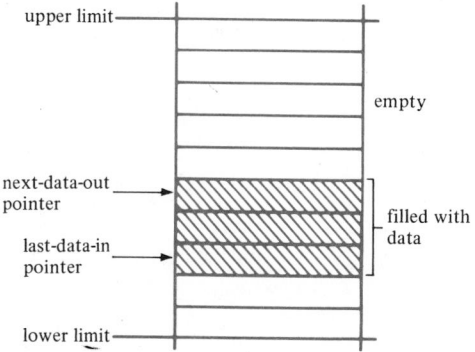

(b) Implementation of a circular queue

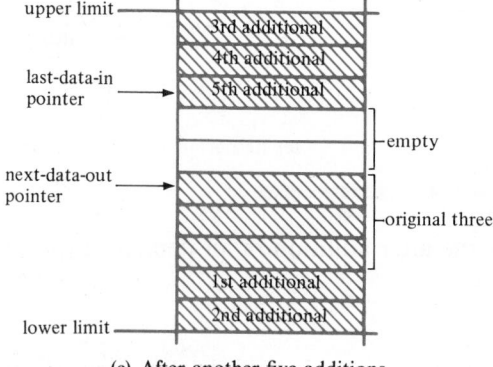

(c) After another five additions

Figure 7–26

A circular queue: (a) concept: (b) implementation; (c) after addition of five more items

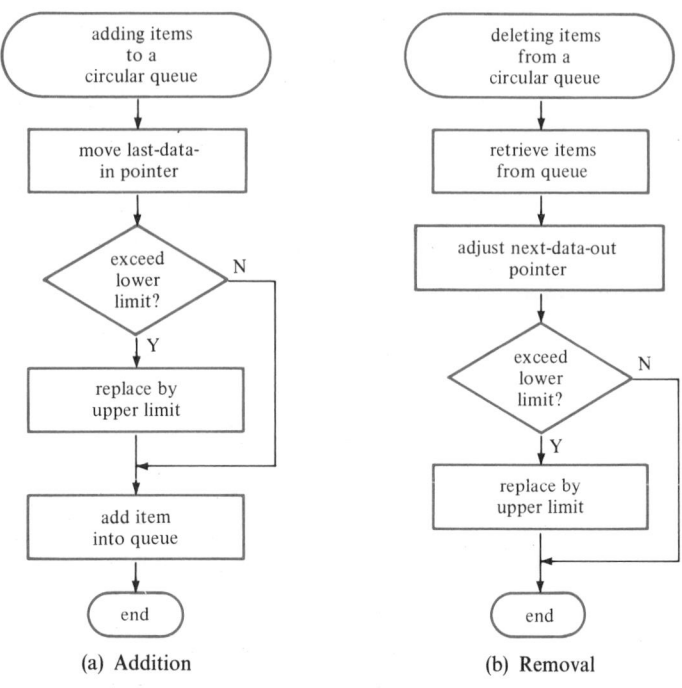

(a) Addition   (b) Removal

Figure 7–27

Flowcharts for: (a) adding items to a circular queue; (b) removing items
from a circular queue

the other. For a dequeue, the data items can enter and exit from either end as
shown in Figure 7–28. A dequeue is maintained by two pointers: the top-of-data
pointer and the bottom-of-data pointer. The memory space between these two
pointers identifies the stored items or, in other words, the contents of the dequeue.

Stack

While a queue maintains data in a first-in, first-out fashion, the *stack* does so in
a last-in, first-out (LIFO) fashion. A queue preserves the order of the data items,
while a stack reverses the order of the data items. A typical stack is shown in
Figure 7–29(a). The bottom of the stack is fixed. As items are entered into the
stack, they are put onto the top of the stack as shown in Figure 7–29(b). As
another data item comes in, it is stacked on top of the last entered item. Hence
the name stack. When the stacked items are retrieved, what is on the top of the
stack is removed first, as shown in Figure 7–29(c). So the order of the data being
removed is reversed when compared to the order of the data being placed on
the stack.

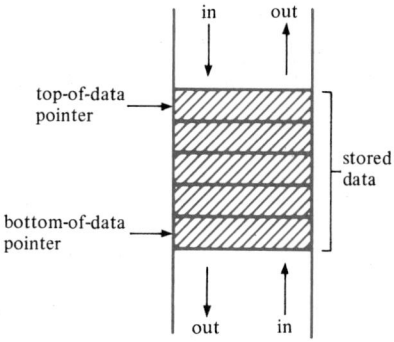

Figure 7–28

A dequeue

Figure 7–29

A stack: (a) concept; (b) after addition of an item; (c) after removal of an item

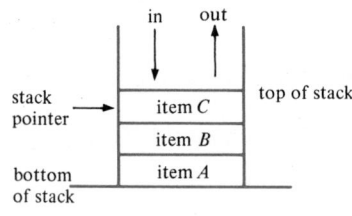

(a) A stack

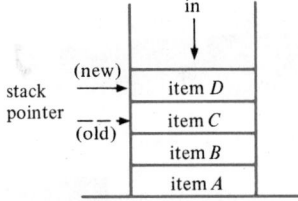

(b) Putting an item onto the stack

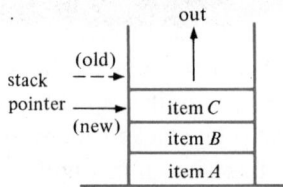

(c) Retrieving an item from the stack

The stack is maintained by a single pointer commonly called the *stack pointer* (SP). As items are stacked, the stack pointer is updated before the items are stored. As items are retrieved, they are retrieved before the stack pointer is changed. The stack pointer is always updated in such a way that it is pointing to the last available item in the stack at all times.

Storing an item onto the stack is called a *PUSH operation*. Similarly, retrieving an item from the stack is called a *POP operation*. If the top of the stack is at a lower address than the bottom of the stack, the PUSH and POP operations can be represented as

PUSH	$((SP) - 1) \leftarrow$ Data
	$(SP) \qquad \leftarrow (SP) - 1$
POP	Data $\leftarrow ((SP))$
	$(SP) \leftarrow (SP) + 1$

On the other hand, if the top of the stack is at a higher address than the bottom of the stack, the stack operations are

PUSH	$((SP) + 1) \leftarrow$ Data
	$(SP) \qquad \leftarrow (SP) + 1$
POP	Data $\leftarrow ((SP))$
	$(SP) \leftarrow (SP) - 1$

When the stack pointer is pointing to the same address as the bottom of the stack, the stack is empty.

Linked List/Tree

While the foregoing data structures are used primarily with homogeneous data items, a tree can be adapted easily for both homogeneous and nonhomogeneous data items. A *tree* is similar to a list of items except that, unlike a list, the data items in a tree are not placed sequentially but are linked together by pointers. No pointers are needed in a list because all items are placed sequentially and consecutively. In a list, data items are assumed to be consecutively placed. In a tree, each data item can be placed anywhere. The relationship between items is preserved by means of pointers. Each data item contains a pointer that indicates its parent and one or more pointers that indicate its children, descendants, or subordinates, as shown in Figure 7–30(a). Hence a tree is also called a *linked list.* The implementation of a linked list is shown in Figure 7–30(b).

A tree begins with the *root node*, which is identified as a node with a null parent pointer. The root leads into *intermediate nodes*. A *binary tree* is one in which each node has only two descendants. A *leaf* or *terminal node* of a tree has no descendants. To traverse a tree from one terminal node to another, one must ascend the tree from the source node to the proper common parent and then descend down to the destination node. For example, to go from node 001 to

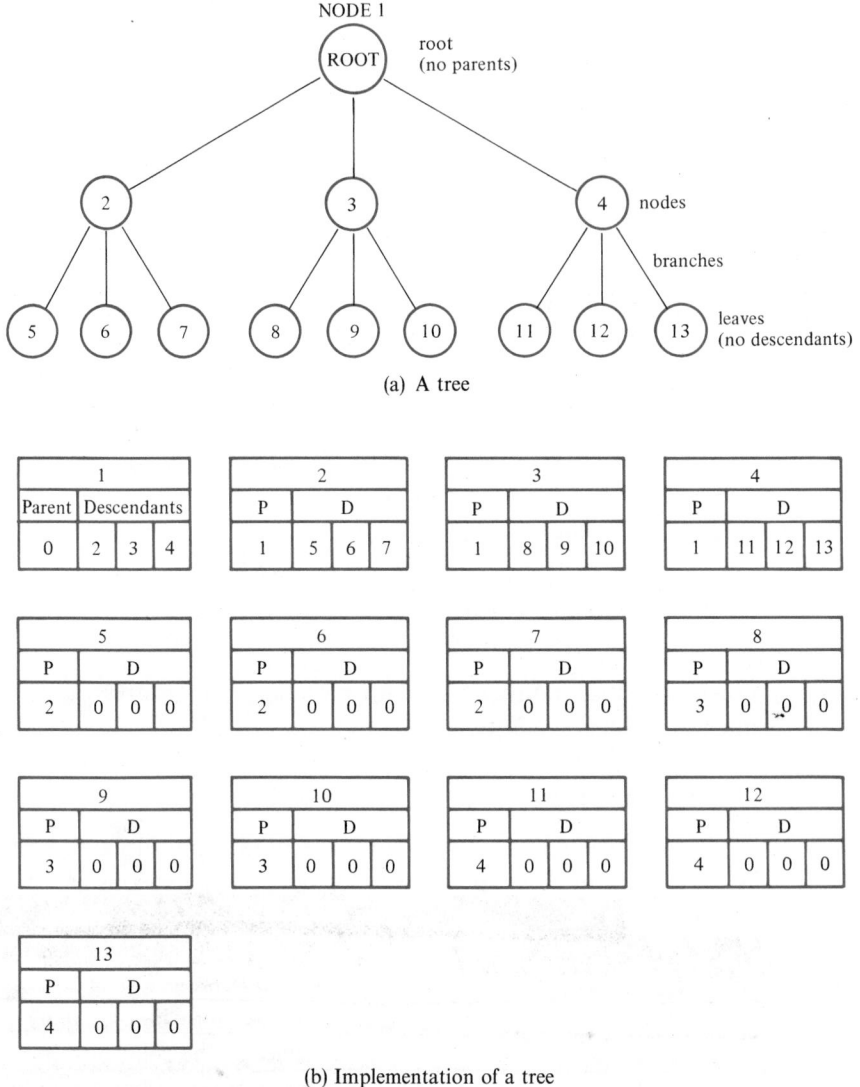

(a) A tree

(b) Implementation of a tree

Figure 7–30
A tree: (a) concept; (b) implementation

node 100 in the binary tree shown in Figure 7–31, one first ascends to node 00*x*, the parent of node 001, then to node 0*xx*, then through the root to node 1*xx*, to its descendant 10*x*, and finally to its descendant 100.

One common use of a tree is to represent algebraic expressions. A binary tree is used because each algebraic operator has two operands. The nodes represent

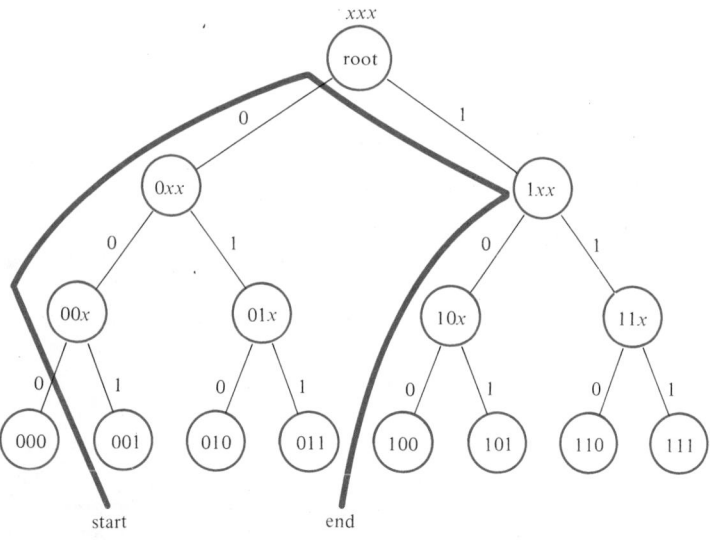

Figure 7–31

Traversing a tree from node 001 to node 100

the operators and the terminal nodes represent the operands. Each level of parentheses requires an additional level of nodes.

Example 7–7

Represent the following algebraic expression by a binary tree:

$$((a + b) * c - d/e + f)/(g * (h - j))$$

The binary tree representation of this expression is shown in Figure 7–32(a).
Note that the order of operations is clearly identified by the different levels of
the tree. No parentheses are required in the final representation.

While the normal algebraic expression is usually given in the infix format—
that is, the operator is placed between the two operands—*reverse Polish notation*
(RPN) is a suffix format in which the operator is placed at the end of the two
operands. Hence the operands precede the operator. The advantage of RPN is
that no parentheses are required to represent any arbitrary algebraic expression.
To convert an expression to RPN form, first represent the expression in the form
of a binary tree. The RPN format results from traversing the tree from left to
right while writing down the descendants before their parents.

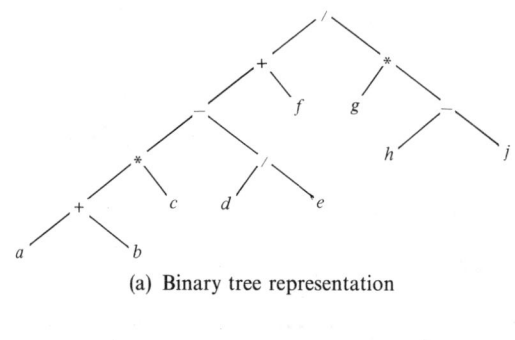

(a) Binary tree representation

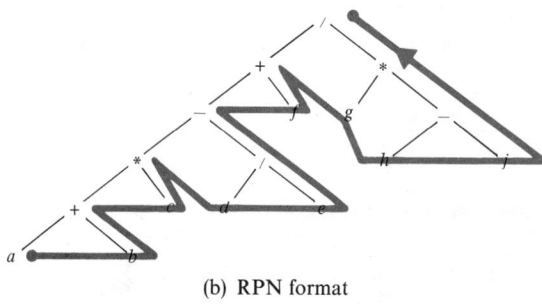

(b) RPN format

Figure 7–32
(a) Representing an algebraic expression by a binary tree; (b) traversing
the tree to obtain the RPN format

Example 7–8

Rewrite the expression in the previous example in RPN form.
By examining the binary tree, the RPN format can be written down immediately. This is shown in Figure 7–32(b).

File

A *file* is a collection of data items with a common name. In general, a file may
be stored in sequential or random access form. A *sequential file* is one that places
records sequentially and consecutively one after another. Hence access to any
record is sequential. Examples of sequential files are files stored on magnetic
tapes and paper tapes. A *random access file* allows direct program access into
any part of the file without prior access to any other part of the file. In a sequential file, the records are placed consecutively. In a random access file, the records

are not placed in any particular order at all. A file is composed mostly of records. Records are the most basic unit in a file. A record may contain simply a single data item, or it may contain a collection of bytes with many different fields each containing different kinds of data types and data formats.

References

1. Ahmad, S. Imtiaz, and Kwok T. Fung. *Introduction to Computer Design and Implementation.* Computer Science Press, Rockville, Md., 1981.
2. Alexandridis, Nikitas A. *Microprocessor System Design Concepts.* Computer Science Press, Rockville, Md., 1984.
3. Baer, Jean-Loup. *Computer Systems Architecture.* Computer Science Press, Potomac, Md., 1980.
4. Booth, Taylor L. *Introduction to Computer Engineering Hardware and Software Design.* John Wiley & Sons, New York, 1984.
5. Camp, R. C., T. A. Smay, and C. J. Triska. *Microprocessor Systems Engineering.* Matrix Publishers, Portland, Ore., 1979.
6. Cannon, Don L. *Fundamentals of Microcomputer Design: System Hardware and Software.* Texas Instruments, Dallas, Tex., 1982.
7. D'Angelo, Henry. *Microcomputer Structures.* Byte Publications, Peterborough, N.H., 1981.
8. Dasgupta, Subrata. *The Design and Description of Computer Architectures.* John Wiley & Sons, New York, 1984.
9. Doty, Keith L. *Fundamentals of Microcomputer Architecture.* Matrix Publishers, Portland, Ore., 1979.
10. Garland, Harry. *Introduction to Microprocessor System Design.* McGraw-Hill, New York, 1979.
11. Gault, James W., and Russell L. Pimmel. *Introduction to Microcomputer-Based Digital Systems.* McGraw-Hill, New York, 1982.
12. Givone, Donald D., and Robert P. Roesser. *Microprocessors/Microcomputers: An Introduction.* McGraw-Hill, New York, 1980.
13. Gorsline, G. W. *Computer Organization: Hardware/Software.* Prentice-Hall, Englewood Cliffs, N.J., 1980.
14. Hamacher, V. Carl, Z. G. Vranesic, and S. G. Zaky. *Computer Organization,* 2d ed. McGraw-Hill, New York, 1984.
15. Hayes, John P. *Computer Architecture and Organization.* McGraw-Hill, New York, 1978.
16. Hayes, John P. *Digital System Design and Microprocessors.* McGraw-Hill, New York, 1984.
17. Heffer, D. E., G. A. King, and D. Keith. *Basic Principles and Practice of Microprocessors.* John Wiley & Sons, New York, 1981.
18. Hill, F. J., and Peterson, G. R. *Digital Logic and Microprocessors.* John Wiley & Sons, New York, 1984.
19. Hitachi Corp. *IC Memories.* Hitachi Corp., San Jose, Calif., 1984.
20. Jermann, William H. *The Structure and Programming of Microcomputers.* Alfred Publishing, Sherman Oaks, Calif., 1982.

21. Johnson, E. L., and M. A. Karim. *Digital Design, A Pragmatic Approach.* PWS Publishers, Boston, 1987.
22. Kline, R. M. *Digital Computer Design.* Prentice-Hall, Englewood Cliffs, N.J., 1977.
23. Kogge, Peter M. *The Structure of Pipelined Computers.* Hemisphere, Washington, D.C., 1981.
24. Kostopoulos, G. K. *Digital Engineering.* John Wiley & Sons, New York, 1975.
25. Lewin, Morton H. *Logic Design and Computer Organization.* Addison-Wesley, Reading, Mass., 1983.
26. Mano, M. Morris. *Digital Logic and Computer Design.* Prentice-Hall, Englewood Cliffs, N.J., 1979.
27. Mano, M. Morris. *Computer System Architecture*, 2d ed. Prentice-Hall, Englewood Cliffs, N.J., 1982.
28. Oleksy, Jerome E., and George B. Rutkowski. *Microprocessor and Digital Computer Technology.* Prentice-Hall, Englewood Cliffs, N.J., 1981.
29. Peatman, J. B. *Microcomputer Based Design.* McGraw-Hill, New York, 1977.
30. Rafiquzzaman, M. *Microcomputer Theory and Applications with the Intel SDK-85.* John Wiley & Sons, New York, 1982.
31. Rafiquzzaman, Mohamed. *Microprocessors and Microcomputer Development Systems: Designing Microprocessor-Based Systems.* Harper & Row, New York, 1984.
32. Ramirez, Edward V. *Microprocessing Fundamentals: Hardware and Software.* McGraw-Hill, New York, 1980.
33. Roth, Charles H., Jr. *Fundamentals of Logic Design*, 3d ed. West Publishing, St. Paul, Minn., 1985.
34. Short, Kenneth L. *Microprocessors and Programmed Logic.* Prentice-Hall, Englewood Cliffs, N.J., 1981.
35. Siewiorek, Daniel P., C. G. Bell, and Allen Newell. *Computer Structures: Principles and Examples.* McGraw-Hill, New York, 1982.
36. Sloan, M. E. *Computer Hardware and Organization*, 2d ed. Science Research Associates, Chicago, 1983.
37. Stone, Harold S., ed. *Introduction to Computer Architecture*, 2d ed. Science Research Associates, Chicago, 1980.
38. Stout, David F. *Microprocessor Applications Manual.* McGraw-Hill, New York, 1982.
39. Tanenbaum, Andrew S. *Structured Computer Organization*, 2d ed. Prentice-Hall, Englewood Cliffs, N.J., 1984.
40. Taub, Herbert. *Digital Circuits and Microprocessors.* McGraw-Hill, New York, 1982.
41. Tennebaum, Aaron M., and Moshe J. Augenstein. *Data Structures Using PASCAL.* Prentice-Hall, Englewood Cliffs, N.J., 1981.
42. Texas Instruments. *Microprocessor/Microcomputers System Design.* McGraw-Hill, New York, 1980.
43. Wakerly, John F. *Microcomputer Architecture and Programming.* John Wiley & Sons, New York, 1981.
44. Ware, W. H. *Digital Computer Technology and Design*, Vol. 1. John Wiley & Sons, New York, 1963.
45. Wiatrowski, Claude A., and Charles H. House. *Logic Circuits and Microcomputer Systems.* McGraw-Hill, New York, 1980.
46. Wilkinson, Barry. *Digital System Design.* Prentice-Hall, Englewood Cliffs, N.J., 1987.
47. Williams, Gerald E. *Digital Technology.* Science Research Associates, Chicago, 1977.
48. Winkel, David, and Franklin Prosser. *The ART of Digital Design.* Prentice-Hall, Englewood Cliffs, N.J., 1980.

Problems

1. With the computer architecture given in Figure 7–11, use RTL format to describe the operations required to sum five elements of an array previously stored in memory.

2. With the computer architecture given in Figure 7–11, use RTL format to describe the operations required to multiply two numbers by the repeated-add method.

3. With the computer architecture given in Figure 7–11, use RTL format to describe the operations required to multiply two numbers by the shift-and-add method.

4. Design the necessary hardware to convert a two-digit BCD number stored in register A into a binary number. Write the sequence of operations required in RTL format.

5. Design the necessary hardware to normalize a floating number. Write the sequence of operations in RTL format.

6. How are loops represented in RTL format?

7. How would you implement loops in hardware? Write the necessary operations in RTL format.

8. What is the difference between ASL and LSL?

9. Devise a circuit to shift more than one bit at a time.

10. Write out the entire sequence of operations needed to execute a complete instruction cycle to add two register operands.

11. Write out the entire sequence of operations needed to execute a complete instruction cycle to add two operands stored in memory locations A and B. Put the result back into memory location C.

12. Assuming an empty stack to begin with, show graphically what would be the contents of the stack after each of the following stack operations:
    (a) PUSH A
    (b) PUSH B
    (c) POP C
    (d) PUSH D
    (e) POP E
    (f) POP F
    (g) PUSH F

13. If the PUSH and POP operations are PUT and GET operations of a queue and if there are already five items in the queue, show graphically the result of the queue if the sequence of PUSHes and POPs in Problem 12 is performed in the queue.

14. Use a binary tree to represent the following arithmetic expressions and rewrite the expressions in RPN form:
    (a) $((a + b) * c)/d$
    (b) $(x * y * z)/(x - y)$
    (c) $(r - s) * (t + u)/(r - u)$

15. Write the algorithm for traversing a tree from one leaf to another.

16. Suppose that we have the architecture shown in Figure P7–1. Show the steps required for the following operations in RTL format:
    (a) The OPCODE fetch
    (b) Summing two operands
    (c) Fetching an operand with direct addressing assuming that the address is located at the word after the OPCODE
    (d) Indexed addressing
    (e) Relative addressing

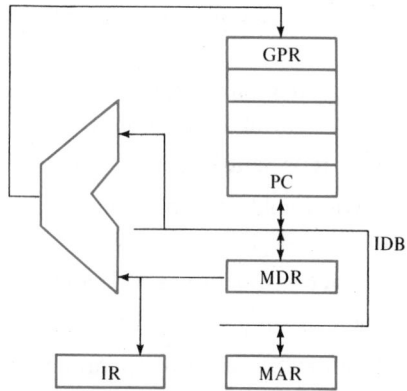

Figure P7–1

Architecture of a hypothetical machine

17. Name a few ways that can be used to clear a register—that is, to zero out a register.

18. Redo the quadratic equation in Example 7–5 using fewer temporary registers.

19. Use RTL format to express a floating-point addition.

20. Use RTL format to express a floating-point normalization.

21. Use RTL format to express a bubble sort.

Memory and I/O

8

In this chapter we discuss the memory and input/output (I/O) aspects of a generic computer system. The memory system is a crucial item in that it is used to store instructions and data. The way the memory is addressed governs the overall performance of the computer system. Concerning the memory, we examine memory hierarchy, various addressing modes, and basic memory organizations. Concerning the input/output facility, we examine the types of I/O, the interrupt system, and the programming techniques of I/O. The latter part of the chapter deals with system organization.

8.1

The Memory

No computer system is complete without memory. Memory is used to store the instructions, constants, data, or any other information pertinent to the proper execution of a computer program. The processor expects instructions and data to be stored in memory. This is called the *stored program concept*. Execution of a program means going through the instructions that are stored in memory sequentially, unless otherwise directed.

Memory Hierarchy

Generally speaking, *memory* refers to any storage medium used for storing information. Since there are many different kinds of storage media, it is common to categorize them according to the memory speed, which in turn determines how the different types of memory are used. This classification forms a memory hierarchy as follows:

1. Cache memory
2. Main (primary) memory
3. Secondary memory
4. Archival memory

Figure 8–1 shows the memory hierarchy according to speed, size, and cost. In

Figure 8–1

Characteristics of the various memory hierarchies

	Memory Type			
	Cache	Main	Secondary	Archival
Technology	TTL	MOS	Magnetic	Paper
Speed (access time)	10^1 ns	10^2 ns	10^3 ns	10^5 ns
Size (capacity)	Small, <8K	Medium, 64K–8M	Large, 300K–1G	Very large, $\to \infty$
Usage	Processor, data buffer	On-line, dynamic, temporary storage	On-line, bulk storage	Archival, off-line, permanent storage
Representative examples	TTL memories	Semiconductor memories	Floppy disk, hard disk, tape	Paper

general, the faster the speed of information access, the higher the cost, and the smaller the size of the memory space. Conversely, the lower the cost of memory, the larger the size, yet the slower the speed.

Cache Memory The *cache memory* is a small section of memory that is usually placed between the processor and the main memory and is used whenever the main memory speed is much slower than the processor speed. The size of the cache memory is usually very small. Its speed, however, is extremely fast and is comparable to the processor speed. It is used as a buffer between the main memory and the processor so that the processor can operate at maximum speed without being otherwise slowed down by the main memory.

Main (Primary) Memory The *main memory* contains the primary storage space that is randomly accessible by the processor. Its size is governed by the intended application area and varies greatly from one installation to another. The primary function of the main memory is to store the instructions and data necessary for the current execution of the processor. Information in the main memory is usually directly accessible by the processor. The width of the words in main memory is usually the same as the internal registers of the processor. The maximum size of the main memory is governed by the number of address bits provided by the processor. If additional memory is needed, several memory extension techniques can be used to increase the size of the memory beyond this limitation.

Secondary Memory The information stored in the main memory is not easily accessible to the programmer and is not suited for storing a lot of information. Secondary storage media are useful for this purpose. *Secondary memory* refers to disks, floppy disks, and magnetic tapes. Typical cost and performance data for these media are listed in Figure 8–2. Fast storage media work in conjunction with the main memory to provide additional bulk storage due to the limited

Figure 8–2

Characteristics of secondary and archival storage media

| Memory Type | Cassette | Secondary | | | Archival |
		Floppy Disk	Hard Disk	Magnetic Tape	Paper
Cost/bit (¢/bit)	10^{-3}	10^{-3}	10^{-3}	10^{-4}	10^{-5}
Capacity (bits)	10^6	10^7	10^9	10^9	$\to \infty$
Access time (s)	10^2	10^1	10^0	10^0	10^2
Transfer rate (bits/s)	10^3	10^4	10^5	10^5	10^2

amount of storage available in the main memory. Information stored in secondary storage is usually not directly accessible by the processor. Special utility routines are needed to retrieve the required section of the data stored in secondary storage and to transfer the information to the main memory for further processing.

Archival Memory *Archival memory* refers to permanent storage media. These include media such as cards and paper printouts. Their use is strictly for the programmer to store computation results. The access time for archival storage is typically extremely slow. The characteristics of an archival memory are also given in Figure 8–2.

A proper treatment of cache memory is beyond the scope of this book. Many references dealing with this subject are listed at the end of this chapter. The main memory is of primary importance to the design of digital systems and will be studied in detail. Secondary storage media are usually related to I/O peripherals, so detailed discussion of secondary storage media is outside the scope of this book. Archival storage media will not be discussed further, so in the remainder of this book, the term "memory" refers to the main memory.

8.2

Addressing Modes

In order to access the memory, an address must be provided. The manner in which the address is specified in an instruction is called the *addressing mode*. A great number of addressing modes are usually made available in a processor. These addressing modes can be categorized into basic and special modes.

Basic addressing modes include:

1. Implied addressing
2. Immediate addressing
3. Register addressing
4. Direct addressing
5. Register indirect addressing
6. Implied register indirect addressing
7. Indexed addressing

Special addressing modes include:

1. Relative addressing
2. Base addressing
3. Page addressing
4. Indirect addressing

For each of these addressing modes, we will present the RTL description of the operations necessary to obtain the operand. In the RTL format supplied with the description of each addressing mode, the parentheses are taken to mean "the contents of." For example, (A) means "the contents of register A," whereas A without the parentheses means "the bit pattern that specifies register A." The effective address, EA, is the final logical address of the operand obtained as specified by the addressing mode.

Basic Addressing Modes

Most computer systems use a number of basic addressing modes to obtain operands from memory; many high-performance processors include additional addressing modes consistent with the design philosophies of the processor. In this section we discuss some of the basic addressing modes found in most processors.

Implied Addressing

EA	None
Operand	(Implied register)

Implied addressing means that the operand is automatically implied according to the context of the instruction and hence does not need to be specified explicitly. Examples of implied addressing are instructions such as STC (set carry) or CMC (complement carry) for the Intel 8085 microprocessor, because the operand—the carry bit of the status register—is implicitly referenced by the OPCODE. The implied addressing mode is shown in Figure 8–3(a).

Immediate Addressing

EA	None
Operand	(Data part of the IR)

or

EA	(PC) + 1
Operand	(Bytes following the OPCODE)

In *immediate addressing*, the operand itself is supplied with the instruction or is stored in memory locations following the OPCODE of the instruction. As part of a multiword instruction, such an operand is often called a *literal*. In computers with large word length, the operand is made part of the instruction and no additional memory access is necessary to get the operand. For microprocessors, the literal is usually stored in the bytes following the OPCODE. Their addresses are specified by the PC and do not have to be specified explicitly. An example of this scheme is the **LXI B,2222H** instruction for the 8085, which loads the literal 2222H, stored after the OPCODE, into the register pair B and C. Illustrations of the immediate addressing mode appear in Figures 8–3(b) and 8–3(c).

Register Addressing

EA   Register number
Operand   (Specified register)

In *register addressing*, or *register direct addressing*, the operand for the instruction is stored in a register file. The only address needed is the register number. No memory access is required, because the operands are already stored internally in the processor. An example of this type of addressing scheme for the 8085 is the MOV B,A instruction, which transfers the contents of register A to register B. Note that both the source and destination operands are identified simply by specifying the names of the associated registers. The register addressing mode is illustrated in Figure 8–3(d).

Figure 8–3

Addressing modes

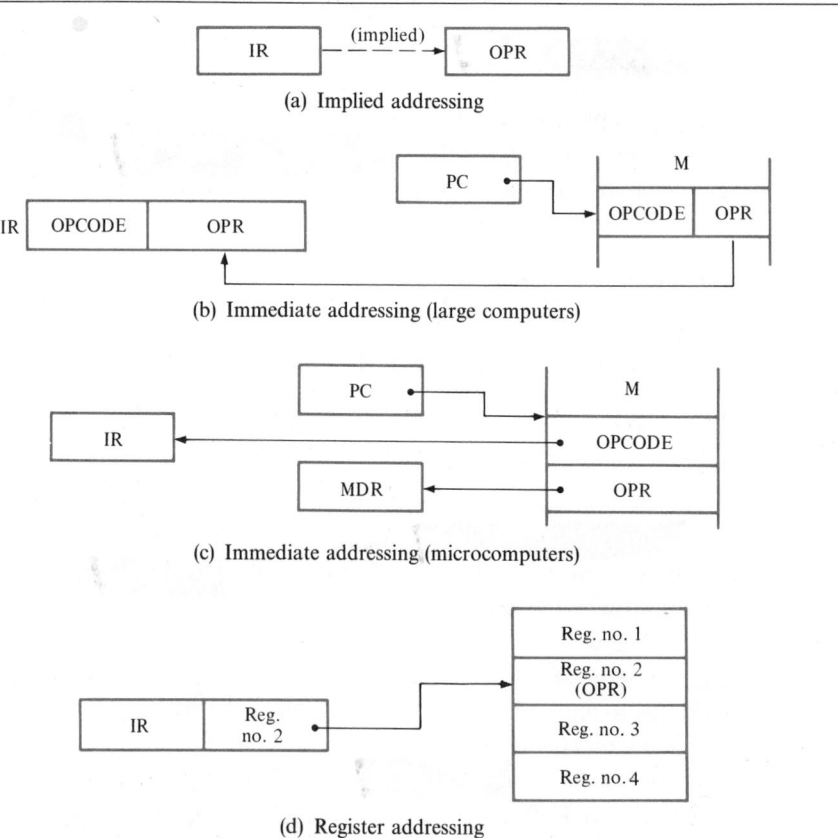

(a) Implied addressing

(b) Immediate addressing (large computers)

(c) Immediate addressing (microcomputers)

(d) Register addressing

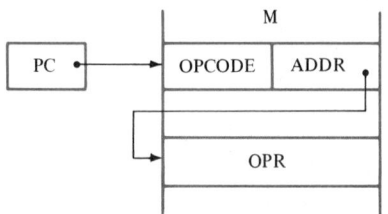

(e) Direct addressing (large computers)

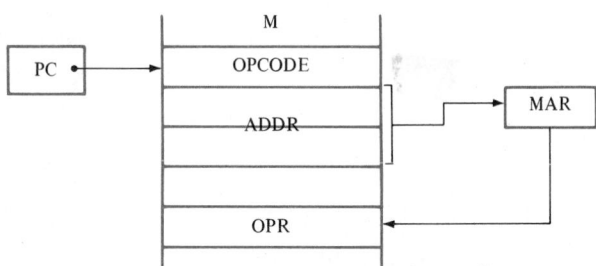

(f) Direct addressing (microcomputers)

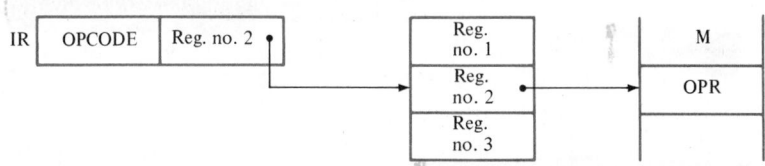

(g) Register indirect addressing

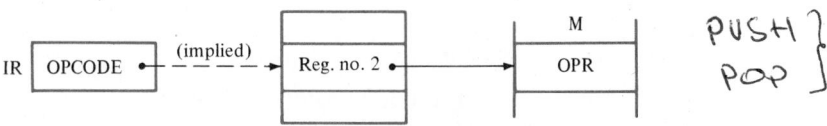

(h) Implied register indirect addressing

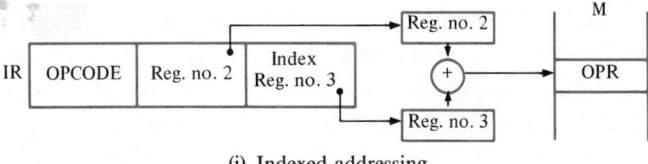

(i) Indexed addressing

Figure 8–3 Continued

Direct Addressing

EA	Address part of the (IR)
Operand	(Address part of the (IR))

or

EA	((PC) + 1)
Operand	((Bytes following the OPCODE))

In *direct addressing*, the memory address of the operand is supplied as part of the instruction. For machines with long word size, the address is actually part of the instruction fetched. For microprocessors, the address is stored in subsequent memory locations. An example of this type of instruction is the LDA 0100H instruction for the 8085, which is used to load the operand stored at location 0100H into accumulator A. The direct addressing mode is shown in Figures 8–3(e) and 8–3(f).

Register Indirect Addressing

EA	(Register number)
Operand	((Specified register))

In *register indirect addressing*, instead of the operand itself being stored in the specified register, it is the address of the desired operand that is stored in the register. When the instruction specifies a certain register, the contents of that register are actually used as an address to access the memory. The contents of the specified memory location give the operand. This is a common type of addressing scheme for microprocessors. An example of this type of instruction is the LDAX B instruction for the 8085, which loads into the accumulator the data whose address is stored in the B and C register pair. The register indirect addressing mode is shown in Figure 8–3(g).

Implied Register Indirect Addressing

EA	(Implied register)
Operand	((Implied register))

Implied register indirect addressing is a special case of register indirect addressing. The former is used when the register is implied by the instruction and does not need to be specified explicitly. For example, the PUSH and POP instructions for the 8085 microprocessor use the implied register indirect addressing mode because the address is always taken from the SP register. The implied register indirect addressing mode is shown in Figure 8–3(h).

Indexed Addressing

EA	(Register number) + (index register)
Operand	((Specified address register) + (index register))

In *indexed addressing,* the address contained in an address register is modified by an offset that is stored in another register commonly called the *index register.* The sum is then used as an address to fetch the operand from memory. This scheme is useful for table and array access. The table address can be specified as part of the instruction or stored in a register. The specific item or subscript of the desired element of the array is left in the index register. The index can be repeatedly incremented or decremented in a loop environment, allowing subsequent items to be accessed in an orderly manner. An example of this addressing scheme is the MOVE (A_n, X_n), A_n instruction. The first operand is the source operand and is obtained by summing the contents of the address register A_n and the contents of the index register X_n. The indexed addressing mode is shown in Figure 8–3(i).

In some processors, there is an auto-increment and/or auto-decrement feature associated with the indexed addressing mode. In the *auto-increment mode,* the index register is automatically incremented either before or after each use. Likewise, in the *auto-decrement mode,* the index register is automatically decremented either before or after each use.

Special Addressing Modes

Many processors have additional or special addressing modes that facilitate ease in implementing language constructs and primitives often used in operating systems and other system programming environments.

Relative Addressing

EA	(PC) + offset
Operand	((PC) + offset)

In *relative addressing,* instead of specifying the complete address as is done in direct addressing, only an offset with respect to the current PC is stored. The reason for this is that the size of an offset for a particular user is often much smaller than that for a complete address, resulting in a much smaller instruction word size. The disadvantage is that an address calculation is necessary for every relative memory reference because the offset must be added to the PC to become an effective address for the operand to be fetched from memory. An example of the relative addressing mode is the BRA LABEL instruction for the Motorola 68000, where the offset between the LABEL and the current PC location is stored in the instruction instead of the absolute address of LABEL. The relative addressing mode is shown in Figure 8–4(a).

Base Addressing

EA	(Base register) + specified address
Operand	((Base register) + specified address)

BRA ±

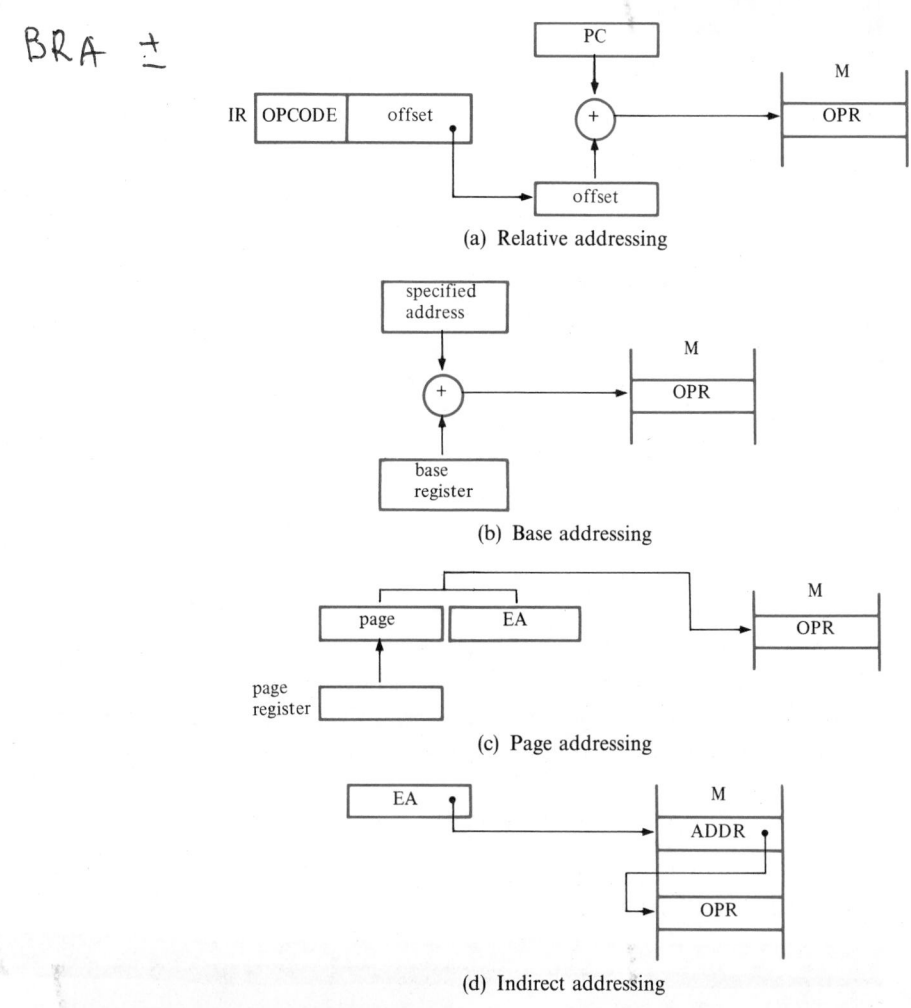

(a) Relative addressing

(b) Base addressing

(c) Page addressing

(d) Indirect addressing

Figure 8–4
Special addressing modes

Base addressing is useful in multiprogramming environments when the programs of more than one user are actively stored in memory at the same time. A memory reference from any user is automatically modified by the proper base address contained in that user's base register. This scheme provides a unique addressing space for each user. The user program is said to be *relocatable* and can reside anywhere in the available memory space. In some processors, base addressing can be used in conjunction with indexed addressing. The base addressing mode is shown in Figure 8–4(b).

Page Addressing

EA	(Page register) + specified address
Operand	((Page register) + specified address)

In some computer systems, it is desirable to extend the width of the address bus beyond the width of the internal address registers. An internally generated address is externally augmented by higher-order address bits. These higher-order bits are called *page numbers*. These higher-order bits effectively divide or segment the total main memory space into pages. Once the page number has been set, the processor needs to deal only with a much smaller address space. This is also a disadvantage, however, because the processor can deal only with operands within that page. The page addressing mode is shown in Figure 8–4(c).

Indirect Addressing

EA	(address)
Operand	((address))

The difference between direct addressing and indirect addressing is that in direct addressing, the address of the operand itself is specified, whereas in *indirect addressing* it is the address of the address of the operand that is specified. Indirect addressing can be combined with other addressing modes to form a class of indirect addressing modes. They are useful for many single accumulator machines that have none or only a limited number of address registers. The indirect addressing mode is illustrated in Figure 8–4(d).

A General Addressing Mode

Some of the addressing schemes mentioned previously can be combined to form additional addressing modes. A general scheme for address generation is shown in Figure 8–5. An address specified as part of the instruction, hence from the IR, or from one of the other designated or implied address registers, is modified by an index in the index register and the specified offset to yield a relative or relocatable address. The relative address is added to the program counter and/or the base register to yield an absolute address. This absolute address is further augmented by the page address to form a complete effective address. If virtual memory is in effect, this absolute address is the logical address to be transformed into an actual physical address for the memory system.

From the above discussion, we see that the various addressing modes provide the mechanism by which a user can specify the effective address of an operand. This effective address is translated into a physical address by the memory segmentation hardware. The actual physical address can now be sent to the memory system for further decoding.

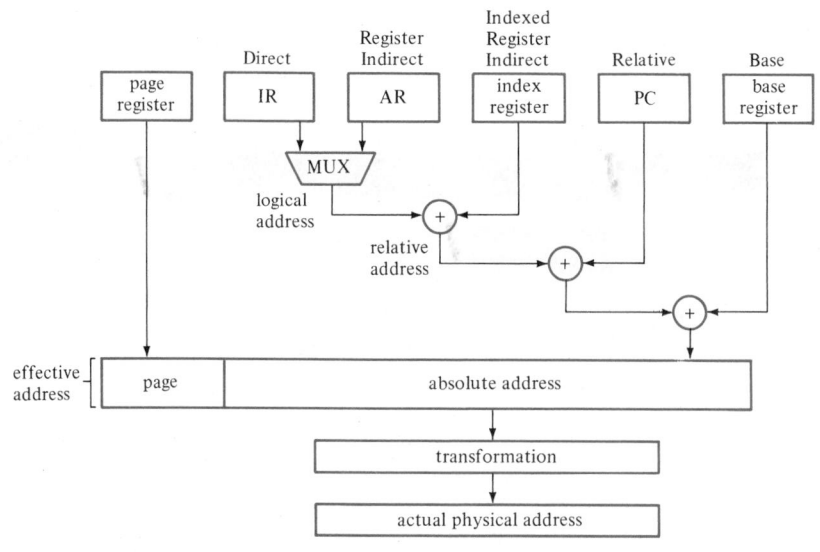

Figure 8-5
A general addressing scheme

Virtual Memory

In some computer systems it is convenient for the user to be able to specify a memory space that is larger than the actual installed memory space. This is called *virtual memory*, because not all the memory space the user may specify is really there in main memory or is actually available to the user. During execution of a program, only the necessary sections of the program are stored in main memory; the rest of the required program memory space is stored in secondary memory.

The operation of a virtual memory system uses the concept of *memory segmentation*. The user's entire virtual memory space is segmented into pages or segments and is stored as pages or segments in secondary memory. A *page* is a fixed block of memory, whereas a *segment* is a variable-size block of memory. We will use the term "section" to refer to either a page or a segment. When the program is initiated, the first program section is brought into memory. When memory references are made, each memory reference is checked to see if it resides in the current section. If so, the logical address is translated into a physical address for immediate memory access. If a reference is made to a memory location outside the present section, the instruction is put on hold until the requested section has been brought into memory from secondary storage. When the user memory space is full, some of the existing sections must be replaced before new sections can be brought in. If any of the sections to be replaced have been modified previously,

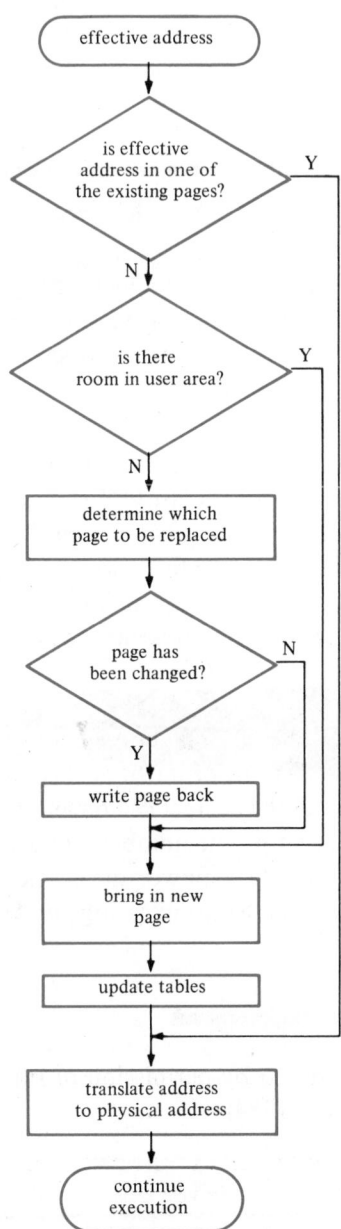

Figure 8–6
Flowchart showing the transformation process from an effective address
to a physical address in a virtual memory system

they must be written back to secondary storage. The choice of which existing section is to be sacrificed is determined by algorithms called *replacement policies*. A flowchart showing the transformation from an effective address to a physical address is shown in Figure 8–6.

8.3

Architecture of the Memory System

The basic data storage unit of a memory system is the *memory cell*, which is able to hold 1 bit of information for an indefinite amount of time for the ROMs and as long as power is applied for the RAMs.

Memory Cells

The memory cell is a simple flip-flop or a bistable multivibrator that can be found in one of two stable states: an ON state and an OFF state. Figure 8–7 shows some typical designs for a memory cell. Reading of the memory cells is accomplished by pulsing the sense lines, which are activated by the decoders. Whenever there is a coincidence in the sense lines, the current state of the memory cell can be detected by sensing the presence or absence of stored charge in the activated memory cell.

Memory components are composed of three major parts: the memory cells, the row decoder, and the column decoder. A typical organization of a memory chip is shown in Figure 8–8. In order to access any of the memory cells, an address is required to identify the correct cell or set of cells. This address is always provided as an input to the chip. The number n of individually addressable memory words or locations is usually a power of 2. The number of address bits required to identify any memory location is $\log_2 n$. According to the organization of the

---

Figure 8–7

Typical memory cells (Reprinted by permission of Intel Corporation, Copyright/Intel Corporation, 1984)

Classification	Dynamic RAM	Static RAM
Example of basic cell circuit	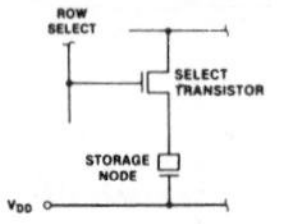	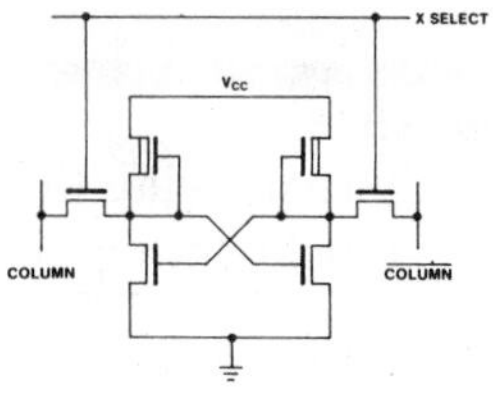

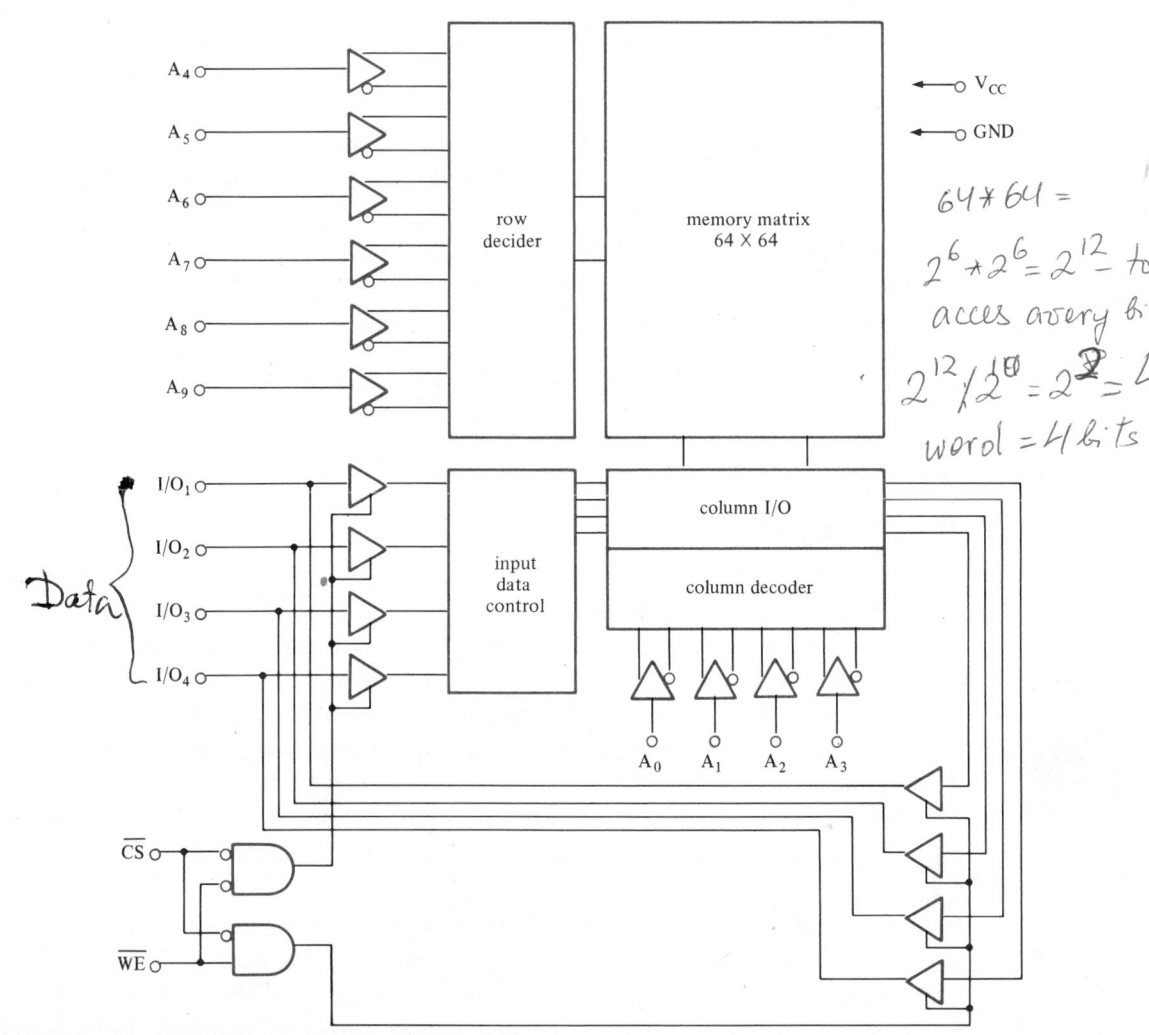

Handwritten notes:
$64 * 64 =$
$2^6 * 2^6 = 2^{12}$ to
acces every bit
$2^{12} / 2^{10} = 2^2 = 4$
word $= 4$ bits

Figure 8-8
Typical organization of a memory chip (Reprinted by permission of Intel
Corporation, Copyright/Intel Corporation, 1979)

memory chip, this memory address is further divided into a row address (x) and
a column address (y). The row address provides the input to the x decoder to
activate one of the x sense lines, and the column address provides the input to the
y decoder to activate one of the y sense lines. The coincidence of both the row
address and the column address is the word being read from or written into.

In addition to the address lines, data lines and control lines are used to control
the operation of the memory chip. Data lines are often bidirectional for RAMs
and output only for ROMs. The control lines include the chip select lines and

the output enable lines. The chip select line is usually active low and governs the overall response of the chip. If the chip is deselected, no data will be read or supplied on the data lines by the memory chip. Data are changed within the chip (a memory write cycle) or are driven onto the bus by the chip (a memory read cycle) only when the chip select is low or activated. The output enable line is used to activate the output drivers. It is also used to indicate whether the current memory operation is a memory read or a memory write operation with respect to the processor. If it is a memory read operation, the output enable is asserted to allow the drivers to supply the data. If it is a memory write operation, the output enable is not asserted so that the processor can drive the bus and supply the data. On some memory chips there is, instead of an output enable line, a write enable line that is functionally opposite to the output enable line. The write enable line is so named because it is an enable signal for the processor to "write" into the chip memory.

Memory Organization

Memory organization is usually given in terms of the number of individually accessible or addressable words in the chip and the word length. For example, a 2114 memory chip with a 1K × 4 organization means that the chip can hold a total of 4096 bits of information divided into 1024 individually addressabie 4-bit words. Hence each address yields a 4-bit word. An individual memory chip is seldom useful or sufficient for most applications. Hence groups of chips are organized together to form the main memory of a computer system according to the word length and amount of memory space required. To accomplish this, the techniques of bit slice and memory segmentation prove useful.

Bit-Slice Organization   The concept of *bit-slice operation* is to have many chips operating together in parallel, each responsible for the operations within a thin slice (a few bits) of the word. Hence, to obtain the word length required, several memory chips are stringed or concatenated together horizontally. For example, four 2114s can produce a memory of 1K × 16 organization. The first 2114 stores the first 4 bits, the next 2114 the next 4 bits, and so forth. This is shown in Figure 8–9(a).

Memory Segmentation   Although the bit-slice organization technique provides the necessary tools for the design of a memory system with any width, the memory segmentation technique gives the necessary tools for the design of a memory system with any depth. In *memory segmentation*, many memory chips are connected together to form the complete memory system. The lower part of the address is usually common to all memory chips. The higher-order address bits are connected to a decoder. The output of the decoder becomes the chip select for each memory chip. Since the data lines of these memory chips are usually connected together,

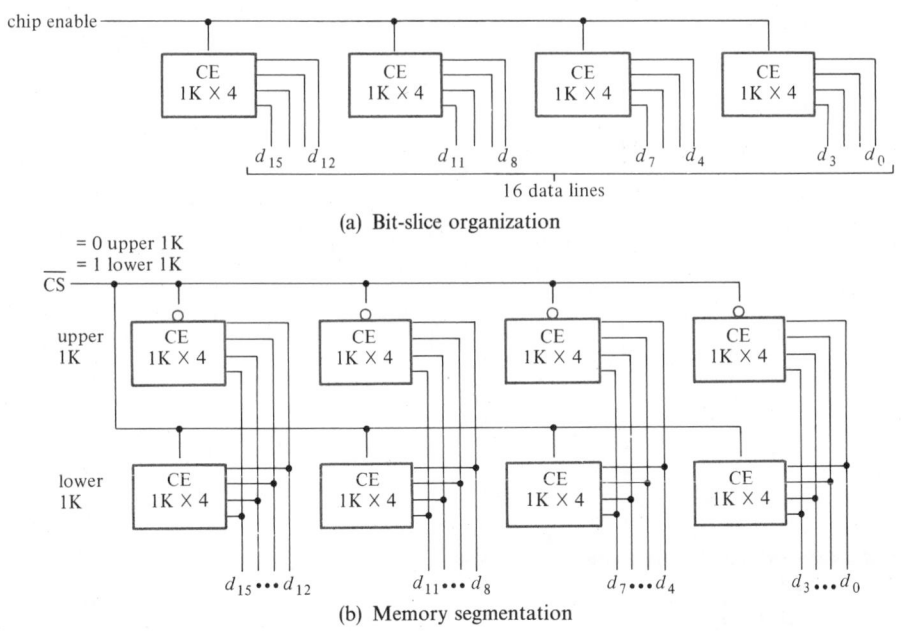

(a) Bit-slice organization

(b) Memory segmentation

Figure 8–9
Bit-slice organization and memory segmentation

it is necessary to arbitrate the memory chips by means of a decoder through higher-order address bits. The use of a decoder ensures that only one of the memory chips is selected at any time.

For example, if more than 1K of memory is required, additional memory chips can be cascaded together vertically. Each set of memory chips is responsible for a segment of the memory space. This is called memory segmentation. Hence, another set of four 2114s added to the original set of four will expand the original 1K × 16 memory organization into 2K × 16, as shown in Figure 8–9(b).

When many memory chips are cascaded together vertically, additional external decoding is needed to arbitrate the data bus so that only one chip is activated at one time. Hence *bus arbitration* is necessary to avoid crashes and to ensure the integrity of information. This arbitration can be accomplished by using the higher-order address to specify and activate the proper set of memory chips by means of the decoder. This is shown in Figure 8–10. The arbitration is done primarily by a decoder connected to the chip select lines of the memory chips. The input of the decoder is fed by the high-order address. The high-order address causes the decoder to choose the proper set of memory chips by asserting only one of the many output lines. The chip select lines of those memory chips that are cascaded horizontally are connected to the same decoder output line. The chip select lines for those memory chips that are cascaded vertically must be connected

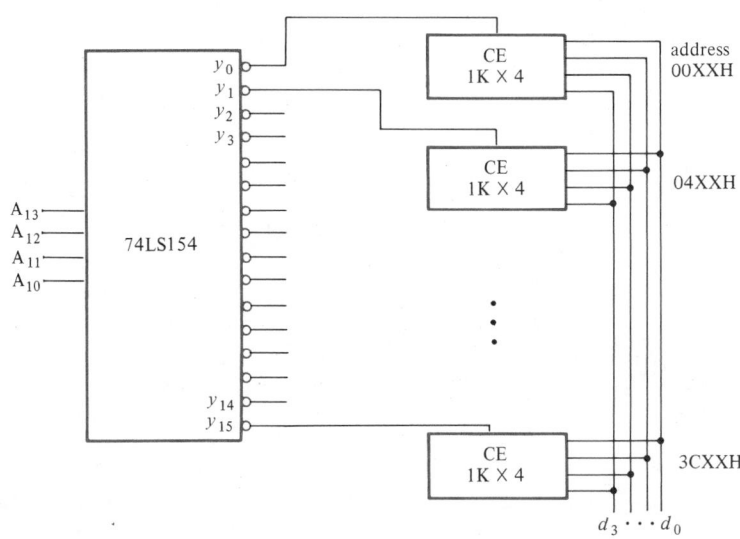

Figure 8–10
Memory decoding

to different decoder output lines. The use of a decoder ensures that only one set of memory chips is activated as depicted by the address, even though the data output lines of all the memory chips cascaded vertically are connected together.

Board/Bank Segmentation Often only a fixed number of memory chips can fit into a specific size of circuit board. If more memory is required, multiple memory boards are used. The portion of memory located on a particular board is called a *memory bank*. The total memory is composed of many memory banks. All banks of memory are usually connected to a *motherboard*. A motherboard is the primary board that provides the same bus connections to all its interfacing boards. Arbitration between the different memory boards is needed. This arbitration is done by a board address, which is usually the most significant bits of the actual memory address. The first thing the memory board looks at is the board address. If the board address is correct, the decoder on that board is activated and data are transferred between the memory board and the motherboard. If the board address is not correct, the decoder is not activated. Thus the present board is effectively isolated from the rest of the system for any further activities. Typically, the comparison of the board addresses is done by a magnitude comparator and a set of user-definable switches or jumpers as shown in Figure 8–11. The switches are set by the user when installing the boards and are used to set the installed memory bank to any starting address.

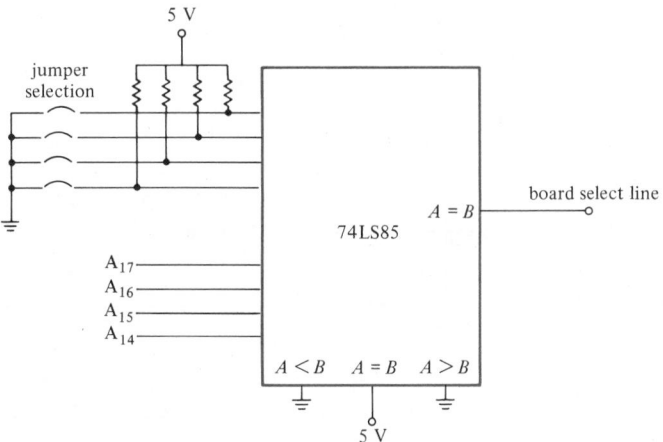

Figure 8–11
Board address selection

In summary, a physical address may be divided into many sections, as shown in Figure 8–12. The most significant part is the board address, which divides the total addressing space into distinct memory banks. Within each memory bank, the next part of the address is applied to the decoder, which selects a particular set of memory chips and deselects the rest. The remaining address bits are supplied to all the memory chips themselves, which in turn separate the address into the row address and the column address for memory cell decoding.

Example 8–1

Design a 64K memory system using 2114 memory chips. Assume that each board can hold only 32 memory chips.

Figure 8–12
Division of address bits

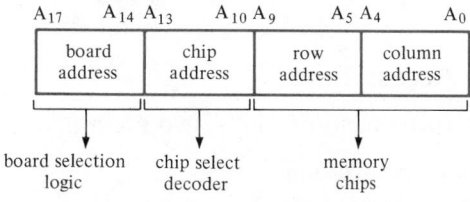

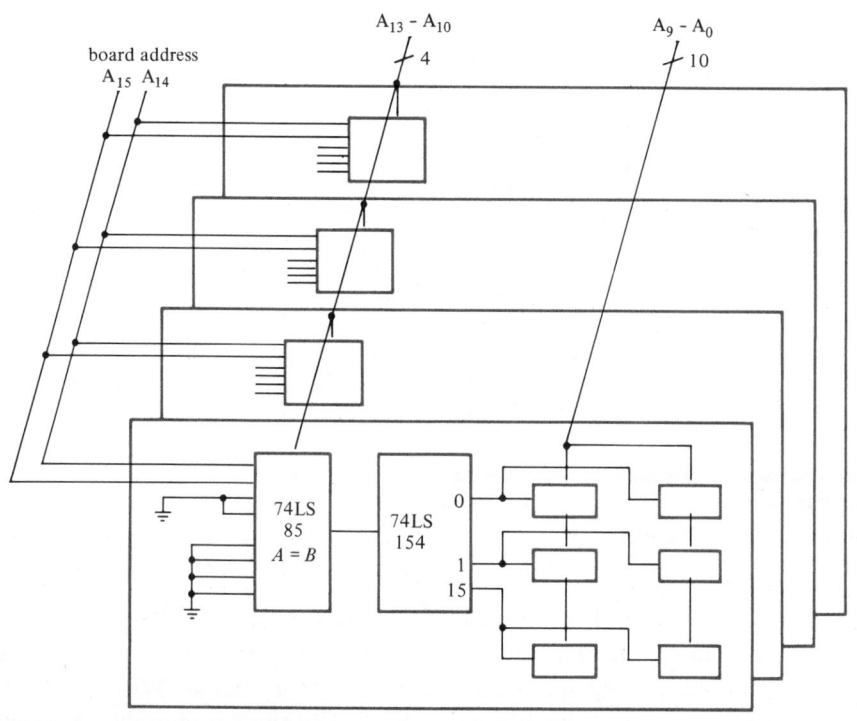

Figure 8-13
A 64K memory system using 2114 memory chips

Since each 2114 memory chip has a 1K × 4 organization, two 2114 chips are required to form 1K bytes. Hence a total of 128 chips are required to form a 64K memory space. If one board can hold only 32 chips, four boards are needed. This means that the two most significant bits of the address, A_{15} and A_{14}, become the board address. On each board, there are altogether 32 chips arranged in pairs. In other words, 16 pairs of memory chips must be decoded. This requires a 4-to-16 decoder such as the 74LS154 decoder. These 16 pairs of chips require the next four address bits, A_{13} to A_{10}, for decoding. The remaining 10 address bits are passed onto each individual chip for row and column address decoding. The complete system is shown in Figure 8–13.

Memory Access Time

In characterizing memory chips, two main parameters are considered:

Memory organization
Maximum access time

We have already discussed memory organization. The second parameter in characterizing memory chips is the maximum memory access time. This is the delay time the chip requires to search through the memory cells for the addressed word. The output can be read after this time. A slightly longer time is sometimes required to write data into memory. This time usually refers to the period beginning with a stable address and ending with the availability of data. Decoding of the memory address occurs during this time. Some typical timings for the 2114 family are:

Type	Access Time
2114 AL-1	100 ns
2114 AL-2	120 ns
2114 AL-3	150 ns
2114 AL-4	200 ns
2114 A-4	200 ns
2114 A-5	250 ns

When choosing memory chips, the memory access time often becomes the overriding factor because an adequate speed must be matched between the memory system and the microprocessor.

8.4
Types of Memory

Memory components are extremely useful for many purposes in addition to storage. There are two basic types of memory:

Read-only memory (ROM)
Read/write memory (commonly referred to as RAM, though the term originally stood for random access memory)

The exact proportion of these two types of memory in different installations may be vastly different depending on applications. A dedicated microprocessor system used in control applications may have mostly ROM memory for program storage and only a little RAM memory for temporary data storage. A general-purpose computer system, on the other hand, may have only a little ROM space for initial program loading and mostly RAM space for program development and execution.

Read-Only Memory (ROM)

The first kind of memory components are ROMs. They are nonvolatile; that is, they retain their information content even after the power is removed. There are many applications for ROMs, some of which include:

(1)   Permanent information storage, including program storage and data storage such as tables and constants
(2)   Code conversion
(3)   Multiple output function generation

(1)   *Permanent data storage* refers to storing programs, data, or any other pertinent information that will not be changed even when the power is turned off. The application is particularly attractive in the consumer market, where the programs of many digital systems are fixed during manufacture and usually require very little change thereafter. Mathematical tables, trigonometric tables, and other system data are also examples of information that usually does not change with time and can be conveniently stored in a ROM.

(2)   Besides permanent data storage, a ROM can be used as a code converter or a function generator. In many applications a translation is required from one type of code to another—for example, from Gray to 8421 BCD or from ASCII to EBCDIC. The attractive part of using a ROM as a translator is that the translation process is very simple. In fact, the process of translation is first accomplished and then permanently stored in a ROM. When a conversion is needed, the information in the ROM can be accessed to provide the converted result immediately. Hence it is possible, by using a ROM, to replace many SSI and MSI chips that would otherwise be required to translate the codes. The use of ROMs thus results in a significant reduction in total chip count.

(3)   Many times, the output signals of a microprocessor may not be directly and perfectly matched to the signals defined for a system bus such as the S-100 bus, the SS-50 bus, or the MULTIBUS. A ROM is an ideal device in this case to match the signals and is often used to bridge the discrepancies by translating the original set of signals to another that matches more closely with the second system.

Types of ROM

As a result of the widespread use of read-only memories, several types of ROMs are now available for different applications:

ROM, or read-only memory

PROM, or programmable read-only memory

EPROM, or erasable programmable read-only memory

EEPROM, or electrically erasable programmable read-only memory

ROM   Typically, ROMs are fabricated with the information already burned into the memory. Sometimes these are called *mask-programmable ROMs*, indicating that the information is programmed by the mask at fabrication time. The information thus encoded cannot be changed at all. ROMs are particularly useful in high-volume consumer products such as games and dedicated microprocessor circuits where the program or data tables are not meant to be changed. This use of ROMs is feasible only for high-volume manufacturers.

PROM   PROMs are used for many applications. PROMs are not programmed during chip fabrication. Rather, a fusible link is provided as a way to hold the information. After fabrication, the information is written into the PROM by a PROM programmer or PROM burner. Programming is accomplished by physically burning the fuse at the desired locations. Hence this type is also called a *field-programmable ROM*. Once the information has been burned into the chip, the information cannot be changed. PROMs are not very popular except for high-volume manufacturer. In principle, the use of a PROM is exactly the same as an EPROM. Some manufacturers also call these production EPROMs. An example of a PROM is the P2764A production EPROM chip.

EPROM   A more versatile and popular type of ROM is the ultraviolet erasable programmable read-only memory or EPROM. Information is programmed into an EPROM by an EPROM programmer or burner. Since it must be possible to change this information, the chip is equipped with a transparent window that exposes the silicon area containing the memory cells to ultraviolet light. If these memory cells are exposed to ultraviolet light for a specified amount of time, the information is erased and new information can then be programmed into the chip. EPROMs are ideal for experimentation and development when the final edition of the program is not yet available.

EEPROM   Lastly, the electrically erasable programmable read-only memory (EEPROM) does not require exposure to ultraviolet light for erasure, but information can be directly and dynamically erased electrically under program control. The main difference between EEPROMs and ordinary RAMs is that when information is written into an EEPROM, the information remains until changed even when the power is off, whereas for RAMs the information remains until changed only as long as power is kept on. Once power has been turned off, RAMs lose all their information content. Electrically erasable programmable read-only memory finds widespread use in equipment that requires user setup. Many times the equipment—for example, a logic analyzer—has a great number of settings to define for a particular environment, which can be cumbersome and time-consuming to do if a user must repeatedly enter these settings every time he or she turns on the machine. A regular EPROM may be used if the user knows what to program into the EPROM. Even this could be a great inconvenience, because the information in an EPROM cannot be changed easily. An EEPROM is very useful in this case, because the equipment can be designed to program the user's setting into its EEPROM and these settings will not be destroyed when the power is turned off. When a new setting is entered, the new information can be written back without the need to physically remove and erase the chip. Other applications for EEPROM are error logs, remote software updates, and diagnosis. Software can be updated automatically by any communication means without the need to send a field engineer to physically remove the old PROMs and replace them with the new ones.

Read/Write Memory (RAM)

The second type of memory is read/write memory, more commonly known as RAM memory. RAMs are volatile memories, because they store information only as long as power is applied. As soon as the power is removed, the information content is lost. The primary use of RAMs is temporary data storage, often called scratchpad memory because the RAM can be written into and later read by the user.

In small dedicated systems such as toys and control applications, the memory of a computer system will be comprised mostly of ROM with very little RAM. In large installations the amount of ROM is usually limited to storing the initialization program. The majority of the memory space is of the RAM type for storing user programs. There are three main types of read/write memory, based on fabrication techniques:

Static memory
Dynamic memory (DRAM)
Pseudo-static memory

Static Memory For *static memory*, the memory cells are simple flip-flops or bistable multivibrators with two stable states so that the cells can hold their state permanently as long as power is applied. Hence these are called static memories (SRAMs). Once information has been written into the memory cells, it remains in the cells and is not destroyed as long as power is on.

Dynamic Memory *Dynamic memory* (DRAM) chips, on the other hand, have much simpler memory cells in which the state of the memory cell is indicated by the presence or absence of stored charge in a capacitor. As time passes, charge leakage occurs and the information is lost even if power is left on. In order to retain the information for an indefinite amount of time, each memory cell must be refreshed periodically by writing the information into the cells again or in essence by recharging the capacitors. Hence memory chips with these kinds of memory cells are called dynamic memories. The main advantage of dynamic memory chips is that the memory cells are much simpler and more compact, thus taking up less space on a chip. This results in a higher-density chip holding more information than a comparable static memory chip with similar chip area.

The main disadvantage of using DRAMs, of course, is that the charge leakage must be controlled in all memory cells by reading and rewriting the information periodically. This is called a *memory refresh process* and is required only for dynamic memories.

Pseudo-Static Memory *Pseudo-static memory* chips are actually dynamic memory. They act like static memory because the refresh circuitry is provided inside the chip, hence requiring no external refreshing circuitry.

8.5
The Input/Output (I/O) Devices

In addition to memory, the processor also communicates with the outside world through its input/output (I/O) devices. Although the use of memory is restricted mostly to program or data storage, the use of I/O is general and is designed for interactions with a variety of devices.

There are several similarities and differences between a memory system and I/O devices. First, an I/O device is unique, just like a memory location. So each I/O device is addressed independently by an I/O address in much the same way as a memory address is. Second, most memory accesses are synchronous and follow a rigid time frame. I/O accesses may be asynchronous and follow a less rigid time frame. Third, it is customary to assume that all memory locations are the same in performance and behavior. Very few I/O devices, on the other hand, are the same. In fact, each I/O device may have its own characteristics and timing constraints.

I/O Programming

Programming I/O is similar in some aspects to programming memory locations in the sense that there are distinct I/O devices that must be individually addressed. For each I/O device, interactions with the processors are either I/O read or I/O write operations. Some computers and microprocessors have special instructions to handle I/O, although this is not absolutely necessary. There are two ways to program I/O operations:

With I/O instructions
With memory instructions

With I/O Instructions It is common with some microprocessors to provide specific I/O instructions to perform I/O. In other words, memory instructions are used only for memory access and I/O instructions are used only for communication with I/O devices. For example, the Intel 8085 microprocessor has many instructions referencing the memory, but only two I/O instructions. These two

I/O instructions are primitive and are not as powerful as the other memory reference instructions.

There are certain trade-offs when performing I/O with I/O instructions. From the programming standpoint, I/O instructions are clearly distinguishable from memory instructions. Hence I/O accesses are clearly separated from memory accesses. On the one hand, this is a definite advantage because of program clarity and readability. Furthermore, special I/O instructions can be designed to better satisfy the requirements of peripheral I/O devices. On the other hand, the disadvantage of this scheme is that I/O instructions are usually less sophisticated in addressing modes than memory instructions, resulting in less efficient programming. From the microprocessor designer's standpoint, separating I/O instructions from memory instructions means that there are more instructions to decode and more hardware is needed to perform them.

With Memory Instructions   Performing I/O with memory instructions can be accomplished by dedicating a section of memory addresses as I/O addresses. Hence this method is often called *memory-mapped I/O*. For a typical microprocessor system the amount of memory space usually ranges from 64K to several million locations, whereas the number of I/O devices is rarely higher than 100. Hence, dedicating the last 100 memory locations, for example, in the total addressable space to I/O addresses is not a great loss at all. Decoding between memory and I/O addresses must be accomplished explicitly by means of external hardware. An example of this type of processor is the 68010. As far as the 68010 is concerned, I/O and memory instructions are the same and are not explicitly differentiated internally within the processor. In fact, there are no specific I/O instructions in the 68010 at all.

There are numerous advantages and disadvantages to memory-mapped I/O. From the programming standpoint, the full power of all the addressing modes in the processor can also be applied to I/O addressing. This is significant in many real-time applications with a large number of I/O devices. However, the programmer must know the memory addresses that are dedicated to I/O addresses so that the I/O devices are not activated inadvertently. From the microprocessor designer's standpoint, memory-mapped I/O simplifies the design of a microprocessor tremendously, because there are now no differences internally between a memory access and an I/O access.

Modes of I/O

When data are transferred and manipulated within the processor, they are always operated upon in parallel. Normally, the input/output facility should also be handled in parallel. However, as the word length begins to grow and the memory or I/O cycle time begins to shrink, timing becomes extremely critical. Many peripheral devices are basically mechanical in nature and operate in seconds rather than in microseconds. This gives rise to two general modes of input and output:

Parallel I/O
Serial I/O

Parallel I/O *Parallel I/O* refers to the form of I/O where all bits of the data path are transferred at the same instant of time as shown in Figure 8–14(a). Such operation usually requires a parallel latch of the same data width to capture the information and a clock to provide the timing for the latch. Once the data have been latched, the processor is free to resume its normal task until another I/O transfer is desired. Parallel I/O is used primarily for two purposes. First, it can be used for single bit control where each bit of the I/O port is used to control a relay, a switch, and so on. Second, parallel I/O is often used for high-speed data transfer, because a whole word or byte of information can be transferred in the same clock pulse.

There are many standards for parallel I/O. For single-bit control-type applications, no formal standards exist because no handshaking is required. Only a buffer is needed between the external signals and the microprocessor to latch

Figure 8–14
(a) Parallel I/O and (b) serial I/O

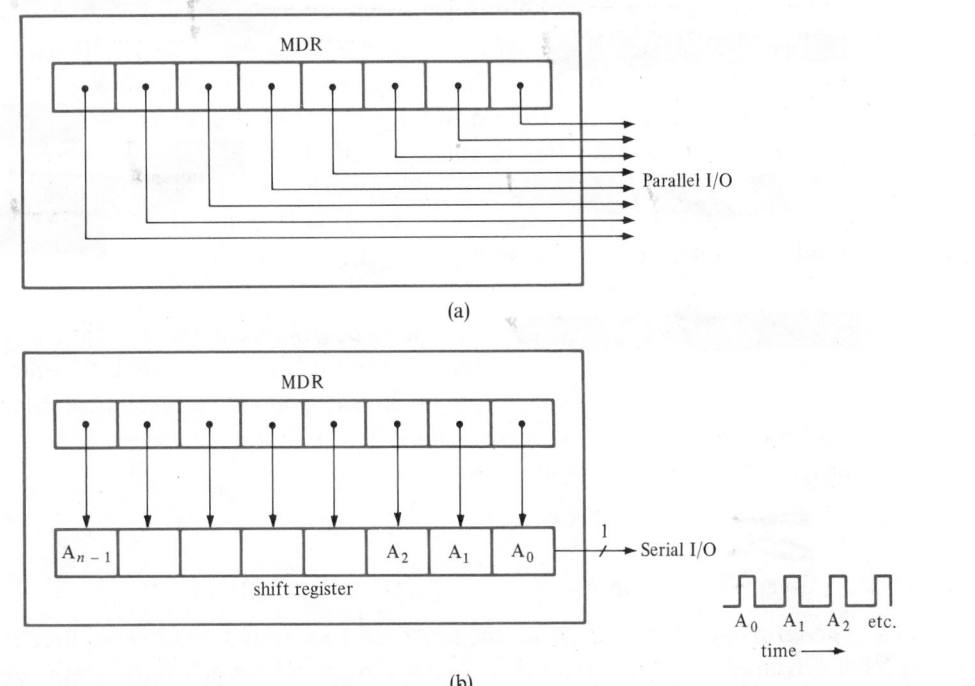

(a)

(b)

and hold the incoming signals and to drive the microprocessor data bus at the requested time. For high-speed data transfer, some type of handshaking is required for time synchronization. One common type of standard between the different parts of the host system in a tightly coupled environment is the definition of the system bus.

Serial I/O *Serial I/O* refers to the form of I/O where the information is transmitted serially and sequentially one bit at a time as shown in Figure 8–14(b). This operation usually requires a parallel-in, serial-out shift register on the sending side and a serial-in, parallel-out shift register on the receiving end. Serial I/O is often used for low-speed and long-distance data transfer such as those between the microprocessor and its terminals, card readers, or other low-speed peripherals.

Serial I/O differs from parallel I/O in that there is only one signal line for data transmission and one signal line for data reception. Only one bit of a word is transmitted at a time. Transmission continues at the specified rate until all the bits of the word have been transmitted or received. The format of serial data transmission may be asynchronous as shown in Figure 8–15(a) or synchronous as shown in Figures 8–15(b), 8–15(c), and 8–15(d).

Asynchronous serial transmission In the asynchronous case, the voltage level on the transmission line is normally high. This is called the *marking position*. When serial transmission begins, the first bit is called the *start bit* and commences when the signal on the line drops. After the start bit, the *character byte* is usually sent from the least significant bit to the most significant bit. The number of bits here is variable and may range from 5 to 8 bits. After the data bits, the *parity bit* follows. The parity bit may be present or absent. If the parity bit is present, it may be in even or odd parity format. After the parity bit, the *stop bits* are sent, with the transmission level returning to the marking level for at least 1, $1\frac{1}{2}$, or 2 bits duration before the next transmission starts.

Synchronous serial transmission In the synchronous case, the data alone are transmitted. No start bit, parity bit, or stop bits are transmitted. When no data are available for transmission, one or more synchronous characters are transmitted as fillers. This way there is always activity in the line. The message transmitted may be in one of three different formats:

Internal monosynchronous
Internal bisynchronous
External synchronous

The term "synchronous" is often shortened to simply "sync." The three formats of synchronous serial transmission differ in the way synchronization is detected and in the number of sync characters received.

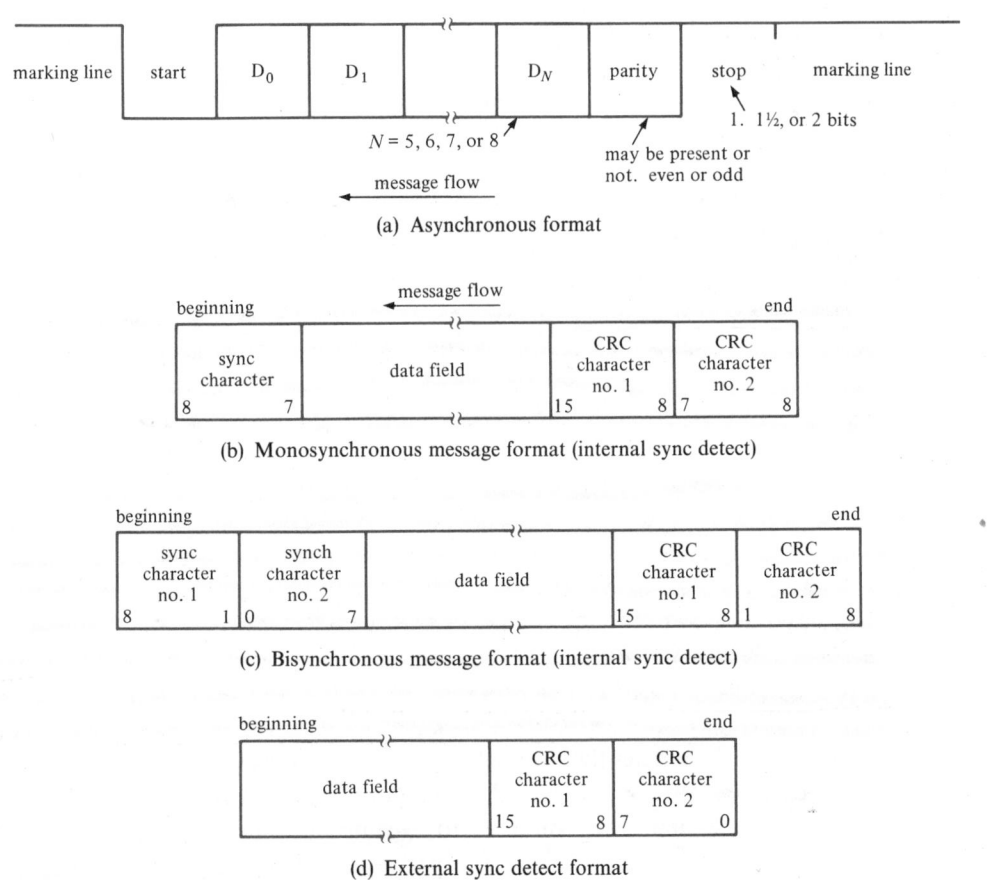

(a) Asynchronous format

(b) Monosynchronous message format (internal sync detect)

(c) Bisynchronous message format (internal sync detect)

(d) External sync detect format

Figure 8–15

Serial I/O formats (Reprinted by permission of Intel Corporation, Copyright/Intel Corporation, 1985)

In the internal monosynchronous case, the serial component device is first loaded with a sync character (8 bits) and the incoming serial line is continuously monitored and compared to the sync character. Once a match is found, subsequent bits are interpreted as data bits of the message until another sync character arrives. Before data transmission or reception, the same sync character must be loaded into both the sending and receiving devices.

In the internal bisynchronous mode, two sync characters rather than one are continuously sent before the data bits of the message. Therefore synchronization occurs when both sync characters are detected in the sequence. For both the monosync and the bisync formats, synchronization is achieved internally by the detection of the sync characters.

In the external synchronous format, no sync characters are sent. Synchronization is handled by other means, and the data reception device is merely told that synchronization has been achieved and is then triggered or enabled to start receiving the incoming bits as data.

8.6

Interrupts

Communication with the processor is bidirectional. Although many times the processor takes an active and dominant role in initiating communications with the I/O peripherals, there are still many situations where a signal must be sent from the I/O peripherals to the processor to initiate further action. This is particularly true for peripheral inputs, for example, when the processor is waiting for user inputs. There is no way to predict when the user inputs will come. We could make the processor wait for the user inputs by checking constantly if the inputs have arrived. Such waiting loops are wasteful, time-consuming, and accomplish little. Another way is to have the processor work on other tasks in the meantime and come back to check for the input status every now and then. This scheme imposes a heavy toll on processor time and programming complexity. If the processor waits too long before checking the status, data arriving at a faster than expected rate may be lost. If the processor checks too often, time is wasted. These situations can be efficiently handled by an interrupt scheme.

An *interrupt process* is a procedure whereby the processor is halted in its normal program sequence and directed to execute another program sequence. An interrupt originates from outside the processor. When a peripheral needs the attention of the processor, an *interrupt request signal* is generated. This interrupt request signal is routed via hardware to the processor. When all the operations related to the current instruction have been completed, this interrupt request signal is polled before the next instruction is initiated. If there is a valid interrupt request pending, the processor halts execution of the next instruction and starts a new sequence of operations called *interrupt servicing*. Programs written for this purpose are called *interrupt servicing routines*. When the interrupt has been serviced, the original programming sequence is activated again and processing continues where the program left off in the original sequence as if nothing had happened.

With an interrupt scheme, the processor does not have to wait for the arrival of user inputs nor to check the input status periodically to determine if the user inputs have arrived or not. The processor merely continues its task in a normal fashion until an external interrupt request arrives. At that point the interrupt is serviced and the processor then proceeds with its normal sequence again.

Interrupt Requests

An interrupt is generated externally by the requesting peripheral. To request an interrupt, a signal must be sent from the I/O peripheral to the processor. This signal must in turn be recognized by the processor. Most processors are designed to have a number of external inputs designated to be interrupt inputs. This means that a transition or a change in voltage level in these inputs indicates an interrupt request.

In a typical computer system there are usually many I/O processors connected to the interrupt system. These interrupt request signals can be routed to the processor in several ways:

Through individual request lines
Through a common request line
Through a daisy chain
Through a priority encoder

These four ways are shown in Figure 8–16.

Through Individual Request Lines   The simplest way to use interrupt request lines is to connect them directly between the peripherals and the processors as shown in Figure 8–16(a). This is possible as long as there are enough slots in the processor to accommodate the needed number of interrupt request lines or channels. An interrupt request is indicated when the interrupt request line is asserted. The peripheral then waits for the interrupt request to be received and acknowledged by the processor. When the interrupt is accepted, an interrupt acknowledge signal is generated by the processor and routed back to the peripheral. The peripheral in turn deactivates the interrupt request line.

The advantage of this scheme is that each peripheral has its own interrupt request line and interrupt acknowledge line. Hence there is no need for communication between peripherals. However, the design on the interrupt part of the processor is more complex since the processor must poll each of the possible interrupt request lines and activate the corresponding interrupt acknowledge line when accepted. There is another problem with this scheme: Since each peripheral basically works independently, it is possible and highly likely that more than one request will arrive at the same time. This means that the processor must have the necessary circuitry to arbitrate the requests and pick the most urgent one to respond to first. This further increases the complexity of the processor design.

The arbitration is usually carried out by a priority system. Two priority systems are prevalent:

Fixed priority
Round robin

In a *fixed-priority system*, each input interrupt request line has a fixed, pre-arranged priority associated with that line. When more than one interrupt occurs simultaneously, the one with the highest interrupt request priority is automatically chosen. This helps to simplify the design of the processor interrupt portion. How-

Figure 8–16

Connection of an interrupt system

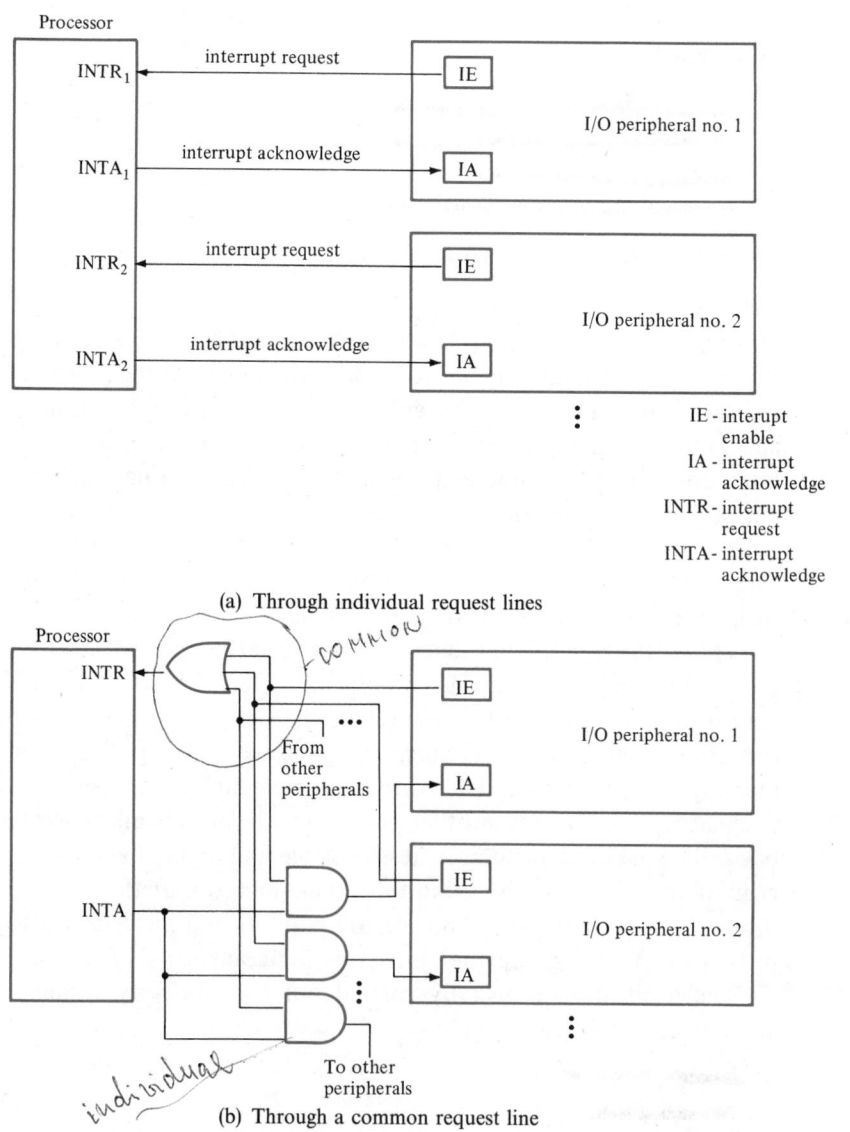

(a) Through individual request lines

IE - interrupt enable
IA - interrupt acknowledge
INTR - interrupt request
INTA - interrupt acknowledge

(b) Through a common request line

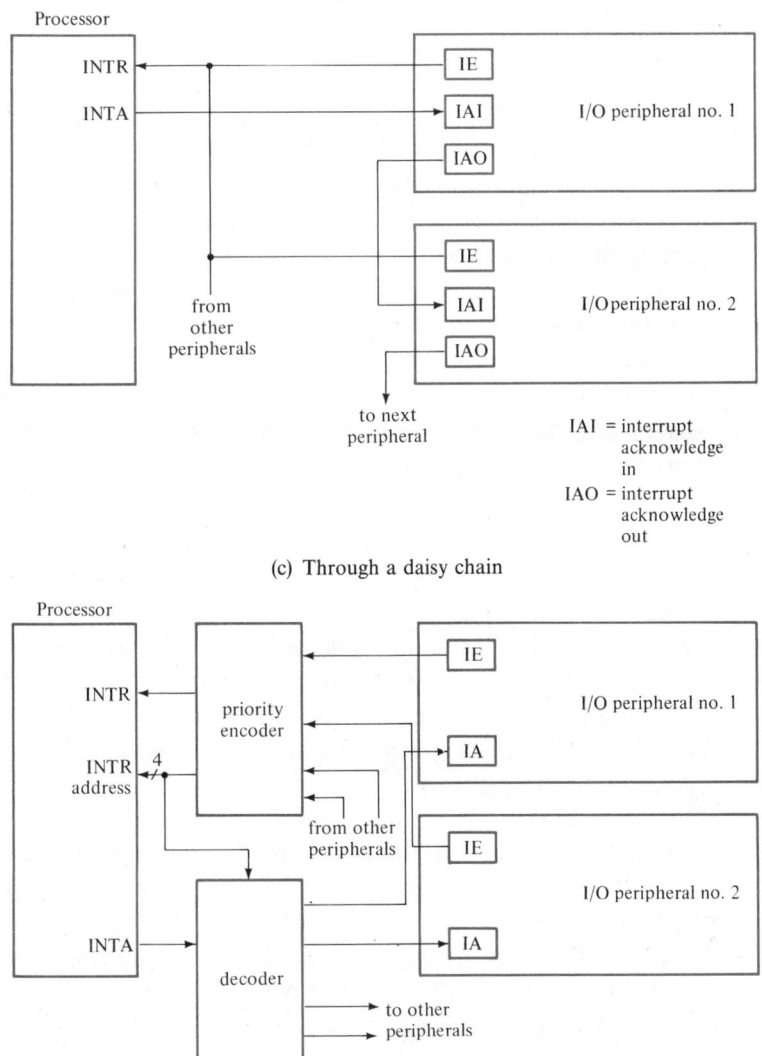

(c) Through a daisy chain

IAI = interrupt
acknowledge
in
IAO = interrupt
acknowledge
out

(d) Through a priority encoder

Figure 8-16 Continued

ever, preferential treatment is always given to those peripherals with top priorities, whereas those with lesser priorities are often delayed. In a *round-robin* system, the processor does not impose an arbitrary priority system on the interrupt request lines. When an interrupt request is active, the one along the chain next to the last serviced interrupt channel is checked first. In this way, all other interrupts are first checked and given a chance before the same interrupt channel is serviced again.

Through a Common Request Line   To provide an individual interrupt request line and an interrupt acknowledge line for each peripheral is simple on the part of the peripheral but cumbersome on the part of the processor. A simpler technique is to provide a single interrupt request line with an interrupt acknowledge line. Since there is only one interrupt request line into the processor, all the individual peripheral interrupt request lines must be ORed together. The interrupt acknowledge signal must then be routed to the proper peripheral. Note that this interrupt acknowledge line cannot be broadcasted to all the peripherals, because only one requesting peripheral can be serviced at one time. An external circuitry implementing the desired priority system is needed to ensure that only one of the interrupt-requesting peripherals receives the acknowledge signal.

Through a Daisy Chain   The use of external circuitry to enforce priority among several interrupt-requesting peripherals is expensive and not well suited for expansion. A more common technique is to daisy chain the peripherals together as shown in Figure 8–16(c). In a *daisy-chain system*, all the interrupt request lines are ORed together and sent to the processor. When the interrupt acknowledge signal is generated by the processor, the signal is routed like a token through the peripherals, one at a time. If a particular peripheral is not one of those requesting interrupts, the token is simply passed along to the next one in the chain. If the peripheral is one of those requesting interrupts, the interrupt acknowledge token is kept by the peripheral and the corresponding interrupt request line is deactivated. The connection of the interrupt acknowledge signal is serial and simple. The priority is fixed and is determined by the location of the priority along the chain. Those peripherals at the front of the chain, close to the processor, have higher priorities, whereas those peripherals at the back of the chain, far away from the processor, have lower priorities.

Through a Priority Encoder   Although the daisy-chain scheme has been used with much success, the use of a priority encoder also simplifies the design of the interrupt system. The *priority encoder* is used to generate an interrupt request signal and to supply the priority level of the associated request to the processor. At the option of the processor, the interrupt may be put on hold if the requesting level is lower than the current processor level, or the interrupt may be serviced immediately if the requesting level is higher than the current processor level. In the latter case the interrupt acknowledge signal can be routed to the requesting peripheral directly through a decoder, because the output of the priority encoder shows which peripheral has the highest priority. This scheme is shown in Figure 8–16(d). It is readily seen that the use of the priority encoder simplifies the design of the interrupt portions of both the processor and the peripherals.

The Interrupt Cycle

When an interrupt request with sufficient priority is detected by the processor, it sends out an interrupt acknowledge signal. At the same time, the processor prepares itself for the interrupt servicing and enters the interrupt cycle instead of the normal instruction fetch-and-execute sequence. The steps for the interrupt cycle are as follows:

and PS

1. The processor saves the contents of the PC in memory for later recall.
2. All further interrupts are disabled.

or here

3. The identity of the interrupting device is ascertained.
4. The address of the interrupt service routine is determined.

PS.

5. The interrupt service routine is executed.
6. The processor retrieves the previous contents of the PC from memory and continues execution of the previous program sequence.

At the end of the instruction cycle, the processor checks to see if there is an interrupt request of a higher priority than the current processor priority level. If so, an interrupt cycle is initiated rather than the instruction fetch cycle. The first task to be performed in the interrupt cycle is to store the contents of the PC in memory so that the execution of the present program can be resumed at the same location at a later time. The status of the other registers should also be saved, but this is left to the interrupt service routines. The next step is to disable all other pending interrupts of lower priority. This is effectively done by raising the processor level to the current priority level, since no interrupt requests of lower priority can interrupt the processor executing at a higher priority. These two steps are done internally in the processor. Next, the processor must determine which peripheral is requesting the interrupt and then determine the starting address of the interrupt service routine. The interrupt service routine is executed next. At the completion of the interrupt service routine, the previous contents of the PC are retrieved from memory so that the processor can return and resume execution of the previous program as if the interrupt had not occurred.

There are several methods that a processor can use to determine which peripheral is the one requesting an interrupt:

Polling
Static vectoring
Dynamic vectoring

Polling   In *polling*, the processor must address each peripheral in turn to ascertain which peripheral has generated the interrupt request. The action of the processor can be shown symbolically in the following pseudo-program:

```
Interrupt_servicing program
IF interrupt pending
THEN
     Store PC
     Disable all further interrupts
     Read status of peripheral #1
     IF status indicates pending interrupt request
     THEN
          Branch to interrupt service routine for peripheral #1
     ELSE
          Read status of peripheral #2
          IF status indicates pending interrupt request
          THEN
               Branch to interrupt service routine for peripheral #2
          ELSE
               .

               .

               .

          END IF
     END IF
END IF
```

This method is used when all the interrupt request lines are ORed together to form a single interrupt request line into the processor. The interrupt priority is now determined by the processor. Typically, those peripherals with critical time constraints, such as timer interrupts or disk interrupts, are assigned higher priorities and are polled first. Peripherals of lower priority are polled last. In some situations, it may be desirable to poll these devices in a round-robin fashion, thus giving the same priority to all the peripherals in the same class. This would be the case when many user terminals are connected to a central computer system with none having a higher priority than the others.

The main advantage of the polling method is that only simple hardware is required. The trade-off is that the process to determine which peripheral is interrupting the processor is done in software. When there are a lot of peripherals, the overhead time consumed in going through the chain interrogating each peripheral in turn can be overwhelming and tedious. For those peripherals with high priorities at the top of the chain, very little time is wasted before they are serviced. But those peripherals with low priorities at the bottom of the chain may be superseded by interrupt requests from those with high priorities.

Static Vectoring The overhead time in the polling system can be eliminated if the processor can determine immediately which peripheral is requesting interrupt, without having to ask each peripheral in turn. This can be done if there are many separate incoming interrupt request lines. If each peripheral has its own interrupt request line, then the processor can readily determine which peripheral is requesting interrupt and can service that peripheral interrupt immediately. In most cases, a table of interrupt addresses is kept in memory. For each incoming interrupt request line, there is a corresponding entry or address of the interrupt service routine for that line. At the time of interrupt, the processor needs only to scan the array of incoming interrupt lines, possibly using a priority encoder. The output of the priority encoder can be used as an offset to the interrupt address table, from which the address of the correct interrupt service routine is obtained immediately. This is shown in Figure 8–17.

The purpose of the *static vectoring* scheme is to eliminate polling at the expense of more interrupt request lines. If the number of peripherals is greater than the number of available interrupt request lines provided by the processor, the peripherals can be grouped together, with each group of peripherals sharing an interrupt request line. Though the processor must still poll, the number of peripherals to poll in each group is substantially smaller than the number of all the peripherals together.

Dynamic Vectoring   In both schemes discussed above, the processor alone determines the priority of the interrupt requests and which interrupt is to be serviced. The interrupting peripheral needs only to generate the interrupt request signal. If each peripheral is now given a unique code—that is, an interrupt address similar to a memory address or an I/O address—then the processor can ask for the interrupt address at the time of interrupt. This unique address can then be used as an offset to the table containing the address of the interrupt service routines. The

Figure 8–17
Static vectoring

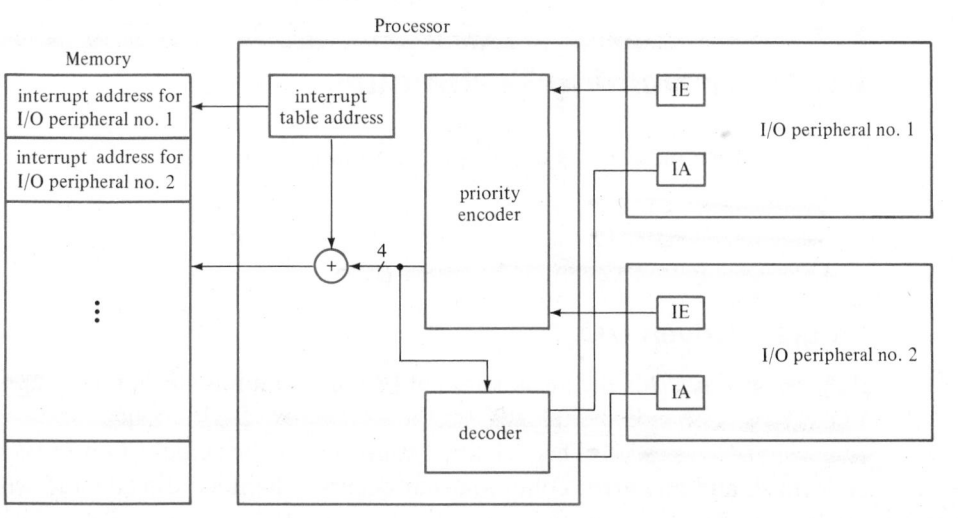

interrupt cycle for fetching the interrupt address is very similar to an ordinary memory fetch cycle. This *dynamic vectoring* scheme has been widely used, particularly in microcomputers.

The Interrupt Service Routine

An *interrupt service routine* is a program written specifically to handle the interrupt request of a particular peripheral. This routine may be written by the system programmer and included as part of the operating system, or the routine may be written by the application programmer. The general framework of an interrupt service routine can be divided into the following steps:

1. Disable any further interrupts.
2. Save the present contents of the registers if they will be destroyed during execution of the service routine.
3. Perform any necessary tasks related to the peripheral.
4. Restore the contents of the registers.
5. Enable the interrupt system.
6. Exit the service routine by restoring the former value of the program counter (PC).

In the interrupt service routine it is necessary to save the contents of any registers that will be destroyed or changed so that, at the end of the interrupt service, the context is unchanged when the processor returns to the previously executed program. The processor context includes the PC, all the registers, the condition codes, and so on. The process of storing the context is called a *context switch*, indicating a switch from the original context of the executing program to another context, that of the interrupt service routine.

8.7
I/O Programming Techniques

There are three main types of I/O programming techniques:

Program-driven I/O
Interrupt-driven I/O
Direct memory access (DMA)-driven I/O

Program-Driven I/O

Program-driven I/O is the most common I/O programming technique. The entire I/O process is handled explicitly by the system or user's application program whenever I/O is needed. There is direct control over the actual I/O process being performed, and any error condition that occurs is handled directly and without delay. The main advantage is program efficiency, because I/O operations can be initiated immediately. The main disadvantage is that the processor is occupied performing I/O operations and is not free to perform other computational tasks. If the I/O operations are fast and of comparable speed as memory access operations, this form of I/O programming is highly desirable. In most practical cases, however, the processor must wait for the data to become available before proceeding. These wait loops are time-consuming and wasteful, because the processor is not performing useful tasks.

An example of program-driven I/O is shown in Figure 8–18. The program computes, outputs data, then computes some more, outputs some more, and so forth. When the program is in the computational stage, no I/O operations are performed. Likewise, when the program is performing I/O, no computations are performed. The time to perform I/O can be very long when slow I/O devices are involved. For example, if the printer is very slow, the program must wait for the printer to finish the current character, the current line, or the current batch of characters before proceeding. In other words, either the I/O device or the processor is active but not both together. These wait loops are wasteful of time and resources and should be avoided at all cost.

Figure 8–18
Program-driven I/O

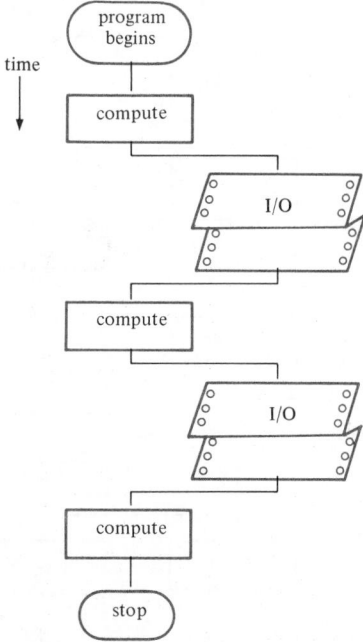

The total time of the program execution is the time to compute and the time to perform the I/O. Note that the time to perform I/O includes the actual clock time for performing the I/O. If the I/O device is slow, so that the processor is always waiting for the I/O device to complete, we call this an *I/O bound condition*. If, on the other hand, the I/O device is always waiting for the computations to be completed, we call this a *compute bound condition*.

Interrupt-Driven I/O

A better way to program the I/O is to use the interrupt service routines to perform the I/O. An interrupt condition causes an abnormal termination in the execution of the current program. The processor stores the current state of the machine and begins to execute a special program called an interrupt service routine. At the end of the interrupt service routine, the processor restores the state of the machine according to the previous program and returns to normal execution of the previous program. In this scheme the processor could be performing useful computations while waiting for the slow I/O devices to finish the job. Hence both the I/O device and the processor appear to be working at the same time. A flowchart for interrupt-driven I/O for the printer problem is shown in Figure 8–19.

Figure 8–19
Interrupt-driven I/O

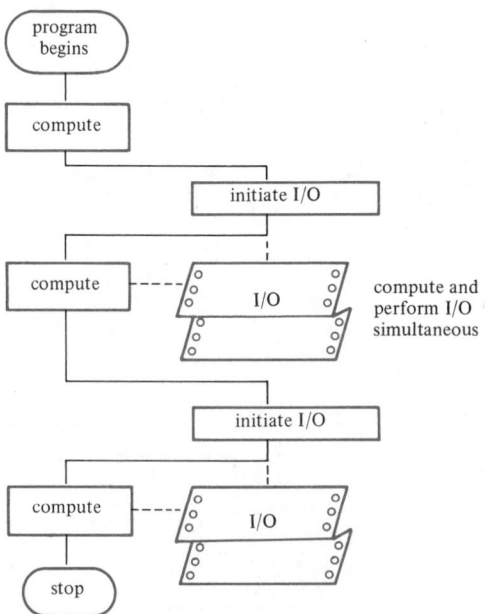

Interrupts are also useful for time synchronization and can be used to signal the occurrence of events. For example, a real-time clock can be used to interrupt the processor at prescribed intervals for timekeeping. Clock interrupts are also used for scheduling real-time events that must occur at periodic intervals. Furthermore, interrupts can be used to signal an event such as a key depression, the arrival of a signal, or the onset of an event.

Direct Memory Access (DMA)-Driven I/O

In *direct memory access (DMA)-driven I/O*, the complete task of I/O operations is given to an external controller to perform. The processor continues its normal operation and is put on hold temporarily when the DMA is actively performing I/O operations. This is shown in Figure 8–20. In terms of programming, this is the most ideal and most efficient type of I/O programming, because an additional controller is keeping track of the I/O while the processor is free to perform other computational and supervisory tasks.

There are various conditions where DMA transfers are most suitable. First, when the CPU is overloaded with too many I/O instructions, the use of DMA may off-load many time-consuming I/O operations from the CPU. Second, when

Figure 8–20
DMA-driven I/O

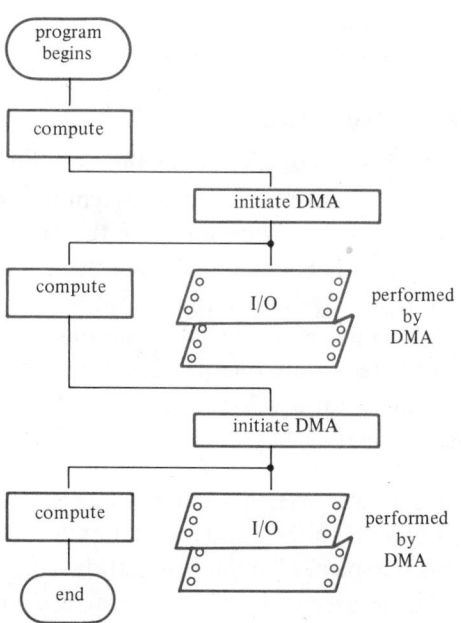

the CPU cannot provide the speed required for data transfer to or from a particular device, a DMA device is often designed that requires much less time to perform a data transfer than the CPU. And third, when the CPU spends too much time in overhead operations for initiating data transfers, a DMA controller can perform many housekeeping tasks and thus free the CPU to perform other duties. Typical applications in which the DMA concept is often used are data transfers between disk/diskette or tape controllers, in memory-to-memory transfers, in block transfers between networking computers, and in multiprocessing environments.

8.8
Microprocessor System Organization

In previous sections we have discussed the basic components of a computer system. In this section we present the concepts of system organization. To have a complete system, one needs more than the processor itself. One also needs adequate memory and peripheral devices for communication and information storage. Increasing demand for computing power and speed has fostered the idea of using more than one processor in the same computer system. This is called a *multiprocessor system*. Other configurations involve connecting different installations together into a network. This is called a *multiple-processor system*. Such innovations have led to three distinct classes of computer systems:

Single-processor systems
Multiprocessor systems
Multiple-processor systems

Single-Processor Systems

The *single-processor system* is by far the simplest and most popular type of system used. A typical single-processor system is shown in Figure 8–21. Basically, the system contains the processor itself, the memory, peripheral controllers, and peripheral devices. The single-processor system is often referred to as a *single-instruction-stream, single-data-stream (SISD) machine*, because there is only one processor performing one instruction on one set of data at any one time.

Although there is only one processor, the system can be connected either to a single user or to an array of users. Hence there are single-user systems and multiple-user systems.

Single-User Systems When a computer system is not shared with others, it is called a *single-user system*, and the computer system is said to be a *dedicated machine*. This is typical of small installations and gave rise to the term "personal computer." Single-user systems, in general, are composed of a terminal, the pro-

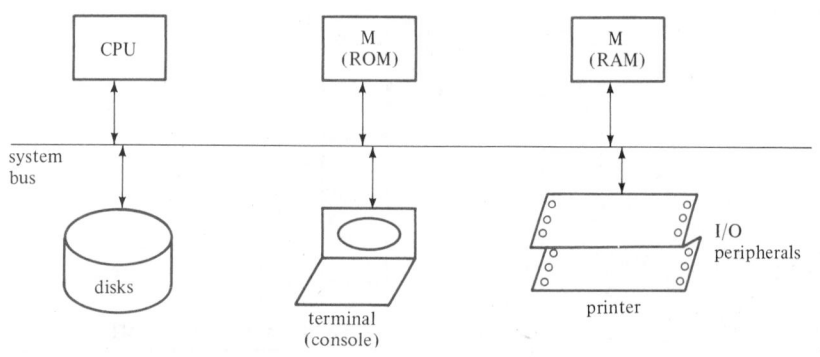

Figure 8–21
Organization of a single-processor system

cessor itself, main memory, and a number of I/O peripherals such as diskettes, hard disks, and printers.

Multiple-User Systems In large installations, a large number of terminals may be connected to the same processor. This is called a *multiple-user* or *multiuser system*. Each terminal represents a user, who gets a small slice of computer time among the others. This is called *time-sharing*. Within the computer system, several programs may be in execution state at the same time. This is called *multi-programming*.

Multiprocessor Systems

In a *multiprocessor system,* several processors work together harmoniously as a single system. When all the processors are the same, the system is called a *homogeneous* multiprocessor system. Where different types of processors are connected together, the system is called a *nonhomogeneous* or *heterogeneous* system. There are three types of multiprocessor systems:

Single-instruction-stream, multiple-data-stream (SIMD) machines
Multiple-instruction-stream, single-data-stream (MISD) machines
Multiple-instruction-stream, multiple-data-stream (MIMD) machines

SIMD Machines One type of multiprocessor system is the SIMD machine, which is more commonly known as a *vector* or *array machine*. A vector machine has many distinct processors. When a large number of additions are to be performed, each processor takes a pair of numbers and performs the operation. All processors then work in time synchronization with one another. A typical block diagram for an array machine is shown in Figure 8–22(a). An array processor is also useful for solving many kinds of numerical problems.

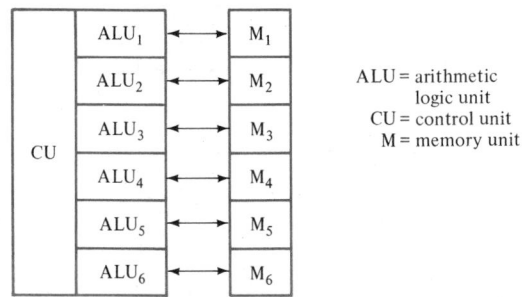

(a) A single-instruction, multiple-data-stream (SIMD) computer

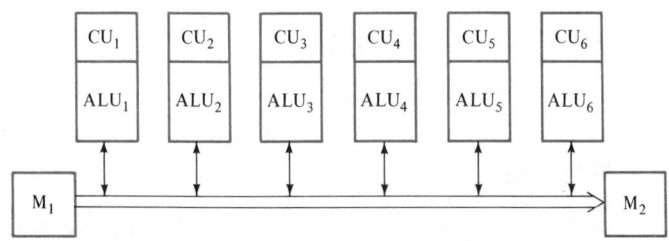

(b) A multiple-instruction, single-data-stream (MISD) computer

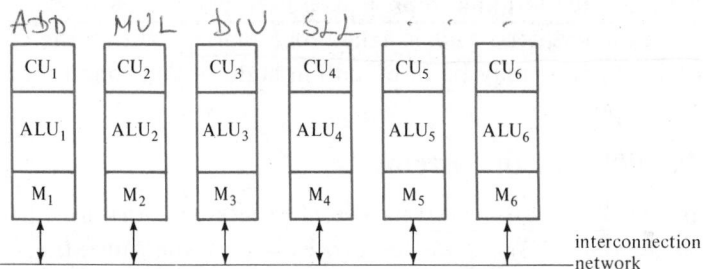

(c) A multiple-instruction, multiple-data-stream (MIMD) computer

Figure 8–22
Multiprocessor systems

MISD Machines   A MISD machine is also called a *pipelined machine*. Pipelined machines have many processors. Each processor operates on a different part of a single data stream. For example, in floating-point additions, the sequence of operations is:

1. Compare the exponents.
2. Align the exponents.
3. Add the mantissas.
4. Normalize the result.

These are distinct steps that must be done one at a time. Each step is like a section of a pipe. The complete operation requires the operands to traverse the whole length of the pipe as shown in Figure 8–22(b). When there are a large number of floating-point additions to be done, the operands are sent to the pipe at the beginning and the results are received at the end of the pipe simultaneously. As the operands traverse down the pipe, they are operated on until the end.

MIMD Machines   In MIMD machines, different processors may work on different tasks at the same time, with the ultimate goal of accomplishing what needs to be completed. A typical MIMD machine is shown in Figure 8–22(c). Examples of MIMD machines are the experimental machines built with a large number of processors based on the data flow concept. There are a few experimental data flow machines, and there are only a few commercially available MIMD machines at this time.

Multiple-Processor Systems

A *multiple-processor system* is a network of single-processor systems connected together for efficient information transfer. Each processor forms a node in the network and is capable of operating by itself. Resources are shared among the nodes as data are passed through the network. A typical block diagram for a multiple-processor system is shown in Figure 8–23.

Figure 8–23
A multiple-processor system

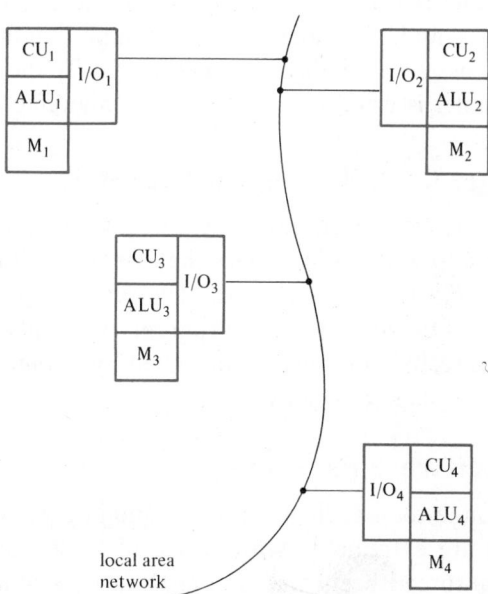

8.9

Microprocessor Software

A complete microprocessor system must include appropriate software for the system to work. Hardware alone is insufficient except in dedicated situations. Many different kinds of software packages are needed, including:

The operating system
The assembler
Language compilers and interpreters
Other software packages

The Operating System

An *operating system* is a program that interacts between the user and the processor. At initial power-up, a simple program stored in ROM is used to load the operating system. At this point, the operating system initializes the rest of the hardware and peripheral devices and prompts the user for commands. When a command is received, it is interpreted and acted upon by linking to the proper program. Any file requests are handled by the file management part of the operating system. When the command has been fully executed, the operating system takes control and initiates interaction with the user again.

The Assembler

An *assembler* is a translation program that reads in an assembly program written by the user and translates it into machine language for execution. Almost all machines use an assembly language to specify the instructions to be performed. An assembler is usually provided by the manufacturer.

Language Compilers and Interpreters

Programs written in the assembly language are hard to read and understand. It is desirable to write in languages that are more similar to common usage. These are called *higher-level languages*. Typical examples are FORTRAN, COBOL, ALGOL, ADA, and BASIC. Programs written in these languages cannot be executed directly, but must be translated by a *compiler* program into assembly or machine language before execution.

Other Software Packages

In addition to tools such as compilers and interpreters, many specialized software packages are extremely useful. Some of these include data base management programs, spreadsheets, and so on. Data base management programs are used

to create, update, sort, search, and maintain a data base. Spreadsheets are used to enter, modify, calculate, and manipulate tables.

References

1. Ahmad, S. Imtiaz, and Kwok T. Fung. *Introduction to Computer Design and Implementation.* Computer Science Press, Rockville, Md., 1981.
2. Alexandridis, Nikitas A. *Microprocessor System Design Concepts.* Computer Science Press, Rockville, Md., 1984.
3. Baer, Jean-Loup. *Computer Systems Architecture.* Computer Science Press, Potomac, Md., 1980.
4. Booth, Taylor L. *Introduction to Computer Engineering Hardware and Software Design.* John Wiley & Sons, New York, 1984.
5. Ciminiera, Luigi, and Adriano Valenzano. *Advanced Microprocessor Architectures.* Addison-Wesley, Reading, Mass., 1987.
6. Clements, Alan. *Microprocessor Systems Design 68000 Hardware, Software, and Interfacing.* PWS Publishers, Boston, 1987.
7. D'Angelo, Henry. *Microcomputer Structures.* Byte Publications, Peterborough, N.H., 1981.
8. Dasgupta, Subrata. *The Design and Description of Computer Architectures.* John Wiley & Sons, New York, 1984.
9. Doty, Keith L. *Fundamentals of Microcomputer Architecture.* Matrix Publishers, Portland, Ore., 1979.
10. Eccles, William J. *Microprocessor Systems: A 16-Bit Approach.* Addison-Wesley, Reading, Mass., 1985.
11. Furht, Borivoje, and Himanshu Parikh. *Microprocessor Interfacing and Communication Using the Intel SDK-85.* Prentice-Hall, Englewood Cliffs, N.J., 1986.
12. Garland, Harry. *Introduction to Microprocessor System Design.* McGraw-Hill, New York, 1979.
13. Gault, James W., and Russell L. Pimmel. *Introduction to Microcomputer-Based Digital Systems.* McGraw-Hill, New York, 1982.
14. Getgen, Lawrence E. *Designing with Microprocessors.* Science Research Associates, Chicago, 1985.
15. Givone, Donald D., and Robert P. Roesser. *Microprocessors/Microcomputers: An Introduction.* McGraw-Hill, New York, 1980.
16. Gorsline, G. W. *Computer Organization: Hardware/Software.* Prentice-Hall, Englewood Cliffs, N.J., 1980.
17. Hamacher, V. Carl, Z. G. Vranesic, and S. G. Zaky. *Computer Organization,* 2d ed. McGraw-Hill, New York, 1984.
18. Hayes, John P. *Computer Architecture and Organization.* McGraw-Hill, New York, 1978.
19. Hayes, John P. *Digital System Design and Microprocessors.* McGraw-Hill, New York, 1984.
20. Heffer, D. E., G. A. King, and D. Keith. *Basic Principles and Practice of Microprocessors.* John Wiley & Sons, New York, 1981.
21. Hill, Frederick J., and Gerald R. Peterson. *Introduction to Switching Theory and Logical Design,* 3d ed. John Wiley & Sons, New York, 1981.

22. Hill, F. J., and Gerald R. Peterson. *Digital Logic and Microprocessors.* John Wiley & Sons, New York, 1984.
23. Hitachi Corp. *IC Memories.* Hitachi Corporation, San Jose, Calif., 1984.
24. Hwang, Kai, and Faye A. Briggs. *Computer Architecture and Parallel Processing.* McGraw-Hill, New York, 1984.
25. Iliffe, J. K. *Advanced Computer Design.* Prentice-Hall International, London, 1982.
26. Intel Corp. *Memory Components Handbook.* Intel Corporation, Santa Clara, Calif., 1984.
27. Jermann, William H. *The Structure and Programming of Microcomputers.* Alfred Publishing, Sherman Oaks, Calif., 1982.
28. Kline, R. M. *Digital Computer Design.* Prentice-Hall, Englewood Cliffs, N.J., 1977.
29. Kostopoulos, G. K. *Digital Engineering.* John Wiley & Sons, New York, 1975.
30. Lewin, Morton H. *Logic Design and Computer Organization.* Addison-Wesley, Reading, Mass., 1983.
31. Mano, M. Morris. *Digital Logic and Computer Design.* Prentice-Hall, Englewood Cliffs, N.J., 1979.
32. Mano, M. Morris. *Computer System Architecture,* 2d ed. Prentice-Hall, Englewood Cliffs, N.J., 1982.
33. Mano, M. Morris. *Digital Design.* Prentice-Hall, Englewood Cliffs, N.J., 1984.
34. O'Connor, Patrick. *Digital and Microprocessor Technology.* Prentice-Hall, Englewood Cliffs, N.J., 1979.
35. Oleksy, Jerome E., and George B. Rutkowski. *Microprocessor and Digital Computer Technology.* Prentice-Hall, Englewood Cliffs, N.J., 1981.
36. Peatman, John B. *Microcomputer Based Design.* McGraw-Hill, New York, 1977.
37. Peatman, John B. *Digital Hardware Design.* McGraw-Hill, New York, 1980.
38. Rafiquzzaman, M. *Microcomputer Theory and Applications with the Intel SDK-85.* John Wiley & Sons, New York, 1982.
39. Rafiquzzaman, Mohamed. *Microprocessors and Microcomputer Development Systems: Designing Microprocessor-Based Systems.* Harper & Row, New York, 1984.
40. Ramirez, Edward V. *Microprocessing Fundamentals: Hardware and Software.* McGraw-Hill, New York, 1980.
41. Richard, R. K. *Arithmetic Operations in Digital Computers.* Van Nostrand, New York, 1955.
42. Rooney, Victor M., and Amin R. Ismail. *Microprocessors and Microcomputers.* Macmillan, New York, 1984.
43. Short, Kenneth L. *Microprocessors and Programmed Logic,* 2d ed. Prentice-Hall, Englewood Cliffs, N.J., 1987.
44. Siewiorek, Daniel P., C. G. Bell, and Allen Newell. *Computer Structures: Principles and Examples.* McGraw-Hill, New York, 1982.
45. Sloan, M. E. *Computer Hardware and Organization,* 2d ed. Science Research Associates, Chicago, 1983.
46. Stone, Harold S., ed. *Introduction to Computer Architecture,* 2d ed. Science Research Associates, Chicago, 1980.
47. Stout, David F. *Microprocessor Applications Manual.* McGraw-Hill, New York, 1982.
48. Tanenbaum, Andrew S. *Structured Computer Organization,* 2d ed. Prentice-Hall, Englewood Cliffs, N.J., 1984.
49. Taub, Herbert. *Digital Circuits and Microprocessors.* McGraw-Hill, New York, 1982.
50. Texas Instruments. *Microprocessor/Microcomputer System Design.* McGraw-Hill, New York, 1980.

51. Wakerly, John F. *Microcomputer Architecture and Programming.* John Wiley & Sons, New York, 1981.

52. Wilkinson, Barry. *Digital System Design.* Prentice-Hall, Englewood Cliffs, N.J., 1987.

Problems

1. How big would the memory space be if there are 8 bits for addressing? What if the number of bits for addresses is increased to 16 bits, 20 bits, 24 bits, or 32 bits?

2. A processor is designed to have lots of memory. How many bits are required for the addresses if the addressing space is
   (a) 1 million bytes
   (b) 1 gigabyte
   (c) 100 gigabytes
   (d) 10 terabytes

3. Which addressing modes do not require any memory access to get the operand? Which addressing modes require only one memory access to get the operand? Which addressing modes require more than one memory access before the operand is obtained?

4. List some uses for the indirect addressing modes.

5. Why is the direct addressing mode less favored than the register indirect addressing mode?

6. Show that the basic and special addressing modes are all degenerate cases of the general addressing mode.

7. Using 2K × 8 memory chips, design a memory of size 32K × 16.

8. Design an address translation device that can map a logical address to a physical address.

9. Most 8-bit microprocessors can address only 64K bytes of memory because 16 bits are allocated for the address. How can this addressing space be expanded to 128K, 256K, or even 1M bytes?

10. Design an expansion memory using 2K × 4 memory chips for the following configurations:
    (a) 2K × 16
    (b) 4K × 16
    (c) 16K × 32

11. Repeat Problem 10 with 1K × 8 memory chips.

12. How can a 16-bit address machine such as the 8085 or the Z-80 be augmented to operate with a memory size that is larger than 64K?

13. Company A hired a super typist who can type 120 words a minute. Assume that the average word length is 12 characters per word. If the transmission is

done serially using asynchronous protocol with two stop bits and one bit for interdigit gap, what is the minimum baud rate? If the transmission is done synchronously, what is the minimum baud rate?

14. A certain portable computer designed by Company XYZ was advertised as having the ability to hold 20 pages of typewritten notes. Assuming that a typewritten page holds 55 lines of 65 characters each, what is the memory capacity of this portable computer?

15. Design a circuit suitable for daisy-chaining the interrupt acknowledge signal from one peripheral to another.

16. Design a scheme to put an interrupt address on the data bus when an interrupt acknowledge signal comes in. Assume that the interrupt acknowledge signal is daisy-chained.

The 8085 Microprocessor

9

Beginning with this chapter, we present the design of VLSI components, more specifically, microprocessors and their supporting chips. The functional block diagram, signals, instruction set, and other features of the 8085 microprocessor are introduced. Simple programs are written showing the use of mnemonics and pseudo-operations. Higher-level language constructs are used extensively to illustrate the assembly programming techniques.

The ability to manufacture the central processing unit (or processor), which is a basic building block of a computer, on one integrated circuit has greatly altered the way of building digital systems. The creation of microprocessors is a result of using metal oxide semiconductor (MOS) technology in very-large-scale integrated (VLSI) circuits. Random-access memory (RAM), read-only memory (ROM), and other types of support hardware such as input/output (I/O) devices have led to the microprocessor-based computer system. These microcomputer systems are far more sophisticated than computer systems of ten to twenty years ago.

In the literature, different terms such as microprocessor, microcomputer, and microcomputer system are sometimes used. The general usage of these terms is as follows:

- *Microprocessor:* A processor implemented on a chip. The chip primarily implements the processing functions and requires additional supporting hardware such as memory chips and external peripheral devices for permanent data storage and communication.
- *Microcomputer:* A complete system with processor, memory, and limited input/output functions on the same chip. Hence microcomputers are also sometimes referred to as *single-chip computers.* These microcomputers are capable of carrying out dedicated control functions with little or no additional supporting chips. Hence they are also sometimes referred to as *microcontrollers.*
- *Microcomputer system:* A complete, stand-alone system that is capable of carrying out its functions with all necessary supporting hardware and peripheral devices.

Microprocessors are used in many modern electronic devices. Some of the most common are personal computers, video games, television sets, microwave ovens, washing machines, stereo receivers, calculators, industrial controllers, automobiles, and many more. We are interested primarily in showing how microprocessors can be used to perform calculations, control functions, and control processes within a digital system.

The microprocessor we choose to examine first in detail is the Intel 8085. The Intel 8085 is an 8-bit microprocessor that succeeds the Intel 8008 and 8080 microprocessors. The main reasons for selecting the Intel 8085 for the present study are that it is popular and easy-to-use, it is readily available from several manufacturers, and it is offered with a rich set of supporting hardware. An 8-bit rather than a 16- or 32-bit microprocessor is chosen because it is easier to understand for those new to the use of microprocessors.

When devices such as the 8085 are made by other manufacturers, this is called *second-sourcing.* An 8085 from another manufacturer is functionally the same as an Intel. For this reason, the name Intel will not be used in the following sections unless specific reference is being made to an Intel component. In this chapter, the terms microprocessor and processor are used interchangeably.

9.1

The 8085 Microprocessor

The 8085 is an 8-bit, general-purpose microprocessor, which is an improvement of Intel's popular 8080 microprocessor. The hardware and pin definitions are different, but the instruction set is essentially the same as the 8080 with two additional instructions to facilitate better interrupt control. The 8085 is easy to use because only a few support chips are needed. Furthermore, only a single 5-V power supply is required.

Internal Organization

The functional block diagram of the 8085 is given in Figure 9–1. The 8085 generates control signals that can be used to select appropriate external devices and functions to perform memory read and write operations. The 8085 can directly address up to 64K bytes of memory and up to 256 different input/output (I/O) ports or locations. The 8085 has excellent interrupt structure and control signals that allow for easy expansion. We will now give a general overview of the internal parts of the 8085.

Internal Registers The 8085 has seven addressable 8-bit registers. Some of these can be treated in pairs for certain types of instructions. The 8085 internal registers are as follows:

1. The accumulator (register A) is a general-purpose register that is used with many instructions, including arithmetic, logic, load, store, and I/O. In fact, almost all data transfers to and from the processor pass through the accumulator.
2. There are six additional general-purpose registers, called B, C, D, E, H, and L. These registers are used for temporary storage of data. With some instructions, the registers are used in pairs—that is, as BC, DE, and HL— to handle 16 bits of data or addresses. Furthermore, the register pair HL functions as a data pointer for a number of instructions with memory references.
3. The flag register contains five 1-bit flags as shown in Figure 9–2, each of which records one aspect of the processor's status information as a result

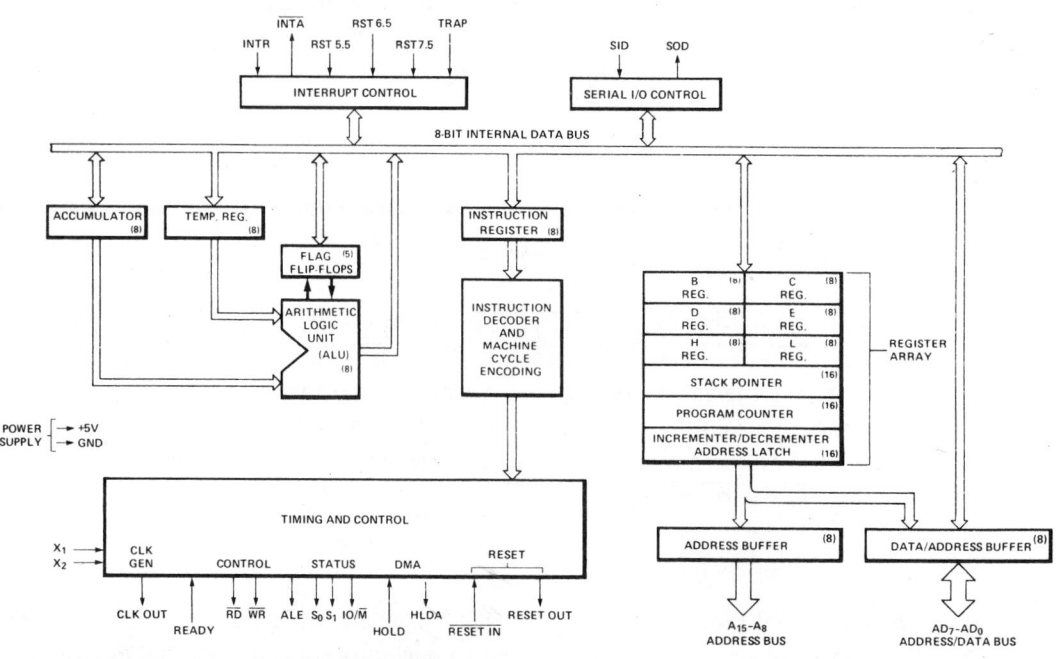

Figure 9-1

Functional block diagram of the 8085 (Reprinted by permission of Intel Corporation, Copyright/Intel Corporation 1979)

Figure 9-2

The status register of the 8085

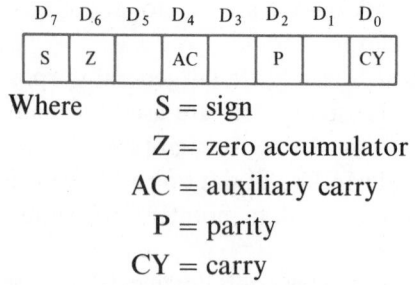

of a previous operation. These flags can be used to control future processor action.

(a) The *carry flag* (CY) indicates that a carry from the most significant bit has been generated from the current operation.

(b) The *auxiliary carry flag* (AC) indicates overflow out of bit 3 of the accumulator. This bit is used primarily for BCD arithmetic.

(c) The *sign flag* (S) indicates the sign of the results in the accumulator (i.e., the most significant bit) for both the arithmetic and logic operations.

(d) The *zero flag* (Z) is set (Z = 1) if the result of certain operations in the accumulator is zero. Otherwise, it is reset (Z = 0). Incrementing and decrementing other registers also affects the zero flag.

(e) The *parity flag* (P) is set (P = 1) if the number of 1 bits in the accumulator is even for certain instructions. Otherwise, it is reset (P = 0). In other words, the P flag can be interpreted as a parity bit generated for odd parity or as a parity checker flag for even parity.

The instructions that affect the flags will be given later. The 8085 has two special 16-bit address registers:

1. The *program counter* (PC) always points to the memory location of the next instruction to be executed.

2. The *stack pointer* (SP) is a special data pointer used to reference a user-defined part of the memory called the *stack*.

Other registers are used for temporary storage to implement various instructions, but they cannot be modified directly under user control. The *temporary register* (T) and the *instruction register* (IR) are two examples. The temporary register is used in arithmetic and logical operations to hold the second operand specified by the instruction. The instruction register is used to hold the operation code of the instruction for further decoding.

Arithmetic Logic Unit (ALU)   The *arithmetic logic unit* (ALU) performs the arithmetic, logical, and rotate shift operations. This is the main data manipulation part of the processor. The 8085 has an accumulator-based architecture. This means that the accumulator is always used as one of the source operands and at the same time the destination operand in dual-operand instructions. In general, the result of an operation is always transferred to the accumulator.

Timing and Control Unit   The first byte of an instruction contains the operation code, which is always transferred into the instruction register. The contents of the instruction register drive the instruction decoder. The outputs of the instruction decoder are timing signals that control the whole operation of the microprocessor, including the hardware parts such as the registers, ALU, data buffers, and address buffers. Hence the output of the instruction decoder together with the internal clock generator produce all the state and machine cycle timing signals.

The internal clock generator is provided in the processor and requires only the addition of an external quartz crystal to establish a timing reference. The clock circuitry generates two nonoverlapping internal clock signals to control the internal timing of the 8085. Other external timing circuits can also be used and are given in the 8085 manuals (see the references at the end of this chapter).

Interrupt Control Unit The *interrupt control unit* is part of the interrupt structure of the 8085. The 8085 has five different interrupt inputs, which can be used to effectively control the operation of the processor. The interrupt structure will be described in detail in the next chapter.

Instruction and Data Formats

Computers of today, no matter how sophisticated, can do only what they are told or instructed to do. A computer program is just a sequence of instructions stored in some type of memory unit. When needed, these instructions are recalled, decoded by the processor, and used to initiate the specified operations. The set of instructions to which the 8085 can respond is permanently fixed at the time the 8085 is designed.

Data Format The data format for the memory attached to the 8085 is organized into 8-bit words (bytes). The total number of memory locations that are directly addressable is $2^{16} = 64K$ (where $K = 1,024_{10} = 2^{10}$), since the 8085 has only 16 address lines. The memory can consist of RAM and ROM in any configuration to fit the appropriate need or application. Data are stored in the form of 8-bit binary integers:

$$\text{data word}$$
$$D_7 D_6 D_5 D_4 D_3 D_2 D_1 D_0$$
$$\text{MSB} \qquad\qquad \text{LSB}$$

The leftmost data bit, D_7, is the most significant bit (MSB), and the rightmost bit, D_0, is the least significant bit (LSB).

Instruction Format The instructions in memory may contain 1, 2, or 3 bytes. They are stored in ascending memory locations for the 2- and 3-byte instructions with the operation code (OPCODE) always being defined as the first byte of each instruction. The number of bytes is a function of the information required by the 8085 to execute the instruction. The form of the instructions is as follows:

1-Byte Instruction

ADDR	Byte 1	$D_7 \cdots D_0$	OPCODE

2-Byte Instruction

ADDR	Byte 1	$D_7 \cdots D_0$	OPCODE
ADDR + 1	Byte 2	$D_7 \cdots D_0$	Data

3-Byte Instruction

ADDR	Byte 1	$D_7 \cdots D_0$	OPCODE
ADDR + 1	Byte 2	$D_7 \cdots D_0$	Data or address (low-order)
ADDR + 2	Byte 3	$D_7 \cdots D_0$	Data or address (high-order)

The OPCODE is the first byte to be obtained from memory during the execution of any instruction. The data or address stored in memory may be 1 or 2 bytes. For 3-byte instructions, the least significant byte is always stored at the memory location one less than the memory location of the most significant byte.

An example of a single-byte instruction is the CMA operation, which is used to complement the contents of the accumulator. Only 1 byte is required to store the OPCODE.

Byte 1	OPCODE	CMA

An example of a 2-byte instruction is the MVI operation, which uses the first byte to store the MVI OPCODE and the second byte to store the immediate data for transfer:

Byte 1	OPCODE	MVI
Byte 2	Data	Data

An example of a 3-byte instruction is the LDA operation, which loads into the accumulator the data whose address is specified in the second and third bytes of the instruction.

Byte 1	OPCODE	LDA
Byte 2	Data or address	Low-order address
Byte 3	Data or address	High-order address

In 3-byte instructions, if data are stored in the bytes following the OPCODE, then the second and third bytes contain the low-order and the high-order parts of the data. If addresses are stored in those bytes, then the second and third bytes contain the low-order address and the high-order address.

Addressing Modes

The mode that specifies how a computer system obtains an operand or the location of data for any instruction is called the *addressing mode*. The data may be located within the processor, the memory, or an I/O device. Some types of addressing modes specify the location of the data entirely within the instruction.

Some specify a register. Others use the contents of a register or memory location to specify the actual location of data. The 8085 does not have a large number of addressing modes, but it has enough to perform most common computations and data manipulation operations. Specifically, the 8085 has four addressing modes: the immediate, direct, register, and register indirect. Some instructions have no operands or do not require a separate operand specification. The latter is called an *implied* addressing mode.

Immediate Mode   Those operations with immediate addressing modes are 2- or 3-byte instructions. The first byte is the OPCODE and the next byte or bytes are the data depending on whether the data width is 8 or 16 bits. In other words, the operands are referenced to the PC. For example, the MVI A, 9 is a 2-byte instruction that is used to move the literal 9 stored in the second byte to register A:

| | Byte 1 | OPCODE | MVI A |
| | Byte 2 | Data | 9 |

The LXI B instruction, on the other hand, is a 3-byte instruction used to move the data stored in the second and third bytes to registers C and B, respectively:

	Byte 1	OPCODE	LXI B
	Byte 2	Data/address	Low-order data/address
	Byte 3	Data/address	High-order data/address

Direct Mode   Those operations with direct addressing modes are 3-byte instructions, with the first byte containing the OPCODE, the second byte the least significant part of the address, and the third byte the most significant part of the address. Some computer manufacturers also call this *absolute addressing*. For example, the LDA instruction is used to load into the accumulator the data whose address is stored in the next 2 bytes of the instruction:

	Byte 1	OPCODE	LDA
	Byte 2	Address	Low-order address
	Byte 3	Address	High-order address

Register Mode   Those operations with register addressing modes are 1-byte instructions, with the instruction byte containing the OPCODE, the source operand, and the destination operand. Both the source and destination of the operands must be register specifications. A group of bits within the OPCODE specifies the register or registers to be used. Instructions with memory references are also included in this address mode. For example, the instruction MOV A,B can be used to move data from register B to register A:

Byte 1   OPCODE   MOV A,B

Register Indirect Mode Those operations with register indirect addressing modes are 1-byte instructions. A group of bits within the OPCODE specifies a register pair that contains the address of the data in memory. An example of this is the MOV A,M instruction, which is used to move to register A the data stored in memory. The address of the particular memory location to be fetched is specified by the HL register pair.

Byte 1   OPCODE   MOV A,M

There are advantages to each of these addressing modes. One-byte instructions may execute or run faster. The direct addressing mode has the advantage of specifying any of the 64K memory locations, whereas the register addressing mode can specify only seven different registers.

9.2
Assembly-Language Programming

The 8085 performs computations and control functions by fetching instructions and data in binary form from memory and input/output devices. The instruction set usually gives both the binary code and the corresponding mnemonic for the OPCODE. A *mnemonic* is a shorthand English description that is far easier for humans to interpret than the binary code. Mnemonics are the basic elements of an assembly language. The writing of codes using these mnemonics is called *assembly-language programming*. Assembly-language programming is similar to other types of programming languages, such as FORTRAN, in that well-defined rules must be followed. The mnemonics used for a particular assembly-language program are, however, applicable only for their specific computer system.

An *assembler* is a program that translates assembly-language mnemonics into a binary or machine-language code. A compiler is a program that translates a higher-level language, such as FORTRAN or COBOL, into a set of binary or machine-language codes that can be executed on a given computer system. An assembler is called a *cross-assembler* if it generates code on one computer system to be executed on a different computer system. Many computer systems have assemblers for the popular microprocessors such as the 8080, 8085, 6800, and others. For uniformity, the mnemonics and assembly program form and symbols used in this book will be the same as the ones used by Intel in describing their 8080/8085 microprocessors.

Assembly-Language Format

There are four fields in an assembly-language program: the label field, the operation code (OPCODE) or mnemonic field, the operand or address field, and the comment field. We will call them the label, OPCODE, operand, and comments for short. The general format is shown below:

Label   OPCODE   Operand   Comments

The general rules for writing a line of assembly code are simple:

1. A colon is required after a label except for the SET and EQU directives discussed later.
2. A space must separate the OPCODE and the operand.
3. A comma separates operands in the operand field.
4. A semicolon is the start of all comments to the end of the line.

A typical line of assembly code is given in the following example.

Example 9–1

Write an assembly code for the MOV instruction.

Label	OPCODE	Operand	Comments
NAME:	MOV	A,B	;Transfer the data from B to A

Label Field   The *label field* contains names that are helpful to identify statements and provide references for program control. The label is any identifier that begins with an alphabetic character and is no more than six characters long. The labels identify a particular address location or a specific data value. In the above example, the label is NAME, which identifies the address of that instruction. If NAME is used elsewhere in the program, then the label represents the address of that particular MOV instruction.

OPCODE Field   The *OPCODE field* specifies the instruction to be performed. Standard mnemonics are used. For communication to the assembler, sometimes assembler directives are also placed in this field. These directives are called *pseudo-operations* and are discussed below.

Operand Field   The *operand field* specifies both address and/or data as needed by the OPCODE. The operand values can be specified in decimal, in binary, or in hexadecimal by appending a D, a B, or an H after the data value, respectively. For example, to represent 15_{10}, one would write:

- In decimal: 15D
- In binary: 00001111B
- In hexadecimal: 0FH

Note that for hexadecimal numbers, if the number starts with the hexadecimal digits A through F, a 0 must be inserted before the hexadecimal number as shown above. In Example 9.1, there are two operands, registers A and B.

Comments Field *Comments* are extremely useful to help understand other people's programs and also your own after a period of time. Comments always follow a semicolon, which can appear anywhere on an assembly line.

Pseudo-Operations

Pseudo-operations, sometimes called *assembler directives*, are not operations that will be executed on the computer. Rather, these pseudo-operations communicate to the assembler concerning the program starting address, the location of the reserve memory space, and other information. There are a great number of pseudo-operations, but we will examine only a few of the most commonly used ones.

Origin (ORG) Directive The first pseudo-operation is the *origin (ORG) directive*. It is used to denote the starting address of the following program segment. The program must be at an address where memory actually exists. The general form is

Label	OPCODE	Operand
	ORG	Expression

END Directive The *END directive* indicates to the assembler the end of the program. The END terminates the assembly process. The general form is

Label	OPCODE	Operand
	END	

EQU Directive The *equate (EQU) directive* assigns the value of the expression to the name in the label field. Note that the name in the label field should not end with a colon. The assembler will equate a 16-bit number to the name when assembled. If only 8 bits are used, the 8 least significant bits will be used. The general form is

Label	OPCODE	Operand
Name	EQU	Expression

Example 9–2

Find the assembled code for the following labels.

The assembled code shows the actual value the assembler assigned to the labels.

Label	OPCODE	Operand	Assembled Code
FIVE	EQU	5D	0005
ONES	EQU	0FFH	00FF
SIXTY	EQU	60D	003C
MEMAD	EQU	1800H	1800

The EQU directive provides an easy representation of constants by means of a name or some other convenient expression. Should the value of the constant be changed, only the EQU statement needs to be changed before reassembling the code.

SET Directive　The *SET directive* is the same as the EQU except that the same name may be used in more than one SET directive. The latest SET directive encountered will be used whenever the name is invoked. In most cases, the EQU directive is used for clarity. The SET directive is used mostly in conditional assembly situations. The general form is

Label	OPCODE	Operand
Name	SET	Expressions or strings

Example 9–3

Set the value of INPORT to 1 initially and set its value to 2 at the end of the program.

Label	OPCODE	Operand
INPORT	SET	1D
INPORT	SET	2D
	END	

Define Byte (DB) Directive　Many different programming examples require the use of tables that are stored in memory. A couple of assembler directives are very useful for storing data and tables into memory. Most assemblers have the capability of doing simple arithmetic calculations such as addition, subtraction, multiplication, and division within the operand.

The *define byte (DB) directive* stores specified data in consecutive memory locations starting with the current setting of the location counter. The general form is

Label	OPCODE	Operand
Optional:	DB	Expressions or strings

The data in the operand field must be separated by commas with a maximum of eight values. The label is optional, but if included, it corresponds to the address of the first data byte. Also, any alphanumerics given between a pair of apostrophes will be converted to their equivalent ASCII code.

Example 9–4

Find the assembled code for the following labels.

Label	OPCODE	Operand	Assembled Code
DATA:	DB	10D,01111110B	0A7E
HERE:	DB	−3H,5*2	FD0A
TEXT:	DB	'HELP'	48454C50

Remember that any time a hexadecimal number is used that starts with a letter (A–F), the number should be preceded with a leading 0 (zero).

Define Word (DW) Directive   The *define word (DW) directive* is used to define 16 bits of data that occupy 2 bytes from memory. The general form is

Label	OPCODE	Operand
Optional:	DW	Expression

The symbols in the expression must have been previously defined. The least significant byte is stored at the first memory location, and the most significant byte is stored at the next memory location. The DW directive is quite often used for a table of addresses.

Example 9–5

Find the assembled code for the following labels.

Label	OPCODE	Operand	Assembled Code
ADDR1:	DW	0F62CH	2CF6
ADDR2:	DW	4H,27D	04001B00
COMP	EQU	67F2H	
ADD:	DW	COMP	F267

Define Storage (DS) Directive   The *define storage (DS) directive* is used to reserve memory locations for data that will be entered when the program is executing. Thus the DS directive should reserve storage only in RAM-type memory. This directive will not initialize any of the memory locations with specific data. The DS directive merely reserves the memory locations. The general form is

Label	OPCODE	Operand
Optional:	DS	Expression

The expression must be defined before the DS directive is encountered. The expression determines the number of bytes of storage to be reserved.

Example 9–6

Reserve 72 memory locations for a TTY buffer and 255 bytes for the analog-to-digital converter.

Label	OPCODE	Operand	Comments
TTYBUF:	DS	72D	;72 bytes are reserved for a ;TTY buffer
ADCTAB:	DS	0FFH	;255 bytes are reserved for ;the data from the analog-to- ;digital converter

---

As was indicated earlier, a number of other Intel 8080/8085 assembler directives are very useful for the more advanced programmer. The reader is referred to the Intel programming reference manual for further discussion.

Assembly-language programming can be developed only by practice and can be enhanced by analyzing well-written programs. Therefore, numerous examples are given in the following sections. Unless otherwise noted, addresses will be in hexadecimal format and will range from 0000H to FFFFH. We will now illustrate the assembly process with a simple program that will fetch data from memory, manipulate the data, and store them back into memory.

Example 9–7

Write a program that will logically complement the contents of memory location 0100H and put the result back into location 0100H.

The assembly program to do this is as follows:

Line No.	Label	OPCODE	Operand	Comments
1		ORG	1000H	;Start at 1000H
2	ADDR	EQU	100H	;Address of location
3	;This is the actual beginning of the program			
4		LDA	ADDR	;Get contents of
5				; location 100
6		CMA		;Complement data
7		STA	ADDR	;Store result
8		END		;End of program

In the above program, the first line of assembly code tells the assembler that the program is to begin at memory location 1000H. Some assemblers have a default starting location such as location 0000H. It is always good practice, however, to specify the program starting location explicitly. The second line defines the identifier ADDR to have a value of 0100H. The use of the EQU directive allows for easy changes of the program if needed later. The name of the identifier goes to the label field. The OPCODE is the pseudo-operation EQU. The value to be equated is the operand. The third line is a comment, since it starts with a semicolon. All information following the semicolon is ignored by the assembler.

The fourth line is the first to produce an executable code. No label is needed here. The operation is a load instruction. The address of the load operations is specified by the identifier ADDR. The fifth line is a comment line. The sixth line contains a single byte instruction with implied addressing. Hence only the OPCODE field is needed. The seventh line is similar to the fourth line. The last line is another pseudo-operation indicating the end of the program.

Program Assembly

After the program has been written, the next step is to assemble the program using the assembler. The assembler converts the assembly program to binary code, which generally is left in hexadecimal format for easy reading. The programmer can also hand-assemble the program.

Example 9–8 ▬▬▬▬▬▬▬▬▬▬▬▬▬▬▬▬▬▬▬▬▬▬▬

Assemble the program in Example 9–7.

All the listings will use one line for each 1-, 2- or 3-byte instruction. The memory address given corresponds to the address of the first byte, that is, OPCODE, and the following bytes will be at consecutive memory locations. The program listing for Example 9–7 is as follows:

Address	Assembled Code	Line No.	Label	OPCODE	Operand
1000		1		ORG	1000H
1000		2	ADDR	EQU	100H
		3	;This is the actual...		
1000	3A 00 01	4		LDA	ADDR
1003		5			;
1003	2F	6		CMA	
1004	32 00 01	7		STA	ADDR
		8		END	

Note that the least significant byte of the address is the second byte of the instruction and the most significant byte is the third byte. This is the address convention used in the 8085. The assembly process will now be explained line by line.

In line 1, the ORG directive set the starting address at 1000H, but being a pseudo-operation, produces no executable code. The second line is also a pseudo-operation that assigns the value 100H to the identifier ADDR. The third line is a comment line, hence no executable code is generated.

The fourth line is actually the first instruction that generates executable code. As will be discussed in Section 9.4, the LDA instruction is a 3-byte instruction. The address of the assembled code starts at 1000H as specified in line 1. The fifth line is a comment. The sixth line contains a single-byte instruction. Only 1 byte is needed to assemble this line. The seventh line is similar to line 4. The last line is another pseudo-operation that generates no executable code: it is there only to signify the end of the program.

When this program is executed starting at location 1000H, an uncertain result will occur when the instruction at location 1007H is fetched. An HLT instruction or a return to the system monitor (to be discussed in Section 9.10) would be appropriate.

9.3

The 8085 Instruction Set

In this section we examine in detail the entire instruction set of the 8085. This will give us a thorough understanding of the software aspects of the 8085 and at the same time allow us to gain experience in assembly-language programming. Each instruction is clearly defined by the operation it performs. Before the instructions are discussed, some notation used for describing the instructions must be presented.

Notation

In describing the instruction set, it is helpful to have some symbols and mnemonics to represent the operations of the instructions.

Symbol	Meaning
A	The accumulator, register A (The terms accumulator and register A are used interchangeably.)
ADDR	A 16-bit address
$DATA_8$	8 bits of data or information
$DATA_{16}$	16 bits of data or information
$BYTE_2$	The second byte (or BYTE TWO) of the instruction
$BYTE_3$	The third byte (or BYTE THREE) of the instruction
PORT	An 8-bit address of an I/O device
R, R_1, R_2	One of the following 8-bit registers: A, B, C, D, E, H, L
SSS	Three bits designating a source register
DDD	Three bits designating a destination register:

DDD or SSS	Register Name
111	A
000	B
001	C
010	D
011	E
100	H
101	L

rp	One of the register pairs: BC, DE, HL [The first register represents the highest-order byte and the second the lowest-order byte. The stack pointer (SP) is also included here.]
RP	The bit pattern designating one of the register pairs:

RP	Register Pair
00	BC
01	DE
10	HL
11	SP

rh	The first (high-order) byte of a designated register pair
rl	The second (low-order) byte of a designated register pair
PC	16-bit program counter
PCH	The higher-order bytes of the PC
PCL	The lower-order bytes of the PC
SP	16-bit stack pointer
SPH	The higher-order bytes of the SP
SPL	The lower-order bytes of the SP

PSW	A term referring to a pseudo-register pair formed with register A and the status register
r_m	Bit m of register r (Bit 7 is on the left and bit 0 is on the right. Bit 7 is considered to be the most significant bit.)
LABEL	A 16-bit address of an instruction
Z	Condition flag for zero ($Z = 1$ if the result of the last instruction that affects the Z flag was 0 in all bits; otherwise $Z = 0$.)
S	Condition flag for the sign ($S = 1$ if the most significant bit in the result of the previous instruction that affects the S flag is 1; otherwise $S = 0$.)
P	Condition flag for the parity ($P = 1$ if the number of bits in the result of the previous instruction that affects the P flag is even; otherwise $P = 0$.)
CY	Condition flag for the carry ($CY = 1$ if the result of the previous instruction that affects the CY flag resulted in a carry or borrow; otherwise $CY = 0$. Note that CY is also defined in the logical instructions.)
AC	Condition flag for the auxiliary carry ($AC = 1$ when there is a carry from bit 3 to bit 4 in the previous instruction that affects the AC flag; otherwise $AC = 0$. This is used for BCD addition.)
( )	The contents of the memory location or register whose address is enclosed within the parentheses.
←	Transfer of data in the direction of the arrow
↔	Exchange of data
∧	Logical AND
∀ or ⊕	Exclusive OR
∨	Inclusive OR
+	Arithmetic addition
−	2's complement subtraction
*	Multiplication
/	Division
n	Restart number in decimal (The restart number must be between 0 and 7.)
NNN	Restart number in binary (The restart number must be between 0 and 7.)

8085 Instructions

The 8085 instructions can be broken down into seven functional types:

1. *The data transfer group* contains instructions that move data between registers and registers, or between registers and memory locations.
2. *The arithmetic operation group* contains arithmetic instructions that add, subtract, increment, or decrement registers and memory locations.

3. *The logical operation group* contains instructions that perform comparisons and logical operations such as AND, OR, Exclusive OR, or complement of data in registers or memory.

4. *The shift/rotate operation group* includes instructions that perform various shifts and rotates on the accumulator.

5. *The stack operation group* includes instructions to perform stack operations and stack maintenance.

6. *The program control group* contains instructions that change the normal sequence of instruction execution through conditional jumps, unconditional jumps, subroutine calls, subroutine returns, and software restarts.

7. *The system control group* contains instructions to perform I/O, interrupts, and other system functions.

The entire instruction set can be quite overwhelming when first examined. The best way to understand the instruction set is first to write programs that will do simple tasks with commonly used instructions, then to practice techniques employing more sophisticated addressing modes and more complex instructions. A summary table of all the instructions according to the order presented here is given in Figure 9–3. For handy reference, a list of those instructions that affect the flags is given in Figure 9–4.

Figure 9–3

Summary of the 8085 instruction set

The Data Transfer Group					Flags				
Instruction	OPCODE	No. of Bytes	No. of T States	Operations	S	Z	AC	P	CY
8-Bit Data Transfer Instructions									
MOV A,A	7F	1	4	(A) ← (A)	(none affected)				
MOV A,B	78	1	4	(A) ← (B)					
MOV A,C	79	1	4	(A) ← (C)					
MOV A,D	7A	1	4	(A) ← (D)					
MOV A,E	7B	1	4	(A) ← (E)					
MOV A,H	7C	1	4	(A) ← (H)					
MOV A,L	7D	1	4	(A) ← (L)					
MOV A,M	7E	1	7	(A) ← ((H)(L))					
MOV B,A	47	1	4	(B) ← (A)					
MOV B,B	40	1	4	(B) ← (B)					
MOV B,C	41	1	4	(B) ← (C)					
MOV B,D	42	1	4	(B) ← (D)					
MOV B,E	43	1	4	(B) ← (E)					
MOV B,H	44	1	4	(B) ← (H)					
MOV B,L	45	1	4	(B) ← (L)					

(continued)

				The Data Transfer Group					
							Flags		
Instruction	OPCODE	No. of Bytes	No. of T States	Operations	S	Z	AC	P	CY
				8-Bit Data Transfer Instructions					
MOV B,M	46	1	7	(B) ← ((H)(L))			(none affected)		
MOV C,A	4F	1	4	(C) ← (A)					
MOV C,B	48	1	4	(C) ← (B)					
MOV C,C	49	1	4	(C) ← (C)					
MOV C,D	4A	1	4	(C) ← (D)					
MOV C,E	4B	1	4	(C) ← (E)					
MOV C,H	4C	1	4	(C) ← (H)					
MOV C,L	4D	1	4	(C) ← (L)					
MOV C,M	4E	1	7	(C) ← ((H)(L))					
MOV D,A	57	1	4	(D) ← (A)					
MOV D,B	50	1	4	(D) ← (B)					
MOV D,C	51	1	4	(D) ← (C)					
MOV D,D	52	1	4	(D) ← (D)					
MOV D,E	53	1	4	(D) ← (E)					
MOV D,H	54	1	4	(D) ← (H)					
MOV D,L	55	1	4	(D) ← (L)					
MOV D,M	56	1	7	(D) ← ((H)(L))					
MOV E,A	5F	1	4	(E) ← (A)					
MOV E,B	58	1	4	(E) ← (B)					
MOV E,C	59	1	4	(E) ← (C)					
MOV E,D	5A	1	4	(E) ← (D)					
MOV E,E	5B	1	4	(E) ← (E)					
MOV E,H	5C	1	4	(E) ← (H)					
MOV E,L	5D	1	4	(E) ← (L)					
MOV E,M	5E	1	7	(E) ← ((H)(L))					
MOV H,A	67	1	4	(H) ← (A)					
MOV H,B	60	1	4	(H) ← (B)					
MOV H,C	61	1	4	(H) ← (C)					
MOV H,D	62	1	4	(H) ← (D)					
MOV H,E	63	1	4	(H) ← (E)					
MOV H,H	64	1	4	(H) ← (H)					
MOV H,L	65	1	4	(H) ← (L)					
MOV H,M	66	1	7	(H) ← ((H)(L))					
MOV L,A	6F	1	4	(L) ← (A)					
MOV L,B	68	1	4	(L) ← (B)					
MOV L,C	69	1	4	(L) ← (C)					
MOV L,D	6A	1	4	(L) ← (D)					
MOV L,E	6B	1	4	(L) ← (E)					
MOV L,H	6C	1	4	(L) ← (H)					
MOV L,L	6D	1	4	(L) ← (L)					
MOV L,M	6E	1	7	(L) ← ((H)(L))					
MOV M,A	77	1	7	((H)(L)) ← (A)					
MOV M,B	70	1	7	((H)(L)) ← (B)					
MOV M,C	71	1	7	((H)(L)) ← (C)					
MOV M,D	72	1	7	((H)(L)) ← (D)					
MOV M,E	73	1	7	((H)(L)) ← (E)					
MOV M,H	74	1	7	((H)(L)) ← (H)					

Figure 9–3 Continued

		No. of Bytes	No. of T States	Operations	Flags				
The Data Transfer Group									
Instruction	OPCODE	No. of Bytes	No. of T States	Operations	S	Z	AC	P	CY
8-Bit Data Transfer Instructions									
MOV M,L	75	1	7	$((H)(L)) \leftarrow (L)$	(none affected)				
MVI A,DATA$_8$	3E	2	7	$(A) \leftarrow (BYTE_2)$					
MVI B,DATA$_8$	06	2	7	$(B) \leftarrow (BYTE_2)$					
MVI C,DATA$_8$	0E	2	7	$(C) \leftarrow (BYTE_2)$					
MVI D,DATA$_8$	16	2	7	$(D) \leftarrow (BYTE_2)$					
MVI E,DATA$_8$	1E	2	7	$(E) \leftarrow (BYTE_2)$					
MVI H,DATA$_8$	26	2	7	$(H) \leftarrow (BYTE_2)$					
MVI L,DATA$_8$	2E	2	7	$(L) \leftarrow (BYTE_2)$					
MVI M,DATA$_8$	36	2	10	$((H)(L)) \leftarrow (BYTE_2)$					
STAX B	02	1	7	$((B)(C)) \leftarrow (A)$					
STAX D	12	1	7	$((D)(E)) \leftarrow (A)$					
LDAX B	0A	1	7	$(A) \leftarrow ((B)(C))$					
LDAX D	1A	1	7	$(A) \leftarrow ((D)(E))$					
STA ADDR	32	3	13	$(ADDR) \leftarrow (A)$					
LDA ADDR	3A	3	13	$(A) \leftarrow (ADDR)$					
16-Bit Data Transfer Instructions									
LXI B,DATA$_{16}$	01	3	10	$(B)(C) \leftarrow (BYTE_3)(BYTE_2)$	(none affected)				
LXI D,DATA$_{16}$	11	3	10	$(D)(E) \leftarrow (BYTE_3)(BYTE_2)$					
LXI H,DATA$_{16}$	21	3	10	$(H)(L) \leftarrow (BYTE_3)(BYTE_2)$					
LXI SP,DATA$_{16}$	31	3	10	$(SP) \leftarrow (BYTE_3)(BYTE_2)$					
SHLD ADDR	22	3	16	$((BYTE_3)(BYTE_2)) \leftarrow (L)$ $((BYTE_3)(BYTE_2) + 1) \leftarrow (H)$					
LHLD ADDR	2A	3	16	$(L) \leftarrow ((BYTE_3), (BYTE_2))$ $(H) \leftarrow ((BYTE_3), (BYTE_2) + 1)$					
XCHG	EB	1	4	$(H) \leftrightarrow (D)$ $(L) \leftrightarrow (E)$					

		No. of Bytes	No. of T States	Operations Performed	Flags				
The Arithmetic Operation Group									
Instruction	OPCODE	No. of Bytes	No. of T States	Operations Performed	S	Z	AC	P	CY
8-Bit Arithmetic Instructions									
ADD A	87	1	4	$(A) \leftarrow (A) + (A)$	X	X	X	X	X
ADD B	80	1	4	$(A) \leftarrow (A) + (B)$	X	X	X	X	X
ADD C	81	1	4	$(A) \leftarrow (A) + (C)$	X	X	X	X	X
ADD D	82	1	4	$(A) \leftarrow (A) + (D)$	X	X	X	X	X
ADD E	83	1	4	$(A) \leftarrow (A) + (E)$	X	X	X	X	X
ADD H	84	1	4	$(A) \leftarrow (A) + (H)$	X	X	X	X	X
ADD L	85	1	4	$(A) \leftarrow (A) + (L)$	X	X	X	X	X
ADD M	86	1	7	$(A) \leftarrow (A) + ((H)(L))$	X	X	X	X	X
ADC A	8F	1	4	$(A) \leftarrow (A) + (A) + (CY)$	X	X	X	X	X
ADC B	88	1	4	$(A) \leftarrow (A) + (B) + (CY)$	X	X	X	X	X
ADC C	89	1	4	$(A) \leftarrow (A) + (C) + (CY)$	X	X	X	X	X

Figure 9–3 Continued

(*continued*)

				The Arithmetic Operation Group					

					Flags				
Instruction	OPCODE	No. of Bytes	No. of T States	Operations Performed	S	Z	AC	P	CY

8-Bit Arithmetic Instructions

Instruction	OPCODE	No. of Bytes	No. of T States	Operations Performed	S	Z	AC	P	CY
ADC D	8A	1	4	$(A) \leftarrow (A) + (D) + (CY)$	X	X	X	X	X
ADC E	8B	1	4	$(A) \leftarrow (A) + (E) + (CY)$	X	X	X	X	X
ADC H	8C	1	4	$(A) \leftarrow (A) + (H) + (CY)$	X	X	X	X	X
ADC L	8D	1	4	$(A) \leftarrow (A) + (L) + (CY)$	X	X	X	X	X
ADC M	8E	1	7	$(A) \leftarrow (A) + ((H)(L)) + (CY)$	X	X	X	X	X
ADI DATA$_8$	C6	2	7	$(A) \leftarrow (A) + (BYTE_2)$	X	X	X	X	X
ACI DATA$_8$	CE	2	7	$(A) \leftarrow (A) + (BYTE_2) + (CY)$	X	X	X	X	X
SUB A	97	1	4	$(A) \leftarrow (A) - (A)$	X	X	X	X	X
SUB B	90	1	4	$(A) \leftarrow (A) - (B)$	X	X	X	X	X
SUB C	91	1	4	$(A) \leftarrow (A) - (C)$	X	X	X	X	X
SUB D	92	1	4	$(A) \leftarrow (A) - (D)$	X	X	X	X	X
SUB E	93	1	4	$(A) \leftarrow (A) - (E)$	X	X	X	X	X
SUB H	94	1	4	$(A) \leftarrow (A) - (H)$	X	X	X	X	X
SUB L	95	1	4	$(A) \leftarrow (A) - (L)$	X	X	X	X	X
SUB M	96	1	7	$(A) \leftarrow (A) - ((H)(L))$	X	X	X	X	X
SBB A	9F	1	4	$(A) \leftarrow (A) - (A) - (CY)$	X	X	X	X	X
SBB B	98	1	4	$(A) \leftarrow (A) - (B) - (CY)$	X	X	X	X	X
SBB C	99	1	4	$(A) \leftarrow (A) - (C) - (CY)$	X	X	X	X	X
SBB D	9A	1	4	$(A) \leftarrow (A) - (D) - (CY)$	X	X	X	X	X
SBB E	9B	1	4	$(A) \leftarrow (A) - (E) - (CY)$	X	X	X	X	X
SBB H	9C	1	4	$(A) \leftarrow (A) - (H) - (CY)$	X	X	X	X	X
SBB L	9D	1	4	$(A) \leftarrow (A) - (L) - (CY)$	X	X	X	X	X
SBB M	9E	1	7	$(A) \leftarrow (A) - ((H)(L)) - (CY)$	X	X	X	X	X
SUI DATA$_8$	D6	2	7	$(A) \leftarrow (A) - (BYTE_2)$	X	X	X	X	X
SBI DATA$_8$	DE	2	7	$(A) \leftarrow (A) - (BYTE_2) - (CY)$	X	X	X	X	X
INR A	3C	1	4	$(A) \leftarrow (A) + 1$	X	X	X	X	
INR B	04	1	4	$(B) \leftarrow (B) + 1$	X	X	X	X	
INR C	0C	1	4	$(C) \leftarrow (C) + 1$	X	X	X	X	
INR D	14	1	4	$(D) \leftarrow (D) + 1$	X	X	X	X	
INR E	1C	1	4	$(E) \leftarrow (E) + 1$	X	X	X	X	
INR H	24	1	4	$(H) \leftarrow (H) + 1$	X	X	X	X	
INR L	2C	1	4	$(L) \leftarrow (L) + 1$	X	X	X	X	
INR M	34	1	10	$((H)(L)) \leftarrow ((H)(L)) + 1$	X	X	X	X	
DCR A	3D	1	4	$(A) \leftarrow (A) - 1$	X	X	X	X	
DCR B	05	1	4	$(B) \leftarrow (B) - 1$	X	X	X	X	
DCR C	0D	1	4	$(C) \leftarrow (C) - 1$	X	X	X	X	
DCR D	15	1	4	$(D) \leftarrow (D) - 1$	X	X	X	X	
DCR E	1D	1	4	$(E) \leftarrow (E) - 1$	X	X	X	X	
DCR H	25	1	4	$(H) \leftarrow (H) - 1$	X	X	X	X	
DCR L	2D	1	4	$(L) \leftarrow (L) - 1$	X	X	X	X	
DCR M	35	1	10	$((H)(L)) \leftarrow ((H)(L)) - 1$	X	X	X	X	

16-Bit Arithmetic Instructions

Instruction	OPCODE	No. of Bytes	No. of T States	Operations Performed	S	Z	AC	P	CY
INX B	03	1	6	$(B)(C) \leftarrow (B)(C) + 1$					
INX D	13	1	6	$(D)(E) \leftarrow (D)(E) + 1$					
INX H	23	1	6	$(H)(L) \leftarrow (H)(L) + 1$					
INX SP	33	1	6	$(SP) \leftarrow (SP) + 1$					
DCX B	0B	1	6	$(B)(C) \leftarrow (B)(C) - 1$					
DCX D	1B	1	6	$(D)(E) \leftarrow (D)(E) - 1$					
DCX H	2B	1	6	$(H)(L) \leftarrow (H)(L) - 1$					

Figure 9–3 Continued

The Arithmetic Operation Group										
								Flags		
Instruction	OPCODE	No. of Bytes	No. of T States	Operations Performed		S	Z	AC	P	CY

16-Bit Arithmetic Instructions

Instruction	OPCODE	No. of Bytes	No. of T States	Operations Performed	S	Z	AC	P	CY
DCX SP	3B	1	6	$(SP) \leftarrow (SP) - 1$					
DAD B	09	1	10	$(H)(L) \leftarrow (H)(L) + (B)(C)$					X
DAD D	19	1	10	$(H)(L) \leftarrow (H)(L) + (D)(E)$					X
DAD H	29	1	10	$(H)(L) \leftarrow (H)(L) + (H)(L)$					X
DAD SP	39	1	10	$(H)(L) \leftarrow (H)(L) + (SP)$					X

Decimal Arithmetic Instructions

Instruction	OPCODE	No. of Bytes	No. of T States	Operations Performed	S	Z	AC	P	CY
DAA	27	1	4	Decimal Adjust Register A	X	X	X	X	X

The Logical Operation Group										
								Flags		
Instruction	OPCODE	No. of Bytes	No. of T States	Operations Performed		S	Z	AC	P	CY

Logical Instructions

Instruction	OPCODE	No. of Bytes	No. of T States	Operations Performed	S	Z	AC	P	CY
ANA A	A7	1	4	$(A) \leftarrow (A) \wedge (A)$	X	X	1	X	0
ANA B	A0	1	4	$(A) \leftarrow (A) \wedge (B)$	X	X	1	X	0
ANA C	A1	1	4	$(A) \leftarrow (A) \wedge (C)$	X	X	1	X	0
ANA D	A2	1	4	$(A) \leftarrow (A) \wedge (D)$	X	X	1	X	0
ANA E	A3	1	4	$(A) \leftarrow (A) \wedge (E)$	X	X	1	X	0
ANA H	A4	1	4	$(A) \leftarrow (A) \wedge (H)$	X	X	1	X	0
ANA L	A5	1	4	$(A) \leftarrow (A) \wedge (L)$	X	X	1	X	0
ANA M	A6	1	7	$(A) \leftarrow (A) \wedge ((H)(L))$	X	X	1	X	0
ANI DATA$_8$	E6	2	7	$(A) \leftarrow (A) \wedge (BYTE_2)$	X	X	1	X	0
XRA A	AF	1	4	$(A) \leftarrow (A) \oplus (A)$	X	X	0	X	0
XRA B	A8	1	4	$(A) \leftarrow (A) \oplus (B)$	X	X	0	X	0
XRA C	A9	1	4	$(A) \leftarrow (A) \oplus (C)$	X	X	0	X	0
XRA D	AA	1	4	$(A) \leftarrow (A) \oplus (D)$	X	X	0	X	0
XRA E	AB	1	4	$(A) \leftarrow (A) \oplus (E)$	X	X	0	X	0
XRA H	AC	1	4	$(A) \leftarrow (A) \oplus (H)$	X	X	0	X	0
XRA L	AD	1	4	$(A) \leftarrow (A) \oplus (L)$	X	X	0	X	0
XRA M	AE	1	7	$(A) \leftarrow (A) \oplus ((H)(L))$	X	X	0	X	0
XRI DATA$_8$	EE	2	7	$(A) \leftarrow (A) \oplus (BYTE_2)$	X	X	0	X	0
ORA A	B7	1	4	$(A) \leftarrow (A) \vee (A)$	X	X	0	X	0
ORA B	B0	1	4	$(A) \leftarrow (A) \vee (B)$	X	X	0	X	0
ORA C	B1	1	4	$(A) \leftarrow (A) \vee (C)$	X	X	0	X	0
ORA D	B2	1	4	$(A) \leftarrow (A) \vee (D)$	X	X	0	X	0
ORA E	B3	1	4	$(A) \leftarrow (A) \vee (E)$	X	X	0	X	0
ORA H	B4	1	4	$(A) \leftarrow (A) \vee (H)$	X	X	0	X	0
ORA L	B5	1	4	$(A) \leftarrow (A) \vee (L)$	X	X	0	X	0
ORA M	B6	1	7	$(A) \leftarrow (A) \vee ((H)(L))$	X	X	0	X	0
ORI DATA$_8$	F6	2	7	$(A) \leftarrow (A) \vee (BYTE_2)$	X	X	0	X	0
CMA	2F	1	4	$(A) \leftarrow (A)'$					
CMC	3F	1	4	$(CY) \leftarrow (CY)'$					X
STC	37	1	4	$(CY) \leftarrow 1$					X

Figure 9–3 Continued

(*continued*)

				The Logical Operation Group					
							Flags		
Instruction	OPCODE	No. of Bytes	No. of T States	Operations Performed	S	Z	AC	P	CY
				Compare Instructions					
CMP A	BF	1	4	$(A) - (A)$	X	X	X	X	X
CMP B	B8	1	4	$(A) - (B)$	X	X	X	X	X
CMP C	B9	1	4	$(A) - (C)$	X	X	X	X	X
CMP D	BA	1	4	$(A) - (D)$	X	X	X	X	X
CMP E	BB	1	4	$(A) - (E)$	X	X	X	X	X
CMP H	BC	1	4	$(A) - (H)$	X	X	X	X	X
CMP L	BD	1	4	$(A) - (L)$	X	X	X	X	X
CMP M	BE	1	7	$(A) - ((H)(L))$	X	X	X	X	X
CPI DATA$_8$	FE	2	7	$(A) - (BYTE_2)$	X	X	X	X	X

				The Shift/Rotate Operation Instructions					
							Flags		
Instruction	OPCODE	No. of Bytes	No. of T States	Operations Performed	S	Z	AC	P	CY
RLC	07	1	4	$(A_{n+1}) \leftarrow (A_n)$ $(A_0) \leftarrow (A_7)$ $(CY) \leftarrow (A_7)$					X
RRC	0F	1	4	$(A_n) \leftarrow (A_{n+1})$ $(A_7) \leftarrow (A_0)$ $(CY) \leftarrow (A_0)$					X
RAL	17	1	4	$(A_{n+1}) \leftarrow (A_n)$ $(A_0) \leftarrow (CY)$ $(CY) \leftarrow (A_7)$					X
RAR	1F	1	4	$(A_n) \leftarrow (A_{n+1})$ $(A_7) \leftarrow (CY)$ $(CY) \leftarrow (A_0)$					X

				The Stack Operation Group					
							Flags		
Instruction	OPCODE	No. of Bytes	No. of T States	Operations Performed	S	Z	AC	P	CY
PUSH B	C5	1	12	$((SP) - 1) \leftarrow (B)$ $((SP) - 2) \leftarrow (C)$ $(SP) \leftarrow (SP) - 2$					
PUSH D	D5	1	12	$((SP) - 1) \leftarrow (D)$ $((SP) - 2) \leftarrow (E)$ $(SP) \leftarrow (SP) - 2$					
PUSH H	E5	1	12	$((SP) - 1) \leftarrow (H)$ $((SP) - 2) \leftarrow (L)$ $(SP) \leftarrow (SP) - 2$					
PUSH PSW	F5	1	12	$((SP) - 1) \leftarrow (A)$ $((SP) - 2) \leftarrow (STATUS\ REGISTER)$ $(SP) \leftarrow (SP) - 2$					

Figure 9-3 Continued

				The Stack Operation Group					
					Flags				
Instruction	OPCODE	No. of Bytes	No. of T States	Operations Performed	S	Z	AC	P	CY
POP B	C1	1	10	(C) ← ((SP))					
				(B) ← ((SP) + 1)					
				(SP) ← (SP) + 2					
POP D	D1	1	10	(E) ← ((SP))					
				(D) ← ((SP) + 1)					
				(SP) ← (SP) + 2					
POP H	E1	1	10	(L) ← ((SP))					
				(H) ← ((SP) + 1)					
				(SP) ← (SP) + 2					
POP PSW	F1	1	10	(STATUS REGISTER) ← ((SP))	X	X	X	X	X
				(A) ← ((SP) + 1)					
				(SP) ← (SP) + 2					
XTHL	E3	1	16	(L) ↔ ((SP))					
				(H) ↔ ((SP) + 1)					
SPHL	F9	1	6	(SP) ← (H)(L)					

				The Program Control Group					
					Flags				
Instruction	OPCODE	No. of Bytes	No. of T States	Operations Performed	S	Z	AC	P	CY
				Conditional and Unconditional Branch Instructions					
JMP ADDR	C3	3	10	(PC) ← (BYTE$_3$)(BYTE$_2$)	(none affected)				
JNZ ADDR	C2	3	7/10	(PC) ← (BYTE$_3$)(BYTE$_2$) IF Z = 0					
JZ ADDR	CA	3	7/10	(PC) ← (BYTE$_3$)(BYTE$_2$) IF Z = 1					
JNC ADDR	D2	3	7/10	(PC) ← (BYTE$_3$)(BYTE$_2$) IF CY = 0					
JC ADDR	DA	3	7/10	(PC) ← (BYTE$_3$)(BYTE$_2$) IF CY = 1					
JPO ADDR	E2	3	7/10	(PC) ← (BYTE$_3$)(BYTE$_2$) IF P = 0					
JPE ADDR	EA	3	7/10	(PC) ← (BYTE$_3$)(BYTE$_2$) IF P = 1					
JP ADDR	F2	3	7/10	(PC) ← (BYTE$_3$)(BYTE$_2$) IF S = 0					
JM ADDR	FA	3	7/10	(PC) ← (BYTE$_3$)(BYTE$_2$) IF S = 1					
PCHL	E9	1	6	(PC) ← (H)(L)					
				Subroutine Call and Return Instructions					
CALL ADDR	CD	3	18	((SP) − 1) ← (PCH)	(none affected)				
				((SP) − 2) ← (PCL)					
				(SP) ← (SP) − 2					
				(PC) ← (BYTE$_3$)(BYTE$_2$)					
CNZ ADDR	C4	3	9/18	((SP) − 1) ← (PCH)					
				((SP) − 2) ← (PCL)					
				(SP) ← (SP) − 2					
				(PC) ← (BYTE$_3$)(BYTE$_2$) IF Z = 0					
CZ ADDR	CC	3	9/18	((SP) − 1) ← (PCH)					
				((SP) − 2) ← (PCL)					
				(SP) ← (SP) − 2					
				(PC) ← (BYTE$_3$)(BYTE$_2$) IF Z = 1					

Figure 9–3 Continued

(continued)

					Flags				
Instruction	OPCODE	No. of Bytes	No. of T States	Operations Performed	S	Z	AC	P	CY

The Program Control Group

Subroutine Call and Return Instructions

Instruction	OPCODE	No. of Bytes	No. of T States	Operations Performed
CNC ADDR	D4	3	9/18	$((SP) - 1) \leftarrow (PCH)$ $((SP) - 2) \leftarrow (PCL)$ $(SP) \leftarrow (SP) - 2$ $(PC) \leftarrow (BYTE_3)(BYTE_2)$ IF CY = 0
CC ADDR	DC	3	9/18	$((SP) - 1) \leftarrow (PCH)$ $((SP) - 2) \leftarrow (PCL)$ $(SP) \leftarrow (SP) - 2$ $(PC) \leftarrow (BYTE_3)(BYTE_2)$ IF CY = 1
CPO ADDR	E4	3	9/18	$((SP) - 1) \leftarrow (PCH)$ $((SP) - 2) \leftarrow (PCL)$ $(SP) \leftarrow (SP) - 2$ $(PC) \leftarrow (BYTE_3)(BYTE_2)$ IF P = 0
CPE ADDR	EC	3	9/18	$((SP) - 1) \leftarrow (PCH)$ $((SP) - 2) \leftarrow (PCL)$ $(SP) \leftarrow (SP) - 2$ $(PC) \leftarrow (BYTE_3)(BYTE_2)$ IF P = 1
CP ADDR	F4	3	9/18	$((SP) - 1) \leftarrow (PCH)$ $((SP) - 2) \leftarrow (PCL)$ $(SP) \leftarrow (SP) - 2$ $(PC) \leftarrow (BYTE_3)(BYTE_2)$ IF S = 0
CM ADDR	FC	3	9/18	$((SP) - 1) \leftarrow (PCH)$ $((SP) - 2) \leftarrow (PCL)$ $(SP) \leftarrow (SP) - 2$ $(PC) \leftarrow (BYTE_3)(BYTE_2)$ IF S = 1
RET	C9	1	10	$(PCL) \leftarrow ((SP))$ $(PCH) \leftarrow ((SP) + 1)$ $(SP) \leftarrow (SP) + 2$
RNZ	C0	1	6/12	$(PCL) \leftarrow ((SP))$ $(PCH) \leftarrow ((SP) + 1)$ $(SP) \leftarrow (SP) + 2$ IF Z = 0
RZ	C8	1	6/12	$(PCL) \leftarrow ((SP))$ $(PCH) \leftarrow ((SP) + 1)$ $(SP) \leftarrow (SP) + 2$ IF Z = 1
RNC	D0	1	6/12	$(PCL) \leftarrow ((SP))$ $(PCH) \leftarrow ((SP) + 1)$ $(SP) \leftarrow (SP) + 2$ IF CY = 0
RC	D8	1	6/12	$(PCL) \leftarrow ((SP))$ $(PCH) \leftarrow ((SP) + 1)$ $(SP) \leftarrow (SP) + 2$ IF CY = 1
RPO	E0	1	6/12	$(PCL) \leftarrow ((SP))$ $(PCH) \leftarrow ((SP) + 1)$ $(SP) \leftarrow (SP) + 2$ IF P = 0
RPE	E8	1	6/12	$(PCL) \leftarrow ((SP))$ $(PCH) \leftarrow ((SP) + 1)$ $(SP) \leftarrow (SP) + 2$ IF P = 1
RP	F0	1	6/12	$(PCL) \leftarrow ((SP))$ $(PCH) \leftarrow ((SP) + 1)$ $(SP) \leftarrow (SP) + 2$ IF S = 0
RM	F8	1	6/12	$(PCL) \leftarrow ((SP))$ $(PCH) \leftarrow ((SP) + 1)$ $(SP) \leftarrow (SP) + 2$ IF S = 1

For CNC ADDR: (none affected)

Figure 9-3 Continued

(continued)

				The Program Control Group					

						Flags			
Instruction	OPCODE	No. of Bytes	No. of T States	Operations Performed	S	Z	AC	P	CY

Software Interrupt Instructions

Instruction	OPCODE	No. of Bytes	No. of T States	Operations Performed	Flags
RST 0	C7	1	12	$((SP) - 1) \leftarrow (PCH)$ $((SP) - 2) \leftarrow (PCL)$ $(SP) \leftarrow (SP) - 2$ $(PC) = 0H$	(none affected)
RST 1	CF	1	12	$((SP) - 1) \leftarrow (PCH)$ $((SP) - 2) \leftarrow (PCL)$ $(SP) \leftarrow (SP) - 2$ $(PC) = 8H$	
RST 2	D7	1	12	$((SP) - 1) \leftarrow (PCH)$ $((SP) - 2) \leftarrow (PCL)$ $(SP) \leftarrow (SP) - 2$ $(PC) = 10H$	
RST 3	DF	1	12	$((SP) - 1) \leftarrow (PCH)$ $((SP) - 2) \leftarrow (PCL)$ $(SP) \leftarrow (SP) - 2$ $(PC) = 18H$	
RST 4	E7	1	12	$((SP) - 1) \leftarrow (PCH)$ $((SP) - 2) \leftarrow (PCL)$ $(SP) \leftarrow (SP) - 2$ $(PC) = 20H$	
RST 5	EF	1	12	$((SP) - 1) \leftarrow (PCH)$ $((SP) - 2) \leftarrow (PCL)$ $(SP) \leftarrow (SP) - 2$ $(PC) = 28H$	
RST 6	F7	1	12	$((SP) - 1) \leftarrow (PCH)$ $((SP) - 2) \leftarrow (PCL)$ $(SP) \leftarrow (SP) - 2$ $(PC) = 30H$	
RST 7	FF	1	12	$((SP) - 1) \leftarrow (PCH)$ $((SP) - 2) \leftarrow (PCL)$ $(SP) \leftarrow (SP) - 2$ $(PC) = 38H$	

				The System Control Group					

Instruction	OPCODE	No. of Bytes	No. of T States	Operations Performed	S	Z	AC	P	CY

I/O Instructions

Instruction	OPCODE	No. of Bytes	No. of T States	Operations Performed	Flags
IN PORT	DB	2	10	$(A) \leftarrow (PORT)$	(none affected)
OUT PORT	D3	2	10	$(PORT) \leftarrow (A)$	

Interrupt-Related Instructions

EI	FB	1	4	ENABLE INTERRUPTS	
DI	F3	1	4	DISABLE INTERRUPTS	
RIM	20	1	4	READ INTERRUPT MASK	
SIM	30	1	4	SET INTERRUPT MASK	

System Instructions

HLT	76	1	4	HALT	
NOP	00	1	4	NO OPERATION	

All mnemonics copyright © Intel Corporation 1976.

Figure 9–3 Continued

Instructions	Status Flags				
	S	Z	AC	P	CY
ACI DATA	X	X	X	X	X
ADC reg	X	X	X	X	X
ADC M	X	X	X	X	X
ADD reg	X	X	X	X	X
ADD M	X	X	X	X	X
ADI DATA	X	X	X	X	X
ANA reg	X	X	1	X	0
ANA M	X	X	1	X	0
ANI DATA	X	X	1	X	0
CMC					X
CMP reg	X	X	X	X	X
CMP M	X	X	X	X	X
CPI DATA	X	X	X	X	X
DAA	X	X	X	X	X
DAD rp					X
DCR reg	X	X	X	X	
DCR M	X	X	X	X	
INR reg	X	X	X	X	
INR M	X	X	X	X	
ORA reg	X	X	0	X	0
ORA M	X	X	0	X	0
ORI DATA	X	X	0	X	0
POP PSW	X	X	X	X	X
RAL					X
RAR					X
RLC					X
RRC					X
SBB reg	X	X	X	X	X
SBB M	X	X	X	X	X
SBI DATA	X	X	X	X	X
STC					X
SUB reg	X	X	X	X	X
SUB M	X	X	X	X	X
SUI DATA	X	X	X	X	X
XRA reg	X	X	0	X	0
XRA M	X	X	0	X	0
XRI DATA	X	X	0	X	0

Figure 9–4
Instructions that affect the condition codes

9.4

The Data Transfer Group

The *data transfer instructions* transfer or move data between registers and registers or between registers and memory. The status flags are not affected by instructions within this group. There are two basic types of data transfers:

8-bit data transfer instructions
16-bit data transfer instructions

8-Bit Data Transfer Instructions

The 8-bit data transfer instructions are listed here:

Instruction	OPCODE	No. of Bytes	Operations Performed	Flags S Z AC P CY
MOV R_2,R_1	—	1	$(R_2) \leftarrow (R_1)$	(none affected)
MOV R_2,M	—	1	$(R_2) \leftarrow ((H)(L))$	
MOV M,R_1	—	1	$((H)(L)) \leftarrow (R_1)$	
MVI $R_2,DATA_8$	—	2	$(R_2) \leftarrow (BYTE_2)$	
MVI $M,DATA_8$	36	2	$((H)(L)) \leftarrow (BYTE_2)$	
STAX B	02	1	$((B)(C)) \leftarrow (A)$	
STAX D	12	1	$((D)(E)) \leftarrow (A)$	
LDAX B	0A	1	$(A) \leftarrow ((B)(C))$	
LDAX D	1A	1	$(A) \leftarrow ((D)(E))$	
STA ADDR	32	3	$(ADDR) \leftarrow (A)$	
LDA ADDR	3A	3	$(A) \leftarrow (ADDR)$	

Example 9–9

Write a program to exchange the contents of registers B and C.

Label	OPCODE	Operand	Comments
	ORG	0H	;Set origin
	MOV	A,B	;Save contents of B in A
	MOV	B,C	;Put contents of C in B
	MOV	C,A	;Restore saved value to C
	HLT		;Stop the 8085
	END		

16-Bit Data Transfer Instructions

The 16-bit data transfer instructions are listed here:

Instruction	OPCODE	No. of Bytes	Operations Performed	Flags S Z AC P CY
LXI B,DATA$_{16}$	01	3	$(B)(C) \leftarrow (BYTE_3)(BYTE_2)$	(none affected)
LXI D,DATA$_{16}$	11	3	$(D)(E) \leftarrow (BYTE_3)(BYTE_2)$	
LXI H,DATA$_{16}$	21	3	$(H)(L) \leftarrow (BYTE_3)(BYTE_2)$	
LXI SP,DATA$_{16}$	31	3	$(SP) \leftarrow (BYTE_3)(BYTE_2)$	
SHLD ADDR	22	3	$((BYTE_3)(BYTE_2)) \leftarrow (L)$ $((BYTE_3)(BYTE_2) + 1) \leftarrow (H)$	
LHLD ADDR	2A	3	$(L) \leftarrow ((BYTE_3), (BYTE_2))$ $(H) \leftarrow ((BYTE_3), (BYTE_2) + 1)$	
XCHG	EB	1	$(H) \leftrightarrow (D)$ $(L) \leftrightarrow (E)$	

Example 9–10

Write a program to read the contents of memory location 100 and put the value into memory location 200.

Label	OPCODE	Operand	Comments
	ORG	0H	
	LDA	100H	;Get value from 100
	STA	200H	;Store in 200
	HLT		
	END		

Example 9–11

Write a program to exchange the contents of registers B and C using the XCHG instruction.

Label	OPCODE	Operand	Comments
	ORG	0H	
	MOV	H,B	;XCHG works only on HL and DE
	MOV	D,C	;Save values into H and D

Label	OPCODE	Operand	Comments
	XCHG		
	MOV	B,H	;Restore values to B and C
	MOV	C,D	
	HLT		
	END		

Example 9-12

Write a program to load data from memory location 100 and store the value in memory location 200 using the register pair H and L.

Label	OPCODE	Operand	Comments
	ORG	0H	
	LXI	H,100H	;Set address
	LXI	D,200H	;Set second address
	MOV	A,M	;Get value from memory
	STAX	D	;Store in location 200
	HLT		
	END		

9.5
The Arithmetic Operation Group

The *arithmetic operation instructions* perform arithmetic operations on data in memory or registers. All instructions affect the flags according to the normal rules unless otherwise indicated. All subtractions are performed in 2's complement arithmetic with $CY = 1$ for a borrow and $CY = 0$ for no borrow. Instructions in this group can be divided into three types:

8-bit arithmetic instructions
16-bit arithmetic instructions
Decimal arithmetic instructions

8-Bit Arithmetic Instructions

The 8-bit arithmetic group includes the majority of the instructions in this group. It provides instructions to perform addition, subtraction, register increment operation, and register decrement operation:

Instruction	OPCODE	No. of Bytes	Operations Performed	Flags S	Z	AC	P	CY
ADD R_1	—	1	$(A) \leftarrow (R_1) + (A)$	X	X	X	X	X
ADD M	86	1	$(A) \leftarrow (A) + ((H)(L))$	X	X	X	X	X
ADC R_1	—	1	$(A) \leftarrow (R_1) + (A) + (CY)$	X	X	X	X	X
ADC M	8E	1	$(A) \leftarrow (A) + ((H)(L)) + (CY)$	X	X	X	X	X
ADI $DATA_8$	C6	2	$(A) \leftarrow (A) + (BYTE_2)$	X	X	X	X	X
ACI $DATA_8$	CE	2	$(A) \leftarrow (A) + (BYTE_2) + (CY)$	X	X	X	X	X
SUB R_1	—	1	$(A) \leftarrow (A) - (R_1)$	X	X	X	X	X
SUB M	96	1	$(A) \leftarrow (A) - ((H)(L))$	X	X	X	X	X
SBB R_1	—	1	$(A) \leftarrow (A) - (R_1) - (CY)$	X	X	X	X	X
SBB M	9E	1	$(A) \leftarrow (A) - ((H)(L)) - (CY)$	X	X	X	X	X
SUI $DATA_8$	D6	2	$(A) \leftarrow (A) - (BYTE_2)$	X	X	X	X	X
SBI $DATA_8$	DE	2	$(A) \leftarrow (A) - (BYTE_2) - (CY)$	X	X	X	X	X
INR R_1	—	1	$(R_1) \leftarrow (R_1) + 1$	X	X	X	X	
INR M	34	1	$((H)(L)) \leftarrow ((H)(L)) + 1$	X	X	X	X	
DCR R_1	—	1	$(R_1) \leftarrow (R_1) - 1$	X	X	X	X	
DCR M	35	1	$((H)(L)) \leftarrow ((H)(L)) - 1$	X	X	X	X	

Example 9–13

Write a program that adds 1 to the contents of memory location 100 and puts the result in location 101. In addition, the program should zero out location 100.

Label	OPCODE	Operand	Comments
	ORG	0H	
	LDA	100H	;Get data in A
	INR	A	;Add 1
	STA	101H	;Store result
	SUB	A	;Clear A
	STA	100H	;Zero location 100
	HLT		;Stop the 8085
	END		

16-Bit Arithmetic Instructions

There are three instructions that deal with 16-bit data, the INX, DCX, and DAD instructions, as shown here:

Instruction	OPCODE	No. of Bytes	Operations Performed	S	Z	AC	P	CY
INX B	03	1	(B)(C) ← (B)(C) + 1					
INX D	13	1	(D)(E) ← (D)(E) + 1					
INX H	23	1	(H)(L) ← (H)(L) + 1					
INX SP	33	1	(SP) ← (SP) + 1					
DCX B	0B	1	(B)(C) ← (B)(C) − 1					
DCX D	1B	1	(D)(E) ← (D)(E) − 1					
DCX H	2B	1	(H)(L) ← (H)(L) − 1					
DCX SP	3B	1	(SP) ← (SP) − 1					
DAD B	09	1	(H)(L) ← (H)(L) + (B)(C)					X
DAD D	19	1	(H)(L) ← (H)(L) + (D)(E)					X
DAD H	29	1	(H)(L) ← (H)(L) + (H)(L)					X
DAD SP	39	1	(H)(L) ← (H)(L) + (SP)					X

Decimal Arithmetic Instructions

There is one instruction specifically designed for decimal arithmetic:

Instruction	OPCODE	No. of Bytes	Operations Performed	S	Z	AC	P	CY
DAA	27	1	Decimal adjust register A	X	X	X	X	X

Example 9–14

Write a program that would add two BCD numbers stored in 0200H and 0201H and store the result in 0202H. Assume that the result will never exceed 99D.

Label	OPCODE	Operand	Comments
	ORG	1000H	
	LXI	H,201H	;Point (HL) to 201H
	LXI	D,200H	;Point (DE) to 200H
	LDAX	D	;Move ((DE)) to A
	ADD	M	;Add ((HL)) to A
	DAA		;BCD correction
	INX	H	;Point (HL) to 202H
	MOV	M,A	;Store result in 202H
	HLT		
	END		

9.6
The Logical Operation Group

The *logical group of instructions* performs logical (Boolean) operations on data in registers and memory. The logical operations are performed in a bitwise manner. In other words, one bit in the source register is operated on the same corresponding bit in the other register. All the flags are generally affected, with the zero, sign, and parity defined according to the standard rules. The CY flag and the AC flag are not affected in any standard manner, but will be covered for each instruction. There are primarily two types of instructions in this group:

1. Instructions to perform logical operations
2. Instructions to do comparisons

Logical Operation Instructions

Logical operations are limited to AND, Exclusive OR, OR, and complements.

Instruction	OPCODE	No. of Bytes	Operations Performed	S	Z	AC	P	CY
ANA R_1	—	1	$(A) \leftarrow (R_1) \wedge (A)$	X	X	1	X	0
ANA M	A6	1	$(A) \leftarrow (A) \wedge ((H)(L))$	X	X	1	X	0
ANI $DATA_8$	E6	2	$(A) \leftarrow (A) \wedge (BYTE_2)$	X	X	1	X	0
XRA R_1	—	1	$(A) \leftarrow (R_1) \oplus (A)$	X	X	0	X	0
XRA M	AE	1	$(A) \leftarrow (A) \oplus ((H)(L))$	X	X	0	X	0
XRI $DATA_8$	EE	2	$(A) \leftarrow (A) \oplus (BYTE_2)$	X	X	0	X	0
ORA R_1	—	1	$(A) \leftarrow (R_1) \vee (A)$	X	X	0	X	0
ORA M	B6	1	$(A) \leftarrow (A) \vee ((H)(L))$	X	X	0	X	0
ORI $DATA_8$	F6	2	$(A) \leftarrow (A) \vee (BYTE_2)$	X	X	0	X	0
CMA	2F	1	$(A) \leftarrow (\overline{A})$					
CMC	3F	1	$(CY) \leftarrow (\overline{CY})$					X
STC	37	1	$(CY) \leftarrow 1$					X

Example 9–15

Write a logical operation program to complement the memory location 250H.

Label	OPCODE	Operand	Comments
	ORG	1000H	
	LDA	250H	;Get value from 250
	CMA		;Complement
	STA	250H	;Store result
	HLT		
	END		

Example 9-16 ▰

Write a logical operation program that zeroes the most significant 4 bits of memory location 100 and puts the result in location 101. Also zero out location 100.

Label	OPCODE	Operand	Comments
	ORG	0H	
	LDA	100H	;Get data in A
	ANI	00001111B	;Mask out bits
	STA	101H	;Store result
	SUB	A	;Clear A
	STA	100H	;Zero location 100
	HLT		;Stop the 8085
	END		

Comparison Instructions

Comparison instructions set the status flags on the status register even though they do not change the contents of the register. Hence they are also included in the logical operations group. The primary purpose of the compare operations is to set the status flags to be used later in conditional branches.

Instruction	OPCODE	No. of Bytes	Operations Performed	Flags				
				S	Z	AC	P	CY
CMP R_1		1	$(A) - (R_1)$	X	X	X	X	X
CMP M	BE	1	$(A) - ((H)(L))$	X	X	X	X	X
CPI DATA$_8$	FE	2	$(A) - (BYTE_2)$	X	X	X	X	X

Examples of comparison instructions will be given in conjunction with the program control instructions presented in Section 9.9.

9.7 ▰

The Shift/Rotate Operation Group

The *shift/rotate group* includes all the operations for shifting and rotating the accumulator. To illustrate the differences among the various shift and rotate

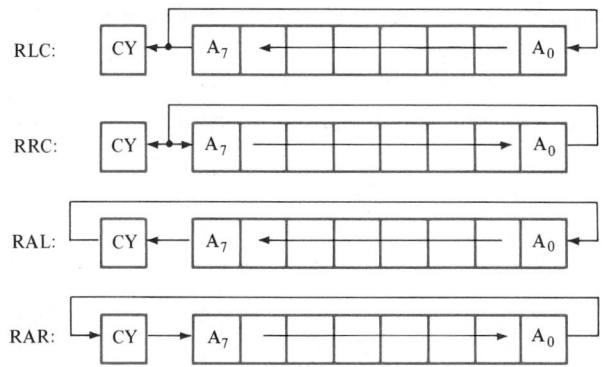

Figure 9–5
Operations of shift and rotate instructions of the 8085

operations, the actions taken by these instructions are shown in Figure 9–5. In-structions in the shift/rotate group are:

Instruction	OPCODE	No. of Bytes	Operations Performed	Flags				
				S	Z	AC	P	CY
RLC	07	1	$(A_{n+1}) \leftarrow (A_n)$ $(A_0) \leftarrow (A_7)$ $(CY) \leftarrow (A_7)$					X
RRC	0F	1	$(A_n) \leftarrow (A_{n+1})$ $(A_7) \leftarrow (A_0)$ $(CY) \leftarrow (A_0)$					X
RAL	17	1	$(A_{n+1}) \leftarrow (A_n)$ $(A_0) \leftarrow (CY)$ $(CY) \leftarrow (A_7)$					X
RAR	1F	1	$(A_n) \leftarrow (A_{n+1})$ $(A_7) \leftarrow (CY)$ $(CY) \leftarrow (A_0)$					X

Example 9-17

Multiply the content of the accumulator by 5. Assume that the binary number in the accumulator is less than 51D. Why must it be so for the algorithm to work?

Label	OPCODE	Operand	Comments
	ORG	1000H	
	MOV	B,A	;Save A
	RLC		;(A) ∗ 2
	RLC		;(A) ∗ 2
	ADD	B	;5 ∗ (A) = 2 ∗ (2 ∗ A) + A
	HLT		
	END		

Example 9-18

Swap the high-order nibble (4 bits) of the accumulator with its low-order nibble.

Label	OPCODE	Operand	Comments
	ORG	1000H	
	RLC		;rotate reg A left
	RLC		;rotate reg A left
	RLC		;rotate reg A left
	RLC		;rotate reg A left
	HLT		
	END		

9.8

The Stack Operation Group

The *stack* concept is an integral part in the operation of the 8085. The stack makes use of the stack pointer. The *stack pointer* is an internal 16-bit register in the 8085 that normally contains the address of RAM in what is referred to as the stack. We will use the higher-order locations of the RAM for our stack, which will require initializing our stack pointer to this area of RAM. Instructions related to the stack pointer are as follows:

Instruction	OPCODE	No. of Bytes	Operations Performed	Flags				
				S	Z	AC	P	CY
PUSH B	C5	1	$((SP) - 1) \leftarrow (B)$ $((SP) - 2) \leftarrow (C)$ $(SP) \quad \leftarrow (SP) - 2$					
PUSH D	D5	1	$((SP) - 1) \leftarrow (D)$ $((SP) - 2) \leftarrow (E)$ $(SP) \quad \leftarrow (SP) - 2$					
PUSH H	E5	1	$((SP) - 1) \leftarrow (H)$ $((SP) - 2) \leftarrow (L)$ $(SP) \quad \leftarrow (SP) - 2$					
PUSH PSW	F5	1	$((SP) - 1) \leftarrow (A)$ $((SP) - 2) \leftarrow (STATUS\ REGISTER)$ $(SP) \quad \leftarrow (SP) - 2$					
POP B	C1	1	$(C) \leftarrow ((SP))$ $(B) \leftarrow ((SP) + 1)$ $(SP) \leftarrow (SP) + 2$					
POP D	D1	1	$(E) \leftarrow ((SP))$ $(D) \leftarrow ((SP) + 1)$ $(SP) \leftarrow (SP) + 2$					
POP H	E1	1	$(L) \leftarrow ((SP))$ $(H) \leftarrow ((SP) + 1)$ $(SP) \leftarrow (SP) + 2$					
POP PSW	F1	1	$(STATUS\ REGISTER) \leftarrow ((SP))$ $(A) \quad \leftarrow ((SP) + 1)$ $(SP) \quad \leftarrow (SP) + 2$	X	X	X	X	X
XTHL	E3	1	$(L) \leftrightarrow ((SP))$ $(H) \leftrightarrow ((SP) + 1)$					
SPHL	F9	1	$(SP) \leftarrow (H)(L)$					

The stack pointer is normally initialized in the first part of any program and then changes only with the various instructions that move information to and from the stack. The programmer must be careful to keep track of the number of times that information is moved to and from the stack. The programmer must ensure that sufficient space is set aside in RAM for the stack so that in the worst case the number of PUSH-type instructions that are executed will not decrement into RAM used for other parts of the program. Note that in the following definition, the term *processor status word* (*PSW*) refers to the register pair consisting of register A and the status register.

Let us look at the PUSH B instruction. This says that the content of register B is moved to a memory location whose address is one less than the content of register SP and the content of register C is moved to memory whose address is two less than the content of register SP; then the SP is decremented by 2. Let us examine the following instruction sequence:

Label	OPCODE	Operand	Comments
	LXI	SP,0900H	
	PUSH	B	

Assume that the above instructions are operating in a system that has 256 bytes of RAM located from 0800H to 08FFH. Also assume that the contents of B and C are (B) = 25H and (C) = 6AH. The above sequence will initialize the SP to 0900H. The PUSH B instruction will move the contents of B to memory location 08FFH and the contents of C to memory location 08FEH. The stack pointer is also decremented by 2, so that the new effective stack pointer contents is 08FEH.

The use of the stack normally implies that for every instruction that moves data into the stack area of RAM, there is a corresponding instruction that moves the data back to an 8085 register and then increments the SP by 2. For the POP B instruction, the contents of the memory location as specified by the contents of the SP are moved to register C. The contents of the memory location that is one more than the contents of the SP are moved to register B, and then the SP is incremented by 2. The following instruction sequence will move the contents of register pair BC to DE:

Label	OPCODE	Operand	Comments
	LXI	SP,0900H	
	PUSH	B	
	POP	D	

The contents of the SP will be 08FEH after the execution of PUSH B and then return to 0900H after POP D is executed. This sequence will transfer the contents of register pair BC to register pair DE, leaving BC unchanged.

Example 9–19

Exchange the contents of the register pair BC with the contents of the register pair HL.

The following program will interchange the contents of register pair BC with the contents of register pair HL:

Label	OPCODE	Operand	Comments
	LXI	SP,0900H	
	PUSH	B	
	PUSH	H	
	POP	B	
	POP	H	
	HLT		
	END		

Note the order of the operands being PUSHed and POPped. What will happen if POP H appears before POP B?

9.9

The Program Control Group

The *program control group* of instructions alters the normal sequential program flow. Instead of the next instruction being executed like the other instructions, a branch instruction may cause an instruction in another section of memory to be executed next. The flags are not affected by any instructions in this group. There are three types of program control instructions:

Conditional and unconditional branches
Subroutine calls and returns
Software interrupts

Conditional and Unconditional Branches

Two kinds of branch instructions are unconditional and conditional transfers. Unconditional transfers simply perform the specified operation on register PC (program counter). Conditional transfers examine the status of one of the four flags (Z, S, P, or CY) to determine if the specified branch is to be executed. Note that there are no conditional branches based on the AC flag. The auxiliary flag is used internally by the DAA instruction only. The conditions are typically shortened as follows, with the three bits, CCC, specifying the condition within the instruction:

	Condition	CCC
NZ	not zero (Z = 0)	000
Z	zero (Z = 1)	001
NC	no carry (CY = 0)	010
C	carry (CY = 1)	011
PO	parity odd (P = 0)	100
PE	parity even (P = 1)	101
P	plus (S = 0)	110
M	minus (S = 1)	111

The instructions will be specified with the word condition, but in actual instructions NZ, Z, etc., are substituted for the condition; that is, the J condition becomes JNZ, JZ, . . . , etc. In Appendix A, two values are given for the number of cycles and states. The first value applies if no jump is made and the second if a jump is made.

Instruction	OPCODE	No. of Bytes	Operations Performed	Flags S Z AC P CY
			Conditional and Unconditional Branch Instructions	
JMP ADDR	C3	3	$(PC) \leftarrow (BYTE_3)(BYTE_2)$	(none affected)
JNZ ADDR	C2	3	$(PC) \leftarrow (BYTE_3)(BYTE_2)$ IF $Z = 0$	
JZ ADDR	CA	3	$(PC) \leftarrow (BYTE_3)(BYTE_2)$ IF $Z = 1$	
JNC ADDR	D2	3	$(PC) \leftarrow (BYTE_3)(BYTE_2)$ IF $CY = 0$	
JC ADDR	DA	3	$(PC) \leftarrow (BYTE_3)(BYTE_2)$ IF $CY = 1$	
JPO ADDR	E2	3	$(PC) \leftarrow (BYTE_3)(BYTE_2)$ IF $P = 0$	
JPE ADDR	EA	3	$(PC) \leftarrow (BYTE_3)(BYTE_2)$ IF $P = 1$	
JP ADDR	F2	3	$(PC) \leftarrow (BYTE_3)(BYTE_2)$ IF $S = 0$	
JM ADDR	FA	3	$(PC) \leftarrow (BYTE_3)(BYTE_2)$ IF $S = 1$	
PCHL	E9	1	$(PC) \leftarrow (H)(L)$	

Example 9–20

Write a program to implement a loop that zeroes out a block of memory from 100H to 1FFH.

Label	OPCODE	Operand	Comments
SIZE	EQU	0100H	
START	EQU	0100H	
	ORG	1000H	
	LXI	D,SIZE	;(DE) = SIZE
	LXI	B,START	;(BC) = START
HERE:	XRA	A	;(A) = 0H
	STAX	B	;((BC)) ← (A)
	INX	B	;(BC) = (BC) + 1
	DCX	D	;(DE) = (DE) − 1
	MOV	A,D	;Is (DE) ZERO
	ORA	E	
	JNZ	HERE	;No, do again
	HLT		
	END		

Example 9-21

Write a program to convert a BCD number into binary.
The program is as follows:

Label	OPCODE	Operand	Comments
	ORG	1000H	;
BCD	EQU	100H	;Location of the BCD #
BINARY	EQU	200H	;Location of the binary #
	LXI	H,BCD	;Get address of BCD #
	MOV	B,M	;Get the BCD #
	LXI	H,BINARY	;Get address of binary #
	XRA	A	;Zero accumulator
HERE:	INR	M	;Increment binary #
	INR	A	;Increment BCD #
	DAA		;
	CMP	B	;Is BCD # the same
	JNZ	HERE	
	HLT		
	END		

The program in Example 9-21 uses a varying amount of time to obtain the binary equivalent depending on the magnitude of the BCD number. Can you write a program that is independent of its magnitude? *Hint*: Use shift and compare techniques.

Example 9-22

Write a program to find the smallest of two numbers stored in location 200H and 201H and place the result in location 202H. Assume that the two numbers are unsigned binary numbers from 00H to 0FFH.
The program is as follows:

Label	OPCODE	Operand	Comments
	ORG	100H	
	LXI	H,200H	;Loads data pointer
	MOV	A,M	;Gets first number
	INX	H	;Increments data pointer
	CMP	M	;Finds smallest
	JC	STOR	;Jump if smallest is in ; register A
	MOV	A,M	;Second is smallest
STOR:	INX	H	;Increments data pointer
	MOV	M,A	;Store smallest
	HLT		;Halt processor
	END		

The program in Example 9–22 makes use of the fact that comparison operations do not alter any registers, only the flags. The next program uses a simple loop to repeat a calculation. The use of loops is very common in most types of assembly-language programs.

Example 9–23 ▰▰▰▰▰▰▰

Write a program to count the number of positive integers that are stored in memory starting at memory location 2001H. Assume that the number of integers is in memory location 2000H and store the number of positive integers in register C. Assume that zero is a positive number.

The program is as follows:

Label	OPCODE	Operand	Comments
	ORG	200H	
	LXI	H,2000H	;Set up data pointer
	MOV	B,M	;Get # of integers
	MVI	C,0D	;Zero reg. C
NEXT:	INX	H	;Increment data pointer
	MOV	A,M	;Get integer
	ANA	A	;Set flags
	JM	HERE	;Check for positive #
	INR	C	;Increment (C) if pos.
HERE:	DCR	B	;Decrement # of integers
	JNZ	NEXT	;Do again
	HLT		
	END		

Loop-type programs typically need a register to keep track of the number of times some operation is performed. Note that if the number at location 2000H is 0, then this program will not work right. How can this be changed?

Software Delay Routine Example   Many times when microprocessors are used to control external hardware that must operate with slow devices, delays need to be built into the software. The delays may be in the range of microseconds, milliseconds, seconds, or even minutes. Most computers have an operation or instruction called a *no operation* (*NOP*) that can be used to insert delay in the microsecond range. The NOP instruction will be discussed in the next section. To obtain delays that are longer, repeated use of NOPs is neither efficient nor appealing. A more convenient way to generate a delay is to preset a register or register pair and count down until it reaches zero. The next example shows the technique.

Example 9–24 ▰▰▰▰▰▰▰▰▰▰▰▰▰▰▰▰▰▰▰▰▰▰▰▰▰▰▰▰▰▰

Write a program that will generate a delay by presetting a register pair and then counting down to zero.

Label	OPCODE	Operand	Comments
	ORG	0H	
DEL	EQU	5000D	
	LXI	D,DEL	;Initialize counter
LOOP:	DCX	D	;Decrement
	MOV	A,D	;Check if (DE) = 0
	ORA	E	;
	JNZ	LOOP	;Jump to loop when (DE) ≠ 0
	HLT		
	END		

The value loaded into the register pair will fix the amount of delay. The most significant part of the delay is equal to the number of times through the loop multiplied by the time needed for the loop. The two instructions MOV A,D and ORA E are needed to determine when (D)(E) = 0, because the decrement and increment register pair instructions do not affect the flags. The number of T states in the loop is

Instruction	No. of T States
DCX D	6
MOV A,D	4
ORA E	4
JNZ LOOP	10 (when jump occurs)

The jump is performed for all cases except the last one through the loop. Assume that the crystal frequency is 5 MHz; the actual delay is then

$$T_{del} = DEL \times (6 + 4 + 4 + 10) \times 2/(5.0 \times 10^6)$$

where DEL represents the number of loops stored in register pair DE.

The actual delay may vary from a minimum of 9.6 μs to 629 ms in 9.6-μs increments, depending on the value of DEL. For DEL = 5000D, the delay is 48 ms. A word of caution: if DEL = 0, the maximum delay is obtained. The checking of the contents of DE after the first decrement causes this to happen. The delay can be increased by doing this routine more than once, or with one

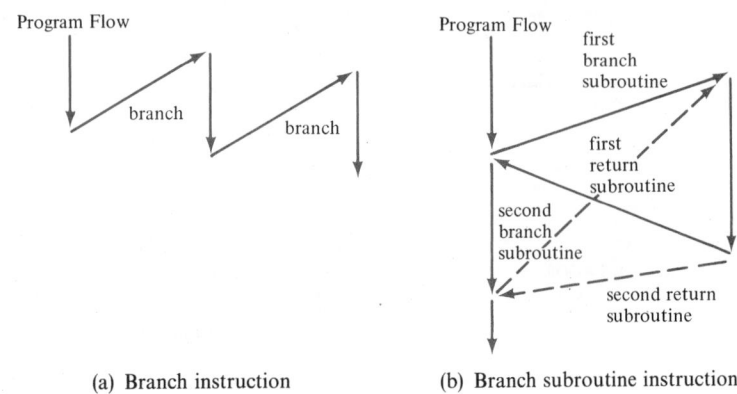

(a) Branch instruction (b) Branch subroutine instruction

Figure 9–6
Branch and branch subroutine instructions

loop inside another loop. These techniques all require the wasteful use of computer time that may be spent performing other tasks. External counters are used in many applications to accomplish longer delays, as will be discussed later.

Subroutine Calls and Returns

Subroutines are used in microprocessor-based systems in the same manner as in any computer system. Assembly-language programs that are used repeatedly are written in the form of a subroutine so that they may be used over and over again. Program control could be changed with the use of a JMP instruction, but the processor would not be able to return to the next instruction following the JMP after the subroutine was completed because the return address is not saved. The CALL instruction is used instead, because the location of the next instruction following the CALL instruction is saved in the stack area of RAM. The difference between a branching instruction (JMP) and a subroutine call instruction (CALL) is shown in Figure 9–6.

The following instructions can be used for subroutine calls and returns. In the case of a conditional subroutine call or return instruction, the next logical instruction is executed when the specified condition is not met.

Subroutine Calls and Returns

Instruction	OPCODE	No. of Bytes	Operations Performed	Flags S Z AC P CY
CALL ADDR	CD	3	$((SP) - 1) \leftarrow (PCH)$ $((SP) - 2) \leftarrow (PCL)$ $(SP) \quad \leftarrow (SP) - 2$ $(PC) \quad \leftarrow (BYTE_3)(BYTE_2)$	(none affected)
CNZ ADDR	C4	3	$((SP) - 1) \leftarrow (PCH)$ $((SP) - 2) \leftarrow (PCL)$ $(SP) \quad \leftarrow (SP) - 2$ $(PC) \quad \leftarrow (BYTE_3)(BYTE_2)$ IF $Z = 0$	
CZ ADDR	CC	3	$((SP) - 1) \leftarrow (PCH)$ $((SP) - 2) \leftarrow (PCL)$ $(SP) \quad \leftarrow (SP) - 2$ $(PC) \quad \leftarrow (BYTE_3)(BYTE_2)$ IF $Z = 1$	
CNC ADDR	D4	3	$((SP) - 1) \leftarrow (PCH)$ $((SP) - 2) \leftarrow (PCL)$ $(SP) \quad \leftarrow (SP) - 2$ $(PC) \quad \leftarrow (BYTE_3)(BYTE_2)$ IF $CY = 0$	
CC ADDR	DC	3	$((SP) - 1) \leftarrow (PCH)$ $((SP) - 2) \leftarrow (PCL)$ $(SP) \quad \leftarrow (SP) - 2$ $(PC) \quad \leftarrow (BYTE_3)(BYTE_2)$ IF $CY = 1$	
CPO ADDR	E4	3	$((SP) - 1) \leftarrow (PCH)$ $((SP) - 2) \leftarrow (PCL)$ $(SP) \quad \leftarrow (SP) - 2$ $(PC) \quad \leftarrow (BYTE_3)(BYTE_2)$ IF $P = 0$	
CPE ADDR	EC	3	$((SP) - 1) \leftarrow (PCH)$ $((SP) - 2) \leftarrow (PCL)$ $(SP) \quad \leftarrow (SP) - 2$ $(PC) \quad \leftarrow (BYTE_3)(BYTE_2)$ IF $P = 1$	
CP ADDR	F4	3	$((SP) - 1) \leftarrow (PCH)$ $((SP) - 2) \leftarrow (PCL)$ $(SP) \quad \leftarrow (SP) - 2$ $(PC) \quad \leftarrow (BYTE_3)(BYTE_2)$ IF $S = 0$	
CM ADDR	FC	3	$((SP) - 1) \leftarrow (PCH)$ $((SP) - 2) \leftarrow (PCL)$ $(SP) \quad \leftarrow (SP) - 2$ $(PC) \quad \leftarrow (BYTE_3)(BYTE_2)$ IF $S = 1$	
RET	C9	1	$(PCL) \leftarrow ((SP))$ $(PCH) \leftarrow ((SP) + 1)$ $(SP) \quad \leftarrow (SP) + 2$	
RNZ	C0	1	$(PCL) \leftarrow ((SP))$ $(PCH) \leftarrow ((SP) + 1)$ $(SP) \quad \leftarrow (SP) + 2$ IF $Z = 0$	
RZ	C8	1	$(PCL) \leftarrow ((SP))$ $(PCH) \leftarrow ((SP) + 1)$ $(SP) \quad \leftarrow (SP) + 2$ IF $Z = 1$	
RNC	D0	1	$(PCL) \leftarrow ((SP))$ $(PCH) \leftarrow ((SP) + 1)$ $(SP) \quad \leftarrow (SP) + 2$ IF $CY = 0$	
RC	D8	1	$(PCL) \leftarrow ((SP))$ $(PCH) \leftarrow ((SP) + 1)$ $(SP) \quad \leftarrow (SP) + 2$ IF $CY = 1$	
RPO	E0	1	$(PCL) \leftarrow ((SP))$ $(PCH) \leftarrow ((SP) + 1)$ $(SP) \quad \leftarrow (SP) + 2$ IF $P = 0$	
RPE	E8	1	$(PCL) \leftarrow ((SP))$ $(PCH) \leftarrow ((SP) + 1)$ $(SP) \quad \leftarrow (SP) + 2$ IF $P = 1$	
RP	F0	1	$(PCL) \leftarrow ((SP))$ $(PCH) \leftarrow ((SP) + 1)$ $(SP) \quad \leftarrow (SP) + 2$ IF $S = 0$	
RM	F8	1	$(PCL) \leftarrow ((SP))$ $(PCH) \leftarrow ((SP) + 1)$ $(SP) \quad \leftarrow (SP) + 2$ IF $S = 1$	

Example 9-25 ▪▪▪▪▪▪▪▪▪

Write a subroutine that will modify bit 7 in register A to ensure that the contents of register A will have even parity when the subroutine is executed. Do not assume that A_7 is 0 or 1.

The subroutine follows.

Label	OPCODE	Operand	Comments
PAR:	ANA	A	;Sets flags
	JPE	DONE	;Jump if parity even
	XRI	10000000B	;Change to even parity
DONE:	RET		;Return

This is a simple subroutine, but it illustrates the way subroutines are constructed. The last executed statement is always the RET. The ANA A instruction is used to set the flags according to the contents of register A without modifying the contents.

In the next example, a program is written to call this subroutine to make sure that a list of 8-bit words will have even parity.

Example 9-26 ▪▪▪▪▪▪▪▪▪

Assume that location 0800H in RAM contains the number of locations to be modified as needed so that the bytes will be read and restored with even parity starting with location 0801H. The subroutine developed in Example 9-25 will be used.

Label	OPCODE	Operand	Comments
	ORG	0H	
	LXI	SP,0900H	;Initialize the SP
	LXI	H,0800H	;Initialize data pointer
	MOV	B,M	;Get count
NEXT:	INX	H	;Update pointer
	MOV	A,M	;Get data
	CALL	PAR	;Modify for even parity
	MOV	M,A	;Restore data
	DCR	B	;Check number of bytes
	JNZ	NEXT	
	⋮		;Go to rest of program
PAR:	ANA	A	;Subroutine starts
	JPE	DONE	;Here
	XRI	10000000B	
DONE:	RET		

The program in Example 9–26 put the subroutine after the main program. Many programmers try to be consistent and place all subroutines at the end of the main program. The CALL PAR instruction will move to the stack the address of the next instruction, which is MOV M,A, and then start executing the subroutine PAR. The next instruction that will use the stack is the RET inside the subroutine. This instruction will move from the stack to the program counter the address of the MOV M,A instruction and thus cause this instruction to be executed after the RET instruction.

The 8085 microprocessor has a number of conditional CALL and RET instructions. These are seldom used, but when appropriate they can reduce the program length and running time. The conditional CALLs and RETs are executed only if the condition is met. For instance, the conditional return RPE (*return on parity even*) could have been used instead of JPE DONE in the subroutine of Example 9–26. Note that then the subroutine would have two return instructions.

The corresponding RET instruction does not have to be the next instruction to affect the stack after the corresponding CALL. For instance, the same number of PUSHes and POPs could be executed in between. During the execution of the subroutine, many other 8085 registers that were used in the main program may be modified by the subroutine. Hence, the programmer must save and restore these registers with PUSH and POP instructions in the subroutine, as is illustrated by the next example.

Example 9–27

The following program is a subroutine that will generate a delay by counting down register pair DE and exiting when it is zero. The PUSH PSW and POP PSW will save and restore the flags and register A because they are modified by the subroutine.

Label	OPCODE	Operand	Comments
	ORG	100H	
DELAY:	PUSH	PSW	;Save A and flags
LOOP:	DCX	D	;Count down
	MOV	A,E	;Test when
	ORA	D	;Zero
	JNZ	LOOP	;Jump if not
	POP	PSW	;Restore A and flags
	RET		

The register pair DE could also be saved and restored if the contents should remain unchanged.

Software Interrupts

Software interrupts are used to abruptly change the normal flow of the program, such as for coroutines or error processing.

Instruction	OPCODE	No. of Bytes	Operations Performed	Flags S Z AC P CY
RST 0	C7	1	$((SP) - 1) \leftarrow (PCH)$ $((SP) - 2) \leftarrow (PCL)$ $(SP) \quad \leftarrow (SP) - 2$ $(PC) = 0H$	(none affected)
RST 1	CF	1	$((SP) - 1) \leftarrow (PCH)$ $((SP) - 2) \leftarrow (PCL)$ $(SP) \quad \leftarrow (SP) - 2$ $(PC) = 8H$	
RST 2	D7	1	$((SP) - 1) \leftarrow (PCH)$ $((SP) - 2) \leftarrow (PCL)$ $(SP) \quad \leftarrow (SP) - 2$ $(PC) = 10H$	
RST 3	DF	1	$((SP) - 1) \leftarrow (PCH)$ $((SP) - 2) \leftarrow (PCL)$ $(SP) \quad \leftarrow (SP) - 2$ $(PC) = 18H$	
RST 4	E7	1	$((SP) - 1) \leftarrow (PCH)$ $((SP) - 2) \leftarrow (PCL)$ $(SP) \quad \leftarrow (SP) - 2$ $(PC) = 20H$	
RST 5	EF	1	$((SP) - 1) \leftarrow (PCH)$ $((SP) - 2) \leftarrow (PCL)$ $(SP) \quad \leftarrow (SP) - 2$ $(PC) = 28H$	
RST 6	F7	1	$((SP) - 1) \leftarrow (PCH)$ $((SP) - 2) \leftarrow (PCL)$ $(SP) \quad \leftarrow (SP) - 2$ $(PC) = 30H$	
RST 7	FF	1	$((SP) - 1) \leftarrow (PCH)$ $((SP) - 2) \leftarrow (PCL)$ $(SP) \quad \leftarrow (SP) - 2$ $(PC) = 38H$	

More information and detailed discussion of software interrupts and their handling are given in the next chapter along with a discussion of hardware interrupt processing.

9.10

The System Control Group

The *system control group* of instructions performs operations related to input/output (I/O), interrupts, and other system functions. The system group can be further divided into three types of instructions:

I/O instructions
Interrupt-related instructions
System instructions

I/O Instructions

I/O instructions are used to communicate with I/O devices that are distinguished from the memory.

Instruction	OPCODE	No. of Bytes	Operations Performed	Flags S Z AC P CY
IN PORT	DB	1	(A) ← (PORT)	(none affected)
OUT PORT	D3	1	(PORT) ← (A)	
RIM	20	1	READ INTERRUPT MASK	
SIM	30	1	SET INTERRUPT MASK	

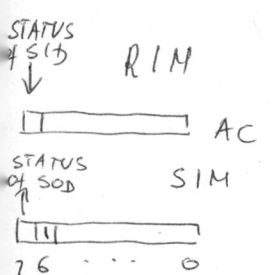

The last two instructions are multipurpose instructions that can be used for both input/output operations and interrupt-related operations.

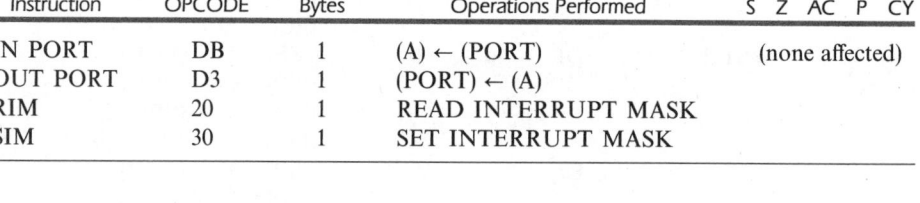

The 8085 has a *serial input data* (SID) line and a *serial output data* (SOD) line. These lines can be used very effectively to communicate with serial devices such as terminals and printers. The functions of these two serial lines are controlled by the RIM and SIM instructions. Whenever the RIM instruction is executed, the status of the SID line is reported in the most significant bit of the accumulator. Likewise, the status of the most significant bit of the accumulator, if the corresponding bit 6 is a 1, is used to modify the SOD line when the SIM instruction is being executed.

Interrupt-Related Instructions

The 8085 has a rich set of instructions dealing with interrupts and interrupt controls.

Instruction	OPCODE	No. of Bytes	Operations Performed	Flags S Z AC P CY
EI	FB	1	ENABLE INTERRUPTS	(none affected)
DI	F3	1	DISABLE INTERRUPTS	
RIM	20	1	READ INTERRUPT MASK	
SIM	30	1	SET INTERRUPT MASK	

The EI instruction enables external events to interrupt the processor through the vectored interrupt lines. The DI instruction disables such privilege so that only the nonmaskable interrupt line can interrupt the processor.

The RIM instruction loads into register A information related to the interrupts and the serial input data (SID) line. After execution, the register contains the following information:

1. Current interrupt mask status for RST 5.5, RST 6.5, and RST 7.5 hardware interrupts in A_0, A_1, and A_2, respectively (1 = mask disabled).
2. Current interrupt enable flag status (1 = interrupts enabled) in A_3 except immediately following a TRAP interrupt.
3. Hardware interrupts pending in A_4, A_5, and A_6 (i.e., signals received but not yet serviced) on the RST 5.5, RST 6.5, and RST 7.5 lines, respectively.
4. Serial input data (SID) line into A_7.

The A_6 bit represents the state of the RST 7.5 flip-flop, not the RST 7.5 input. The contents of register A after the execution of the RIM instruction is

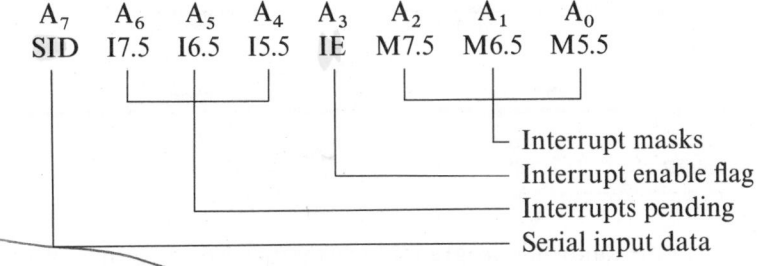

The execution of the SIM instruction uses the contents of register A (which must be previously loaded) to perform the following functions:

1. Programs the RST 5.5, RST 6.5, and RST 7.5 interrupt masks according to A_0, A_1, and A_2, respectively. To program the masks, first make $A_3 = 1$. Then set A_0, A_1, and A_2 to 0 to enable interrupts and 1 to disable interrupts. Next execute the SIM instruction. If $A_3 = 0$ and SIM is executed, the status of the masks will not change.

2. Resets the edge-triggered RST 7.5 input flip-flop. If $A_4 = 1$ and the SIM instruction is executed, then the RST 7.5 input flip-flop is reset. The RST 7.5 input flip-flop is also reset in response to a RESET IN or an internal processor acknowledgment of an RST 7.5 interrupt.

3. Loads the SOD output flip-flop. If $A_6 = 1$ and the SIM instruction is executed, then A_7 is loaded into the SOD output flip-flop and thus appears on the SOD output line. If $A_6 = 0$ and SIM is executed, then the SOD output flip-flop is not changed. SOD is always reset by the RESET IN signal.

The contents of register A before the execution of SIM to control the 8085 is specified as follows:

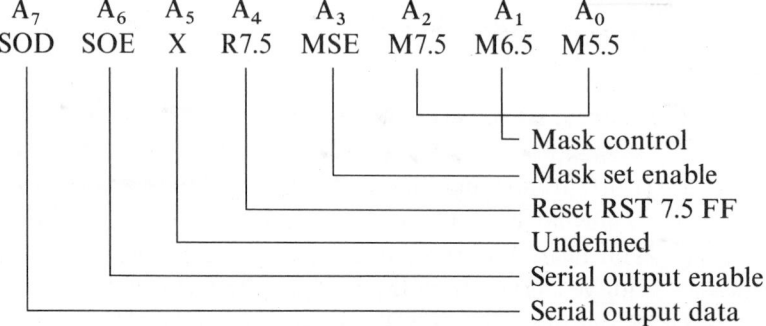

$$\begin{array}{cccccccc} A_7 & A_6 & A_5 & A_4 & A_3 & A_2 & A_1 & A_0 \\ \text{SOD} & \text{SOE} & \text{X} & \text{R7.5} & \text{MSE} & \text{M7.5} & \text{M6.5} & \text{M5.5} \end{array}$$

Mask control
Mask set enable
Reset RST 7.5 FF
Undefined
Serial output enable
Serial output data

System Instructions

There are two system instructions, the HLT and the NOP instructions.

Instruction	OPCODE	No. of Bytes	Operations Performed	Flags S Z AC P CY
HLT	76	1	HALT	(not affected)
NOP	00	1	NO OPERATION	

In the following programming example, we will write a program that will put out a square wave on the SOD.

Example 9–28

Write a program that will generate the highest-frequency square wave on the SOD line on the 8085. Also determine its frequency.

The SOD line is controlled by making bit $A_6 = 1$, $A_7 =$ data in register A, and then executing a SIM instruction. To put out a square wave of maximum frequency requires complementing A_7 and executing SIMs repeatedly in an endless loop. The program is as follows:

Label	OPCODE	Operand	Comments
	ORG	0H	;ROM memory at location 0000H
	MVI	B,10000000B	;Set up comp mask $B_7 = 1$
	MVI	A,01000000B	;$A_6 = 1$
	LXI	H,LOOP	;Initialize HL
LOOP:	SIM		;Output data on SOD
	XRA	B	;Complement A_7 only
	PCHL		;Jump to Loop
	HLT		;Halt program
	END		

There are two unconditional jump instructions, JMP and PCHL. The first takes 10 T states and the second 6 T states to execute. When speed is important, the second jump is better. The LXI H,LOOP instruction loads the appropriate address to jump to in register pair HL. Many times the PCHL jump cannot be used if register pair HL is being used for something else. The frequency can be calculated by counting the number of T states in the loop only:

Instruction	No. of T States
SIM	4
XRA	4
PCHL	6

The loop must be executed twice for each complete cycle of the square wave, and also the frequency of the internal clock is one-half the crystal frequency. Then the period for the square wave is

$$t_{square} = 2 * (4 + 4 + 6) * 2/5.0 \text{ MHz} = 11.2 \ \mu s$$

Assuming that the crystal frequency is 5 MHz, the output frequency is

$$F_{square} = 89.29 \text{ kHz}$$

This frequency could be increased by using a higher-frequency crystal, but this example should illustrate that very-high-frequency signals cannot be generated under program control with a microprocessor.

The next example makes use of the SID in a simple case. We will look at more involved problems in the next chapter. The execution of the RIM instruction will read the contents of the SID into A_7 of register A.

Instruction		Code	Bytes	T States 8085A	T States 8080A	Machine Cycles
ACI	DATA	CE data	2	7	7	F R
ADC	REG	1000 1SSS	1	4	4	F
ADC	M	8E	1	7	7	F R
ADD	REG	1000 0SSS	1	4	4	F
ADD	M	86	1	7	7	F R
ADI	DATA	C6 data	2	7	7	F R
ANA	REG	1010 0SSS	1	4	4	F
ANA	M	A6	1	7	7	F R
ANI	DATA	E6 data	2	7	7	F R
CALL	LABEL	CD addr	3	18	17	S R R W W*
CC	LABEL	DC addr	3	9/18	11/17	S R•/S R R W W*
CM	LABEL	FC addr	3	9/18	11/17	S R•/S R R W W*
CMA		2F	1	4	4	F
CMC		3F	1	4	4	F
CMP	REG	1011 1SSS	1	4	4	F
CMP	M	BE	1	7	7	F R
CNC	LABEL	D4 addr	3	9/18	11/17	S R•/S R R W W*
CNZ	LABEL	C4 addr	3	9/18	11/17	S R•/S R R W W*
CP	LABEL	F4 addr	3	9/18	11/17	S R•/S R R W W*
CPE	LABEL	EC addr	3	9/18	11/17	S R•/S R R W W*
CPI	DATA	FE data	2	7	7	F R
CPO	LABEL	E4 addr	3	9/18	11/17	S R•/S R R W W*
CZ	LABEL	CC addr	3	9/18	11/17	S R•/S R R W W*
DAA		27	1	4	4	F
DAD	RP	00RP 1001	1	10	10	F B B
DCR	REG	00SS S101	1	4	5	F*
DCR	M	35	1	10	10	F R W
DCX	RP	00RP 1011	1	6	5	S*
DI		F3	1	4	4	F
EI		FB	1	4	4	F
HLT		76	1	5	7	F B
IN	PORT	DB data	2	10	10	F R I
INR	REG	00SS S100	1	4	5	F*
INR	M	34	1	10	10	F R W
INX	RP	00RP 0011	1	6	5	S*
JC	LABEL	DA addr	3	7/10	10	F R/F R R†
JM	LABEL	FA addr	3	7/10	10	F R/F R R†
JMP	LABEL	C3 addr	3	10	10	F R R
JNC	LABEL	D2 addr	3	7/10	10	F R/F R R†
JNZ	LABEL	C2 addr	3	7/10	10	F R/F R R†
JP	LABEL	F2 addr	3	7/10	10	F R/F R R†
JPE	LABEL	EA addr	3	7/10	10	F R/F R R†
JPO	LABEL	E2 addr	3	7/10	10	F R/F R R†
JZ	LABEL	CA addr	3	7/10	10	F R/F R R†
LDA	ADDR	3A addr	3	13	13	F R R R
LDAX	RP	000X 1010	1	7	7	F R
LHLD	ADDR	2A addr	3	16	16	F R R R R

Instruction		Code	Bytes	T States 8085A	T States 8080A	Machine Cycles
LXI	RP,DATA16	00RP 0001 data16	3	10	10	F R R
MOV	REG,REG	01DD DSSS	1	4	5	F*
MOV	M,REG	0111 0SSS	1	7	7	F W
MOV	REG,M	01DD D110	1	7	7	F R
MVI	REG,DATA	00DD D110 data	2	7	7	F R
MVI	M,DATA	36 data	2	10	10	F R W
NOP		00	1	4	4	F
ORA	REG	1011 0SSS	1	4	4	F
ORA	M	B6	1	7	7	F R
ORI	DATA	F6 data	2	7	7	F R
OUT	PORT	D3 data	2	10	10	F R O
PCHL		E9	1	6	5	S*
POP	RP	11RP 0001	1	10	10	F R R
PUSH	RP	11RP 0101	1	12	11	S W W*
RAL		17	1	4	4	F
RAR		1F	1	4	4	F
RC		D8	1	6/12	5/11	S/S R R*
RET		C9	1	10	10	F R R
RIM (8085 only)		20	1	4	–	F
RLC		07	1	4	4	F
RM		F8	1	6/12	5/11	S/S R R*
RNC		D0	1	6/12	5/11	S/S R R*
RNZ		C0	1	6/12	5/11	S/S R R*
RP		F0	1	6/12	5/11	S/S R R*
RPE		E8	1	6/12	5/11	S/S R R*
RPO		E0	1	6/12	5/11	S/S R R*
RRC		0F	1	4	4	F
RST	N	11XX X111	1	12	11	S W W*
RZ		C8	1	6/12	5/11	S/S R R*
SBB	REG	1001 1SSS	1	4	4	F
SBB	M	9E	1	7	7	F R
SBI	DATA	DE data	2	7	7	F R
SHLD	ADDR	22 addr	3	16	16	F R R W W
SIM (8085 only)		30	1	4	–	F
SPHL		F9	1	6	5	S*
STA	ADDR	32 addr	3	13	13	F R R W
STAX	RP	000X 0010	1	7	7	F W
STC		37	1	4	4	F
SUB	REG	1001 0SSS	1	4	4	F
SUB	M	96	1	7	7	F R
SUI	DATA	D6 data	2	7	7	F R
XCHG		EB	1	4	4	F
XRA	REG	1010 1SSS	1	4	4	F
XRA	M	AE	1	7	7	F R
XRI	DATA	EE data	2	7	7	F R
XTHL		E3	1	16	18	F R R W W

Machine cycle types:

F	Four clock period instr fetch	DDD	Binary digits identifying a destination register	B = 000, C = 001, D = 010 Memory = 110
S	Six clock period instr fetch	SSS	Binary digits identifying a source register	E = 011, H = 100, L = 101 A = 111
R	Memory read	RP	Register Pair	BC = 00, HL = 10
I	I/O read			DE = 01, SP = 11
W	Memory write			
O	I/O write			
B	Bus idle			
X	Variable or optional binary digit			

*Five clock period instruction fetch with 8080A.

†The longer machine cycle sequence applies regardless of condition evaluation with 8080A.

•An extra READ cycle (R) will occur for this condition with 8080A.

Figure 9–7

Index of the 8085 instruction set (Reprinted by permission of Intel Corporation, Copyright/Intel Corporation 1979)

Example 9–29 ▰▰▰▰▰▰▰▰▰▰▰▰▰▰▰▰

Write a program that will read the SID input and output the complement on the SOD line.

The program is as follows:

Label	OPCODE	Operand	Comments
	ORG	0H	
	MVI	B,10000000B	:Mask to control SOD
HERE:	RIM		;Get data
	RAL		;Data into CY
	CMC		;Complements data
	MOV	A,B	;Initialize reg. A
	RAR		;A_7 = data, A_6 = 1, A_3 = 0
	SIM		;Output comp. data
	JMP	HERE	;Do again
	END		

Register A must have $A_6 = 1$ and $A_3 = 0$ if the SOD signal is to be affected and no change is to occur in the interrupt structure. This will be a very slow inverter, but this example shows how the rotates and carry can be used effectively to manipulate data.

To aid in the process of assembling programs, a short summary of the OPCODEs in alphabetical order and their corresponding binary codes are given in Figure 9–7. As an example of assembling a program, one of the examples is assembled and shown in Example 9–30.

Example 9–30 ▰▰▰▰▰▰▰▰▰▰▰▰▰▰▰▰

Assemble the program in Example 9–29. *22*
The solution is as follows:

Address	Assembled Code	Label	OPCODE	Operand
			ORG	100H
0100	21 00 02		LXI	H,200H
0103	7E		MOV	A,M
0104	23		INX	H
0105	BE		CMP	M
0106	DA 0A 01		JC	STOR
0109	7E		MOV	A,M
010A	23	STOR:	INX	H
010B	77		MOV	M,A
010C	76		HLT	
			END	

This assembly requires a forward reference on the JC STOR instruction. Two bytes are left blank until the address of STOR is found.

9.11

Implementation of Higher-Level Language Constructs

In the previous sections we examined the complete instruction set of the 8085 and illustrated how the assembly language can be used in numerous short examples. In this section we concentrate on using the 8085 instruction set to demonstrate the implementation of higher-level language constructs. These language constructs are fundamental and can be found in most higher-level languages. For the sake of clarity, we will use mostly the syntax of FORTRAN and PASCAL in the following sections. Specific language constructs that we will examine here include:

> The logical IF statement
> The arithmetic IF statement
> The WHILE statement
> The DO loop statement
> The computed GOTO statement
> The CASE statement

The Logical IF Statement

The *logical IF statement* requires a simple implementation when using the 8085 instructions. A typical logical IF statement takes the following form:

```
IF condition true
   THEN first group of statements
   ELSE second group of statements
```

In the operation of the logical IF statement, the specified condition is first tested. If the condition is true, then the first group of statements is performed. If the condition is false, then the second group of statements is performed. At the end of either group of statements, the next statement is performed.

The condition to be tested is usually a comparison of arithmetic variables, such as LT, LE, GT, GE, EQ, NE, or logical operations of logical variables.

Example 9-31

Implement the following logical IF statement using the 8085 instruction set:

```
IF (X.LT.Y) THEN GOTO 100 ELSE GOTO 200
```

Assume that X and Y are the actual labels or addresses of the variables X and Y. The above construct can be implemented by first comparing X with Y and branching to the address of statement 100 (S100) or the address of statement 200 (S200) based on the result of the comparison. This is shown below:

```
            LXI   H,Y
            LDA   X
            CMP   M
            JM    S200
    ;
    ;First group of statements starts here.
    S100:    .

             .

             .
            JMP   CONT
    ;
    ;Second group of statements starts here.
    S200:    .

             .

             .
    ;
    ;Common continuation point for both groups of statements.

    CONT:    .           ;Program continues here
             .           ; with the next statement
             .
```

Note that the GOTO statements can represent blocks of statements denoted by BEGIN and END in some structured languages such as PASCAL. The implementation, however, is the same as shown above.

Other types of arithmetic comparisons can be treated in the same way. This has been left for the programmer as an exercise. More complicated expressions for the condition to be tested can be programmed in a like manner.

The Arithmetic IF Statement

The *arithmetic IF statement* is used for many FORTRAN programs and takes the following form:

```
IF (IX-IY) 100,200,300
```

The arithmetic IF statement executes in the following manner. First, the integer arithmetic operation within the pair of parentheses is carried out. The result is compared to zero. If the result is less than zero, statement 100 is executed next. If the result is equal to zero, statement 200 is executed next. If the result is greater than zero, statement 300 is executed next.

Example 9-32

Implement the arithmetic IF statement "IF (IX − IY) 100,200,300."

Assume that IX and IY are the actual labels or addresses for the FORTRAN variables IX and IY. The above arithmetic IF statement can be implemented very easily in 8085 assembly language:

```
          LXI   H,IY
          LDA   IX
          SUB   M
          JZ    S200
          JP    S300
      ;
      ;If result is less than zero, continue here.
      S100:    .

               .

               .
          JMP   CONT
      ;
      ;If result is zero, continue here.
      S200:    .

               .

               .
          JMP   CONT
      ;
      ;If result is greater than zero, continue here.
      S300:    .

               .

               .
      ;
      ;Common return
      CONT:    .

               .

               .
```

The variables IX and IY are assumed to be integers. Slightly more involved computational procedures result if these variables are in floating-point representation.

The WHILE Statement

The *WHILE statement* is a very powerful feature and has the following form:

```
WHILE condition DO ...  ENDWHILE
```

The execution of the WHILE statement is as follows. The condition is tested at the beginning of the block. If the condition is met—i.e., true—the block of statements is executed. If the condition is not met—i.e., false—then the next statement is executed.

The WHILE statement can be implemented in a loop fashion with the test at the beginning to provide an exit out of the loop when the condition is not met.

Example 9-33 ■■■■■■■■■■■■■■■■■■■■■■■■■■■■■■■

Implement the following WHILE loop:

WHILE J<10 DO ... ENDWHILE

Let J be the address of the variable *J*. The above WHILE loop requires the condition that the variable *J* must be less than 10 in order for the loop to be executed. This is implemented as shown below:

```
LOOP:       LDA   J
            SUI   10D
            CMA
            JP    CONT
;
;WHILE loop starts here.
            .

            .

            .
            JMP   LOOP
;
;Next statement starts here.
CONT:       .

            .

            .
            END
```

Note that a reverse logic is used at the beginning of the loop after 10 is subtracted from the present value of *J*. A JM instruction cannot be used alone to terminate the loop, because the JM instruction does not respond to the situation when the value of *J* is exactly 10. In other words, zero is always counted as a positive number.

The DO Loop Statement

Looping is fundamental in all computer programs. In addition to the WHILE loop discussed above, a completely specified DO loop is often used in higher-level language programs. The *DO loop* takes the following form:

```
        DO 100 L = I,J,K
        .

        .

        .
    100 CONTINUE
```

The DO loop statement shows that the enclosed statements are executed first, with the index L having a value of I. At the end of the loop, the index L is incremented by the value of K and compared to the terminating value of J. If the value of the index L is less than that of J, the loop is repeated; otherwise, the following statement is executed next.

There are many possible versions of DO loop implementations. Variations include pre-loop or post-loop testing. Some compilers allow I, J, and K to be signed floating-point numbers. The structures of these various versions are very similar.

Example 9–34 ▰▰▰▰▰▰▰▰▰▰▰▰▰▰▰▰▰▰▰▰▰▰▰▰▰▰▰▰

Implement the above DO loop according to standard FORTRAN convention with post-loop testing and integer representations on the index and loop limits.

Let I, J, K, and L be the labels or addresses for the variables I, J, K, and L, respectively. The above DO loop construct can be implemented as follows:

```
;
;Pre-loop set-up
;
            LDA   I
            STA   L
;
;This is the top of the loop.
LOOP:       .
            .
            .
;
;This is the post-loop test.
            LXI   H,L
            LDA   K
            ADD   M
            MOV   M,A
            LDA   J
            CMP   M
            JP    LOOP
;
;End of DO loop, program continues here.
            .
            .
            .
```

If pre-loop testing is desired, the loop test section can be moved to the top of the loop with a reverse test. The exact implementation is left to the programmer as an exercise.

The Computed GOTO Statement

The *computed GOTO statement* is often used in menus where the user is given a list of options and the answer is polled to see which option is selected. The form of the computer GOTO is

```
GOTO (100,200,300,400,500),J
```

If the value of *J* is 1, then the next statement to be executed is 100. If the value is 2, or 3, or 4, or 5, then the next statement to be executed is 200, or 300, or 400, or 500, respectively. If the value of *J* is less than 1 or greater than 5, in this case, then the next statement following the computed GOTO is executed.

Example 9–35 ▮▬▬▬▬▬▬▬▬▬▬▬▬▬▬▬▬▬▬▬▬▬▬▬▬▬▬▬

Implement the computed GOTO statement "GOTO (100,200,300,400,500), J)."

Assume that J is the label for the variable *J*, and assume that S100, S200, S300, S400, and S500 are the labels for the corresponding statements. Then the computed GOTO can be implemented as follows:

```
            LDA    J
            JM     S000
            JZ     S000
            DCR    A
            JZ     S100
            DCR    A
            JZ     S200
            DCR    A
            JZ     S300
            DCR    A
            JZ     S400
            DCR    A
            JZ     S500
    ;
    ;Invalid index value comes here.
    S000:      .
               .
               .
    ;If J = 1, program continues here.
    S100:      .
               .
               .
    ;If J = 2, program continues here.
    S200:      .
               .
               .
```

```
;If J = 3, program continues here.
S300:     .
          .
          .
;If J = 4, program continues here.
S400:     .
          .
          .
;If J = 5, program continues here.
S500:     .
          .
          .
```

The CASE Statement

The *CASE statement* is a more refined construct but similar in principle to the computed GOTO statement. The form of the CASE statement is

```
CASE      J
1         BEGIN ... END
5         BEGIN ... END
.         .
.         .
.         .
END CASE
```

In the CASE statement, the index, J in this example, can take on any finite value as listed on the left-hand side. If the value of J matches that on the left, the associated block of statements is executed. If no value matches that of J, the statement after the END CASE statement is executed.

In many installations, a range of values can also be specified on the left-hand side. Furthermore, the same value can be specified in different rows.

Example 9-36

Assume that the value of J is in integer format only. Implement the above CASE statement.

Let J be the label or address that represents the variable J, the CASE statement can be realized as follows:

```
LDA   J
CPI   1D
JZ    S100
CPI   5D
JZ    S200
```

```
              .
              .
              .
        JMP   ENDCASE
;
;Individual blocks of statements go here.
S100:   .
              .
              .
        JMP   ENDCASE
;
S200:   .
              .
              .
        JMP   ENDCASE
;
        .
        .
        .
;
;End of the CASE statement
ENDCASE:  .
              .
              .
```

The above implementation is very similar to the computed GOTO statement. The CPI instruction is used instead of the DCR instruction to make the comparison easier, since any value can be used on the left. Note that the above implementation does not allow the same value of J to appear in more than one row. To relax this requirement, a slight rearrangement of the tests is required. This is shown below:

```
        LDA   J
;
;First test
        CPI   1D
        JNZ   NOT1
;
;Statements for the first row go here.
              .
              .
              .
;
;Second test
NOT1:   CPI   5D
        JNZ   NOT5
;
;Statements for the second row go here.
              .
              .
              .
```

```
;
;Next tests and statements
NOT5:    .

         .

         .

;
;After the last test and the last set of statements,
; come here.
ENDCASE:  .

         .

         .
```

The above implementation allows the value of *J* to be tested against all possible values for every row, thus allowing the same value of *J* to cause execution in several rows.

References

1. D'Angelo, Henry. *Microcomputer Structures*. Byte Publications, Peterborough, N.H., 1981.
2. Dasgupta, Subrata. *The Design and Description of Computer Architectures*. John Wiley & Sons, New York, 1984.
3. Doty, Keith L. *Fundamentals of Microcomputer Architecture*. Matrix Publishers, Portland, Ore., 1979.
4. Friedman, Arthur D. *Fundamentals of Logic Design and Switching Theory*. Computer Science Press, Rockville, Md., 1986.
5. Furht, Borivoje, and Himanshu Parikh. *Microprocessor Interfacing and Communication Using the Intel SDK-85*. Prentice-Hall, Englewood Cliffs, N.J., 1986.
6. Garland, Harry. *Introduction to Microprocessor System Design*. McGraw-Hill, New York, 1979.
7. Gault, James W., and Russell L. Pimmel. *Introduction to Microcomputer-Based Digital Systems*. McGraw-Hill, New York, 1982.
8. Getgen, Lawrence E. *Designing with Microprocessors*. Science Research Associates, Chicago, 1985.
9. Givone, Donald D., and Robert P. Roesser. *Microprocessors/Microcomputers: An Introduction*. McGraw-Hill, New York, 1980.
10. Gorsline, G. W. *Computer Organization: Hardware/Software*. Prentice-Hall, Englewood Cliffs, N.J., 1980.
11. Hamacher, V. Carl, Z. G. Vranesic, and S. G. Zaky. *Computer Organization*, 2d. ed. McGraw-Hill, New York, 1984.
12. Hayes, John P. *Computer Architecture and Organization*. McGraw-Hill, New York, 1978.
13. Hayes, John P. *Digital System Design and Microprocessors*. McGraw-Hill, New York, 1984.
14. Iliffe, J. K. *Advanced Computer Design*. Prentice-Hall International, London, 1982.
15. Intel Corp. *Intel MCS-85 User's Manual*. Intel Corporation, Santa Clara, Calif., 1978.
16. Intel Corp. *Intel 8080/8085 Assembly Language Programming*. Intel Corporation, Santa Clara, Calif., 1979.

17. Jermann, William H. *The Structure and Programming of Microcomputers.* Alfred Publishing, Sherman Oaks, Calif., 1982.
18. Johnson, E. L., and M. A. Karim. *Digital Design, A Pragmatic Approach.* PWS Publishers, Boston, 1987.
19. Kline, R. M. *Digital Computer Design.* Prentice-Hall, Englewood Cliffs, N.J., 1977.
20. Kostopoulos, G. K. *Digital Engineering.* John Wiley & Sons, New York, 1975.
21. Leventhal, L. *8080/8085 Assembly Language Programming.* Osborne/McGraw-Hill, Berkeley, Calif., 1978.
22. Lewin, Morton H. *Logic Design and Computer Organization.* Addison-Wesley, Reading, Mass., 1983.
23. Liu, Yu-Cheng, and Glenn A. Gibson. *Microcomputer Systems: The 8086/8088 Family Architecture, Programming, and Design,* 2d ed. Prentice-Hall, Englewood Cliffs, N.J., 1986.
24. Mano, M. Morris. *Digital Logic and Computer Design.* Prentice-Hall, Englewood Cliffs, N.J., 1979.
25. Mano, M. Morris. *Computer System Architecture,* 2d ed. Prentice-Hall, Englewood Cliffs, N.J., 1982.
26. Mano, M. Morris. *Digital Design.* Prentice-Hall, Englewood Cliffs, N.J., 1984.
27. O'Connor, Patrick. *Digital and Microprocessor Technology.* Prentice-Hall, Englewood Cliffs, N.J., 1979.
28. Peatman, J. B. *Microcomputer Based Design.* McGraw-Hill, New York, 1977.
29. Peatman, John B. *Digital Hardware Design.* McGraw-Hill, New York, 1980.
30. Rafiquzzaman, M. *Microcomputer Theory and Applications with the Intel SDK-85.* John Wiley & Sons, New York, 1982.
31. Rafiquzzaman, Mohamed. *Microprocessors and Microcomputer Development Systems: Designing Microprocessor-Based Systems.* Harper & Row, New York, 1984.
32. Ramirez, Edward V. *Microprocessing Fundamentals: Hardware and Software.* McGraw-Hill, New York, 1980.
33. Rooney, Victor M., and Amin R. Ismail. *Microprocessors and Microcomputers.* Macmillan, New York, 1984.
34. Roth, Charles H., Jr. *Fundamentals of Logic Design,* 3d ed. West Publishing Co., St. Paul, Minn., 1985.
35. Short, Kenneth L. *Microprocessors and Programmed Logic,* 2d ed. Prentice-Hall, Englewood Cliffs, N.J., 1987.
36. Taub, Herbert. *Digital Circuits and Microprocessors.* McGraw-Hill, New York, 1982.
37. Ware, W. H. *Digital Computer Technology and Design,* Vol. 1. John Wiley & Sons, New York, 1963.
38. Wilkinson, Barry. *Digital System Design.* Prentice-Hall, Englewood Cliffs, N.J., 1987.
39. Williams, Gerald E. *Digital Technology.* Science Research Associates, Chicago, 1977.
40. Winkel, David, and Franklin Prosser. *The ART of Digital Design.* Prentice-Hall, Englewood Cliffs, N.J., 1980.

Problems

1. Write a program that will find the larger of two numbers in locations 1000 and 1001 and put the result in location 1001.

2. Write a program that finds the 2's complement of a 16-bit number stored in the register pair BC and leaves the result in the register pair DE.

3. Write a program that will interchange the following bits in register A:

$$A_7 \leftrightarrow A_0$$
$$A_6 \leftrightarrow A_1$$
$$A_5 \leftrightarrow A_2$$
$$A_4 \leftrightarrow A_3$$

4. List all the addressable 8- and 16-bit registers in the 8085.

5. Write a program that will add two 24-bit numbers (3 bytes). Assume that locations 100, 101, and 102 contain the three bytes making up one number and locations 200, 201, and 202 contain the other. Put the result in locations 300, 301, and 302. Assume that the least significant byte is in location 100, 200, and 300 and the most significant byte is in location 102, 202, and 302.

6. Assemble the program in Problem 5.

7. Assemble the program in Problem 2.

8. Write a program that finds the largest number in a block starting at location 100 with the length of the block in location FF, stores the maximum number in location 200, and assumes that the numbers are all unsigned binary numbers.

9. Assume that you have an ASCII character in register A in bits A_6 to A_0. Write a program that will modify bit A_7 as needed to ensure that the total character has even parity. Leave the result in register A.

10. The hexadecimal numbers 0 through 9 and A through F many times need to be converted to ASCII code to print on a terminal. Assume that you have a hexadecimal number 0 through F in the four least significant bits of register B and that the four most significant bits are not equal to zero. Convert this hexadecimal number into the equivalent ASCII code in register C. For example:

$$(B) = 09$$
$$(C) = 39$$

11. Write a program similar to the program in Problem 10 that will convert an ASCII character in register C into a hexadecimal character in register B. You may assume that the ASCII character is from 30 to 39 and 41 to 46 only.

12. Write a program that will convert a BCD in register A into the equivalent binary number with the result in register B. (*Hint*: One technique is to count down in register A in BCD while the other register counts up in binary.)

13. Write a program that puts out a square wave of 50 percent duty cycle on SOD on the 8085 at a frequency of 1 kHz.

14. Write a program that tests if a 2's complement number in register A is positive but not zero. Jump to location 1800H if $(A) > 0$, and to location 1900H if $(A) \le 0$.

15. Write a program to multiply two 8-bit numbers by the shift-and-add method.

16. Write a program to convert a BCD number to binary by the shifting method. The BCD number is shifting from the least significant bit to the most significant

bit of the binary number. After every shifting operation, the BCD digit is checked. If the BCD digit is larger than or equal to 8, then 3 must be subtracted from that BCD digit.

17. Write a program to convert a binary number to BCD representation using the shifting method. The most significant bit of the binary number is shifted into the least significant bit of the BCD number. However, before each BCD digit is shifted, 3 must be added to any BCD digit that is larger than or equal to 5.

18. Write a program to convert a two-digit BCD number into binary by interrogating a conversion table.

19. Write a program to compare two blocks of memory when only the size of the blocks and the starting addresses are given.

20. Write a program to check a string to see if it is a palindrome. A palindrome is one that produces the same string when read backwards.

The 8085 Microprocessor-Based System

10

In this chapter we present the details concerning the hardware aspects of the Intel 8085. We examine the pin configuration and the signals required to interface the 8085 to its surrounding environment. In the second half of the chapter we examine a minimum microprocessor configuration that is typically used in many microprocessor-controlled devices. The different types of supporting hardware that are attached to the ports to control inputs and outputs in simple input/output, handshaking, and interrupt-driven environments are also discussed. The hardware and software needed to implement a digital clock are included to illustrate the design concepts.

In the previous chapter we presented in detail a programming model of the Intel 8085 microprocessor. In this chapter we discuss the hardware characteristics of the 8085 and the techniques used in interfacing the processor with the outside world. We also examine the support chips that have been designed specifically for the 8085. These are the 8355/8755 and the 8156/8155 chips designed to connect external hardware to the 8085. Numerous examples illustrate the use of the support hardware and the methodology of interfacing to a microprocessor.

10.1

Hardware Characteristics of the 8085

The functional block diagram given in the previous chapter shows all information necessary for the user to program the 8085; however, it is insufficient to specify the characteristics of the 8085 for the purpose of interfacing. The hardware characteristics refer to the signal definitions, the polarity, the timing, and the voltage levels of the signals. The pin-outs of the 8085 microprocessor are shown in Figure 10–1.

Signal Definitions of the 8085

The signals on the pins of the 8085 have a number of different functions. Many of them are related to one another as a group, as shown in Figure 10–1. The fol-

Figure 10–1

Pin-out diagram of the 8085 (Reprinted by permission of Intel Corporation, Copyright/Intel Corporation 1979)

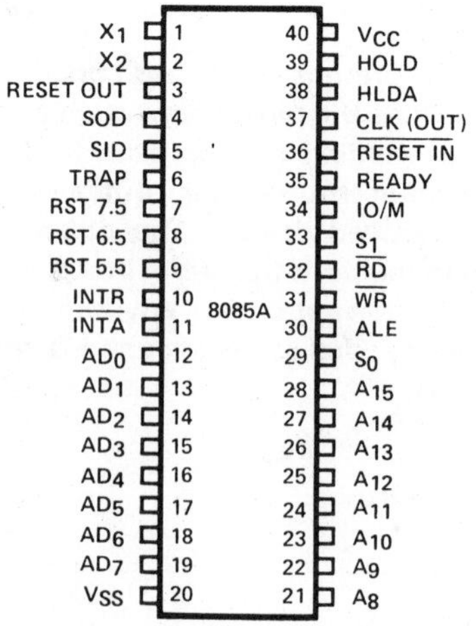

X_1	1	40	V_{CC}
X_2	2	39	HOLD
RESET OUT	3	38	HLDA
SOD	4	37	CLK (OUT)
SID	5	36	$\overline{\text{RESET IN}}$
TRAP	6	35	READY
RST 7.5	7	34	$\text{IO}/\overline{\text{M}}$
RST 6.5	8	33	S_1
RST 5.5	9	32	$\overline{\text{RD}}$
INTR	10	31	$\overline{\text{WR}}$
$\overline{\text{INTA}}$	11	30	ALE
AD_0	12	29	S_0
AD_1	13	28	A_{15}
AD_2	14	27	A_{14}
AD_3	15	26	A_{13}
AD_4	16	25	A_{12}
AD_5	17	24	A_{11}
AD_6	18	23	A_{10}
AD_7	19	22	A_9
V_{SS}	20	21	A_8

(8085A)

lowing list gives a description of the function of each pin, the notation of the signal (given in capital letters), its name, and the input/output (I/O) type or direction. The notations of the directions for the I/O are as follows:

I   Input
O   Output
I/O   Bidirectional
OT   Tri-state output

Signal	Name	Type	Function
1. *Parallel address/data bus*			
AD_0–AD_7	Address/data	I/O	These eight lines serve a dual function. At the beginning of a data transfer cycle, these lines carry the low-order 8 bits of the address bus or the I/O location. At the end of a data transfer cycle, these lines carry the 8 bits of data between the microprocessor and the memory (or I/O).
A_8—A_{15}	Address bus	OT	These are the high-order 8 bits of the address. They are used to specify a memory (or I/O) location.
2. *Serial data lines*			
SID	Serial input	I	This signal is used for reading in a single bit from an external device to the 8085.
SOD	Serial output	O	This signal is used for outputting a single bit from the 8085 to an external device.
3. *Memory and I/O control lines*			
ALE	Address latch	OT	This signal specifies when a memory or I/O port address is valid on the address lines.
$\overline{RD}$	Read control	OT	This signal goes low when information is read from memory or an I/O device into the 8085.
$\overline{WR}$	Write control	OT	This signal goes low when information is written to memory or an I/O device from the 8085.
$IO/\overline{M}$	I/O or memory	OT	This signal is high when I/O devices are addressed and low when memory is addressed.

(continued)

Signal	Name	Type	Function
S_0, S_1	State	O	These two signals indicate the state of the 8085.
READY	Request	I	This signal can be used to slow down the 8085 to obtain extra time when communicating with slower devices.
HOLD	Hold request	I	This signal is used for direct memory access request by putting a high on this line.
HLDA	Hold	O	This signal acknowledges the request from a direct memory access device in response to its activity on the HOLD line.

4. Interrupt control lines

Signal	Name	Type	Function
TRAP	Nonmaskable interrupt	I	This signal is a vectored interrupt input that cannot be disabled internally in the 8085.
RST 5.5 RST 6.5 RST 7.5	Interrupt inputs	I	These three signals are vectored interrupt inputs that can be disabled internally in the 8085.
INTR	Interrupt request	I	This signal is a nonvectored interrupt input that can be disabled internally in the 8085.
$\overline{\text{INTA}}$	Interrupt acknowledge	O	This signal indicates that an interrupt request on the INTR input is granted.

5. System lines

Signal	Name	Type	Function
$\overline{\text{RESET IN}}$	Reset	I	A low on this signal causes the 8085 to be reset. Program execution begins at address 0000H after this input goes high.
RESET OUT	Reset output	O	This signal can be used to reset all peripherals attached to the 8085. The RESET OUT is high in response to a low-level input at the $\overline{\text{RESET IN}}$.
X_1, X_2	Crystal	I	A crystal or an external clock can be attached to these inputs for the 8085 timing reference. A resistor-capacitor timing circuit or other resonant circuits can also be used.
CLK OUT	Clock output	O	This is the clock output, which is one-half the frequency of the X_1 or X_2 input. It is used for external timing on other devices.
V_{cc}, V_{ss}	Power, ground	I	+5 V is applied to V_{cc}, and ground is applied to V_{ss}.

The expanded descriptions of these signals with detailed explanations will follow as they are referenced.

Timing Characteristics of the 8085

In this section we concentrate on the hardware characteristics of the 8085. Timing of the control signals is a major concern here. For the 8085 to operate as specified, all timing constraints must be met. Hence a thorough understanding of the relationship of the 8085 signals themselves and their respective timings is imperative in order to successfully interface other devices to the 8085.

Internal Clock Generator An internal clock generator is provided with the processor, and the addition of an external quartz crystal is required to establish timing reference. A suitable crystal for the standard 8085A must be parallel resonant at a fundamental frequency of 6.25 MHz or less (10 MHz or less for the 8085A-2 version). It is also possible to drive the X_1 input directly with an external clock. A Schmitt trigger is used interchangeably as an oscillator and an input conditioner, depending on whether a crystal or an external source is used. The clock circuitry generates two nonoverlapping internal clock signals, ϕ_1 and ϕ_2, which together control the internal timing of the 8085. It is important to note that the frequency of ϕ_1 and ϕ_2 is one-half the crystal or external frequency. The internal clock signal is also available externally on the CLK OUT pin, which is a buffered version of ϕ_1.

8085 Cycles The execution of an 8085 program consists of a sequence of read and write operations that transfer data between the 8085 and memory or I/O peripheral, based on the address or device selected. These read and write operations are the only communication between the 8085 and the other components in the system and are all that is necessary to execute any instruction or program. There are three levels of cycles in the 8085:

1. Instruction cycle
2. Machine cycle
3. Clock cycle

An *instruction cycle* is composed of one or more machine cycles. A *machine cycle* is in turn composed of four or more clock cycles. Therefore the execution of an instruction cycle is accomplished by executing a specific set of machine cycles, and a machine cycle is accomplished by executing a specific sequence of events corresponding to that machine cycle in synchrony with the clock.

Instruction Cycles To an assembly-language programmer, the 8085 operates on an instruction level, i.e., one instruction at a time. An expanded view of an instruction is shown in Figure 10–2. The total time to completely fetch and execute

STA FFAA$_H$

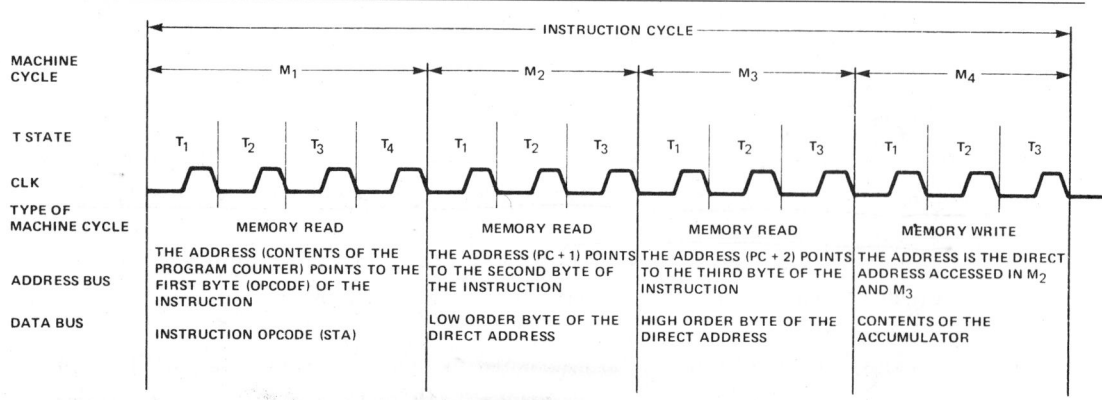

Figure 10–2

Relationship among instruction cycles, machine cycles, and T states for
the STA instruction (Reprinted by permission of Intel Corporation,
Copyright/Intel Corporation 1979)

an instruction is called an *instruction cycle*. Within an instruction cycle, there are
several machine cycles. Each *machine cycle* is associated with a memory (or I/O)
operation. For each instruction, there is at least one memory operation needed
to fetch the OPCODE from memory. There may be more memory (or I/O) cycles
depending on the type of instruction and its addressing mode. Within each memory cycle are states, called T states. Each T state corresponds to a particular phase
of the memory (or I/O) cycle. There is one clock cycle for every T state. The number of T states in a particular machine cycle is determined by the type of machine
cycle, and, in turn, the number of machine cycles is determined by the particular
instruction being executed.

Machine Cycles   The sequence of execution for any instruction consists of a
series of machine cycles whose nature of operation and subsequent timing are
determined by the first byte, normally called OPCODE, of the instruction that
is stored in memory. Each instruction takes up at least one machine cycle. Some
instructions may take as many as five machine cycles. There are altogether seven
different types of machine cycles. These are the only allowable machine cycles
executed by the 8085. In other words, the 8085 at any time is executing one of
these seven machine cycles:

1. OPCODE fetch
2. Memory read
3. Memory write
4. I/O read
5. I/O write

6. Interrupt acknowledge
7. Bus idle

An OPCODE fetch cycle is used to fetch the OPCODE from memory. It is similar to a memory read cycle except for two things. First, the information fetched is always the OPCODE; second, additional T states may be inserted for instruction execution. A memory read cycle is used to retrieve operands for instruction execution, whereas a memory write cycle is used to store the resultant operand back to the memory. The I/O read and I/O write cycles are used for inputting and outputting operands in the I/O address space rather than the memory space. The interrupt acknowledge cycle is used to acquire the interrupt vector when an interrupt occurs. The bus idle cycle indicates that the 8085 has been halted. These seven types of machine cycles can be externally determined and decoded by the state (values) of the three status lines IO/$\overline{M}$, S_0, and S_1, and the three control signals $\overline{RD}$, $\overline{WR}$, and $\overline{INTA}$ as shown in Figure 10–3.

Clock Cycles   A machine cycle is further divided into clock cycles called T states. The duration of one T state is the same as one clock period of ϕ_1 (or ϕ_2). Remember that this is one-half the frequency of the crystal attached to pins X_1 and X_2. There are 10 different types of T states as shown in Figure 10–4. There are six normal states (T_1 to T_6) and four special states (T_{Reset}, T_{Wait}, T_{Hold}, and T_{Halt}).

Most machine cycles consist of three T states, except the OPCODE fetch cycle, which normally has either four or six T states. The actual number of T states required to perform any instruction, however, depends on the instruction

Figure 10–3

Signals for external decoding of various machine cycles (Reprinted by permission of Intel Corporation, Copyright/Intel Corporation 1979)

Machine Cycle		IO/$\overline{M}$	S_1	S_0	$\overline{RD}$	$\overline{WR}$	$\overline{INTA}$
		Status			**Control**		
OPCODE fetch	(OF)	0	1	1	0	1	1
Memory read	(MR)	0	1	0	0	1	1
Memory write	(MW)	0	0	1	1	0	1
I/O read	(IOR)	1	1	0	0	1	1
I/O write	(IOW)	1	0	1	1	0	1
INTR acknowledge	(INA)	1	1	1	1	1	0
Bus idle	(BI) DAD	0	1	0	1	1	1
	INA(RST/TRAP)	1	1	1	1	1	1
	HALT	TS	0	0	TS	TS	1

TS—output tristated

Tri-state

Machine State	Status and Buses				Control		
	S_1, S_0	IO/$\overline{M}$	A_8–A_{15}	AD_0–AD_7	$\overline{RD}, \overline{WR}$	$\overline{INTA}$	ALE
T_1	X	X	X	X	1	1	1
T_2	X	X	X	X	X	X	0
T_{Wait}	X	X	X	X	X	X	0
T_3	X	X	X	X	X	X	0
T_4	1	0	X	TS	1	1	0
T_5	1	0	X	TS	1	1	0
T_6	1	0	X	TS	1	1	0
T_{Reset}	X	TS	TS	TS	TS	1	0
T_{Halt}	0	TS	TS	TS	TS	1	0
T_{Hold}	X	TS	TS	TS	TS	1	0

TS—output tristated
X—unspecified

Figure 10–4

T states of the 8085 (Reprinted by permission of Intel Corporation, Copyright/Intel Corporation 1979)

being executed and on the number of WAIT and HOLD states inserted into each machine cycle.

The six normal T states, T_1–T_6, represent a normal time progression of events, e.g., from T_1 to T_3 for a three T-state machine cycle. The other four T states represent special action that occurs in response to certain inputs or conditions.

T State	Condition to Enter
T_{Reset}	This state is entered in response to $\overline{RESET}$ $\overline{IN}$ being set at 0. It is used primarily to reset the 8085.
T_{Wait}	This state is entered in response to READY = 1. It is used to temporarily suspend further operations on the 8085 to wait for slower devices.
T_{Hold}	This state is entered in response to HOLD = 1. This is used for direct memory access devices to suspend the operations of the 8085.
T_{Halt}	This state is entered when the HLT instruction is executed. It is used to halt the 8085. A reset or an interrupt is needed for the 8085 to exit the halt state.

The state transition diagram in Figure 10–5 illustrates how the 8085 proceeds through each machine cycle. The state transition diagram shows when the READY, HOLD, and interrupt signals (TRAP, RST 7.5, RST 6.5, RST 5.5, and INTR) are sampled and how they modify the basic instruction sequence.

Starting from reset, the machine progresses from T_{Reset} to T_1. If no halt condition exists, T_2 is entered. At this point, the READY line is sampled for possible

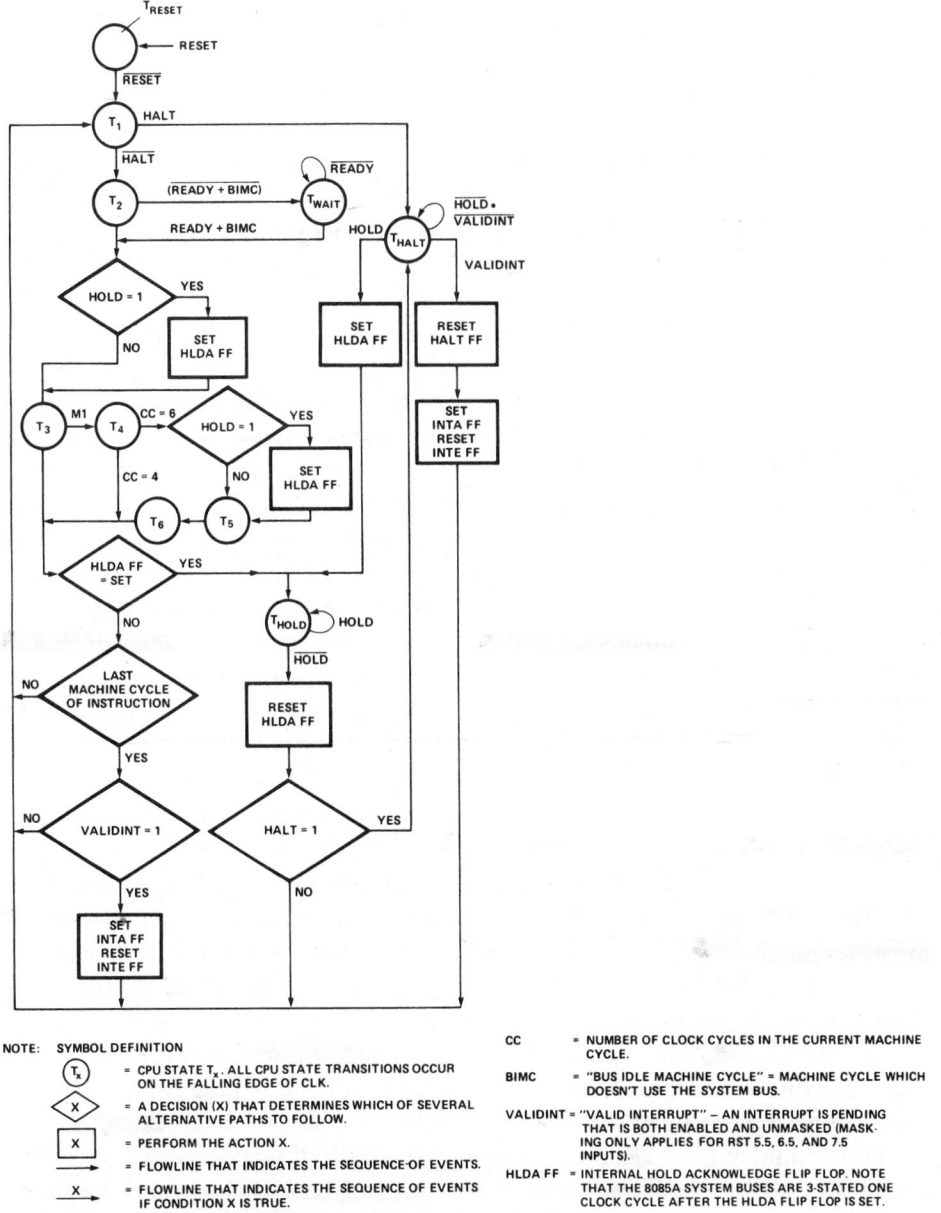

NOTE: SYMBOL DEFINITION

T_x = CPU STATE T_x. ALL CPU STATE TRANSITIONS OCCUR ON THE FALLING EDGE OF CLK.

X = A DECISION (X) THAT DETERMINES WHICH OF SEVERAL ALTERNATIVE PATHS TO FOLLOW.

X = PERFORM THE ACTION X.

→ = FLOWLINE THAT INDICATES THE SEQUENCE OF EVENTS.

$\xrightarrow{X}$ = FLOWLINE THAT INDICATES THE SEQUENCE OF EVENTS IF CONDITION X IS TRUE.

CC = NUMBER OF CLOCK CYCLES IN THE CURRENT MACHINE CYCLE.

BIMC = "BUS IDLE MACHINE CYCLE" = MACHINE CYCLE WHICH DOESN'T USE THE SYSTEM BUS.

VALIDINT = "VALID INTERRUPT" – AN INTERRUPT IS PENDING THAT IS BOTH ENABLED AND UNMASKED (MASKING ONLY APPLIES FOR RST 5.5, 6.5, AND 7.5 INPUTS).

HLDA FF = INTERNAL HOLD ACKNOWLEDGE FLIP FLOP. NOTE THAT THE 8085A SYSTEM BUSES ARE 3-STATED ONE CLOCK CYCLE AFTER THE HLDA FLIP FLOP IS SET.

Figure 10–5

State transition diagram for a machine cycle in the 8085 (Reprinted by permission of Intel Corporation, Copyright/Intel Corporation 1979)

insertion of T_{Wait} states. Coming either from T_2 or out from T_{Wait}, the HOLD condition is sampled before proceeding to T_3 and T_4 if necessary. If T_5 and T_6 are required, the HOLD condition is once again sampled. At this point, the present machine cycle is completed and the whole process is repeated until the last machine cycle of the instruction is completed. At the end of the instruction, the HOLD condition and the interrupt condition are sampled. If neither condition is active, the normal OPCODE fetch cycle is executed again. If either the HOLD or interrupt condition is active, then appropriate actions are taken.

The 8085 Timing Diagrams

The timing diagrams presented in this section are very useful in showing the relative timing between different signals. Refer to the manufacturer's specifications for exact values in timing. A high value on a signal line implies a logic 1 and a low a logic 0. Some of the signal lines, such as the address lines, represent a group of signals. In this case, the timing diagram is trying to represent a general specification as to the interval when the signals must be stable. The signals are stable when the two lines do not intersect during a T state. The signals change, or may be changed, during those T states when the two lines cross. Also, when the two lines form a single dotted line, they are said to be in the *tri-state mode*. When a change on one signal causes a subsequent change in another signal, the change on the first signal is identified with a circle with an arrow pointing to the change in the other signal.

The OPCODE Fetch Cycle   The instruction execution will always start with an OPCODE fetch, which may be 4 to 6 states. Referring to Figure 10–6, an instruction cycle begins with the machine cycle that fetches the OPCODE. This is called the *OPCODE fetch* (OF) *cycle*. The OF cycle is identified by setting $\text{IO}/\overline{\text{M}} = 0$, $S_1 = 1$, and $S_0 = 1$. This implies a read operation from a memory location to obtain the OPCODE. The timing for the OF cycle of the DCX instruction is shown in Figure 10–6.

During the T_1 state, the 8085 sends out the 16-bit address contained in the program counter along with the control signal that identifies the type of machine cycle. The higher-order byte (PCH) is placed on the address lines A_8-A_{15}, where it will stay until at least T_4. The low-order byte (PCL) is placed on the AD_0-AD_7 lines for only the duration of the T_1 state. These lines must be read during state T_1 at the falling edge of ALE into an external register. This register is a part of the normal 8085 device family. Otherwise, an 8-bit register must be used to hold the lower-order byte of the address through the T_3 state. This is necessary because the AD_0-AD_7 lines are used for data transfer during the T_2 and T_3 states. After the status signals and address have been sent out and the AD_0-AD_7 drivers have been disabled, the 8085 drives the $\overline{\text{RD}}$ line low to enable the addressed memory

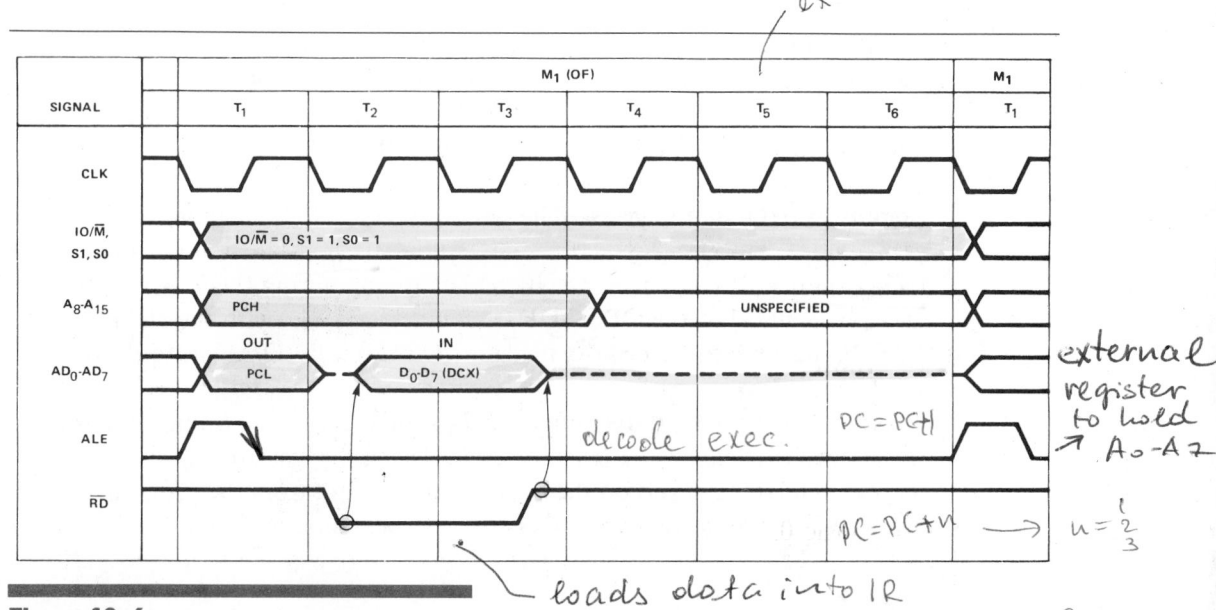

_ex_

decode | exec.

PC = PCH

loads data into IR

PC = PC+u → u = 2/3

external register to hold A₀-A₇

DCX Rp
Rp = Rp-1

Figure 10-6

Timing diagram for an OPCODE fetch cycle for the DCX instruction
(Reprinted by permission of Intel Corporation, Copyright/Intel
Corporation 1979)

device. The assertion of the $\overline{RD}$ line signals external devices to latch the address
before it is removed from the AD lines. At the same time, the memory will use the
$\overline{RD}$ line to start preparing the data for transfer to the processor.

During T_3, sufficient time has elapsed for the memory to gain access to its
internal data storage. The memory device now drives the AD bus with the data.
The 8085 will respond by loading the data from the AD_0–AD_7 lines into the instruction register and then deactivate the $\overline{RD}$ line. At this point the memory access
is complete and the OPCODE of the instruction is obtained. Since this is an OF
cycle, additional T states are necessary.

During T_4, the 8085 will decode the OPCODE in the instruction register and
decide whether to enter the T_5 state or start a new machine cycle. In this case,
since the instruction fetched is the DCX instruction, the 8085 will enter into the
T_5 and T_6 states before returning to T_1 for another machine cycle. This instruction decrements a register pair during the T_5 and T_6 states that will affect the
A_8–A_{15} address lines. The T_5 state is primarily for execution of the instruction.
Any data transfer within the processor and any arithmetic or logical operations
are all accomplished during this time. This is true as long as there are no memory
operands; i.e., all operands are internally stored in registers. During the T_6 state,
the program counter is incremented and general preparatory work is done to get
the processor ready for the next machine cycle.

not 9

The timing diagram presented in Figure 10–6 is for the OPCODE fetch machine cycle. A number of instructions are composed of the OPCODE fetch machine cycle only. Other types of instructions require one or more read and/or write machine cycles following the fetching of the OPCODE.

Memory (or I/O) Read Cycles The timing diagrams for a memory read (MR) or I/O read (IOR) machine cycle, the first without a T_{Wait} state and the second with a T_{Wait} state, are given in Figure 10–7. The timing during the T_1 to T_3 states is basically identical to the OPCODE fetch machine cycle, with the exception that the status bits indicate a memory read (or I/O read) with $IO/\overline{M} = 0$ (or $IO/\overline{M} = 1$), $S_1 = 1$, and $S_0 = 0$.

Note that the next T state after the T_3 state is normally T_1, not T_4, as is the case in the OPCODE fetch machine cycle. This is always true for all other types of cycles. The signal levels on the address lines are not from the program counter anymore, but may have a variety of sources depending on the instruction. Also,

Figure 10–7

Timing diagrams for memory (or I/O) read machine cycles (Reprinted by permission of Intel Corporation, Copyright/Intel Corporation 1979)

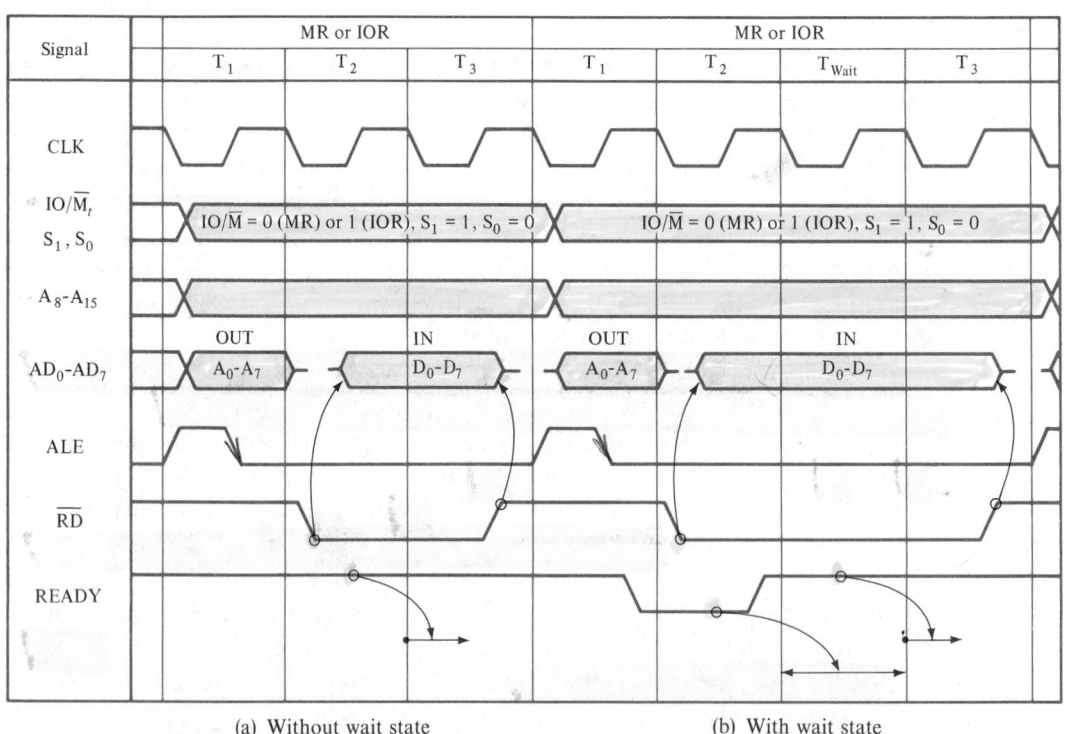

(a) Without wait state   (b) With wait state

the destination for the data is no longer the instruction register but some other appropriate register.

Wait States The READY input may be used by external devices to slow down the 8085 to accommodate slower memory or external devices. The control of the READY signal is performed by examining the proper status and address lines during the T_2 state. In this manner, the READY signal may be used to insert T_{Wait} states for slow I/O devices or possibly slow portions of memory. This is shown in Figure 10–7(b) for the memory (or I/O) read cycle. The T_{Wait} state is added only after the T_2 states when the READY line is sampled during the T_2 state and is seen to be active. This is also apparent from the state transition diagram in Figure 10–5.

Memory (or I/O) Write Cycles The timing diagrams for the memory write (MW) or I/O write (IOW) machine cycles are given with and without the T_{Wait} state in Figure 10–8. The timing diagrams for the MW or IOW are very similar to those

Figure 10–8

Timing diagrams for memory (or I/O) write machine cycles (Reprinted by permission of Intel Corporation, Copyright/Intel Corporation 1979)

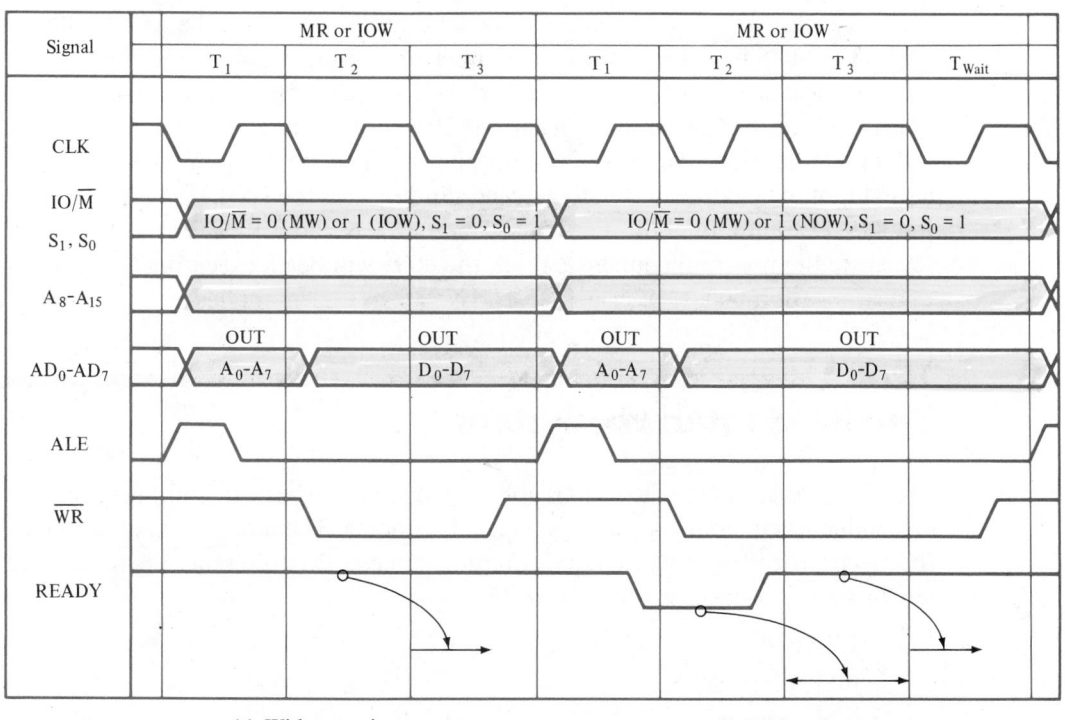

(a) Without wait state   (b) With wait state

for the MR or IOR cycle. The differences are that $S_1 = 0$ and $S_0 = 1$ and that it is the $\overline{\text{WR}}$ line that goes low instead of the $\overline{\text{RD}}$ line.

The address used in the IOR and IOW machine cycles comes from the second byte of the IN (input) and OUT (output) instructions, respectively. This address is called the *port address*. It is very important to keep in mind that the same port address is also duplicated on both the AD_0–AD_7 lines and the address lines A_8–A_{15} during the execution of these two instructions.

Power-On Reset

The 8085 uses a special substrate bias generator to increase speed by creating a negative voltage that biases the substrate. The circuit employs an oscillator and a charge pump, which require a certain amount of time to stabilize after power is initially applied. Hence the 8085 is not guaranteed to work until 10 ms after V_{cc} reaches 4.75 V. A simple RC network should be used to keep $\overline{\text{RESET IN}}$ low until 10 ms have elapsed after V_{cc} reaches 4.75 V.

The $\overline{\text{RESET IN}}$ line is latched every time CLK OUT = 1. The latched signal is recognized by the 8085 during the CLK OUT = 1 period of the next T state. If it is low, the 8085 will make RESET OUT = 1 and enter the T_{Halt} for the next T state. $\overline{\text{RESET IN}}$ should be kept low for a minimum of three clock periods to ensure proper synchronization of the 8085. When $\overline{\text{RESET IN}} = 1$, the 8085 will then enter the T_1 state of the next machine cycle, which will be an OPCODE fetch. The 8085 will clear and set a number of internal registers and flip-flops in response to the $\overline{\text{RESET IN}}$ being set to 0. Those that are reset include the program counter, the instruction register, the interrupt flip-flop, the RST 7.5 flip-flop, the TRAP flip-flop, the SOD flip-flop, the machine state flip-flop, the machine cycle flip-flop, and any internally latched flip-flops for the HOLD, the INTR, and the READY lines. In addition, the RST 5.5, RST 6.5, and RST 7.5 masks will be set. Since the program counter is reset, the 8085 will begin executing the instruction at location 0000H after a $\overline{\text{RESET IN}} = 0$ has been applied.

10.2

The 8085 Interrupt System

Most computer systems use interrupts to reduce the overhead computer time to determine when input/output devices need servicing. The interrupt system signals the microprocessor when new information is needed or present. There are five hardware interrupt inputs on the 8085:

1. TRAP
2. RST 7.5
3. RST 6.5

4. RST 5.5

5. INTR

They are listed in descending order of priority if all are enabled and requesting service at the same time.

The 8085 Hardware Interrupts

The RST 5.5, RST 6.5, and RST 7.5 hardware interrupts are maskable through the use of the SIM instruction, which enables or disables these interrupts by clearing or setting the corresponding mask flags based on data in the accumulator. The status of the interrupt mask previously set by a SIM instruction may be read by executing a RIM instruction, which loads the masks into register A. The RIM instruction also loads the interrupt enable flag status and the pending RST 5.5, RST 6.5, and RST 7.5 interrupts into the accumulator.

The interrupt structure can be explained with the aid of Figure 10–9. Enabling the RST 5.5, RST 6.5, and RST 7.5 interrupts requires the corresponding interrupt mask flip-flops M5.5, M6.5, and M7.5 to be 0 and the interrupt enable flip-flop to be 1 as shown in Figure 10–9. Note that the outputs of the AND gates 2, 3, and 4 symbolically imply that the 8085 will jump to the vector locations (addresses) indicated when the output of the AND gate is 1.

The TRAP Interrupt   The TRAP interrupt is not maskable internally by the 8085. It is activated with a positive transition and a sustained high level on the input line. The TRAP input should be held high until the interrupt is recognized. The internal hardware of the 8085 will reset the flip-flop associated with the TRAP when a TRAP interrupt is recognized. Thus, if the TRAP input is held high, a second interrupt will not be recognized until the TRAP input goes low and then high again.

The RST 7.5 Interrupt   The RST 7.5 interrupt is a positive transition-sensitive input. This implies that the input does not have to be held high until it is recognized. An internal D flip-flop is used to sense this transition. This D flip-flop will be cleared when the RST 7.5 interrupt is acknowledged and also with the SIM instruction with $A_4 = 1$ in register A as follows:

Label	OPCODE	Operand	Comments
	MVI	A,00001101B	
	SIM		
	EI		

Note that the interrupt input positive pulse can be used only on the RST 7.5 and none of the others, because they must be held high until the inputs are recognized.

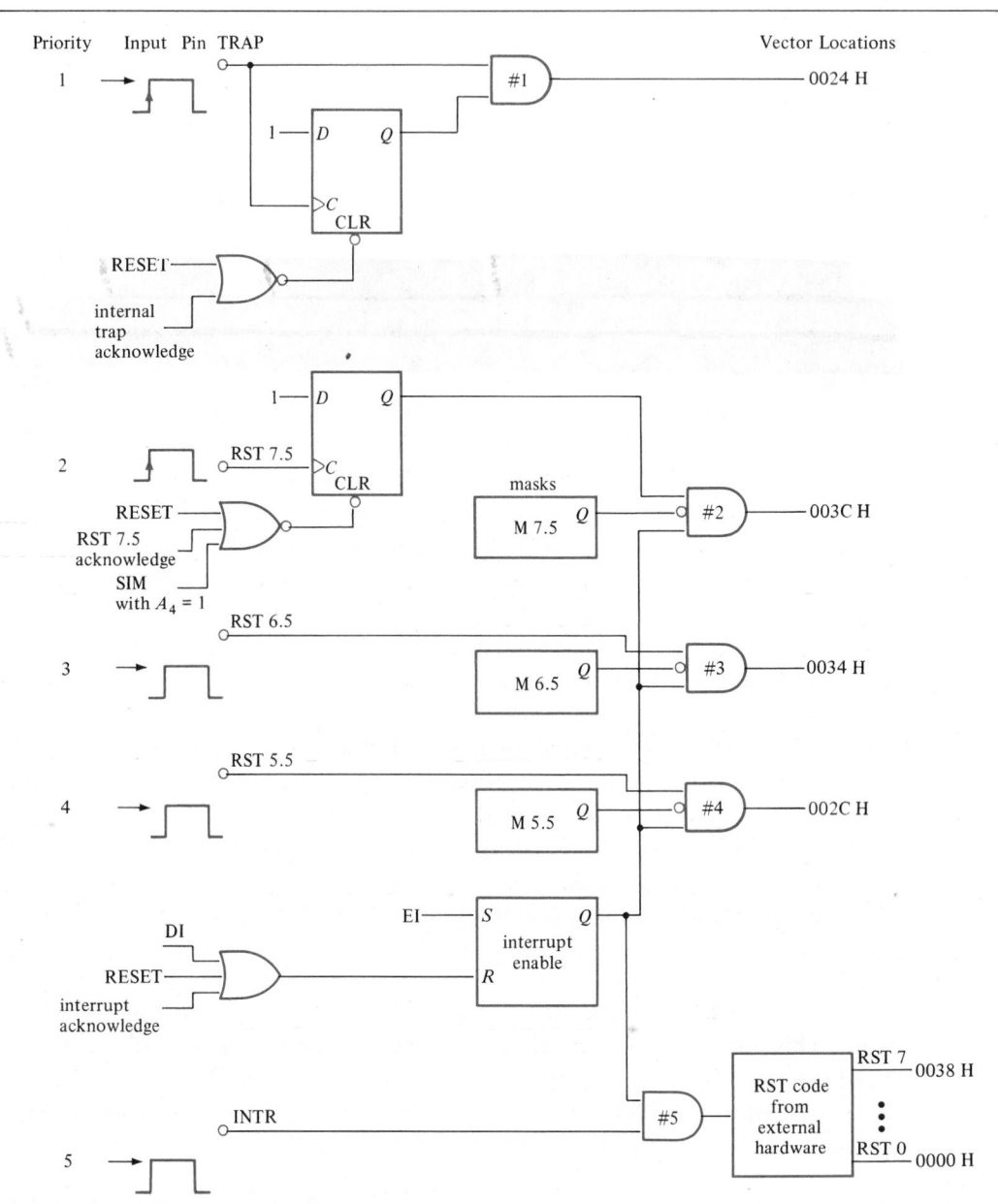

Figure 10–9
The 8085 interrupt structure

The RST 6.5 Interrupt   The RST 6.5 interrupt is a level-sensitive input and can be enabled by the following instructions:

Label	OPCODE	Operand	Comments
	MVI	A,00001010B	
	SIM		
	EI		

The MVI and SIM reset the mask M6.5 flip-flop. Bit 3 represents the mask set enable bit, which must be set to 1 to change the masks. The EI instruction sets the interrupt enable flip-flop and then the output of AND gate 3 becomes 1 when RST 6.5 is high. When this occurs, the 8085 pushes the contents of the program counter on the stack and then jumps to address 0034H. Note that the interrupt enable flip-flop is cleared when any interrupt is recognized. Thus all maskable interrupts are disabled. Hence, after every interrupt is recognized, the programmer must make sure the EI instruction is executed to enable further interrupts.

The sampling of all interrupts occurs on the falling edge of the 8085 clock, one cycle before the end of the instruction in which the interrupt is activated. To be recognized, a valid interrupt must occur at least 160 ns before the falling clock transition. Also, these interrupt inputs (except RST 7.5) must be held until they are recognized. This implies that they must be held for the longest instruction that can be executed on the 8085, namely, a CALL instruction. This assumes that no WAIT or HOLD is currently effective.

The reset signal that appears in Figure 10–9 is generated in response to the RESET IN signal on the 8085. This signal resets the flip-flops on the TRAP and RST 7.5 circuitry, and in addition the interrupt enable flip-flop. The circuitry is not shown, but the reset also sets all mask flip-flops.

The RST 5.5 Interrupt   As indicated in Figure 10–9, the RST 5.5 interrupt is level-sensitive. It operates in the same manner as the RST 6.5 interrupt.

The INTR Interrupt   The INTR is a maskable interrupt with no individual mask assigned other than the interrupt enable flip-flop. When the INTR line is activated, a restart code is expected to be placed on the data bus by the requesting device and the interrupt service routine is taken according to the restart code received.

The 8085 Software Restarts

The RST n-type instruction is similar in operation to a hardware interrupt except that it is invoked by an instruction rather than by an external interrupt. The execution of this instruction is similar to interrupt processing with the program counter being loaded with 8 times n. Since n can be only 0, 1, 2, 3, 4, 5, 6, or 7, only eight different locations can be reached with this instruction, which is a disadvantage. However, the advantages are that only 1 byte is needed for the instruction, and it executes faster. When the RST is used, the subroutines must start at the locations given in Figure 10–10. The interrupt structure of the 8085 is basically similar to the RST instructions, but the subroutines are called by external signals driving the interrupt inputs.

Figure 10–10 shows the addresses for both the software restarts and the hardware interrupts. The naming of the interrupts as RST 5.5, etc., becomes apparent when one observes the address to which the jumps are performed in relation to the RST 5, RST 6 instructions. The RST 5 instruction and RST 5.5 interrupt work in a very similar manner and are basically subroutine calls to different addresses, with the first being activated by software control and the second by external hardware control.

The interrupt structure on the 8085 is referred to as a *vectored interrupt system*, because every interrupt is directed or vectored to a specific memory address that contains either the program or a jump to the address of the program to service the interrupt when it occurs. The interfacing of interrupt hardware and software routines will be discussed later.

Interrupt Programming

A few comments are in order concerning interrupt programming. First, it is the responsibility of the programmer and the hardware designer to ensure that the interrupt inputs are returned low after each interrupt is recognized and then

Figure 10–10
Restart and interrupt addresses of the 8085

Instruction or Interrupt	Hexadecimal Address
RST 0	0000
RST 1	0008
RST 2	0010
RST 3	0018
RST 4	0020
TRAP	0024
RST 5	0028
RST 5.5	002C
RST 6	0030
RST 6.5	0034
RST 7	0038
RST 7.5	003C

brought high only when the next interrupt is required. The mechanism for changing the mask flip-flops (M5.5, M6.5, and M7.5) is not shown in Figure 10–9. The masks are all set to 1 when the $\overline{\text{RESET IN}}$ on the 8085 is driven low. This reset will automatically disable the RST 5.5, RST 6.5, and RST 7.5 interrupt inputs.

Example 10–1 ▮▬▬▬▬▬▬▬▬▬▬▬▬▬▬▬▬▬▬▬▬▬▬▬▬▬▮

Write a program that will enable the M7.5 and M5.5 interrupts and disable the M6.5 interrupt.

Label	OPCODE	Operand	Comments
	MVI	A,00001010B	
	SIM		
	EI		

The current state of the masks and the interrupt enable flip-flops can be determined by executing a RIM and then examining register A. Any pending interrupts can also be read into register A, thus providing useful information when more than one interrupt is used.

10.3
An 8085 Minimum System

A microprocessor-based system requires additional external memory and input/output integrated circuits in order to function properly. Usually, both ROM for program storage and RAM for temporary scratchpad storage are needed. The addition of the 8156/8155 and the 8355/8755 to the 8085 microprocessor will create a minimum microcomputer system. A possible 8085 minimum system with most of the interconnections shown is given in Figure 10–11. The hardware of the minimum system in Figure 10–11 will be used to start writing more programs with simple types of input/output. Additional hardware will be added to this system to illustrate other techniques. Then only the circuits indicating changes will be shown.

The 8156/8155 and 8355/8755 are designed as part of a compatible family with the 8085, and they make possible a system with very few integrated circuits. The programs that we will write can be considered to reside in the ROM memory of the 8355 or in the programmable memory of the 8755. The latter requires a PROM programmer to store a binary program. The 8355 would need to be programmed at the factory as the integrated circuit was being manufactured. The 2K of ROM memory can be addressed in the system of Figure 10–11 at addresses

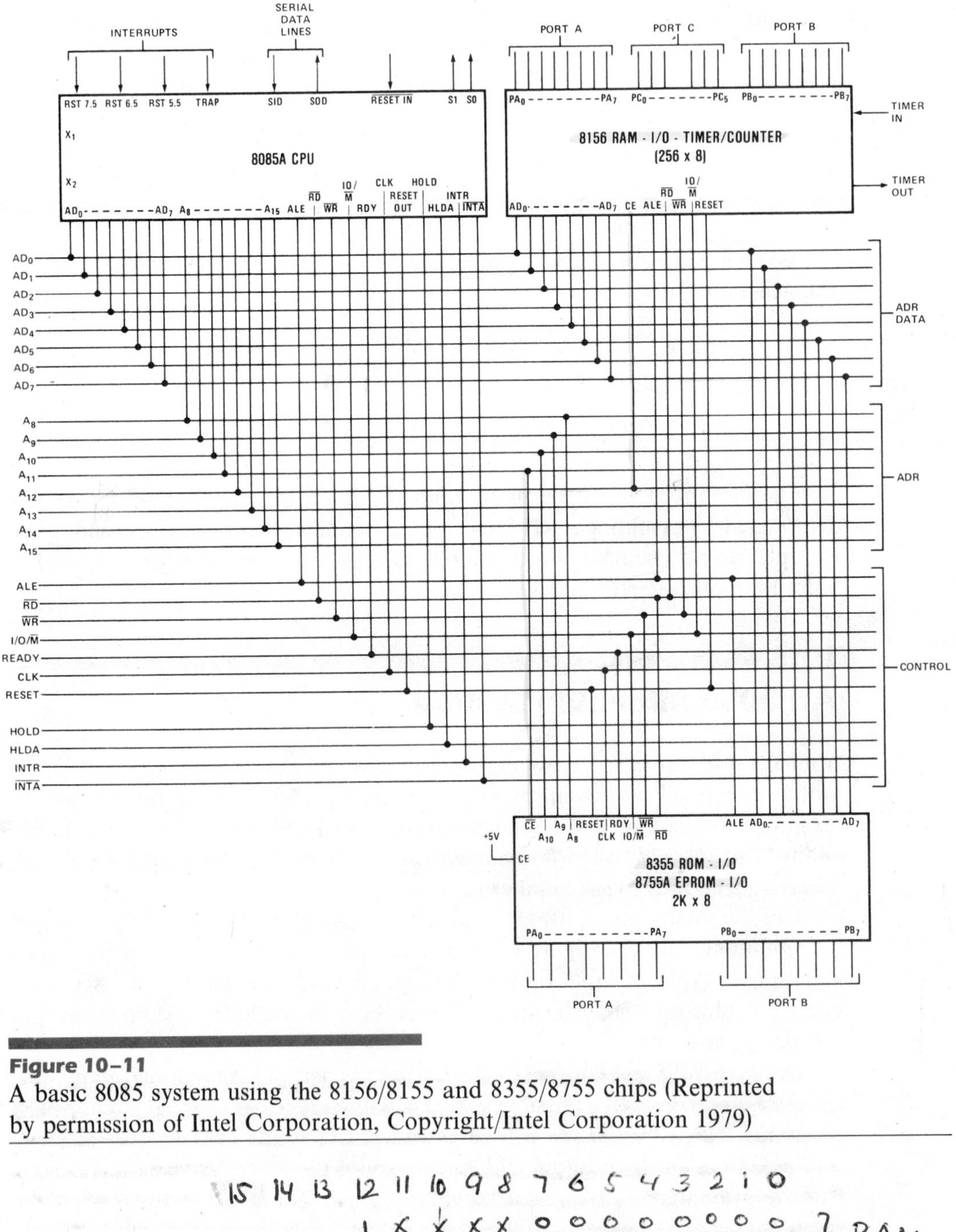

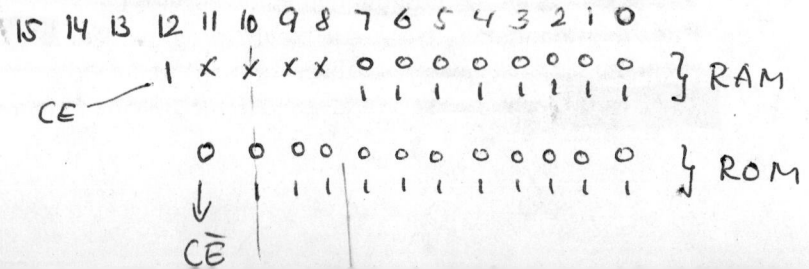

Figure 10–11
A basic 8085 system using the 8156/8155 and 8355/8755 chips (Reprinted by permission of Intel Corporation, Copyright/Intel Corporation 1979)

15 14 13 12 11 10 9 8 7 6 5 4 3 2 1 0

1 X X X X 0 0 0 0 0 0 0 0 } RAM
1 1 1 1 1 1 1 1

CE

0 0 0 0 0 0 0 0 0 0 0 } ROM
1 1 1 1 1 1 1 1 1 1 1

CE

0000H to 07FFH. The 256 words of RAM can be addressed at locations 1800H to 18FFH. A word of caution is that other addresses can also be used, because our system does not have full memory decoding. To prevent the multiple-address problem, more decoder-type chips would be needed in the system. This situation will be discussed later.

The ports on both the 8156/8155 and the 8355/8755 in Figure 10–11 have specific port numbers. The port numbers will be determined as we further examine these integrated circuits in the following sections. The $\overline{\text{RESET IN}}$ circuitry will cause the 8085 to start executing the program stored at memory address 0000H when power is initially applied. The programs must be written so that proper execution will start at that address. Also, remember that the RESET OUT = 1 will reset the 8156 and 8355/8755. The pull-up resistors on the RDY, $\overline{\text{WR}}$, and $\overline{\text{RD}}$ lines will be needed if direct memory access (DMA) is added to the system.

10.4
The 8355/8755 ROM I/O Chip

The 8355 and the 8755 are very similar. The difference lies in the fact that the 8355 uses a mask-programmable ROM and the 8755 an EPROM. The 8355/8755 is an NMOS chip packaged in a 40-pin DIP; it contains $2K \times 8$-bit words of ROM (or EPROM) and two 8-bit I/O ports. Each bit on the two I/O ports can be programmed individually as an input or output. The chip uses a single 5-V power supply, and all inputs and outputs are TTL-compatible. This chip is compatible with the 8085.

Signal Definitions of the 8355/8755

A simplified block diagram of the 8355/8755 is given in Figure 10–12. The definitions of the signals on the pins are listed below.

Signal	Name	Type	Function
1. *Address/data bus*			
AD_{0-7}	Address/data lines	I/O	This is the bidirectional address/data bus that can be connected directly to the AD_{0-7} of the 8085.
A_{8-10}	Address lines	I	These are the high-order bits of the ROM address and are normally connected to the address lines A_8–A_{10} of the 8085.
2. *Parallel input/output lines*			
PA_{0-7}	I/O port A	I/O	These are general-purpose I/O pins collectively identified as port A that are controlled by the data direction register for port A.

(continued)

Signal	Name	Type	Function
PB_{0-7}	I/O port B	I/O	These are general-purpose I/O pins collectively identified as port B. Its function is the same as PA_{0-7}.
3. *Memory and I/O control lines*			
ALE	Address latch enable	I	The signals on AD_{0-7}, $IO/\bar{M}$, A_{8-10}, CE_2, and $\overline{CE}_1$ are latched at the $1 \rightarrow 0$ transition of ALE. It is normally connected directly to the ALE line of the 8085.
$\overline{CE}_1$, CE_2	Chip enable	I	The 8355 chip is enabled (selected) only when $\overline{CE}_1 = 0$ and $CE_2 = 1$.
$IO/\bar{M}$	IO or memory	I	If the latched $IO/\bar{M} = 1$ and $\overline{RD} = 0$, then the data on AD_{0-7} come from an I/O port. If $IO/\bar{M} = 0$, the data on AD_{0-7} come from the ROM.
$\overline{IOR}$	I/O read	I	Input/output reads and performs same function as $\overline{RD}$ and $IO/\bar{M}$.
$\overline{IOW}$	I/O write	I	If $\overline{CE}_1 = 0$, $CE_2 = 1$, and $\overline{IOW} = 0$, the ports or data direction registers are written to from AD_{0-7}. $IO/\bar{M}$ is ignored.
$\overline{RD}$	Read	I	If $\overline{CE}_1 = 0$, $CE_2 = 1$, and $\overline{RD} = 0$, then the AD_{0-7} lines are driven from either the ROM or an I/O port.
4. *System control lines*			
CLK	Clock	I	This is the clock input that controls the READY line for slow devices.
READY	Ready line	O	The READY signal may be used to insert wait states into the 8085 cycle.
RESET	Reset	I	The RESET input that is high causes both port A and port B to be inputs.
V_{cc}, V_{ss}	Power, ground	I	$+5$ V supply and ground, respectively.

The 8755 differs from the 8355 in that the 8755 uses an EPROM instead of a mask-programmable ROM. An unused pin on the 8355 is committed to V_{DD} of $+25$ V for programming the EPROM on the 8755. Also, the $\overline{CE}_1$ pin on the 8355 is called a $PROG/\overline{CE}_1$ on the 8755. These two pins are used to program the 8755. It is important to mention that once the mask-programmable ROM on the 8355 or the EPROM on the 8755 is programmed, the use of the 8755 is identical to the 8355 in both memory reference and input/output programming.

Operation of the 8355/8755

The main items to be considered in the operation of the 8355/8755 are memory operations of the ROM/EPROM and the operations of the I/O ports.

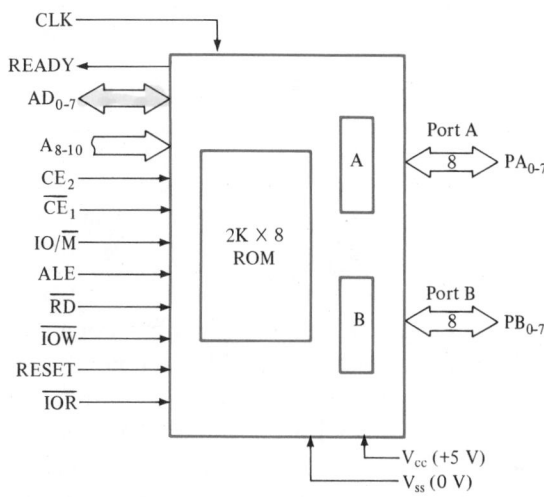

Figure 10–12

A simplified block diagram of the 8355 chip (Reprinted by permission of
Intel Corporation, Copyright/Intel Corporation 1979)

Memory Operations Let the 8355/8755 be connected to the 8085 in the configuration shown in Figure 10–11. If the address line A_{11} of the 8085 is connected to the $\overline{CE}_1$ of the 8355/8755 and the address lines A_{13}–A_{15} are not used,
then the ROM could be addressed at any one of the following eight 2K memory
segments. The address line A_{12} when equal to 1 selects the 8156, so we cannot
use any address that activates two memory devices.

$$0000 \text{ to } 07\text{FF}$$
$$2000 \text{ to } 27\text{FF}$$
$$4000 \text{ to } 47\text{FF}$$
$$\vdots \quad \vdots \quad \vdots$$
$$\text{E}000 \text{ to } \text{E}7\text{FF}$$

We will always use the address range from 0000H to 07FFH for clarity.

I/O Port Operations The input/output section is controlled by the values on
AD_0 and AD_1 when $\overline{CE}_1 = 0$ and $CE_2 = 1$ during an $\overline{RD} = 0$ or $\overline{IOW} = 0$. Two
8-bit data direction registers (DDRs) in the 8355/8755, one for each port, deter-
mine the input/output status of each pin in the corresponding ports. A 0 in bit i
of DDR A causes PA_i to be an input, a 1 makes PA_i an output. The same holds
true for DDR B. This makes it possible to make any combination of the 16 port

A and port B lines to be inputs or outputs. The values on the AD_0 and AD_1 lines when $ALE = 1$ determine the port or data direction register as follows:

AD_1	AD_0	Selection
0	0	Port A
0	1	Port B
1	0	Port A Data Direction Register (DDR A)
1	1	Port B Data Direction Register (DDR B)

The logic diagram that shows the control of a pin in port A is given in Figure 10–13. This logic is repeated for all lines in port A and port B. The logic diagram also shows that even though a line is an output, the value in the output latch can be read into the 8085. Furthermore, whenever a line is an input, writing to the output latch will not affect the signal at the pin because of the presence of the tri-state gate. Note that the hardware RESET causes the DDR latch to be 0 and the line to be an input. The actual output will change on the rising edge of $\overline{IOW}$. Also, the data direction registers cannot be read from; they can only be written into by the processor.

Figure 10–13
Equivalent circuit for the internal control of one pin on the 8355 (Reprinted by permission of Intel Corporation, Copyright/Intel Corporation 1979)

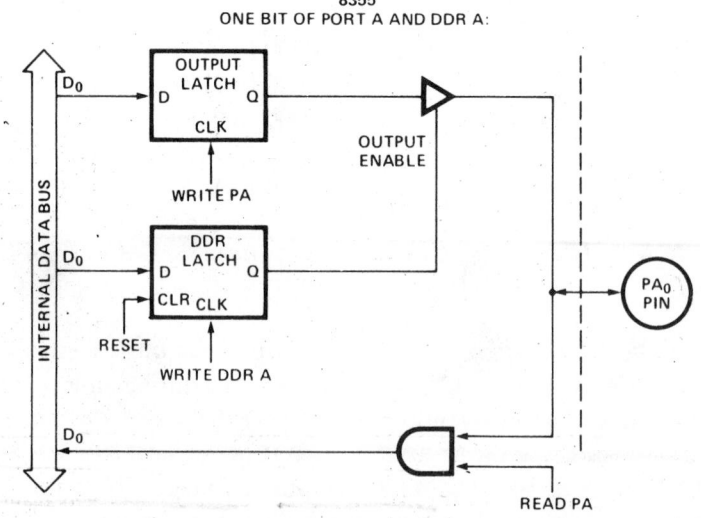

The ports and the DDRs in the 8355/8755 as shown in the minimum system configuration can be controlled when address line $A_{11} = 0$, with AD_1 and AD_0 previously defined, with an IN port or OUT port instruction from the 8085. Whenever an IN or OUT instruction is executed with the corresponding 8-bit port address, the 8-bit port address is duplicated on AD_0–AD_7 and address lines A_8–A_{15}. The port address occurs on AD_{0-7} when ALE = 1. The value on AD_0 is duplicated on A_8, AD_1 on A_9, etc. Therefore, the address of the ports and the data direction registers on the 8355/8755 as shown in the minimum system configuration are selected as follows:

Port	Selection
00	Port A 8355/8755
01	Port B 8355/8755
02	Port A DDR A 8355/8755
03	Port B DDR B 8355/8755

These addresses are not unique, but they are the ones that we will use in our programs.

Programming Examples with the 8355/8755

Several programming examples using the 8355/8755 will now be given. We will assume that the 8355/8755 chip is connected to the 8085 as shown previously in the minimum system configuration. Therefore all the port addresses and memory addresses hold.

Example 10-2

Write a program to configure port A on the 8355/8755 in the minimum system configuration as an output. Furthermore, the output on port A should count from 00 to FF in binary and infinitely repeat the sequence again.

The first operation in the program is to configure port A as an output port. Then data that increase in a binary fashion are sent to port A. The program follows.

Label	OPCODE	Operands	Comments
	ORG	0H	
	MVI	A,11111111B	;Makes all lines
	OUT	02H	; on port A outputs
	SUB	A	;Zeroes reg. A
LOOP:	OUT	00H	;Outputs to port A
	INR	A	;Increments in binary
	JMP	LOOP	;Do again
	END		

The number of T states in the loop is 24, and with a 5.0-MHz crystal the basic period of the MSB on output PA_7 is *— half of the freq*

$$T_{period} = 2^8 \times (24) \times 2/(5 \times 10^6) = 2.45 \text{ ms}$$
$$f_{period} = 406.9 \text{ Hz}$$

The output frequency could be increased by using the PCHL form of jump.

Example 10–3

Assume that the 7400 quad NAND gates are connected to port A and port B as shown in Figure 10–14. Write a program that will logically test the four NAND gates for all input combinations. If the chip is okay, output a 1 on PB_4; if not, output a 0.

The first operation is to configure port A as an output, PB_0–PB_3 as inputs, and PB_4 as an output. All other unused pins will be configured as inputs. The inputs to the NAND gates are connected to the output of the 8355/8755. Likewise, the outputs of the NAND gates are connected to the inputs of the 8355/8755.

Figure 10–14

Circuit diagram for Example 10–3 with 7400 quad NANDs connected to the 8355/8755 chip

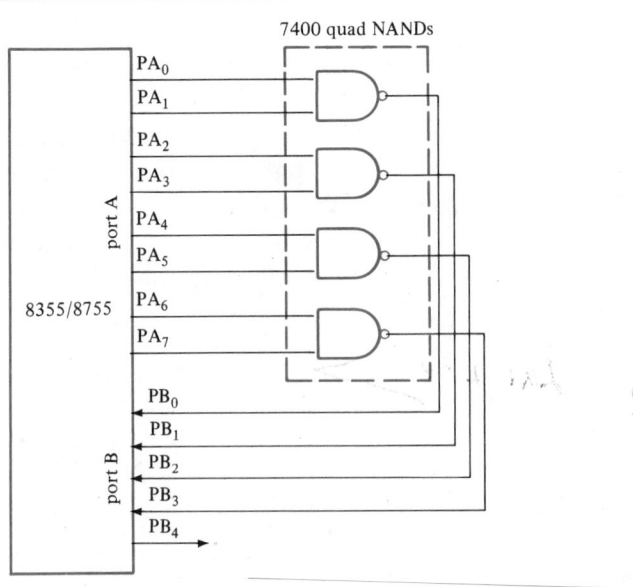

To carry out the test of the NAND gates, supply the same input patterns to each NAND gate simultaneously. The outputs of the NAND gates are then read and verified. If the output pattern of the NAND gates passes the tests, we will make $PB_4 = 1$; otherwise $PB_4 = 0$.

Label	OPCODE	Operand	Comments
	ORG	0H	
	MVI	A,11111111B	;Configure ports with
	OUT	02H	; Port A output
	MVI	A,00010000B	;PB0 — 3 inputs
	OUT	03H	;PB4 — output
	SUB	A	
	OUT	01H	;PB4 = 0
;			
	OUT	00H	;Tests for 00 on each NAND
	IN	01H	;Read in outputs
	ANI	0FH	;Mask off 4 MSB
	CPI	0FH	;Compares with all 1 outputs
	JNZ	BAD	;Jump if bad
;			
	MVI	A,01010101B	;Test for 01 inputs
	OUT	00H	;Make sure
	IN	01H	; all outputs are 1
	ANI	0FH	
	CPI	0FH	
	JNZ	BAD	;Jump if bad
;			
	MVI	A,10101010B	;Test for 10 inputs
	OUT	00H	;Make sure
	IN	01H	; all outputs are 1
	ANI	0FH	
	CPI	0FH	
	JNZ	BAD	;Jump if bad
;			
	MVI	A,11111111B	;Test for 11 inputs
	OUT	00H	;Make sure
	IN	01H	; all outputs are 0
	ANI	0FH	
	CPI	00H	
	JNZ	BAD	;Jump if bad
;			
	MVI	A,00010000B	;Chip okay
	OUT	01H	;PB4 = 1
	HLT		
;			
BAD:	SUB	A	;Chip bad
	OUT	01H	;PB4 = 0
	HLT		
	END		

The above examples showed simple programming using port A and port B. The next example shows the use of the delay routine.

Example 10–4

Assume that a seven-segment LED display is connected to port B on the 8355/8755 as shown in Figure 10–15(a). Write a program that will display the hexadecimal numbers in sequence from 0 to F, repeating with approximately a one-half second display period for each digit.

The 7406 are open-collector TTL inverters capable of sinking 16 mA. The resistance value R is calculated in Figure 10–15(b) assuming that 10 mA of current is needed to light an individual red LED segment with a 1.7-V drop across the LED and a 0.2-V collector-to-emitter saturation voltage. The diagram shows that when $PB_0 = 1$, the corresponding segment of the LED is on. Likewise, if $PB_0 = 0$, the segment is off. Thus, to turn on any segment, the appropriate PB

Figure 10–15

Circuit diagram for Example 10–4 showing the interfacing between a seven-segment display and the 8355 chip

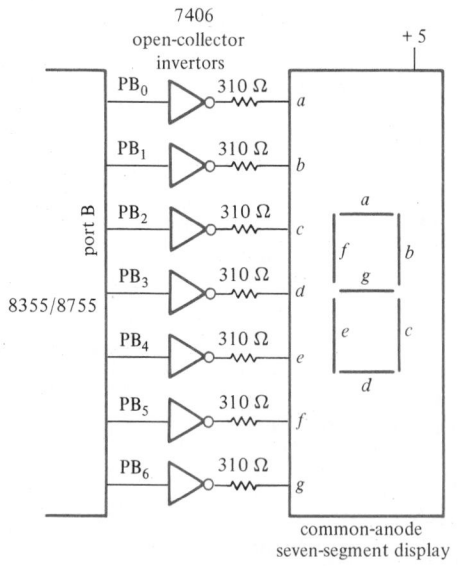

$$R = \frac{V_{sop} - V_{LED} - V_{CE}}{I_{LED}}$$

$$= \frac{5 - 1.7 - 0.2}{10 \text{ mA}} = 310\ \Omega$$

(b) Individual segment

(a) Seven-segment display connections

line should be 1 and otherwise 0. The letters *a* through *g* on the seven-segment displays in the indicated order are used almost universally by manufacturers.

The hexadecimal numbers can be shown on the seven-segment displays using a combination of upper- and lowercase letters for the digits A through F as indicated in Figure 10–16. The digits are formed in the normal manner, taking care to ensure that the 6 and the lowercase *b* are not the same. The program will be written using a table look-up to light the correct segments and then delaying a one-half second before displaying the next digit.

Figure 10–16
Seven-segment decode table for hexadecimal numbers

seven-segment display	PB 7	6 *g*	5 *f*	4 *e*	3 *d*	2 *c*	1 *b*	0 *a*	corresponding byte
	0	0	1	1	1	1	1	1	3 F
	0	0	0	0	0	1	1	0	0 6
	0	1	0	1	1	0	1	1	5 B
	0	1	0	0	1	1	1	1	4 F
	0	1	1	0	0	1	1	0	6 6
	0	1	1	0	1	1	0	1	6 D
	0	1	1	1	1	1	0	1	7 D
	0	0	0	0	0	1	1	1	0 7
	0	1	1	1	1	1	1	1	7 F
	0	1	1	0	1	1	1	1	6 F
	0	1	1	1	0	1	1	1	7 7
	0	1	1	1	1	1	0	0	7 C
	0	0	1	1	1	0	0	1	3 9
	0	1	0	1	1	1	1	0	5 E
	0	1	1	1	1	0	0	1	7 9
	0	1	1	1	0	0	0	1	7 1

Label	OPCODE	Operand	Comments
	ORG	0H	
DECTAB:	DB	3FH, 06H, 5BH, 4FH, 66H, 6DH, 7DH, 07H	
	DB	7FH, 6FH, 77H, 7CH, 39H, 5EH, 79H, 71H	
DEL	EQU	51083D	
;			
START:	MVI	A,01111111B	;PB$_0$–PB$_6$ outputs
	OUT	03H	
DOZERO:	LXI	H,DECTAB	;Initialize HL to start of table
	MVI	B,16D	;Initialize counter
NEXT:	MOV	A,M	;Get data for 7 segment
	OUT	01H	;Display data
	LXI	D,DEL	;Delay one-half second
LOOP:	DCX	D	
	MOV	A,D	
	ORA	E	
	JNZ	LOOP	
	DCR	B	;Test if start at zero again
	JZ	DOZERO	
	INX	H	;Increment to next digit
	JMP	NEXT	
	HLT		
	END		

This program shows how the DB directive allows us to define a table with the address loaded into the register pair HL based on the initial label. The program must begin execution at the location with the START label.

10.5

The 8156/8155 RAM I/O Chip

The 8156/8155 chip contains 256 bytes of RAM, three parallel I/O ports, and a programmable timer. The 8156/8155 is also specially designed to operate with the 8085.

Signal Definitions of the 8156/8155

A simplified block diagram is given in Figure 10–17. The 8155 is identical to the 8156 except that the chip enable pin is active high on the 8156 and active low (that is, $\overline{CE}$ is used) on the 8155.

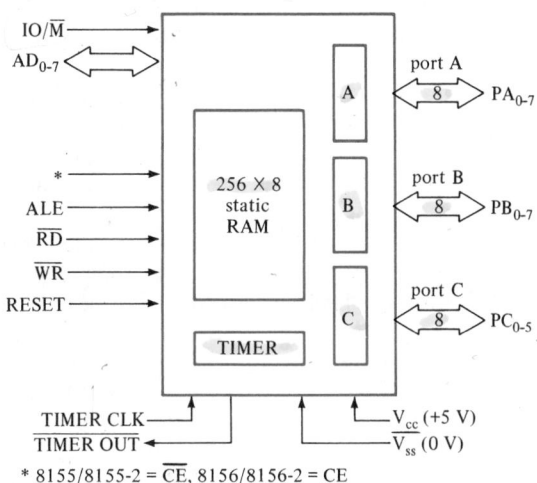

Figure 10–17

A simplified block diagram for the 8156/8155 chip (Reprinted by permission of Intel Corporation, Copyright/Intel Corporation 1979)

A brief description of the signals and their corresponding symbols are given here:

Signal	Name	Type	Function
1. *Address/data bus*			
AD_{0-7}	Address/data	I/O	These are normally connected to the AD_0–AD_7 lines of the 8085 and are used in the normal manner.
2. *Parallel I/O lines*			
PA_{0-7}	Port A	I/O	Port A, with eight general-purpose I/O lines.
PB_{0-7}	Port B	I/O	Port B, with eight general-purpose I/O lines.
PC_{0-5}	Port C	I/O	Port C, with six general-purpose I/O lines. The lines can also be used as control signals for port A and port B.
3. *Memory and I/O control lines*			
CE/$\overline{CE}$	Chip enable	I	The 8156 is activated whenever CE = 1. On the other hand, the 8155 is activated whenever $\overline{CE}$ = 0.

(continued)

Signal	Name	Type	Function
RD	Read control	I	A low on this pin with CE = 1 for the 8156 (or $\overline{CE} = 0$ for the 8155) will cause a read from the RAM locations, the status registers, and the I/O ports, depending on IO/$\overline{M}$.
WR	Write control	I	A low on this pin with CE = 1 for the 8156 (or $\overline{CE} = 0$ for the 8155) will cause a write to the RAM locations, the command registers, and the I/O ports.
ALE	Address latch enable	I	This control signal will cause a reading of the AD_0–AD_7 lines, CE, and IO/$\overline{M}$.
IO/$\overline{M}$	I/O or memory	I	If IO/$\overline{M}$ is low, then memory is selected. If IO/$\overline{M}$ is high, then the I/O ports and the command registers are selected.
4. Timer lines			
TIMER IN	Timer clock	I	Input to the timer/counter.
$\overline{\text{TIMER OUT}}$	Timer output	O	Output of the timer/counter.
5. System control lines			
RESET	Reset	I	High on this line resets 8156/8155 and puts all ports into the input mode.
V_{cc}, V_{ss}	Power	I	+5 V power supply and ground, respectively.

Operation of the 8156/8155

To examine the operation of the 8156/8155, we also connect the chip to the 8085 shown earlier in the minimum system configuration.

Memory Addresses   The RAM on the 8156/8155 in the minimum system is selected when the address line $A_{12} = 1$ and the IO/$\overline{M}$ = 0 during MEMORY READ and MEMORY WRITE instructions. The remaining address lines A_{8-10} and A_{13-15} are not connected to the 8156/8155, so these are ignored. We will assume that they are zero and thus we will address the 256 bytes of RAM in the 8156/8155 as memory locations 1800 to 18FF. Note that the address line $A_{11} = 1$ selects the 8156/8155, whereas $A_{11} = 0$ selects the 8355/8755. (ROM)
    (RAM)

I/O Port Addresses   All the lines of port A in the 8156/8155 can be collectively configured as all inputs or all outputs. The direction of a single line cannot be programmed individually as is the case for the 8355/8755. The same situation applies to port B. The lines in port C can be programmed in one of several different modes, discussed later.

The programming of the 8156/8155 chip for I/O is much more involved than the 8355/8755. The 8156/8155 has a command register and a status register for controlling and storing the status information of various signals. The 8156/8155 uses the multiplexed address/data lines AD_2, AD_1, and AD_0 with $IO/\overline{M} = 1$ and $CE = 1$ to address various I/O ports and internal registers. This is shown below:

AD_2	AD_1	AD_0	Selection
0	0	0	Status/command register
0	0	1	I/O port A
0	1	0	I/O port B
0	1	1	I/O or control port C
1	0	0	Counter/timer register, low-order byte
1	0	1	Counter/timer register, high-order byte

The status and command registers are decoded using the same pattern. As far as the 8085 is concerned, they are differentiated, in that the command register can only be written to and the status register can only be read from the 8085. Thus, after the command register is written into, it cannot be read from later.

The 8085 communicates with the timer and the I/O ports of the 8156/8155 by means of the IN and OUT instructions. Recall that the 8-bit port address is duplicated on the address lines A_8–A_{15} and AD_0–AD_7. Now the minimum system uses the address line A_{12} as the chip enable. Since the AD_0, AD_1, and AD_2 lines are duplicated on the address lines A_8, A_9, and A_{10}, the four lines A_{8-10} and A_{12} (and AD_0–AD_2, AD_4) are used to address the timer and I/O ports. The address line $A_{11} = 1$ to select the 8156/8155. The three most significant bits A_{13}–A_{15} (and AD_5–AD_7) are not used, and as before, we will specify them as zeroes. The port designations used for the 8085 system in the minimum system configuration are

Port	Selection
0 0	Port A of the 8355/8755
0 1	Port B of the 8355/8755
0 2	Port A—DDR A of the 8355/8755
0 3	Port B—DDR B of the 8355/8755
1 8	Status/command register of the 8156/8155
1 9	Port A of the 8156/8155
1 A	Port B of the 8156/8155
1 B	Port C of the 8156/8155
1 C	Counter/timer register, low-order byte of the 8156/8155
1 D	Counter/timer register, high-order byte of the 8156/8155

ROM

RAM

The Command Register   The command register consists of 8 bits whose values determine the configuration and operation of the timer and I/O ports. The bit assignment of the command register is given in Figure 10–18. The two least significant bits (0 and 1) program port A and port B with all lines as inputs or outputs in each port. Bits 2 and 3 control port C to be inputs or outputs or the handshaking mode. These two bits have ALT 1, ALT 2, ALT 3, and ALT 4 designations. ALT stands for alternative, and the four alternatives are spelled out in Figure 10–19. The use of the alternatives for port C will be discussed later. Bits 4 and 5 are used to enable the interrupt structure for port A and port B. Bits 6 and 7 are used to start, stop, and reconfigure the timer. The command register will be programmed by loading the appropriate values into register A and then executing an OUT 18H instruction.

The Status Register The status register contains 7 bits, with six for the status of the ports and one for the status of the timer. The bit assignment of the status register is given in Figure 10–20. The status register can be read only by executing

Figure 10–18

Bit assignment in the command register of the 8156/8155 (Reprinted by permission of Intel Corporation, Copyright/Intel Corporation 1979)

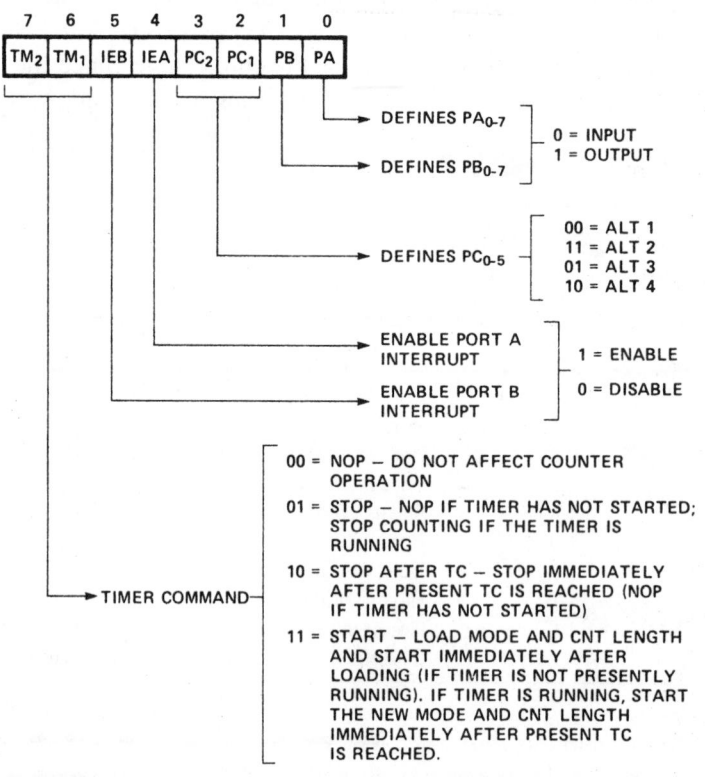

Pin	ALT 1	ALT 2	ALT 3	ALT 4
PC_0	Input port	Output port	A INTR (port A interrupt)	A INTR (port A interrupt)
PC_1	Input port	Output port	A BF (port A buffer full)	A BF (port A buffer full)
PC_2	Input port	Output port	A $\overline{STB}$ (port $\overline{A}$ strobe)	A $\overline{STB}$ (port $\overline{A}$ strobe)
PC_3	Input port	Output port	Output port	B INTR (port B interrupt)
PC_4	Input port	Output port	Output port	B BF (port B buffer full)
PC_5	Input port	Output port	Output port	B $\overline{STB}$ (port $\overline{B}$ strobe)

Figure 10–19

Port assignments for the different ALT modes in the 8156/8155 (Reprinted by permission of Intel Corporation, Copyright/Intel Corporation 1979)

an **IN 18H** instruction, which will move the contents of the status register into register A in the 8085.

The I/O Ports The logic diagram for an individual line on port A or port B is given in Figure 10–21. Each individual line can be programmed for the output mode, the simple input mode, or the strobed input mode that operates in conjunction with a data storage latch. This output latch is always cleared when the

Figure 10–20

Bit assignment in the status register of the 8156/8155 (Reprinted by permission of Intel Corporation, Copyright/Intel Corporation 1979)

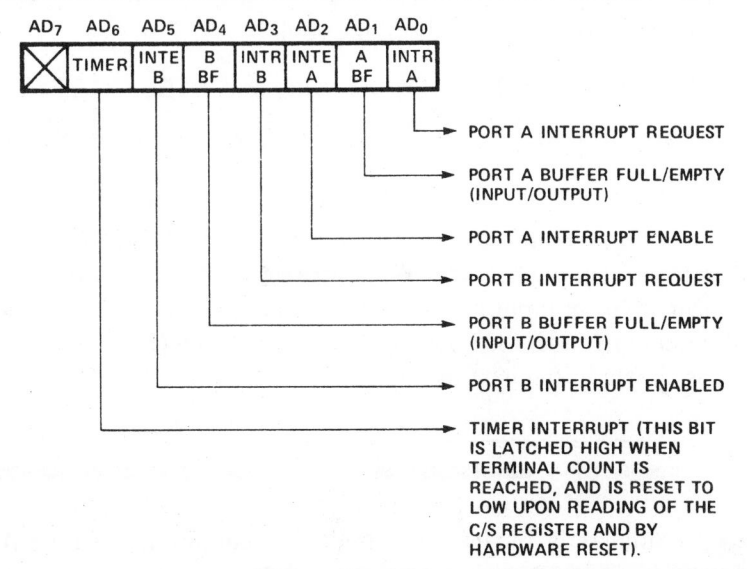

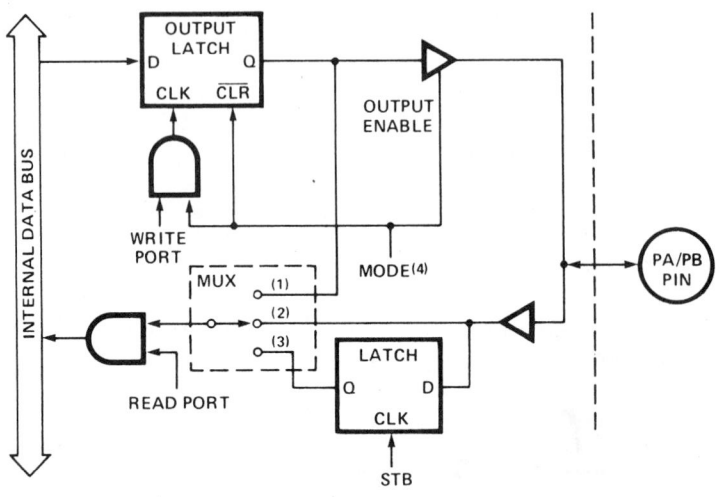

Figure 10-21
Equivalent logic diagram of a port in the 8156/8155 (Reprinted by permission of Intel Corporation, Copyright/Intel Corporation 1979)

line is an input. Hence, when changing any line from input to output, the output is always zero initially. Furthermore, when the line is in the output mode, the value in the output latch can be read into the 8085 with an IN instruction. Port A and port B can be in the basic input/output mode or can be in the strobed (handshaking) mode in conjunction with port C lines. These various techniques are illustrated in the following examples.

Programming Techniques with the 8156/8155

In this section we present a few programming examples using the 8156/8155 chip.

I/O Programming Without Handshaking The 8156/8155 can be used for basic programming when **ALT 1** and **ALT 2** modes are used with port C. This is the same type of input/output programming that was used in the input/output programming of the 8355/8755. The command register can be used to set the mode, as shown in the next example.

Example 10-5

Simulate a binary-up counter that counts up on port A on the 8156/8155 when a $0 \rightarrow 1$ transition occurs on PC_0.

This problem requires making port C an input and port A an output. Since port B is unused, we will make it an input. The recognition of the $0 \to 1$ transition on PC_0 requires the storage of the previous value that was read in so that we can compare it with the present value. We will use register B to store the previous value and register C to provide a temporary working copy of the present PC_0 value. The past value is stored in its complement form. The counter should increment when

$$\overline{PC}_{0(old)} \cdot PC_{0(new)} = 1$$

The program is as follows:

Label	OPCODE	Operand	Comments
	ORG	0H	
	MVI	A,00000001B	; Configure ports
			; Port A as output
	OUT	18H	; Port B as input
			; Port C ALT 1 —
			;No timer or interrupts
	MVI	B,00H	;Initialize $\overline{PC}_0$(old) = 0
HERE:	IN	1BH	;Read PC_0
	MOV	C,A	;Temporary copy of PC_0
	ANA	B	;AND with PC_0(old)
	ANI	00000001B	;Check A0 only — flags set
	MOV	A,C	;Complement old value
	CMA		;and
	MOV	B,A	;Store in reg. B
	JZ	HERE	;Jump if not $0 \to 1$ transition
	IN	19H	;Read in counter state
	INR	A	;Increment counter
	OUT	19H	;Output incremented value
	JMP	HERE	
	END		

(handwritten annotations: "command register Fig 10-18" next to command register lines; "port select p.561" next to IN 1BH; "carry = 0" next to ANI line; "port A" next to IN 19H)

The solution of the problem above requires recognizing the $0 \to 1$ transition. The program, however, can be made simpler if handshaking is used.

I/O Programming with Handshaking

The basic idea with input/output programming using handshaking is to effectively communicate between the computer and the external device by means of additional control signals. The handshaking process on inputting data requires that the external device tell the computer that new data are available and the computer responds by indicating that the data were read. The handshaking operation on outputting data is similar. The handshaking operation ensures that data transfer takes place in an orderly manner and ensures that the computer and the external device are synchronized together.

The handshaking mode on the 8156/8155 for the **ALT 3** or **ALT 4** mode uses the pins in port C for communication control. The appropriate interrupt enable bits must be programmed in the command register. To begin with, we will not use the 8085 interrupt structure. The contents of the status register is first sampled and then data are transferred accordingly. The PC_{0-2} lines are used with handshaking with port A while in **ALT 3** or **ALT 4**, and PC_{3-5} lines are used with handshaking with port B while in **ALT 4** only. The meaning and condition for the handshaking lines are as follows:

Pin	Signal	Type	Function
PC_0	A INTR	O	Port A interrupt in ALT 3 or ALT 4 mode with INTE A = 1 in command register, same as bit 0 of status register (INTR A) and may be connected to interrupt on 8085.
PC_1	A BF	O	Port A buffer full in ALT 3 or ALT 4 mode and is the same as bit 1 of status register (A BF). It indicates to the external device when the input or output buffer is full.
PC_2	$\overline{\text{A STB}}$	I	Port A strobe in ALT 3 or ALT 4 mode. When port A is configured as an input port, the external device may pulse this line low only when the buffer full line is low ($PC_1 = 0$) to load new data. When port A is configured as an output port, the external device may pulse this line low only when the buffer full line is high ($PC_1 = 1$) to indicate that the data at the output of port A were read by the external device.
PC_3	B INTR	O	Port B interrupt in ALT 4 mode only with INTE B = 1 in command register. Same as bit 3 in status register (INTR B). It may be connected to an interrupt on 8085.
PC_4	B BF	O	Port B buffer full in ALT 4 mode and is the same as bit 4 of status register (B BF). This signal indicates to the external device when input or output buffer is full.
PC_5	$\overline{\text{B STB}}$	I	Port B strobe in ALT 4 mode and is the same in usage as PC_2 ($\overline{\text{A STB}}$).

The control signals are initialized as follows when port C is programmed to ALT 3 or ALT 4:

Control	Input Mode	Output Mode
BF	Low	Low
INTR	Low	High
STB	Input control	Input control

The next examples show how the command and status registers are used to co-ordinate the data transfer between the 8085 and an external device through the 8156/8155.

Example 10–6 ▰▰▰▰▰▰▰

An external device will present data to the 8156/8155 on port A. After the data have been read by the 8085, the external device will continue to supply new data. To do this, the external device needs only to monitor A **BF** (PC$_1$); when A BF = 0, the external device may change the data on port A and then pulse $\overline{\text{A STB}}$ low. The circuit diagram is shown in Figure 10–22.

The first operation is to configure the 8156/8155 by storing the appropriate contents in the 8156/8155 command register. Assume that the data will be stored at consecutive locations starting at 0800H. We will not check for an upper limit on our memory locations. The program is as follows:

Label	OPCODE	Operand	Comments
	ORG	0H	
	MVI	A,00010100B	;A interrupt enable, ALT 3
	OUT	18H	;Port A and B inputs
			;Implies A BF = 0
			; and A INTR = 0
	LXI	H,1800H	;Initialize data pointer
CHECK:	IN	18H	;Read status register
	RAR		;CY = A INTR, 1 = new data
	JNC	CHECK	;Jump if A INTR = 0
	IN	19H	;Read port A
	MOV	M,A	;Store data
	INX	H	;Increment data pointer
	JMP	CHECK	;Repeat
	END		

(handwritten annotation near Operand heading: "B input / A input")

▰▰▰▰▰▰▰▰▰▰▰▰▰▰▰▰▰

Figure 10–22

Circuit diagram for Example 10–6 showing the connection of control lines for input handshaking between the 8156 and an external device

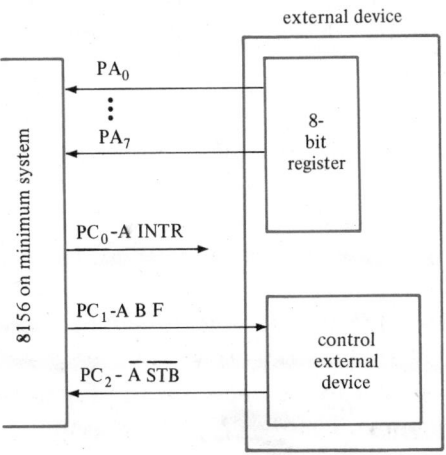

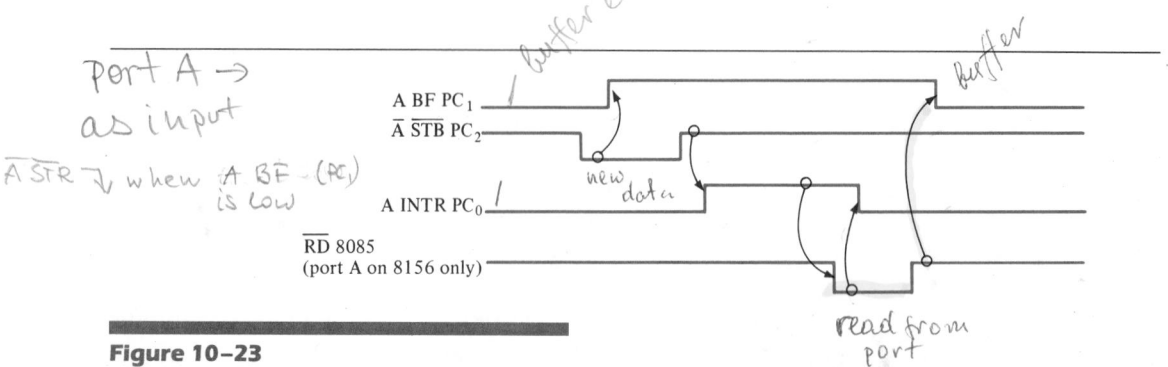

port A →
as input

Ā STR ↓ when A BF - (PC₁)
is Low

buffer empty

new data

buffer

read from port

Figure 10–23

Timing diagram for the input handshaking protocol (Reprinted by permission of Intel Corporation, Copyright/Intel Corporation 1979)

The timing diagram for the input handshaking is given in Figure 10–23. What is shown is rather typical for most types of input routines. The external device may pulse the $\overline{\text{A STB}}$ line low if A BF = 0. The low level on $\overline{\text{A STB}}$ causes A BF = 1, which implies that the data have been loaded into the input latch on the 8156/8155 (see Figure 10–23). When $\overline{\text{A STB}}$ goes high, this causes the A INTR to go high, which is all a function of the hardware in the 8156/8155. The programming loop,

Label	OPCODE	Operand	Comments
	IN	18H	
	RAR		
	JNC	CHECK	

is used to determine when A INTR (which has the same value as INTR PC0) in the status register goes to 1. When A INTR = 1, then IN 19H is executed, which pulses the $\overline{\text{RD}}$ line to be low for a read from port A of the 8156/8155. The $\overline{\text{RD}}$ line being low causes the A INTR bit in the status register to be cleared and thus A INTR (PC0) to go low. The $\overline{\text{RD}}$ line going high causes the A BF line to go low, and the timing cycle is thus complete. Orderly transfer of data can take place when the external device and the microprocessor system operate in the hand-shaking mode.

Example 10-7 illustrates the handshaking when data are output to port B.

Example 10-7 ▬▬▬▬▬▬▬▬

V

The 8156/8155 will be used to simulate an 8-bit binary counter using the $\overline{\text{B STB}}$ signal as the clock input and port B as the counter output. The circuit diagram is similar to the one shown in Figure 10–22 except we will use port B instead for output. The $\overline{\text{B STB}}$ line may be pulsed low only when the B BF line is

1. The contents of the command register must be initialized to make port **B** an output, to enable the ALT 4 mode, and to enable the interrupt.

The program is given here:

Label	OPCODE	Operand	Comments
	ORG	0H	
	MVI	A,00101010B	;B int enabled, ALT 4
	OUT	18H	;Port A input & Port B
			; output imply BF = 0 and
			; INTR B = 1
TEST:	IN	18H	;Read status register
	XRI	00001000B	;Check INTR B
	JZ	TEST	;Jump if INTR B = 0
	IN	1AH	;Read counter = Port B
	INR	A	;Increment counter
	OUT	1AH	;Output, Port B
	JMP	TEST	;Repeat cycle
	END		

(handwritten: PB = out / PA = in)

The output handshaking timing diagram is given in Figure 10–24, which is typical for most types of output routines. The B INTR line is initialized at a high value, which is the same as the third bit in the status register. The program segment

Label	OPCODE	Operand	Comments
	IN	18H	
	XRI	00001000B	
	JZ	TEST	

is used to test the B INTR bit. When B INTR = 1, the program will execute the OUT 0AH instruction, which generates the $\overline{\text{WR}}$ pulse on the timing diagram. The falling transition on this pulse causes B INTR to go low and the rising transition causes B BF to go high, which, for this example, means that the output was incremented. The external control may pulse the $\overline{\text{B STB}}$ low whenever B BF = 1,

Figure 10–24

Timing diagram for the output handshaking protocol (Reprinted by permission of Intel Corporation, Copyright/Intel Corporation 1979)

(handwritten: B port for output / B STB ↓ when B BF (PC4) is 1)

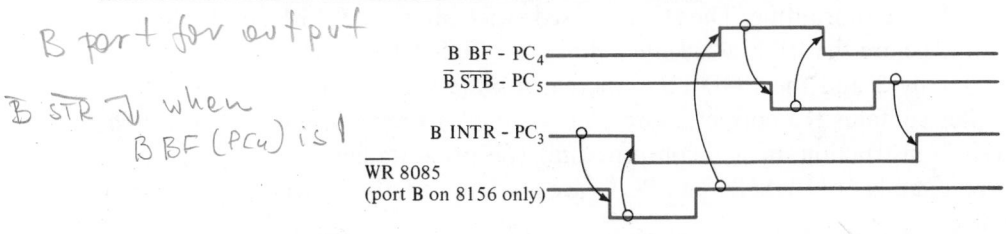

which clears B BF and then sets the B INTR line high. This, in effect, asks for the output to be incremented again, and the cycle is completed. Thus an orderly transfer of data can take place.

Example 10–8 ▬▬▬▬▬▬▬▬▬▬▬▬▬▬▬▬▬▬▬▬▬▬▬

Write a program that generates a number of positive pulses on the PA_0 output of port A on the 8156/8155 that is equal to the content of register D. The pulses should be approximately 500 ms apart and ensure that PA_1 through PA_7 are zero. Use the DELAY subroutine given in the previous chapter, and assume the minimum system configuration.

Label	OPCODE	Operand	Comments
	ORG	0H	
	LXI	SP,1900H	;Initialize SP
	MVI	A,00000001B	;Configure 8156/8155
	OUT	18H	;Port A output
	SUB	A	;Clear A
	OUT	19H	;$PA_0 = 0$
HERE:	INR	A	;
	OUT	19H	;$PA_0 = 1$
	SUB	A	;
	OUT	19H	;$PA_0 = 0$
	DCR	D	;Check for number
	JZ	DONE	; of pulses
	PUSH	D	;Save pulse count
	LXI	D,52083D	;Constant for .5MS
	CALL	DELAY	; delay
	POP	D	;Restore pulse count
	JMP	HERE	
DONE:	.		

(Rest of the program)

This program together with the DELAY subroutine show that the saving and restoring operations of the registers can be done in the subroutine and/or before entering the subroutine. The stack is used twice after the CALL inside the subroutine before the RET is executed. Since the PUSH PSW decrements the stack pointer by 2 and the POP PSW increments the stack pointer by 2, the stack pointer contains the correct address in the stack to retrieve the address for the POP D instruction in the main program. The programmer must make sure the correct number of PUSHes and POPs are used for orderly sequencing of program control.

Interrupt Programming with the 8156/8155

The following examples involve the use of the interrupt structure on the 8156/8155.

Example 10-9

Assume that a one-shot is connected to the RST 7.5 input on the 8085 as shown in Figure 10–25. The one-shot produces a 1-μs positive pulse whenever the switch is depressed. Assume that the switch is debounced. Write a program so that the output on port B of the 8156/8155 will count down by one every time the switch is depressed.

The program is as follows:

Label	OPCODE	Operand	Comments
	ORG	0H	
	LXI	SP,1900H	;Initialize the stack
			;pointer
	MVI	A,00000010B	;Configure port B
	OUT	18H	;as output on 8156/8155
	MVI	A,00011011B	;Clears int flip-flop
	SIM		;and enables RST 7.5
	EI		;Enable interrupts
	HLT		;Wait for interrupt
	ORG	003CH	;Interrupt routine
	IN	1AH	;Read port B
	DCR	A	;Decrement and
	OUT	1AH	;output the count
	EI		;Enable interrupts
	RET		;Return to main program
	END		

Figure 10–25

Circuit diagram for Example 10–9, showing an interrupt-driven counter

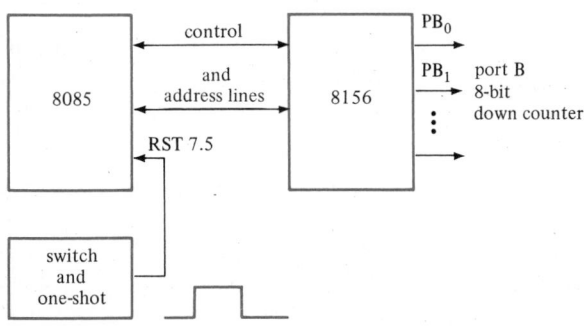

The HLT instruction can be used on the 8085 while waiting for interrupts. Most systems will perform other calculations and control functions so as not to waste the computation and control capabilities while waiting for the interrupts. With most of the examples discussed in this chapter, we are more interested in showing the mechanics of the hardware and the software in the interrupt mechanism than in what other computations the computer can perform at the same time.

The interrupt service routine was positioned by the ORG statement to occur at the address to which the RST 7.5 would jump when an interrupt occurs. The EI instruction is used to set the interrupt enable flip-flop because the interrupt had already reset the flip-flop. The EI instruction will be enabled immediately following the instruction that allows the 8085 to return to the HLT instruction before the next interrupt will be recognized. The EI instruction is normally put just before the RET instruction in any interrupt-driven subroutine.

Example 10–10

The 8156/8155 will be used to simulate an 8-bit binary counter using the B $\overline{STB}$ signal as the clock input and port B as the counter output. The circuit diagram will be almost the same as in Figure 10–25, which was used with the previous example.

The only difference for this example is to connect the B INTR (PC_3) directly to the RST 6.5 input. Instead of checking the B INTR bit in the status register of the 8156/8155 to determine its status, the program will wait for an interrupt because the signal on the B INTR pin is also connected to the RST 6.5 pin. The program will now be written in an interrupt-driven manner.

Label	OPCODE	Operand	Comments
	ORG	0H	
	MVI	A,00101010B	;B int enable, ALT 4
	OUT	18H	;Port A input,
			;Port B output
			;Implies B BF = 0
			;and B INTR = 1
	MVI	A,00001101B	;Enable RST 6.5
	SIM		;interrupt
	EI		;Enable interrupts
	HLT		;Wait
			;
	ORG	0034H	;RST 6.5 interrupt address
	IN	1AH	;Increment output
	INR	A	;when interrupt
	OUT	1AH	;occurs
	EI		;Enable interrupts
	RET		;Return
	END		

The timing diagram of the output handshaking is the same as shown in Figure 10–24. The $\overline{\text{WR}}$ pulse is generated in response to the interrupt routine. The interrupt routine is used in contrast to the continuous loop, which tests the bit in the status register of the 8156/8155 as was done before.

The 8156/8155 Counter/Timer

The timer on the 8156/8155 is a 14-bit down counter that counts the TIMER IN pulses and provides either a single or recurring square wave or pulse on the $\overline{\text{TIMER OUT}}$ pin. The addresses for the timer section of the 8156/8155 for our minimum system were 1CH and 1DH for the low-order and high-order bytes, respectively. To program the timer, first the COUNT REGISTER is loaded one byte at a time by selecting the two addresses. Bits 0–13 of the count register will specify the length of the next count, and bits 14–15 will specify the timer output mode as given in Figure 10–26(a). The low-order 8 bits of the count length are addressed when AD_2, AD_1, AD_0 = 100. The mode and the six most significant bits are addressed when AD_2, AD_1, AD_0 = 101. The $\overline{\text{TIMER OUT}}$ waveforms are given in Figure 10–26(b). The two mode bits M_2 and M_1 determine the type of waveform that is generated, which can be a single square pulse, a continuous square pulse, a single pulse on terminal count, or continuous pulses.

The value loaded into the count length register can have any value from 0002H through 3FFFH. The square wave will have the first half-cycle of the square wave output high one count longer than the second half-cycle when the count length is odd.

Bits 6 and 7 in the command register are used to start and stop the counter. There are four commands to choose from:

Bit 7 TM$_2$	Bit 6 TM$_1$	Operation of the Counter
0	0	NOP This selection does not affect counter operation.
0	1	STOP This selection does not affect the timer if it has not been started and will stop the timer from counting if it is running.
1	0	STOP AFTER TC This selection will stop the timer immediately after the present terminal count is reached. The same selection will not affect the timer if it has not been started.
1	1	START This selection will load the mode, load the count length, and start the timer immediately after loading if timer is not presently running. If the timer is running, start the new mode and count length immediately after the present terminal count is reached.

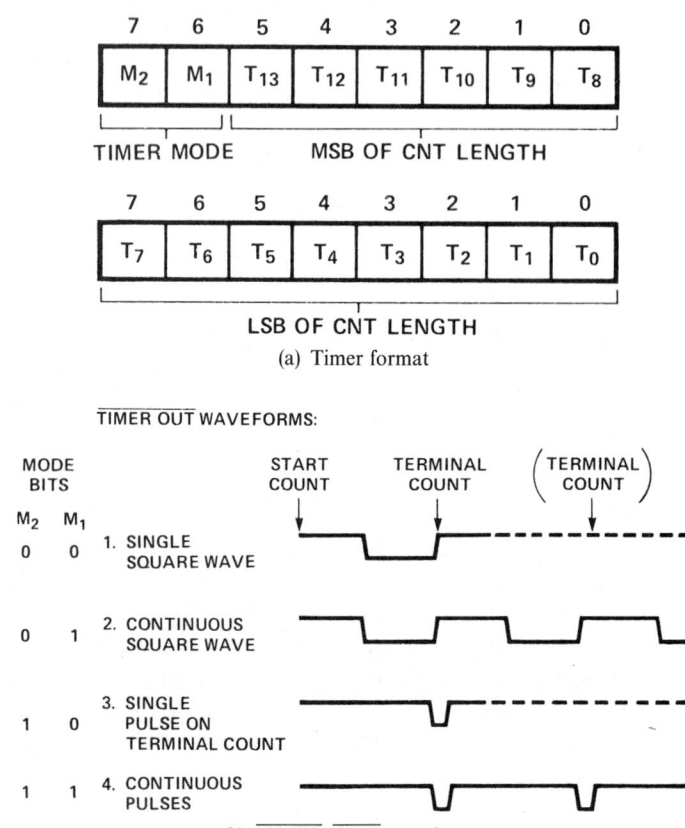

(a) Timer format

(b) $\overline{\text{TIMER}}$ $\overline{\text{OUT}}$ waveforms

Figure 10–26

The format and modes of the timer in the 8156/8155 chip (Reprinted by permission of Intel Corporation, Copyright/Intel Corporation 1979)

Note that while the counter is counting, a new count and a new mode can be loaded into the registers. Before the new count and mode will be used by the counter, a new START command must be issued. This applies even though you may only want to change the count and use the previous mode. The next example shows how the counter/timer can be used.

Example 10–11

Write a program for the minimum system that will generate a continuous square wave on the $\overline{\text{TIMER}}$ $\overline{\text{OUT}}$ of 1 kHz.

The register must be loaded with the correct mode and length. The TIMER IN will be connected to the CLK on the 8085, which is a 2.5-MHz signal. This signal must be divided by 2,500 to generate a 1-kHz signal. Note that $2,500D = 9C4H$.

Label	OPCODE	Operand	Comments
	ORG	0H	
	MVI	A,0C4H	;Store low order byte
	OUT	1CH	; in register
	MVI	A,49H	;Mode = 01 and count is
	OUT	1DH	; MSB = 09H
	MVI	A,1100000B	;Start counter
	OUT	18H	; and generate square wave
	HLT		
	END		

This program will affect the command register, and care should be taken not to change the port definitions in normal cases.

10.6

Microprocessor Systems that Use the 8085

In this section we want to put the different components together to make a working system. There are many variations in system requirements, and a circuit designer must gather the proper components to form a system that fits a particular application.

An 8085 Minimum System

In the above sections we examined a basic 8085 system and its corresponding components. In order for this basic system to work, a few additional components must be added. A complete and operable system is shown in Figure 10–27. This system is often referred to as the minimum system, because it contains the minimum number of chips.

The minimum system shown in Figure 10–27 differs from the basic system shown previously in that the latter has a few additional connections. A manual reset is added with the proper resistor and capacitor values chosen to satisfy the reset timing. A manual reset is useful to reset the system without the need to remove and reconnect the power. A choice of normal I/O or memory-mapped I/O is provided by means of a set of jumpers. The IO/$\overline{\text{M}}$ line on the supporting chips is connected to either the IO/$\overline{\text{M}}$ of the 8085 for normal I/O or the address line A_{15} for memory-mapped I/O. Another jumper on the READY line of the 8085 allows the user to choose the option of having one T_{wait} state. Note the

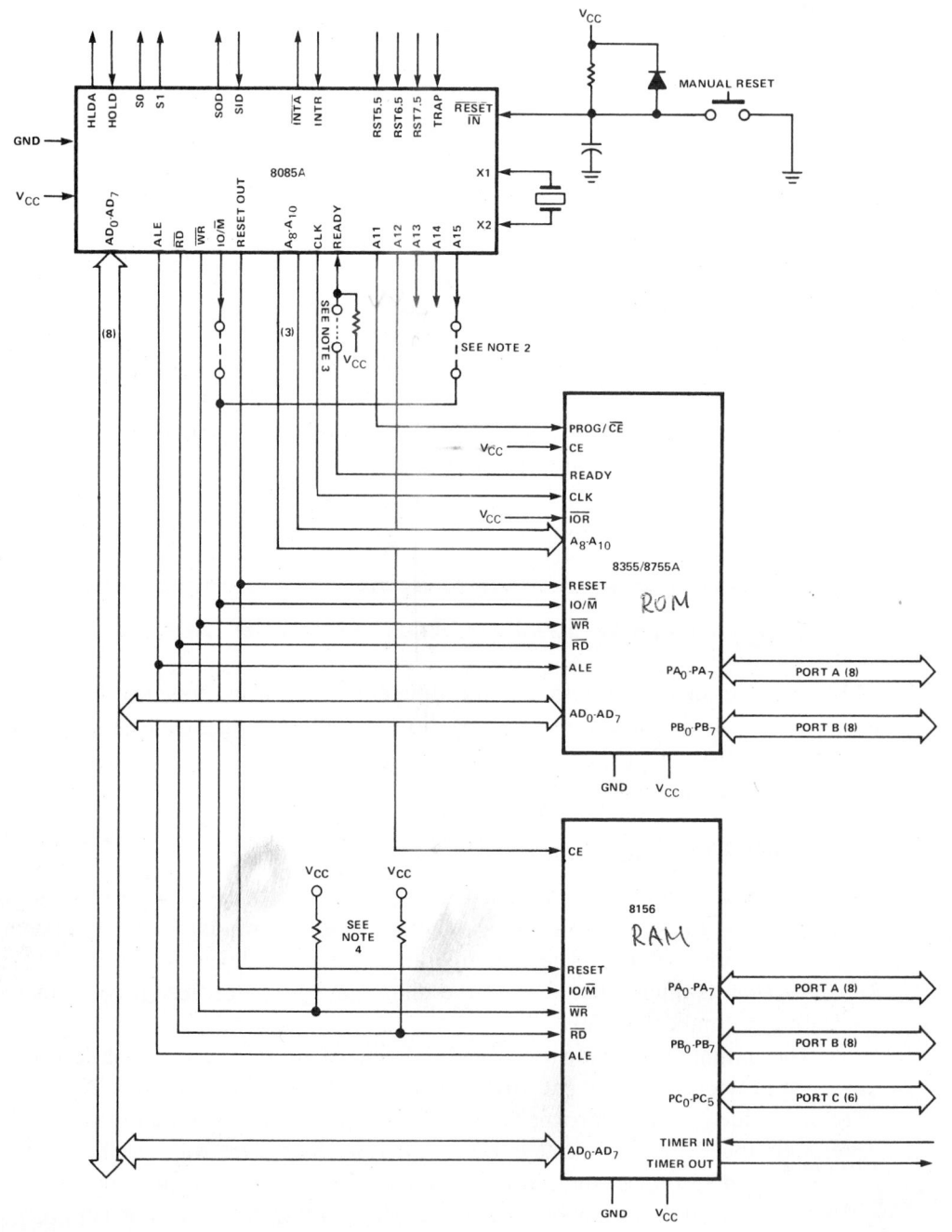

Figure 10–27

A complete 8085 minimum system (Reprinted by permission of Intel
Corporation, Copyright/Intel Corporation 1979)

use of pull-up resistors on the control lines $\overline{RD}$ and $\overline{WR}$. Whenever a control line is connected to more than one load, it is advisable to provide external pull-up capability.

An 8085 Expanded System

The minimum system in Figure 10–28 does not have unique addresses for each peripheral device. Namely, the 8355/8755 ROM can be addressed whenever the address line $A_{11} = 0$ and address lines A_{12} through A_{15} are ignored. For instance, a load from any of the following addresses:

$$0000H$$
$$2000H$$
$$4000H$$
$$\vdots$$
$$E000H$$

will load from the first byte of ROM memory. Likewise, the RAM portion of the 8156/8155 will be addressed for any memory instruction when address line $A_{12} = 1$ and address lines A_8 to A_{10} and A_{13} to A_{15} are don't cares. We will also make $A_{11} = 1$ so as not to select the 8355/8755. Thus the first byte in RAM of the 8156/8155 can be addressed at the following locations:

1800H, 1900H, 1A00H, 1B00H, 1C00H, 1D00H, 1E00H, 1F00H
3800H, 3900H, 3A00H, 3B00H, 3C00H, 3D00H, 3E00H, 3F00H
$\vdots$ $\vdots$ $\vdots$ $\vdots$ $\vdots$ $\vdots$ $\vdots$ $\vdots$
D800H, D900H, DA00H, DB00H, DC00H, DD00H, DE00H, DF00H
F800H, F900H, FA00H, FB00H, FC00H, FD00H, FE00H, FF00H

This multiple use of addresses does not cause a problem in our minimum system; however, with expansion we will have difficulty. A means of adding additional memory using the CE and $\overline{CE}$ on the 8355/8755 and the CE on the 8156/8155 is shown in Figure 10–28. The 8355/8755 can be addressed when the CE = 1 and $\overline{CE} = 0$. In this way, the second 8355/8755 can be addressed when $A_{12} = 1$ and $A_{13} = 0$. The address range for the memory on each support chip in the expanded system is now given, where any unused address line is assumed to be zero and no two chips are activated simultaneously.

Address Device	Storage Bytes
0000–07FFH 8355/8755 (#1)	2K ROM
1800–1FFFH 8355/8755 (#2)	2K ROM
2800–28FFH 8355/8755 (#3)	2K ROM
4800–48FFH 8156/8155 (#1)	256 RAM
8800–88FFH 8156/8155 (#2)	256 RAM

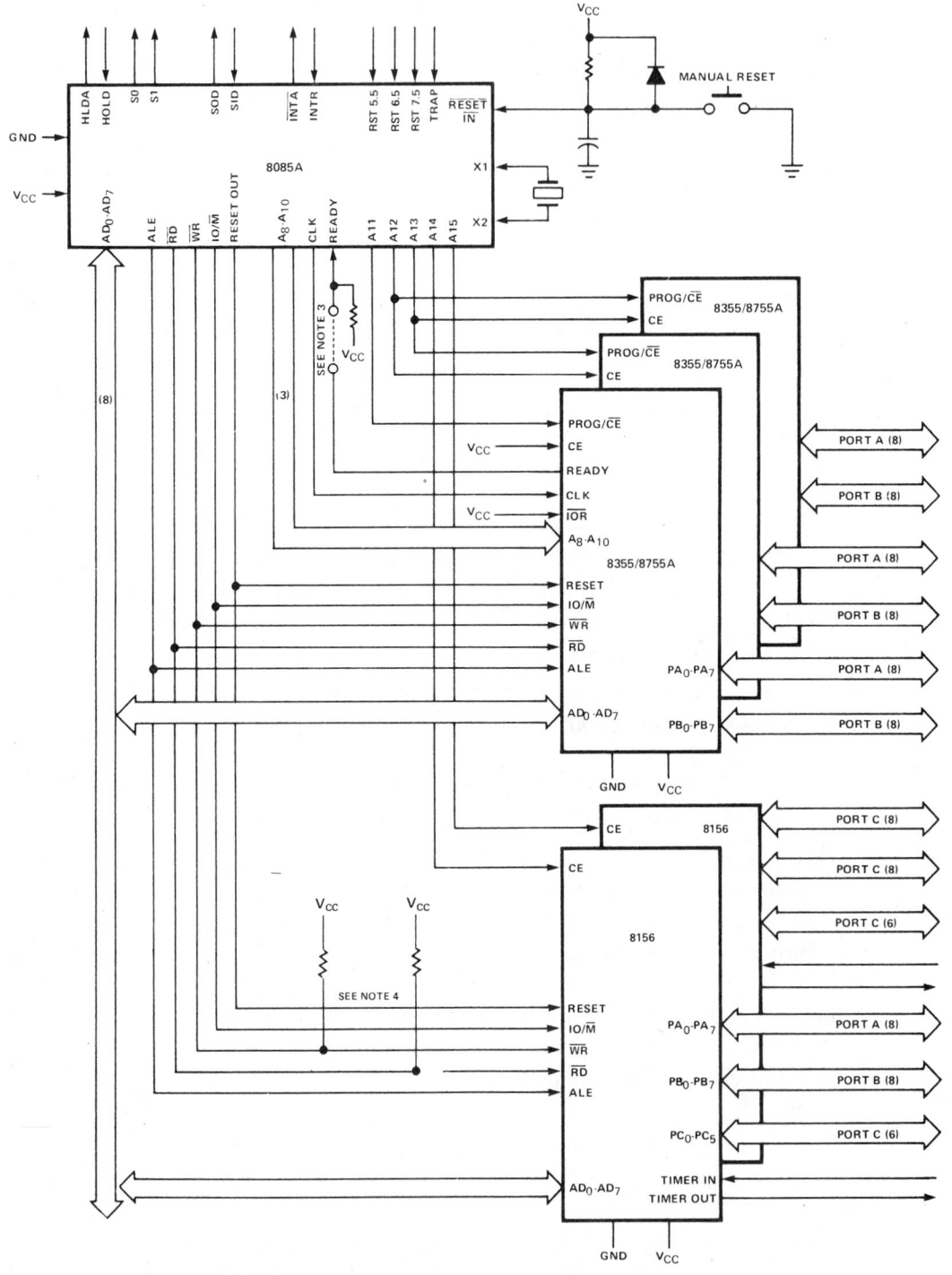

Figure 10–28

An expanded 8085 system with no additional decoding (Reprinted by permission of Intel Corporation, Copyright/Intel Corporation 1979)

This memory expansion is not fully decoded, because two memory devices could be activated simultaneously. A write operation such as

    STA C800H

would store information in both 8156s. Likewise, a read operation such as

    LDA C800H

would try to read from both 8156 chips. Uncertain system behavior will result, since both devices are trying to drive the data bus at the same time. In small systems with small amounts of memory that are not fully decoded, the programmer must make sure that two devices are not selected simultaneously.

The designer must use additional decoding chips in systems where more memory is required. In Figure 10–29, the two 74LS154s, which are 4-to-16 decoders, are used to separate the memory into 2K blocks. The $\overline{CE}$ outputs can be connected directly to chip enable inputs that are active low and that represent

Figure 10–29

Full memory decoding into 2K blocks using two 74LS154 decoders

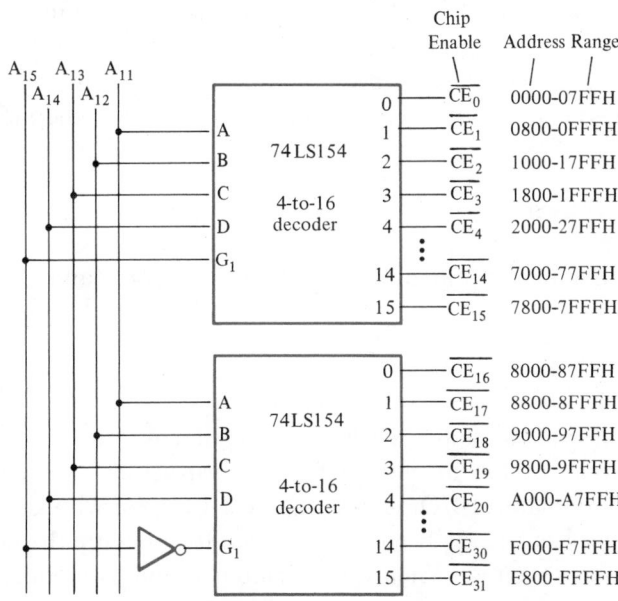

2K memory blocks. The 8355/8755 represents 2K of ROM and can thus be connected directly to these chip enables. The 8156 (or 8155 with active low chip enable) can be connected with an inverter to the $\overline{CE}$ but will have only 256 bytes of RAM. This will waste 1,792 (2K − 256) bytes of address space. Unfortunately, the 8156/8155 uses the address lines A_8, A_9, and A_{10} for selection of internal registers, so no further division of the address space is possible. It should be pointed out that very few microprocessor control applications need the full 64K address space. Usually, 2K to 4K of memory is sufficient for most control applications, and full decoding is not usually necessary.

Memory-Mapped I/O

The 8085 microprocessor normally uses a separate set of instructions and addresses when communicating with peripheral devices. The IN and OUT instructions are distinguished to the devices when the 8085 drives the control line $IO/\overline{M} = 1$. A number of microprocessors do not use any special I/O instructions. Any peripheral I/O-type chips are connected to the processor as if they were memory devices. The communication with the I/O devices is then accomplished using standard memory reference instructions. This is called *memory-mapped I/O*.

The 8085 system can also be hardwired into the normal memory address range by connecting the $IO/\overline{M}$ line on the devices to a memory address line from the 8085 instead of the $IO/\overline{M}$ on the 8085. This can be accomplished in our minimum system in Figure 10–27 by connecting the $IO/\overline{M}$ on the 8156/8155 to the address line A_{15} on the 8085. Then, using the I/O portion of the 8156/8155, the address line A_{15} is set to 1 and, using the memory, the address line A_{15} is set to 0. The address line A_{12} must be 1 in our system to select the 8156/8155. An instruction such as

```
STA 9800
```

would move the contents of register A to the command register, and an instruction such as

```
LDA 9800
```

would load the status register into register A.

Some of the memory reference instructions are helpful when used with memory-mapped I/O versus standard I/O. We will assume that the command register has already been programmed to configure the ports as input or output ports as required. When the minimum system is modified for memory-mapped I/O, the registers and ports in the 8156/8155 can be addressed with the following addresses:

Address	Selection
9800H	Command/status
9801H	Port A
9802H	Port B
9803H	Port C
9804H	Lower byte, counter
9805H	Higher byte, counter

In the following examples, we will assume the same port addresses used earlier for the standard I/O in our minimum system.

Example 10–12

Show how to increment port A using standard and memory-mapped I/O.

The two sequences below will increment port A (assuming that port A is configured as an output).

Standard I/O		Memory-Mapped I/O	
IN	19H	LXI	H,9801 H
INR	A	INR	M
OUT	19H		

Example 10–13

Show how to do arithmetic using standard and memory-mapped I/O.

The two sequences below will subtract the input data at port B from register A. Assume that data are already loaded into register A.

Standard I/O		Memory-Mapped I/O	
MOV	B,A	LXI	H,9802H
IN 1	AH	SUB	M
MOV	C,A		
MOV	A,B		
SUB	C		

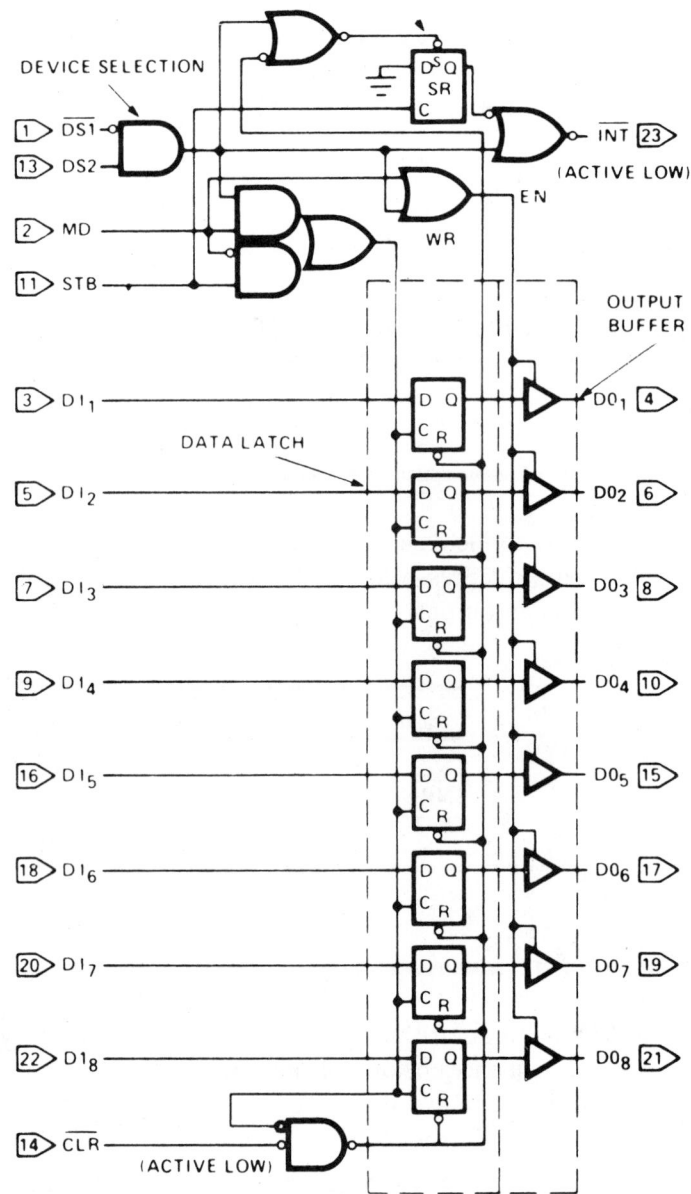

Figure 10–30

The equivalent block diagram of the 8212 chip (Reprinted by permission
of Intel Corporation, Copyright/Intel Corporation 1979)

Example 10–14 ▪▪▪▪▪▪▪▪▪▪▪▪▪▪▪▪▪▪▪▪▪▪▪▪▪▪▪▪▪

Show how to output data using standard and memory-mapped I/O.

The two sequences below will move information from register B to output port A. Assume that the data in register A must not be changed.

Standard I/O		Memory-Mapped I/O	
PUSH	PSW	LXI	H,9801H
MOV	A,B	MOV	M,B
OUT	19H		
POP	PSW		

One of the main problems with standard I/O is that all of the information must pass through register A and will hence destroy whatever information is already there. Other examples can easily be generated that will read the data from the ports directly and perform logical AND, OR, or Exclusive OR operations on the data directly. Thus many designers prefer memory-mapped I/O. It should also be noted that the memory address space for conventional memory is reduced when using this technique.

10.7 ▬▬▬▬▬▬▬▬▬▬▬▬▬▬▬▬▬▬▬▬▬▬▬▬▬▬▬▬

The 8212 I/O Port

The Intel 8212 input/output port is an 8-bit register and tri-statable output buffer with an asynchronous clear. It is also equipped with an interrupt request pin for handshaking. It is useful in many applications.

Hardware Characteristics of the 8212

As a latch, the 8212 can be synchronized to read the input data when available. As a register, it can retain the input values. As an output buffer, it can be configured to drive the output lines on demand. The equivalent block diagram is shown in Figure 10-30. The pin definitions and the signal names for the 8212 are given on the following page.

Signal	Name	Type	Function
1. *Data input lines*			
$DI_1–DI_8$	Data input	I	These eight data input lines are latching D flip-flops. The outputs Q of the flip-flops follow the inputs when the clock input (C) is high. Hence latching of input data occurs on the negative edge of C.
2. *Data output lines*			
$DO_1–DO_8$	Data output	O	These eight output lines are the outputs of the latch and are enabled together by the internal enable (EN) line. When the output is disabled, they remain in a high impedance state.
3. *Control lines*			
$\overline{DS1}$, DS2	Device select lines	I	These two device select lines function together for selecting the 8212. When the 8212 is deselected, no data from the input lines are read or latched. Note that $\overline{DS1}$ is active low and DS2 is active high. When the device is selected, the service request flip-flop is automatically set asynchronously.
MD	Mode control	I	The mode control works in conjunction with the $\overline{DS1}$, DS2, and STB lines to control the output functions. When MD is high, the source of data output is either the data latch or the input lines. When MD is low, the output is determined depending on the other control lines.
STB	Strobe	I	The strobe input can be considered as a clock input if MD is low.
$\overline{CLR}$	Clear	I	The clear line asynchronously clears the internal latch and sets the service request flip-flop.
$\overline{INT}$	Interrupt request	O	The interrupt request line can be used to indicate to the processor any activity on the device.

Operation of the 8212

The operation of the 8212 is completely specified by the four control lines $\overline{DS1}$, DS2, MD, and STB. The source of the output data with respect to the input is given in Figure 10–31. Depending on the application, not all control lines need to be used. The general philosophy of the control lines is that $\overline{DS1}$ and DS2 control the output buffer, the STB controls the input like a clock, and the mode

STB	MD	(DS_1, DS_2)	Data out Equals
0	0	0	3-state
1	0	0	3-state
0	1	0	Data latch
1	1	0	Data latch
0	0	1	Data latch
1	0	1	Data in
0	1	1	Data in
1	1	1	Data in

Figure 10–31

The source of output data with respect to the control lines (Reprinted by permission of Intel Corporation, Copyright/Intel Corporation 1979)

determines whether the device is in the input mode (MD = 0) or in the output mode (MD = 1). In the input mode, the output is determined by the device select lines and the source of the clock is the STB. In the output mode, the output is determined by the STB and the source of data by the device select lines.

The interrupting function of $\overline{\text{INT}}$ is given in Figure 10–32. The INT line is asserted (active low) when the device is being selected or when the STB line goes low.

Figure 10–32

The operation of the $\overline{\text{INT}}$ line as a function of the input conditions (Reprinted by permission of Intel Corporation, Copyright/Intel Corporation 1979)

CLR	$(\overline{DS}_1 \cdot DS_2)$	STB	*SR	INT
0	0	0	1	1
0	1	0	1	0
1	1	⌐_	0	0
1	1	0	1	0
1	0	0	1	1
1	1	⌐_	1	0

* internal SR flip-flop

Application of the 8212

The 8212 can be used in a variety of ways. Some of the more common applications include:

Gated buffer
Input port
Output port
Bidirectional port
Status or address latch

Gated Buffer   A *gated buffer* is a transparent latch. In other words, by connecting MD to ground and the STB to high, the device is set in the input mode. This is shown in Figure 10–33. Hence both data input and output are determined by the device select lines. When the device is selected, data presented at the data input lines pass through the latch onto the output lines directly. When the device is not selected, the output buffers are deactivated, causing the device to go into the tri-state condition. This is why it is called a *transparent* latch.

Input Port   The 8212 can be easily configured as an input port. This is shown in Figure 10–34(a). The MD is grounded to set the 8212 in the input mode.

Figure 10–33

Using the 8212 as a gated buffer (Reprinted by permission of Intel Corporation, Copyright/Intel Corporation 1979)

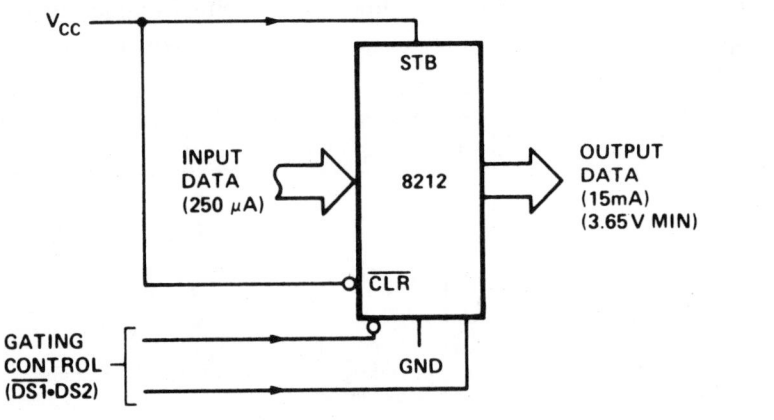

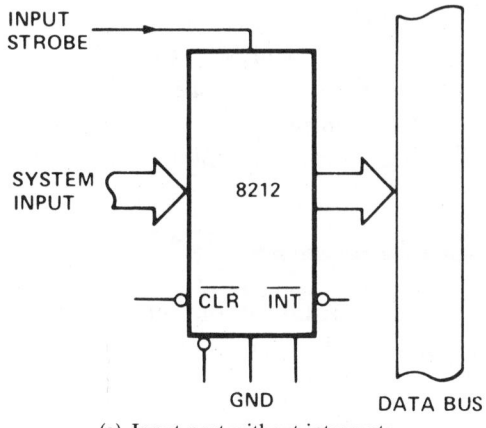

(a) Input port without interrupts

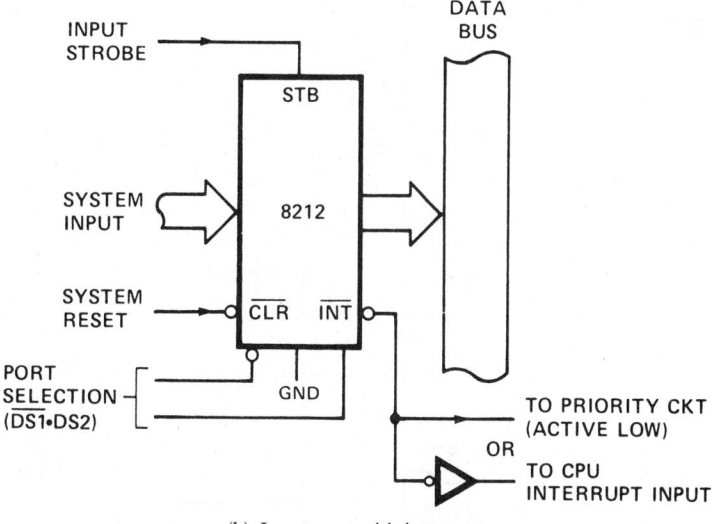

(b) Input port with interrupts

Figure 10–34

Using the 8212 as an input port (a) without and (b) with interrupts
(Reprinted by permission of Intel Corporation, Copyright/Intel
Corporation 1979)

When input data are ready, the STB is used to read the data into the latches. The
output is determined by the device select lines. A special case of an input port
is an interrupting input port as shown in Figure 10–34(b). The $\overline{INT}$ line is used
to interrupt the processor when data have been read and are ready for transfer
into the processor for further processing.

Output Port The 8212 can also be configured as an output port with the MD line tied high to set the device in the output mode. This is shown in Figure 10–35(a). When data are ready for output from the processor, data are read when the device is selected. The state of the output is determined by the STB, which is now acting as an output strobe line. The output buffers are synchronized to an external device and will drive the output lines when strobed.

Figure 10–35
Using the 8212 as an output port (a) without and (b) with handshaking (Reprinted by permission of Intel Corporation, Copyright/Intel Corporation 1979)

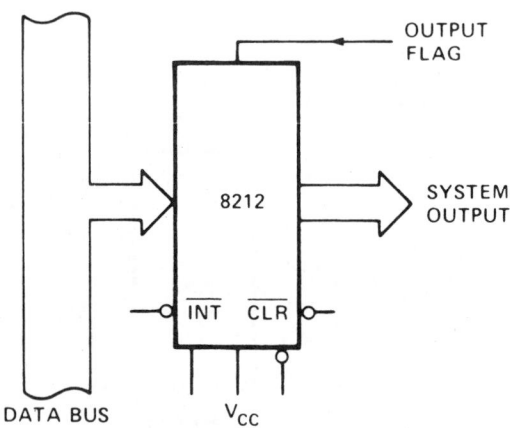

(a) Output port without handshaking

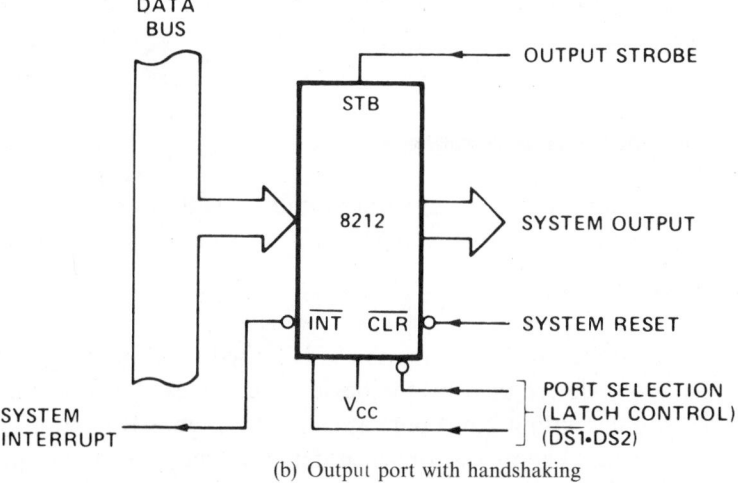

(b) Output port with handshaking

With the added use of the INT and $\overline{\text{CLR}}$ lines, the output can support some simple handshaking as shown in Figure 10–35(b). When the system is ready for output, the data are latched into the 8212 by means of the device select lines. Data can now be read by an external device whenever the output strobe connected to the STB line is pulsed. On the negative edge of the STB line, $\overline{\text{INT}}$ is asserted and is used as an interrupt to the processor signifying that data have been read by an external device and that the output process is completed. The $\overline{\text{CLR}}$ line can be used as a system reset line.

Bidirectional Port When two 8212s are connected together back to back as shown in Figure 10–36, we have a bidirectional port capable of driving data

Figure 10–36

Using the 8212 as a bidirectional port (Reprinted by permission of Intel Corporation, Copyright/Intel Corporation 1979)

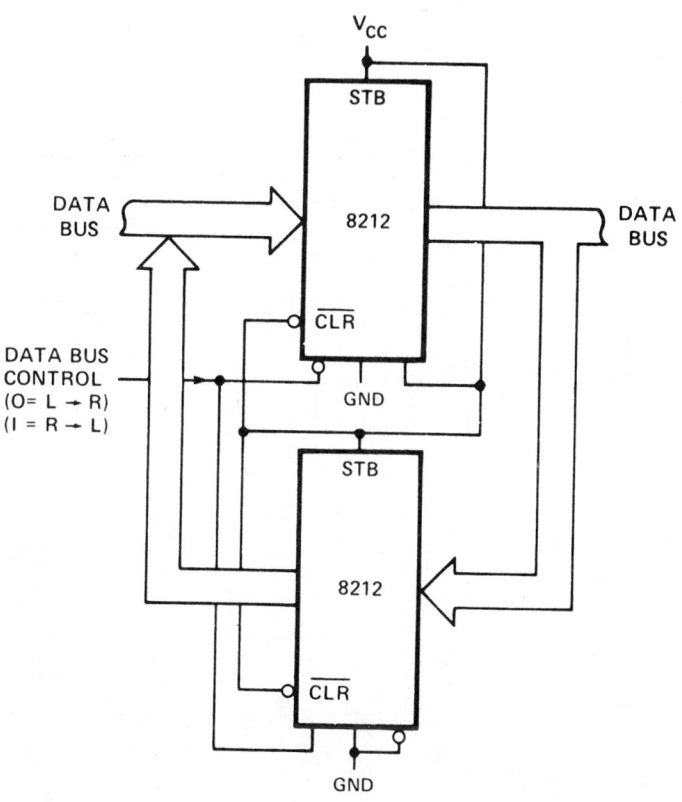

in either direction. The direction control and its complement may be connected to the device select lines as shown in Figure 10–36, or alternatively, the direction control may be connected to the $\overline{DS1}$ line on one and the DS2 on another so that only one of the 8212s is active at any one time.

Status or Address Latch   Whenever a device has multiplexed lines, the 8212 can be used to demultiplex the signals. For example, in the 8080 processor, the status signals are multiplexed. In the 8085 system, the address and data are multiplexed. The demultiplexing circuit is given in Figure 10–37. The address latch enable (ALE) on the 8085 is used to select the 8212, which is configured in the output mode so that the output is always enabled.

Figure 10–37

Using the 8212 as a status or address label (Reprinted by permission of Intel Corporation, Copyright/Intel Corporation 1979)

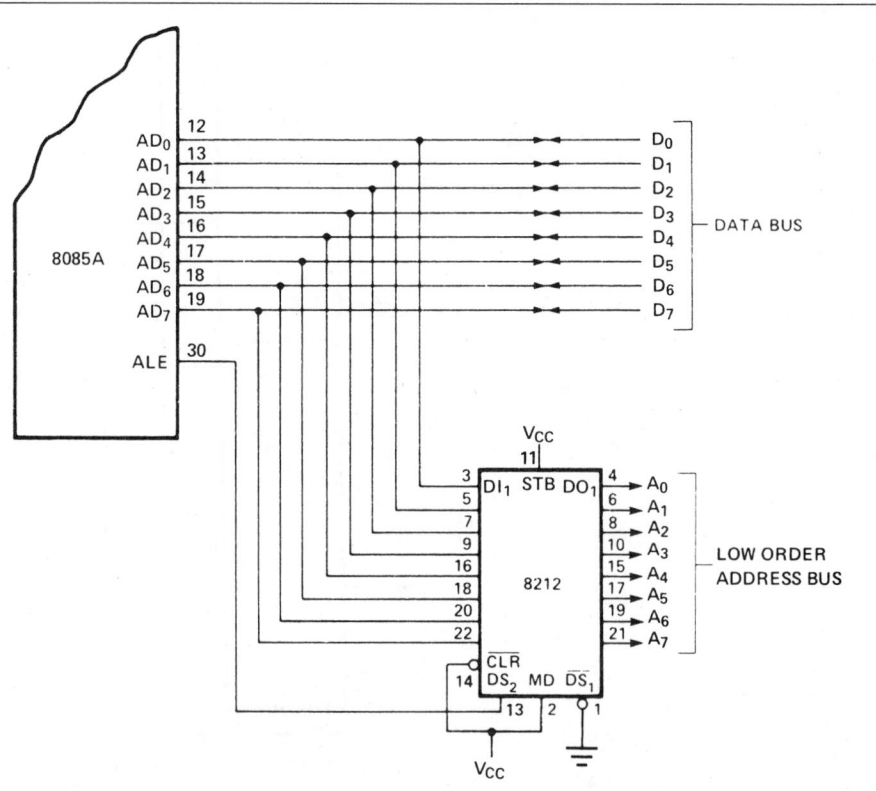

10.8

An 8085 System with Demultiplexed Bus

The 8085 bus structure multiplexes the lower-order 8 bits of address and data on the AD_0–AD_7 lines. The standard 8085 peripheral devices (such as the 8156/8155 and 8355/8755) perform the demultiplex operation that requires the reading and holding of the lower-order 8 address bits, which are placed on the AD_0–AD_7 lines first and then followed by data. Most conventional memories require that all address lines be held constant throughout the read and write cycles. The circuit in Figure 10–38 uses an octal latch to store the low-order address bits. The address appears on the AD_0–AD_7 lines on the falling transition

Figure 10–38

Demultiplexing the address and data bus of the 8085

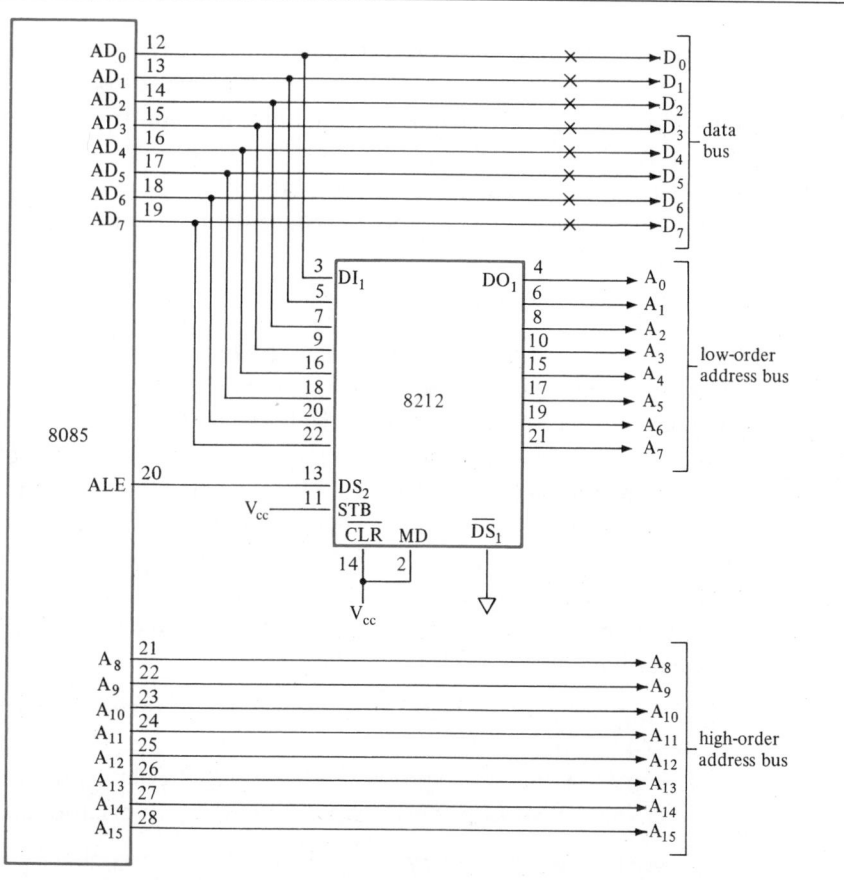

of the ALE and can be used to strobe the address into the 8212. In this configuration, all 16 address lines are available, with the least significant 8 bits on the 8212 and the most significant 8 bits directly from the 8085.

The conventional RAM chips such as the 2114 (1K × 4) and the 6116 (2K × 8) can now be interfaced quite easily to the 8085. Likewise, PROMs such as the 2716, 2732, 2764, and 27128 can be connected directly for ROM storage.

10.9

A Design Example

In this section we present a detailed example showing how a complete microprocessor system is developed. We will use as an example a digital clock.

Example 10–15

Use the minimum system to simulate a clock that can be set.

The clock will use seven-segment multiplexed displays, which implies that only one display is illuminated at a time in a continuous time sequence at a rate above the normal flicker frequency (normally above 30 Hz is sufficient). Then the displays will appear to be on continuously. This is the conventional technique used with clocks and calculators to reduce the pin count from integrated circuits.

Hardware Operation of the Clock

The circuitry in our clock is given in Figure 10–39, which uses six Hewlett-Packard 0.3-in. seven-segment displays (5082–7610). These displays are red and are packaged as a common anode display in a 14-pin package. They can provide a reasonable intensity with an average of 5 mA per segment. An individual seven-segment display will be on for approximately one-sixth of the time. This implies that 30 mA must flow when the individual segment is on for normal intensity. The outputs of the 7447 BCD-to-seven-segment decoder can sink up to 40 mA with an output voltage of 0.3 V typical.

The normal sequence to generate a multiplexed display for displaying the decimal digits of the clock is as follows:

1. Apply the BCD code of the least significant digit to the inputs of the 7447 chip.
2. Turn on the corresponding IRFD9120 hexfet transistor, which is a p-channel VMOS power transistor made by International.

These transistors are very simple to operate as a switch with any MOS circuits for three reasons: (1) their extremely high input impedance on the gate (greater

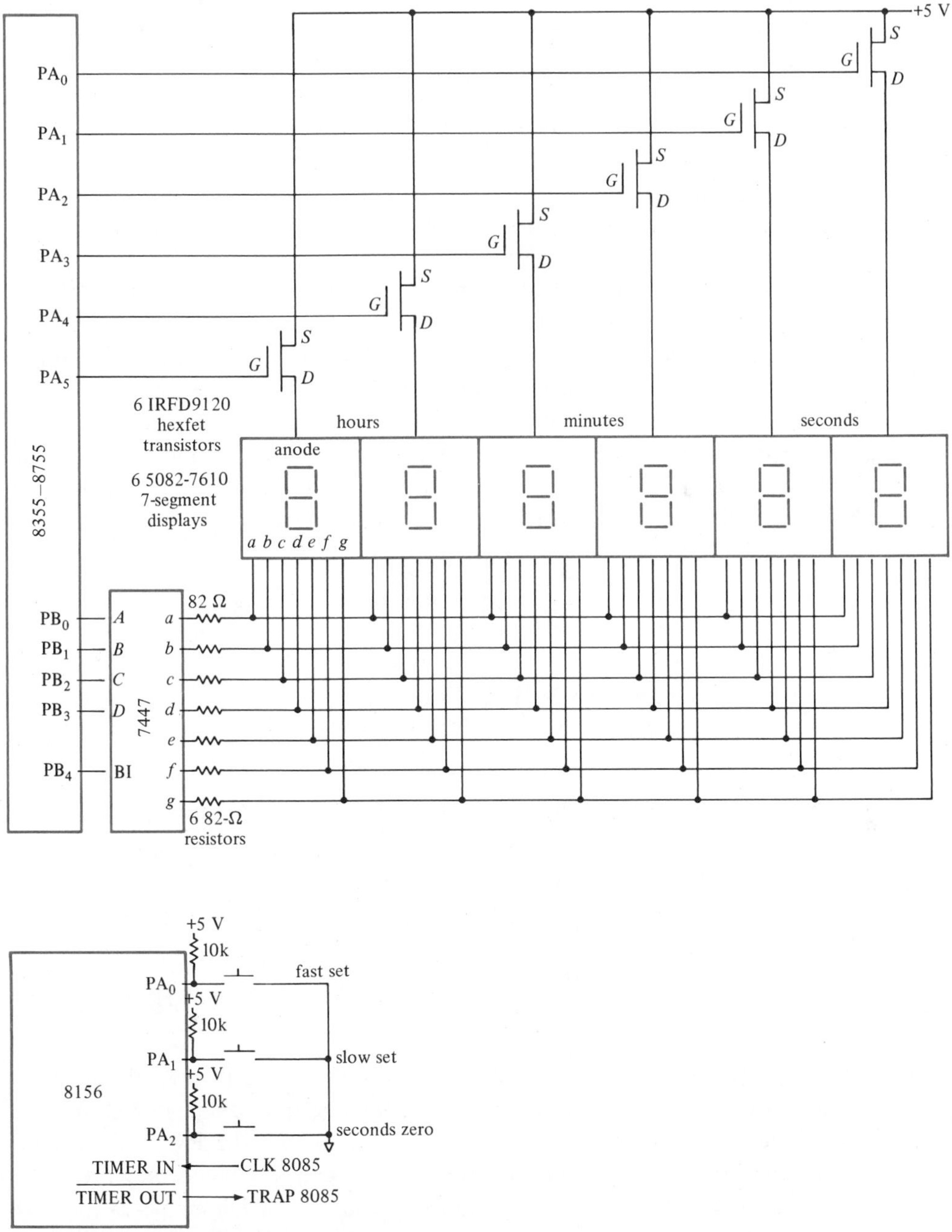

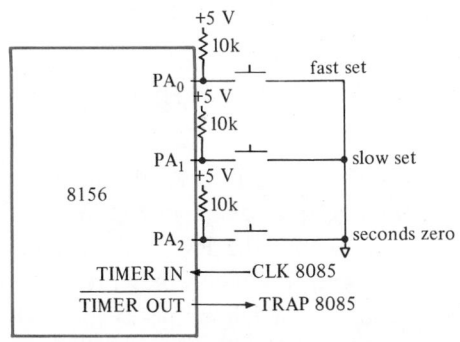

Figure 10–39
Circuit diagram of a clock for the design example

than 40 MΩ), (2) their low on resistance from source to drain (approximately 0.5 Ω), and (3) no external biasing resistors are needed. The VMOS transistors are turned on when the gate is negative with respect to the source, so a low on the PA_0 line turns the VMOS on and thus drives the least significant digit of the segment. A corresponding high on the PA_0 turns the digit off.

The procedure to control the display is to place the units of the seconds on PB_0 to PB_3 and let $PA_0 = 0$ and PA_1 to $PA_5 = 1$ for a time S. Next, place the tens of the seconds on PB_0 to PB_3 and let $PA_1 = 0$ and PA_0, PA_2 to $PA_5 = 1$ for a time S, etc. We must ensure that $6 * S < \frac{1}{60}$ second so that the display will return to the first one in time to be above the flicker frequency. Thus $S < 2.77$ ms.

A common unit for S is 1 ms or 1 kHz for a number of multiplexed display systems. The main reason for multiplexing displays is to reduce the number of pins and to share some of the external electronics. This technique is used on virtually all clock and calculator-type chips. The Intel 8279 is an example of a display and keyboard interface chip that performs in this fashion. The resistance value on the outputs of the 7447 will determine the current, and hence the brightness, flowing in each segment. The relationship is

$$R = \frac{V_{ss} - V_{seg} - V_{SD} - V_{sat}}{n \times I_{seg}}$$

where

V_{ss} = supply voltage (5 V)

V_{seg} = forward voltage drop for diode (~ 2.0 V)

V_{SD} = voltage drop, source to drain (~ 0.1 V)

V_{sat} = saturation output voltage on $a, b, c, \ldots, g$ (0.3 V)

n = number of displays (six displays)

I_{seg} = average segment current (5 mA)

Thus,

$$R = \frac{5 - 2.0 - 0.1 - 0.3}{6 \times 5 \text{ mA}} = 86.87 \ \Omega \qquad \text{(Use 82 }\Omega\text{)}$$

The 8156/8155 on the minimum system is connected to several pushbuttons to produce a fast set, slow set, and a seconds zero as given in Figure 10–39. The switches are wired with a pull-up resistor so that the PA_0, PA_1, and PA_2 inputs are high when the buttons are released and low when they are pressed. The clock hours display should advance at the rate of 1 hour per second when the fast set is depressed. The clock minutes display should advance at the rate of 1 minute per second when the slow set is depressed, with no carry to the hours display. The seconds display should be zeroed whenever the seconds zero button is depressed and remain at zero until released. Furthermore, the minutes and hours displays should also be held at their current values. Otherwise the clock should count in the normal sequence in a 12-hour format.

The TIMER IN and $\overline{\text{TIMER OUT}}$ are connected to the CLK and TRAP lines on the 8085. The interrupt on the TRAP input will be used to provide the basic timing information in the 8085.

Software Programming of the Clock

The software is written with a number of subroutines that are called from the main program. The first subroutine is used to increment the memory locations and to prepare the data for display. Two tables will be used. The first consists of three memory locations, which contain the seconds, minutes, and hours, and will start at the address called SEC. The second table has 12 entries, where the first entry is the code to activate the seconds unit display on port A (00111110) followed by the units of the seconds. Then the code to activate the tens of seconds on the display (00111101) followed by the units of minutes is given, and so on. The first entry starts with SECD.

The RAM memory will be used for the stack and contain storage for the time and the time data that are also modified for display purposes with the mask for turning on the correct display. The reserved RAM with locations specified are given next.

Label	OPCODE	Operand	Comments
	ORG	1800H	
SEC:	DS	1	;Initialize clock RAM
MIN:	DS	1	;Save space for clock
			; values
HOUR:	DS	1	
			;
SECD:	DS	4	;Save space for each
			; clock digit and
			; corresponding mask
MIND:	DS	4	
HOURD:	DS	4	
REF:	DS	2	;Refresh address
MOD:	DS	2	;Divide

The three locations SEC, MIN, and HOUR are used for the current time. The next four locations starting with SECD are: the mask to turn on the units of the seconds; the units of the seconds; the mask for the tens of seconds; and tens of seconds. MIND specifies the same data string for the minutes display. The memory locations SECD, MIND, and HOURD specify the first addresses of the seconds, minutes, and hours portion of the display RAM.

The first subroutine increments the memory location specified by register pair HL. The maximum count is specified in register B. The register pair DE contains the address of the units corresponding to HL.

Label	OPCODE	Operand	Comments

;This subroutine increments the memory location specified
; by register pair HL with the outflow count in register B.
;The data are prepared for displaying at the
; address specified by register pair DE
;(C) = 0 for no carry, and (C) = 1 for carry
;

Label	OPCODE	Operand	Comments
ADVAN:	MVI	C,0H	;Set C for no carry
	MOV	A,M	
	INR	A	;Get data and increment
	DAA		; in BCD
	CMP	B	;Store if no carry
	JNZ	STORE	; to next units
	MVI	C,01H	;Reg C = 1 with carry
	MOV	A,B	;Check if hours display
	CPI	13H	
	MVI	A,00H	;Zero A without flag
			; changes
	JNZ	STORE	;Jump if not hours
	INR	A	;Set to 1 for hours
STORE:	MOV	M,A	;Store result
	PUSH	B	;Save BC
	MOV	C,A	;Copy A
	ANI	0FH	;Set 4 MSB = 0
	STAX	D	;Store units position
	INX	D	;Point to
	INX	D	;Tens position
	MOV	A,C	;Get data
	RAR		;Put 4 MSB in 4 LSB
	RAR		
	RAR		
	RAR		
	ANI	0FH	;Set 4 MSB = 0
	MOV	C,A	;Save A temp
	MOV	A,B	;Check if hours
	CPI	13H	
	MOV	A,C	;Restore units position
	JNZ	SAVE	;
	CPI	0H	;Check tens of hours for
	JNZ	SAVE	;Zero and then blank
			; leading digit
	ORI	00010000B	;Blank for tens of hours
SAVE:	STAX	D	;Store tens
	POP	B	;Restore BC
	RET		

The REF byte in the table stands for the memory location that contains the address of the data for the next seven-segment display to be illuminated. The MOD is a 2-byte memory location that will be counted down every time the 8085 is interrupted. When the count reaches zero, the time will be incremented by 1 s. The interrupts will occur every millisecond, so the initial count will have to be set to 1000 decimal. The contents of all these RAM locations must be initialized with the subroutine in ROM that follows. This will be executed on power-up.

Label	OPCODE	Operand	Comments
INIAT:	SUB	A	;Initialize seconds and
	LHLD	SEC	; minutes to zero
	MOV	M,A	
	INX	H	
	MOV	M,A	;
;			
	INX	H	;Initialize hours
	MVI	A,12H	; to twelve
	MOV	M,A	
;			
	INX	H	
	MVI	M,11111110B	;Mask for units of second
	INX	H	
	MVI	M,0H	;Zero units of second ; in the display
;			
	INX	H	
	MVI	M,11111101B	;Mask for tens of second
	INX	H	
	MVI	M,0H	;Zero tens of seconds ; display
;			
	INX	H	
	MVI	M,11111011B	;Mask for units of minutes
	INX	H	
	MVI	M,0H	;Zero units of minutes ; display
;			
	INX	H	
	MVI	M,11110111B	;Mask for tens of minutes
	INX	H	
	MVI	M,0H	;Zero tens of minutes ; display
;			
	INX	H	
	MVI	M,11101111B	;Mask for units of hours
	INX	H	
	MVI	M,02H	;Set units of hours to two
;			
	INX	H	
	MVI	M,11011111B	;Mask for tens of hours
	INX	H	
	MVI	M,01H	;Set tens of hours to one

(continued)

Label	OPCODE	Operand	Comments
	LXI	H,SECD	;Load HL with address
	SHLD	REF	; of SECD to store in REF
	LXI	H,1000D	;Initialize MOD to 1000
	SHLD	MOD	
	RET		

Several subroutines are written to take care of the time settings when the buttons Fast Set, Slow Set, and Seconds Zero are pressed. The first subroutine will handle the Fast Set button.

Label	OPCODE	Operand	Comments
FAST:	LXI	H,HOUR	;Load HL with hour address
	LXI	D,HOURD+1	;Load DE with hour units
			; address
	MVI	B,13H	;Hour overflow
	CALL	ADVAN	;Advance hours
	RET		

The next subroutine handles the Slow Set button by incrementing the minutes when the Slow Set button is pressed.

Label	OPCODE	Operand	Comments
SLOW:	LXI	H,MIN	;Load HL with minutes address
	LXI	D,MIND+1	;Load DE with minute units address
	MVI	B,60H	;Minute overflow
	CALL	ADVAN	;Advance minutes
	RET		

The next subroutine handles the Seconds Zero button by zeroing out the seconds display.

Label	OPCODE	Operand	Comments
SECOND:	SUB	A	;Zero register A
	LXI	H,SEC	;Zero seconds
	MOV	M,A	
	LXI	H,SECD+1	;Zero second units
	MOV	M,A	
	INX	H	
	INX	H	
	MOV	M,A	;Zero tens of seconds
	RET		

The setting of the clock is handled differently than the normal keeping of time. The following subroutine will test if any of the clock-setting buttons are depressed and then set the clock accordingly. The Fast Set will be checked first, then the Slow Set, and finally the Seconds Zero. The first button that is depressed will be determined and then the proper action taken. All other depressed buttons will be ignored. The carry will be set if any setting takes place in the subroutine.

Label	OPCODE	Operand	Comments
SET:	IN	19H	;Read buttons
	RAR		;Check fast set
	JC	SETS	
	CALL	FAST	;Set hours if pressed
	STC		
	RET		
;			
SETS:	RAR		;Check slow set
	JC	SETZ	
	CALL	SLOW	;Set minutes if pressed
	STC		
	RET		
;			
SETZ:	RAR		;Check seconds zero
	CMC		
	RNC		;Return on no carry with
			; (CY) = 0
	CALL	SECOND	;Zero seconds if pressed
	STC		
	RET		

The next subroutine will turn on the seven-segment display specified by the address stored at memory location REF. Furthermore, the 2 bytes at REF will be incremented for displaying the next digit. The contents must be restored to the memory location needed to display the units of the seconds after the tens of the hours have been displayed.

Label	OPCODE	Operand	Comments
DISP:	MVI	A,00010000B	;Blank displays while
	OUT	01H	;changing data
	LHLD	REF	;Get address of data
	MOV	A,M	;Get mask data
	OUT	00H	;Turn on FET
	INX	H	
	MOV	A,M	;Get BCD code
	OUT	01H	;Display digit
	INX	H	
	LXI	D,REF	;Test if past tens of hours
	MOV	A,D	; display address
	CMP	H	
	JNZ	HERE	
	MOV	A,E	
	CMP	L	
	JNZ	HERE	
	LXI	H,SECD	;Reload to units of seconds
HERE:	SHLD	REF	;Store address of next display
	RET		

The 8085 will be interrupted once every millisecond. For timekeeping purposes, we will decrement a memory location that has an initial value of 1000D.

When the contents equal zero, we will increment the clock or perform the necessary time-setting functions. For this we use two memory locations, because 1000 decimal is larger than what can be stored in 1 byte. The 2-byte location is a memory location MOD as defined earlier. The subroutine returns with the Zero Flag set when time out occurs. At this point, the location MOD is restored to 1000D.

Label	OPCODE	Operand	Comments
COUNT:	LHLD	MOD	;Load HL with count
	DCX	H	;Decrement count
	MOV	A,H	;Tests when (HL) = 0
	ORA	L	
	JNZ	CONT	;Jump if not zero
	LXI	H,1000D	;Restore count
CONT:	SHLD	MOD	
	RET		

The 8085 is interrupted with the TRAP input every millisecond. This interrupt is used to turn on the next seven-segment display for refresh purposes. At the same time, a counter is decremented to provide other timekeeping functions. The interrupt-driven routine must be placed at location 24H as follows:

Label	OPCODE	Operand	Comments
	ORG	24H	
INTR:	CALL	DISP	;Refresh 7 segment displays
	CALL	COUNT	;Divide by 1000D
	RNZ		;Return if not zero
	CALL	SET	;Call time setting routines
	RC		;Return if any set button is pressed
	LXI	H,SEC	;Increment second
	LXI	D,SECD+1	; count and display
	MVI	B,60H	
	CALL	ADVAN	;When timed out
	MOV	A,C	;Check if carry from
	RAR		;Seconds to minutes
	RNC		;Return if not
	LXI	H,MIN	;Increment minutes
	LXI	D,MIND+1	; count and display
	MVI	B,60H	;
	CALL	ADVAN	;When timed out
	MOV	A,C	;Check if carry from
	RAR		;Minutes to hours
	RNC		;Return if not
	LXI	H,HOUR	;Increment hours
	LXI	D,HOURD+1	; count and display
	MVI	B,13H	
	CALL	ADVAN	;When timed out
	RET		;Return and wait for
			; another interrupt

The initialization routine must be positioned starting at location 0000H, which configures the ports and sets up the counter in the 8156/8155 to interrupt the 8085 every millisecond. The routine follows:

Label	OPCODE	Operand	Comments
	ORG	0H	
	MVI	A,0FFH	;Port A on 8355/8755 outputs
	OUT	02H	
	MVI	A,1FH	;PB_{0-4} outputs
	OUT	03H	;and PB_{5-7} inputs
	LXI	SP,1900H	;Initialize stack
	MVI	A,0C4H	;Load count register
	OUT	1CH	; for square wave and
	MVI	49H	;Divide by 2500D = 9C4H
	OUT	1DH	
	MVI	A,1100000B	;Start counter and
	OUT	18H	; Port A inputs on the 8156/8155
	CALL	INIAT	;Initialize the RAM
	HLT		;Wait for interrupts

All of the subroutines may be placed one after the other following the interrupt routine. The order is not important. The section of the program (INTR) containing the RAM information must occur at location 1800H, the first location of the RAM. Place this after all subroutines. The END statement must follow the last part of the program.

References

1. Baer, Jean-Loup. *Computer Systems Architecture*. Computer Science Press, Potomac, Md., 1980.
2. Becher, William D. *Logical Design Using Integrated Circuits*. Hayden Book Company, Rochelle Park, N.J., 1977.
3. Cannon, Don L. *Fundamentals of Microcomputer Design: System Hardware and Software*. Texas Instruments, Dallas, Tex., 1982.
4. D'Angelo, Henry. *Microcomputer Structures*. Byte Publications, Peterborough, N. H., 1981.
5. Furht, Borivoje, and Himanshu Parikh. *Microprocessor Interfacing and Communication Using the Intel SDK-85*. Prentice-Hall, Englewood Cliffs, N.J., 1986.
6. Garland, Harry. *Introduction to Microprocessor System Design*. McGraw-Hill, New York, 1979.
7. Getgen, Lawrence E. *Designing with Microprocessors*. Science Research Associates, Chicago, 1985.
8. Givone, Donald D., and Robert P. Roesser. *Microprocessors/Microcomputers: An Introduction*. McGraw-Hill, New York, 1980.

9. Hamacher, V. Carl, Z. G. Vranesic, and S. G. Zaky. *Computer Organization*, 2d ed. McGraw-Hill, New York, 1984.
10. Hayes, John P. *Computer Architecture and Organization*. McGraw-Hill, New York, 1978.
11. Hayes, John P. *Digital System Design and Microprocessors*. McGraw-Hill, New York, 1984.
12. Intel Corp. *Intel MCS-85 User's Manual*. Intel Corporation, Santa Clara, Calif., 1978.
13. Intel Corp. *Intel 8080/8085 Assembly Language Programming*. Intel Corporation, Santa Clara, Calif., 1979.
14. Intel Corp. *Memory Components Handbook*. Intel Corporation, Santa Clara, Calif., 1984.
15. Jermann, William H. *The Structure and Programming of Microcomputers*. Alfred Publishing, Sherman Oaks, Calif., 1982.
16. Kline, R. M. *Digital Computer Design*. Prentice-Hall, Englewood Cliffs, N.J., 1977.
17. Leventhal, L. *8080/8085 Assembly Language Programming*. Osborne/McGraw-Hill, Berkeley, Calif., 1978.
18. Mano, M. Morris. *Digital Logic and Computer Design*. Prentice-Hall, Englewood Cliffs, N.J., 1979.
19. Mano, M. Morris. *Computer System Architecture*, 2d ed. Prentice-Hall, Englewood Cliffs, N.J., 1982.
20. National Semiconductor Corp. *CMOS Data Book*. National Semiconductor Corporation, Santa Clara, Calif., 1981.
21. National Semiconductor Corp. *Logic Data Book*. National Semiconductor Corporation, Santa Clara, Calif., 1981.
22. National Semiconductor Corp. *PAL Data Book*. National Semiconductor Corporation, Santa Clara, Calif., 1982.
23. O'Connor, Patrick. *Digital and Microprocessor Technology*. Prentice-Hall, Englewood Cliffs, N.J., 1979.
24. Oleksy, Jerome E., and George B. Rutkowski. *Microprocessor and Digital Computer Technology*. Prentice-Hall, Englewood Cliffs, N.J., 1981.
25. Rafiquzzaman, Mohamed. *Microcomputer Theory and Applications with the Intel SDK-85*. John Wiley & Sons, New York, 1982.
26. Rafiquzzaman, Mohamed. *Microprocessors and Microcomputer Development Systems: Designing a Microprocessor-Based System*. Harper & Row, New York, 1984.
27. Rooney, Victor M., and Amin R. Ismail. *Microprocessors and Microcomputers*. Macmillan, New York, 1984.
28. Short, Kenneth L. *Microprocessors and Programmed Logic*, 2d ed. Prentice-Hall, Englewood Cliffs, N.J., 1987.
29. Signetics Corp. *Signetics Logic—TTL Data Manual*. Signetics Corporation, Sunnyvale, Calif., 1978.
30. Sloan, M. E. *Computer Hardware and Organization*, 2d ed. Science Research Associates, Chicago, 1983.
31. Stone, Harold S., ed. *Introduction to Computer Architecture*, 2d ed. Science Research Associates, Chicago, 1980.
32. Texas Instruments. *The TTL Data Book for Design Engineers*, 2d ed. Texas Instruments, Dallas, Tex., 1976.
33. Wilkinson, Barry. *Digital System Design*. Prentice-Hall, Englewood Cliffs, N.J., 1987.
34. Williams, Gerald E. *Digital Technology*. Science Research Associates, Chicago, 1977.
35. Winkel, David, and Franklin Prosser. *The ART of Digital Design*. Prentice-Hall, Englewood Cliffs, N.J., 1980.

Problems

1. Write a program that will make a 16-bit up counter on ports A and B of the 8355 in the minimum system. Assume that PA_0 is the LSB up to PB_7, which is the MSB. Configure the ports and make the output count up as fast as possible. Determine the frequency of the MSB on PB_7.

2. Connect a 74LS10 three-input NAND gate to ports A and B of the 8355/8755 in the minimum system as follows:

Port Pin	IC Pin	
PA_0	1—1A input	gate 1
PA_1	2—1B input	
PA_2	13—1C input	
PA_3	3—2A input	gate 2
PA_4	4—2B input	
PA_5	5—2C input	
PA_6	9—3A input	gate 3
PA_7	10—3B input	
PB_0	11—3C input	
PB_1	12—1Y	output gate 1
PB_2	6—2Y	output gate 2
PB_3	8—3Y	output gate 3

Configure the ports properly and write a program that will completely test the 74LS10, putting out a 1 on PB_4 if bad and a 0 if good.

3. Assume that you are going to model a 74LS163 counter with the minimum system in the following manner. Assume that the following pins on ports A, B, and C of the 8156 are to simulate the 74LS163.

Pin	Function
PA_0	QA
PA_1	QB
PA_2	QC
PA_3	QD
PA_4	RCO (ripple carry output)
PB_0	A
PB_1	B
PB_2	C
PB_3	D
PB_4	Clock
PB_5	Clear
PB_6	Load
PC_0	Enable T
PC_1	Enable P

Write a program such that the designated outputs model the 74LS163 as closely as possible.

4. An analog comparator has two analog inputs normally designated with + and − such that the digital output is 1 if the voltage on the + input is greater than the voltage on the − input. Otherwise the output is 0. The symbol is given in Figure P10–1(a), which is basically a specially designed operational amplifier. Assume that the comparator is connected to the port A of the 8355 in Figure P10–1(b). The LED should be turned on when the input analog voltage exceeds 3 V continuously for more than 3 s and off when the analog voltage is less than 3 V continuously for more than 5 s.

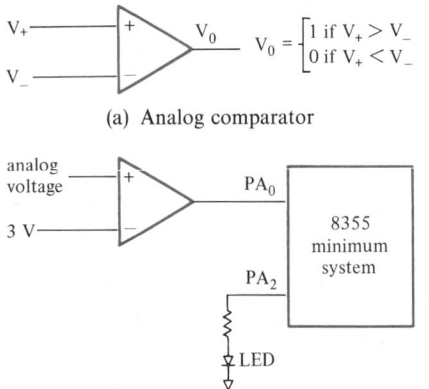

(a) Analog comparator

(b) Analog comparator connected to minimum system

Figure P10–1

Analog comparison system with an LED indicator

5. A digital-to-analog converter (DAC) converts a digital input code into an equivalent analog output voltage. The number of bits varies but can range from 4 to 16 bits, with 8- and 10-bit converters being the most common. The outputs for most integrated circuit DACs range from 0 to 5, 0 to 10, and −5 to 5 V. Current-output DACs are also commercially available. The output voltage is typically proportional to a reference voltage called V_{ref}. Unipolar converters have an output voltage swing that is only positive or negative. The analog output for a 4-bit unipolar converter for two binary codes is shown in Figure P10–2. Note that the maximum output voltage is 1 LSB less than V_{ref} ($10.000 - 0.625 = 9.375$ V). This is true of all types of DACs. The binary code for a different number of bits is a straightforward extension of the 4-bit case. Assume that a unipolar 8-bit DAC with a $V_{ref} = 10.0$ V is connected to port A of the 8156 in the minimum system with the LSB of the DAC connected to PA_0 and the MSB to PA_7. Write a program that configures the port and that generates a stair-step approximation to a periodic ramp that goes from 0.0 V to 9.9609375 V with 1 LSB increments and then repeats on the output of the DAC. Calculate the maximum frequency.

Input Binary Code	Output	
Straight Complementary	Fraction of V_{ref}	V_0 with $V_{ref} = 10$
0000 1111	0/16	0.000
0001 1110	1/16	0.625
0010 1101	2/16	1.250
0011 1100	3/16	1.875
0100 1011	4/16	2.500
0101 1010	5/16	3.125
0110 1001	6/16	3.750
0111 1000	7/16	4.375
1000 0111	8/16	5.000
1001 0110	9/16	5.625
1010 0101	10/16	6.250
1011 0100	11/16	6.875
1100 0011	12/16	7.500
1101 0010	13/16	8.125
1110 0001	14/16	8.750
1110 0000	15/16	9.375

Figure P10–2
Analog-to-digital converter values

6. Repeat Problem 5 but generate a sinusoidal output with a 5.0-V DC component. Use 256 points per cycle to approximate the sinusoid, using the maximum swing possible. Calculate the frequency of the sinusoid.

7. A bipolar DAC output produces both positive and negative voltages on the output. A unipolar DAC together with a summing amplifier can produce a bipolar DAC as shown in Figure P10–3. Assume that the unipolar DAC has a straight binary code. Determine the code for the output V_{DAC}.

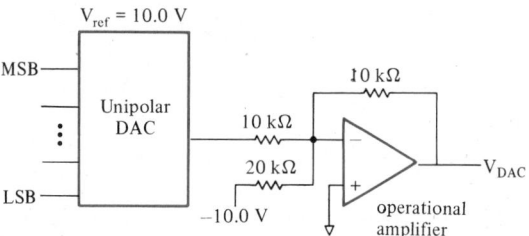

Figure P10–3
A bipolar digital-to-analog converter

8. Repeat Problem 6 by eliminating the DC component using the bipolar DAC of Figure P10–3.

9. An analog-to-digital converter (ADC) converts an analog signal to an equivalent digital value. A number of ADCs use a DAC, a comparator, and other logic to produce the necessary conversions. The circuit in Figure P10–3, with the aid of the minimum system microprocessor, can simulate several types of converters. A counter-type ADC starts with the digital code on the DAC of all zeroes. (Assume that the bipolar DAC is the same as in Figure P10–2.) Note that all zeroes = 00H implies that $V_{DAC} = 0$. The comparator compares V_{DAC} and V_{AS}, which is an analog input signal. Assume that $-5.0 \leq V_{AS} \leq 5$. If $V_{DAC} \leq V_{AS}$, the digital equivalent is 00H. If not, continue to increment the output of port A until $V_{DAC} \leq V_{AS}$. The digital code at this point is equivalent to V_{AS}. Write a program for the 8085 that implements this algorithm. How long does it take for the conversion?

10. The successive approximation ADC can use the same hardware as in Figure P10–4. The successive approximation ADC applies the digital code in a different sequence. The first code that is applied generates $V_{DAC} = 0$ V, which is 80H on the DAC inputs. This is the halfway point between -5 and 5 V. If $V_{AS} \geq 0$, the halfway point between 0 and 5 V is applied, namely 40H. If $V_{AS} < 0$, the halfway point between -5 and 0 V is applied, namely C0H. The code applied will always split the remaining voltage range in half and is continued until all bits are determined. This is equivalent to the binary search algorithm. Note that the MSB is determined first, then the second MSB is determined, etc. Write a program for the 8085 that uses the successive approximation technique to convert the V_{AS} into a digital equivalent. How long does the conversion take?

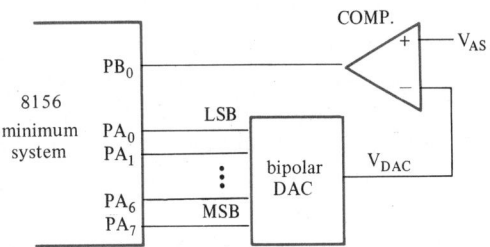

Figure P10–4
Connection of a digital-to-analog converter and a comparator to the 8156

11. Modify the circuit in Figure P10–4 to include another input called start of conversion (SOC) to PB_7 and an output called end of conversion (EOC), which is PC_0. A conversion is started whenever a $0 \rightarrow 1 \rightarrow 0$ pulse occurs on SOC. When SOC goes $0 \rightarrow 1$, make EOC = 0 and then, when the conversion is finished, let EOC = 1. Most converters have these two signals, which control the conversion process and indicate when it is finished. Write a program for the 8085 that performs the successive approximate ADC with the addition of these two signals.

12. The circuit in Figure P10–5 is called an R-2R DAC and is commonly used in IC technology. Assume that $a_0, a_1, \ldots, a_n$ correspond to the digital inputs and that when a_i is 0, the switch is connected to ground and when a_i is 1, the switch is connected to V_{ref}. The switches are typically transistor switches and can operate in the nanosecond range. Show that:

$$V_{DAC} = \frac{-V_{ref}R_f}{3 \cdot 2^n \cdot 2R} (a_n 2^{n-1} + a_{n-1} 2^{n-2} + \cdots + a_1 2 + a_0)$$

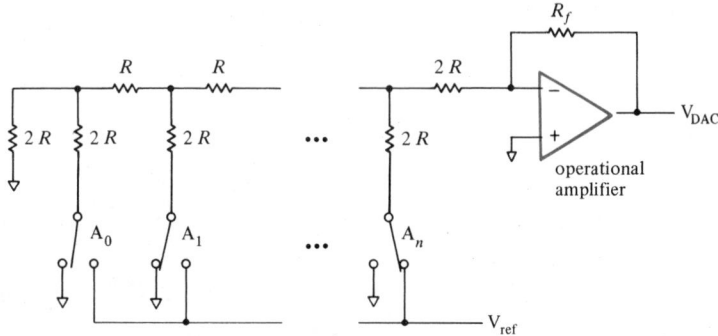

Figure P10–5
R-2R network for a digital-to-analog converter

The 68010 Microprocessor

11

In this chapter we examine another microprocessor, the Motorola 68010 microprocessor. The 68010 contains many advanced features that are not found in earlier designs. We present the programming aspects of the processor in this chapter and leave the hardware details of the 68010 to the next chapter. After a brief introduction to the 68010 processor, we present the addressing modes and briefly discuss its instruction set.

In previous chapters we have examined in some detail a useful microprocessor, the Intel 8085, and its supporting chips. In this chapter we examine a more powerful microprocessor, the Motorola 68010. Differences between the two will be emphasized.

11.1

The 68010 Microprocessor

Since the announcement of the Intel 4004, there have been three major generations in microprocessor development. The 8085 studied so far is often referred to as a first-generation microprocessor, and the 68010 can be considered a second-generation microprocessor.

Internal Organization

Whereas the 8085 processor was first announced and put to use in the middle of the 1970s, the 68010 was announced at the beginning of the 1980s. The 68010 was designed with several specific goals in mind. In particular, the 68010 is designed so that language directives can be easily and directly implemented. To achieve this, a consistent structure for assembly instructions is maintained for almost all instruction types. Instead of channeling everything through the accumulator, a set of data registers is used to hold temporary data operands. Hence a dual-operand structure is used for each instruction. To facilitate system programming, a set of powerful addressing modes is available. For the most part, the addressing mode of each operand is independent.

To facilitate implementation of language constructs, three types of registers are available, the data registers, address registers, and index registers. In multiprogramming environments, the machine can be operated in either the supervisory or the user mode. Certain privileged instructions are not available in the user mode. Separation of the user stack pointer and supervisory stack pointer is also helpful to preserve the integrity of the user and the supervisory program. In particular, reentrant programs can be constructed easily with special instructions for linking and unlinking data pointers. Modular programs can be designed with relocatable codes using the available addressing modes. An elaborate exception processing environment is also provided to handle software traps and hardware interrupts.

In terms of hardware, the 68010 is equipped with vectored interrupts to speed the interrupt handling process. A special trace mode causes the microprocessor to flush all registers at the end of every instruction cycle to facilitate program development and debugging. Some of the major differences between the 8085 and the 68010 can be tabulated as follows:

	8085	68010
External characteristics:		
Chip size	40 pins	64 pins
External data register size	8 bits	16 bits
External address path size	16 bits	24 bits
Number of interrupts	4 lines	1 line
Internal characteristics:		
Internal data register size	8/16 bits	8/16/32 bits
Internal address register size	16 bits	32 bits
Number of data registers	1–7	8
Number of address registers	2–3	9
Other characteristics:		
Generation	First	Second
Multiprocessing capability	Possible	Yes

The 68010 is a popular choice among the 68000 series offered by Motorola. Besides the 68010, there are also the 68000, 68008, and the 68020. Some of the differences among the various types are shown in Figure 11–1. As can be seen from Figure 11–1, there is very little internal difference between the 68000, 68008, and 68010. All functions that are in the 68000 and 68008 are also in the 68010. Externally, the data path in the 68008 is byte-oriented, whereas both the 68000 and the 68010 are word-oriented. For this reason, the 68010 is chosen to represent the three processors.

Figure 11–1

Characteristics of the 68000, 68008, 68010, and 68020 microprocessors

	68000	68008	68010	68020
Number of data registers	8	8	8	8
Number of address registers	10	10	11	11
Width of data registers	32	32	32	32
Width of address registers	32	32	32	32
Width of external data path	16	8	16	32
Width of external address path	24	20	24	32
Special features	—	8-bit bus	Virtual memory	32-bit addressing

Internal Registers   There are many internal registers in the 68010, as shown in Figure 11–2(a), specifically:

Eight 32-bit data registers

Seven 32-bit address registers

Two 32-bit stack pointers

One 32-bit program counter

One 16-bit status register

All the data registers and address registers are 32 bits long. Both the data registers and the address registers can be used as index registers. In addition, there is a 16-bit status register with a system byte and a user byte for system and user information, respectively.

There are only seven address registers, because the eighth one is always taken to be the stack pointer. There are also two stack pointers, though only one is used at any one time. When the processor is running in the user mode, the user stack pointer is used. When the processor is running in the supervisory mode, the supervisor stack pointer is used. This distinction between the two stack pointers provides a clear separation between the user's memory space and the supervisor's memory space, thus preserving data integrity for both the user and the supervisor.

In the status register, there are seven flags and a 3-bit interrupt mask in the status register. These are shown in Figure 11–2(b). The status register is divided into two parts: the supervisory part and the user part. A user may only access the user part of the status register. Only the supervisor has access to both parts of the status register. In the supervisory part, the *trace flag* T is used to set the processor in the trace mode. When the trace mode is active, all internal registers in the processor are stacked into memory after every instruction. The *supervisory flag* S indicates whether the processor is in the supervisory mode or in the user mode. Many privileged instructions can be executed only when the processor is in the supervisory mode. The *interrupt mask* I_0, I_1, I_2 indicates the interrupt level of the processor. Only interrupts with higher priorities are allowed to interrupt the processor. There are five flags in the user part of the status register. The *extend flag* X is used specifically for multiprecision arithmetic. The *sign flag* N carries the sign of the result of the last operation. The *zero flag* Z indicates if the result of the operation is zero. The *overflow flag* V indicates an overflow condition. The *carry flag* C shows that there is a carry-out from the most significant bit of the operand.

Stacks

The concept of stacks has been discussed in previous chapters. The use of the stack is crucial to the operation of the 68010. The stack may be affected either explicitly using stack instructions or implicitly by means of certain instructions such as calls to subroutines. A stack can be created by means of either the stack

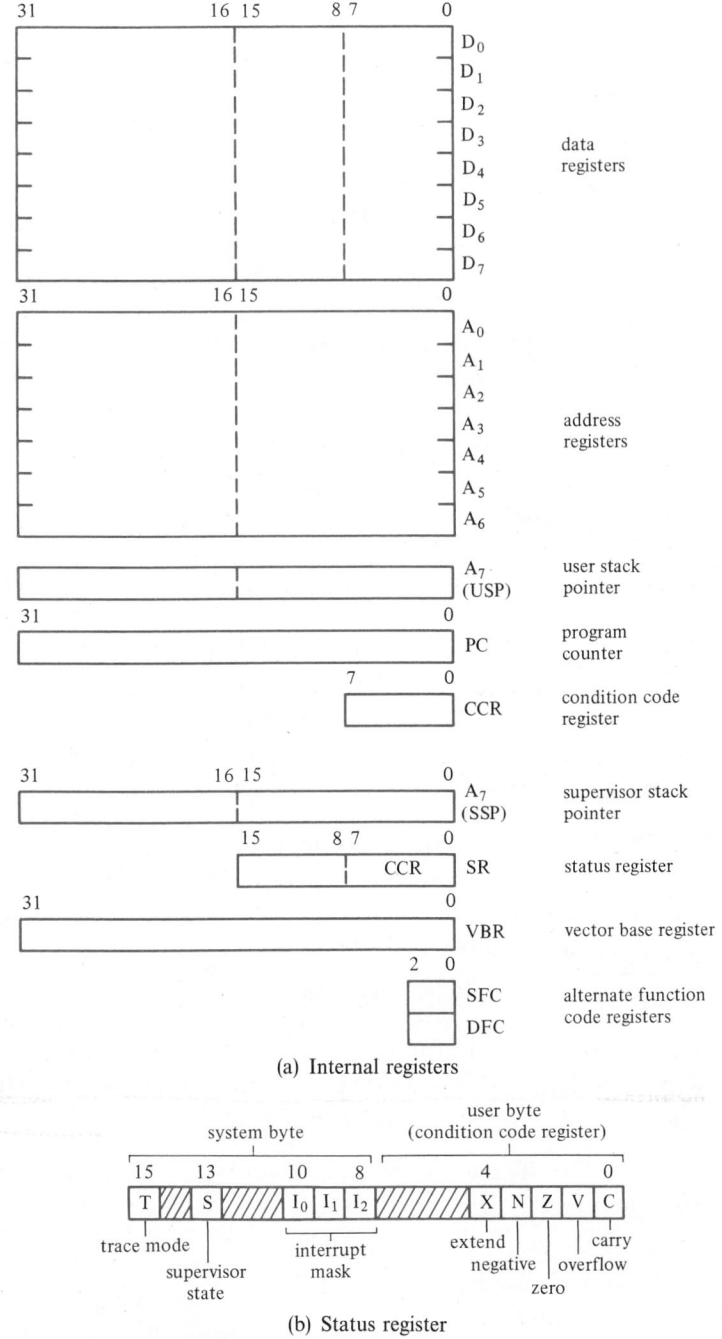

(a) Internal registers

(b) Status register

Figure 11–2

(a) Internal registers and (b) status register of the 68010 (Reprinted with permission of Motorola Inc.)

pointer or any user-specified address register. There are two stacks maintained in the 68010:

A system stack
A user stack

If the stack is referred to by means of the system stack pointer, it is called a *system stack*. If the stack is referred to by any other address register, it is called a *user stack*.

System Stack   The system stack actually refers to one of two stacks. If the system is in the supervisory mode, the stack is referenced by the *supervisor stack pointer* (SSP). If the system is in the user mode, the stack is referenced by the *user stack pointer* (USP). Either the SSP or the USP is actually one of the address registers (A_7). The system stack is always created with the bottom of the stack in high memory and the top of the stack in low memory. This means that when items are pushed onto the stack, the pre-decrement addressing mode, −(SP), is used. Similarly, as items are retrieved from the stack, the post-increment addressing mode, (SP)+, is used.

The system stack is used for storing the PC and other system data during subroutine calls and exception processing. During subroutine calls, only the PC is pushed onto the system stack. Return from subroutine is merely a POP operation to retrieve the return address. During exception processing with traps and interrupts, both the PC and other information are stored onto the system stack.

User Stack   The user stack can be created by any of the address registers (A_0 to A_6). Any of these address registers can be used for stack operations with the proper addressing modes, as seen later. User stacks are created and maintained by the user at all times and at the user's convenience.

Data and Instruction Formats

The 68010 has a powerful set of instructions comprised of a total of 56 instruction types. Each instruction type can operate on a variety of five main data types. The operands of an instruction can be specified in one of the 14 addressing modes.

Data Format   There are five main data types supported and used by the 68010 processor:

Bit: 1-bit
BCD digit: 4-bit
Integer: 8-bit
Word: 16-bit
Long word: 32-bit

Figure 11–3 shows how the different data types are organized in memory for the 68000 and the 68010. The organization of the data types in the 68008 stored in

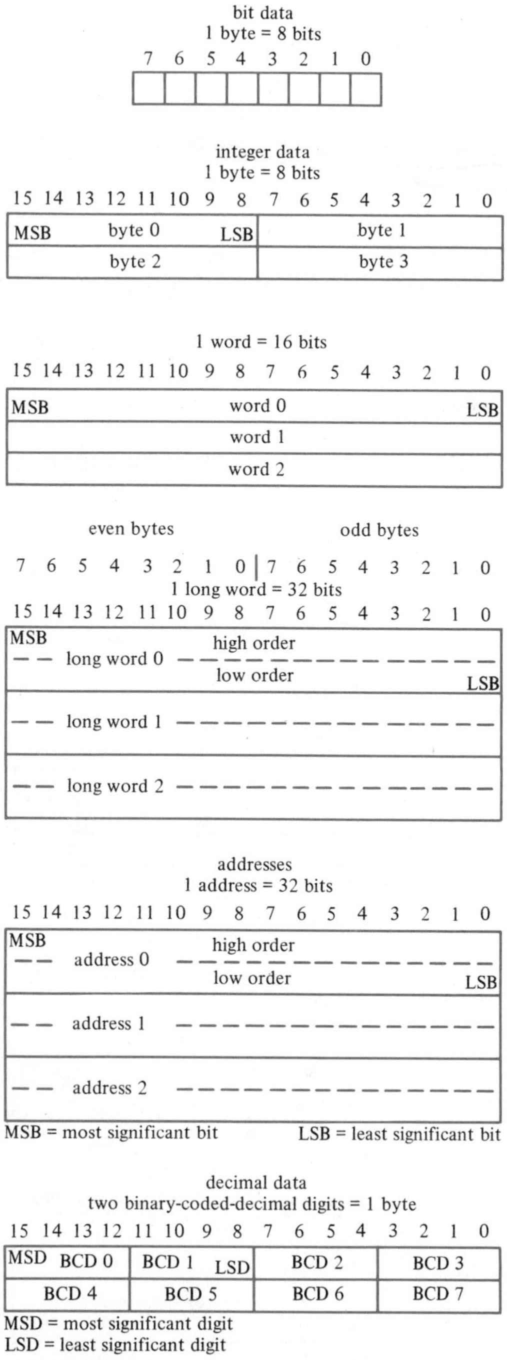

bit data
1 byte = 8 bits

```
 7  6  5  4  3  2  1  0
┌──┬──┬──┬──┬──┬──┬──┬──┐
│  │  │  │  │  │  │  │  │
└──┴──┴──┴──┴──┴──┴──┴──┘
```

integer data
1 byte = 8 bits

```
15 14 13 12 11 10  9  8  7  6  5  4  3  2  1  0
┌─────────────────────────┬─────────────────────────┐
│MSB    byte 0        LSB  │        byte 1           │
├─────────────────────────┼─────────────────────────┤
│         byte 2          │        byte 3           │
└─────────────────────────┴─────────────────────────┘
```

1 word = 16 bits

```
15 14 13 12 11 10  9  8  7  6  5  4  3  2  1  0
┌───────────────────────────────────────────────┐
│MSB              word 0                     LSB │
├───────────────────────────────────────────────┤
│                 word 1                         │
├───────────────────────────────────────────────┤
│                 word 2                         │
└───────────────────────────────────────────────┘
```

even bytes   odd bytes

```
 7  6  5  4  3  2  1  0 │ 7  6  5  4  3  2  1  0
```
1 long word = 32 bits
```
15 14 13 12 11 10  9  8  7  6  5  4  3  2  1  0
┌───────────────────────────────────────────────┐
│MSB              high order                      │
│── long word 0 ─────────────────────────────── │
│                 low order                  LSB │
├───────────────────────────────────────────────┤
│── long word 1 ─────────────────────────────── │
├───────────────────────────────────────────────┤
│── long word 2 ─────────────────────────────── │
└───────────────────────────────────────────────┘
```

addresses
1 address = 32 bits

```
15 14 13 12 11 10  9  8  7  6  5  4  3  2  1  0
┌───────────────────────────────────────────────┐
│MSB              high order                      │
│── address 0 ───────────────────────────────── │
│                 low order                  LSB │
├───────────────────────────────────────────────┤
│── address 1 ───────────────────────────────── │
├───────────────────────────────────────────────┤
│── address 2 ───────────────────────────────── │
└───────────────────────────────────────────────┘
```

MSB = most significant bit   LSB = least significant bit

decimal data
two binary-coded-decimal digits = 1 byte

```
15 14 13 12 11 10  9  8  7  6  5  4  3  2  1  0
┌──────────┬──────────┬──────────┬──────────┐
│MSD BCD 0 │ BCD 1 LSD│  BCD 2   │  BCD 3   │
├──────────┼──────────┼──────────┼──────────┤
│  BCD 4   │  BCD 5   │  BCD 6   │  BCD 7   │
└──────────┴──────────┴──────────┴──────────┘
```

MSD = most significant digit
LSD = least significant digit

Figure 11–3

Organization of data types in memory for the 68000 and 68010
(Reprinted with permission of Motorola Inc.)

the internal registers is the same as the 68010. However, the organization of the external memory is byte-oriented, as shown in Figure 11–4.

All addresses used in the 68010 are 32 bits long, although only 24 bits are used externally. They are stored in the memory with the most significant part first followed by the least significant part, as shown in Figure 11–4. Only word or long word data types are valid for addresses. If a particular address is a word size address, the address is sign-extended to 32 bits long before further manipulation.

Figure 11–4

Organization of data types in memory for the 68008 (Reprinted with permission of Motorola Inc.)

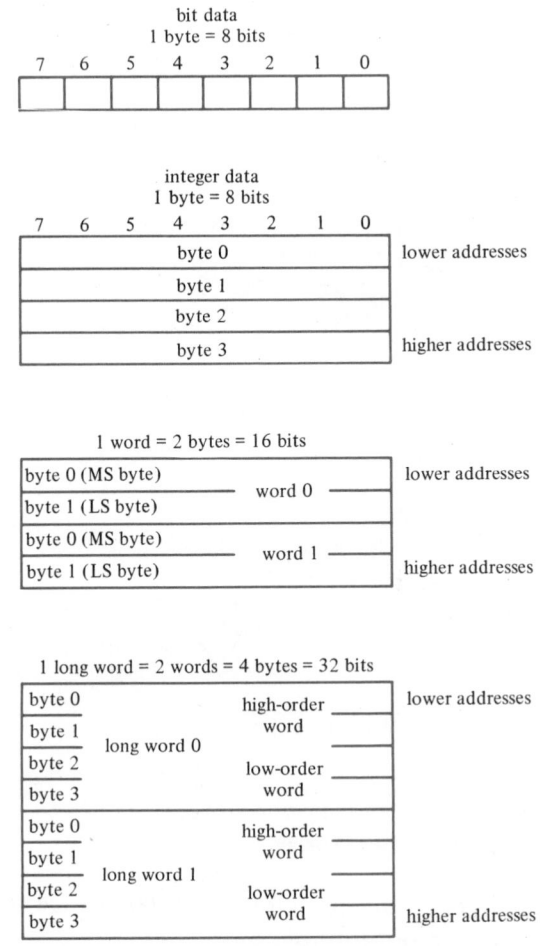

Since the memory address is usually addressed in terms of bytes, there are rules for accessing the memory:

1. Word and long word types of memory accesses must be made from an even address in memory.
2. Byte accesses can be made from either an even or an odd address.
3. Operation codes must be accessed on an even address.

The above rules are enforced because the 68010 is primarily word-oriented when accessing the memory. Only a few byte-oriented instructions require byte-size data. The bulk of the instructions and all operation codes (OPCODEs) are stored in even word boundaries.

Instruction Format The length of a 68010 instruction can range from 1 to 5 words as shown in Figure 11–5, depending on the instruction and its addressing schemes. The first word will always be the instruction operation code, including the necessary addressing modes and the register selections. The next one or two words, called *extension words*, contain the immediate operand if any. The next one or two words are the source effective address if any. The last one or two words are the destination effective address if any.

All memory references are differentiated into one of two categories: program references or data references. A *program reference* refers to a memory reference in the program space that may be read only. A *data reference* refers to a memory reference in the user data space that must necessarily be read/write memory. The

Figure 11–5

Instruction formats of the 68010 (Reprinted with permission of Motorola Inc.)

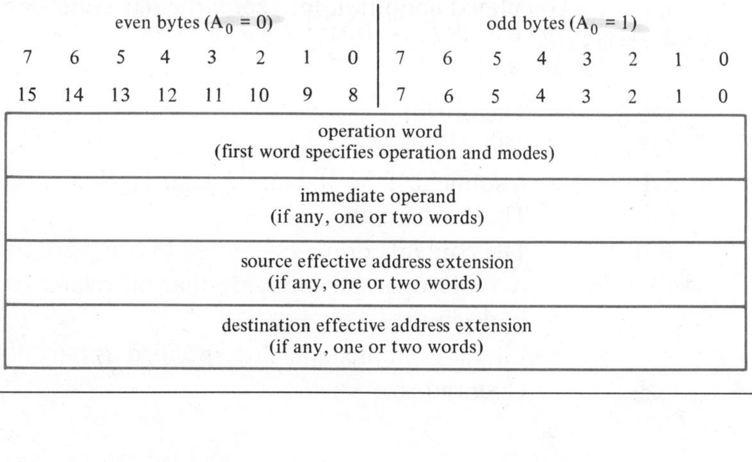

even bytes ($A_0 = 0$)								odd bytes ($A_0 = 1$)							
7	6	5	4	3	2	1	0	7	6	5	4	3	2	1	0
15	14	13	12	11	10	9	8	7	6	5	4	3	2	1	0

operation word (first word specifies operation and modes)
immediate operand (if any, one or two words)
source effective address extension (if any, one or two words)
destination effective address extension (if any, one or two words)

purpose of this type of memory reference separation is useful for memory protection and preservation of the integrity of supervisory memory.

Notation

The first word of the instruction is always the operation code. It is comprised of three sections, a 10-bit operation code, a 3-bit addressing mode, and a 3-bit register select. In the description of the various instructions, it is handy to use a uniform notation. The following notation is used when describing the 68010 (courtesy of Motorola Inc.).

Register Definitions

An	Address register
Dn	Data register
Rn	Any register, data or address
Xn	An index register that can be either a data or an address register
PC	Program counter
SR	Status register
CCR	The condition code part of the status register
SP	The active stack pointer (either user stack pointer or supervisor stack pointer, depending on the status of the machine)
USP	User stack pointer
SSP	Supervisor stack pointer
VBR	Vector base register
SFC	Alternate function code source register
DFC	Alternate function code destination register
X	Extend condition code or bit 4 of SR
N	Sign condition code or bit 3 of SR
Z	Zero condition code or bit 2 of SR
V	Overflow condition code or bit 1 of the SR
C	Carry condition code or bit 0 of the SR
U	Undefined condition for the particular condition code involved

Operand Definitions

OPR	The operand
OPR(i)	The ith bit of the operand
OPR(i:j)	A subfield of the operand specified by bit i through bit j inclusively
( )	The contents of
(OPR)	The contents of the referenced location specified as OPR
OPR10	A BCD operand implying that all related operations are done in decimal
#	Literal or immediate data specified as part of the instruction
SZ	Operand size

Addressing Mode Specifications

EA	The effective address of the operand
d	A 2's complement displacement
Source	Source effective address
Destination	Destination effective address
(Source)	Source operand or the contents of the source effective address
(Dest)	Destination operand or the contents of the destination effective address
Word 2	Second extension word address
Word 3	Third extension word address
Vector	Exception vector location
Dn	Data register direct. The operand is in the specified data register
An	Address register direct. The operand is in the specified address register
(An)	Address register indirect. The operand is in the memory location specified by the address register A_n
(An)+	Address register indirect with post-increment. The operand is in the memory location specified by the address register A_n that is being incremented after being referenced
−(An)	Address register indirect with pre-decrement. The operand is in the memory location specified by the address register A_n that is being decremented before it is used
d(An)	Address register indirect with displacement. The operand is in the memory location whose address is in the specified address register modified by the displacement
d(An,Xn)	Address register indirect with index. The operand is in the memory location whose address is in the specified address register modified by the displacement and the specified index register
Abs.W	Absolute short address. The operand address is stored in the first extension word giving a 16-bit address
Abs.L	Absolute long address. The operand address is stored in the first and second extension words giving a complete 32-bit address
d(PC)	Program counter with displacement. The operand address is given by modifying the PC with the displacement
d(PC,Xn)	Program counter with displacement and index. The operand address is given by the program counter modified by the displacement and the contents of the index register
Imm	Immediate addressing. The operand itself is given in the extension word/words

Operators

←	The left-hand side is replaced by the right-hand side
↔	The left-hand side is exchanged with the right-hand side
+	Addition
−	Subtraction
*	Multiplication
/	Division
∧	Logical AND operation
∨	Logical Inclusive OR operation
⊕	Logical Exclusive OR operation
<	Relational test that is true if the left operand is less than the right operand
>	Relational test that is true if the left operand is greater than the right operand
'	Logical complement
IF cc THEN ... ELSE ...	Relational test based on the condition cc. When the condition is true, the first set of operations is performed. When the condition fails, the second set of operations is performed

11.2
Addressing Modes

Since the 68010 is basically a two-address machine, most instructions have a source address and a destination address. In cases where a particular operator requires two operands, one of the operands is both the source and the destination operand. The effective address can be computed from one of the many addressing modes using one of the addressing registers. There are many addressing modes possible in the 68010. These are generally categorized into four main classes:

Register direct mode
Memory addressing mode
Special addressing mode
Implicit reference mode

A general format of the OPCODE with single-operand specification is composed of the OPCODE and 6 specification bits that describe the operand's effective address. The first 3 bits specify the addressing mode, while the last 3 bits specify the desired register number. We will examine each of these addressing modes in detail in the following sections. For each addressing mode, we will give the

Addressing Mode	Mode	Register	Addressing Categories				Assembler Syntax
			Data	Memory	Control	Alterable	
Data register direct	000	Reg. no.	X	—	—	X	Dn
Address register direct	001	Reg. no.	—	—	—	X	An
Address register indirect	010	Reg. no.	X	X	X	X	(An)
Address register indirect with post-increment	011	Reg. no.	X	X	—	X	(An)+
Address register indirect with pre-decrement	100	Reg. no.	X	X	—	X	−(An)
Address register indirect with displacement	101	Reg. no.	X	X	X	X	d(An)
Address register indirect with index	110	Reg. no.	X	X	X	X	d(An, IX)
Absolute short	111	000	X	X	X	X	xxx.W
Absolute long	111	001	X	X	X	X	xxx.L
Program counter with displacement	111	010	X	X	X	—	d(PC)
Program counter with index	111	011	X	X	X	—	d(PC, IX)
Immediate	111	100	X	X	—	—	#xxx

Figure 11–6

Relationship of addressing modes and memory reference categories
(Reprinted with permission of Motorola Inc.)

EA generation technique in RTL and the assembler syntax. A summary table listing all the addressing modes, the corresponding mode specification, and the register numbers is given in Figure 11–6.

Register Direct Mode

The *register direct addressing mode* means that the instruction operand is in one of the data or address registers specified. There are only two register direct modes:

1. Data register direct
2. Address register direct

These two addressing modes are illustrated in Figure 11–7.

Data Register Direct

Generation	EA = Dn
Assembler syntax	Dn

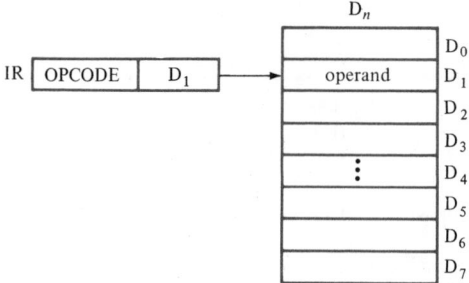

(a) Data register direct

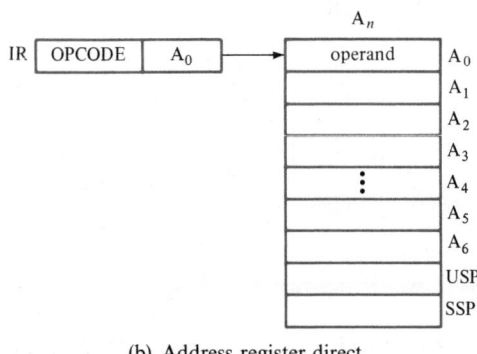

(b) Address register direct

Figure 11-7
Register direct addressing modes

In the data register direct mode, the operand is stored in one of the data registers. Specification of the proper data register is all that is needed. This specification requires a total of 3 bits to indicate one of the eight data registers. The data register direct addressing mode is useful for manipulation of temporary data stored in the data registers and provides the fastest execution rate because no memory references are required.

Address Register Direct

Generation $\qquad$ EA = An
Assembler syntax $\qquad$ An

In the address register direct mode, the operand is in one of the address registers. Specification of the proper address register requires 3 bits in the same way as the data register direct mode. The address register direct mode is used for address manipulation when implementing data structures. Typical examples are subscript calculations and manipulation of pointers in a linked list.

Memory Addressing Mode

In the *memory addressing mode*, the effective address of a memory operand is specified by the contents of one of the address registers. In essence, this is the register indirect addressing scheme, because the address of the required operand is in one of the specified address registers. The address register can be optionally pre-decremented or post-incremented before use. Furthermore, the effective address in the address register can be modified by an offset and/or the contents of another index register. There are five memory addressing modes:

1. Address register indirect
2. Address register indirect with post-increment
3. Address register indirect with pre-decrement
4. Address register indirect with displacement
5. Address register indirect with index

Note that all the above addressing modes use only the address registers and not the data registers for address specification. These five addressing modes are illustrated in Figure 11–8.

Address Register Indirect

Generation	EA = (An)
Assembler syntax	(An)

In the address register indirect mode, the effective address is contained in the specified address register. The address in the specified address register remains unchanged. The address register indirect mode is useful for getting memory operands. The address may be dynamic and may be manipulated as required because the address is stored in the address register.

Address Register Indirect with Post-Increment

Generation	EA = (An)
	(An) = (An) + SZ
Assembler syntax	(An)+

In the address register indirect with post-increment mode, the address of the operand is contained in the specified address register like the address register indirect mode. However, after the operand address is used, the address in the specified address register is incremented by the operand length. If the instruction specifies the size of the operand to be a byte, or a word, or a long word, the address in the address register will be incremented by 1, 2, or 4, respectively. When

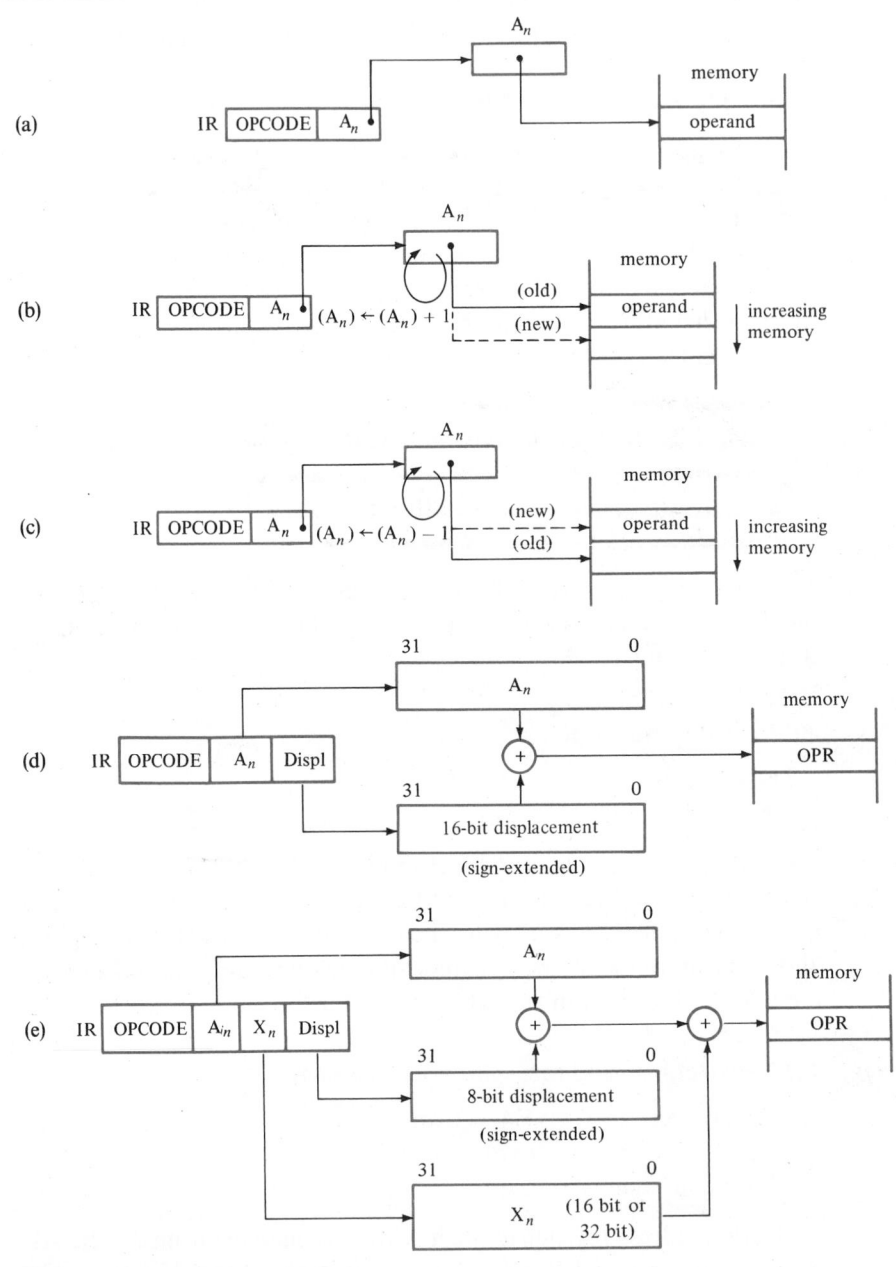

Figure 11–8
Memory addressing modes: (a) address register indirect; (b) address
register indirect with post-increment; (c) address register indirect with
pre-decrement; (d) address register indirect with displacement; and (e)
address register indirect with index

loading or storing items in a list, it is helpful to use the address register indirect with post-increment mode so that the address is always pointing to the next available data in the list. The post-increment mode is used when the top of the list is located at a lower address.

Address Register Indirect with Pre-Decrement

Generation	$(An) = (An) - SZ$
	$EA = (An)$
Assembler syntax	$-(An)$

The address register indirect with pre-decrement mode is similar to the address register indirect with post-increment except that in the pre-decrement case the address register is decremented before the address is used in the specified register instead of the address being incremented after it is used in the post-increment case. The amount to be incremented is according to the size specified. The address register indirect with pre-decrement mode is useful also for loading and storing items to and from a list. In this case, the pre-decrement mode is used when the top of the list is at a higher address or when the direction of access is reversed.

Address Register Indirect with Displacement

Generation	$EA = (An) + d_{16}$
Assembler syntax	$d_{16}(An)$

In the address register indirect with displacement mode, the address contained in the specified address register is first modified by a 16-bit sign-extended displacement before being used as an effective address to fetch the operand. This address mode requires an additional 16-bit word following the instruction word to hold the 16-bit displacement. This addressing mode is often used for referencing subscripted variables.

Address Register Indirect with Index

Generation	$EA = (An) + (Xn) + d_8$
Assembler syntax	$d_8(An,Xn.W)$ or $d_8(An,Xn.L)$

In the address register indirect with index mode, the address in the specified address register is first modified by the 8-bit sign-extended displacement and the sign-extended contents of the index register before being used as an effective address to fetch the operand. The length of the specified index register may be 16 bits (word) or 32 bits (long word) long. This addressing mode requires a second word of storage to hold the specification of the index register and the displacement. The format of this second word is shown in Figure 11–9. The D/A bit represents data or address register and the W/L bit word or long word size.

	Even Byte								Odd Byte						
7	6	5	4	3	2	1	0	7	6	5	4	3	2	1	0
15	14	13	12	11	10	9	8	7	6	5	4	3	2	1	0
D/A	register		W/L	0	0	0		displacement integer							

Figure 11-9

Format of the second extension word when an index register is specified
(Reprinted with permission of Motorola Inc.)

Special Addressing Mode

In addition to the register direct mode and memory addressing mode, there are numerous special addressing modes, including:

Absolute short address
Absolute long address
Program counter with displacement
Program counter with index
Immediate data

These five special addressing modes are illustrated in Figure 11-10. Note that the absolute and immediate addressing modes do not require explicit specification of address registers.

Absolute Short Address

Generation	EA (given in the extension)
Assembler syntax	xxxx.W

In the absolute short addressing mode, the effective address of the required operand is stored in the next extension word. This 16-bit address is sign-extended before it is used to fetch the operand. The absolute short addressing mode provides direct addressing of any memory operands located at the lowest 32K or the highest 32K of the memory space, because only 16 bits of address can be specified.

Absolute Long Address

Generation	EA (given in the next two extension words)
Assembler syntax	xxxxxxxx.L

In the absolute long address, the effective address is specified in full (32 bits) in the next two extension words. The most significant part of the address is placed in the first extension word, whereas the least significant part of the address is placed in the second extension word. This addressing mode provides memory reference anywhere in the total addressing space, because the full 32-bit address

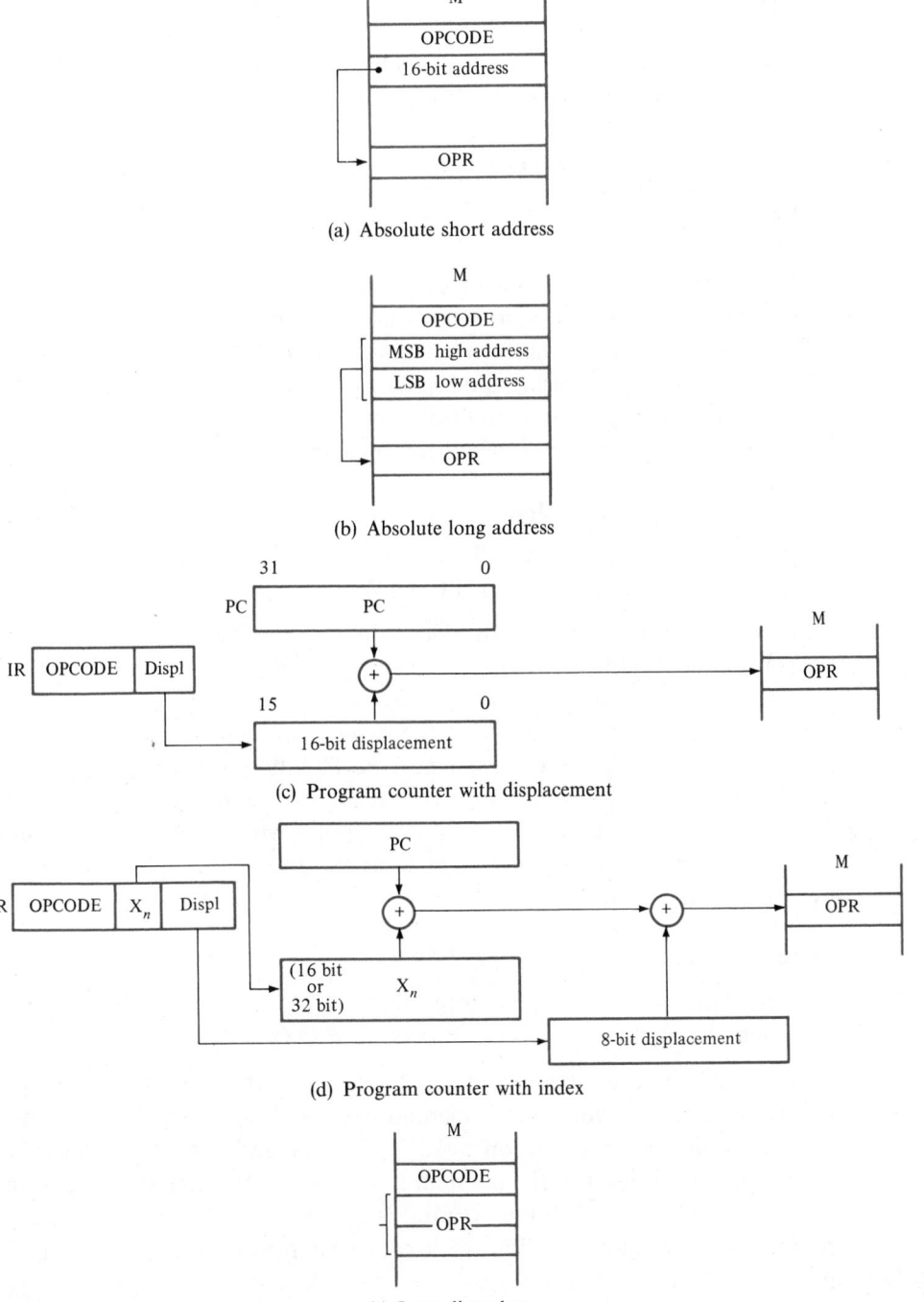

(a) Absolute short address

(b) Absolute long address

(c) Program counter with displacement

(d) Program counter with index

(e) Immediate data

Figure 11–10
Special addressing modes

can be specified at the expense of an additional extension word when compared to the absolute short addressing mode.

Program Counter with Displacement

Generation $\qquad$ EA $= (PC) + d_{16}$
Assembler syntax $\qquad$ LABEL(PC)

In the program counter with displacement addressing mode, the effective address is taken to be the sum of the current address contained in the PC and the sign-extended 16-bit displacement contained in the extension word. Although the absolute short addressing mode addresses only the lowest and highest 64K pages, the program counter with displacement mode allows addressing anywhere within 32K on either side of the current program counter location.

Program Counter with Index

Generation $\qquad$ EA $= (PC) + (Xn) + d_8$
Assembler syntax $\qquad$ LABEL(PC,Xn.W) or LABEL(PC,Xn.L)

In the program counter with index addressing mode, the effective address is the sum of three things: the current value of the PC, the sign-extended 8-bit displacement, and the sign-extended address in the index register. Both the specifications for the index register and the 8-bit displacement are supplied in the extension word. The index register may be 16 bits or 32 bits long. The format of the extension word is the same as that for the address register indirect with index mode shown earlier. The program counter with index addressing mode is very useful for writing relocatable code as the PC or the index register can be used as a base register.

Immediate Data

Generation $\qquad$ Operand (given)
Assembler syntax $\qquad$ #xx or #xxxx or #xxxxxxxx

The immediate addressing mode implies that the operand itself is in the following extension words. The operand may be 8 bits, 16 bits, or 32 bits long, occupying one or two extension words. If the operand is byte size, then the next extension word is used. If the operand is word size, then the next extension contains the word required. If the operand is 32 bits long, then the most significant part of the operand is stored in the first extension word and the least significant part in the second extension word.

Implicit Reference Mode

In the *implicit reference mode*, no specific reference is needed because the affected register is implied by the instruction itself. Registers affected by these types of instructions include the program counter, the system stack pointer, the user stack pointer, and the status register.

Categories of Memory References

There are four major addressing categories used to describe the operand address:

1. Data
2. Memory
3. Alterable
4. Control

The relations among these addressing categories and their associated addressing modes are shown in Figure 11–6.

A *data reference* refers to a register or memory reference that contains data. A data reference is valid for all addressing modes except for the address register direct mode. A *memory reference* refers to operands that reside in memory only. This addressing mode is clearly available to all references except the data register direct and address register direct modes, because in the latter two modes the operands are stored in the specified registers and no memory references are required. An *alterable reference* denotes operands that can be read or changed at the end of the instruction. This would be the case if the operand serves as both a source and a destination operand. In many applications, references to the program memory may not be alterable because the program is stored in a read-only memory. A *control reference* refers to operands that do not have an associated size, hence are not likely to be a datum that carries data-type information. A control reference contains special information to augment instructions.

11.3

Assembly-Language Programming

Programming the 68010 is made easy by a powerful assembly-language assembler and other utility programs. In this section we provide some basic information for the assembler.

Assembly-Language Format

Most assemblers operate on a line-by-line basis. The basic assembler format is composed of four fields for each line of code:

1. Label
2. OPCODE
3. Operand
4. Comments

The layout for these four fields is as follows:

[LABEL]   OPCODE   [OPERAND][, OPERAND]   [*COMMENTS]

The square brackets indicate optional items.

Label Field The *label field* is the first field in the assembly line when present. A label that begins at column 1 of the line must be terminated by a space or a colon and a space. A label that does not begin at column 1 must be terminated by a colon and a space. The label is optional. When the label is present, it represents the address for that program segment, the value placed at the operand field, or the name of macros depending on the usage decoded by what is inserted in the OPCODE field. If only the label is given, the label is assumed to represent the current address location. A label is composed of up to eight significant characters, including A–Z inclusive, 0–9 inclusive, "$", ".", and "_". A label, however, must begin with an alphabetic character or a period. There are two kinds of labels:

Class 1 labels
Class 2 labels

Class 1 labels are often used to indicate mnemonics and macro calls. As such, the value of class 1 labels is absolute and can be used only in the operation field. *Class 2 labels* are used primarily for address labels. Address labels may take on one of two possible attributes. If the address label is defined in a relocatable section, the value of the address for the label is relative to the beginning of the section. If the label is defined in an absolute section, then the value assigned to the label is absolute.

OPCODE Field The *operation field*, sometimes called the *OPCODE field*, contains the specification for the instruction and the size of the operand. The format is the mnemonic followed by the operand size specification, which is separated by a period. In addition to instruction mnemonics, the OPCODE field can also contain assembler directives or macro calls. The latter two types are also called pseudo-operations.

There are many sizes of operand, as mentioned before. Four of them require explicit specification:

B   Byte size for 8-bit data
W   Word size for 16-bit data (default)
L   Long word size for 32-bit data
S   Short displacement for jump instructions only

Operand Field The *operand field* contains specifications for the necessary operands required in the execution of the present instruction. There may be none, one, or many operands, depending on the instruction. For those instructions that require only one operand, the first operand is usually the source operand, whereas the second is the destination operand. For those instructions requiring two operands, the first operand is always the source operand, whereas the second is both the source and the destination operand.

Comments Field   There are two ways to include comments in the assembly program. If the assembly line begins with a star (*), then the whole line is taken to be comments and the rest of the line after the star is ignored for assembly. The second way to include comments is to put them in the last field of the line separated by at least a space after the operation field or the operand field. Comments are strongly encouraged in all assembly programs for readability.

Expressions

There are certain notations used in the assembler. When identifying literal information, special characters are used to indicate whether they are octal, binary, decimal, or hexadecimal numbers, or characters:

Symbol	Meaning	Example
$	Hexadecimal digits	$FF89
(none)	Decimal digits	2345
@	Octal digits	@0377
%	Binary digits	%00110011
' '	Character string	'HELP'

Expressions are basic to 68010 assembly programming. They are often used in addressing calculation and manipulation in the operand field. When expressions are used to indicate addresses, certain constraints apply that govern the attributes of the resulting value. There are a number of arithmetic and logical operations that can be performed in expressions. Their symbols are

    & Logical AND
    ! Logical OR
    « Left shift
    » Right shift
    + Arithmetic addition
    − Arithmetic subtraction or unary minus
    * Arithmetic multiplication
    / Arithmetic division

Since symbols and labels can be absolute or relocatable, combinations of these labels may be categorized into three main types:

Absolute
Simple relocatable
Complex relocatable

Mode	Invocation	Comments
1. Register direct	An Dn	
2. Memory address		
(a) Simple indirect	(An)	
(b) Pre-decrement	−(An)	
(c) Post-increment	(An)+	
(d) Indirect with displacement (16-bit)	⟨absolute⟩(An) ⟨complex⟩(An)	

(e) Indirect with index
    (16- or 32-bit) plus
    displacement (8-bit)
	⟨absolute⟩(An,Xn)	

3. Special address		
(a) PC with displacement (16-bit)	⟨simple⟩	Expression is an address (not a displacement) which must be backward, within current relocatable section.
	⟨absolute⟩(PC) ⟨simple⟩(PC) ⟨complex⟩(PC)	Forced PC-relative. Must fit within 16-bit signed field; resolved at assembly or link time.

(b) PC with index
    (16- or 32-bit) plus
    displacement (8-bit)
	⟨simple⟩(Xn)	Expression is an address that must be backward, within current relocatable section.

	⟨absolute⟩(PC,Ri) ⟨simple⟩(PC,Ri)	Forced PC-relative; expression must be within current program section.
(c) Absolute (16- or 32-bit)	⟨absolute⟩ ⟨complex⟩ ⟨simple⟩	Expression must be forward reference or not in current program section.
(d) Immediate (8-, 16-, or 32-bit)	#⟨absolute⟩ #⟨simple⟩ #⟨complex⟩	
4. Implicit PC reference		Invoked by conditional branch (Bcc) or DBcc instruction; the effective address is a displacement from the PC; the displacement is either 8 or 16 bits.

Figure 11–11

Allowable expression types for the different addressing modes (Reprinted with permission of Motorola Inc.)

An *absolute expression* results when expressions are comprised solely of absolute labels and quantities. Absolute expressions also result from differences of two relocatable variables with the same origin. A *simple relocatable expression* results when only one relocatable variable remains. In other words, the result is a single offset with respect to the beginning of a single section. A *complex relocatable expression* is one that involves many offsets to many different sections. Different addressing modes generate different types of expressions. Some addressing modes may only resolve a certain kind of expression. Figure 11–11 shows the allowable expression types for the different addressing modes.

Assembly Control Directives

There are three primary groups of pseudo-operations in the 68010 assembler:

1. Assembly control directives
2. Symbol definition directives
3. Data and storage definition directives

Assembly control directives are pseudo-operations that inform the assembler concerning the way the subsequent code should be assembled.

Origin (ORG) Directive

```
ORG ⟨expression⟩ ⟨comments⟩
```

The **ORG** directive communicates to the assembler the absolute address of the code that follows the directive. All subsequent statements are assembled with absolute memory references starting at the specified address. The expression must be absolute and must not contain any unresolved forward references.

SECTION Directive

```
⟨name⟩ SECTION ⟨number⟩
```

The **SECTION** directive is used to denote a relocatable code segment. The number identifies the section number and must be less than 16, hence a total of 16 relocatable segments is possible. The name of the section is optional.

END Directive

```
END ⟨start address⟩
```

The **END** directive indicates to the assembler the end of the current program segment. All subsequent statements are ignored. If desired for another segment at a later time, the starting address can be specified with the END directive. The starting address for the END directive is optional.

INCLUDE Directive

```
INCLUDE <file.specification>
```

The INCLUDE directive is used to insert statements from another file into the current location. The specification of the file to be inserted follows the directive. The INCLUDE directive is often used when multiple copies of symbol definitions are needed in many different programs.

Symbol Definition Directives

The next group of assembler directives provides the way to define symbols. Defined symbols are meaningful within the current program segment unless otherwise specified.

The EQU Directive

```
<label> EQU <expression> <comments>
```

The EQU directive is used to assign values to class 2 symbols. The expression to the right of the EQU directive is evaluated and the resultant value is assigned to the label. No complex relocatable expressions or expressions containing forward references are allowed. Once the label has been assigned a value, this value cannot be changed.

The SET Directive

```
<label> SET <expression> <comments>
```

The SET directive also assigns the value of the expression to the label in much the same way as the EQU directive does except for one important difference: The same label can be reassigned to another value at any time anywhere in the program.

The REG Directive

```
<label> REG <register list> <comments>
```

The REG directive translates a list of registers identified on the <register list> location and assigned a 16-bit representation to the label in register list mask format. The label can then be used in a MOVEM instruction, which is discussed in Section 11.5.

Data and Storage Definition Directives

The next group of directives deals with directives that direct the assembler to reserve memory locations for data storage.

The DC Directive

```
<label> DC.B <operand(s)>
<label> DC.W <operand(s)>
<label> DC.L <operand(s)>
```

The DC directive defines constants in memory. The size specifies the length of the operand. If .B is specified, 1 byte of memory is allocated at a time and is usually used for character representation. If .W is used, the allocation will begin at a word boundary. The last word reserved will be zero-filled when only an odd number of bytes are used. If .L is used, the allocation will also begin on a word boundary with the last long word (4 bytes) being zero-filled when not used. Labels are optional.

The DS Directive

```
<label> DS.B <operand>
<label> DS.W <operand>
<label> DS.L <operand>
```

The DS directive reserves the specified number of memory locations. When the operand is zero, the DS directive can be used to align the next memory address on a word boundary. The labels are optional.

The DCB Directive

```
<label> DCB.B <length>, <value> <comment>
<label> DCB.W <length>, <value> <comment>
<label> DCB.L <length>, <value> <comment>
```

The DCB directive is similar to the DS directive in that both allocate the specified number of memory locations. The DCB differs from the DS directive, however, in that the DCB directive not only allocates but also assigns a constant value to every one of the allocated memory locations, whereas the DS directive merely allocates the memory locations but does not assign values. Labels and comments are optional.

11.4
The 68010 Instruction Set

The instruction set of the 68010 is rather powerful and can be divided into eight major groups. All but the last group of privileged instructions are available to the application user. The privileged instructions are reserved for system programming and are executed only when the processor is in the supervisory mode. The

Addressing Mode

Mnemonic Operation	Size	Addr. Mode	Dn	#	An	#	(An)	#	(An)+	#	-(An)	#	(An)	#	d(An)	#	d(An, Xi)	#	Abs W	#	Abs L	#	d(PC)	#	d(PC, Xi)	#	s Immed d SR/CC	#	Opcode Bit Pattern 1111 11 / 5432 1098	7654 3210	Boolean	Condition Codes X N Z V C
ABCD Add Digits	B	s—Dn s—-(An)	6	2							19	2																	1100 RRR1	0000 Orrr	d10+s10+X—d	* U * U *
ADD Add Binary	B/W L W L	d—Dn s—Dn d—Dn s—Dn d—An	4	2	ADDA 2* 4 ADDA		13 8 22 13	2 2 2 2	13 8 22 14	2 2 2 2	15 10 24 16	2 2 2 2	17 12 26 18	4 4 4 4	19 14 28 20	4 4 4 4	17 12 26 18	4 4 4 4	21 16 30 22	6 6 6 6	12 18	4 4	14 20	4 4	8 14	4 6	1101 DDD0 1101 DDD0 1101 DDD1 1101 DDD1 1101 10ee	SSEE EEEE 1Oee EEEE SSEE eeee 1Oee EEEE 1Oee eeee	d+Dn—Dn Dn+s—Dn d+Dn—Dn Dn+s—Dn	* * * * *
ADDA Add Address	W L	d—An d—An	8 8	2 2	ADDA ADDA		14 16	2 2	14 16	2 2	16 18	2 2	18 21	4 4	20 23	4 4	18 21	4 4	22 25	6 6	16 18	4 4	18 20	4 4	12 14	4 4	1101 AAA1 1101 AAA1	11ee ecce 11ee ecce	An+s—An	— — — — —
ADDI Add Immed	B/W L	d—Dn d—Dn	8 16	6 8			17 30	4 6	17 30	4 6	19 32	4 6	21 34	6 8	23 36	6 8	21 34	6 8	25 38	8 10							0000 0110	SSEE EEEE	d+#—d	* * * * *
ADDQ Add Quick	B/W L	d—Dn d—An	4 8	2 2	2* 2	2 2	13 22	2 2	13 22	2 2	15 24	2 2	17 26	4 4	19 28	4 4	17 26	4 4	21 30	6 6							0101 QQQ0	SSEE EEEE	d+#—d	* * * * *
ADDX Add Multi-precision	B/W L	s—Dn s—-(An) s—Dn s—-(An)	4 8	2 2							19 32	2 4															1101 RRR1 1101 RRR1 1101 RRR1 1101 RRR1	SS00 Orrr SS00 1rrr 1000 Orrr 1000 1rrr	d+s+X—d	* * * * *
AND Logical And	B/W L	d—Dn s—Dn d—Dn s—Dn	4 8	2 2			13 22	2 2	13 22	2 2	15 24	2 2	17 26	4 4	19 28	4 4	17 26	4 4	21 30	6 6	12 18	4 4	14 20	4 4	8 14	4 6	1100 DDD0 1100 DDD0 1100 DDD1 1100 DDD1	SSEE EEEE 1Oee EEEE SSEE EEEE 1Oee EEEE	d and Dn—Dn Dn and s—Dn d and Dn—Dn Dn and s—Dn	* * * 0 0
ANDI And Immed	B/W L	s—Imm s—Imm	8 16	4 4			17 30	4 6					17 30	4 6					21 34	4 6					8 20	8 20	0000 0010	SSEE EEEE	d and #—d	* * * 0 0
ASL, ASR Arithmetic Shift	B/W L	count—Dn count=#1 8 count—Dn count=#1 8	6+2n 6+2n 8+2n 8+2n	2 2 2 2							19	2															1110 rrrf 1110 QQQf 1110 rrrf 1110 QQQf	SS10 ODDD SS00 ODDD 1010 ODDD 1000 ODDD		* * * * *
BCHG Test and Change	B L	bit#—Dn bit#—Imm bit#—Dn bit#—Imm	<8 <12	2 4			13 17	2* 4	13 17	2 4	15 19	2 4	17 21	4* 6	19 23	4* 6	17 21	4 6	21 25	6* 8							0000 rrr1 0000 1000	01EE EEEE 01EE EEEE	'bit#' of d—Z, 'bit#' of d—'bit#' of d	— — * — —
BCLR Test and Clear	B L	bit#—Dn bit#—Imm bit#—Dn bit#—Imm	<10 <14	2 4			13 17	2 4	13 17	2 4	15 19	2 4	17 21	4 6	19 23	4 6	17 21	4 6	21 25	6 8							0000 rrr1 0000 1000	10EE EEEE 10EE EEEE	'bit#' of d—Z, 0—'bit#' of d	— — * — —
BSET Test and Set	B L	bit#—Dn bit#—Imm bit#—Dn bit#—Imm	<8 <12	2 4			13 17	2 4	13 17	2 4	15 19	2 4	17 21	4 6	19 23	4 6	17 21	4 6	21 25	6 8							0000 rrr1 0000 1000	11EE EEEE 11EE EEEE	'bit#' of d—Z, 1—'bit#' of d	— — * — —
BTST Bit Test	B L	bit#—Dn bit#—Imm bit#—Dn bit#—Imm	<6 <10	2 4			8 12	2 4	10 14	2 4	8 12	2 4	12 16	4 6	14 18	4 6	12 16	4 6	16 20	6 8							0000 rrr1 0000 1000	00EE EEEE 00EE EEEE	'bit#' of d—Z	— — * — —

CHK	W				< 43	← trap →			< 47		< 47		< 49		< 51		< 53		< 51		< 55		< 51		< 53		< 47	0100 DDD1	10ec eeee	If Dn < 0, or	– • 0 0 0
Check Reg-		d – Dn	s –	2				2		2		2		4		4		4		6		4		4		4				Dn > (bound),	
ister Against			(bound)		8′	← no →			12′		12′		14′		16′		18′		16′		20		16′		18′		12′			then trap	
Bounds						trap																									
CLR	B/W		d –	2	4			2	13	2	13	2	15	4	17	4	19	4	17	6	21							0100 0010	SSEE EEEE	0 → d	– 0 1 0 0
Clear Operand	L		d –	2	6			2	22	2	22	2	24	4	26	4	28	4	26	6	30										
CMP	B/W	d – Dn	s –	2	4	2*	4	2	8	2	8	2	10	4	12	4	14	4	12	6	16	4	12	4	14	4	8	1011 DDD0	SSee eeee	Dn – s	– • • • •
Compare	L	d – Dn	s –	2	6	2	6	2	14	2	14	2	16	4	18	4	20	4	18	6	22	4	18	4	20	6	14				
Binary																															
CMPA	W	d – An	s –	2	6	2	6	2	10	2	10	2	12	4	14	4	16	4	12	6	18	4	14	4	16	4	10	1011 AAA0	1lee eeee	An – s	– • • • •
Compare	L	d – An	s –	2	6	2	6	2	14	2	14	2	16	4	18	4	20	4	18	6	22	4	18	4	20	6	14	1011 AAA1	1lee eeee		
Address																															
CMPI	B/W	s – Imm	d –	4	8	CMPA		4	12	4	12	4	14	6	16	6	18	6	16	8	20							0000 1100	SSEE EEEE	d – #	– • • • •
Compare Imm.	L	s – Imm	d –	6	14	CMPA		6	20	6	20	6	22	8	24	8	26	8	24	10	28										
CMPM	B/W	s – (An) +	d –							2	12																	1011 RRR1	SS00 1rrr	d – s	– • • • •
Compare	L	s – (An) +	d –							2	20																				
Memory																															
DIVS	W	d – Dn	s –	2	< 158			2	< 162	2	< 162	2	< 164	4	< 166	4	< 168	4	< 166	6	< 170	4	< 166	4	< 168	4	< 162	1000 DDD1	1lee eeee	Dn32/s16 →	– • • • 0
Divide Signed																														Dn(r:q)	
DIVU	W	d – Dn	s –	2	< 140			2	< 144	2	< 144	2	< 146	4	< 148	4	< 150	4	< 148	6	< 152	4	< 148	4	< 150	4	< 144	1000 DDD0	1lee eeee	Dn32/s16 →	– • • • 0
Divide																														Dn(r:q)	
Unsigned																															
EOR	B/W	s – Dn	d –	2	4			2	13	2	13	2	15	4	17	4	19	4	17	6	21							1011 rrr1	SSEE EEEE	d ⊕ Dn → d	– • • 0 0
Exclusive OR	L	s – Dn	d –	2	8			2	22	2	22	2	24	4	26	4	28	4	26	6	30										
Logical																															
EORI	B/W	s – Imm	d –	4	8			4	17	4	17	4	19	6	21	6	23	6	21	8	25					4	20	0000 1010	SSEE EEEE	d ⊕ # → d	– • • 0 0
Exclusive OR	L	s – Imm	d –	6	16			6	30	6	30	6	32	8	34	8	36	8	34	10	38										
Immediate																															
EXG	L	s – Dn	d –	2	6	2	6																					1100 DDD1	0100 0DDD	s ↔ d	– – – – –
Exchange		s – An	d –	2	6	2	6																					1100 AAA1	0100 1AAA		
Registers																												1100 DDD1	1000 1AAA		
EXT	W		d –	2	4																							0100 1000	1000 0DDD	bit 7 → bit 8-15	
Sign Extend	L		d –	2	4																							0100 1000	1100 0DDD	bit 15 → bit 16-31	– • • 0 0
LEA	L	d – An	s –					2	4					4	8	4	12	4	8	6	12	4	8	4	12			0100 AAA1	1lee eeee	s → An	– – – – –
Load Effect-																															
ive Address																															
LINK		disp – Imm	s –			4	18																					0100 1110	0101 0AAA	An → – (SP)	– – – – –
Link and																														SP → An	
Allocate																														SP + disp → SP	
LSL, LSR	B/W	count – Dn	d –	2	6 + 2n																							1110 rrrf	SS10 1DDD		• • • 0 •
Logical Shift		count – #1-8	d –	2	6 + 2n																							1110 QQQf	SS00 1DDD	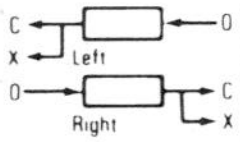	
	L	count – Dn	d –	2	8 + 2n																							1110 rrrf	1010 1DDD		
		count – #1-8	d –	2	8 + 2n																							1110 QQQf	1000 1DDD		
Memory	W	count – 1	d –					2*	13	2*	13	2*	15	4*	17	4*	19	4*	17	6*	21							1110 001f	11EE EEEE		
MOVE	B/W	s – Dn	d –	2	4	MOVEA		2	9	2	9	2	9	4	13	4	15	4	13	6	17							00SS RRRM	MMee eeee	s → d	– • • 0 0
Move Data		s – An	s –	2	4	MOVEA		2	9	2	9	2	9	4	13	4	15	4	13	6	17										
		s – (An)	d –	2	8	MOVEA		2	13	2	13	2	13	4	17	4	19	4	17	6	21										
		s – (An) +	d –	2	8	MOVEA		2	13	2	13	2	13	4	17	4	19	4	17	6	21										
		s – – (An)	d –	2	10	MOVEA		2	15	2	15	2	15	4	19	4	21	4	19	6	23										
		s – d(An)	d –	4	12	MOVEA		4	17	4	17	4	17	6	21	6	23	6	21	8	25										

Figure 11–12

Summary table showing all the instructions of the 68010 (in alphabetical order) (A: Address register number; C: Test condition, D: Data register number; e: Source effective address; E: Destination effective address; f: Direction, 0–Right, 1–Left; M: Destination EA mode; P: Displacement; Q: Quick immediate data; r: Source register; R: Destination register; S: Size, 00–B, 01–W, 10–L, 11–another operation; V: Vector number). (Reprinted with permission of Motorola Inc.)

(continued)

Mnemonic Operation	Size	Addr. Mode		Dn		An		(An)		(An)+		(An)		d(An)		d(An, Xi)		Abs.W		Abs.L		d(PC)		d(PC, Xi)		s = Immed d = SR/CC		Opcode Bit Pattern 1111 11 5432 1098 7654 3210	Boolean	Condition Codes X N Z V C
				#	~	#	~	#	~	#	~	#	~	#	~	#	~	#	~	#	~	#	~	#	~	#	~			
		s-d(An,X)	d-	4	14	MOVEA		4	19	4	19	4	19	6	23	6	25	6	23	8	27									
		s-Abs.W	d-	4	12	MOVEA		4	17	4	17	4	17	6	21	6	23	6	21	8	25									
		s-Abs.L	d-	6	16	MOVEA		6	21	4	21	6	21	8	25	8	27	8	25	10	29									
		s-d(PC)	d-	4	12	MOVEA		4	17	4	17	4	17	6	21	6	23	6	21	8	25									
		s-d(PC,X)	d-	4	14	MOVEA		4	19	4	19	4	19	6	3	6	5	6	3	8	27									
		s-Imm	d-	4	8	MOVEA		4	13	4	13	4	13	6	17	6	19	6	17	8	21									
	L	s-Dn	d-	2	4	MOVEA		2	14	2	14	2	13	4	18	4	2C	4	18	6	22									
		s-An	d-	2	4	MOVEA		2	14	2	14	2	13	4	18	4	2U	4	18	6	22									
		s-(An)	d-	2	12	MOVEA		2	22	2	22	2	22	4	26	4	28	4	26	6	30									
		s-(An)+	d-	2	12	MOVEA		2	22	2	22	2	22	4	26	4	28	4	26	6	30									
		s- -(An)	d-	2	14	MOVEA		2	24	2	24	2	24	4	28	4	30	4	28	6	32									
		s-d(An)	d-	4	16	MOVEA		4	26	4	26	4	26	6	30	6	32	6	30	8	34									
		s-d(An,X)	d-	4	18	MOVEA		4	28	4	28	4	28	6	32	6	34	6	32	8	36									
		s-Abs.W	d-	4	16	MOVEA		4	26	4	26	4	26	6	30	6	32	6	30	8	34									
		s-Abs.L	d-	6	20	MOVEA		6	30	4	30	6	30	8	34	8	36	8	34	10	38									
		s-d(PC)	d-	4	16	MOVEA		4	26	4	26	4	26	6	30	6	32	6	30	8	34									
		s-d(PC,X)	d-	4	18	MOVEA		4	28	4	28	4	28	6	32	6	34	6	32	8	36									
		s-Imm	d-	6	12	MOVEA		6	22	6	22	6	22	8	26	8	28	8	26	10	30									
MOVE	W	d-CCR	s-	2	12			2	16	2	16	2	18	4	20	4	22	4	20	6	24	4	20	4	22	4	16	0100 0100 11ee eeee	s→CCR	* * * * *
Move to Condition Codes																														
MOVE	W	d-SR	s-	2	12			2	16	2	16	2	18	4	20	4	22	4	20	6	24	4	20	4	22	4	16	0100 0110 11ee eeee	s→SR	* * * * *
Move to/from Status Reg.		s-SR	d-	2	6			2	13	2	13	2	15	4	17	4	19	4	17	6	21							0100 0000 11EE EEEE	SR→d	– – – – –
MOVE	L	s-USP	d-			2	4																					0100 1110 0110 1AAA	USP→An	– – – – –
Move to/from User SP (A7)		d-USP	s-			2	4																					0100 1110 0110 0AAA	An→USP	
MOVEA	W	d-An	s-	2	4	2	4	2	8	2	8	2	10	4	12	4	14	4	12	6	16	4	12	4	14	4	8	0001 AAA0 01ee eeee	s→An	– – – – –
Move Address	L	d-An	s-	2	4	2	4	2	12	2	12	2	14	4	16	4	18	4	16	6	20	4	16	4	18	6	12	0010 AAA0 01ee eeee		
MOVEM	W	s-Xn	d-					4	8+5n			4	8+5n	6	12+5n	6	14+5n	6	12+5n	8	16+5n							0100 1000 10EE EEEE a7 a0 d7 d0†	Xn→d	– – – – –
Move Multiple Registers		d-Xn	s-					4	12+4n	4	12+4n			6	16+4n	6	18+4n	6	16+4n	8	20+4n	6	16+4n	6	18+4n			0100 1100 10ee eeee a7 a0 d7 d0	s→Xn	
	L	s-Xn	d-					4	8+10n			4	8+10n	6	12+10n	6	14+10n	6	12+10n	8	16+10n							0100 1000 11EE EEEE a7 a0 d7 d0†	Xn→d	
		d-Xn	s-					4	12+8n	4	12+8n			6	16+8n	6	18+8n	6	16+8n	8	20+8n	6	16+8n	6	18+8n			0100 110C 11ee eeee a7 a0 d7 d0	s→Xn	
MOVEP	W	s-Dn	d-											4	18													0000 DDD1 1000 1AAA	Dn→d by bytes	– – – – –
Move		s-d(An)	d-	4	16																							0000 DDD1 0000 1AAA	s→Dn by bytes	
Peripheral	L	s-Dn	d-											4	28													0000 DDD1 1100 1AAA	Dn→d by bytes	
		s-d(An)	d-	4	24																							0000 DDD1 0100 1AAA	s→Dn by bytes	
MOVEQ	L	s-Imm8	d-	2	4																							0111 DDD0 QQQQ QQQQ	#→Dn	– * * 0 0
Move Quick																														
MULS	W	d-Dn	s-	2	<70			2	<74	2	<74	2	<76	4	<78	4	<80	4	<78	6	<82	4	<78	4	<80	4	<74	1100 DDD1 11ee eeee	Dn×s→Dn	– * * 0 0
Multiply Signed																														
MULU	W	d-Dn	s-	2	<70			2	<74	2	<74	2	<76	4	<78	4	<80	4	<78	6	<82	4	<78	4	<80	4	<74	1100 DDD0 11ee eeee	Dn×s→Dn	– * * 0 0
Multiply Unsigned																														
NBCD	B		d-	2	6			2	13	2	13	2	15	4	17	4	19	4	17	6	21							0100 1000 00EE EEEE	0-d10-X→d	* U * U *
Negate Digit																														

Addressing Mode

Instruction	Operands	Operation	Encoding	Flags
NEG Negate Binary	B/W d-; L d-	0−d→d	0100 0100 SSEE EEEE	• • • • •
NEGX Negate Multi-precision	L d-	0−d−X→d	0100 0000 SSEE EEEE	• • • • •
NOT Logical Complement	B/W d-; L d-	ʹd→d	0100 0110 SSEE EEEE	− • • 0 0
OR Inclusive OR Logical	B/W d=Dn; s=Dn; d=Dn; s=Dn; L d=Dn; s=Dn	d<or>Dn→Dn; Dn<or>s→Dn; d<or>Dn→Dn; Dn<or>s→Dn	1000 DDD1 SSEE EEEE; 1000 DDD0 SSEE EEEE; 1000 DDD1 SSEE EEEE; 1000 DDD0 SSEE EEEE	− • • 0 0
ORI OR Immediate Logical	B/W d=Imm; L s=Imm	d<or>#→d	0000 0000 SSEE EEEE	− • • 0 0
PEA Push Effective Address	L s-	s→−(SP)	0100 1000 01ee eeee	− − − − −
ROR, ROL Rotate without X	B/W count=Dn; count=#1 8; s=Dn d-; count=#1 8 d-; W count=1 d-	(rotate right/left)	1110 rrrf SS11 1DDD; 1110 rrrf SS01 1DDD; 1110 rrrf 1011 1DDD; 1110 rrrf SS01 1DDD; 1110 011f 11EE EEEE	− • • 0 •
ROXR, ROXL Rotate through X	B/W count=Dn; count=#18 d-; count=#18 d-; W count=1 d-	(rotate through X)	1110 rrrf SS11 0DDD; 1110 rrrf SS01 0DDD; 1110 rrrf 1011 0DDD; 1110 010f 11EE EEEE	• • • 0 •
SBCD Subtract digits	B s=Dn d-; s=−(An) d-	d10−s10−X→d	1000 RRR1 0000 0rrr; 1000 RRR1 0000 1rrr	• U • U •
Scc Set Conditionally	B cc d-	If cc true, 1's→d; Else, 0's→d	0101 CCCC 11EE EEEE	− − − − −
SUB Subtract Binary	B/W d=Dn; d=Dn; L d=Dn; d=Dn	d−Dn→Dn; Dn−s→Dn; d−Dn→Dn; Dn−s→Dn	1001 DDD1 SSEE EEEE; 1001 DDD0 SSEE EEEE; 1001 DDD1 SSEE EEEE; 1001 DDD0 SSEE EEEE	• • • • •
SUBA Subtract Address	W d=An; L d=An	Dn−s→An; An−s→An	1001 AAA0 11ee eeee; 1001 AAA1 11ee eeee	− − − − −
SUBI Subtract Immediate	B/W d=Imm; L s=Imm	d−#→d	0000 0100 SSEE EEEE	• • • • •
SUBQ Subtract Quick	B/W d=Imm3; L s=Imm3	d−#→d	0101 QQQ1 SSEE EEEE	• • • • •
SUBX Subtract Multiprecision	B/W s=Dn d-; L s=−(An) d-	d−s−X→d	1001 RRR1 SS00 0rrr; 1001 RRR1 SS00 1rrr	• • • • •
SWAP Swap Register Halves	W d-	Dn[31:16]↔Dn[15:0]	0100 1000 0100 0DDD	− • • 0 0

Figure 11-12 Continued

(continued)

Mnemonic / Operation	Size	Addr Mode	Dn	An	(An)	(An)+	-(An)	d(An)	d(An, Xi)	Abs W	Abs L	d(PC)	d(PC, Xi)	s Immd / d SR/CC	Opcode Bit Pattern (1111 11 5432 1098 / 7654 3210)	Boolean	Condition Codes (X N Z V C)
TAS Test and Set Operand	B	d-	4	-	15	15	17	19	21	19	23				0100 1010 11EE EEEE	test d→cc, 1→bit 7 of d	- · · 0 0
TST Test	B/W	d-	4		8	8	10	12	14	12	16				0100 1010 1SEE EEEE	test d→cc	- · · 0 0
	L	d-	4		12	12	14	16	18	16	20						
UNLK Unlink				12											0100 1110 0101 1AAA	An→SP, (SP)+→An	- - - - -

Program Control Instructions

Mnemonic / Operation	Size	Addr Mode	Dn	An	(An)	(An)+	-(An)	d(An)	d(An, Xi)	Abs W	Abs L	d(PC)	d(PC, Xi)	s Immd / d SR/CC	Opcode Bit Pattern	Boolean	Condition Codes (X N Z V C)
Bcc Branch Conditionally	B	disp-	{10 / 12' / 14'}										bra taken / bra not taken	2 / 8' / 10	0110 CCCC PPPP PPPP	If cc true, PC+disp→PC	- - - - -
	W	disp-											bra taken / bra not taken	14' / 2 / 4			
BRA Branch Always	B	disp-													0110 0000 PPPP PPPP	PC+disp→PC	- - - - -
	W	disp-															
BSR Branch to Subroutine	B	disp-												20	0110 0001 PPPP PPPP	PC→-(SP), PC+disp→PC	- - - - -
	W	disp-imm												20			
DBcc Decrement Counter, & Branch Until Condition True or Count = -1	W	counter-													0101 CCCC 1100 1DDD	If cc false, Dn-1→Dn & if Dn ≠ -1, PC+disp→PC. Else, NOP	- - - - -
JMP Jump to		d-			8			10	14	10	12	10	14		0100 1110 11EE EEEE	d→PC	- - - - -
JSR Jump to Subroutine		d-			18			20	24	20	22	20	24		0100 1110 10EE EEEE	PC→-(SP), d→PC	- - - - -
NOP No Operation															0100 1110 0111 0001	none	- - - - -
RESET Reset External Devices			132												0100 1110 0111 0000	assert RESET pin	- - - - -
RTE Return from Exception			20												0100 1110 0111 0011	(SP)+→SR, (SP)+→PC	· · · · ·
RTR Return from Subroutine/Restore CC			20												0100 1110 0111 0111	(SP)+→CC, (SP)+→PC	· · · · ·
RTS Return from Subroutine			16												0100 1110 0111 0101	(SP)+→PC	- - - - -
STOP Load SR/Stop			37											4	0100 1110 0111 0010	#→SR, Wait for Interrupt	· · · · ·
TRAP Trap			37 / 4												0100 1110 0100 VVVV	PC→-(SP), SR→-(SP), (Vector)→PC	- - - - -
TRAPV Trap if Overflow Set			37 / 4												0100 1110 0111 0110	If V=1, then PC→-(SP), SR→-(SP), (TRAPV vector)→PC. else NOP	- - - - -

Branch sub-table (for Bcc / DBcc):

cc	Counter	Branch
false		yes
true		no
false		no
	≠ -1	
	≠ -1	
	expired	

Figure 11-12 Continued

eight groups of instructions available in the 68010 are listed below:

Data movement group
Integer arithmetic group
Logical operation group
Shift/rotate operation group
Bit manipulation group
BCD arithmetic group
Program control group
System control group

We will examine each of these instructions in detail in the following sections. For each instruction, we will present the assembler format, the name, the RTL description of the operations involved, and the addressing category. A more detailed description of the instructions is given in Appendix B. A summary table showing all the instructions in alphabetical order is shown in Figure 11–12.

11.5

The Data Movement Group

The data movement group of instructions deals with movement of data among data registers, address registers, memory, and other special internal registers. The data movement instructions are very powerful and allow many forms of specification for both the source and destination operands. There are five categories of data movement instructions for:

Data movement instructions for data registers
Data movement instructions for address registers
Data movement instructions for the status register
Data movement instructions for special registers
Special register movement instructions

Data Movement Instructions for Data Registers

The first category of data movement instructions involves the data registers. The basic operation allows movement of data operands of any size from any source to almost any alterable destination.

MOVE EA,EA   (Move data from source to destination)
    Operation: (Dest) ← (Source)
    Source addressing category: All
    Destination addressing category: Data alterable

MOVEQ #Data,Dn   (Move quick)
   Operation: (Dest) ← Immediate data
   Addressing category: None
MOVEM ⟨register list⟩,EA   (Move multiple registers)
   Operation: (Dest's) ← (Rn's)
   Addressing category: Control alterable or pre-decrement
MOVEM EA,⟨register list⟩   (Move multiple registers)
   Operation: (Rn's) ← (Source's)
   Addressing category: Control or post-increment
MOVEP Dn,d(An)   (Move peripheral data)
   Operation: (Dest) ← (Source)
   Addressing category: None
MOVEP d(An),Dn   (Move peripheral data)
   Operation: (Dest) ← (Source)
   Addressing category: None

In reality, there is only one general data movement instruction, the MOVE instruction. The other two instructions, the MOVEA (next section) and the MOVEQ instructions, are variants of the MOVE instruction. The MOVE instruction cannot alter address registers. Hence, if the destination is an address register, MOVEA is used. The MOVEA instruction is presented in the next section. If the source operand is a literal that is less than 8 bits long, MOVEQ can be used with much faster execution time because the literal is actually stored in the OPCODE, thus requiring no additional memory cycles to fetch the operand.

The MOVEM instruction allows storing and retrieving the contents of any or all of the data and address registers in one instruction. This is particularly useful in subroutine calls and in context-change environments. In the post-increment and control modes, the order of registers saved is D_0 to D_7, then A_0 to A_7. However, the order for the pre-decrement mode is just the reverse: starting with A_7 to A_0 and then D_7 to D_0. The MOVEP instruction is similar to input/output instructions, especially with byte-oriented peripherals, and allows input or output of up to 4 bytes of data in one instruction.

Data Movement Instructions for Address Registers

The instructions presented in the last section deal mostly with data registers. Address registers are handled in somewhat different ways than data registers. Although byte operations are allowed with operations using data registers, byte operations are not allowed for instructions dealing with address registers.

MOVEA EA,An   (Move address)
   Operation: (Dest) ← (Source)
   Addressing category: All
LEA EA,An   (Load effective address)
   Operation: (An) ← Source EA
   Addressing category: Control

PEA EA   (Push effective address)
    Operation: $(-(SP)) \leftarrow$ Source EA
    Addressing category: Control
LINK An, #d   (Link and allocate)
    Operation: $((SP)-) \leftarrow (An)$
    $(An) \leftarrow (SP)$
    $(SP) \leftarrow (SP) + d$
    Addressing category: None
UNLK,An   (Unlink)
    Operation: $(SP) \leftarrow (An)$
    $An \leftarrow ((SP)+)$
    Addressing category: None

The MOVEA instruction is similar to the general MOVE instruction except that the destination is always taken to be one of the address registers. The LEA and PEA instructions are used for storing or retrieving address parameters in subroutines and procedure calls to or from address registers. The LINK and UNLK instruction pair allows a powerful mechanism to allocate space for local variables. The LINK and UNLK instructions are also useful for linked list manipulations. The use of the LINK and UNLK instructions will be discussed in detail in Section 11.14.

Data Movement Instructions for the Status Register

There are a number of MOVE instructions dealing with the status registers. Those instructions that affect only the user part of the status registers are common and available to the user; that is, they are nonprivileged. Those instructions that affect the complete status register, both the user part and the supervisory part, are privileged instructions and must be executed only while the processor is in the supervisory mode. Otherwise, a TRAP is generated.

MOVE CCR,EA   (Move from the condition code register)
    Operation: $(Dest) \leftarrow (CCR)$
    Addressing category: Data alterable
MOVE EA,CCR   (Move to condition codes)
    Operation: $(CCR) \leftarrow (Source)$
    Addressing category: Data
MOVE SR,EA   (Move from the status register)
    Operation: If supervisory state then $(Dest) \leftarrow (SR)$
                       else TRAP
    Addressing category: Data alterable
MOVE EA,SR   (Move to the status register)
    Operation: If supervisory state then $(SR) \leftarrow (Source)$
                       else TRAP
    Addressing category: Data

The MOVE from CCR and MOVE to CCR instructions allow manipulations of the condition codes but the form is the same as the general MOVE instruction. The MOVE from SR and MOVE to SR instructions allow manipulations of the complete status register and therefore are privileged instructions.

Data Movement Instructions for Special Registers

There are a number of special registers in the 68010. Most of these are useful for basic machine operations. Hence the manipulation of the data in these special registers is privileged and is restricted to system programs being executed when the processor is in the supervisory mode.

MOVE USP,An   (Move user stack pointer)
    Operation: If supervisory state then (An) ← (USP)
                         else TRAP
    Addressing category: Data alterable
MOVE An,USP   (Move user stack pointer)
    Operation: If supervisory state then (USP) ← (An)
                         else TRAP
    Addressing category: None
MOVEC Rc,Rn   (Move to/from control register)
    Operation: If supervisory state then (Rn) ← (Rc)
                         else TRAP
    Addressing category: None
MOVEC Rn,Rc   (Move to/from control register)
    Operation: If supervisory state then (Rc) ← (Rn)
                         else TRAP
    Addressing category: None
MOVES Rn,EA   (Move from address space)
    Operation: If supervisory state then (Dest DFC) ← (Rn)
                         else TRAP
    Addressing category: Alterable memory
MOVES EA,Rn   (Move to address space)
    Operation: If supervisory state then (Rn) ← (Source SFC)
                         else TRAP
    Addressing category: Alterable memory

For context changes, it is helpful for the supervisor to save and restore the user stack pointer. This is accomplished by the MOVE from USP and MOVE to USP instructions. The reading and restoring of other special registers are accomplished in the same way.

Special Register Movement Instructions

The last group of data movement instructions provides data interchange within the register (SWAP) or between registers (EXG).

EXG Rx,Ry   (Exchange registers)
   Operation: (Ry) ↔ (Rx)
   Addressing category: None
SWAP Dn   (Swap register halves)
   Operation: (Dn)(15:0) ↔ (Dn)(31:16)
   Addressing category: None

Note that the EXG instruction also allows data interchange between data registers and address registers.

Programming Examples for the Data Movement Group

Data movement instructions are basic to the operation of any program. In this section we examine several examples using data movement instructions. Examples dealing with the LINK and UNLK instructions are presented and discussed in more detail later in the section dealing with parameter passing in subroutines and procedure calls.

Example 11–1

Write a program to move 5 words from memory into data registers D_0 through D_4.

The program requires sending the address of the values to an address register and the actual transfer of data.

```
Label      OPCODE       Operand                  Comments
*
* THIS PROGRAM TRANSFERS FIVE MEMORY LOCATIONS
* INTO DATA REGISTERS
*
           ORG          $1000
           LEA          TABLE(PC),A0    *LOAD TABLE ADDRESS
           MOVE         (A0)+,D0        *TRANSFER DATA FROM MEMORY
           MOVE         (A0)+,D1        *ADDRESS INCREMENTED AFTER
           MOVE         (A0)+,D2        *EACH LOAD
           MOVE         (A0)+,D3
           MOVE         (A0)+,D4
             .
             .
             .
TABLE      DS 5
           END
```

Example 11-2 ▬▬▬▬▬▬▬▬▬▬▬▬▬▬▬▬▬▬▬▬▬▬▬▬▬▬▬▬▬

Write a program to clear all condition codes.

To clear the condition codes, we will zero out a data register and move the contents to the condition codes. An alternative is to use the MOVE to CCR instruction with immediate addressing for the source operand.

```
Label       OPCODE      Operand                 Comments
*
* THIS PROGRAM CLEARS ALL CONDITION CODES
*
            ORG         $1000
            MOVEQ       #0,D0       *ZERO DATA REGISTER
            MOVE        D0,CCR      *TRANSFER TO CCR
            END
```

Example 11-3 ▬▬▬▬▬▬▬▬▬▬▬▬▬▬▬▬▬▬▬▬▬▬▬▬▬▬▬▬▬

Simulate the following MOVEM instruction with regular MOVE instructions:

```
MOVEM D1-D4/A2-A3,-(A6)
```

In the above MOVEM instruction, the data registers D_1 through D_4 and the address registers A_2 and A_3 are stored in the stack whose address is pointed to by address register A_6 in a pre-decrement mode. To simulate this instruction, we will need a MOVE instruction for each register.

```
Label       OPCODE      Operand                     Comments
*
* THIS PROGRAM SIMULATES A MOVEM INSTRUCTION
*
            ORG         $1000
            MOVE        A3,-(A6)    *ADDRESS REGISTERS ARE FIRST TO
            MOVE        A2,-(A6)    *BE STORED
            MOVE        D4,-(A6)    *DATA REGISTERS ARE LAST TO BE
            MOVE        D3,-(A6)    *STORED IN POST-DECREMENT
            MOVE        D2,-(A6)
            MOVE        D1,-(A6)
            END
```

Example 11-4

Convert a hexadecimal digit into its ASCII, decimal, and Gray code equivalents.

This is an example of a table look-up application. The value of the datum received becomes the index of a table, and the corresponding item from the table is read. The program is as follows:

```
Label      OPCODE          Operand                    Comments
*
* USING LOOKUP TABLES, THE ASCII, DECIMAL,
* AND GRAY CODE VALUES OF A HEX DIGIT
* IN D0 ARE STORED AT LOCATIONS TO WHICH
* A1 IS POINTING. A0 POINTS TO THE SET
* OF LOOKUP TABLES
*
           ORG             $1000
           LEA             ASCTABLE(PC),A0            *A0 POINTS TO ASCTABLE
           MOVE.B          0(A0,D0),(A1)+             *ASCII VALUE GOES TO TEMP 1
           MOVE.B          $10(A0,D0),(A1)+           *BCD VALUE GOES TO TEMP 2
           MOVE.B          $20(A0,D0),(A1)+           *GRAY VALUE GOES TO TEMP 3
            .
            .
            .
ASCTABLE DC.B $30,$31,$32,$33,$34,$35,$36,$37,$38,$39
         DC.B $41,$42,$43,$44,$45,$46
BCDTABLE DC.B 0,1,2,3,4,5,6,7,8,9,$10,$11,$12,$13,$14
GRCTABLE DC.B 0,1,3,2,6,7,5,4,$C,$D,$F,$E,$A,$B,9,8
         END
```

(Reprinted with permission of Motorola Inc.)

11.6

The Integer Arithmetic Group

The integer arithmetic group of instructions provides extensive support for data manipulation operations. There are four categories of integer arithmetic instructions:

Dual-operand integer arithmetic instructions
Single-operand integer arithmetic instructions
General comparison instructions
Special comparison instructions

Dual-Operand Integer Arithmetic Instructions

Most integer arithmetic instructions have two operands. Since only two operands can be specified, the destination of the operation is taken to be one of the source operands. Hence the second specified operand is both the source and destination operand. The main operations included in this category are arithmetic addition, subtraction, multiplication, and division.

ADD EA,Dn   (Add binary)
   Operation: (Dest) ← (Source) + (Dest)
   Addressing category: All

ADD Dn,EA   (Add binary)
   Operation: (Dest) ← (Source) + (Dest)
   Addressing category: Alterable memory

ADDA EA,An   (Add address)
   Operation: (Dest) ← (Source) + (Dest)
   Addressing category: All

ADDI #Data,EA   (Add immediate)
   Operation: (Dest) ← Immediate data + (Dest)
   Addressing category: Data alterable

ADDQ #Data,EA   (Add quick)
   Operation: (Dest) ← Immediate data + (Dest)
   Addressing category: Alterable

ADDX Dy,Dx   (Add extended)
   Operation: (Dest) ← (Source) + (Dest) + (X)
   Addressing category: None

ADDX −(Ay),−(Ax)   (Add extended)
   Operation: (Dest) ← (Source) + (Dest) + (X)
   Addressing category: None

SUB EA,Dn   (Subtract binary)
   Operation: (Dest) ← (Dest) − (Source)
   Addressing category: All

SUB Dn,EA   (Subtract binary)
   Operation: (Dest) ← (Dest) − (Source)
   Addressing category: Alterable memory

SUBA EA,An   (Subtract address)
   Operation: (Dest) ← (Dest) − (Source)
   Addressing category: All

SUBI #Data,EA   (Subtract immediate)
   Operation: (Dest) ← (Dest) − Immediate data
   Addressing category: Data alterable

SUBQ #Data,EA   (Subtract quick)
   Operation: (Dest) ← (Dest) − Immediate data
   Addressing category: Alterable

SUBX Dy,Dx   (Subtract with extend)
   Operation: (Dest) ← (Dest) − (Source) − (X)
   Addressing category: None

SUBX −(Ay),−(Ax)   (Subtract with extend)
    Operation: (Dest) ← (Dest) − (Source) − (X)
    Addressing category: None
MULS EA,Dn   (Signed multiply)
    Operation: (Dest) ← (Source) * (Dest)
    Addressing category: Data
MULU EA,Dn   (Unsigned multiply)
    Operation: (Dest) ← (Source) * (Dest)
    Addressing category: Data
DIVS EA,Dn   (Signed divide)
    Operation: (Dest) ← (Dest)/(Source)
    Addressing category: Data
DIVU EA,Dn   (Unsigned divide)
    Operation: (Dest) ← (Dest)/(Source)
    Addressing category: Data

There are several variants of the ADD and SUB instructions. Take the ADD instruction as an example. Besides the general ADD instruction for adding any operand to one of the data operands and storing the result either in memory or in the specified data register, there are four additional variants. The ADDA instruction is used when the destination is an address register. The ADDI instruction is used when one of the source operands is a literal. The ADDQ instruction is used when one of the source operands is a literal and is also less than 8 bits. The ADDX instruction is used when the contents of the extend bit X is also added into the sum. The ADDX instruction is used when manipulating multiple-precision numbers. The same variations are provided for subtraction.

For both multiplication and division, a signed operation and an unsigned operation are provided. The signed operation is used for single-precision arithmetic, whereas the unsigned version is used primarily for multiple-precision arithmetic and for unsigned numbers.

Single-Operand Integer Arithmetic Instructions

The single-operand instructions include operations that change the form of a single operand. There are four instructions in this category.

NEG EA   (Negate)
    Operation: (Dest) ← 0 − (Dest)
    Addressing category: Data alterable
NEGX EA   (Negate with extend)
    Operation: (Dest) ← 0 − (Dest) − (X)
    Addressing category: Data alterable
EXT Dn   (Sign extend)
    Operation: (Dest) ← Sign-extended (Dest)
    Addressing category: None

CLR EA   (Clear an operand)
    Operation: (Dest) ← 0
    Addressing category: Data alterable

The difference between the NEG and the NEGX instructions is the extend bit. The instruction NEG is used for single-precision numbers, whereas NEGX is for multiple-precision numbers. The EXT instruction allows conversions from byte to word quantities or from word to long word quantities. The CLR instruction is used for clearing or zeroing any register operands or memory operands.

General Comparison Instructions

Another category in the integer arithmetic group is comparison instructions. These are put here because most of the comparisons are performed arithmetically by subtraction in 2's complement form. The result of the subtraction is not stored anywhere; however, the condition codes are changed according to the result of the subtraction.

CMP EA,Dn   (Compare)
    Operation: (Dest) − (Source)
    Addressing category: All
CMPA EA,An   (Compare address)
    Operation: (Dest) − (Source)
    Addressing category: All
CMPI #Data,EA   (Compare immediate)
    Operation: (Dest) − Immediate data
    Addressing category: Data alterable
CMPM (Ay)+,(Ax)+   (Compare memory)
    Operation: (Dest) − (Source)
    Addressing category: None

The CMP instruction is done by subtracting the destination operand from the source operand and by altering the condition codes according to the result of the subtraction. Similar to the ADD and SUB instructions, the CMP instruction also has three variants. The CMPA is used whenever the destination operand is in the address register. The CMPI instruction is used when the source operand is a literal. The CMPM instruction is used whenever both operands, source and destination, are in memory. In this case, only the address register indirect with post-increment addressing mode can be used.

Special Comparison Instructions

Special comparison instructions are operations related to the condition codes or other operand-checking functions. There are four special comparison instructions.

TST EA   (Test an operand)
    Operation: (CCR) ← (Dest) Tested
    Addressing category: Data alterable

Scc EA   (Set according to condition)
  Operation: If condition true then (Dest) ← 1's
           else (Dest) ← 0's
  Addressing category: Data alterable
TAS EA   (Test and set an operand)
  Operation: (CCR) ← (Dest) Tested
    (Dest(7)) ← 1
  Addressing category: Data alterable
CHK EA,Dn   (Check register against bounds)
  Operation: If (Dn) < 0 or (Dn) > (EA) then TRAP
  Addressing category: Data

The TST instruction merely sets the condition codes according to the current contents of the specified operand. The TST instruction differs from the CMP instruction in that no subtraction is made. In essence, the TST instruction is the same as the CMP instruction with the destination operand set to zero.

The Scc instruction is used to set the values of logical variables based on either an arithmetic comparison or a logical operation. A list of conditions and their encoding is given in Figure 11–13.

The TAS instruction is used to test and modify, in one indivisible instruction, the specified bit in the operand specified by the effective address. This instruction

Figure 11–13

Branching conditions and their encoding bits (Reprinted with permission of Motorola Inc.)

Mnemonic	Condition	Encoding	Test
T	True	0000	1
F	False	0001	0
HI	High	0010	$\bar{C} \cdot \bar{Z}$
LS	Low or same	0011	$C + Z$
CC (HS)	Carry clear	0100	$\bar{C}$
CS (LO)	Carry set	0101	C
NE	Not equal	0110	$\bar{Z}$
EQ	Equal	0111	Z
VC	Overflow clear	1000	$\bar{V}$
VS	Overflow set	1001	V
PL	Plus	1010	$\bar{N}$
MI	Minus	1011	N
GE	Greater or equal	1100	$N \cdot V + \bar{N} \cdot \bar{V}$
LT	Less than	1101	$N \cdot \bar{V} + \bar{N} \cdot V$
GT	Greater than	1110	$N \cdot V \cdot \bar{Z} + \bar{N} \cdot \bar{V} \cdot \bar{Z}$
LE	Less or equal	1111	$Z + N \cdot \bar{V} + \bar{N} \cdot V$

is needed in the implementation of semaphores or critical sections where data integrity must be guaranteed.

The CHK instruction can be used to check the bounds of the contents of the identified data register against the specified operand. If the specified operand is an address, then the CHK instruction can be used to detect references outside the user's area. If the specified operand is an index, then the CHK instruction can be used to check the limits of array subscripts. A trap is generated if the contents of the data register are outside the limits of the specified operand.

Programming Examples for the Integer Arithmetic Group

In this section we present several examples dealing with the use of integer arithmetic instructions. We also look at some examples dealing with comparisons.

Example 11-5

Write a program to add five numbers stored in memory starting at location TABLE. Store the result at the end of the table.

The solution of this example is similar to that given earlier. Instead of using the MOVE instruction, we can use the ADD instruction here.

```
        Label       OPCODE      Operand                     Comments
        *
        * THIS PROGRAM ADDS FIVE MEMORY OPERANDS
        * THE RESULT IS STORED AT THE END OF THE TABLE
        *
                    ORG         $1000
                    LEA         TABLE(PC),A0        *GET ADDRESS OF TABLE
                    MOVE        (A0)+,D0            *GET FIRST OPERAND
                    ADD         (A0)+,D0            *ADD TO REST OF OPERANDS
                    ADD         (A0)+,D0
                    ADD         (A0)+,D0
                    ADD         (A0)+,D0
                    MOVE        D0,(A0)             *STORE RESULT
                      .
                      .
                      .
        TABLE       DS 6
                    END
```

Example 11–6 ▰▰▰▰▰▰▰

Write a program to multiply an unsigned 32-bit multiplicand by an unsigned 16-bit multiplier.

When unsigned numbers are used, the arithmetic of multiplication for multiprecision operand is rather simple. Assume that the 32-bit multiplicand is in D_0, and the 16-bit multiplier is in D_1. The most significant bits of the 48-bit result should be stored in D_2 and the least significant bits in D_3.

Label	OPCODE	Operand	Comments
*			

```
* THIS PROGRAM MULTIPLIES A 32-BIT MULTIPLICAND IN D0
* WITH A 16-BIT MULTIPLIER IN D1
        ORG      $1000
        CLR.L    D2          *CLEAR DATA REGISTER
        MOVE.W   D0,D2       *GET LOWER HALF OF MULTIPLICAND
        MULU     D1,D2       *MULTIPLY
        MOVE.L   D0,D4       *GET UPPER HALF OF MULTIPLICAND
        CLR.W    D4          *STRIP OFF LOWER HALF
        SWAP     D4
        MULU     D1,D4       *MULTIPLY
*
* COMBINE THE TWO HALVES OF RESULTS
*
        CLR.L    D3
        MOVE.W   D4,D3       *GET LOWER PART
        SWAP     D3
        ADD.L    D3,D2       *LOWER PART OF RESULT
        CLR.L    D3
        SWAP     D4
        ADDX.W   D4,D3       *UPPER PART OF RESULT
        END
```

Example 11–7 ▰▰▰▰▰▰▰

Assume that there is a two-dimensional array with dimensions stored in memory locations A and B. Write a subscript-checking routine to check if the unknown subscripts X and Y are within bounds.

To check if a subscript is out of bounds, first store the bounds in memory. The subscript can then be put into a data register and checked against the contents of the specified memory operands.

Label	OPCODE	Operand	Comments
*			

* THIS PROGRAM CHECKS THE LIMITS OF A SUBSCRIPT

Label	OPCODE	Operand	Comments
*			
	ORG	$1000	
	MOVE	X(PC),D0	*GET FIRST SUBSCRIPT
	CHK	A(PC),D0	*CHECK
	MOVE	Y(PC),D0	*GET SECOND SUBSCRIPT
	CHK	B(PC),D0	*CHECK
	.		
	.		
	.		
X	DS	1	
Y	DS	1	
A	DS	1	
B	DS	1	
	END		

Example 11–8

Demonstrate the use of the Scc instruction.

The Scc instruction sets the value of a logical variable for later testing and actions. In the following example, the equal flag is stored in D_2 and the positive flag in D_3. These flags are set according to the operand stored in D_1.

Label	OPCODE	Operand	Comments
*			

* THE SCC INSTRUCTION ALLOWS THE USER
* TO REMEMBER THE RESULT (FROM THE CCR)
* OF AN INSTRUCTION AND ACT UPON IT LATER

Label	OPCODE	Operand	Comments
*			
	ORG	$2000	
	CLR.W	D0	
	CLR.W	D1	
	CMP.W	D0,D1	
	SEQ.B	D2	
	ADDQ.W	#1,D0	
	SEQ.B	D3	
	END		

(Reprinted with permission of Motorola Inc.)

11.7

The Logical Operation Group

The logical operation group provides a complete set of instructions for performing logic operations. Most of the common logic operations are available in the 68010. There are three types of logic operations:

Dual-operand logic instructions for data registers
Dual-operand logic instructions for the status register
Single-operand logic instructions

These are now examined in more detail.

Dual-Operand Logic Instructions for Data Registers

Most of the dual-operand logic instructions use a data register as one of the operands.

AND EA,Dn   (Logical AND)
    Operation: (Dest) ← (Source) ∧ (Dest)
    Addressing category: Data
AND Dn,EA   (Logical AND)
    Operation: (Dest) ← (Source) ∧ (Dest)
    Addressing category: Alterable memory
ANDI #Data,EA   (AND immediate)
    Operation: (Dest) ← Immediate data ∧ (Dest)
    Addressing category: Data alterable
OR EA,Dn   (Logical OR)
    Operation: (Dest) ← (Source) ∨ (Dest)
    Addressing category: Data alterable
OR Dn,EA   (Logical OR)
    Operation: (Dest) ← (Source) ∨ (Dest)
    Addressing category: Alterable memory
ORI #Data,EA   (Inclusive OR immediate)
    Operation: (Dest) ← Immediate data ∨ (Dest)
    Addressing category: Data alterable
EOR Dn,EA   (Exclusive OR logical)
    Operation: (Dest) ← (Source) ⊕ (Dest)
    Addressing category: Data alterable
EORI #Data,EA   (Exclusive OR immediate)
    Operation: (Dest) ← Immediate data ⊕ (Dest)
    Addressing category: Data alterable

Three principal logic operations are provided. These are the AND, OR, and Exclusive OR operations. For all logic instructions, one of the operands must be either a data register or immediate data. Note that for the EOR instruction, there is no EOR EA,Dn form. Why? There are other logic instructions dealing with the condition codes and the status register. These instructions are presented later in the system group.

Dual-Operand Logic Instructions for the Status Register

In the next group of logic operations, the instructions all have the condition code register or the status register as their destination operand.

ANDI #Data,CCR   (AND immediate to condition codes)
    Operation: (CCR) ← (CCR) ∧ (Source)
    Addressing category: None
ORI #Data,CCR   (Inclusive OR immediate to condition codes)
    Operation: (CCR) ← (Source) ∨ (CCR)
    Addressing category: None
EORI #Data,CCR   (Exclusive OR immediate to condition codes)
    Operation: (CCR) ← (CCR) ⊕ (Source)
    Addressing category: None
ANDI #Data,SR   (AND immediate to the status register)
    Operation: If supervisory state then (SR) ← (SR) ∧ (Source)
                                   else TRAP
    Addressing category: None
ORI #Data,SR   (Inclusive OR immediate to the status register)
    Operation: If supervisory state then (SR) ← (SR) ∨ (Source)
                                    else TRAP
    Addressing category: None
EORI #Data,SR   (Exclusive OR immediate to the status register)
    Operation: If supervisory state then (SR) ← (Source) ⊕ (SR)
                                   else TRAP
    Addressing category: None

As before, those operations dealing only with the condition code register are nonprivileged instructions, whereas those operations dealing with the complete status register are privileged instructions.

Single-Operand Logic Instructions

There is only one single-operand logic instruction. This is the NOT instruction as shown below.

NOT EA   (Logical complement)
    Operation: (Dest) ← $\overline{\text{(Dest)}}$
    Addressing category: Data alterable

The NOT operation takes the logical complement of the specified operand. Its function is similar to its arithmetic counterpart of the unary minus.

Programming Examples for the Logical Operation Group

In this section we examine a few examples dealing with logic instructions.

Example 11-9 ▰▰▰▰▰▰▰▰▰▰▰▰▰▰▰▰▰▰▰▰▰▰▰▰▰▰▰▰▰▰▰▰▰▰▰▰▰▰

Evaluate the following logic expression assuming that A, B, C, D, and E are addresses of arithmetic variables.

`((A-B)/(C-D).EQ.E).AND.(A.NE.B)`

The above expression can be evaluated using both arithmetic and logical operations.

Label	OPCODE	Operand	Comments
*			
* THIS PROGRAM EVALUATES THE LOGICAL EXPRESSION			
* GIVEN ABOVE			
*			
	ORG	$1000	
	MOVE	A(PC),D0	*GET A
	SUB	B(PC),D0	*A − B
	MOVE	C(PC),D1	*GET C
	SUB	D(PC),D1	*C − D
	DIVS	D1,D0	*(A − B)/(C − D)
	CMP	E(PC),D0	
	SEQ	D0	
	MOVE	A(PC),D1	*COMPARE A AND B
	CMP	B(PC),D1	
	SNE	D1	
	AND	D1,D0	*RESULT IN D0
	.		
	.		
	.		
A	DS	1	
B	DS	1	
C	DS	1	
D	DS	1	
E	DS	1	
	END		

11.8
The Shift/Rotate Operation Group

The shift/rotate group of instructions provides both shift and rotate instructions. In addition, both logical and arithmetic shifts are implemented. We can separate instructions in this group into:

Shift instructions
Rotate instructions

To further understand the difference between the various shift instructions and rotate instructions, these instructions are illustrated in Figure 11–14.

Figure 11–14

Shift and rotate instructions (Reprinted with permission of Motorola Inc.)

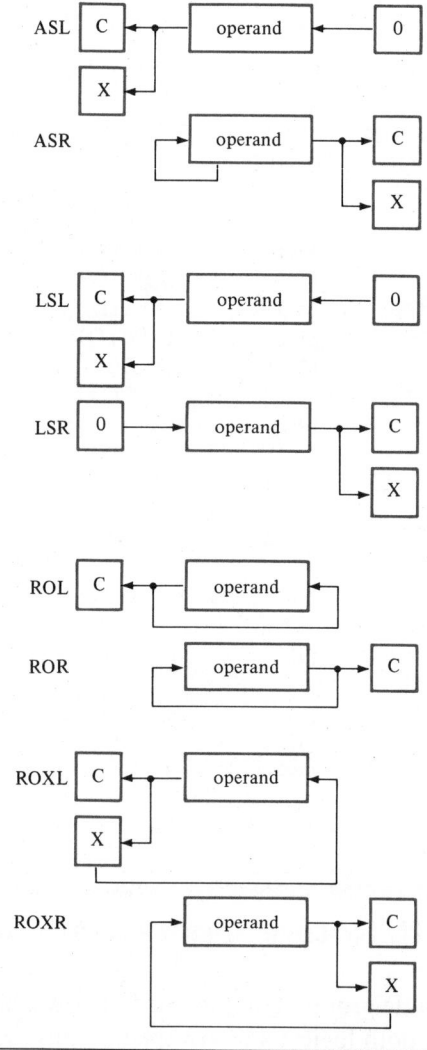

Shift Instructions

There are two kinds of shifts: arithmetic shifts and logical shifts. In an arithmetic shift, the most significant bit is considered as a sign bit, and the sign of the number is thus preserved. In a logical shift, the most significant bit is not considered as a sign bit.

ASL Dx,Dy   (Arithmetic shift)
    Operation: (Dest) ← (Dest) shifted by ⟨count⟩ in Dx
    Addressing category: None

ASL #Data,Dy   (Arithmetic shift)
    Operation: (Dest) ← (Dest) shifted by ⟨count⟩ in #Data
    Addressing category: None

ASL EA   (Arithmetic shift)
    Operation: (Dest) ← (Dest) shifted by 1 bit
    Addressing category: Memory alterable

ASR Dx,Dy   (Arithmetic shift)
    Operation: (Dest) ← (Dest) shifted by ⟨count⟩ in Dx
    Addressing category: None

ASR #Data,Dy   (Arithmetic shift)
    Operation: (Dest) ← (Dest) shifted by ⟨count⟩ in #Data
    Addressing category: None

ASR EA   (Arithmetic shift)
    Operation: (Dest) ← (Dest) shifted by 1 bit
    Addressing category: Memory alterable

LSL Dx,Dy   (Logical shift)
    Operation: (Dest) ← (Dest) shifted by ⟨count⟩ in Dx
    Addressing category: None

LSL #Data,Dy   (Logical shift)
    Operation: (Dest) ← (Dest) shifted by ⟨count⟩ in #Data
    Addressing category: None

LSL EA   (Logical shift)
    Operation: (Dest) ← (Dest) shifted by 1 bit
    Addressing category: Memory alterable

LSR Dx,Dy   (Logical shift)
    Operation: (Dest) ← (Dest) shifted by ⟨count⟩ in Dx
    Addressing category: None

LSR #Data,Dy   (Logical shift)
    Operation: (Dest) ← (Dest) shifted by ⟨count⟩ in #Data
    Addressing category: None

LSR EA   (Logical shift)
    Operation: (Dest) ← (Dest) shifted by 1 bit
    Addressing category: Memory alterable

There are arithmetic shifts and logical shifts. Multiple shifts in a single instruction are possible if the operand is stored in a data register. Depending on the number of shifts required, the shift count can be specified statically as part of the OPCODE or dynamically in another data register. The latter case is very useful, especially when the number of shifts is dependent on the data and cannot be determined during program development. In addition to shifting data in a data register, it is also possible to shift memory operands. However, only one bit can be shifted with each instruction.

Rotate Instructions

In the rotate instructions, the extend bit can be included or excluded in the shift path. Whatever is rotated out of one end of the operand is shifted into the other end of the operand. This is also sometimes called a circular shift.

ROL Dx,Dy   (Rotate without extend)
    Operation: (Dest) ← (Dest) rotated by ⟨count⟩ in Dx
    Addressing category: None

ROL #Data,Dy   (Rotate without extend)
    Operation: (Dest) ← (Dest) rotated by ⟨count⟩ in #Data
    Addressing category: None

ROL EA   (Rotate without extend)
    Operation: (Dest) ← (Dest) rotated by 1 bit
    Addressing category: Alterable memory

ROR Dx,Dy   (Rotate without extend)
    Operation: (Dest) ← (Dest) rotated by ⟨count⟩ in Dx
    Addressing category: None

ROR #Data,Dy   (Rotate without extend)
    Operation: (Dest) ← (Dest) rotated by ⟨count⟩ in #Data
    Addressing category: None

ROR EA   (Rotate without extend)
    Operation: (Dest) ← (Dest) rotated by 1 bit
    Addressing category: Alterable memory

ROXL Dx,Dy   (Rotate with extend)
    Operation: (Dest, X) ← (Dest, X) rotated by ⟨count⟩ in Dx
    Addressing category: None

ROXL #Data,Dy   (Rotate with extend)
    Operation: (Dest, X) ← (Dest, X) rotated by ⟨count⟩ in #Data
    Addressing category: None

ROXL EA   (Rotate with extend)
    Operation: (Dest, X) ← (Dest, X) rotated by 1 bit
    Addressing category: Alterable memory

ROXR Dx,Dy   (Rotate with extend)
    Operation: (X, Dest) ← (X, Dest) rotated by ⟨count⟩ in Dx
    Addressing category: None

ROXR #Data,Dy   (Rotate with extend)
    Operation: (X, Dest) ← (X, Dest) rotated by ⟨count⟩ in #Data
    Addressing category: None

ROXR EA   (Rotate with extend)
    Operation: (X, Dest) ← (X, Dest) rotated by 1 bit
    Addressing category: Alterable memory

The rotate instructions are very similar to the shift instructions. The difference is that in rotate instructions, the shifted-out bits are recycled back either through the X bit or not through the X bit.

Programming Examples for the Shift/Rotate Operation Group

It is interesting to examine the trade-offs between shifting in memory and shifting by means of a data register. This is done in the next example.

Example 11–10 ■■■■■■■■■■■■■■■■■■■■■■■■■

Find the minimum shift length so that shifting by a data register is faster than shifting in memory.

The solution can be obtained by calculating the number of clock cycles required between shifting in memory and shifting by a data register. The program to shift an operand in memory is as follows:

```
ROL (A0)          13 clock cycles
ROL (A0)          13 clock cycles
    .
    .
    .
```

The above statements are repeated N times for an N-bit shift. The total number of clock cycles required is $N_m = 13 * N$.

The program to shift by moving the data into a data register is as follows:

```
MOVE (A0),D1      8 clock cycles
ROL #N,D1         6 + 2N clock cycles
MOVE D1,(A0)      9 clock cycles
```

The total number of clock cycles is $N_d = 23 + 2 * N$.

To compare, we wish to find out when is it feasible to bring data into a data register for shifting. For $N_d < N_m$, N must be greater than 2.1. Hence the break point for the shift count when N_m is greater than N_d is 3. In other words, if the shift count N is 2 or less, then a memory shift is faster. If N is 3 or larger, then bringing the operand into the data register before shifting is faster. (Reprinted with permission of Motorola Inc.)

Example 11–11 ■■■■■■■■■■■■■■■■■■■■■■■■■

Convert a long word with eight BCD digits into eight ASCII digits and store them starting at location ASCADDR.

In this example, eight BCD digits are stored in one memory location, X. Each group of 4 bits must be shifted out to yield a single BCD digit to form an ASCII character.

Label	OPCODE	Operand	Comments
*			
* THIS PROGRAM CONVERTS 8 BCD DIGITS INTO THEIR			
* ASCII EQUIVALENTS			
*			
	ORG	$1000	
BCDADDR	DS	1	
ASCADDR	DS.B	8	
	LEA	ASCADDR(PC),A1	*SET ADDRESSES
	LEA	BCDADDR(PC),A0	
*			
	MOVE.B	(A0),D0	*GET FIRST BYTE WITH
	LSR	#4,D0	*2 DIGITS, CHANGE
	ANDI.B	#$30,D0	*LEFT DIGIT INTO ASCII
	MOVE.B	D0,(A1)+	
*			
	MOVE.B	(A0)+,D0	*GET BYTE AGAIN AND
	ANDI.B	#$30,D0	*INC ADDRESS, CHANGE
	MOVE.B	D0,(A1)+	*RIGHT DIGIT INTO ASCII
*			
	MOVE.B	(A0),D0	*GET SECOND BYTE
	LSR	#4,D0	*CHANGE LEFT DIGIT
	ANDI.B	#$30,D0	
	MOVE.B	D0,(A1)+	
*			
	MOVE.B	(A0)+,D0	
	ANDI.B	#$30,D0	*CHANGE RIGHT DIGIT
	MOVE.B	D0,(A1)+	
*			
	MOVE.B	(A0),D0	*GET THIRD BYTE
	LSR	#4,D0	*CHANGE LEFT DIGIT
	ANDI.B	#$30,D0	
	MOVE.B	D0,(A1)+	
*			
	MOVE.B	(A0)+,D0	
	ANDI.B	#$30,D0	*CHANGE RIGHT DIGIT
	MOVE.B	D0,(A1)+	
*			
	MOVE.B	(A0),D0	*GET FOURTH BYTE
	LSR	#4,D0	*CHANGE LEFT DIGIT
	ANDI.B	#$30,D0	
	MOVE.B	D0,(A1)+	
*			
	MOVE.B	(A0)+,D0	
	ANDI.B	#$30,D0	*CHANGE RIGHT DIGIT
	MOVE.B	D0,(A1)+	
	END		

Example 11-12 ▰▰▰▰▰▰▰▰▰▰▰▰▰▰▰▰▰▰▰▰▰▰▰▰▰▰▰▰▰▰▰▰▰▰▰▰▰▰

Convert a positive four-digit BCD number into its binary equivalent. The BCD number is stored in BCDTMP, and the binary number is stored in BINTMP.

This number conversion can be accomplished simply by multiplying each successive BCD digit or decade by 10. This is shown below.

```
     Label      OPCODE       Operand                      Comments
*
* THIS PROGRAM CONVERTS A BCD NUMBER INTO BINARY
*
                ORG          $1000
BCDTMP          DS.W         1
BINTMP          DS.L         1
                MOVE.W       #10,D0              *LOAD MULTIPLIER
                LEA          BCDTMP(PC),A0       *GET ADDRESSES
*               LEA          BINTMP(PC),A1
                MOVE.B       (A0),D1             *GET FIRST BYTE WITH 2 DIGITS
                LSR.B        #4,D1               *GET FIRST DIGIT
                MULU         D0,D1               *MULTIPLY BY 10
                MOVE.B       (A0)+,D2            *GET SECOND DIGIT
                AND.B        #$0F,D2
                ADD          D2,D1               *ADD TO SUM
                MULU         D0,D1               *MULTIPLY BY 10
*
                MOVE.B       (A0),D2             *GET SECOND BYTE
                LSR.B        #4,D2               *GET FIRST DIGIT
                ADD.B        D2,D1               *ADD TO SUM
                MULU         D0,D1               *MULTIPLY BY 10
                MOVE.B       (A0)+,D2            *GET LAST DIGIT
                AND.B        #$0F,D2
                ADD          D2,D1               *ADD TO SUM
                MOVE         D1,(A1)             *STORE RESULT
                END
```

11.9 ▰▰

The Bit Manipulation Group

The bit manipulation group of instructions allows testing and operation on individual bits within the specified operand. In general, the specification of a particular bit is by the address of a word or long word operand and the bit position within the operand. The bit position is usually specified by the source operand, whereas the address of the operand is specified by the destination operand.

Bit Manipulation Instructions

There are four general types of bit manipulation instructions: bit change, bit clear, bit set, and bit test instructions. These are listed below.

BCHG Dn,EA   (Test a bit and change)

Operation: $(Z) \leftarrow \overline{(Dest(Dn))}$

$(Dest(i)) \leftarrow \overline{(Dest(i))}$

Addressing category: Data alterable

BCHG #Data,EA   (Test a bit and change)

Operation: $(Z) \leftarrow \overline{(Dest(\#Data))}$

$(Dest(i)) \leftarrow \overline{(Dest(i))}$

Addressing category: Data alterable

BCLR Dn,EA   (Test a bit and clear)

Operation: $(Z) \leftarrow \overline{(Dest(Dn))}$

$(Dest(i)) \leftarrow 0$

Addressing category: Data alterable

BCLR #Data,EA   (Test a bit and clear)

Operation: $(Z) \leftarrow \overline{(Dest(\#Data))}$

$(Dest(i)) \leftarrow 0$

Addressing category: Data alterable

BSET Dn,EA   (Test a bit and set)

Operation: $(Z) \leftarrow \overline{(Dest(Dn))}$

$(Dest(i)) \leftarrow 1$

Addressing category: Data alterable

BSET #Data,EA   (Test a bit and set)

Operation: $(Z) \leftarrow \overline{(Dest(\#Data))}$

$(Dest(i)) \leftarrow 1$

Addressing category: Data alterable

BTST Dn,EA   (Test a bit)

Operation: $(Z) \leftarrow \overline{(Dest(Dn))}$

Addressing category: Data

BTST #Data,EA   (Test a bit)

Operation: $(Z) \leftarrow (Dest(\#Data))$

Addressing category: Data

The bit manipulation instructions allow bit operations on any register or memory operand. The operand to be tested is specified by the effective address. The specification of the particular bit within the operand to be tested is stored either as data in a data register or as immediate data. The former case allows dynamic bit testing, whereas the latter allows static bit testing. The results of all the bit-testing instructions are always stored in the Z flag. This result can be used later in a number of ways to control program flow.

Programming Example for the Bit Manipulation Group

In this section we demonstrate the use of bit manipulation instructions.

Example 11-13 ▰▰▰▰▰▰▰▰▰▰▰▰▰▰▰▰▰▰▰▰▰▰▰▰▰▰▰▰▰▰▰

Determine the sign of a memory operand stored in location ADDR. Store the sign as a logical variable in D_0.

Many arithmetic routines require determination of the sign so that the proper sign of the resultant operand can be set accordingly. This requires the sign of the operands to be read and stored. The program to do this is as follows:

```
Label    OPCODE      Operand                  Comments
*
* THIS PROGRAM TESTS THE SIGN OF THE OPERAND
*
         ORG         $1000
         BTST        #7,ADDR(PC)      *TEST OPERAND IN MEMORY
         SZ          D0               *SET LOGICAL VARIABLE
         END
```

11.10 ▰▰▰▰▰▰▰▰▰▰▰▰▰▰▰▰▰▰▰▰▰▰▰▰▰▰▰▰▰▰▰▰▰

The BCD Arithmetic Group

The BCD arithmetic group of instructions is designed to support operands that represent decimal numbers in binary-coded decimal (BCD) format. Similar to the binary case, there are two types of BCD arithmetic operations:

Dual-operand BCD instructions
Single-operand BCD instructions

Dual-Operand BCD Instructions

There are two main operations dealing with BCD arithmetic, addition and subtraction. No multiplication and division operations are provided for BCD arithmetic.

ABCD Dy,Dx   (Add decimal with extend)
    Operation: (Dest) ← (Source)$_{10}$ + (Dest)$_{10}$ + (X)
    Addressing category: None
ABCD −(Ay),−(Ax)   (Add decimal with extend)
    Operation: (Dest) ← (Source)$_{10}$ + (Dest)$_{10}$ + (X)
    Addressing category: None
SBCD Dy,Dx   (Subtract decimal with extend)
    Operation: (Dest) ← (Dest)$_{10}$ − (Source)$_{10}$ − (X)
    Addressing category: None
SBCD −(Ay),−(Ax)   (Subtract decimal with extend)
    Operation: (Dest) ← (Dest)$_{10}$ − (Source)$_{10}$ − (X)
    Addressing category: None

All of the above instructions are BCD operations that require two operands. Note that all additions and subtractions are performed taking into consideration the contents of extend flag X. Hence care is needed to zero out the extend bit before proceeding with the computations.

Single-Operand BCD Instructions

There is only one single-operand BCD instruction.

NBCD EA   (Negate decimal with extend)
    Operation: (Dest) ← 0 − (Dest)$_{10}$ − (X)
    Addressing category: Data alterable

The NBCD instruction is similar to the unary minus operator for binary arithmetic. Note that the operation also takes into consideration the contents of extend bit X.

Programming Example for the BCD Arithmetic Group

Here we present an example involving the use of BCD arithmetic operations.

Example 11–14

Write a program to add two four-digit BCD numbers together. The BCD numbers are stored in BCD$_1$ and BCD$_2$.

The solution is rather simple, since BCD addition automatically takes into consideration the extend flag. Care must be exercised, however, to zero out the extend flag before the addition.

Label	OPCODE	Operand	Comments
*			
* THIS PROGRAM ADDS TWO BCD NUMBERS			
*			
	ORG	$1000	
BCD1	DS	1	
BCD2	DS	1	
BCDSUM	DS	1	
	MOVE	BCD1(PC),D0	*GET FIRST OPERAND
	MOVE	BCD2(PC),D1	*GET SECOND OPERAND
	LEA	BCDSUM(PC),A0	*SET DESTINATION ADDRESS
	MOVE	#$10,CCR	*CLEAR CONDITION CODES
	ABCD	D0,D1	*ADD
	MOVE	D1,(A0)	*STORE SUM
	END		

11.11

The Program Control Group

The program control instructions direct the program flow. There are three main types of instructions in this group:

Unconditional branches
Conditional branches
Subroutine calls and returns

Unconditional Branches

There are two unconditional branches: a branch (BRA) instruction and a jump (JMP) instruction. The branch instruction is for short branches with less than 16-bit displacements relative to the current PC location, whereas the jump instruction allows a full 32-bit change in the PC.

BRA LABEL   (Branch always)
    Operation: $(PC) \leftarrow (PC) + d$
    Addressing category: None
JMP EA   (Jump)
    Operation: $(PC) \leftarrow Dest$
    Addressing category: Control

Both the BRA and JMP instructions change the contents of the PC by the effective address. The BRA instruction is used for relative addressing with respect to the PC and can be used for position-independent programs. The only limitation to the BRA instruction is that the displacement cannot be larger than 16 bits. Although the JUMP instruction takes slightly longer to execute, it allows a branch anywhere within the 32-bit addressing space.

Conditional Branches

In conditional branches, a condition is tested before the specified action is taken. If the condition is not satisfied, then the next logical instruction is executed. A list of conditions and their encoding formats has already been given in Figure 11–13. Instructions to perform conditional branches are as follows:

Bcc LABEL   (Branch conditionally)
    Operation: If condition true then $(PC) \leftarrow (PC) + d$
    Addressing category: None
DBcc Dn,LABEL   (Test condition, decrement, and branch)
    Operation: If condition false then $\{(Dn) \leftarrow (Dn) - 1$
                       If $(Dn) \neq -1$
                             then $(PC) \leftarrow (PC) + d\}$
                   else $(PC) \leftarrow (PC) + 2$
    Addressing category: None

Conditional branches are very useful. There are 16 conditions that can be tested. Note that the conditional branch instruction is limited to a 16-bit displacement. There are no conditional jump instructions that provide the full 32-bit jump. The DB instruction is very useful and is equivalent to the LOOP-WHILE-CONDITION-TRUE operation. It decrements the counter to keep track of the loop index. Additionally, the specified testing condition is polled every time, so the looping continues only as long as the test condition is false.

Subroutine Calls and Returns

Subroutine or procedure calls differ from the branching instructions in that the return address is stored onto the stack before the branching operation takes place. This provides for a proper return to the calling routine. The instructions for handling subroutine calls are as follows:

BSR LABEL   (Branch to subroutine)
    Operation: $(-(SP)) \leftarrow (PC)$
              $(PC) \leftarrow (PC) + d$
    Addressing category: None
JSR EA   (Jump to subroutine)
    Operation: $(-(SP)) \leftarrow (PC)$
              $(PC) \leftarrow$ Dest
    Addressing category: Control
RTD #d   (Return and de-allocate parameters)
    Operation: $(PC) \leftarrow ((SP)+)$
              $(SP) \leftarrow (SP) + d$
    Addressing category: None
RTR   (Return and restore condition codes)
    Operation: $(CC) \leftarrow ((SP)+)$
              $(PC) \leftarrow ((SP)+)$
    Addressing category: None
RTS   (Return from subroutine)
    Operation: $(PC) \leftarrow ((SP)+)$
    Addressing category: None

Similar to the branching instructions, there are the BSR and JSR instructions for subroutine calls. There are three ways to return from a subroutine. The RTS instruction is the basic one, which merely returns from a subroutine by retrieving the return address from the stack. The RTR instruction not only returns from the subroutine but also restores the condition codes. Note that a subroutine itself is responsible for storing the condition codes if the RTR instruction is to be used. The RTD instruction is used to return from a subroutine and to de-allocate stack space. Note also that the subroutine itself is also responsible for allocating the stack space in the first place.

Examples of Program Control Instructions

Most programs require many program control statements. We present a number of programming examples dealing with conditional and unconditional branches.

Example 11–15 ▮▮▮▮▮▮▮▮▮▮▮▮▮▮▮▮▮▮▮

Implement a block move operation using conditional branch instructions. Both the source and destination of the block to be moved reside in memory.

The block move operation can be implemented simply by setting the source address, the destination address, and the block count. This is shown as follows.

```
     Label        OPCODE        Operand                  Comments
*
* MOVE A BLOCK OF DATA FROM
* SOURCE TO DESTINATION. D0 CONTAINS
* THE NUMBER OF LONG WORDS TO
* BE MOVED. A0 CONTAINS THE SOURCE
* ADDRESS AND A1 CONTAINS THE
* DESTINATION ADDRESS
*
                  ORG           $2000
     AGAIN        MOVE.L        (A0)+,(A1)+      *MOVE DATA
                  SUBQ.W        #1,D0            *DECREMENT COUNT
                  BNE           AGAIN            *CONDITIONAL BRANCH
                    .
                    .
                    .
                  RTS
                  END
```

(Reprinted with permission of Motorola Inc.)

Example 11–16 ▮▮▮▮▮▮▮▮▮▮▮▮▮▮▮▮▮▮▮

Convert a positive BCD number stored in memory location BCD into binary, and store the result in memory location BINARY.

Since no multiplication operations are available in BCD, we will solve this problem by decrementing the BCD number and by incrementing the binary number.

Label	OPCODE	Operand	Comments
*			
* THIS PROGRAM CONVERTS A POSITIVE BCD NUMBER INTO BINARY			
* THE BCD NUMBER IS STORED IN MEMORY LOCATION BCD			
* THE BINARY NUMBER IS TO BE STORED IN MEMORY LOCATION			
* CALLED BINARY			
*			
	ORG	$1000	
BCD	DS.B	1	
BINARY	DS.L	1	
	LEA	BINARY(PC),A0	*SET ADDRESS
	MOVE.B	BCD(PC),D0	*GET OPERAND
	MOVEQ	#−1,D1	*SET −1 FOR DECREMENT IN BCD
	CLR.L	(A0)	*CLEAR DEST
LOOP	ABCD	D1,D0	*DECRE BCD
	BZ	DONE	*IF ZERO THEN DONE
	ADDQ.L	#1,(A0)	*INCRE BINARY
	BRA	LOOP	*LOOP AGAIN
DONE	.		
	.		
	.		
	END		

Example 11–17

Find the maximum number in an array of numbers so that the largest number is in the lowest memory location.

Finding the largest number and storing it in the lowest memory location is in essence an integral part of the bubble sort. To implement this a single loop is used to compare consecutive numbers stored in a table.

Label	OPCODE	Operand	Comments
*			
* SEQUENCES A STRING OF WORDS SUCH THAT			
* THE LARGEST NUMBER IS IN THE LOWEST MEMORY			
* LOCATION. A0 POINTS TO THE BEGINNING OF THE			
* STRING, AND A1 TO THE END			
*			
	ORG	$2000	
	MOVEM.L	A2/D0/D1,−(SP)	
	MOVE.L	A0,A2	*SAVE BEGINNING ADDRESS
BGNAGN	MOVE.L	A2,A0	*BEGINNING ADDRESS → A0
NXTPR	CMPM.W	(A0)+,(A0)+	*COMPARE SEQUENTIAL WORDS
	BHI.S	EXCHNG	*IF DW > SW THEN DO EXCHNG

Label	OPCODE	Operand	Comments
	SUBQ.L	#2,A0	*DEC POINTER BY 2
	CMP.L	A0,A1	*IF NOT END OF STRING
	BNE	NXTPR	*THEN CHECK NEXT WORD
	MOVEM.L	(SP)+,D0/D1/A2	*ELSE DONE
	.		
	.		
	.		
	RTS		
EXCHNG	MOVE.W	−(A0),D0	*DW → D0
	MOVE.W	−(A0),D1	*SW → D1
	MOVEM.W	D0/D1,(A0)	*STORED IN REVERSE ORDER
	BRA	BGNAGN	*START OVER
	END		

(Reprinted with permission of Motorola Inc.)

Example 11–18 ▰▰▰▰▰▰▰▰

Implement a sorting algorithm by insertion sort where a new element is added to a sorted array.

The insertion sort requires comparison of the new element with the old array values. When the place for the new element is found, all the rest of the array elements must be shifted down to make room for the new element. The program to perform this is as follows.

Label	OPCODE	Operand	Comments
*			
* A NEW NUMBER IS INSERTED INTO			
* THE PROPER PLACE OF A SEQUENCED			
* LIST. THE NUMBER TO BE INSERTED			
* IS IN D0. THE LOWEST ADDRESS OF			
* THE LIST IS IN A0. THE HIGHEST			
* ADDRESS+1 IS IN A1. THE LARGEST			
* DATA IS IN THE LOWEST ADDRESS,			
* AND DATA IS MOVED DOWN TO MAKE			
* ROOM FOR THE NEW WORD			
*			
	ORG	$2000	
	CMP.W	(A0),D0	*IF NEW > MAX. STR WORD
	BCC.S	FINAL	*THEN DO FINAL; ELSE
MVDWN	MOVE.W	(A0)+,−4(A0)	*MOVE.W DOWN NEXT WORD
	CMP.W	(A0),D0	*IF NEW > THIS WORD
	BCC.S	FINAL	*THEN DO FINAL; ELSE
	CMP.W	A0,A1	*IF NOT END OF STRING
	BNE	MVDWN	*THEN DO MVDWN; ELSE

(continued)

Label	OPCODE	Operand	Comments
FINAL	MOVE.W	D0,−(A0)	*INSERT NEW IN STRING
			*DONE
	.		
	.		
	.		
	RTS		
	END		

(Reprinted with permission of Motorola Inc.)

Example 11–19

Write a program to compare two strings for similarities.

The string comparison program can be done by comparing successive words of two strings. This is shown below.

```
       Label      OPCODE      Operand              Comments
*
* TWO STRINGS ARE COMPARED FOR
* SAMENESS. A0 POINTS TO THE BEGIN-
* NING OF THE FIRST STRING AND
* A1 TO THE BEGINNING OF THE SECOND.
* THE NUMBER OF LONG WORDS TO BE COMPARED
* IS IN D1.
*
              ORG         $2000
ENTRY         ORI         #$04,CCR      *FORCE Z = 1
              BRA.S       SKIP          *IS NUMBER IN D1 = 0?
AGAIN         CMPM.L      (A0)+,(A1)+   *CMPAR-STR1 WITH STR2
SKIP          DBNE        D1,AGAIN      *WHILE Z = 1 & D1 ≠ −1 DO AGAIN
NOCOMP        NOP
                 .
                 .
                 .
              END
```

(Reprinted with permission of Motorola Inc.)

Example 11–20

Given a 16-bit word, determine whether it is a palindrome or not. A palindrome is a word that reads the same from either end.

The palindrome can be detected by setting up two address pointers and comparing the bit obtained from both pointers for symmetry. This is shown below.

Label	OPCODE	Operand	Comments

```
*
* A 16-BIT WORD POINTED TO BY A0
* IS CHECKED FOR SYMMETRY, I.E.,
* THE WORD READS THE SAME LEFT
* TO RIGHT AS IT READS RIGHT TO
* LEFT. IF IT IS DETERMINED TO BE
* A PALINDROME, THE BYTE TO WHICH
* THE STACK POINTER IS POINTING
* AT END WILL BE ALL 1'S
*
                ORG         $2000
PALCHK          MOVEM.L     D0/D1/D2,-(SP)
                MOVEQ.L     #7,D2           *7 → COUNTER
                MOVE.W      (A0),D0         *TEST WORD → D0.W
AGAIN           LSR.W       #1,D0           *LSB,D0.W → X,CCR
                ROXL.W      #1,D1           *X,CCR → LSB,D1.W
                DBF         D2,AGAIN        *DO THIS LOOP 8 TIMES
                CMP.B       D0,D1           *TST WORD = MIRROR IMAGE
                                            *OF LSB
                SEQ         16(SP)          *THN $FF → 16(SP)
                MOVEM.L     (SP)+,D0/D1/D2  *ELSE 00 → 16(SP)
                .
                .

                .
                RTS
                END
```

(Reprinted with permission of Motorola Inc.)

11.12

The System Control Group

The system control group of instructions pertains to the operation of the microprocessor itself. The bulk of the instructions are privileged instructions that can be executed only in the supervisory mode. We will divide this group of instructions into two categories:

Nonprivileged instructions
Privileged instructions

Nonprivileged Instructions

There are a number of nonprivileged system control instructions:

NOP   (No operation)
   Operation: None
   Addressing category: None

ILLEGAL   (Illegal instruction)
   Operation: $(-(SSP)) \leftarrow (PC)$
                $(-(SSP)) \leftarrow (SR)$
   Addressing category: None

TRAP #Vector   (Trap)
   Operation: $(-(SSP)) \leftarrow (PC)$
                $(-(SSP)) \leftarrow (SR)$
                $(PC) \leftarrow (Vector)$
   Addressing category: None

TRAPV   (Trap on overflow)
   Operation: If V then TRAP
   Addressing category: None

The NOP instruction is useful for timing and delay loops. The ILLEGAL instruction is used to cause a software TRAP. Other nonimplemented OPCODEs also cause TRAP conditions but are subject to future revisions and changes. The ILLEGAL instruction will always cause a TRAP condition regardless of future changes. The TRAP instruction allows the user to make system calls to the supervisor. The restart address is taken from the vector table. The TRAPV instruction is a specialized conditional TRAP instruction that is activated whenever the overflow condition is set.

Privileged Instructions

All operations dealing with the complete 16-bit status register are privileged instructions, because only the supervisor should have access to the system part, the most significant byte, of the status register. Remember that the least significant byte of the status register is commonly called the condition code register and is accessible when the processor is in both the user and supervisory modes.

RTE   (Return from exception)
   Operation: If supervisory state then $\{(SR) \leftarrow ((SP)+)$
                             $(PC) \leftarrow ((SP)+)$
                             If $((SP)+) =$ long format
                                 then full restore$\}$
                           else TRAP
   Addressing category: None

RESET   (Reset external devices)
   Operation: If supervisory state then assert RESET line
                           else TRAP
   Addressing category: None

STOP #Data (Load status register and stop)
    Operation: If supervisory state then (SR) ← Immediate data
                                       else TRAP
  Addressing category: None

The RTE instruction is used to return from exception processing. The RESET causes a software-generated RESET condition to synchronize all peripherals. The STOP instruction halts the processor until the occurrence of a hardware reset or an interrupt condition.

11.13
Position-Independent Programs

The instruction set of the 68010 is specially designed for writing position-independent programs. This is particularly important for multiprogramming environments. A position-independent program is one that is relocatable; that is, the code is written in such a way that it can be executed anywhere in the program space.

Characteristics of a Position-Independent Program

A position-independent program possesses the property that there are no absolute memory references in the code segment. If the program is written in such a way that it can be executed when placed anywhere in memory, there cannot be any absolute references used for specifying program branches or data. All addresses and memory references must be relative with respect to the PC or an address register. If the memory addresses are made with respect to the PC, the following addressing modes can be used:

    d(PC)   Program counter relative
    d(PC,Xn)   Program counter relative with index

If the memory addresses are made with respect to an address register, then the following addressing modes can be used:

    (An)   Address register indirect
    (An)+   Address register indirect with post-increment
    −(An)   Address register indirect with pre-decrement
    d(An)   Address register indirect with displacement
    d(An,Xn)   Address register indirect with index

The address in the address register can be set by the parent program or by means of the LEA instruction using the PC relative addressing modes.

Examples of Position-Independent Programs

Many programs can be written in relocatable form simply by using one of the above-listed addressing modes. Consider the problem of adding 10 numbers in memory by means of a loop. One can symbolically represent the operations in the following FORTRAN segments:

```
        DIMENSION A(10)
        SUM = 0.
        DO 100 I = 1,10
        SUM = SUM + A(I)
100     CONTINUE
```

We will look at two different ways of implementing the above segment: a position-dependent way and a position-independent way.

Example 11–21

Implement the FORTRAN segment listed above in a position-dependent manner.

When the program is position-dependent, absolute references can be used as follows:

Label	OPCODE	Operand	Comments
*			
* A POSITION DEPENDENT PROGRAM			
*			
	ORG	$1000	*ABSOLUTE ORIGIN
SUM	DS	1	
A	DS	10	
	CLR	SUM	*CLEAR TOTAL
	LEA	A,A0	*GET ADDRESS
	MOVEQ	#10,D0	*SET COUNTER
LOOP	ADD	(A0)+,D1	*ADD
	DBRA	D0,LOOP	*CHECK LOOP CONDITION
	MOVE	D	*STORE RESULT
	END		

Note that in this program, the addresses stored in instructions CLR, LEA, and MOVE are all absolute addresses. Hence this code is not relocatable to any other origin except wherever specified.

Example 11–22

Implement the same FORTRAN segment using position-independent coding techniques.

Label	OPCODE	Operand	Comments
*			
* A POSITION INDEPENDENT PROGRAM			
*			
* DATA SEGMENT			
*			
	SECTION	1	*RELOCATABLE SEGMENT
SUM	DS	1	
A	DS	10	
	CLR	SUM(PC)	*CLEAR TOTAL
	LEA	A(PC),A0	*GET ADDRESS
	MOVEQ	#10,D0	*SET COUNTER
LOOP	ADD	(A0)+,D1	*ADD
	DBRA	D0,LOOP	*CHECK LOOP CONDITION
	MOVE	D(PC)	*STORE RESULT
	END		

Note the differences between the position-dependent and the position-independent programs. In the latter program, all addresses are relative addresses with respect to the current value of the program counter. Hence, if this section is relocated to another addressing space, the program will execute correctly with the right address pointing to the right data.

11.14
Data Structure and Language Construct Implementations

In the discussion of the 8085 microprocessor, we have already reviewed the implementation of different language constructs. However, the instruction set of the 8085 is not designed for implementing many data structures, such as linked lists, trees, etc. In this section we illustrate the use of the 68010 instruction set in implementing different language constructs and various operations with data structures.

Specifically, we examine the following data structures and language constructs:

Stacks
Queues
Linked lists
Loops
Parameter passing and space allocation for local variables in subroutine calls
Semaphores

Stacks

Stacks, especially user stacks, can be created to expand in either of two directions: toward low memory or toward high memory. If the stack expands toward low memory, with the top of the stack in low memory and the bottom of the stack in high memory, then the following pair of addressing modes is used to PUSH and POP items from the stack:

PUSH   −(An)
POP   (An)+

This is the same as the system stack. On the other hand, if it is more desirable to expand the user stack toward high memory, then the following pair of addressing modes can be used:

PUSH   (An)+
POP   −(An)

The operation of these two types of user stacks is shown in Figure 11–15.

Queues

Whereas stacks are last-in, first-out structures, queues are first-in, first-out structures. Queues are not used by the 68010 in its system operation, though they can be easily implemented by the addressing modes. Just like stacks, queues can be

Figure 11–15

Implementation of stacks in memory (Reprinted with permission of Motorola Inc.)

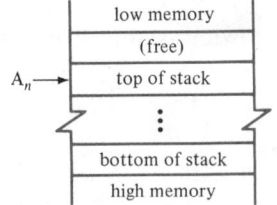

(a) Stack expands toward low memory

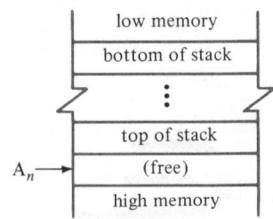

(b) Stack expands toward high memory

created to expand toward higher or lower memory as shown in Figure 11–16. If the queue expands toward high memory as shown in Figure 11–16(a), the following pair of addressing modes can be used for the GET and PUT pointers:

GET pointer   (An)+
PUT pointer   (An)+

The PUT pointer is always pointing to the next available free location that can be used to store any incoming items, whereas the GET pointer is always pointing to the next item to be retrieved.

If the queue expands toward the low memory as shown in Figure 11–16(b), the following pair of addressing modes can be used:

GET pointer   −(An)
PUT pointer   −(An)

Figure 11–16
Implementation of queues in memory (Reprinted with permission of Motorola Inc.)

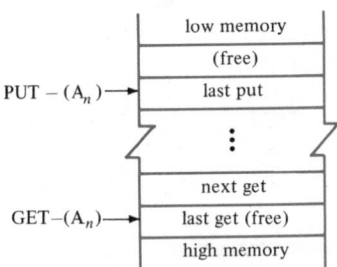

(a) Queue expands toward low memory

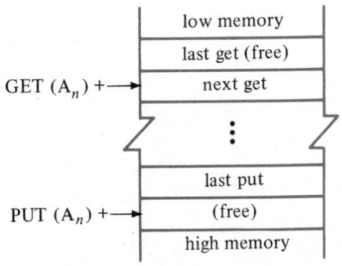

(b) Queue expands toward high memory

Note that in this case the PUT pointer always points to the last item stored in the queue, whereas the GET pointer always points to the last item retrieved from the queue.

Linked Lists

The addressing modes of the 68010 made it easy to manipulate linked list operations. The most difficult part of implementing a linked list is address calculations when traversing a tree. One must determine where the parent address is and where the child addresses are for any particular data item.

Example 11–23

Implement the binary tree shown in Figure 11–17.

In addition to information related to a node, every node in a binary tree has three addresses for linking purposes. These addresses are pointers to the parent and the two children. A possible implementation is shown below.

Label	OPCODE	Operand	Comments
*			
* IMPLEMENTATION OF A BINARY TREE			
*			
	SECTION	1	
NODEXX	DC.L	0	*PARENT ADDRESS
	DC.L	NODE0X	*CHILD ADDRESS
	DC.L	NODE1X	*CHILD ADDRESS
	DS	10	*ADDITIONAL NODE INFO

Figure 11–17
Implementation of a linked list

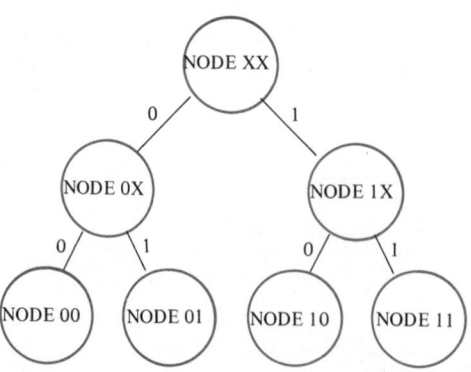

Label	OPCODE	Operand	Comments
*			
NODE0X	DC.L	NODEXX	*PARENT ADDRESS
	DC.L	NODE00	*CHILD ADDRESS
	DC.L	NODE01	*CHILD ADDRESS
	DS	10	*ADDITIONAL NODE INFO
*			
NODE1X	DC.L	NODEXX	*PARENT ADDRESS
	DC.L	NODE10	*CHILD ADDRESS
	DC.L	NODE11	*CHILD ADDRESS
	DS	10	*ADDITIONAL NODE INFO
*			
NODE00	DC.L	NODE0X	*PARENT ADDRESS
	DC.L	0	*CHILD ADDRESS
	DC.L	0	*CHILD ADDRESS
	DS	10	*ADDITIONAL NODE INFO
*			
NODE01	DC.L	NODE0X	*PARENT ADDRESS
	DC.L	0	*CHILD ADDRESS
	DC.L	0	*CHILD ADDRESS
	DS	10	*ADDITIONAL NODE INFO
*			
NODE10	DC.L	NODE1X	*PARENT ADDRESS
	DC.L	0	*CHILD ADDRESS
	DC.L	0	*CHILD ADDRESS
	DS	10	*ADDITIONAL NODE INFO
*			
NODE11	DC.L	NODE1X	*PARENT ADDRESS
	DC.L	0	*CHILD ADDRESS
	DC.L	0	*CHILD ADDRESS
	DS	10	*ADDITIONAL NODE INFO
	END		

To traverse the tree, one must ascend to the node's parent or descend to one of the node's descendants. This is shown below.

Example 11–24 ▰▰▰▰▰▰▰▰▰▰▰▰▰▰▰▰▰▰▰▰▰▰▰▰▰▰▰▰▰

Assume that the address pointer is pointing to a particular node. Write a program to change the address pointer to point to the node's parent.

Assume that the address pointer A_0 is pointing to the beginning of a particular node. The next address is the address of its parent. Hence this offset can be accomplished simply by means of the address register indirect with displacement addressing mode.

Label	OPCODE	Operand	Comments
*			
* ROUTINE TO ASCEND THE BINARY TREE			
*			
	SECTION	1	
ASCEND	LEA	$0(A0),A0	*LOAD PARENT ADDRESS
	RTS		
	END		

Example 11–25 ▬▬▬▬▬▬▬▬▬▬▬▬▬▬▬▬▬▬▬▬▬▬▬▬

Assume that the address pointer is pointing to a particular node. Write a program to change the address pointer to point to one of the node's descendants, depending on the value of the logical variable stored in D_0.

The proper descendant can be chosen by storing the offset in an index register. This is shown below.

Label	OPCODE	Operand	Comments
*			
* THIS ROUTINE FINDS THE PROPER DESCENDANT			
*			
	SECTION	1	
DESCEND	LEA	$1(A0,D0),A0	*LOAD DESCENDANT ADDRESS
	RTS		
	END		

Note that in this program the offset of the descendant table is taken care of by the displacement. The correct descendant is chosen by the contents of the data register.

Loops

In the instruction set of the 68010, several powerful instructions are provided for looping control. These include the usual conditional branching instructions and two decrement and conditional branching instructions. Execution of the DBcc instruction, for example, causes not only the specified condition to be tested for possible abnormal loop termination, but also the counter to be decremented and tested for normal loop termination. We will illustrate its use by implementing the WHILE loop and the DO loop constructs.

Example 11-26 ▮▮▮▮▮▮▮▮▮▮▮▮▮▮▮▮▮▮▮▮▮▮▮▮▮

Implement the following WHILE loop:

```
WHILE J < 10 DO ... END DO
```

Assume that JADDR is the address of the variable *J*. The WHILE loop can be easily implemented as follows:

Label	OPCODE	Operand	Comments
*			
* TEST FOR WHILE LOOP IS AT THE BEGINNING			
*			
LOOP	MOVE	JADDR(PC),D0	
	SUBQ	#$10,D0	
	BGE	ENDLOOP	
*			
* WHILE LOOP GOES HERE			
*			
	.		
	.		
	.		
	BRA	LOOP	
*			
* END OF THE LOOP			
*			
ENDLOOP	.		
	.		
	.		
	END		

Example 11-27 ▮▮▮▮▮▮▮▮▮▮▮▮▮▮▮▮▮▮▮▮▮▮▮▮▮

Implement the following FORTRAN DO loop:

```
DO 100 L=I,J,K
      .
      .
      .
100   CONTINUE
```

Assume that the addresses for the variables *I*, *J*, *K*, and *L* are IADDR, JADDR, KADDR, and LADDR, respectively. The loop can be implemented using a pre-loop test or a post-loop test. In accordance with FORTRAN definition, we will use the post-loop test to implement this DO loop.

Label	OPCODE	Operand	Comments
*			
* IMPLEMENTATION OF A FORTRAN DO LOOP			
*			
	ORG	$1000	
	LEA	LADDR(PC),A0	*SET ADDRESS OF INDEX
	MOVE.L	IADDR(PC),D0	*GET INITIAL VALUE
* LOOP BEGINS HERE			
*			
LOOP	MOVE.L	D0,(A0)	*SET LOOP INDEX
	.		
	.		
	.		
	ADD.L	KADDR(PC),D0	*ADD INCREMENT
	CMP	JADDR(PC),D0	*CHECK WITH ENDING
	BLT	LOOP	
	END		

Parameter Passing and Space Allocation for Local Variables in Subroutine Calls

Parameter passing and space allocation for local variables in subroutine calls are not easily done using the 8085 instruction set because the subroutine call statement uses the stack pointer to store the return address. Hence a fair amount of stack manipulation is required when the parameters are also stored in the stack. In the instruction set of the 68010, the tasks of passing parameters and allocating space for local variables are made easy by means of two instructions, the LINK and UNLK instructions.

Example 11–28

Demonstrate the parameter-passing capability in subroutine call instructions. Parameter passing in subroutine calls can be accomplished by means of a frame pointer in addition to the stack pointer. The relation between the FORTRAN codes and the assembler program is given in Figure 11–18. The relationship between the stack pointer and the memory is given in Figure 11–19. (Note: This example is adapted from Example 11–29 which is reprinted with permission of Motorola Inc.)

High-Level Source Code	Compiler-Generated Code
A: PROCEDURE;	A:
.	.
.	.
.	.
CALL B (C);	PEA ⟨EA⟩ OF C
.	JSR B
.	LEA 4(SP),SP
.	.
END {A} ;	.
B: PROCEDURE (X);	
.	B: LEA −4(SP),A0
.	.
.	.
END {B} ;	.
	RTS

Figure 11–18

An assembler program with its FORTRAN equivalent for parameters passing in subroutine calls (Reprinted with permission of Motorola Inc.)

```
        Label      OPCODE      Operand                    Comments
*
* PROGRAM TO DEMONSTRATE PARAMETER PASSING IN SUBROUTINES
*
            ORG       $1000
            LEA       $2000,A3        *SET FRAME POINTER
            LEA       $1FF0,SP        *SET STACK POINTER
*
* START OF PROCA
*
PROCA       .
            .
            .
            PEA       −6(A3)          *PASS PARAMETER
            JSR       PROCB(PC)       *CALL PROC. B
            LEA       4(SP),SP        *RESTORE STACK
            .
            .
            .
            RTS
*
* START OF PROCB
```

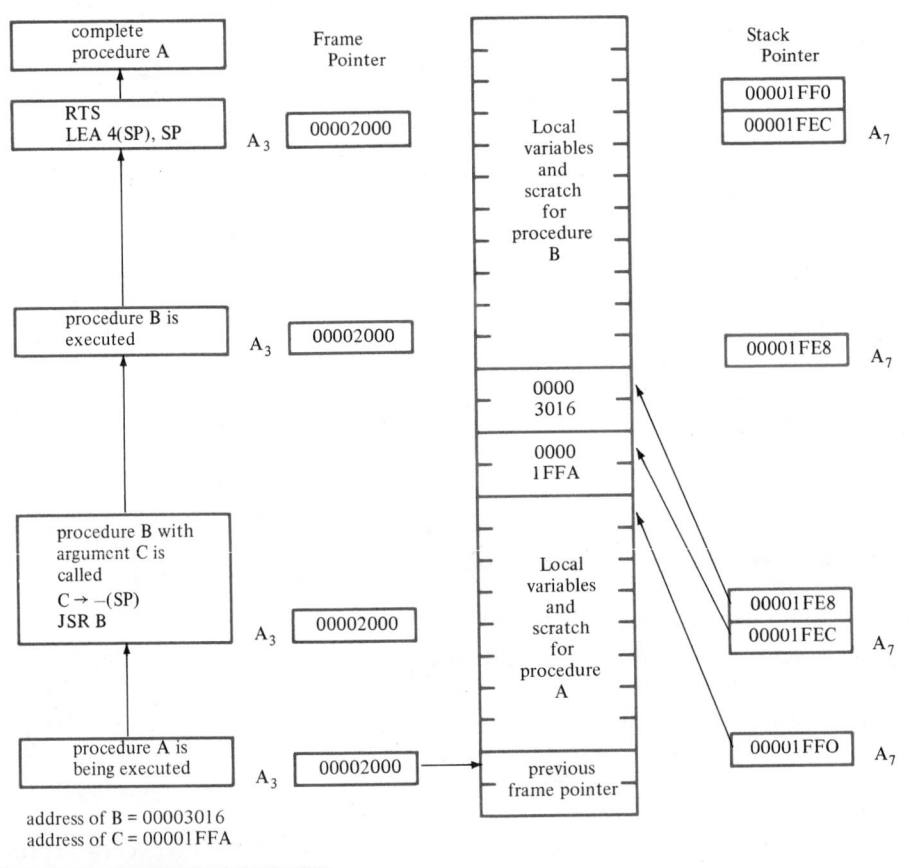

Figure 11-19
Use of the stack pointer and the frame pointer in subroutine calls
(Adapted from Figure 11-20)

Label	OPCODE	Operand	Comments
*			
PROCB	LEA	−4(SP),A0	*GET ADDRESS OF PARAMETER
	.		
	.		
	.		
	RTS		
	END		

The above example shows how parameters can be passed from one routine
to another. In many higher-level languages, subroutines may require storage for

local variables. The LINK and UNLK instructions can be used effectively for this purpose as shown in the following example.

Example 11–29 ▬▬▬▬▬▬▬▬

Demonstrate how a subroutine can allocate stack space for local variables. By means of an address register, the address of the reserved stack space can be stored and freely used by the subroutine. The relationship between the frame pointer A_3 and the stack pointer to the memory is shown in Figure 11–20.

Figure 11–20

How the LINK and UNLK instructions can be used for local variables
(Reprinted with permission of Motorola Inc.)

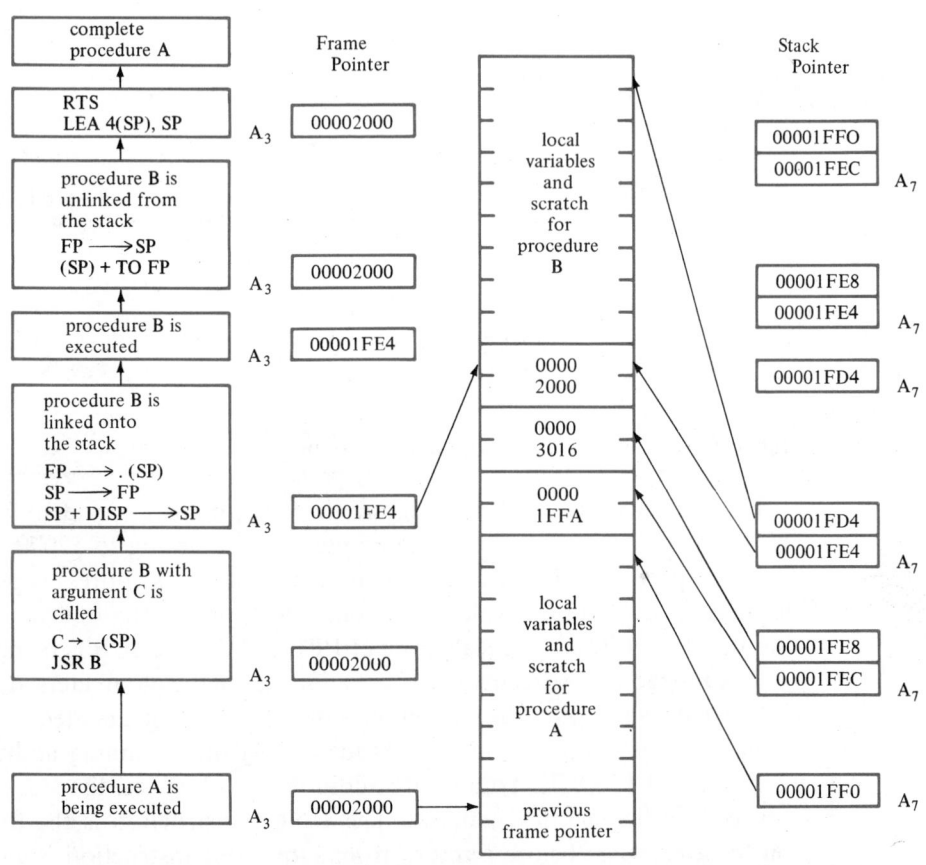

Label	OPCODE	Operand	Comments
*			

```
* PROGRAM TO DEMONSTRATE PARAMETER PASSING
* AND STACK SPACE ALLOCATION FOR LOCAL VARIABLES
*
                ORG       $3000
                LEA       $2000,A3
                LEA       $1FF0,SP
*
* START OF PROCA
*
PROCA           .
                .
                .
                PEA       -6(A3)        *PASS PARAMETER
                JSR       PROCB(PC)     *CALL PROC. B
                LEA       4(SP),SP      *RESTORE STACK
                .
                .
                .
                UNLK      A3
                RTS
*
* START OF PROCB
*
PROCB           LEA       -4(SP),A0     *GET PARAMETERS
                LINK      A3,#-$10      *GET SOME RAM
                .
                .
                UNLK      A3            *RELEASE RAM
                RTS
                END
```

(Reprinted with permission of Motorola Inc.)

The activity of the stack pointer and the frame pointer is evident from Figure 11–20. At the beginning, the frame pointer A_3 is pointing to $2000, identifying the stack space for PROCA. To pass the argument B, its address is pushed onto the stack. The subroutine call is executed, which pushes the return address also onto the stack within the local space of PROCA. As PROCB is being executed, a LINK instruction is executed after the addresses of the parameters are retrieved. The current frame pointer is pushed onto the stack and the updated stack pointer becomes the new frame pointer, thus identifying the beginning address for the work space of PROCB. To allocate additional work space, the stack pointer is displaced by the specified amount. The UNLK instruction is the first instruction to be executed before a return from subroutine instruction is encountered. The UNLK instruction merely reverses the operation of the LINK instruction.

Semaphores

In multiprogramming or multi-user environments, semaphores or critical sections are required to ensure data integrity. A *semaphore* operation requires the testing of the contents of a common operand and at the same time changing the contents of that very operand in one indivisible operation. There are no such provisions in the instruction set of the 8085. However, it is possible to implement this using the 68010 instruction set.

There are primarily four instructions that can be used effectively for implementation of semaphores:

TAS	Test and set
BCLR	Bit test and clear
BCHG	Bit test and change
BSET	Bit test and set

All of the above instructions test and change the contents of the destination operand in the same operation. The TAS instruction changes the contents of the most significant bit of the destination operand as a logical variable. The 3-bit manipulation instructions change the specified bit in memory.

Example 11–30 ▊▊▊▊▊▊▊▊▊▊▊▊▊▊▊▊▊▊▊▊▊▊▊▊▊▊▊▊▊▊▊▊

Write a program that would interrogate the IN USE flag for a certain I/O device. If the flag is 0, the device is idle and can be allocated for any process to use. If the flag is 1, the device is in use and any interrogating process must wait.

Assume that the flag is stored in memory location FLAG. Any interrogating device must be able to read the true contents of the flag and act upon them before being interrupted by other processes that may be contending for the same device

```
        Label      OPCODE      Operand         Comments
    *
    * INTERROGATION OF A SEMAPHORE
    *
                   SECTION     1
    PROCESS        LEA         FLAG(PC),A0     *GET FLAG ADDRESS
                   TAS         (A0)            *TEST AND SET FLAG
                   BNE         PROCESS         *CHECK STATUS
                    .
                    .
                    .
                   END
```

In this program the address of the flag is first loaded into address register A_0. To interrogate the contents of the flag, rather than using a MOVE or other load instructions, the TAS instruction is used. If the flag is 0, the TAS will set the FLAG. If, after the TAS instruction is executed, the routine is interrupted at this point, no other process will be able to use the device because the flag has already been set. Hence the device has been granted to the requesting process through the TAS instruction.

References

1. Camp, R. C., T. A. Smay, and C. J. Triska. *Microprocessor Systems Engineering.* Matrix Publishers, Portland, Ore., 1979.
2. Cannon, Don L. *Fundamentals of Microcomputer Design: System Hardware and Software.* Texas Instruments, Dallas, Tex., 1982.
3. Ciminiera, Luigi, and Adriano Valenzano. *Advanced Microprocessor Architectures.* Addison-Wesley, Reading, Mass., 1987.
4. Clements, Alan. *Microprocessor Systems Design: 68000 Hardware, Software, and Interfacing.* PWS Publishers, Boston, 1987.
5. Eccles, William J. *Microprocessor Systems: A 16-Bit Approach.* Addison-Wesley, Reading, Mass., 1985.
6. Garland, Harry. *Introduction to Microprocessor System Design.* McGraw-Hill, New York, 1979.
7. Gorsline, G. W. *Computer Organization: Hardware/Software.* Prentice-Hall, Englewood Cliffs, N.J., 1980.
8. Hayes, John P. *Digital System Design and Microprocessors.* McGraw-Hill, New York, 1984.
9. Heffer, D. E., G. A. King, and D. Keith. *Basic Principles and Practices of Microprocessors.* John Wiley & Sons, New York, 1981.
10. Johnson, E. L., and M. A. Karim. *Digital Design, A Pragmatic Approach.* PWS Publishers, Boston, 1987.
11. Lewin, Morton H. *Logic Design and Computer Organization.* Addison-Wesley, Reading, Mass., 1983.
12. Mano, M. Morris. *Digital Logic and Computer Design.* Prentice-Hall, Englewood Cliffs, N.J., 1979.
13. Mano, M. Morris. *Computer System Architecture,* 2d ed. Prentice-Hall, Englewood Cliffs, N.J., 1982.
14. Mano, M. Morris. *Digital Design.* Prentice-Hall, Englewood Cliffs, N.J., 1984.
15. Motorola, Inc. *M68000: 16/32-Bit Microprocessor, Programmer's Reference Manual,* 4th ed. Prentice-Hall, Englewood Cliffs, N.J., 1984.
16. Motorola, Inc. *MC68000: 16-Bit Microprocessor.* Motorola Semiconductor Products Literature ADI-814, Austin, Tex.
17. Motorola, Inc. *MC68000: 16-Bit Microprocessor.* Motorola Semiconductor Products Literature ADI-839, Austin, Tex.
18. Motorola, Inc. *MC68000 Technical Training Course Notes.* Motorola Semiconductor Products Literature, Austin, Tex.
19. Motorola, Inc. *MC68000 Family Resident Structured Assembler Reference Manual.* Motorola Semiconductor Products Literature M68KMASM/D7, Austin, Tex.
20. Motorola, Inc. *MC68010: 16-Bit Microprocessor.* Motorola Semiconductor Products Literature ADI-942, Austin, Tex.

21. Motorola, Inc. *MC68030: Enhanced 32-Bit Microprocessor User's Manual.* Motorola Semiconductor Products Literature MC68030UM/AD, Austin, Tex.

22. National Semiconductor Corp. *Logic Data Book.* National Semiconductor Corporation, Santa Clara, Calif., 1981.

23. Ramirez, Edward V. *Microprocessing Fundamentals: Hardware and Software.* McGraw-Hill, New York, 1980.

24. Siewiorek, Daniel P., C. G. Bell, and Allen Newell. *Computer Structures: Principles and Examples.* McGraw-Hill, New York, 1982.

25. Signetics Corp. *Signetics Logic—TTL Data Manual.* Signetics Corporation, Sunnyvale, Calif., 1978.

26. Sloan, M. E. *Computer Hardware and Organization,* 2d ed. Science Research Associates, Chicago, 1983.

27. Taub, Herbert. *Digital Circuits and Microprocessors.* McGraw-Hill, New York, 1982.

28. Texas Instruments, Inc. *The TTL Data Book for Design Engineers,* 2d ed. Texas Instruments, Dallas, Tex., 1976.

29. Wakerly, John F. *Microcomputer Architecture and Programming.* John Wiley & Sons, New York, 1981.

30. Wilcox, Alan D. *68000 Microcomputer Systems Designing and Troubleshooting.* Prentice-Hall, Englewood Cliffs, N.J., 1987.

31. Wilkinson, Barry. *Digital System Design.* Prentice-Hall, Englewood Cliffs, N.J., 1987.

32. Williams, Gerald E. *Digital Technology.* Science Research Associates, Chicago, 1977.

33. Winkel, David, and Franklin Prosser. *The ART of Digital Design.* Prentice-Hall, Englewood Cliffs, N.J., 1980.

Problems

1. What are the differences between the three data definition or storage directives? List their uses.

2. Are there any differences between the ASL and LSL instructions?

3. Implement the equivalent of a SWAP using rotate instructions.

4. Give some examples where shifting in memory could be useful.

5. Write a program to convert a hexadecimal number from ASCII to binary.

6. Implement a two-dimensional array with m rows and n columns. If the specified element has subscripts j and k, calculate the address of the specified element. Don't forget to check the bounds of the subscript in both directions.

7. Convert a four-digit decimal number in ASCII to its binary equivalent.

8. Perform a 32-bit × 32-bit unsigned multiplication by the shift-and-add method.

9. Perform a 32-bit × 32-bit unsigned multiplication by the MULU instruction.

10. The picture of a 256 × 256 display is stored in memory beginning at location DISPLAY. Write a program to draw a square with dimensions 100 × 100 on the display. The width of the edges of the square must be 3 rasters wide.

11. Multiply two four-digit BCD numbers together.

12. In BASIC, all integer variables are represented internally in 15-digit BCD. Write a set of routines to add and subtract BASIC variables.

13. Write a position-independent routine that implements the logical IF statement.

14. Write a position-independent routine that implements the arithmetic IF statement.

15. Why do we need the MOVE USP,Dn instruction? How does it differ from the MOVE SP,Dn or MOVE A7,Dn instruction?

16. Many subroutines such as MIN and MAX should be written in such a way that a variable number of arguments may be passed to the subroutine. How would you handle the passing of a variable number of arguments?

17. What are the advantages of storing local variables on the stack using LINK instructions?

The 68010 Microprocessor-Based System

12

In this chapter we discuss the hardware characteristics and interfacing techniques of the 68010. We first discuss the signal characteristics of the 68010 and their use. Then detailed timing analyses are presented showing both the asynchronous nature and the synchronous nature of the system bus. The processing states are next introduced, with special attention to the concept of exception processing. The processor spends considerable time in exception processing. For an applications-oriented user, the processor will spend most of its time in the user state. For a system-oriented programmer, the processor may spend as much time under the supervisory mode with system calls and exception processing as under the user mode. Finally, design techniques based on the 68010 are discussed.

12.1

Hardware Characteristics of the 68010

The Motorola 68010 microprocessor is packaged in a 64-pin dual-in-line package as shown in Figure 12–1. There are therefore a total of 64 signals coming out of the package. With a 24-bit address bus and a 16-bit data bus, a significant number of pins remains for the control bus, so signal multiplexing is not necessary. Eliminating the need for signal multiplexing or multiple signal definitions on the same pins tends to ease the system design and reduce total chip count for a complete microprocessor-based system.

Signal Description of the 68010

The input and output signals on the 68010 chip can be logically divided into groups according to their functions and relationships with one another. There are seven groups as shown in Figure 12–2. These seven groups include:

1. The data bus
2. The address bus

Figure 12–1

The footprint of the 68010 microprocessor (Reprinted with permission of Motorola Inc.)

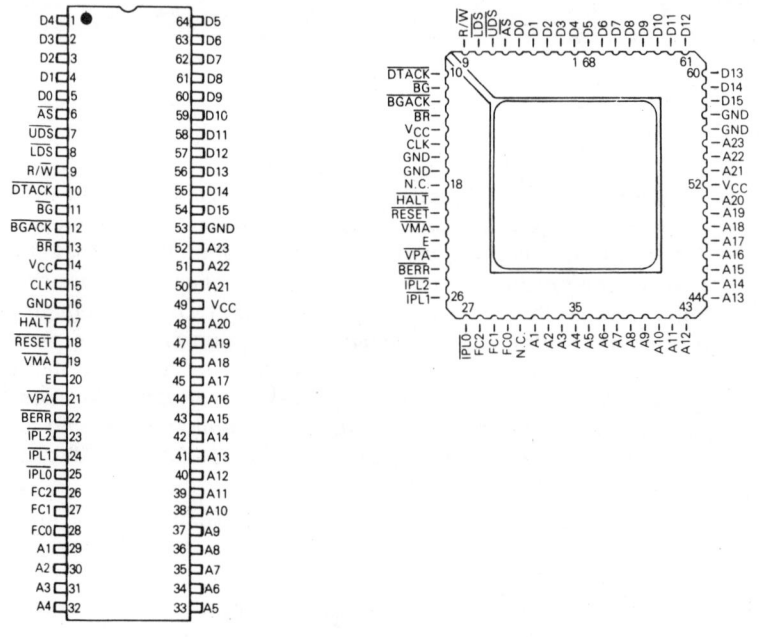

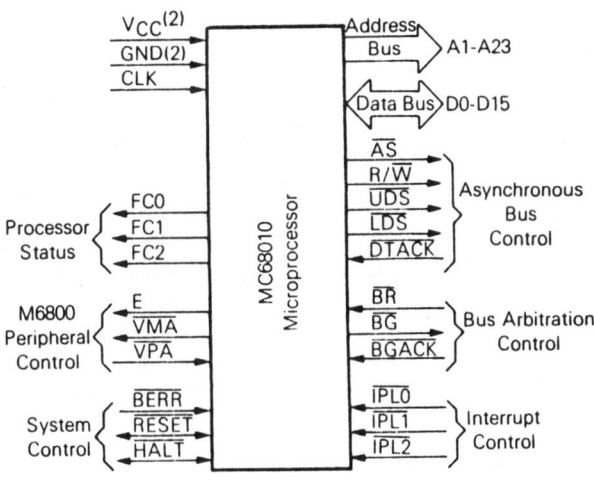

Figure 12–2

Functional grouping of pins on the 68010 (Reprinted with permission of Motorola Inc.)

3. The asynchronous bus control
4. The bus arbitration control
5. The interrupt control
6. The peripheral control
7. The system control

The definition and function of each of these signals is listed below.

Signal	Name	Type	Function
1. *Data bus*			
D_0-D_{15}	Data bus	I/O	This is a bidirectional tri-statable data bus for transferring data into and out of the micro-processor. The sequence of steps for word transfers is asynchronously related to the $\overline{\text{AS}}$, R/$\overline{\text{W}}$, $\overline{\text{DTACK}}$, $\overline{\text{UDS}}$, and $\overline{\text{LDS}}$.
2. *Address bus*			
A_1-A_{23}	Address bus	O	This is a 23-bit, output-only tri-statable address bus in normal operations. The lower 3 bits are also used to indicate the level of interrupt during an interrupt cycle. The lowest address bit, A_0, which indicates the odd and even bytes, is represented by a pair of control signals called $\overline{\text{UDS}}$ and $\overline{\text{LDS}}$.

(continued)

Signal	Name	Type	Function
3. Asynchronous bus control			
$\overline{\text{AS}}$	Address strobe	O	This signal indicates that a valid address has been put on the address bus.
R/$\overline{\text{W}}$	Read/$\overline{\text{Write}}$	O	This signal indicates the direction of the data bus.
$\overline{\text{UDS}}$	Upper data strobe	O	This signal indicates that the upper byte of the word is used and is used for either input or output of the upper byte.
$\overline{\text{LDS}}$	Lower data strobe	O	This signal indicates that the lower byte of the word is used and is used for either input or output of the lower byte.
$\overline{\text{DTACK}}$	Data transfer acknowledge	I	This input signal from the memory or an I/O device indicates that data transfer is completed and the microprocessor is now free to terminate the current read or write cycle.
4. Bus arbitration control			
$\overline{\text{BR}}$	Bus request	I	For bus arbitration, an input on this signal constitutes a request to use the bus.
$\overline{\text{BG}}$	Bus grant	O	This signal indicates that the microprocessor is ready to grant the use of the system bus and release the use of the bus at the end of the current bus cycle.
$\overline{\text{BGACK}}$	Bus grant acknowledge	I	The assertion of this signal by the interrupt requesting device indicates that the device is in the bus master mode and the microprocessor should remain in an inactive state.
5. Interrupt control			
$\overline{\text{IPL}}_0$–$\overline{\text{IPL}}_2$	Interrupt control	I	These input lines indicate the encoded priority of the interrupting device. A logic low on all three lines indicates that no interrupts are pending. A logic high on all three lines indicates a nonmaskable interrupt is pending.
6. Peripheral control			
E	Enable signal	O	An enable signal to 6800 peripheral devices. It simulates and functions as a clock signal provided by the 6800 microprocessor.
$\overline{\text{VMA}}$	Valid memory address	O	This output signal indicates that a valid address is put on the address bus and the 68010 is internally synchronized to the E signal.
$\overline{\text{VPA}}$	Valid peripheral address	I	This input signal indicates that a 6800 device is being addressed and is ready for data transfer.

Signal	Name	Type	Function
7. *System control*			
FC_0–FC_2	Processor status	O	These are output signals that indicate whether the processor is in the user mode or in the supervisory mode, and whether the memory reference is of data type or program type.
$\overline{BERR}$	Bus error	I	This input signal indicates that a bus error has occurred and the microprocessor should either reexecute the current bus cycle or perform exception processing.
$\overline{RESET}$	Reset	I/O	This signal is asserted as an input to reset the system and as an output signal to reset peripheral devices.
$\overline{HALT}$	Halt	I/O	This signal can be used as an input by peripheral devices to halt the 68010, which will subsequently float all tri-state lines and relinquish all other control lines. The $\overline{HALT}$ line may also be driven by the processor as an output signal to indicate that the processor has stopped.
CLK	Clock	I	A TTL-compatible clock input for the processor.
V_{cc}	Power	I	This is the 5-V power supply pin.
GND	Ground	I	This is the ground pin.

These 64 signal and control lines are divided into functional groups as shown in Figure 12–2. We will discuss the function of each pin in association with its corresponding group in the following sections.

Internal Organization

In addition to data registers and address registers, major components in the 68010 processor include three arithmetic logic units and an instruction prefetch queue.

Arithmetic Logic Units The three 16-bit arithmetic logic units are used primarily for data and address calculations. All three units are designed to work in parallel. Among the three, one special arithmetic logic unit is the main workhorse and is used primarily for data manipulations. The other two arithmetic units are used together for address calculations, which are always 32 bits long. Therefore, within the same time unit, a 16-bit data operation and a 32-bit address calculation can be made. This arrangement of parallel operations contributes an important factor in increasing the processing speed of the 68010.

Instruction Prefetch Queue   The architecture of the 68010 is of the von Neumann type. As such, the processor operates with the usual instruction fetch and execute cycles. To increase the speed of processing, the processor is internally equipped with an instruction prefetch queue. The prefetch queue increases the processing speed in two ways.

For branch operations, the instruction prefetch logic tries to fetch the next instruction if a branch is to take place or if a branch is not to take place. In this case, as soon as the branch condition can be determined, instruction decode for the next instruction can proceed immediately because the next instruction for either a branch or no-branch condition is already available. Another way the prefetch queue is useful in increasing the processing speed is in loop conditions. When a loop is made with loopable instructions, it is possible for all the instructions of the whole loop to be stored inside the queue. If this is so, the processor needs only to fetch operands from memory, so no instruction fetches are necessary.

External Operation

A block diagram of the basic 68010 system is shown in Figure 12–3. There are two types of bus transfer in the 68010:

Asynchronous bus transfer
Synchronous bus transfer

The asynchronous bus is used primarily for interfacing with all 68000-compatible devices and is specially designed to accommodate a variety of devices with very different timing characteristics. Hence the bus is designed in an asynchronous manner. On the other end, the synchronous bus is made specifically for interfacing with 6800 peripherals that are synchronous in nature. Hence a synchronizing clock is included as one of the control lines and is used for handshaking.

Figure 12–3

Basic system diagram for the 68010 (Reprinted with permission of Motorola Inc.)

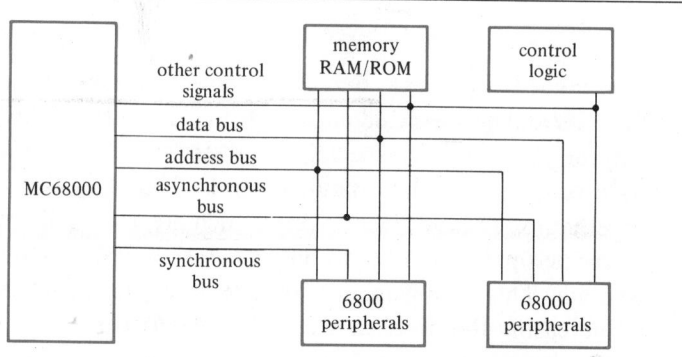

Most data transfer associated with the 68010 is made through the asynchronous bus. Though internally all signals are synchronized with the system clock, external devices do not need to be synchronized with the processor clock. Wait states are inserted automatically when the response from the requested device using the asynchronous bus is slow or delayed. The asynchronous nature of the bus makes interface to the 68010 simple. One pitfall is that a watchdog timer must be installed so that the processor does not wait indefinitely for some nonexistent device to respond. Since the bus is asynchronous, arbitration of the bus for other DMA devices is also asynchronous, relying wholly on handshaking signals. In order to be compatible with existing peripheral chips, special control signals are provided in the 68010 to facilitate synchronous data transfer by means of a system clock. An input on the $\overline{VPA}$ line indicates that the current data transfer cycle is synchronous.

For memory protection and other interfacing purposes, there are three output signal lines that collectively indicate the status of the processor. These are the FC_0, FC_1, and FC_2 lines. They indicate to external circuitry whether the current memory cycle is an interrupt cycle or a normal memory cycle. Furthermore, they indicate whether the requested memory cycle comes from a user's program or from the system program.

12.2
The Asynchronous Bus of the 68010

Most of the data transfer in the 68010 is accomplished by means of the asynchronous bus. By being asynchronous, a great variety of devices can be connected to the 68010. However, handshaking signals are also required to ensure an orderly transfer of information. For the 8085, all the bus transfers are strictly synchronous and are governed by the system clock generated by the microprocessor. The 68010, on the other hand, operates in an asynchronous bus mode. This means that external signals are internally synchronized to the internal clock by the processor itself. The signals involved with the asynchronous bus are listed below:

D_0–D_{15}	The data bus
A_1–A_{23}	The address bus
$\overline{AS}$	The address strobe
$R/\overline{W}$	The read/not write line
$\overline{LDS}$	The lower data strobe
$\overline{UDS}$	The upper data strobe
$\overline{DTACK}$	The data acknowledge line

A memory operation may be byte-oriented, word (2 bytes)-oriented, or long word (4 bytes)-oriented. Since the external data bus is 16 bits wide, it is possible to read in a total of 2 bytes of information with each data transfer. For byte operations, it is necessary to specify an even byte or an odd byte according to whether the byte is located with an even address or an odd address, respectively. For word operations, the operands fetched always have even addresses with an upper byte and a lower byte. All 16 bits of the data bus are used. For long word operands, two word operands are fetched on even address boundaries. The memory map according to the 68010 is shown in Figure 12–4.

Note that the memory address is specified in terms of bytes. A word or a long word always falls on an even boundary. Since there are only 23 address lines, namely A_1 to A_{23}, the least significant bit of the address is carried by the information shown in the $\overline{\text{LDS}}$ and $\overline{\text{UDS}}$ lines. When $A_0 = 0$, then $\overline{\text{LDS}} = 1$ and $\overline{\text{UDS}} = 0$. When $A_0 = 1$, then $\overline{\text{LDS}} = 0$ and $\overline{\text{UDS}} = 1$. When the memory operand is either word size or long word size, A_0 is always zero and so are $\overline{\text{UDS}}$ and

Figure 12–4

Memory map organization for the 68010 (Reprinted with permission of Motorola Inc.)

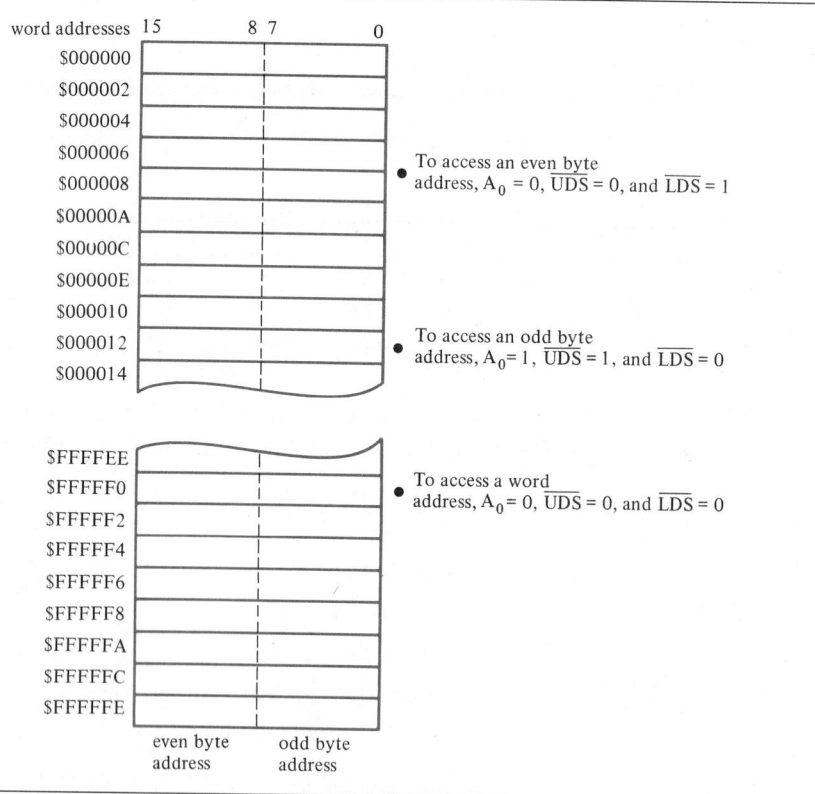

To access an even byte address, $A_0 = 0$, $\overline{\text{UDS}} = 0$, and $\overline{\text{LDS}} = 1$

To access an odd byte address, $A_0 = 1$, $\overline{\text{UDS}} = 1$, and $\overline{\text{LDS}} = 0$

To access a word address, $A_0 = 0$, $\overline{\text{UDS}} = 0$, and $\overline{\text{LDS}} = 0$

$\overline{UDS}$	$\overline{LDS}$	R/$\overline{W}$	D_8-D_{15}	D_0-D_7
0	0	0	Valid write data	Valid write data
0	0	1	Valid read data	Valid read data
0	1	0	Valid write data	Same as D_8-D_{15}
0	1	1	Valid read data	Invalid data
1	0	0	Same as D_0-D_7	Valid write data
1	0	1	Invalid data	Valid read data
1	1	0	Invalid data	Invalid data
1	1	1	Invalid data	Invalid data

Figure 12–5

Functional relations of the $\overline{UDS}$, $\overline{LDS}$, and R/$\overline{W}$ lines for memory references (Reprinted with permission of Motorola Inc.)

$\overline{LDS}$. Figure 12–5 shows the functional relationship of the $\overline{LDS}$, $\overline{UDS}$, and R$\overline{W}$ lines.

Timing Analysis of the Asynchronous Bus

Since the bus is asynchronous, timing is of utmost importance. The most frequent use of the asynchronous bus is for memory cycles. There are no explicit input/output (I/O) instructions in the 68010. Memory operations and I/O operations are decoded externally. In other words, the I/O space is memory-mapped. Hence all bus transfers, including both memory and I/O transfers, are referred to as memory cycles. Three kinds of memory cycles are possible:

The read cycle
The write cycle
The read-modify-write cycle

The Read Cycle A flowchart of a typical memory read cycle is shown in Figure 12–6. Four steps are necessary to complete a memory read operation:

1. The bus master addresses the device
2. The bus slave inputs the data
3. The bus master acquires the data
4. The bus slave terminates the cycle

A read cycle begins with the processor or the bus master addressing the device. This process is accomplished by the processor placing the address on the address bus and asserting the control lines R/$\overline{W}$, FC_0-FC_2, $\overline{AS}$, $\overline{UDS}$, and $\overline{LDS}$. The R/$\overline{W}$ line indicates the direction of the data transfer. The FC_0-FC_2 lines

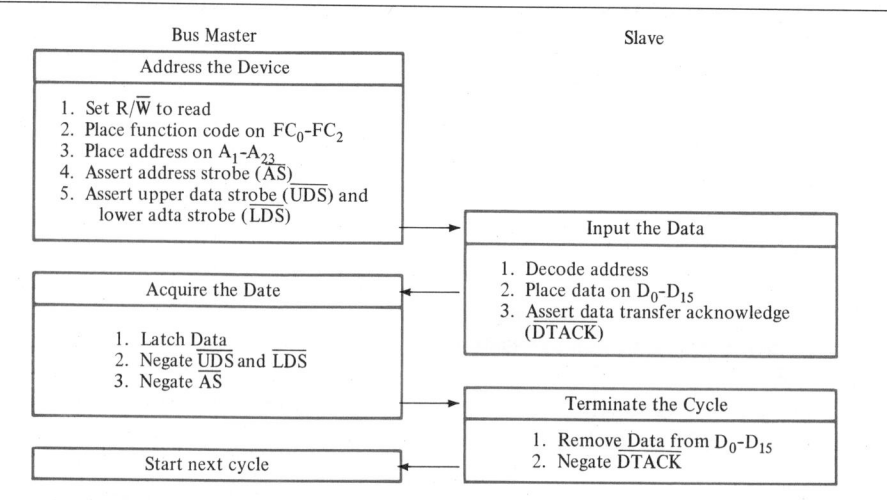

Figure 12–6

Flowchart for a memory read cycle (Reprinted with permission of Motorola Inc.)

indicate whether it is a user or supervisory mode for the hardware memory protection circuit and also whether the particular memory reference is of data or of program type. The assertion of the $\overline{AS}$ line indicates that there is a valid address on the address lines. The $\overline{UDS}$ and $\overline{LDS}$ pair indicates whether it is a byte or word operand that is being addressed, and whether it is the upper or the lower byte in the case of a byte operand. The next step of a read cycle is the slave's response by putting the data on the bus. At this point the slave, which can be either the memory or the peripheral device used in memory-mapped input/output, decodes the address and places the data on the data bus. The $\overline{DTACK}$ is then asserted to indicate valid data on the data bus. The third step of a read cycle is the acceptance of data by the bus master. After the bus master has latched the data on the data bus, the signals $\overline{UDS}$, $\overline{LDS}$, and $\overline{AS}$ are deactivated. The slave, on the other hand, monitors the $\overline{AS}$ line. When the $\overline{AS}$ line is deactivated, the slave responds by removing the data from the data bus and deactivating the $\overline{DTACK}$ line. This is the last step of the data transfer. The de-assertion of the $\overline{DTACK}$ line signifies the completion of the current data transfer cycle.

A timing diagram for the read cycle is shown in Figure 12–7(a). A normal read cycle will take four clock cycles (eight half cycles). For slow devices, wait states are inserted internally when the $\overline{DTACK}$ remains high a short time after the bus master asserts all its control signals. The memory cycle will complete in a normal manner with one and a half clock cycles after the $\overline{DTACK}$ is asserted as shown in Figure 12–7(b).

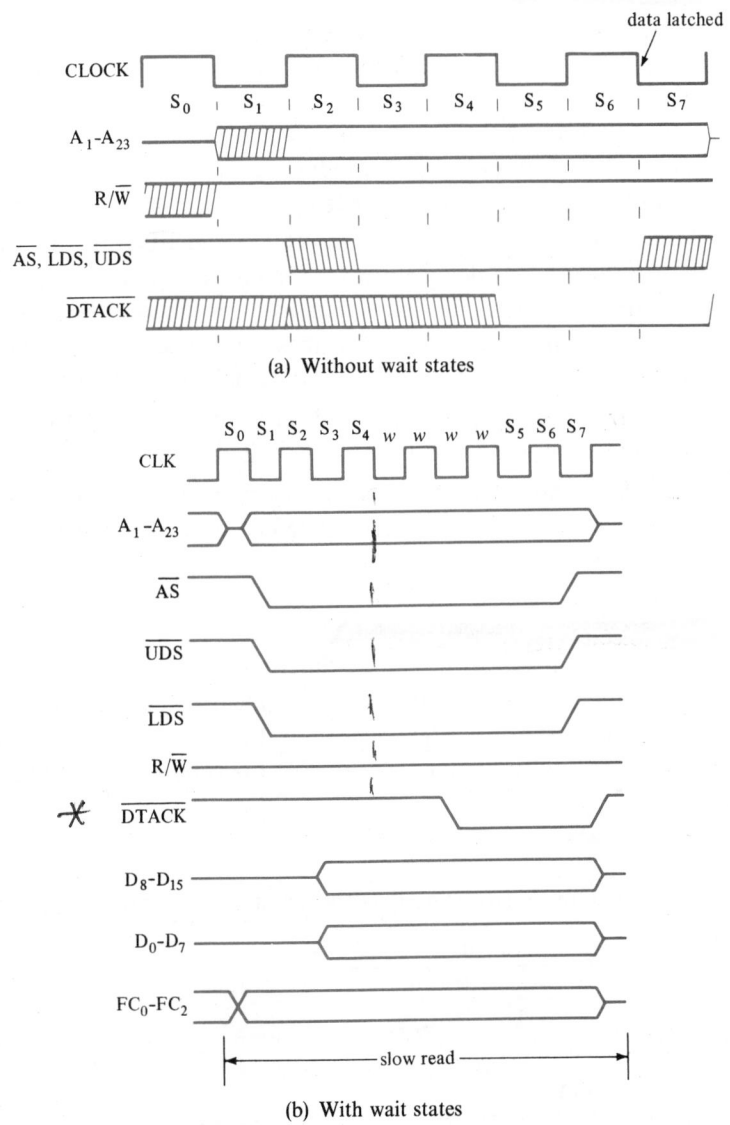

Figure 12–7
Timing diagram for a read cycle (a) without wait states; (b) with wait states (Reprinted with permission of Motorola Inc.)

It is possible to transfer a single byte instead of transferring a whole word. In a byte transfer, the assertion of $\overline{UDS}$ or $\overline{LDS}$ indicates which half of the word is to be transferred. Hence $\overline{UDS}$ and $\overline{LDS}$ can be used to select the proper memory chips to transfer the requested byte. All other signals and steps of the data transfer remain the same.

The Write Cycle A memory write cycle is similar in principle to a memory read cycle. A flowchart of a write cycle is shown in Figure 12–8, and a timing diagram is shown in Figure 12–9. The same four steps as the read cycle are required for the write cycle.

In the first step, the bus master addresses the device by asserting the various control lines. In addition, the data are placed on the data bus. One difference between a read cycle and a write cycle is that the $R/\overline{W}$ line remains low for the duration of a write cycle. Note that $\overline{UDS}$ and $\overline{LDS}$ are asserted in state S_4 rather than state S_2. This delay provides the necessary timing to indicate valid data on the data bus. Therefore data are placed on the data bus at the same time as the assertion of the $\overline{LDS}$ and $\overline{UDS}$ lines. In the second step, the slave decodes the supplied address and latches the data for storage. After the data are latched, $\overline{DTACK}$ is asserted. In the third step, the assertion of $\overline{DTACK}$ by the slave causes a response by the bus master. The bus master de-asserts the control lines $\overline{UDS}$, $\overline{LDS}$, and $\overline{AS}$ and removes the data from the data bus. In the last step, the slave responds by de-asserting the $\overline{DTACK}$ line, thus terminating the complete write memory cycle.

The Read-Modify-Write Cycle In order to facilitate multiprocessing and to achieve synchronization among different processors, it is necessary to be able to test and change a particular memory location used as a semaphore. This is accomplished by using a *test and set* (TAS) instruction. The execution of the TAS instruction requires that the memory operand in question is read and changed

Figure 12–8

Flowchart for a write cycle (Reprinted with permission of Motorola Inc.)

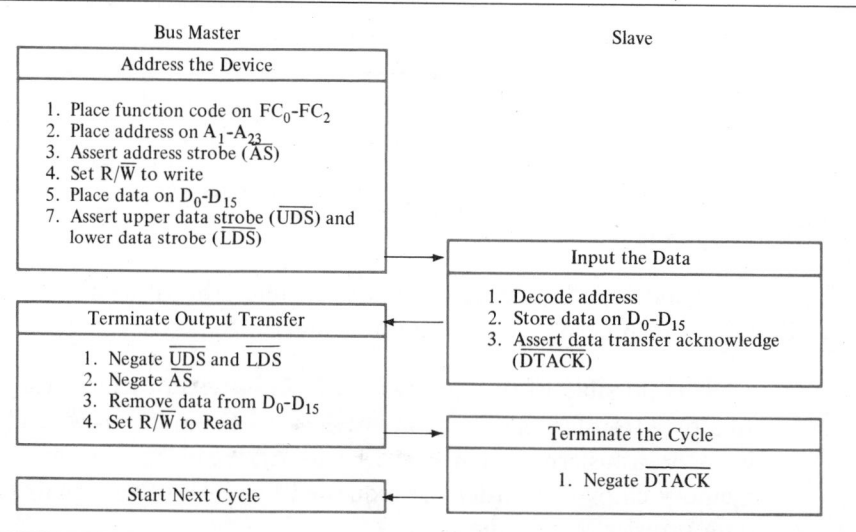

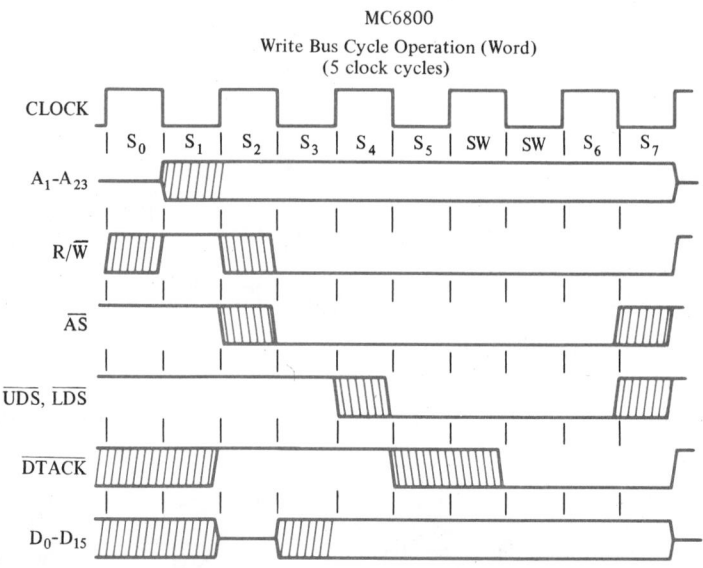

Figure 12–9

Timing diagram for a write cycle (Reprinted with permission of Motorola Inc.)

before the instruction is completed. This is done by a read-modify-write cycle. This operation is similar in principle to a read cycle followed immediately by another write cycle at the same location except that both parts of the same cycle are indivisible. A flowchart of the operations involved in a read-modify-write cycle is shown in Figure 12–10, and a typical timing diagram is shown in Figure 12–11. There are nine steps in a read-modify-write cycle:

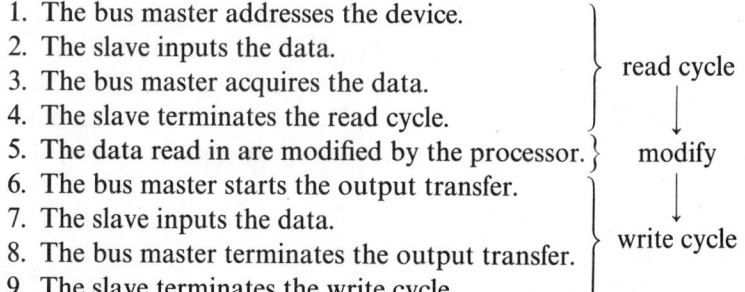

1. The bus master addresses the device.
2. The slave inputs the data.
3. The bus master acquires the data.
4. The slave terminates the read cycle.
5. The data read in are modified by the processor.
6. The bus master starts the output transfer.
7. The slave inputs the data.
8. The bus master terminates the output transfer.
9. The slave terminates the write cycle.

The first four steps are identical to a read memory cycle, with the bus master addressing the device, the slave inputting the data, the bus master acquiring the data, and the slave terminating the read cycle. The last four steps are similar to a write memory cycle, with the bus master starting the output transfer, the slave

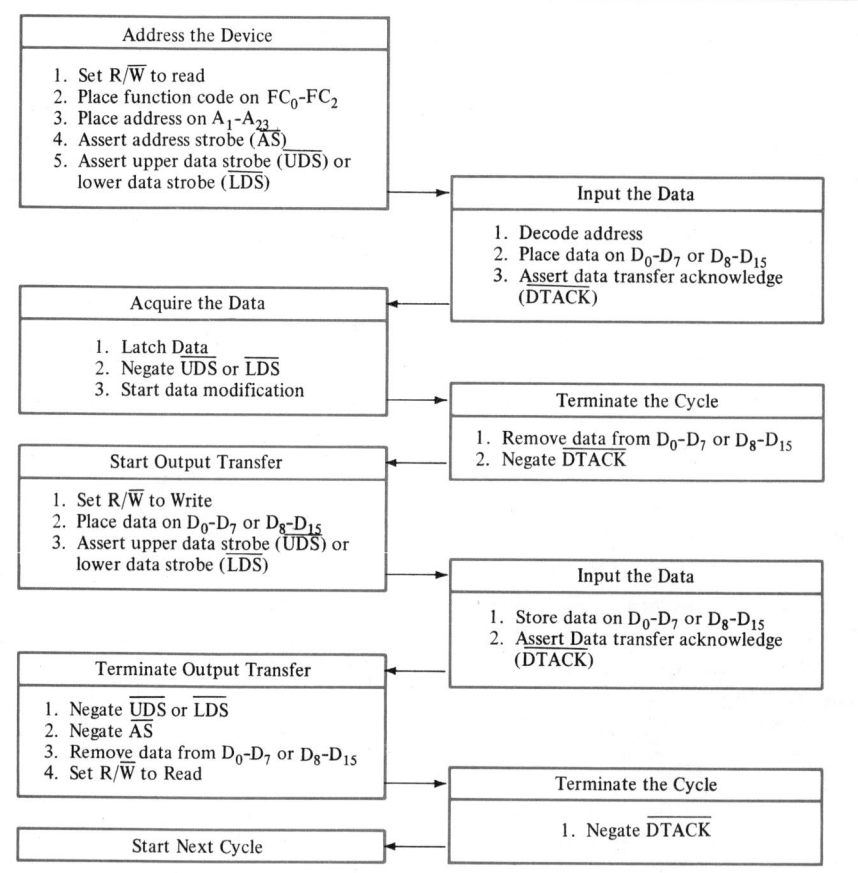

Figure 12–10

Flowchart for a read-modify-write cycle (Reprinted with permission of Motorola Inc.)

inputting the data, the bus master terminating the output transfer, and the slave terminating the write cycle. Note that although the address remains stable throughout the cycle, the $\overline{AS}$ does not because the assertion and de-assertion of the $\overline{AS}$ line are integral parts of the handshaking procedure for the read and write cycles.

Arbitration of the Asynchronous Bus

To facilitate interfacing peripherals with the 68010 and to provide for multiprocessing configuration, bus arbitration is handled by two input control lines and one output control line. These lines are used to arbitrate the use of the bus between the microprocessor, which is normally the bus master, and the rest of the

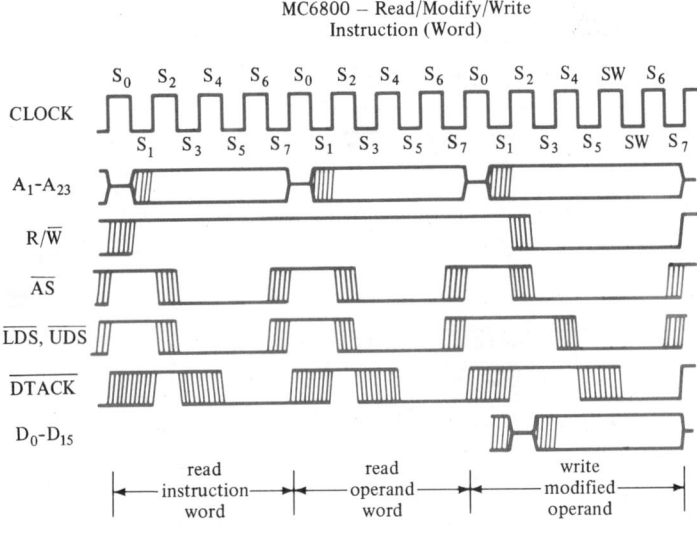

Figure 12–11

Timing diagram for a read-modify-write cycle (Reprinted with permission of Motorola Inc.)

peripherals. The control lines used for the arbitration of the asynchronous bus are

$\overline{BR}$	The bus request line
$\overline{BG}$	The bus grant line
$\overline{BGACK}$	The bus grant acknowledge line

When a peripheral device such as a DMA device desires to use the bus, the sequence of events as outlined in Figure 12–12 must occur. The timing diagram is shown in Figure 12–13. There are five steps a requesting DMA device must go through:

1. The DMA device requests the bus.
2. The processor grants the bus arbitration.
3. The DMA device acknowledges the bus master status.
4. The processor terminates the arbitration.
5. The DMA device operates as the current bus master.

Any DMA device that desires to use the bus may do so by making a request to the processor. This is the first step and is done by asserting the $\overline{BR}$ line. A wired-OR logic is usually used if there is more than one DMA device connected to the system bus. When the bus request is detected by the processor, the bus use is granted to the requesting device at state 6. This is the second step. At that

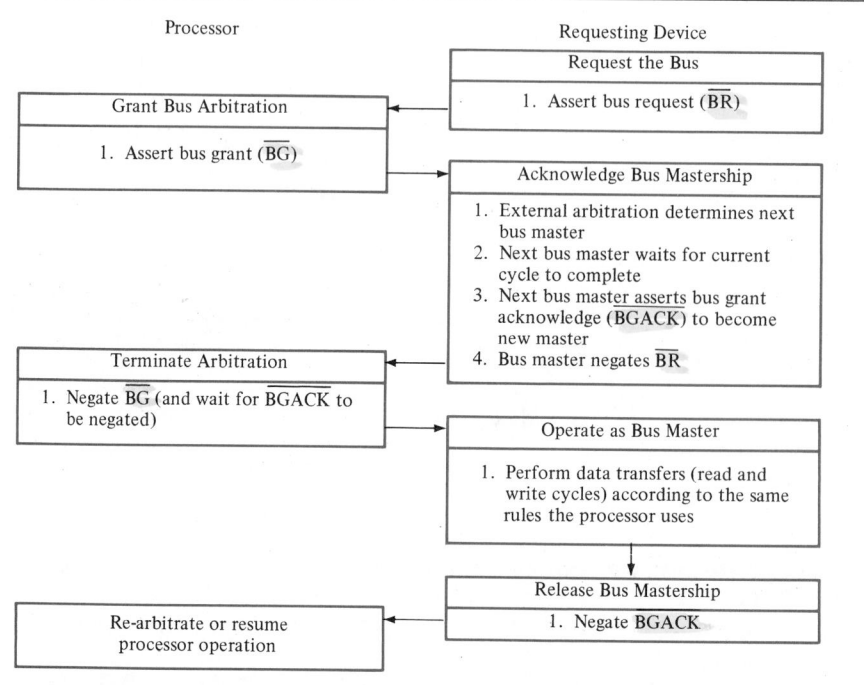

Figure 12–12
Flowchart for a bus arbitration cycle (Reprinted with permission of
Motorola Inc.)

time, the $\overline{BG}$ line is asserted to indicate that the external requesting device may now proceed.

When the $\overline{BG}$ line is asserted, several techniques can be used to route the $\overline{BG}$ line to the requesting device. If more than one device is making requests, an arbitration logic is required to pass the $\overline{BG}$ line to the one with the highest priority. A priority decoder can be used effectively in such situations. Another simple technique is called *daisy chaining*, as shown in Figure 12–14. In daisy chaining, all request lines of the DMA devices are wired-OR together. However, the $\overline{BG}$ line goes only from one to another, hence the name daisy chaining. When the bus grant signal comes to a device through the $\overline{BG}$-in control line, if the device is not the one making the DMA request, then the $\overline{BG}$ line is passed on through the $\overline{BG}$-out signal to the next device. If the device is the one making the request, then the $\overline{BG}$ signal is not passed on to the next device. Hence, keeping the $\overline{BG}$ signal is the same as keeping the bus usage privilege. The priority of a daisy-chain setup is therefore fixed. Whoever gets the $\overline{BG}$ line first has the highest priority. Whoever is connected to the $\overline{BG}$-output line of a device has a priority lower than that device.

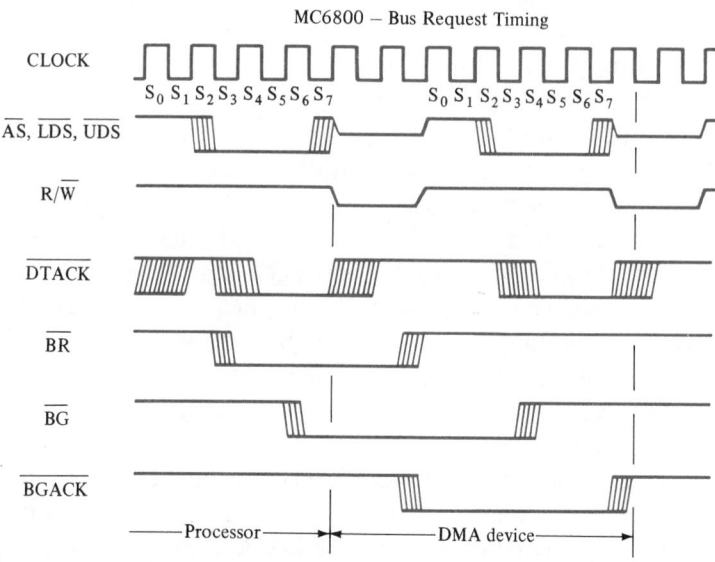

Figure 12–13

Timing analysis of bus request (Reprinted with permission of Motorola Inc.)

Figure 12–14

Sample circuit for bus arbitration by the daisy-chaining technique (Reprinted with permission of Motorola Inc.)

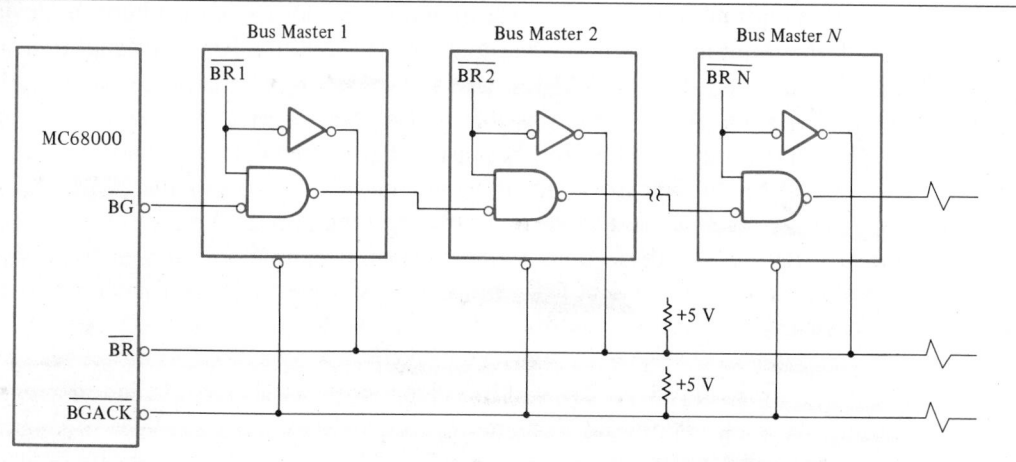

The third step in the DMA request-grant process is the slave acknowledging the bus master status. With the proper external arbitration logic, the device that made the DMA request and has the highest priority gets the $\overline{BG}$ signal. The selected device that receives the $\overline{BG}$ signal becomes the next bus master. The next bus master then monitors the $\overline{AS}$ line and watches for the end of the current processor bus cycle. When $\overline{AS}$ is deactivated at the end of the bus cycle, the next bus master asserts $\overline{BGACK}$ and deactivates the $\overline{BR}$ line.

The assertion of the $\overline{BGACK}$ line indicates that a new bus master has been found. Hence the processor, after detecting the presence of the new bus master via the assertion of the $\overline{BGACK}$ line, deactivates the $\overline{BG}$ line and becomes idle while all the time monitoring for the negation of the $\overline{BGACK}$ line. This is the fourth step. As long as the $\overline{BGACK}$ is asserted, the processor will remain in an idle state until the $\overline{BGACK}$ is deactivated. In the meantime, the DMA requesting device has full use of the bus, acting as the bus master and providing all the control signals needed for the operation of the bus. The DMA device may monopolize the bus for as long as desired by keeping the $\overline{BGACK}$ line low.

To terminate bus usage by the DMA device, only two steps are needed. First, the DMA device deactivates the $\overline{BGACK}$ line. The de-assertion action signals to the processor the end of the current data transfer session by the DMA device. At this point, if another $\overline{BR}$ is active, the process of bus arbitration again convenes from the beginning. If the $\overline{BR}$ line is not active, the processor resumes its normal processing.

Bus Error

There are two problems associated with the asynchronous nature of the bus. First, it is possible that the processor is addressing some nonexistent device. As a result, the cycle is never completed because the processor is continuously waiting for a termination signal, which never materializes because of the phantom device. The same thing may happen when a spurious signal inadvertently makes a bus request. Since no slaves would respond to the $\overline{BG}$ signal, the processor is stalled at the $\overline{BG}$ state. Second, it is possible that a DMA device requests the bus and is granted the use of the bus, but because of hardware failure or other connection problems, the bus release signal through the de-assertion of the $\overline{BGACK}$ line never comes back to the processor. In this case the processor will remain on hold forever. Because of these potential problems, a simple solution is to provide separate circuitry for a watchdog timer that supplies a bus error signal to the processor.

Since the bus is asynchronous, external circuitry is needed to detect any possible bus error conditions. Typical bus errors such as peripheral time-outs could occur as a result of nonexistent memory, a nonexistent peripheral device, a signal fault, or a malfunctioning controller. These error conditions must be detected externally and reported back to the 68010 for special error handling procedures.

BERR	HALT	Resulting Operation
0	0	Bus error—Rerun current cycle
0	1	Bus error—Exception processing
1	0	Single bus cycle operation
1	1	Normal operation

Figure 12–15

Processor operation in relation to the $\overline{\text{BERR}}$ and $\overline{\text{HALT}}$ inputs (Reprinted with permission of Motorola Inc.)

When a bus error is detected, two options are possible, as shown in Figure 12–15, depending on the status of the input control lines $\overline{\text{BERR}}$ and $\overline{\text{HALT}}$. First, the 68010 may be forced to enter an exception processing mode for special error handling; or second, the 68010 can be made to rerun the last instruction provided it is not a TAS instruction. (The TAS instruction should not be rerun because it is possible that the rerun will give a different result.) Exception error processing is selected when the $\overline{\text{BERR}}$ line is asserted without the $\overline{\text{HALT}}$ line being asserted. The rerun option is selected when both the $\overline{\text{BERR}}$ and $\overline{\text{HALT}}$ lines are asserted.

If the exception processing option is selected, the 68010 will perform in a way similar to interrupt processing. Specifically, the following steps are taken:

1. Save the program counter and the status register in the SSP area.
2. Save the error information in the SSP.
3. Execute the bus error handler from the address stored in the bus error vector table.

During the stack saving operations, if another bus error is detected, the processor will be halted. This is called the double bus fault condition. In this condition, the 68010 can be restarted only with a hardware reset.

If the rerun option is selected, the 68010 remains in a halt state until the HALT signal is removed. Then the 68010 will rerun the previous cycle using the same function codes and the same data. This option is extremely useful and is often used by the memory management controller to facilitate the implementation of virtual memory. A bus fault during the rerun phase is not considered a double bus fault and will not halt the processor.

12.3
The Synchronous Bus of the 68010

For compatibility with other MC6800 or synchronous peripherals, the 68010 is also designed to interact with synchronous bus peripherals. This interface is accomplished by including three signal lines that are common to all MC6800 or other synchronous peripherals:

1. The enable (E) line
2. The valid memory address ($\overline{\text{VMA}}$) line
3. The valid peripheral address ($\overline{\text{VPA}}$) line

For the synchronous bus, the E line acts as a system clock signal. Assertion of the $\overline{\text{VPA}}$ line by a valid synchronous peripheral identifies the current memory cycle to be synchronous.

Timing Analysis of the Synchronous Bus

The flowchart of a typical peripheral transfer is shown in Figure 12–16 and the timing diagram in Figure 12–17. There are five steps in the sequence of events:

1. The processor initiates the cycle.
2. The slave defines the peripheral transfer cycle.
3. The processor synchronizes with the E line.
4. The slave transfers the data.
5. The processor terminates the cycle.

To begin a peripheral transfer cycle, the processor begins with a normal read or write cycle by placing the address and/or the data on the bus and by asserting the $\overline{\text{AS}}$, $\overline{\text{LDS}}$, or $\overline{\text{UDS}}$ line. This is the first step. Immediately, the slave, which

Figure 12–16

Flowchart for a synchronous transfer (Reprinted with permission of Motorola Inc.)

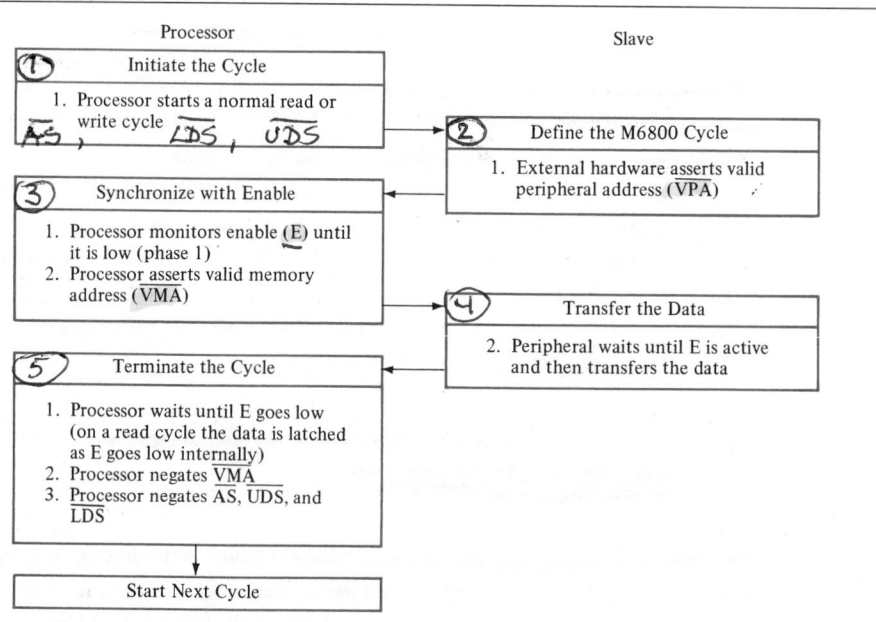

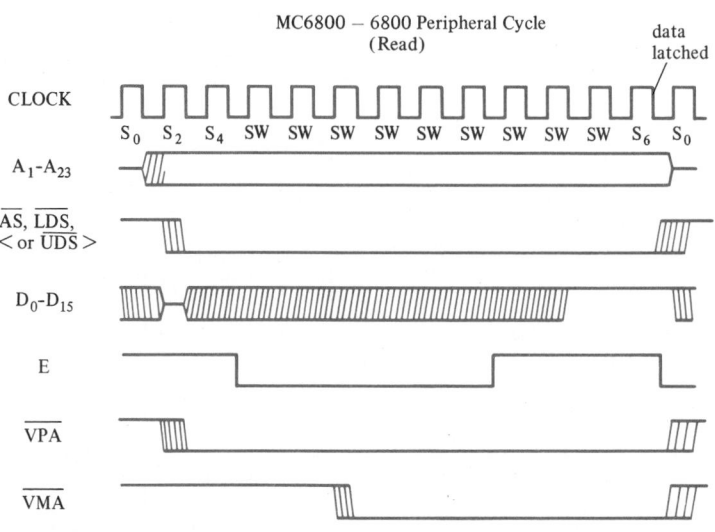

Figure 12–17

Timing diagram for a synchronous transfer (Reprinted with permission of Motorola Inc.)

(2) is the peripheral device, asserts the $\overline{\text{VPA}}$ line. This signifies to the processor that this is a synchronous bus transfer.

(3) When the processor detects the signal of a synchronous transfer via the assertion of the $\overline{\text{VPA}}$ line, the processor will internally synchronize itself with the negative edge of the E line. After a predetermined delay time, the $\overline{\text{VMA}}$ line is asserted low. This is the third step, which is the synchronization process on the processor side. The $\overline{\text{VMA}}$ line is used to indicate to the synchronous devices that a valid address has been placed on the address bus and that all subsequent timing is synchronized to the assertion of the $\overline{\text{VMA}}$ line.

(4) In the fourth step, the slave will wait until the E line is asserted high before data transfer is performed. Hence the E signal can also be used as an output control. In the last step, the processor terminates the cycle when the E line is (5) asserted low again. If it is a memory read cycle, data are latched at the negative edge of the E line. If it is a write cycle, the processor removes the data at the negative edge of the E line. The processor then completes the termination process by deactivating the $\overline{\text{VMA}}$, $\overline{\text{AS}}$, $\overline{\text{UDS}}$, and $\overline{\text{LDS}}$ lines.

Design of a Synchronous Interface

The primary component used for interfacing devices through the synchronous bus is a latch. The latch is strobed with the data when available and is set to release the data to or from the processor at the requested time. A block diagram of the synchronous interface is shown in Figure 12–18.

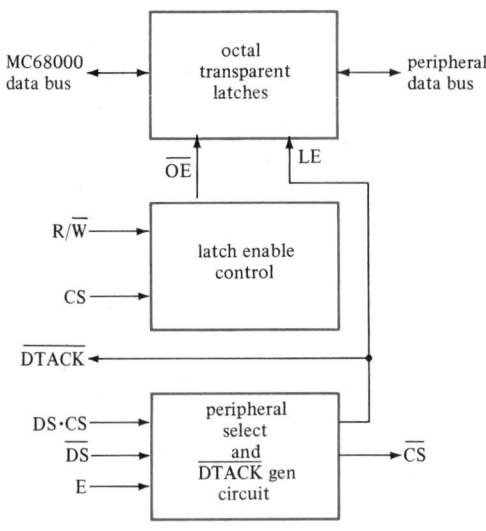

Figure 12–18
Block diagram for a synchronous interface (Reprinted with permission of Motorola Inc.)

On one side of the latch, the bus is connected to the 68010 data bus. On the other side of the latch, the bus is connected to the peripheral data bus. The latch is driven by a latch enable (LE) line and an output enable ($\overline{OE}$) line. When a data output cycle occurs, the latch enable line must be synchronously pulsed to receive the data from the 68010 while the output is synchronized to the peripheral receiving the data. Similarly, during a data input cycle, the LE signal is pulsed when data are available and the $\overline{OE}$ line is synchronized to the 68010 timing signals.

The $\overline{OE}$ line is generated by examining the peripheral chip select and the 68010 R/$\overline{W}$ line. The peripheral chip select determines whether the corresponding device is selected or not. The 68010 R/$\overline{W}$ line indicates the direction of data flow. The LE signal, on the other hand, is generated by a combination of several control signals, including the peripheral chip select, the 68010 data strobe ($\overline{UDS}$ or $\overline{LDS}$) line, and the 68010 E clock.

Example 12–1

Design an interface for a typical device using the synchronous bus.

Since the data bus is bidirectional, two sets of transparent latches are required. The output enable circuit can be generated by simple combinational logic. The latch enable circuit can be generated by flip-flops clocked by the E signal. A typical interface circuit is shown in Figure 12–19.

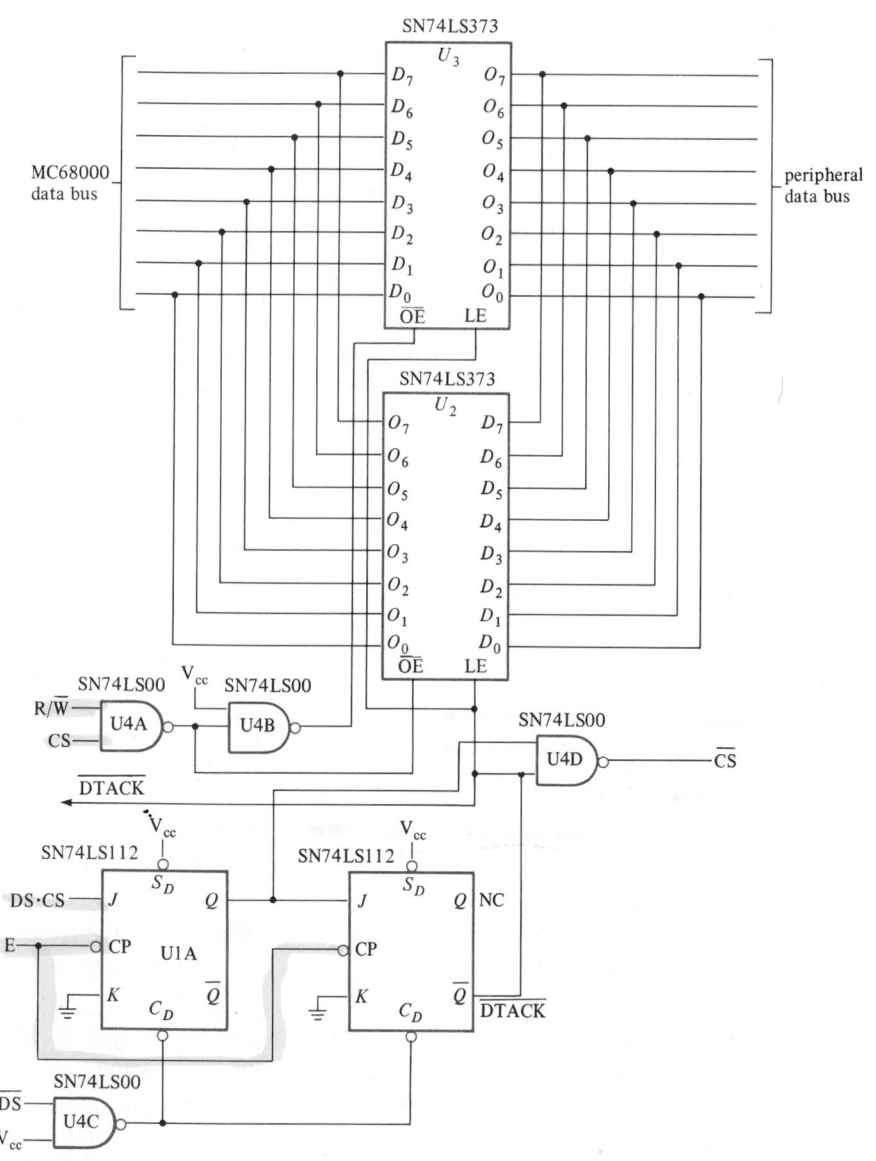

Figure 12–19

Typical interface circuit for synchronous transfer (Reprinted with permission of Motorola Inc.)

Note that each latch requires a different output enable signal. The R/$\overline{\text{W}}$ line is first gated with the peripheral chip select line. If the chip is not selected, the output of the U4A NAND gate will always be high, putting the latches into a read mode. If the chip is selected, a read cycle will cause the output of gate U4A to be low, thus enabling the U2 latch to be active in transferring data to the processor. An inversion of this signal by U4B will enable the outgoing U3 latch and disable the incoming U2 latch.

For the latch enable circuitry, the peripheral chip select signal generated through address decoding is ANDed with the data strobe DS and routed to the J input of the JK flip-flop U1A. The flip-flop is gated by the E clock for synchronization. The output of U1A provides the actual device chip select line. The second flip-flop provides the $\overline{\text{DTACK}}$ signal at the next clock cycle, which can be used for enabling the latches. Both flip-flops are cleared at the de-assertion of the data strobe.

12.4

Processor Status

The 68010 may be operated in any one of the following three processing states:

Normal processing state
Exception processing state
Halted state

During the *normal processing state,* the 68010 is actively making memory references and executing instructions. This is the normal state for the processor. There are two special occasions that could cause the processor to deviate from its normal instruction fetch and execute routines. First, it is possible that upon execution of a **STOP** instruction, the processor is stopped with no further memory references. At this stopped state, the processor can be restarted by external interrupts. This stopped state is still considered a part of the normal processing state. Second, the processor may be found in a looped mode, where all required instructions are stored in the instruction prefetch queue so that no instruction fetches are required. All subsequent memory references are operand fetches only. Under the looped condition, the processor is still considered to be in the normal processing state.

The *exception processing state* can be entered with either internal or external conditions. Exception processing is generated internally after the execution of trap instructions, when the processor is in a trace mode, and when other internally generated exception conditions occur. Externally, the presence of an interrupt, a bus error, or a reset will cause exception processing. Exception processing is used extensively for communication between the user and the supervisor, for input/

output operations, and for other system operations. Virtual memory systems also use exception processing to detect page faults.

The *halted state* indicates that a catastrophic problem in the system, such as a double bus error, has occurred and has been detected. Only an external hardware reset is able to restart the processor. The halted state is different from a stopped state. The former causes the processor to cease completely because of hardware problems or errors. The latter causes the processor to remain idle for power conservation until external conditions require further actions by the processor.

12.5
Normal Processing State

During normal processing, the processor may be found in one of two states:

User state
Supervisory state

This separation is mandated for easier programming, especially for system programs. The primary difference between the supervisory state and the user state is in the use of the stack and the execution of privileged instructions.

The distinction between the user state and the supervisory state is indicated internally by the S bit on the status register. If a program is running under the supervisor mode, the S bit can be interrogated. If a program is running under the user mode, any privileged instructions including the interrogation of the status register will cause a trap to occur. Externally, all memory references are tagged as being a user- or a supervisor-generated reference through the function codes FC_0 through FC_3 as shown in Figure 12–20.

Figure 12–20
Relationship between function code and processor state (Reprinted with permission of Motorola Inc.)

FC_2	FC_1	FC_0	State	Mode
0	0	0	Reserved	User
0	0	1	Data space	User
0	1	0	Program space	User
0	1	1	Reserved	User
1	0	0	Reserved	Supervisor
1	0	1	Data space	Supervisor
1	1	0	Program space	Supervisor
1	1	1	Interrupt acknowledge	Supervisor

When the processor is in the user state, it cannot change to the supervisory state except by exception processing. When the processor is in the supervisory state, it can change to the user state at any time simply by writing to the status register.

User State

In the user state, all use of the stack pointer and any stack references automatically refer to the user stack pointer (USP). Furthermore, in the user state, privileged instructions cannot be executed. Any attempt to execute privileged instructions will cause a TRAP condition to occur.

Internal register usage and external memory references are restricted when the processor is running in the user state. Internally, some of the registers, such as the supervisor part of the status register, the source function code (SFC) register, the destination function code (DFC) register, the vector base register (VBR), and the supervisor stack pointer (SSP), are not available and cannot be used at all. Externally, because the state of the processor is reflected in the function codes, hardware memory protection circuits can be designed to recognize and therefore lock out memory references that are outside the prescribed user memory boundaries. In addition to restricted access to registers and memory space, a program can enter into the supervisory state only in a controlled manner through exception processing.

Supervisory State

In the supervisory state, the processor has unlimited access internally to all programmable registers and externally to all memory references as long as allowed by the hardware protection circuits. All stack operations in the supervisor state use the supervisor stack pointer instead of the user stack pointer (USP). However, the USP is still available to the supervisor for storage and manipulation during context environment changes.

The supervisory state is used primarily for two main purposes: system programming and exception processing. In system programming environments, the supervisory state can be used to implement systems calls, resource allocation routines, and other operating system-related routines. In exception processing, trap, error conditions, and interrupt conditions are handled by the supervisor.

Changing Between States

The distinction between the user state and the supervisory state is made to facilitate memory protection. Hence, on one hand, all changes from the user state to the supervisory state must be controlled. On the other hand, changes from the supervisory state to the user state are common and should not require special control or monitoring.

Changing from User State to Supervisory State It is sometimes helpful for a user program to access supervisory routines or to communicate with the operating system. This requires a change from the user state to the supervisory state. There are three ways whereby one may change from the user state to the supervisory state:

1. By executing a TRAP instruction
2. By executing a conditional TRAP instruction
3. By external interrupts

Among the three methods of entering the supervisory state, the execution of a TRAP instruction is a guaranteed method. The TRAP instruction provides 16 vector entry points into the system and can be effectively considered system calls. Since the execution of a TRAP instruction causes exception processing (discussed in the next section), the contents of both the program counter and the status register are saved onto the system stack before processing continues at the supervisory state. Upon return from exception processing, the original status register is reloaded and the user program proceeds at the user state.

In addition to the TRAP instruction, there are other conditional TRAP instructions, such as TRAPV and CHK, that will generate exception processing at the supervisory level. These error conditions would normally be handled by error routines provided by the system routines. All external interrupts cause exception processing in the supervisory state. Hence any occurrence of external interrupts will change the processor level from user state to supervisory state. More details on this are given later in the section dealing with interrupt processing.

Changing from Supervisory State to User State A change from supervisory state to user state is needed when the operating system is transferring control to the user program. This change is simple and is accomplished merely by resetting the S bit of the status register. Since the processor is already in the supervisory state, it has available all the instructions, including the privileged instruction, to change the status register.

12.6
Exception Processing State

Exception processing occurs during times of interrupts, traps, tracing, and bus errors. An exception occurs when something unexpected occurs, whether internally or externally generated. Internally, an exception may be generated by address errors, an illegal instruction, a divide check, a trap instruction, etc. Externally, an exception may be generated by interrupts, bus errors, bus requests, etc.

Exception Processing

When a condition calls for exception processing, the processor follows a sequence of four events:

1. Save the status register.
2. Determine the vector number.
3. Save the processor context.
4. Create a new context for exception processing.

When an exception occurs, the first event is to make a temporary copy of the status register so that all user's information may be preserved. The S bit is forced to be 1 to put the processor into the supervisory mode. Tracing, if active, is deactivated. The interrupt priority mask is also updated.

In the second step, the vector number is obtained from the data base and the address of the exception vector is determined. A vector number is a two-word (except for reset) offset from which the vector address of the exception routine is calculated and fetched. Each exception generates a unique vector number. In the case of externally generated exceptions, this vector number is supplied to the lower byte of the data bus. The vector address is obtained by multiplying the vector number by 4 bytes and adding the product as an offset to the contents of the vector base register. The resultant address becomes an entry in the vector table, which is 512 bytes long, in the supervisory data space (SD). The only exception to this is the reset vector, which always lies in the supervisory program space (SP). An example of the vector table is shown in Figure 12–21.

In the third step, the current processor context, including at least the status register and the program counter, is saved in memory. Additional information is also saved depending on the exception. When an exception has occurred, the current processor context must be saved on the stack. The amount of information stored depends on the exception and may be as short as 4 words or as long as 29 words. The basic information stored on the stack is:

1. The status register
2. The high word of the program counter
3. The low word of the program counter
4. The special format code and the vector offset
5. Other information

Most exceptions use the short format, but the reset, address error, and bus error exceptions use the long format to completely define the processor context. An

Vector Number(s)	Offset			Assignment
	Dec	Hex	Space	
0	0	000	SP	Reset initial SSP
1	4	004	SP	Reset initial PC
2	8	008	SD	Bus error
3	12	00C	SD	Address error
4	16	010	SD	Illegal instruction
5	20	014	SD	Zero divide
6	24	018	SD	CHK instruction
7	28	01C	SD	TRAPV instruction
8	32	020	SD	Privilege violation
9	36	024	SD	Trace
10	40	028	SD	Line 1010 emulator
11	44	02C	SD	Line 1111 emulator
12*	48	030	SD	(Unassigned, reserved)
13*	52	034	SD	(Unassigned, reserved)
14	56	038	SD	Format error
15	60	03C	SD	Uninitialized interrupt vector

Vector Number(s)	Offset			Assignment
	Dec	Hex	Space	
16–23*	64	04C	SD	(Unassigned, reserved)
	95	05F		—
24	96	060	SD	Spurious interrupt
25	100	064	SD	Level 1 interrupt autovector
26	104	068	SD	Level 2 interrupt autovector
27	108	06C	SD	Level 3 interrupt autovector
28	112	070	SD	Level 4 interrupt autovector
29	116	074	SD	Level 5 interrupt autovector
30	120	078	SD	Level 6 interrupt autovector
31	124	07C	SD	Level 7 interrupt autovector
32–47	128	080	SD	TRAP instruction vectors
	191	0BF		—
48–63*	192	0C0	SD	(Unassigned, reserved)
	255	0FF		—
64–255	256	100	SD	User interrupt vectors

* Vector numbers 12, 13, 16–23, and 48–63 are reserved for future enhancements by Motorola. No user peripheral devices should be assigned to these numbers.

Figure 12–21

Typical vector table for the 68010 (Reprinted with permission of Motorola Inc.)

example of the long format is shown in Figure 12–22(a). Only the first 4 words are used for the short format. The information shown in the special status word is given in Figure 12–22(b). In the final step, a new processor context is obtained, and the processor continues normal processing with the new context.

Types of Exceptions

Exception conditions, both internally and externally generated, can be roughly categorized into three main groups as shown in Figure 12–23:

1. Group 0 exceptions
2. Group 1 exceptions
3. Group 2 exceptions

Group 0 exceptions have the highest priority. When they occur, the current instruction is aborted, and exception processing convenes in less than two clock cycles. Group 1 exceptions, when they occur, are honored after the end of the current instruction. Group 2 exceptions have the lowest priority. They are generated as a result of the execution of the current instruction. When multiple exception conditions occur, they are handled in an orderly manner according to the priorities.

```
            15                                              0
SP ──▶  ┌──────────────────────────────────────────────┐
        │               status register                 │
        ├──────────────────────────────────────────────┤
        │            program counter (high)             │
        ├──────────────────────────────────────────────┤
        │            program counter (low)              │
        ├───────────┬──────────────────────────────────┤
        │  1 0 0 0  │          vector offset            │
        ├───────────┴──────────────────────────────────┤
        │             special status word               │
        ├──────────────────────────────────────────────┤
        │             fault address (high)              │
        ├──────────────────────────────────────────────┤
        │             fault address (low)               │
        ├──────────────────────────────────────────────┤
        │              (unused, reserved)               │
        ├──────────────────────────────────────────────┤
        │              data output buffer               │
        ├──────────────────────────────────────────────┤
        │              (unused, reserved)               │
        ├──────────────────────────────────────────────┤
        │               data input butter               │
        ├──────────────────────────────────────────────┤
        │              (unused, reserved)               │
        ├──────────────────────────────────────────────┤
        │            instruction input buffer           │
        ├──────────────────────────────────────────────┤
        │          internal information, 16 words       │
        └──────────────────────────────────────────────┘
```

Note: The stack pointer is decremented by 29 words, although only 26 words of information are actually written to memory. The three additional words are reserved for future use by Motorola.

(a) Exception stack order

```
 15   14   13   12   11   10    9    8   7 - 3   2   1   0
┌────┬────┬────┬────┬────┬────┬────┬────┬──────┬──────────┐
│ RR │ •  │ IF │ DF │ RM │ HB │ BY │ RW │  •   │ FC₂-FC₀  │
└────┴────┴────┴────┴────┴────┴────┴────┴──────┴──────────┘
```

RR	Rerun flag. 0 = processor rerun (default), 1 = software rerun
IF	Instruction fetch to the instruction input buffer
DF	Data fetch to the data input buffer
RM	Read modify/write cycle
HB	High-byte transfer from the data output buffer or to the data input buffer
BY	Byte transfer flag. HB selects the high or low byte of the transfer register. If BY is clear, the transfer is word
RW	Read/write flag. 0 = write. 1 = read
FC	The function code used during the faulted access
•	These bits are reserved for future use by Motorola and will be zero when written by the MC68010

(b) Special status word format

Figure 12–22

(a) Exception stack order (bus and address error); (b) special status word format (Reprinted with permission of Motorola Inc.)

Group 0 Exceptions

Conditions for group 0 exceptions include reset, address error, and bus error. They have the highest priority among the other groups. Within group 0 itself, the priority of these conditions is

1. Hardware reset
2. Address error
3. Bus error

Group	Exception	Processing
0	Reset Address error Bus error	Exception processing begins within two clock cycles
1	Trace Interrupt Illegal Privilege	Exception processing begins before the next instruction
2	TRAP, TRAPV, CHK, Zero divide Format error	Exception processing is started by normal instruction execution

Figure 12–23

Groups of exception conditions (Reprinted with permission of Motorola Inc.)

Occurrence of exceptions in Group 0 preempts the current instruction to complete. Exception processing usually occurs after two clock cycles. Each of these exceptions within the group requires special attention. They are discussed individually here.

Hardware Reset Exception The hardware reset input generates the highest exception level. Its main function is for system initialization and recovery. Upon reset, all former processing context is lost. All contents of internal registers are assumed random. For this reason, the vector base register is always set to 0, and the vector number is internally generated to be 0 also. The processor is set at the supervisory state, tracing is deactivated, and the processor priority mask is set at the highest level. The first 2 words at location 0 of the supervisory program space are assumed to be the initial supervisory stack pointer, and the next 2 words to be the initial program counter. Normal instruction processing in the supervisory state then proceeds at the address pointed to by the initial program counter. An external reset timing pulse is also provided for resetting other peripherals. Note that the reset exception can be generated only by a hardware reset and is not generated by the RESET instruction. The RESET instruction merely asserts the RESET signal line for peripheral synchronization and does not cause total system initialization.

Address Error Exception Address error occurs when the processor is attempting to access a word that is not at an even word boundary. Exception processing occurs immediately, without the completion of the instruction. Address error exception processing is similar to bus error exception conditions. The long format is used to write information out to the memory. Exception handling can use software to emulate the instruction and return to normal processing by the RTE instruction.

Bus Error Exception Bus error occurs when external logic has detected a bus error and notified the processor through the bus error signal. When the bus error signal is received, the processor stops whatever it is doing and begins exception processing. For a bus error, 29 bytes of stack locations are used to store the information inside the processor. If the exception is a single bus error, processing continues at the vector entry table with the offset of $008. If the exception is a double bus fault, the processor merely halts.

In a virtual memory environment, the single bus error can be used to detect page faults. Sufficient internal information has been loaded onto the stack so that the present state of the processor before the page fault can be restored. A bus error then can be used as a signal to the operating system that a new page is needed. After this page has been brought into memory, an RTE instruction can be executed to continue processing. Alternatively, a bus error can be analyzed internally by a bus fault service routine. When the fault is located, it can be corrected by any corrective actions in software.

Group 1 Exceptions

Group 1 exceptions include trace exceptions, external interrupts, illegal instructions, and privilege violation conditions. These exceptions are of second priority. Within the group, the priority is:

1. Trace
2. External interrupt
3. Uninitialized interrupt
4. Spurious interrupt
5. Illegal instruction
6. Unimplemented instruction
7. Privilege violation

The occurrence of exceptions in group 1 will force the processor to enter into an exception processing mode only after completion of the current instruction. These exceptions that are part of group 1 are now discussed individually in more detail.

Trace Exception To facilitate program debugging, a trace mode is provided. The trace mode is activated by the T bit in the supervisory part of the status register. When the trace mode is active, an exception is forced automatically after the execution of every complete instruction so that the debugging program can take over to monitor progress. The trace exception will not occur upon occurrence of other exception conditions such as bus errors, address errors, etc. Instruction exceptions such as TRAP-generating instructions are honored before the trace exception, but the interrupt exception is not honored until after the trace exception processing.

External Interrupt Exception   An external interrupt request is made by encoding the interrupt request level on the interrupt request lines. If the interrupt request level is lower than the current processor priority, the processing of the interrupt request is postponed. This is called *maskable* interrupt processing. If the interrupt is of a higher priority, the interrupt request will be honored at the end of the current instruction cycle. The processor is set to the supervisory mode, the tracing is suppressed, and the priority level is set to the interrupting priority level. The processor then requests the vector number from the interrupting device through the lower byte of the data bus. The usual procedure for exception then follows. A *nonmaskable* interrupt can be effected by generating an interrupt at the highest priority level, that is, priority level 7. Any interrupts at this level will be honored immediately.

Uninitialized Interrupt Exception   If the vector register during an interrupt acknowledge cycle has not been initialized because of a programming error, the processor will internally generate vector 15 as an uninitialized interrupt and continue exception processing.

Spurious Interrupt Exception   A spurious interrupt occurs when, during the interrupt acknowledge cycle, no interrupting device responds or an error condition occurs in the bus signal. The condition is considered a spurious interrupt rather than a bus error.

Illegal Instruction Exception   Illegal instructions occur when the first word of an instruction fetched is not an operation code. Leaving room for future expansion, there are at least three operation codes that will always cause an illegal instruction exception. These are 4AFA and 4AFB, which are reserved for Motorola products, and 4AFC, which is reserved for others. There are also eight breakpoints, 4848 to 484F, that will initiate and execute a breakpoint bus cycle before exception processing is entered. These are software breakpoints useful for synchronizing other external devices.

Unimplemented Instruction Exception   Unimplemented instruction exception processing is useful for emulation. Any operation code word patterns in the form of AXXX or FXXX are unimplemented instructions. Unimplemented instructions are handled separately from illegal instructions, because unimplemented instructions can be used to emulate hardware devices such as the floating-point processor or non-native codes targeted for another machine.

Privilege Violation Exception There are a number of privileged instructions. Some of these include the ANDI to SR, EORI to SR, MOVE to SR, MOVE from SR, MOVEC, MOVES, MOVE USP, ORI to SR, RESET, RTE, and STOP instructions. Any attempt to execute these instructions in the user state will cause an exception.

Group 2 Exceptions

The last group of exceptions has the lowest priority and includes traps and processing checks. These exceptions are caused by the instructions themselves and may also occur as a result of the execution of an ordinary instruction during the normal processing of instructions. Since only one instruction can be executed at one time, there is no priority among exceptions within this group. Instruction traps are generated by an abnormal termination of an instruction. Three instructions are designed to cause traps. These are the TRAP, TRAPV, and CHK instructions. These instructions are used for communication between the user in the user processing state and the supervisory processing state.

12.7

Interrupt Processing

There are seven levels of interrupts. The current interrupt level of the processor is indicated by the 3-bit mask in the processor status register. All external interrupts are inhibited unless they have a priority that is higher than or equal to the current processor interrupt level. In other words, level 7 is the highest interrupt level. Of all the interrupt levels, level 0 is not recognized as a valid interrupt input. Levels 1 through 6 are maskable interrupts in the sense that they may or may not be acknowledged depending on the current processor level. If the current processor level is higher than the requesting interrupt level, the pending interrupt request will not be acknowledged. However, input interrupt level 7 is handled differently. A level 7 interrupt is nonmaskable and is always acknowledged as long as it is held stable until the beginning of the IACK cycle. Note that the level 7 interrupt is edge-triggered. This means that the input interrupt level must return to a lower level before another level 7 interrupt is acknowledged.

When a legitimate interrupt is present, a sequence of events occurs to transfer the processor from its normal processing sequence to initiate interrupt processing. These steps are shown in Figure 12–24. After a valid interrupt is recognized, the processor will commence interrupt processing, which can be divided into three phases:

1. Interrupt vector acquisition cycle
2. Context save cycle
3. Interrupt processing cycle

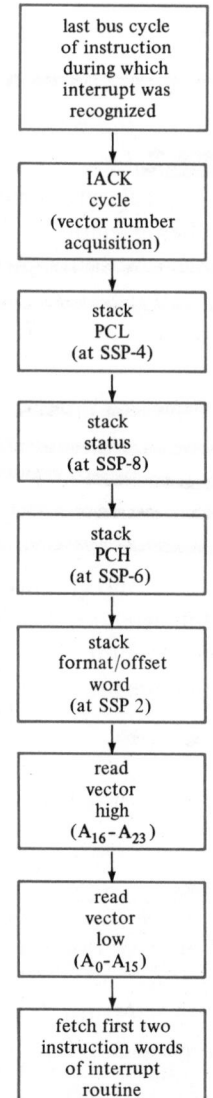

last bus cycle
of instruction
during which
interrupt was
recognized

IACK
cycle
(vector number
acquisition)

stack
PCL
(at SSP-4)

stack
status
(at SSP-8)

stack
PCH
(at SSP-6)

stack
format/offset
word
(at SSP 2)

read
vector
high
$(A_{16}-A_{23})$

read
vector
low
(A_0-A_{15})

fetch first two
instruction words
of interrupt
routine

Note: SSP refers to the value
of the supervisor stack pointer
before the interrupt occurs.

Figure 12–24

Interrupt processing sequence (Reprinted with permission of
Motorola Inc.)

These steps are now examined below. The first step is of primary importance among the three steps.

Interrupt Vector Acquisition Cycle

There are several pins on the 68010 that are dedicated for interrupt control. These are the $\overline{IPL0}-\overline{IPL2}$ lines. These interrupt lines are encoded. When these three lines indicate a logical zero, that is, level 0, it is taken to mean that no interrupt requests are being made.

When the incoming interrupt request has a higher priority than the current processor level, the first thing to occur is the interrupt vector acquisition cycle. A flowchart of the sequence of events is shown in Figure 12–25. There are five steps in this cycle:

1. The device requests the interrupt.
2. The processor grants the interrupt.
3. The device provides the vector number.
4. The processor acquires the vector number.
5. The device releases the data transfer.

The first step of the interrupt process begins with the interrupting device requesting the interrupt by encoding the interrupt request level on the interrupt request lines. Whenever the interrupt level is not 0, an interrupt request is being made. Since there is only one interrupt level input, all possible interrupting devices

Figure 12–25

Flowchart for vector acquisition (Reprinted with permission of Motorola Inc.)

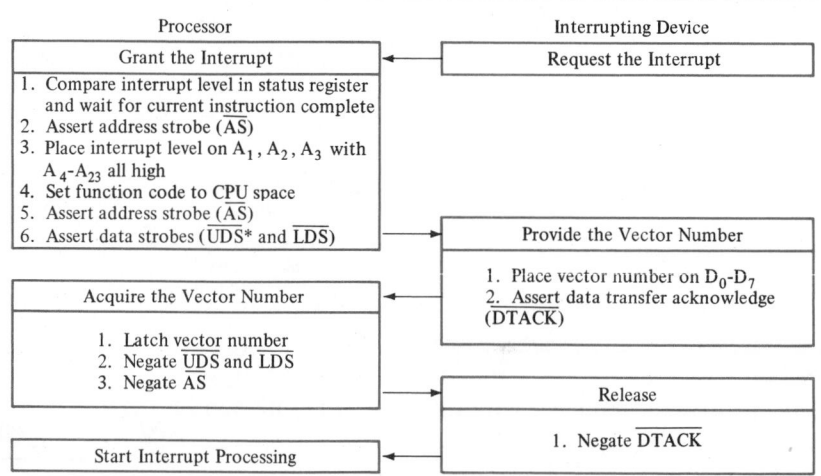

*Although a vector number is 1 byte, both data strobes are asserted due to the microcode used for exception processing. The processor does not recognize anything on data lines D_8 through D_{15}.

must be externally prioritized so that the highest interrupt level is reflected on these input lines.

In the second step, the processor recognizes an interrupt request being made and starts the interrupt processing sequence when the interrupt request level is higher than the current processor interrupt level. Since interrupt processing is of group 1 exception type, no action is taken until the execution of the current instruction is complete. The processor then enters into an IACK cycle. The timing diagram of the IACK cycle is shown in Figure 12–26. In the IACK cycle, the processor asserts the $\overline{AS}$, $\overline{UDS}$, and $\overline{LDS}$ lines with the interrupt level in the A_1, A_2, and A_3 address lines. All function codes are set to 1's to reflect an interrupt acknowledge condition.

Figure 12–26

Timing diagram for an interrupt acknowledge cycle (Reprinted with permission of Motorola Inc.)

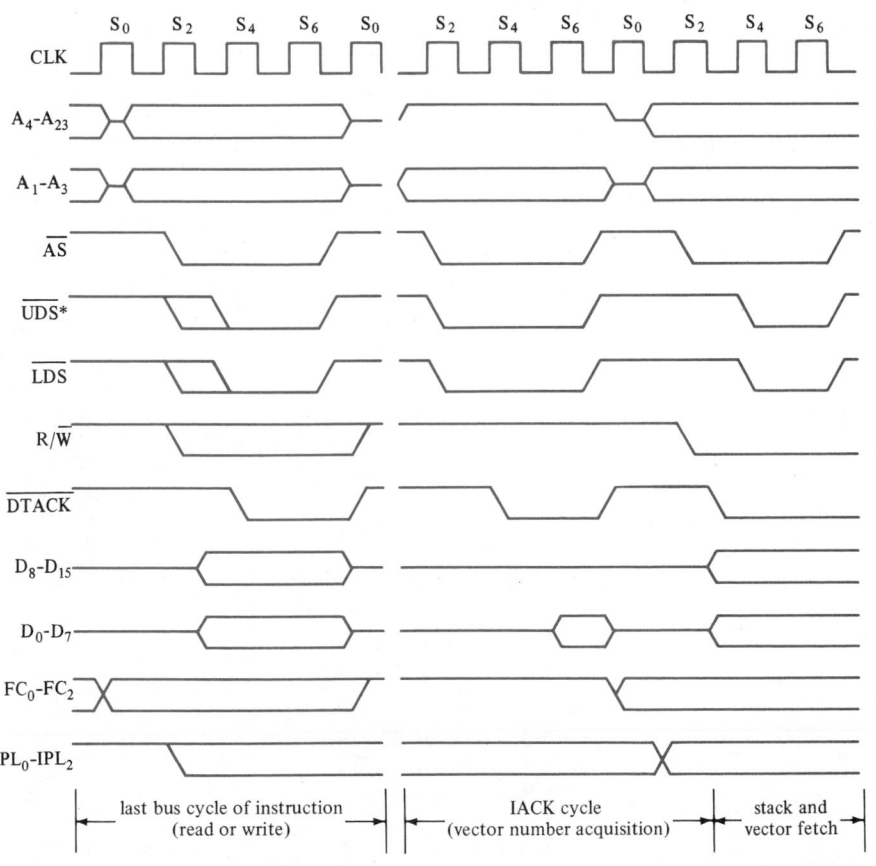

The purpose of the IACK cycle is to request the interrupting device for the vector number, so in the third step the device merely places the vector number on the lower byte of the data bus and asserts the $\overline{\text{DTACK}}$ line. This is an asynchronous data transfer operation. When the IACK cycle comes, external decoding must be used to determine which of the requesting devices must supply the vector number. This may be done either by a priority decoding/encoding scheme or by a daisy-chaining technique.

In the priority encoding/decoding scheme, individual requests come from the various devices. These requesting lines are encoded by a priority encoder to yield a single interrupt request code for the processor. When the interrupt acknowledge cycle comes via the assertion of the $\overline{\text{AS}}$, FC_0, FC_1, and FC_2 lines, a decoder is

Figure 12–27

A priority encoding/decoding scheme for interrupts (Reprinted with permission of Motorola Inc.)

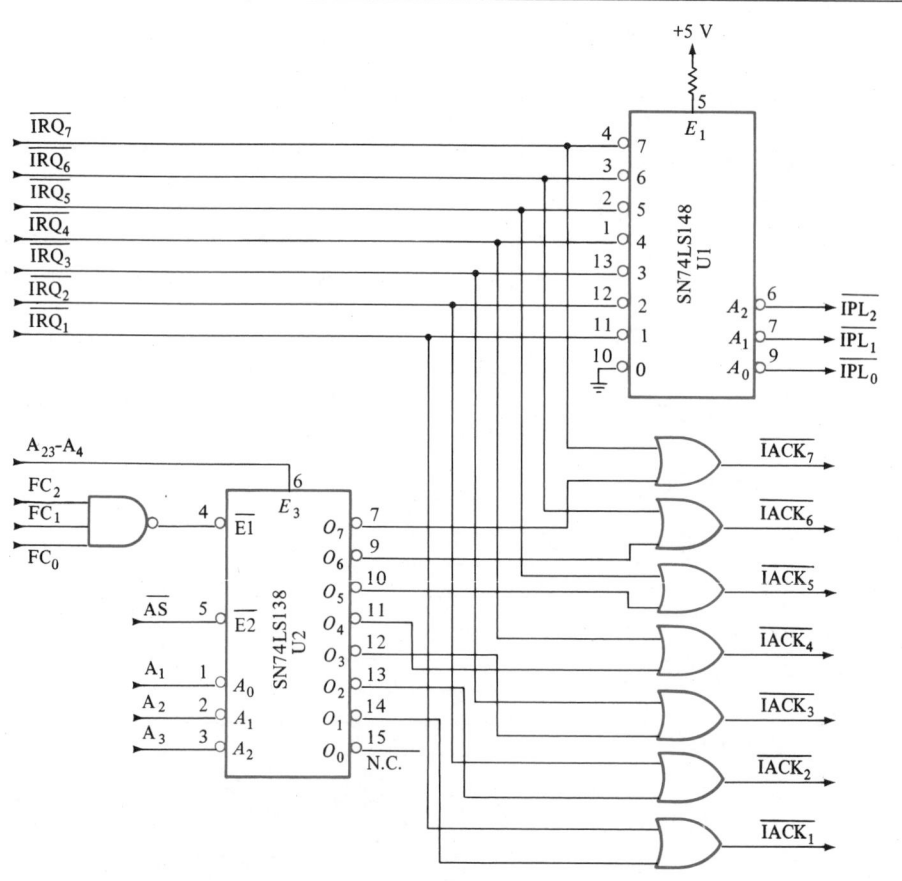

needed to generate the individual acknowledge lines. This is shown in Figure 12–27. This scheme is particularly attractive when only a small number of interrupt devices are present.

Another method for routing the interrupt acknowledge signal to the requesting device is by means of the daisy-chaining technique. The interrupt request line that carries the interrupt acknowledge "token" is passed from one device to another. Whoever makes the request keeps the token and supplies the vector number. This is shown in Figure 12–28. Sometimes, when a large number of interrupt devices must be accommodated, both methods are used together to provide the proper connection of all the interrupting devices to the processor.

In the fourth step, the processor latches the vector number supplied by the device and de-asserts the $\overline{\text{UDS}}$, $\overline{\text{LDS}}$, and $\overline{\text{AS}}$ lines. At this point, the acquired vector number is internally multiplied by 4 and added to the contents of the vector base register to form the interrupt vector. To finish the process of asynchronous data transfer, the device deactivates $\overline{\text{DTACK}}$ after sensing the de-assertion of $\overline{\text{AS}}$. Now the processor is ready to start interrupt processing.

Context Save Cycle

After the acquisition of the interrupt vector number, the processor is able to calculate the exception vector and locate the address of the interrupt processing routine. In the interrupt processing routine, the next logical thing to do is to store

Figure 12–28

A daisy-chaining method for interrupts (Reprinted with permission of Motorola Inc.)

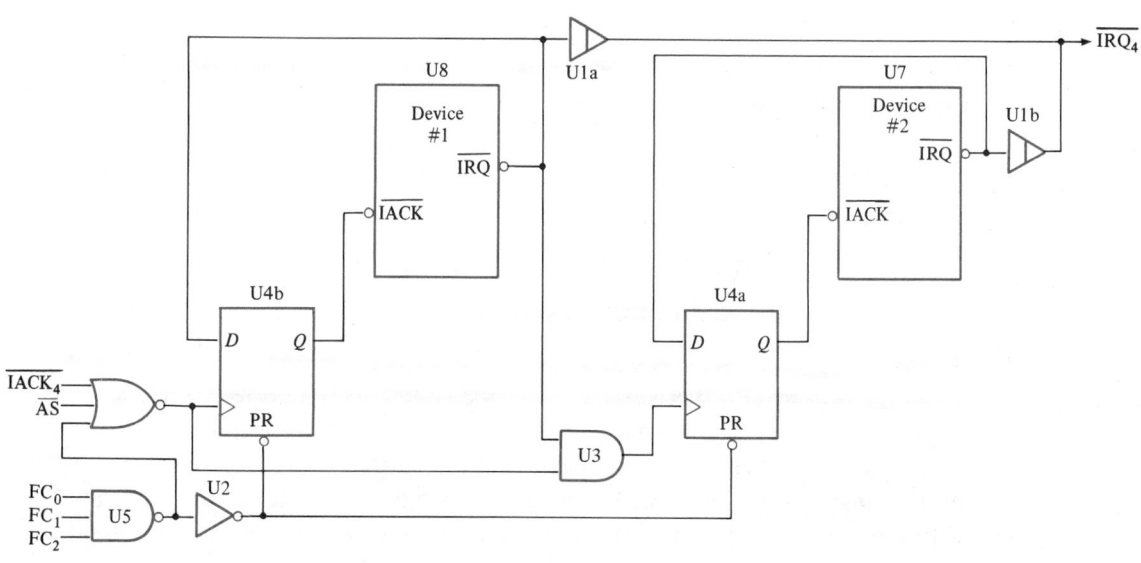

the current context. Since interrupt processing is part of exception processing, the information being saved in memory according to the supervisor stack pointer is as follows:

1. Status register
2. High order of program counter
3. Low order of program counter
4. Stack format and offset word

The above information is indeed the first 4 words of the exception short format. Since interrupt processing is not a group 0 exception, the long format is not used.

Interrupt Processing Cycle

The final step in interrupt processing is to use the previously computed vector address to obtain the first two instruction words of the interrupt routine. After storing the current context information, the vector number is used to locate the offset to the vector table from which the full 32-bit address of the interrupt routine is read. With this address, the first 2 words of the interrupt routine are read into the processor. However, before the instructions in the interrupt routine are executed, the interrupt control pins are sampled once more for the presence of a valid interrupt. If there is a valid interrupt pending that is of higher priority, the first instruction of the original interrupt routine will not be executed; rather, a new interrupt sequence will be initiated. After the new interrupt has been recognized and serviced, an RTE instruction will return the processor to the original interrupt routine and continue the service routine for the original interrupt.

12.8

Virtual Memory

The 68010 is designed to support the concepts of virtual memory. In this section we present some implementation and design techniques concerning virtual memory.

Concept of Virtual Memory Systems

The concept of virtual memory was developed mainly for systems that have a physical memory size much smaller than the maximum addressing space. From a system designer's point of view, when a processor has a large addressing space, it is neither practical nor possible to fill the entire addressing space. From a programmer's point of view, however, it is highly desirable to write programs without needing to worry about the actual amount of space available. Besides,

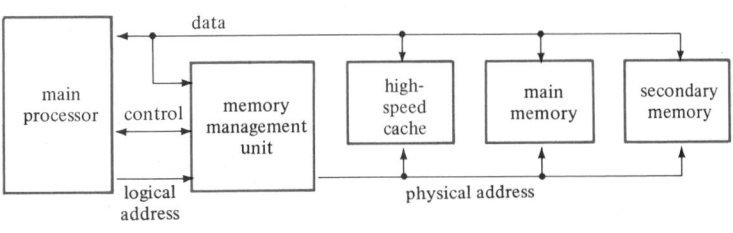

Figure 12–29

A memory hierarchy system

the actual amount of space is often subject to change. Hence the implementation of a virtual memory system is highly desirable.

In a virtual memory system, the processor generates logical addresses for data fetches. Without virtual memory support, this logical address becomes the actual physical address supplied to the memory for fetches. When the addressing space is larger than the actual memory available, the total required memory space to execute a particular program may be kept in a backing store, which is typically high-speed disk storage. When a particular memory block is needed, the memory block is down-loaded from the backing store to the main memory. This arrangement is shown in Figure 12–29.

Though the size of memory addressing space available to the programmer is limited only by the maximum number of addressing bits, the size of the actual addressing space in the main memory is usually limited to what is installed. Hence it is necessary to detect addresses that are outside the current available memory. When such an illegal address is detected, the program flow must be temporarily stopped until the situation has been rectified by bringing into memory from secondary storage the memory block or page that contains the requested memory reference. A translation by the memory management unit is also needed to convert the requested logical address to an actual physical address for memory fetches. The complete process is called a virtual memory system.

Implementation of a Virtual Memory System

To implement a virtual memory system, it is necessary to perform a series of steps on each memory reference. A flowchart of the steps is shown in Figure 12–30. The steps are:

1. Check the logical address by examining the page availability table to see if the requested page is currently available in memory. If so, proceed to Step 5.
2. If the current page is not available, stop the processing and save all pertinent information.

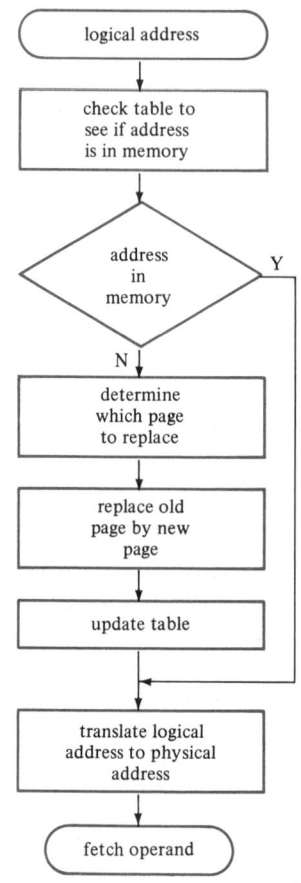

Figure 12–30

Flowchart for a virtual memory system

3. Replace an existing page with the newly requested page and update the page availability table.
4. Restore all previously saved information to continue program execution.
5. Convert the logical address to a physical address by examining the page availability table for the correct page address.
6. Fetch the operand from memory.

The first step, checking for the availability of the requested page, is not performed implicitly by the 68010 processor but is performed automatically by the memory management unit (MMU). The 68010 processor is, however, equipped to communicate with the memory management unit on the status of the check through the $\overline{\text{BERR}}$ line. If the current memory page is in memory as detected by the MMU, then no page fault occurs. The logical address is then translated as described in Step 5. If a page fault is detected, the $\overline{\text{BERR}}$ line is asserted to signify to the processor that a page fault is present.

Upon detecting activity on the $\overline{\text{BERR}}$ line, the processor immediately enters exception processing. The next logical step is to save all pertinent information so that the same context is restored after rectification of the access fault. A total of 52 bytes of information are stored on the system stack. If another bus fault is detected during this period, the processor will enter a halted state. The processor must halt because exception processing cannot proceed any further.

The next step is to replace an existing page with the newly requested page. Different replacement policies can be used. Some of these have been presented in Chapter 7. After determining which page has the lowest priority, the page is written back to secondary memory if the page has been modified. The newly requested page is then retrieved from memory and the page availability table is subsequently updated. The tasks described in this step are accomplished through exception processing by the processor in the supervisory state in response to the detection of a bus fault by the MMU.

After the newly requested page has been retrieved from memory, all pertinent information must be restored. The machine must be restored to exactly the same context as prior to the page fault before program execution can continue. The fifth step is to proceed with program execution by translating the logical address into an actual physical address by the MMU. The translated address is then supplied to the memory in the last step for operand fetches.

Processing Bus Faults

When a bus fault is detected, normal program execution is temporarily stopped and the page containing the requested reference is retrieved from backup storage. At this point, there are two ways to proceed:

Instruction restart
Instruction continuation

In the first method, the entire instruction is executed again from the beginning. In the second method, only the last microinstruction is reexecuted. There are advantages and disadvantages to both methods.

Instruction Restart Instruction restart is the simplest method to implement, because the only information to be saved is the program counter or the instruction register. Hence, after the requested page has been retrieved, decoding of the OPCODE commences again from the beginning, and the instruction is reexecuted from the beginning.

There are several drawbacks to the instruction restart method. It is necessary for the processor to make a copy of the original values of all affected registers for the purpose of recovery in the case of a bus fault. This may be done at the beginning of the instruction by making a copy of all the original values for a full recovery. Alternatively, the final values of all affected registers can be temporarily stored so that all original values are preserved, and all final values are not written back until the end of the instruction when no bus faults are detected. Another technique is to tag all affected registers in such a way that the processor can back-trace its steps to the original values. All these techniques are rather cumbersome. If each instruction performs only a single task, then restoration is simple. If each instruction performs a complex series of tasks, such as the MOVEM or the TAS instruction, then keeping a backup copy of all the registers and restoring them tends to be tedious and difficult. Another drawback to instruction restart involves input/output operations. It is somewhat easier to restore memory contents, but it is almost impossible to restore input/output peripherals, especially when peripheral status is involved and has been changed.

Instruction Continuation The instruction restart is easy to implement but provides an incomplete solution to context restoration for processors with a complex set of instructions. An alternative to the instruction restart solution is the instruction continuation method. In this method, the instruction is continued at the microinstruction level at the point where the page fault occurred.

If a certain instruction requires a number of microinstructions to complete, the completed microinstructions need not be executed again. However, in order for the context to be completely restored, in addition to the information in all user-accessible registers, all internal information, including the state of the processor and the contents of all internal registers, must be stored for a full recovery. The immediate drawback of this method is that a lot of information must be saved and restored, resulting in a much longer latency time than with the instruction restart method. This accounts for the long stack format used by the 68010 system.

12.9
A Design Example

In this section we discuss design techniques for microprocessor-based systems using the 68010. For a minimum system, the usual configuration consists of the processor, RAMs, ROMs, serial input/output, parallel input/output, and other supporting chips. Additional hardware can be added in a similar manner for a high-performance system or for an expanded system. We will concentrate on a particular minimum system configuration for simplicity. The same system design can be used effectively for many other expanded systems.

Because of the use of the asynchronous bus, it is rather easy to connect memory chips and peripherals to the system bus. Timing is critical only when a high-performance condition with no wait states is desired. The design of a minimum system can be accomplished in three steps:

1. Design the processor support and its system bus
2. Design RAM and ROM memory
3. Design connection of I/O peripherals

The first step sets the general architecture of the system. The second step fills in the memory portion of the system. The total amount of memory desired is simply the choice of the type of memory chips and the number of memory chips with their chip select lines connected to the decoders. The last step illustrates the connection of I/O peripherals.

Designing the Processor Support and Its System Bus

The 68010 processor is extremely easy to interface. Since the data bus is not multiplexed with the address bus, no external latch is necessary to demultiplex the address/data bus as is the case for the 8085. All three components of the system bus—the address bus, the data bus, and the control bus—come directly from or go to the processor.

The Address Bus There are 23 address lines, which are the A_1 to A_{23} address bits. Each must be connected to a buffer/driver, as each will be connected to a lot of memory chips and peripheral chips. It is advisable to connect even the higher-order address bits to drivers before connecting them to memory address decoders and peripheral I/O decoders. If the address bus is connected to read-only memory (ROM) and static read/write memory (SRAM), only address decoders connected to the higher-order bits are necessary. If the address bus is connected to dynamic read/write memory (DRAM), all the address bits must be connected to multiplexers or a DRAM controller for multiplexing the column address and the row address. The least significant address bit is reflected by $\overline{UDS}$ for the upper data byte and $\overline{LDS}$ for the lower data byte. The separation of the upper and lower data bytes is necessary because some instructions require only byte-size operands. Hence it is necessary to determine whether the operand is byte size or word size, and whether it is the even byte or the odd byte that is being addressed.

By monitoring the processor status lines, it is possible to tell the category of the current memory reference. Memory protection is then possible by separating supervisor references from user references and by separating program references from data references. A simple memory protection circuit is shown in Figure 12–31. In this circuit the supervisor memory space is protected against user references, yet the supervisor can address any area in the user area.

The Data Bus The data bus is bidirectional. Depending on the loading of the data bus, bidirectional latches/drivers may be required for the data bus. The direc-

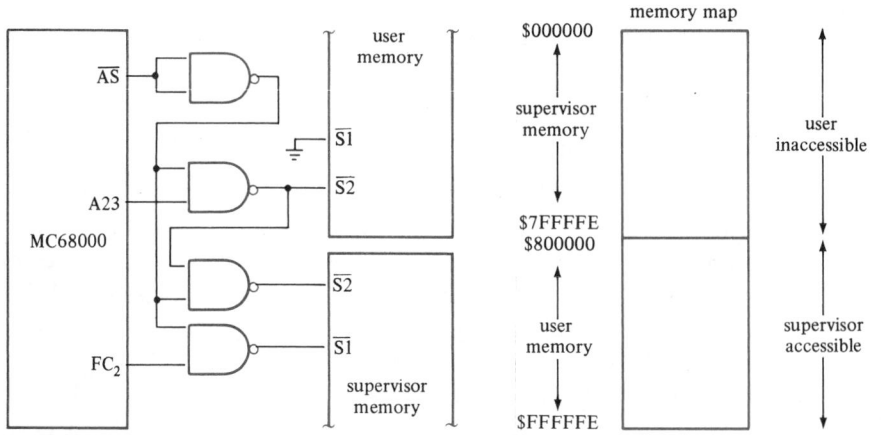

Figure 12–31
A memory protection circuit (Reprinted with permission of
Motorola Inc.)

tion of the data bus is given by the R/$\overline{\text{W}}$ line. The latches may be enabled through
decoding using simple combinational logic involving the $\overline{\text{AS}}$, R/$\overline{\text{W}}$, and the processor status lines.

The Control Bus In previous sections we discussed most of the external connections for specific groups of control lines. We will therefore only summarize those
discussions below. The interrupt priority lines are usually connected to the output of a priority decoder that prioritizes the incoming interrupt requests. The
interrupt acknowledge signal is often daisy-chained in order to route the acknowledge signal to the proper interrupt requesting device.

The asynchronous bus is controlled by the $\overline{\text{AS}}$ and $\overline{\text{DTACK}}$ lines. Since the
bus is asynchronous, as long as $\overline{\text{DTACK}}$ is not asserted, wait states are automatically generated internally. Hence it is an effective means to take advantage
of this property to implement single-step operations by manually generating the
$\overline{\text{DTACK}}$ signal. This way of single stepping will cause the processor to wait for
the assertion of the $\overline{\text{DTACK}}$ line in the middle of every memory cycle. For arbitration of the asynchronous bus, DMA requesting devices can be connected to
the $\overline{\text{BR}}$, $\overline{\text{BG}}$, and $\overline{\text{BGACK}}$ lines. Synchronous peripherals can be connected to the
synchronous bus by asserting the $\overline{\text{VPA}}$ line. The synchronous clock is taken from
the E control signal. For an expanded system, a large number of I/O peripherals
can be connected to the synchronous bus. For memory management, the $\overline{\text{BERR}}$
and $\overline{\text{HALT}}$ lines are used to indicate page faults and to force the processor into
exception processing.

Designing the RAM and ROM Memory

The design of the memory portion is rather straightforward. The lower-order address lines are connected to the ROMs and the SRAMs; they may also be connected to multiplexers and DRAM controllers in order to multiplex the column and row address if DRAMs are used. The higher-order address lines are connected to decoders for proper address decoding. The data lines of the memory chips can be connected to the data bus with or without transceivers.

The most important aspect of designing the memory part is the explicit generation of the $\overline{\text{DTACK}}$ signal. Neither the processor nor the memory chips provide the $\overline{\text{DTACK}}$ signal. It must be externally generated. Since the bus is asynchronous, wait states are generated automatically until the assertion of the $\overline{\text{DTACK}}$ signal. This is the primary difference between designing a memory system for an asynchronous bus versus a synchronous bus. From the timing diagrams, it can be seen that the $\overline{\text{DTACK}}$ signal is sampled at S_4. If $\overline{\text{DTACK}}$ is asserted, then no wait states are generated, and the processor continues with the rest of the states for the remaining part of the memory cycle. If the memory chips are fast enough and no wait states are desired, a counter can be used to count the elapsed time of four states and to assert the $\overline{\text{DTACK}}$ line afterwards. The count values are chosen to provide the appropriate number of clock cycles to accommodate the access time of the memory chips selected.

Designing the Connection of I/O Peripherals

Input/output peripherals can be connected to either the asynchronous bus or the synchronous bus. Connection to the asynchronous bus is by means of the $\overline{\text{AS}}$ and $\overline{\text{DTACK}}$ lines. Connection to the synchronous bus is by means of the $\overline{\text{VPA}}$, E, and $\overline{\text{VMA}}$ lines.

If connection is made to the asynchronous bus, I/O peripherals are connected in a similar way as memory chips. Most 68010-compatible peripheral chips will generate the proper $\overline{\text{DTACK}}$ signal for connection to the $\overline{\text{DTACK}}$ line of the processor. A real danger exists when the processor makes an I/O reference to a nonexistent peripheral. In this case the processor may wait forever for the nonexistent peripheral to respond. The solution to this problem is to install a time-out feature for the detection of bus errors. A counter clocked by the E signal is used to time the response on the $\overline{\text{DTACK}}$ line. If the terminal count is obtained on the down counter, a time-out condition is present and the $\overline{\text{BERR}}$ line of the processor is asserted to indicate the condition.

A Typical System Design

A typical design example for a system based on the 68000 is given in Figure 12–32. No significant changes are required to upgrade the system shown to one based on the 68010.

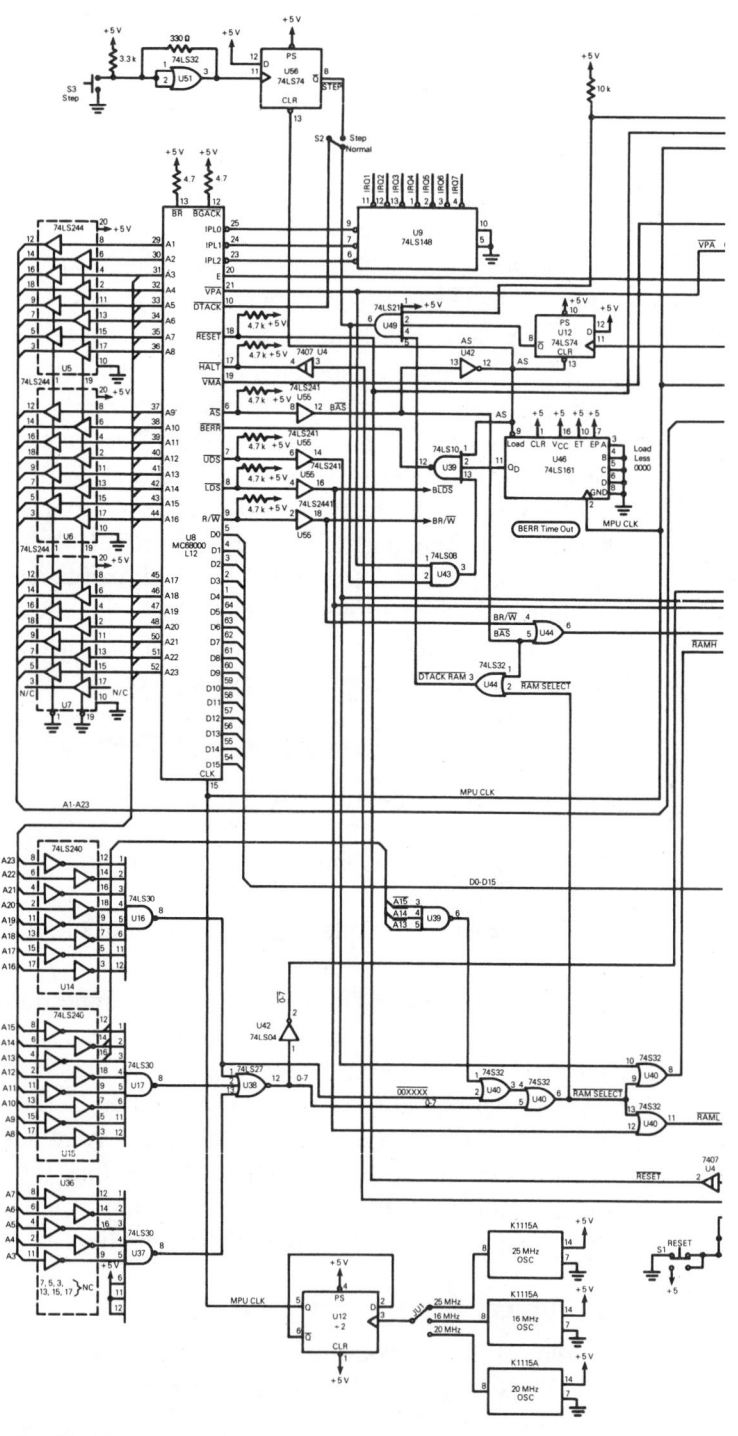

Figure 12-32 A sample design using a 68000 processor (Reprinted with permission of Motorola Inc.)

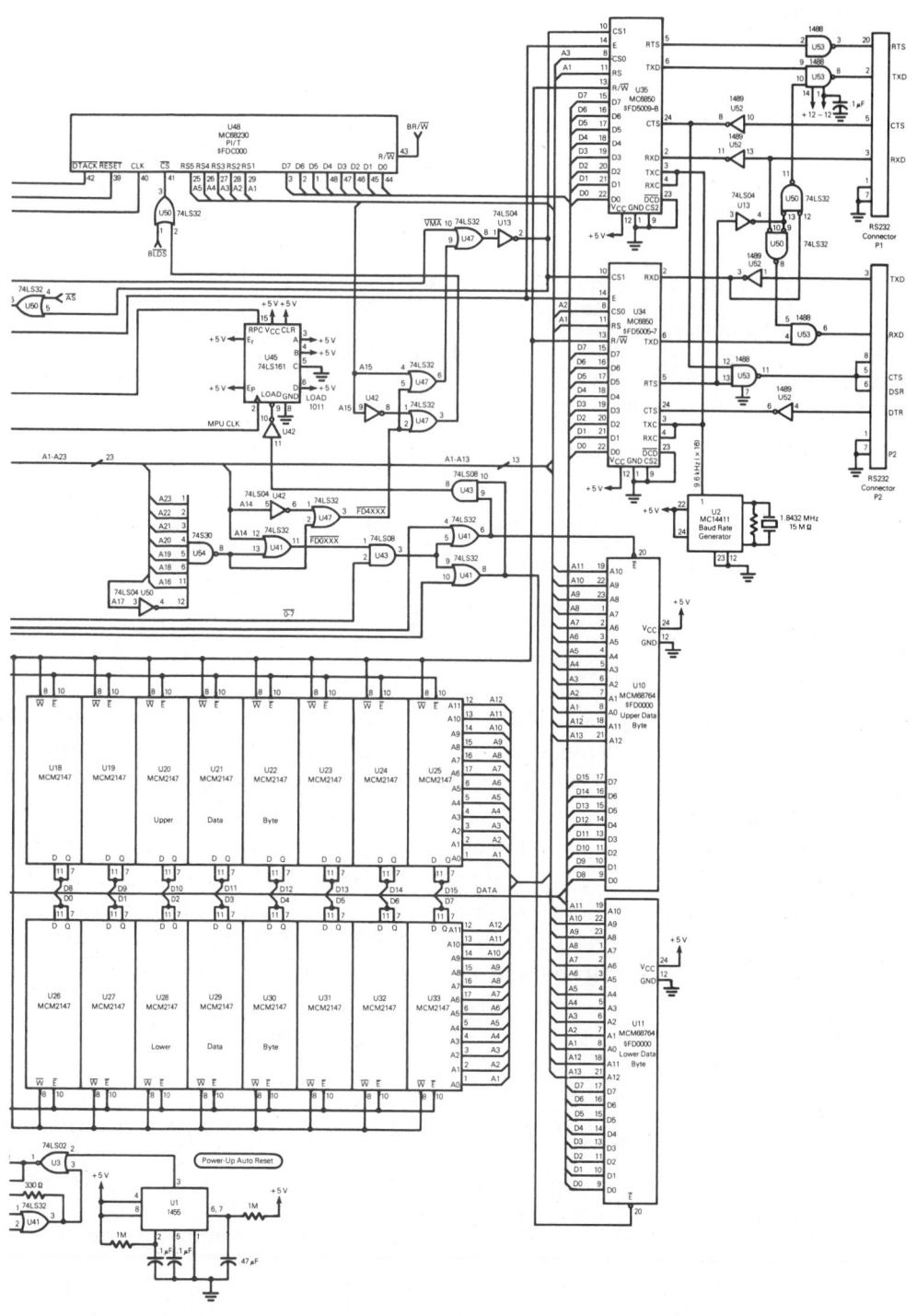

Figure 12-32 Continued

(continued)

List of ICs Used in MC68000L 12 System with No Wait States

IC Number	Part Number	IC Number	Part Number
U1	MC1455 timing circuit	U30, U31, U32, U33	
U2	MC14411 baud rate generator	U34, U35	MC6850 asynchronous
U3	74LS02 quad NOR gate		communication interface adapter
U4	7407 hex buffer with open	U38	74S27 3-input NOR gate
	collector outputs	U39	74S10 3-input NAND gate
U5, U6, U7	74LS244 octal line driver	U40, U41, U44, U47	74S32 quad OR gate
	noninverting outputs	U50, U51	74LS32 quad OR gate
U8	MC68000 16-bit microprocessor	U43	74S08 quad AND gate
U9	74LS148 8-to-3 encoder	U45, U46	74LS161 sync. 4-bit counter
U10, U11	MCM68764 8K × 8 EPROM	U48	MC68230 parallel interface/timer
U12	74S74 dual D latch	U49	74LS21 dual AND gate
U13, U42	74S04 hex inverter	U52	MC1489 level shifter/line driver
U14, U15, U36	74S240 octal buffer inverted	U53	MC1488 level shifter/line driver
	outputs	U55	74S241 octal buffer noninverting
U16, U17, U37, U54	74S30 8-input NAND gate	U56	74LS74 dual D latch
U18, U19, U20, U21	MCM2147 4K × 1 static RAM	U57	24-MHz oscillator
U22, U23, U24, U25	70-ns access time	U58	16-MHz oscillator
U26, U27, U28, U29			

Figure 12-32   Continued

By examining the timing diagrams for the read and write cycles, it can be seen that the maximum memory access time is determined by the time when $\overline{\text{LDS}}$ and $\overline{\text{UDS}}$ are asserted. Using a 25-MHz, clock for the processor, the internal clock period is 40 ns per cycle or 20 ns per state. The assertion of $\overline{\text{LDS}}$ and $\overline{\text{UDS}}$ comes during S_2 and goes during S_7, assuming that no wait states are generated. Therefore there are fewer than four states or 80 ns available from the assertion of $\overline{\text{LDS}}$ and $\overline{\text{UDS}}$. This period includes the address decoding time and memory access time. If the chosen memory system is not fast enough, the processor can be forced to insert additional wait states by changing the count used to assert the $\overline{\text{DTACK}}$ line.

References

1. Baer, Jean-Loup. *Computer Systems Architecture.* Computer Science Press, Potomac, Md., 1980.
2. Ciminiera, Luigi, and Adriano Valenzano. *Advanced Microprocessor Architectures.* Addison-Wesley, Reading, Mass., 1987.
3. Clements, Alan. *Microprocessor Systems Design: 68000 Hardware, Software, and Interfacing.* PWS Publishers, Boston, 1987.
4. Eccles, William J. *Microprocessor Systems: A 16-bit Approach.* Addison-Wesley, Reading, Mass., 1985.
5. Garland, Harry. *Introduction to Microprocessor System Design.* McGraw-Hill, New York, 1979.

6. Gorsline, G. W. *Computer Organization: Hardware/Software.* Prentice-Hall, Englewoods Cliffs, N.J., 1980.

7. Hamacher, V. Carl, Z. G. Vranesic, and S. G. Zaky. *Computer Organization*, 2d ed. McGraw-Hill, New York, 1984.

8. Hayes, John P. *Computer Architecture and Organization.* McGraw-Hill, New York, 1978.

9. Hayes, John P. *Digital System Design and Microprocessors.* McGraw-Hill, New York, 1984.

10. Hill, F. J., and G. R. Peterson. *Digital Logic and Microprocessors.* John Wiley & Sons, New York, 1984.

11. Hitachi Corp. *IC Memories.* Hitachi Corporation., San Jose, Calif., 1984.

12. Jermann, William H. *The Structure and Programming of Microcomputers.* Alfred Publishing Co., Sherman Oaks, Calif., 1982.

13. Kline, R. M. *Digital Computer Design.* Prentice-Hall, Englewood Cliffs, N.J., 1977.

14. Mano, M. Morris. *Digital Logic and Computer Design.* Prentice-Hall, Englewood Cliffs, N.J., 1979.

15. Mano, M. Morris. *Computer System Architecture*, 2d ed. Prentice-Hall, Englewood Cliffs, N.J., 1982.

16. Mano, M. Morris, *Digital Design.* Prentice-Hall, Englewood Cliffs, N.J., 1984.

17. Motorola, Inc. *MC68000: 16/32-Bit Microprocessor, Programmer's Reference Manual,* 4th ed. Prentice-Hall, Englewood Cliffs, N.J., 1984.

18. Motorola, Inc. *MC68000: 16-Bit Microprocessor.* Motorola Semiconductor Products Literature ADI-814, Austin, Tex.

19. Motorola, Inc. *MC68000: 16-Bit Microprocessor.* Motorola Semiconductor Products Literature ADI-839, Austin, Tex.

20. Motorola, Inc. *MC68000 Technical Training Course Notes.* Motorola Semiconductor Products Literature, Austin, Tex.

21. Motorola, Inc. *MC68000 Family Resident Structured Assembler Reference Manual.* Motorola Semiconductor Products Literature M68KMASM/D7, Austin, Tex.

22. Motorola, Inc. *MC68010: 16-Bit Microprocessor.* Motorola Semiconductor Products Literature ADI-942, Austin, Tex.

23. Motorola, Inc. *MC68030: Enhanced 32-Bit Microprocessor User's Manual.* Motorola Semiconductor Products Literature MC68030UM/AD, Austin, Tex.

24. National Semiconductor Corp. *Logic Data Book.* National Semiconductor Corporation, Santa Clara, Calif., 1981.

25. Oleksy, Jerome E., and George B. Rutkowski. *Microprocessor and Digital Computer Technology.* Prentice-Hall, Englewood Cliffs, N.J., 1981.

26. Ramirez, Edward V. *Microprocessing Fundamentals: Hardware and Software.* McGraw-Hill, New York, 1980.

27. Short, Kenneth L. *Microprocessors and Programmed Logic.* Prentice-Hall, Englewood Cliffs, N.J., 1981.

28. Siewiorek, Daniel P., C. G. Bell, and Allen Newell. *Computer Structures: Principles and Examples.* McGraw-Hill, New York, 1982.

29. Sloan, M. E. *Computer Hardware and Organization,* 2d ed. Science Research Associates, Chicago, 1983.

30. Tanenbaum, Andrew S. *Structured Computer Organization,* 2d ed. Prentice-Hall, Englewood Cliffs, N.J., 1984.

31. Wakerly, John F. *Microcomputer Architecture and Programming.* John Wiley & Sons, New York, 1981.

32. Ware, W. H. *Digital Computer Technology and Design*, Vol. 1. John Wiley & Sons, New York, 1963.
33. Wilcox, Alan D. *68000 Microcomputer Systems Designing and Troubleshooting.* Prentice-Hall, Englewood Cliffs, N.J., 1987.
34. Wilkinson, Barry. *Digital System Design.* Prentice-Hall, Englewood Cliffs, N.J., 1987.

Problems

1. Design a simple memory circuit to interface the 68010 to memory components. Remember that the memory must respond to byte-oriented operations also. Use the asynchronous data transfer technique.

2. Design a simple memory circuit to interface the 68010 to memory components. Remember that the memory must respond to byte-oriented operations also. Use the synchronous data transfer technique.

3. Design a simple memory circuit to interface the 68010 to byte-oriented memory components.

4. Design a simple memory protection circuit that separates user memory space from supervisor memory space.

5. Design a complete daisy-chaining circuit for five DMA requesting devices.

6. Design a complete circuit for five DMA requesting devices using the priority encoding/decoding scheme.

Appendix A
The 8085 Instruction Set

A.1 Notation
A.2 The 8085 Instructions

A.1
Notation

In describing the 8085 instruction set, it is helpful to use symbols and mnemonics to better represent and understand the operations of the instructions. We will use the Intel convention for the symbols and mnemonics.

Symbol	Meaning
A	The accumulator or register A (The terms accumulator and register A are used interchangeably.)
ADDR	A 16-bit address
$DATA_8$	8 bits of data or information
$DATA_{16}$	16 bits of data or information
$BYTE_2$	The second byte of the instruction
$BYTE_3$	The third byte of the instruction
PORT	An 8-bit address of an I/O device
r, r_1, r_2	One of the following 8-bit registers: A, B, C, D, E, H, or L
SSS	3 bits designating a source register

DDD	3 bits designating a destination register:

DDD or SSS	Register Name
111	A
000	B
001	C
010	D
011	E
100	H
101	L

rp   One of the register pairs BC, DE, or HL [The first register representing the highest-order byte and the second the lowest-order byte. The stack pointer (SP) is also included here.]

RP   The bit pattern designating one of the register pairs:

RP	Register Pair
00	BC
01	DE
10	HL
11	SP

rh   The first (high-order) register of a designated register pair

rl   The second (low-order) register of a designated register pair

PC   16-bit program counter

PCH   The higher-order bytes of the PC

PCL   The lower-order bytes of the PC

SP   16-bit stack pointer

SPH   The higher-order bytes of the SP

SPL   The lower-order bytes of the SP

PSW   A term referring to a pseudo-register pair formed with register A and the status register

r_m Bit m of register r (Bit 7 is on the left and bit 0 on the right. Bit 7 is considered to be the most significant bit.)

LABEL   A 16-bit address of an instruction

Z   Condition flag for zero (Z = 1 if result of last instruction that affects the Z flag was 0 in all bits; otherwise Z = 0.)

S   Condition flag for the sign (S = 1 if the most significant bit in the result of the previous instruction that affects the S flag is 1; otherwise S = 0.)

P   Condition flag for the parity (P = 1 if the number of bits in the result of the previous instruction that affects the P flag is even; otherwise P = 0.)

CY   Condition flag for the carry (CY = 1 if the result of the previous instruction that affects the CY flag resulted in a carry or borrow; otherwise CY = 0. Note: CY is also defined in logical instructions.)

AC   Condition flag for the auxiliary carry (AC = 1 when there is a carry from bit 3 to bit 4 in the previous instruction that affects the AC flag; otherwise AC = 0. This is used for BCD addition.)

( )   The contents of the memory location or register whose address is enclosed within the parentheses

←   Transfer of data in the direction of the arrow

↔   Exchange of data

∧   Logical AND

$\forall$ or $\oplus$ Exclusive OR

∨   Inclusive OR

+   Arithmetic addition

−   2's complement subtraction

*   Multiplication

/   Division

n Restart number in decimal (The restart number must be between 0 and 7.)

NNN   Restart number in binary (The restart number must be between 0 and 7.)

cc   The condition code:

	Condition	CCC
NZ	Not zero ($Z = 0$)	000
Z	Zero ($Z = 1$)	001
NC	No carry ($CY = 0$)	010
C	Carry ($CY = 1$)	011
PO	Parity odd ($P = 0$)	100
PE	Parity even ($P = 1$)	101
P	Plus ($S = 0$)	110
M	Minus ($S = 1$)	111

A.2
The 8085 Instructions

The instructions for the 8085 microprocessor are listed alphabetically below and described in a standardized form. The first line gives the instruction mnemonic and the operands, which are copyrighted © 1976 by Intel Corporation. The mnemonics are in a one-to-one correspondence with the binary bit patterns of the OPCODE of each instruction. The name of the instruction is enclosed in parentheses following the mnemonic. The second line contains a symbolic description of what the instruction does. On the next line the binary code of the instruction is given, with the bytes separated by commas in 2- and 3-byte instructions. The addressing mode is presented next. The number of machine cycles, the number of T states, and the flags that are changed by the instruction are given on the last line. A narrative description of the operation performed by the instruction follows. Note that the program counter is always incremented by 1, 2, or 3 bytes, depending on the instruction. This will not be shown for each instruction, but will be assumed.

ACI DATA$_8$ (Add immediate data with carry to register A)

Operation: $(A) \leftarrow (A) + (BYTE_2) + (CY)$
Binary code: 11001110, BYTE$_2$
Addressing mode: Immediate
Cycles: 2; states: 7
Flags: Z, S, P, CY, AC

The contents of BYTE$_2$ of the instruction are added to register A, taking into account the carry, and the result is stored in register A.

ADC M   (Add immediate data with carry to register A)

Operation: $(A) \leftarrow (A) + ((H)(L)) + (CY)$
Binary code: 10001110
Addressing mode: Register indirect
Cycles: 2; states: 7
Flags: Z, S, P, CY, AC

The contents of the memory location specified by register pair HL and the contents of the carry are added to register A. The result is stored in register A.

ADC r   (Add register with carry to register A)

Operation: $(A) \leftarrow (A) + (r) + (CY)$
Binary code: 10001SSS
Addressing mode: Register
Cycles: 1; states: 4
Flags: Z, S, P, CY, AC

The contents of register r and the contents of the carry are added to register A. The result is stored in register A.

ADD M   (Add memory to register A)

Operation: $(A) \leftarrow (A) + ((H)(L))$
Binary code: 10000110
Addressing mode: Register indirect
Cycles: 2; states: 7
Flags: Z, S, P, CY, AC

The contents of the memory location specified by register pair HL are added to register A, and the result is stored in register A.

ADD r   (Add register to register A)

Operation: $(A) \leftarrow (A) + (r)$
Binary code: 10000SSS
Addressing mode: Register
Cycles: 1; states: 4
Flags: Z, S, P, CY, AC

The contents of register r are added to the contents of register A, and the result is stored in register A.

ADI DATA$_8$ (Add immediate data to register A)

Operation: $(A) \leftarrow (A) + (BYTE_2)$
Binary code: 11000110, BYTE$_2$
Addressing mode: Immediate
Cycles: 2; states: 7
Flags: Z, S, P, CY, AC

The contents of BYTE$_2$ of the instruction are added to register A, and the result is stored in register A.

ANA M   (AND memory to register A)

Operation: $(A) \leftarrow (A) \wedge ((H)(L))$
Binary code: 10100110
Addressing mode: Register indirect
Cycles: 2; states: 7
Flags: Z, S, P, CY, AC

The contents of the memory location specified by register pair HL are logically ANDed with register A, and the result is stored in register A. Note that the CY flag is always reset, and the AC flag is set after the execution of this instruction.

ANA r   (AND register to register A)

Operation: $(A) \leftarrow (A) \wedge (r)$
Binary code: 10100SSS
Addressing mode: Register
Cycles: 1; states: 4
Flags: Z, S, P, CY, AC

The contents of register r are logically ANDed with the contents of register A, and the result is stored in register A. The CY flag is cleared (CY = 0) and the AC flag is set (AC = 1).

ANI DATA$_8$ (AND immediate data to register A)

Operation: $(A) \leftarrow (A) \wedge (BYTE_2)$
Binary code: 11100110, BYTE$_2$
Addressing mode: Immediate
Cycles: 2; states: 7
Flags: Z, S, P, CY, AC

BYTE$_2$ of the instruction is logically ANDed with register A, and the result is stored in register A. Note that the CY flag is always reset, and the AC flag is set after the execution of this instruction.

CALL ADDR   (Unconditional call)

Operation: ((SP) − 1) ← (PCH)
           ((SP) − 2) ← (PCL)
           (SP)   ← (SP) − 2
           (PC)   ← (BYTE$_3$)(BYTE$_2$)
Binary code: 11001101, BYTE$_2$, BYTE$_3$
Addressing mode: Immediate, register indirect
Cycles: 5; states: 18
Flags: None affected

The high-order and low-order bytes of the program counter of the next instruction are moved to the addresses that are 1 and 2 less than what is specified by the SP (stack pointer), respectively. The contents of the SP are decremented by 2, and control is transferred to the instruction whose address is specified in BYTE$_3$ and BYTE$_2$ of the instruction.

Ccc ADDR   (Conditional call)

Operation: IF condition THEN BEGIN
                        ((SP) − 1) ← (PCH)
                        ((SP) − 2) ← (PCL)
                        (SP)   ← (SP) − 2
                        (PC)   ← (BYTE$_3$)(BYTE$_2$)
Binary code: 11CCC100, BYTE$_2$, BYTE$_3$
Addressing mode: Immediate, register indirect
Cycles: 2/5; states: 9/18
Flags: None affected

If the specified conditions are true, the actions of the CALL instruction are performed; otherwise, control continues sequentially.

CMA   (Complement register A)

Operation: (A) ← ($\bar{A}$)
Binary code: 00101111
Addressing mode: None
Cycles: 1; states: 4
Flags: None affected

The contents of register A are complemented; no flags are affected.

CMC   (Complement carry)

Operation: (CY) ← $\overline{(CY)}$
Binary code: 00111111
Addressing mode: None
Cycles: 1; states: 4
Flags: CY

The CY flag is complemented. No other flags are affected.

CMP M   (Compare memory to register A)

Operation: $(A) - ((H)(L))$
Binary code: 10111110
Addressing mode: Register indirect
Cycles: 2; states: 7
Flags: Z, S, P, CY, AC

The contents of the memory location specified by register pair HL are subtracted from register A, and register A remains unchanged. The flags are set according to the results of the subtraction. Furthermore, the Z flag is set if the contents of register A are equal to the contents of the specified register, and the CY flag is set if the contents of A are less than the contents of the specified register.

CMP r   (Compare register to register A)

Operation: $(A) - (r)$
Binary code: 10111SSS
Addressing mode: Register
Cycles: 1; states: 4
Flags: Z, S, P, CY, AC

The contents of register r are subtracted from register A, and register A remains unchanged. The flags are set according to the results of the subtraction. Furthermore, the Z flag is set if the contents of register A are equal to the contents of the specified register, and the CY flag is set if the contents of A are less than the contents of the specified register.

CPI DATA$_8$ (Compare immediate data to register A)

Operation: $(A) - (BYTE_2)$
Binary code: 11111110, $BYTE_2$
Addressing mode: Immediate
Cycles: 2; states: 7
Flags: Z, S, P, CY, AC

The contents of $BYTE_2$ of the instruction are subtracted from register A, and register A remains unchanged. The flags are set according to the results of the subtraction. Furthermore, the Z flag is set if the contents of register A are equal to the contents of the specified register, and the CY flag is set if the contents of A are less than the contents of the specified register.

DAA   (Decimal adjust accumulator)

Operation: $(A) \leftarrow$ Decimal adjusted (A)
Binary code: 00100111
Addressing mode: None
Cycles: 1; states: 4
Flags: Z, S, P, CY, AC

The 8-bit number in the accumulator is adjusted to form two 4-bit BCD digits by the following process:

1. If the contents of the least significant 4 bits in register A are greater than 9 or if AC = 1, then 6 is added to register A to correct for the decimal carry in the least significant digit.
2. If the value of the most significant 4 bits of register A are now greater than 9 or if the CY flag is set, then 6 is added to the most significant 4 bits of register A. All flags are affected.

DAD rp   (Add register pair to register pair HL)

Operation: (H)(L) ← (H)(L) + (rh)(rl)
Binary code: 00RP1001
Addressing mode: Register
Cycles: 3; states: 10
Flags: CY

The contents of register pair rp are added to the contents of register pair HL, and the result is stored in register pair HL. Register pair HL is always one of the source operands and is always the destination operand. Only the CY flag is affected. CY = 1 when carry occurs from the most significant bit during the DAD instruction; otherwise CY = 0.

DCR M   (Decrement memory)

Operation: ((H)(L)) ← ((H)(L)) − 1
Binary code: 00110101
Addressing mode: Register indirect
Cycles: 3; states: 10
Flags: Z, S, P, AC

The contents of the memory location specified by register pair HL are decremented by 1. All flags are affected except CY.

DCR r   (Decrement register)

Operation: (r) ← (r) − 1
Binary code: 00DDD101
Addressing mode: Register
Cycles: 1; states: 4
Flags: Z, S, P, AC

The contents of register r are decremented by 1. All flags except CY are affected.

DCX rp   (Decrement register pair)

Operation: (rh)(rl) ← (rh)(rl) − 1
Binary code: 00RP1011
Addressing mode: Register
Cycles: 1; states: 6
Flags: None affected

The contents of register pair rp are decremented by 1. No flags are affected.

DI   (Disable interrupts)

    Operation: Disable internal interrupt flip-flop
    Binary code: 11110011
    Addressing mode: None
    Cycles: 1; states: 5
    Flags: None affected

The interrupt system is disabled immediately following the execution of the DI instruction. Furthermore, no interrupts are allowed during the execution of the DI instruction.

EI   (Enable interrupts)

    Operation: Enable internal interrupt flip-flop
    Binary code: 11111011
    Addressing mode: None
    Cycles: 1; states: 4
    Flags: None affected

The interrupt system is enabled following the execution of the next instruction. Furthermore, no interrupts are allowed during the execution of the EI instruction.

HLT   (Halt)

    Operation: Halt the processor
    Binary code: 01110110
    Addressing mode: None
    Cycles: 1; states: 5
    Flags: None affected

The processor is stopped and the registers and flags are not affected. A second ALE is generated during the execution of HLT to strobe out the halt cycle status information.

IN port   (Input to register A)

    Operation: (A) ← (port)
    Binary code: 11011011, $BYTE_2$
    Addressing mode: Direct
    Cycles: 3; states: 10
    Flags: None affected

The contents of the input port specified in $BYTE_2$ of the instruction are moved to register A.

INR M   (Increment memory)

    Operation: ((H)(L)) ← ((H)(L)) + 1
    Binary code: 00110100
    Addressing mode: Register indirect
    Cycles: 3; states: 10
    Flags: Z, S, P, AC

The contents of the memory location specified by the register pair HL are incremented by 1. All flags except CY are affected.

INR r   (Increment register)

    Operation: $(r) \leftarrow (r) + 1$
    Binary code: 00DDD100
    Addressing mode: Register
    Cycles: 1; states: 4
    Flags: Z, S, P, AC

The contents of register r are incremented by 1. All flags except CY are affected.

INX rp   (Increment register pair)

    Operation: $(rh)(rl) \leftarrow (rh)(rl) + 1$
    Binary code: 00RP0011
    Addressing mode: Register
    Cycles: 1; states: 6
    Flags: None affected

The contents of register pair rp are incremented by 1. No flags are affected.

Jcc ADDR   (Conditional jump)

    Operation: IF condition THEN $(PC) \leftarrow (BYTE_3)(BYTE_2)$
    Binary code: 11CCC010, $BYTE_2$, $BYTE_3$
    Addressing mode: Immediate
    Cycles: 2/3; states: 7/10
    Flags: None affected

If the specified condition is true, control is transferred to the instruction specified by the address contained in $BYTE_3$ and $BYTE_2$ of the instruction; otherwise the next instruction is executed.

JMP ADDR   (Unconditional jump)

    Operation: $(PC) \leftarrow (BYTE_3)(BYTE_2)$
    Binary code: 11000011, $BYTE_2$, $BYTE_3$
    Addressing mode: Immediate
    Cycles: 3; states: 10
    Flags: None affected

Control is transferred to the instruction specified by the address contained in $BYTE_3$ and $BYTE_2$ of the instruction.

LDA ADDR   (Load accumulator direct)

    Operation: $(A) \leftarrow ((BYTE_3)(BYTE_2))$
    Binary code: 00111010, $BYTE_2$, $BYTE_3$
    Addressing mode: Direct
    Cycles: 4; states: 13
    Flags: None affected

The contents of the memory location, whose address is specified in $BYTE_2$ and $BYTE_3$ of the instruction, are moved to register A.

LDAX rp   (Load accumulator indirect)

Operation: $(A) \leftarrow ((rp))$
Binary code: 00RP1010
Addressing mode: Register indirect
Cycles: 2; states: 7
Flags: None affected

The contents of the memory location that is specified by the register pair rp are moved to register A. Only the register pairs BC and DE may be specified.

LHLD ADDR   (Load H and L direct)

Operation: $(L) \leftarrow ((BYTE_3)(BYTE_2))$
              $(H) \leftarrow ((BYTE_3)(BYTE_2) + 1)$
Binary code: 00101010, $BYTE_2$, $BYTE_3$
Addressing mode: Direct
Cycles: 5; states: 16
Flags: None affected

The contents of the memory location specified by $BYTE_2$ and $BYTE_3$ in the instruction are moved to register L, and the contents of the next memory location are moved to register H.

LXI rp,$DATA_{16}$ (Load register pair immediate)

Operation: $(rh) \leftarrow (BYTE_3)$
              $(rl) \leftarrow (BYTE_2)$
Binary code: 00RP0001, $BYTE_2$, $BYTE_3$
Addressing mode: Register, immediate
Cycles: 3; states: 10
Flags: None affected

The contents of $BYTE_3$ and $BYTE_2$ of the instruction are moved into the high-order and low-order registers, respectively, of register pair rp. This instruction is very useful for loading addresses directly into the specified register pair.

MOV M,r   (Move register to memory)

Operation: $((H)(L)) \leftarrow (r)$
Binary code: 01110SSS
Addressing mode: Register indirect, register
Cycles: 2; states: 7
Flags: None affected

The contents of register r are moved to the memory location specified by the register pair H and L. Because there are only seven source registers—A, B, C, D, E, H, and L—the bit pattern 110 for SSS is not used. It is reserved for register indirect instructions using register pair H and L.

MOV r,M   (Move memory to register)

Operation: (r) ← ((H)(L))
Binary code: 01DDD110
Addressing mode: Register, register indirect
Cycles: 2; states: 7
Flags: None

The contents of the memory location that is specified by the register pair H and L are moved to register r.

MOV r_1,r_2 (Move register to register)

Operation: $(r_1) ← (r_2)$
Binary code: 01DDDSSS
Addressing mode: Register, register
Cycles: 1; states: 4
Flags: None affected

The contents of register r_2 are moved to register r_1. The binary patterns for the DDD (destination register) and SSS (source register) have already been given. For example, the instruction MOV A,B means a transfer of the contents of register B to register A and has a binary code of 01 111 000, where 111 represents destination register A and 000 represents source register B.

MVI M,DATA$_8$ (Move to memory immediate)

Operation: ((H)(L)) ← (BYTE$_2$)
Binary code: 00110110, BYTE$_2$
Addressing mode: Register indirect, immediate
Cycles: 3; states: 10
Flags: None affected

The value of BYTE$_2$ of the instructions is moved to the memory location specified by the register pair H and L.

MVI r,DATA$_8$ (Move immediate data to register)

Operation: (r) ← (BYTE$_2$)
Binary code: 00DDD110, BYTE$_2$
Addressing mode: Register, immediate
Cycles: 2; states: 7
Flags: None affected

The 8-bit value of BYTE$_2$ of the instruction is moved to register r. This instruction is often used for loading constants and literals into the designated register.

NOP   (No operation)

Operation: No operation
Binary code: 00000000

Addressing mode: None
Cycles: 1; states: 4
Flags: None affected

No operation is performed; and the registers and flags are not affected.

ORA M   (OR memory to register A)

Operation: (A) ← (A) ∨ ((H)(L))
Binary code: 10110110
Addressing mode: Register indirect
Cycles: 2; states: 7
Flags: Z, S, P, CY, AC

The contents of the memory location specified by register pair HL are logically ORed with the contents of register A, and the result is stored in register A. Both the CY flag and the AC flag are reset after the execution of this instruction.

ORA r   (OR register r to register A)

Operation: (A) ← (A) ∨ (r)
Binary code: 10110SSS
Addressing mode: Register
Cycles: 1; states: 4
Flags: Z, S, P, CY, AC

The contents of register r are ORed with the contents of register A, and the result is stored in register A. Both the CY flag and the AC flag are reset after the execution of this instruction.

ORI DATA$_8$ (OR immediate to register A)

Operation: (A) ← (A) ∨ (BYTE$_2$)
Binary code: 11110110, BYTE$_2$
Addressing mode: Immediate
Cycles: 2; states: 7
Flags: Z, S, P, CY, AC

BYTE$_2$ of the instruction is logically ORed with register A, and the result is stored in register A. Both the CY flag and the AC flag are reset after the execution of this instruction.

OUT port   (Output from register A)

Operation: (Port) ← (A)
Binary code: 11010011, BYTE$_2$
Addressing mode: Direct
Cycles: 3; states: 10
Flags: None affected

The contents of register A is moved to the output port specified in BYTE$_2$ of the instruction.

PCHL   (Jump based on register pair HL)

Operation: (PCH) ← (H)
            (PCL) ← (L)
Binary code: 11101001
Addressing mode: Register
Cycles: 1; states: 6
Flags: None affected

The contents of register pair HL are transferred to the PC. The next instruction executed will be at the address specified in register pair HL.

POP PSW   (Pop processor status word)

Operation: (Status register) ← ((SP))
            (A)   ← ((SP) + 1)
            (SP)   ← (SP) + 2
Binary code: 11110001
Addressing mode: Register indirect
Cycles: 3; states: 10
Flags: Z, S, P, CY, AC

The contents of the memory location specified by the contents of SP and the contents of SP plus 1 are transferred to the flag register and register A, respectively. The SP is incremented by 2.

POP rp   (Pop register pair)

Operation: (rl) ← ((SP))
            (rh) ← ((SP) + 1)
            (SP) ← (SP) + 2
Binary code: 11RP0001
Addressing mode: Register indirect
Cycles: 3; states: 10
Flags: None affected

The contents of the memory location specified by the contents of SP and the contents of SP plus 1 are transferred to the low- and high-order bytes of register pair rp, respectively. The SP is incremented by 2.

PUSH PSW   (Push processor status word)

Operation: ((SP) − 1) ← (A)
            ((SP) − 2) ← (Status register)
            (SP)   ← (SP) − 2
Binary code: 11110101
Addressing mode: Register indirect
Cycles: 3; states: 12
Flags: None

The contents of A and the flags are transferred to memory at the address specified at 1 and 2 less than the contents of SP, respectively. The SP is decremented by 2.

PUSH rp   (Push register pair)

Operation: ((SP) − 1) ← (rh)
            ((SP) − 2) ← (rl)
            (SP)   ← (SP) − 2
Binary code: 11RP0101
Addressing mode: Register indirect
Cycles: 3; states: 12
Flags: None affected

The contents of the high- and low-order bytes of register pair rp are transferred to memory at the address specified at 1 and 2 less than the contents of
SP, respectively. The contents of SP is also decremented by 2. Note that rp may
not be specified as SP.

Rcc   (Conditional return)

Operation: IF condition THEN BEGIN
                            (PCL) ← ((SP))
                            (PCH) ← ((SP) + 1)
                            (SP) ← (SP) + 2
                         END
Binary code: 11CCC000
Addressing mode: Register indirect
Cycles: 1/3; states: 6/12
Flags: None affected

If the specified condition is true, the actions of the RET instruction are performed; otherwise, control continues sequentially.

RAL   (Rotate register A left through carry)

Operation: $(A_{n+1}) ← (A_n)$
            $(CY)   ← (A_7)$
            $(A_0)   ← (CY)$
Binary code: 00010111
Addressing mode: None
Cycles: 1; states: 4
Flags: CY

The contents of register A are rotated left one position through the CY flag.
The low-order bit in register A is set equal to CY, and the CY flag is set according
to the high-order bit.

RAR   (Rotate register A right through carry)

Operation: $(A_n) ← (A_{n+1})$
            $(CY) ← (A_0)$
            $(A_7) ← (CY)$
Binary code: 00011111
Cycles: 2; states: 4
Flags: CY

The contents of register A are rotated right one position through the CY flag. The high-order bit is set to the CY flag, and the CY flag is set to the low-order bit.

RET   (Unconditional return)

Operation: $(PCL) \leftarrow ((SP))$
$(PCH) \leftarrow ((SP) + 1)$
$(SP) \quad \leftarrow (SP) + 2$
Binary code: 11001001
Addressing mode: Register indirect
Cycles: 3; states: 10
Flags: None affected

The contents of the memory locations specified by the contents of register SP and the contents plus 1 of register SP are moved to the low-order and high-order bytes of register PC, respectively. The contents of register SP are incremented by 2.

RIM   (Read interrupt masks)

Operation: $(A) \leftarrow$ SID and Interrupt mask
Binary code: 00100000
Addressing mode: None
Cycles: 1; states: 4
Flags: None affected

The RIM instruction loads into register A information related to the interrupts and the serial input data (SID) line. These data contain the following information:

1. Current interrupt mask status for RST 5.5, RST 6.5, and RST 7.5 hardware interrupts in A_0, A_1, and A_2, respectively (1 = mask disabled)
2. Current interrupt enable flag status (1 = interrupts enabled) in A_3 except immediately following a TRAP interrupt
3. Hardware interrupts pending in A_4, A_5, and A_6 (i.e., signals received but not yet serviced) on the RST 5.5, RST 6.5, and RST 7.5 lines, respectively
4. Serial input data (SID) line into A_7

The A_6 bit represents the state of the RST 7.5 flip-flop, not the RST 7.5 input. The contents of register A after the execution of the RIM instruction are

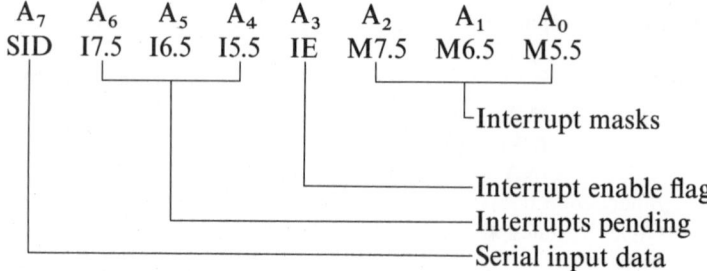

RLC   (Rotate register A left)

Operation: $(A_{n+1}) \leftarrow (A_n)$
            $(A_0) \quad \leftarrow (A_7)$
            $(A_7) \quad \leftarrow (CY)$
Binary code: 00000111
Addressing mode: None
Cycles: 1; states: 4
Flags: CY

The contents of register A are rotated left one position. The low-order bit and CY are both set to the most significant bit in register A.

RRC   (Rotate register A right)

Operation: $(A_n) \quad \leftarrow (A_{n+1})$
            $(A_7) \leftarrow (A_0)$
            $(CY) \leftarrow (A_0)$
Binary code: 00001111
Addressing mode: None
Cycles: 1; states: 4
Flags: CY

The contents of register A are rotated right one position. The high-order bit and CY are both set to the least significant bit in register A.

RST n   (Restart)

Operation $((SP) - 1) \leftarrow (PCH)$
            $((SP) - 2) \leftarrow (PCL)$
            $(SP) \qquad \leftarrow (SP) - 2$
            $(PC) \qquad \leftarrow 8 * (NNN)$
Binary code: 11NNN111
Addressing mode: Register indirect
Cycles: 3; states: 12
Flags: None affected

The high-order and low-order bytes of the program counter of the next instruction are moved to the address that is 1 and 2 less than is specified by the SP, respectively. The contents of the SP are decremented by 2, and control is transferred to the instruction that is eight times the contents of NNN.

SBB M   (Subtract memory with borrow from register A)

Operation: $(A) \leftarrow (A) - ((H)(L)) - (CY)$
Binary code: 10011110
Addressing mode: Register indirect
Cycles: 2; states: 7
Flags: Z, S, P, CY, AC

The contents of the memory location specified by register pair HL and the contents of CY are both subtracted from register A, and the result is stored in register A.

SBB r   (Subtract register with borrow from register A)

Operation: $(A) \leftarrow (A) - (r) - (CY)$
Binary code: 10011SSS
Addressing mode: Register
Cycles: 1; states: 4
Flags: Z, S, P, CY, AC

The contents of register r and the contents of CY are both subtracted from register A, and the result is stored in register A.

SBI DATA$_8$ (Subtract immediate data with borrow from register A)

Operation: $(A) \leftarrow (A) - (BYTE_2) - (CY)$
Binary code: 11011110, BYTE$_2$
Addressing mode: Immediate
Cycles: 2; states: 7
Flags: Z, S, P, CY, AC

The contents of BYTE$_2$ of the instruction and the contents of CY are both subtracted from register A and the result is stored in register A.

SHLD ADDR   (Store H and L direct)

Operation: $((BYTE_3)(BYTE_2)) \leftarrow (L)$
$\qquad\qquad ((BYTE_3)(BYTE_2) + 1) \leftarrow (H)$
Binary code: 00100010, BYTE$_2$, BYTE$_3$
Addressing mode: Direct
Cycles: 5; states: 16
Flags: None affected

The contents of register L are moved to the memory location specified by BYTE$_2$ and BYTE$_3$ of the instruction, while the contents of register H are moved to the next memory location.

SIM   (Set interrupt masks)

Operation: SOD and interrupt masks $\leftarrow (A)$
Binary code: 00110000
Addressing mode: None
Cycles: 1; states: 4
Flags: None

The execution of the SIM instruction uses the contents of register A (which must be previously loaded) to perform the following functions:

1. Program the RST 5.5, RST 6.5, and RST 7.5 interrupt masks according to A_0, A_1, and A_2, respectively. To program the masks, first make $A_3 = 1$. Then set A_0, A_1, and A_2 to 0 to enable interrupts and 1 to disable interrupts. Next execute the SIM instruction. If $A_3 = 0$ and SIM is executed, the status of the masks will not change.
2. Reset the edge-triggered RST 7.5 input flip-flop. If $A_4 = 1$ and the SIM instruction is executed, then the RST 7.5 input flip-flop is reset. The RST

7.5 input flip-flop is also reset in response to a RESET IN or an internal processor acknowledge to an RST 7.5 interrupt.

3. Load the SOD output flip-flop. If $A_6 = 1$ and the SIM instruction is executed, then A_7 is loaded into the SOD output flip-flop and thus appears on the SOD output line. If $A_6 = 0$ and SIM is executed, then the SOD output flip-flop is not changed. SOD is always reset by the RESET IN signal.

The contents of register A before the execution of SIM to control the 8085 are specified as follows:

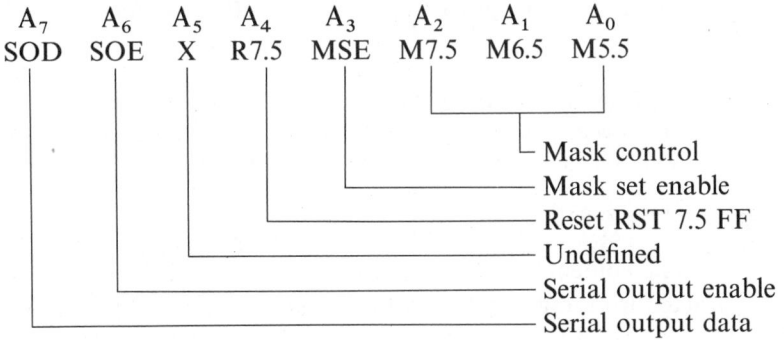

SPHL   (Move register pair HL to stack pointer)

    Operation: $(SP) \leftarrow (H)(L)$
    Binary code: 11111001
    Addressing mode: Register
    Cycles: 1; states: 6
    Flags: None affected

The contents of register pair HL are moved to register SP. The previous contents of the stack pointer are lost.

STA ADDR   (Store accumulator direct)

    Operation: $((BYTE_3)(BYTE_2)) \leftarrow (A)$
    Binary code: 00110010, $BYTE_2$, $BYTE_3$
    Addressing mode: Direct
    Cycles: 4; states: 13
    Flags: None affected

The contents of the accumulator are moved to the memory location specified by $BYTE_2$ and $BYTE_3$ of the instruction. Both the LDA and STA instructions provide the means to load and store the contents of the accumulator directly from and to memory.

STAX rp   (Store accumulator indirect)

    Operation: $((rp)) \leftarrow (A)$
    Binary code: 00RP0010
    Addressing mode: Register indirect
    Cycles: 2; states: 7
    Flags: None affected

The contents of register A are moved to the memory location that is specified by the register pair rp. Only register pairs BC and DE may be specified.

STC (Set carry)

Operation: (CY) ← 1
Binary code: 00110111
Addressing mode: None
Cycles: 1; states: 4
Flags: CY

The CY flag is set to 1 and no other flags are affected.

SUB M (Subtract memory from register A)

Operation: (A) ← (A) − ((H)(L))
Binary code: 10010110
Addressing mode: Register indirect
Cycles: 2; states: 7
Flags: Z, S, P, CY, AC

The contents of the memory location specified by register pair HL are subtracted from register A, and the result is stored in register A.

SUB r (Subtract register r from register A)

Operation: (A) ← (A) − (r)
Binary code: 10010SSS
Addressing mode: Register
Cycles: 1; states: 4
Flags: Z, S, P, CY, AC

The contents of register r are subtracted from the contents of register A, and the result is stored in register A.

SUI DATA$_8$ (Subtract immediate data from register A)

Operation: (A) ← (A) − (BYTE$_2$)
Binary code: 11010110, BYTE$_2$
Addressing mode: Immediate
Cycles: 2; states: 7
Flags: Z, S, P, CY, AC

BYTE$_2$ of the instruction is subtracted from the contents of register A, and the result is stored in register A.

XCHG (Exchange H and L with D and E)

Operation: (H) ↔ (D)
         (L) ↔ (E)
Binary code: 11101011
Addressing mode: Register
Cycles: 1; states: 4
Flags: None affected

The contents of registers H and L are exchanged with the contents of registers D and E. Since the memory instructions only assume that HL is the address pointer, this instruction can be used effectively to exchange the two address pointers DE and HL when register pair DE is also used as an address pointer.

XRA M   (Exclusive OR memory to register A)

Operation: $(A) \leftarrow (A) \oplus ((H)(L))$
Binary code: 10101110
Addressing mode: Register indirect
Cycles: 2; states: 7
Flags: Z, S, P, CY, AC

The contents of the memory location specified by register pair HL are logically Exclusive ORed with the contents of register A, and the result is stored in register A. Both the CY flag and the AC flag are reset after the execution of this instruction.

XRA r   (Exclusive OR register to register A)

Operation: $(A) \leftarrow (A) \oplus (r)$
Binary code: 10101SSS
Addressing mode: Register
Cycles: 1; states: 4
Flags: Z, S, P, CY, AC

The contents of register r are logically Exclusive ORed with the contents of register A, and the result is stored in register A. Both the CY flag and the AC flag are reset after the execution of this instruction.

XRI DATA$_8$ (Exclusive OR immediate to register A)

Operation: $(A) \leftarrow (A) \oplus (BYTE_2)$
Binary code: 11101110, BYTE$_2$
Addressing mode: Immediate
Cycles: 2; states: 7
Flags: Z, S, P, CY, AC

BYTE$_2$ of the instruction is logically Exclusive ORed with the contents of register A, and the result is stored in register A. Both the CY flag and the AC flag are reset after the execution of this instruction.

XTHL   (Exchange top of stack with register pair HL)

Operation: $(L) \leftrightarrow ((SP))$
          $(H) \leftrightarrow ((SP) + 1)$
Binary code: 11100011
Addressing mode: Register indirect
Cycles: 5; states: 16
Flags: None affected

The contents of registers L and H are exchanged with the contents of the memory location specified by the contents of SP and the contents of SP plus 1, respectively.

Appendix B
The 68010 Instruction Set

B.1
Notation

The first word of the instruction is always the operation code. It is comprised of three sections, a 10-bit operation code, a 3-bit addressing mode, and a 3-bit register select. The operation code can also include the register number, the data size, etc. For uniformity, the following notation is adopted when describing 68010 instructions.

Register Definitions

An	Address register
Dn	Data register
Rn	Any register, data or address
Xn	An index register that can be either a data or an address register
PC	Program counter
SR	Status register
CCR	The condition code part of the status register
SP	The active stack pointer (either user stack pointer or supervisor stack pointer, depending on the status of the machine)
USP	User stack pointer
SSP	Supervisor stack pointer
VBR	Vector base register

SFC   Alternate function code source register
DFC   Alternate function code destination register
X   Extend condition code or bit 4 of SR
N   Sign condition code or bit 3 of SR
Z   Zero condition code or bit 2 of SR
V   Overflow condition code or bit 1 of the SR
C   Carry condition code or bit 0 of the SR
U   Undefined condition for the particular condition code involved

Operand Definitions

OPR	The operand
OPR(i)	The ith bit of the operand
OPR(i:j)	A subfield of the operand specified by bit i through bit j inclusive
( )	The contents of
(OPR)	The contents of the referenced location specified as OPR
OPR_{10}	A BCD operand implying that all related operations are done in decimal
#	Literal or immediate data specified as part of the instruction
SZ	Operand size

Addressing Mode Specifications

EA	The effective address of the operand
d	A 2's complement displacement
Source	Source effective address
Destination or Dest	Destination effective address
(Source)	Source operand (The source operand is the contents of the source effective address.)
(Dest)	Destination operand (The destination operand is the contents of the destination effective address.)
Word 2	Second extension word address
Word 3	Third extension word address
Vector	Exception vector location
Dn	Data register direct (The operand is in the specified data register.)
An	Address register direct (The operand is in the specified address register.)
(An)	Address register indirect (The operand is in the memory location specified by the address register A_n.)
(An)+	Address register indirect with post-increment (The operand is in the memory location specified by the

	address register A_n, which is being incremented after being referenced.)
$-(An)$	Address register indirect with pre-decrement (The operand is in the memory location specified by the address register A_n, which is being decremented before it is used.)
$d(An)$	Address register indirect with displacement (The operand is in the memory location whose address is in the specified address register modified by the displacement.)
$d(An,Xn)$	Address register indirect with index (The operand is in the memory location whose address is in the specified address register modified by the displacement and the specified index register.)
Abs.W	Absolute short address (The operand address is stored in the first extension word giving a 16-bit address.)
Abs.L	Absolute long address (The operand address is stored in the first and second extension words giving a complete 32-bit address.)
$d(PC)$	Program counter with displacement (The operand address is given by modifying the PC with the displacement.)
$d(PC,Xn)$	Program counter with displacement and index (The operand address is given by the program counter modified by the displacement and the contents of the index register.)
Imm	Immediate addressing (The operand itself is given in the extension word/words.)

Operators

$\leftarrow$	The left-hand side is replaced by the right-hand side
$\leftrightarrow$	The left-hand side is exchanged with the right-hand side
$+$	Addition
$-$	Subtraction
$*$	Multiplication
$/$	Division
$\wedge$	Logical AND operation
$\vee$	Logical Inclusive OR operation
$\oplus$	Logical Exclusive OR operation

<	Relational test that is true if the left operand is less than the right operand
>	Relational test that is true if the left operand is greater than the right operand
$-$	Logical complement
IF cc THEN ... ELSE ...	Relational test based on the condition cc (When the condition is true, the first set of operations is performed. When the condition fails, the second set of operations is performed.)

OPCODE Definitions

MMMRRR	Bit pattern of the effective address of either the source or the destination operand
RRX	Bit pattern of a data or address register
RRY	Bit pattern of another data or address register
A	A 1-bit pattern indicating whether the operand is in a data register or in memory
S	A 1-bit pattern of the operand size
SS	A 2-bit pattern of the operand size
D	A 1-bit pattern indicating the direction of shifting
cc	Branching condition:

Mnemonic	Condition	Encoding
T	True	0000
F	False	0001
HI	High	0010
LS	Low or same	0011
CC(HS)	Carry clear	0100
CS(LO)	Carry set	0101
NE	Not equal	0110
EQ	Equal	0111
VC	Overflow clear	1000
VS	Overflow set	1001
PL	Plus	1010
MI	Minus	1011
GE	Greater or equal	1100
LT	Less than	1101
GT	Greater than	1110
LE	Less or equal	1111

E	A 1-bit pattern showing whether the count is stored in a data register or in the OPCODE
DDDDDDDD	An 8-bit displacement

B.2

The 68010 Instructions

In the following sections we examine the 68010 instructions in detail. For each instruction, we present the assembler format, the name, the RTL description, the binary code, the available source/destination addressing modes, the addressing categories, the allowable operand size, and the flags that are affected. A brief description of the instruction follows the standard format.

ABCD Dy,Dx   (Add decimal with extend)
ABCD −(Ay),−(Ax)

    Operation: (Dest) ← (Source)$_{10}$ + (Dest)$_{10}$ + (X)
    Binary code: 1100 RRX 10000 A RRY
    Source addressing mode: Dn or −(An)
    Destination addressing mode: Dn or −(An)
    Addressing category: None
    Size: Byte
    Flags: X, N = U, Z, V = U, C

This instruction performs BCD addition taking into consideration the content of the extend bit. The operands can be stored either in the data registers (RRX and RRY) or in memory as specified by address registers RRX and RRY. The selection between data register operands or memory operands is determined by bit A in the binary code. If the operands are in memory, the pre-decrement memory address mode is assumed.

ADD EA,Dn   (Add binary)

    Operation: (Dest) ← (Source) + (Dest)
    Binary code: 1101 RRX 0 SS MMMRRR
    Source addressing mode: Dn, An, (An), (An)+, −(An), d(An), d(An,Xn), Abs.W, Abs.L,
                  d(PC), d(PC,Xn), Imm
    Destination addressing mode: Dn
    Addressing category: All
    Size: Byte, word, long
    Flags: X, N, Z, V, C

ADD Dn,EA   (Add binary)

    Operation: (Dest) ← (Source) + (Dest)
    Binary code: 1101 RRX 1 SS MMMRRR
    Source addressing mode: Dn
    Destination addressing mode: (An), (An)+, −(An), d(An), d(An,Xn), Abs.W, Abs.L
    Addressing category: Alterable memory
    Size: Byte, word, long
    Flags: X, N, Z, V, C

This instruction performs a signed 2's complement addition between the two operands. One of the operands must be stored in a data register. The other operand, which may be a source or destination operand, may be stored in memory.

Whether the destination operand is in the data register or in memory is determined by bit 8 in the OPCODE. The size of the operands is given by SS. All flags are affected by the ADD instruction. The ADD instruction has several variants. When the destination is an address register, ADDA is used. When the source operand is a literal, ADDI is used. When the literal is between the range 1 to 8, ADDQ is used.

ADDA EA,An   (Add address)

    Operation: (Dest) ← (Source) + (Dest)
    Binary code: 1101 RRX S 11 MMMRRR
    Source addressing mode: Dn, An, (An), (An)+, −(An), d(An), d(An,Xn), Abs.W, Abs.L,
            d(PC), d(PC,Xn), Imm
    Destination addressing mode: An
    Addressing category: All
    Size: Word, long
    Flags: None affected

The ADDA instruction is a variant of the ADD instruction and is used whenever the destination operand is stored in the address register. The ADDA instruction is used to perform signed 2's complement addition if the destination is an address register. The source operand can be anywhere. Only word and long word operations are allowed.

ADDI #Data,EA   (Add immediate)

    Operation: (Dest) ← Immediate data + (Dest)
    Binary code: 0000 0110 SS MMMRRR, Word 1, Word 2
    Source addressing mode: Imm
    Destination addressing mode: Dn, (An), (An)+, −(An), d(An), d(An,Xn), Abs.W, Abs.L
    Addressing category: Data alterable
    Size: Byte, word, long
    Flags: X, N, Z, V, C

This instruction is a variant of the ADD instruction and is used when one of the source operands is a literal. The ADDI instruction adds the contents of the destination operand to the immediate data contained in word 1 and word 2. The result is stored in the specified destination operand. If the size of the operation is specified to be byte size, the source operand is taken to be the lower half of word 2. If the operation is word size, the source operand is taken to be both halves of word 2. If the operation is long word size, the source operand is contained in word 2 and word 3, with word 2 containing the most significant part and word 3 the least significant part. In the latter case, two words are required after the OPCODE.

ADDQ #Data,EA   (Add quick)

    Operation: (Dest) ← Immediate data + (Dest)
    Binary code: 0101 DDD 0 SS MMMRRR
    Source addressing mode: Imm

Destination addressing mode: Dn, An, (An), (An)+, −(An), d(An), d(An,Xn),
Abs.W, Abs.L
Addressing category: Alterable
Size: Byte, word, long
Flags: X, N, Z, V, C

The ADDQ instruction is a variant of the ADD instruction. The ADDQ instruction is used when the source operand is a literal and is in the range 1 to 8. The immediate data are stored as part of the OPCODE in bits DDD. Hence word 2 and word 3 are not required. In essence, the ADDQ instruction is similar to the common increment instruction but is more powerful than the increment instruction in that the amount to be incremented can be specified to be between 1 and 8.

ADDX Dy,Dx   (Add extended)
ADDX −(Ay),−(Ax)

Operation: (Dest) ← (Source) + (Dest) + (X)
Binary code: 1101 RRX 1 SS 00 A RRY
Source addressing mode: Dn or −(An)
Destination addressing mode: Dn or −(An)
Addressing category: None
Size: Byte, word, long
Flags: N, Z, V, C, X

This instruction is similar to the ADD instruction except that the contents of the extend bit (X) are also summed when performing the signed 2's complement addition. This instruction is useful for implementing multiple-precision addition. Similar to the ADD instruction, the operands can only be stored either in data registers or in memory as determined by bit A. If memory operations are specified, the address register pre-decrement mode is assumed.

AND EA,Dn   (Logical AND)

Operation: (Dest) ← (Source) ∧ (Dest)
Binary code: 1100 RRX 0 SS MMMRRR
Source addressing mode: Dn, (An), (An)+, −(An), d(An,Xn), Abs.W, Abs.L,
d(PC), d(PC,Xn), Imm
Destination addressing mode: Dn
Addressing category: Data
Size: Byte, word, long
Flags: N, Z, V = 0, C = 0

AND Dn,EA   (Logical AND)

Operation: (Dest) ← (Source) ∧ (Dest)
Binary code: 1100 RRX 1 SS MMMRRR
Source addressing mode: Dn
Destination addressing mode: (An), (An)+, −(An), d(An), d(An,Xn), Abs.W, Abs.L
Addressing category: Alterable memory
Size: Byte, word, long
Flags: N, Z, V = 0, C = 0

The AND instruction performs a logical bit-by-bit AND operation on the operands. One of the operands must be in the data register, whereas the other is specified by the effective address. Whether the result is to be stored in a data register or in memory as specified by the effective address is determined by bit 8 in the OPCODE. The AND instruction has one variant form. If the source operand is a literal, then the ANDI instruction is used.

ANDI #Data,EA   (AND immediate)

Operation: (Dest) ← Immediate data ∧ (Dest)
Binary code: 0000 0010 SS MMMRRR, Word 2, Word 3
Source addressing mode: Imm
Destination addressing mode: Dn, (An), (An)+, −(An), d(An), d(An,Xn), Abs.W, Abs.L
Addressing category: Data alterable
Size: Byte, word, long
Flags: N, Z, V = 0, C = 0

The ANDI is a variant of the AND instruction and is used whenever one of the source operands is immediate data. The ANDI instruction performs a logical AND operation between the immediate data supplied and the destination operand as specified by the effective address. If the size of the operation is specified to be byte size, then the lower half of word 2 is used as the operand. If the operation is word size, both halves of word 2 are used. If the operation is long word size, then both word 2 and word 3 are used.

ANDI #Data,CCR   (AND immediate to condition codes)

Operation: (CCR) ← (CCR) ∧ (Source)
Binary code: 0000 0010 0011 1100, Word 2
Source addressing mode: Imm
Destination addressing mode: Implied
Addressing category: None
Size: Byte
Flags: X, N, Z, V, C

The immediate data stored in the lower half of word 2 is logically ANDed with the condition code register. This instruction can be used to change or set the condition codes, that is, the user part of the status register only. The relation between the bits of the immediate data and the flags is

Flag	Bit
C	0
V	1
Z	2
N	3
X	4

ANDI #Data,SR   (AND immediate to the status register)
                                                 (Privileged instruction)

> Operation: IF supervisor state
>   THEN (SR) ← (SR) ∧ (Source)
>   ELSE TRAP
> Binary code: 0000 0010 0111 1100, Word 2
> Source addressing mode: Imm
> Destination addressing mode: Implied
> Addressing category: None
> Size: Word
> Flags: X, N, Z, V, C

This instruction changes both the user part and the supervisor part of the status register. Hence, it is a privileged instruction. The immediate data supplied are logically ANDed with all 16 bits of the status register, and the result is left in the status register. This instruction must be executed when the processor is in the supervisory state, else a TRAP condition is generated. The flags affected by this instruction are

Flag	Bit
C	0
V	1
Z	2
N	3
X	4
Interrupt	8–10
S	13
T	15

ASL Dx,Dy   (Arithmetic shift)
ASL #Data,Dy
ASR Dx,Dy
ASR #Data,Dy

> Operation: (Dest) ← (Dest) Shifted by ⟨count⟩
> Binary code: 1110 RRX D SS E 00 RRR
> Addressing mode: Dn
> Addressing category: None
> Size: Byte, word, long
> Flags: X, N, Z, V, C

ASL EA   (Arithmetic shift)
ASR EA

> Operation: (Dest) ← (Dest) Shifted by 1 bit
> Binary code: 1110 000 D 11 MMMRRR
> Addressing mode: (An), (An)+, −(An), d(An), d(An,Xn), Abs.W, Abs.L
> Addressing category: Memory alterable
> Size: Word
> Flags: X, N, Z, V, C

The ASL and ASR instructions can be used to shift the operand to the left or to the right, respectively, as specified by bit 8 (i.e., bit D) of the OPCODE. The operand may be stored in a data register or in memory. If the operand is shifted to the left, the last bit shifted out from the most significant side is stored in both the X and the C flags, and the operand is zero-filled on the right. If the operand is shifted to the right, the last bit shifted out on the right is also stored in the X and C flags, and the operand is sign-extended on the left.

If the operand is stored in a data register, any number of shifts in either direction is possible, and the shift count is stored in a data register. If the shift count is between the range 0 to 7, the shift count may be stored as part of the OPCODE as determined by bit E. In the latter case the shift count is supplied as immediate data in the source operand position.

If the operand is stored in memory, only one bit can be shifted at a time, and the size of the operand is always taken to be word size. A number of addressing modes are available for the effective address specification.

Bcc LABEL   (Branch conditionally)

 Operation: IF condition true
     THEN (PC) ← (PC) + d
 Binary code: 0110 CCCC DDDDDDDD, Word 2
 Addressing mode: None
 Addressing category: None
 Size: Byte, word
 Flags: None

Branching occurs when the specified condition (*cc*) is true. The new address is specified as a 2's complement displacement from the current PC value that is 2 bytes beyond the current instruction address. If the new address is within an 8-bit displacement, then the 8-bit displacement is specified as part of the OPCODE. If the relative distance is larger than 8 bits but smaller than 16 bits, then the 8-bit displacement field DDDDDDDD will be zero and the 16-bit displacement is taken from word 2. All branching conditions are allowed except for the T and the F conditions. See Section B.1 for a complete list of branching conditions.

BCHG Dn,EA   (Test a bit and change)

 Operation: (Z) ← $\overline{(\text{Dest}(i))}$
    (Dest(i)) ← $\overline{(\text{Dest}(i))}$
 Binary code: 0000 RRX 101 MMMRRR
 Addressing mode: Dn, (An), (An)+, −(An), d(An), d(An,Xn), Abs.W, Abs.L
 Addressing category: Data alterable
 Size: Byte, long
 Flags: Z

BCHG #Data,EA   (Test a bit and change)

    Operation: $(Z) \leftarrow \overline{(Dest(i))}$

                  $(Dest(i)) \leftarrow \overline{(Dest(i))}$

    Binary code: 0000 1000 01 MMMRRR, Word 2

    Addressing mode: Dn, (An), (An)+, −(An), d(An), d(An,Xn), Abs.W, Abs.L

    Addressing category: Data alterable

    Size: Byte, long

    Flags: Z

Within the duration of a single instruction, the specified bit (bit i) of the given operand is tested, changed, and written back to its destination. The specified bit is tested by storing the complement of the bit content in the Z register. The specified bit is changed by storing the complement of the bit content back. If the bit to be tested is in the data register, any one of the 32 bits can be tested. If the bit is in memory, then one of the 8 bits (lower byte) of the memory word can be tested. The bit specification can be performed statically by storing the bit number as immediate data in word 2 or dynamically by storing the bit number in a data register.

BCLR Dn,EA   (Test a bit and clear)

    Operation: $(Z) \leftarrow \overline{(Dest(i))}$

               $(Dest(i)) \leftarrow 0$

    Binary code: 0000 RRX 110 MMMRRR

    Addressing mode: Dn, (An), (An)+, −(An), d(An), d(An,Xn), Abs.W, Abs.L

    Addressing category: Data alterable

    Size: Byte, long

    Flags: Z

BCLR #Data,EA   (Test a bit and clear)

    Operation: $(Z) \leftarrow \overline{(Dest(i))}$

               $(Dest(i)) \leftarrow 0$

    Binary code: 0000 1000 10 MMMRRR, Word 2

    Addressing mode: Dn, (An), (An)+, −(An), d(An), d(An,Xn), Abs.W, Abs.L

    Addressing category: Data alterable

    Size: Byte, long

    Flags: Z

The operation performed in the BCLR instruction is similar to that in the BCHG instruction except that instead of complementing the bit and writing it back, the specified bit is cleared before writing back. If the bit to be tested is in the data register, any one of the 32 bits can be tested. If the bit is in memory, then one of the 8 bits (lower byte) of the memory word can be tested. The bit specification can be performed statically by storing the bit number as immediate data in word 2 or dynamically by storing the bit number in a data register.

BRA LABEL   (Branch always)

Operation: (PC) ← (PC) + d
Binary code: 0110 0000 DDDDDDDD, Word 2
Addressing mode: None
Addressing category: None
Size: Byte, word
Flags: None affected

This instruction differs from the Bcc instruction in that branching always occurs. The new address is specified as a 2's complement displacement from the current PC value that is 2 bytes beyond the current instruction address. If the new address is within an 8-bit displacement, then the 8-bit displacement is specified as part of the OPCODE. If the relative distance is larger than 8 bits but smaller than 16 bits, then the 8-bit displacement field DDDDDDDD will be zero and the 16-bit displacement is taken from word 2.

BSET Dn,EA   (Test a bit and set)

Operation: (Z)   ← $\overline{(Dest(i))}$
            (Dest(i)) ← 1
Binary code: 0000 RRR 111 MMMRRR
Addressing mode: Dn, (An), (An)+, −(An), d(An), d(An,Xn), Abs.W, Abs.L
Addressing category: Data alterable
Size: Byte, long
Flags: Z

BSET #Data,EA   (Test a bit and set)

Operation: (Z)   ← $\overline{(Dest(i))}$
            (Dest(i)) ← 1
Binary code: 0000 1000 11 MMMRRR, Word 2
Addressing mode: Dn, (An), (An)+, −(An), d(An), d(An,Xn), Abs.W, Abs.L
Addressing category: Data alterable
Size: Byte, long
Flags: Z

The operation performed in the BSET instruction is exactly the same as the BCHG instruction except that instead of complementing the bit and writing it back, the specified bit is set before writing back. If the bit to be tested is in the data register, any one of the 32 bits can be tested. If the bit is in memory, then one of the 8 bits (lower byte) of the memory word can be tested. The bit specification can be performed statically by storing the bit number as immediate data in word 2 or dynamically by storing the bit number in a data register.

BSR LABEL   (Branch to subroutine)

Operation: (−(SP)) ← (PC)
            (PC)   ← (PC) + d
Binary code: 0110 0001 DDDDDDDD, Word 2
Addressing mode: None
Addressing category: None
Size: Byte, word
Flags: None affected

This instruction differs from the BRA instruction in that not only does branching occur but also the current PC value is stored in the stack for return linkage. The new address is specified as a 2's complement displacement from the current PC value that is 2 bytes beyond the current instruction address. If the new address is within an 8-bit displacement, then the 8-bit displacement is specified as part of the OPCODE. If the relative distance is larger than 8 bits but smaller than 16 bits, then the 8-bit displacement field DDDDDDDD will be zero and the 16-bit displacement is taken from word 2.

BTST Dn,EA   (Test a bit)

    Operation: $(Z) \leftarrow \overline{(Dest(i))}$
    Binary code: 0000 RRX 100 MMMRRR
    Addressing mode: Dn, (An), (An)+, −(An), d(An), d(An,Xn), Abs.W, Abs.L, d(PC),
               d(PC,Xn), Imm
    Addressing category: Data
    Size: Byte, long
    Flags: Z

BTST #Data,EA   (Test a bit)

    Operation: $(Z) \leftarrow \overline{(Dest(i))}$
    Binary code: 0000 1000 00 MMMRRR, Word 2
    Addressing mode: Dn, (An), (An)+, −(An), d(An), d(An,Xn), Abs.W, Abs.L, d(PC),
               d(PC,Xn), Imm
    Addressing category: Data
    Size: Byte, long
    Flags: Z

The operation performed in the BTST instruction is exactly the same as the BCHG instruction except that the destination operand is not changed at all. If the bit to be tested is in the data register, any one of the 32 bits can be tested. If the bit is in memory, then one of the 8 bits (lower byte) of the memory word can be tested. The bit specification can be performed statically by storing the bit number as immediate data in the least significant byte of word 2 or dynamically by storing the bit number in a data register.

CHK EA,Dn   (Check register against bounds)

    Operation: IF (Dn) < 0 or (Dn) > (Source)
              THEN TRAP
    Binary code: 0100 RRX 110 MMMRRR
    Addressing mode: Dn, (An), (An)+, −(An), d(An), d(An,Xn), Abs.W, Abs.L, d(PC),
               d(PC,Xn), Imm
    Addressing category: Data
    Size: Word
    Flags: N, Z = U, V = U, C = U

The CHK instruction is used to check the content of a data register against a specified bound. If the data register value is less than zero or greater than the given bound, a TRAP condition is generated. This instruction is useful for checking subscripts or checking if the user address is within the user area.

CLR EA (Clear an operand)

    Operation: (Dest) ← 0
    Binary code: 0100 0010 SS MMMRRR
    Addressing mode: Dn, (An), (An)+, −(An), d(An), d(An,Xn), Abs.W, Abs.L
    Addressing category: Data alterable
    Size: Byte, word, long
    Flags: N = 0, Z = 1, V = 0, C = 0

The destination operand is cleared. The size of the operation may be specified to be byte, word, or long. The operand may be in a data register or in memory. If the operand size is byte-oriented, the other part of the register or memory will not be cleared or changed at all.

CMP EA,Dn (Compare)

    Operation: (Dest) − (Source)
    Binary code: 1011 RRX 0 SS MMMRRR
    Source addressing mode: Dn, An, (An), (An)+, −(An), d(An), d(An,Xn), Abs.W, Abs.L,
                   d(PC), d(PC,Xn), Imm
    Destination addressing mode: Dn
    Addressing category: All
    Size: Byte, word, long
    Flags: N, Z, V, C

The CMP instruction is used to compare the source operand with the destination operand. Neither of the operands is changed. The result of the subtraction is stored in the status register for later use and testing. The CMP instruction has several variants. When the destination specified is an address register, CMPA is used. When the source is a literal, CMPI is used. When both operands are in memory, CMPM is used.

CMPA EA,An (Compare address)

    Operation: (Dest) − (Source)
    Binary code: 1011 RRX SS 1 MMMRRR
    Source addressing mode: Dn, An, (An), (An)+, −(An), d(An), d(An,Xn), Abs.W, Abs.L,
                   d(PC), d(PC,Xn), Imm
    Destination addressing mode: An
    Addressing category: All
    Size: Word, long
    Flags: N, Z, V, C

This instruction is a variant of the CMP instruction and is used whenever the destination is an address register. Only word size or long word size operations are allowed on address registers. The CMPA instruction is used to compare the source operand with the destination operand. Neither of the operands is changed. The result of the subtraction is stored in the status register for later use and testing.

CMPI #Data,EA   (Compare immediate)

    Operation: (Dest) − Immediate data
    Binary code: 0000 1100 SS MMMRRR, Word 1, Word 2
    Source addressing mode: Imm
    Destination addressing mode: Dn, (An), (An)+, −(An), d(An), d(An,Xn), Abs.W, Abs.L
    Addressing category: Data alterable
    Size: Byte, word, long
    Flags: N, Z, V, C

The CMPI instruction is a variant of the CMP instruction and is used whenever the source operand is a literal. The CMPI instruction is used to compare the source operand with the destination operand. Neither of the operands is changed. The result of the subtraction is stored in the status register for later use and testing. The CMPI instruction may be operated with byte, word, or long word size data. If the operand size is specified to be byte size, then the lower half of word 2 is taken to be the source operand. If the operand is word size, both halves of word 2 are taken to be the source operand. If the operand is long word size, then both word 2 and word 3 are taken together as the source operand.

CMPM (Ay)+,(Ax)+   (Compare memory)

    Operation: (Dest) − (Source)
    Binary code: 1011 RRX 1 SS 001 RRY
    Source addressing mode: (An)+
    Destination addressing mode: (An)+
    Addressing category: None
    Size: Byte, word, long
    Flags: N, Z, V, C

The CMPM instruction is a variant of the CMP instruction and is used whenever both operands are in memory. The OPCODE specifies the two address registers, and the operands are obtained by means of these address registers in a post-increment addressing mode. The CMPM instruction is used to compare the source operand with the destination operand. Neither of the operands is changed. The result of the subtraction is stored in the status register for later use and testing.

DBcc Dn,LABEL   (Test condition, decrement, and branch)

    Operation: IF condition false
              THEN BEGIN
                  (Dn) ← (Dn) − 1
                  IF (Dn) ≠ −1
                      THEN (PC) ← (PC) + d
                  END
               ELSE (PC) ← (PC) + 2
    Binary code: 0101 CCCC 11001 RRR, Word 2
    Addressing mode: Dn
    Addressing category: None
    Size: Word
    Flags: None affected

This is a looping instruction that will test, decrement, and branch within a single instruction cycle. The instruction begins by testing the specified termination condition. If the terminating condition is true, no operation is performed. Program execution continues with the following instruction as pointed to by the program counter. If the terminating condition is false, the counter stored in the specified data register is decremented. If the resultant count is -1, then a normal loop termination is valid and execution continues at the next instruction. If termination count has not been reached, then the specified branch address is taken. This instruction is useful when multiple termination conditions may occur. A normal terminating condition is encountered usually at the end of loop operations. An abnormal terminating condition occurs when unusual environments are present, such as a divide by zero, etc. This abnormal terminating condition can be specified as the test condition. Any one of the 16 testing conditions can be specified. The branch address is specified by a 16-bit displacement with respect to the current PC. This displacement is computed from the given LABEL.

DIVS EA,Dn   (Signed divide)

Operation: (Dest) ← (Dest)/(Source)
Binary code: 1000 RRX 111 MMMRRR
Source addressing mode: Dn, (An), (An)+, −(An), d(An), d(An,Xn), Abs.W, Abs.L, d(PC), d(PC,Xn), Imm
Destination addressing mode: Dn
Addressing category: Data
Size: Word
Flags: N, Z, V, C = 0

This instruction performs a signed binary division. The dividend is a long operand (32 bits) specified by the destination operand. The divisor is a word operand (16 bits) specified by the source operand. The result is stored in the destination, with the remainder in the most significant word and the quotient in the least significant word. If the divisor is zero, a TRAP condition results. If the quotient cannot be represented by a 16-bit signed integer, an overflow condition occurs. The overflow condition is detected before the division process, and no operands are destroyed when overflow occurs.

DIVU EA,Dn   (Unsigned divide)

Operation: (Dest) ← (Dest)/(Source)
Binary code: 1000 RRX 011 MMMRRR
Source addressing mode: Dn, (An), (An)+, −(An), d(An), d(An,Xn), Abs.W, Abs.L, d(PC), d(PC,Xn), Imm
Destination addressing mode: Dn
Addressing category: Data
Size: Word
Flags: N, Z, V, C = 0

The DIVU instruction performs an unsigned binary division. The dividend is a long operand (32 bits) specified by the destination operand. The divisor is a word operand (16 bits) specified by the source operand. The result is stored in the destination, with the remainder in the most significant word and the quotient in the least significant word. If the divisor is zero, a TRAP condition results. If the quotient cannot be represented by a 16-bit signed integer, an overflow condition occurs. The overflow condition is detected before the division process, and no operands are destroyed when overflow occurs.

EOR Dn,EA   (Exclusive OR logical)

    Operation: (Dest) ← (Source) ⊕ (Dest)
    Binary code: 1011 RRX 1 SS MMMRRR
    Source addressing mode: Dn
    Destination addressing mode: Dn, (An), (An)+, −(An), d(An), d(An,Xn), Abs.W, Abs.L
    Addressing category: Data alterable
    Size: Byte, word, long
    Flags: N, Z, V = 0, C = 0

The EOR instruction performs the logical Exclusive OR operation between the operands. One of the source operands must be in a data register. The destination operand, however, can be stored either in a data register or in memory. A variation of the EOR instruction is the EORI instruction, which is used when the source operand is a literal.

EORI #Data,EA   (Exclusive OR immediate)

    Operation: (Dest) ← Immediate data ⊕ (Dest)
    Binary code: 0000 1010 SS MMMRRR, Word 2, Word 3
    Source addressing mode: Imm
    Destination addressing mode: Dn, (An), (An)+, −(An), d(An), d(An,Xn), Abs.W, Abs.L
    Addressing category: Data alterable
    Size: Byte, word, long
    Flags: N, Z, V = 0, C = 0

The EORI instruction is a variant of the EOR instruction and is used whenever the source operand is a literal stored in word 2 and/or word 3. The EORI instruction may be operated with byte, word, or long word size data. If the operand size is specified to be byte size, then the lower half of word 2 is taken to be the source operand. If the operand is word size, both halves of word 2 are taken to be the source operand. If the operand is long word size, then both word 2 and word 3 are taken together to be the source operand.

EORI #Data,CCR   (Exclusive OR immediate to condition codes)

    Operation: (CCR) ← (CCR) ⊕ (Source)
    Binary code: 0000 1010 0011 1100, Word 2
    Source addressing mode: Imm
    Destination addressing mode: Implied
    Addressing category: None
    Size: Byte
    Flags: X, N, Z, V, C

The EORI to CCR instruction performs a logical Exclusive OR operation between the literal data, the source operand, and the contents of the CCR. Only the lower byte of word 2 is used. This instruction affects only the user part of the status register. The flags affected by this instruction are

Flag	Bit
C	0
V	1
Z	2
N	3
X	4

EORI #Data,SR   (Exclusive OR immediate to the status register)
                                                                   (Privileged instruction)

Operation: IF supervisor state
            THEN (SR) ← (Source) ⊕ (SR)
            ELSE TRAP
Binary code: 0000 1010 0111 1100, Word 2
Source addressing mode: Imm
Destination addressing mode: Implied
Addressing category: None
Size: Word
Flags: X, N, Z, V, C

The EORI to SR instruction performs a logical Exclusive OR operation between the literal and the contents of the SR. All 16 bits of word 2 are used to change the status register. Hence this is a privileged instruction that can be executed only when the processor is in the supervisory state. Any other processor state will cause a trap. The flags affected by this instruction are

Flag	Bit
C	0
V	1
Z	2
N	3
X	4
Interrupt	8–10
S	13
T	15

EXG Rx,Ry   (Exchange registers)

Operation: (Ry) ↔ (Rx)
Binary code: 1100 RRX 1 FFFFF RRY
Source addressing mode: Dn or An
Destination addressing mode: Dn or An

Addressing category: None
Size: Long
Flags: None affected

The contents of the two specified registers are exchanged. Either register can be a data register or an address register. Transfer between a data register and an address register is also allowed.

The 5-bit operation mode FFFFF determines the type of transfer to be performed. There are three possible values for FFFFF:

01000   Data registers exchange

01001   Address registers exchange

10001   Data register and address register exchange

EXT Dn   (Sign extend)

Operation: (Dest) ← Sign-extended (Dest)
Binary code: 0100 100 0 SS 000 RRX
Addressing mode: Dn
Addressing category: None
Size: Word, long
Flags: N, Z, V = 0, C = 0

The sign of the specified data register is extended by this instruction. If the operand size is specified to be word size, then the byte information in the data register is sign-extended to be word size. If the operand is specified to be long word size, then the sign of the word operand is extended to be long word size. A sign extension operation is performed only with data in the data registers.

ILLEGAL   (Illegal instruction)

Operation: (−(SSP)) ← (PC)
                (−(SSP)) ← (SR)
                (PC)   ← (Illegal instruction vector)
Binary code: 0100 1010 1111 1100
Addressing mode: None
Addressing category: None
Size: Unsized
Flags: None affected

This is the only valid instruction that will always cause the processor to enter into exception processing. Other reserved bit patterns may also cause exception processing, though they are reserved for future uses and hence are subject to changes.

JMP EA   (Jump)

Operation: (PC) ← Destination EA
Binary code: 0100 1110 11 MMMRRR
Addressing mode: (An), d(An), d(An,Xn), Abs.W, Abs.L, d(PC), d(PC,Xn)
Addressing category: Control
Size: Unsized
Flags: None affected

The address specified by the destination is loaded into the PC and is executed next. This instruction differs from the BRA instruction in that only 16-bit displacement from the PC is allowed in the BRA instruction. The JMP instruction, on the other hand, allows a true 32-bit address to be loaded into the PC, causing a true jump into any random location within the allowable addressing space of the processor.

JSR EA   (Jump to subroutine)

Operation: $(-(SP)) \leftarrow (PC)$
$\qquad (PC) \qquad \leftarrow$ Destination EA
Binary code: 0100 1110 10 MMMRRR
Addressing mode: (An), d(An), d(An,Xn), Abs.W, Abs.L, d(PC), d(PC,Xn)
Addressing category: Control
Size: Unsized
Flags: None affected

The JSR instruction is similar to the JMP instruction except that before the branching operation takes place, the current PC is first stored in the stack. The JSR instruction allows a branch anywhere within the allowable addressing space, whereas the BSR instruction allows a branch only within a 16-bit displacement from the current PC.

LEA EA,An   (Load effective address)

Operation: An $\leftarrow$ Source EA
Binary code: 0100 RRX 111 MMMRRR
Addressing mode: (An), d(An), d(An,Xn), Abs.W, Abs.L, d(PC), d(PC,Xn)
Addressing category: Control
Size: Long
Flags: None affected

The destination specified by the effective address is loaded into the address register by this instruction. This is one way that an address can be loaded into the address register for further manipulation.

LINK An,#d   (Link and allocate)

Operation: $(-(SP)) \leftarrow (An)$
$\qquad (An) \qquad \leftarrow (SP)$
$\qquad (SP) \qquad \leftarrow (SP) + d$
Binary code: 0100 1110 01010 RRR, Word 2
Addressing mode: An
Addressing category: None
Size: Unsized
Flags: None affected

In the LINK instruction, the specified address register is first stored onto the stack. The contents of the current stack pointer is then stored in the specified address register. Finally, the stack pointer is modified by the offset amount as specified by the instruction. The LINK and UNLK instructions form a pair to provide support to many applications, such as argument passing between sub-routines and stack space allocation for local variables in subroutines.

LSL Dx,Dy   (Logical shift)
LSL #Data,Dy
LSR Dx,Dy
LSR #Data,Dy

    Operation: (Dest) ← (Dest) Shifted by ⟨count⟩
    Binary code: 1110 RRX D SS E 01 RRR
    Addressing mode: Dn
    Addressing category: None
    Size: Byte, word, long
    Flags: X, N, Z, V = 0, C

LSL EA   (Logical shift)
LSR EA

    Operation: (Dest) ← (Dest) Shifted by 1 bit
    Binary code: 1110 001 D 11 MMMRRR
    Addressing mode: (An), (An)+, −(An), d(An), d(An,Xn), Abs.W, Abs.L
    Addressing category: Memory alterable
    Size: Byte, word, long
    Flags: X, N, Z, V = 0, C

The LSL and LSR instructions can be used to logically shift the operand to the left or to the right. The direction is determined by bit D, bit 8, of the OPCODE. The operand may be stored in a data register or in memory. If the operand is shifted to the left, the last bit shifted out from the most significant side is stored in both the X and the C flags, and the operand is zero-filled on the right. If the operand is shifted to the right, the last bit shifted out on the right is also stored in the X and C flags, and zero is inserted on the left.

If the operand is stored in a data register, any number of shifts in either direction is possible when the shift count is stored in a data register. If the shift count is in the range 0 to 7, the shift count may be stored as part of the OPCODE as determined by bit E.

If the operand is stored in memory, only one bit can be shifted at a time, and the size of the operand is always taken to be word size. A number of addressing modes are available for the effective address specification.

MOVE CCR,EA   (Move from the condition code register)

    Operation: (Dest) ← (CCR)
    Binary code: 0100 0010 11 MMMRRR
    Source addressing mode: Implied
    Destination addressing mode: Dn, (An), (An)+, −(An), d(An), d(An,Xn), Abs.W, Abs.L
    Addressing category: Data alterable
    Size: Word
    Flags: None affected

The contents of the lower half of the status register that contains the condition codes are transferred to the destination. The upper byte of the destination is zero-filled. The MOVE from the CCR instruction is not a privileged instruction. Note that this instruction is a word instruction, whereas AND to CCR, OR to CCR, and EOR to CCR are byte operations. The flags are placed in the following bit positions:

Flag	Bit
C	0
V	1
Z	2
N	3
X	4

MOVE EA,CCR   (Move to condition codes)

Operation: (CCR) ← (Source)
Binary code: 0100 0100 11 MMMRRR
Source addressing mode: Dn, (An), (An)+, −(An), d(An), d(An,Xn), Abs.W, Abs.L, d(PC), d(PC,Xn), Imm
Destination addressing mode: Implied
Addressing category: Data
Size: Word
Flags: X, N, Z, V, C

The contents of the source operand is loaded into the condition code part or the lower part of the status register. This is not a privileged instruction. Note that this instruction is a word instruction, whereas AND to CCR, OR to CCR, and EOR to CCR are byte operations. The flags affected by this instruction are

Flag	Bit
C	0
V	1
Z	2
N	3
X	4

MOVE EA,EA   (Move data from source to destination)

Operation: (Dest) ← (Source)
Binary code: 00 SS RRRMMM MMMRRR
Source addressing mode: Dn, An, (An), (An)+, −(An), d(An), d(An,X), Abs.W, Abs.L, d(PC), d(PC,Xn), Imm

Destination addressing mode: Dn, (An), (An)+, −(An), d(An), d(An,X), Abs.W, Abs.L
Source addressing category: All
Destination addressing category: Data alterable
Size: Byte, word, long
Flags: N, Z, V = 0, C = 0

The contents of the source effective address is moved and becomes the contents of the destination effective address. The operation may be done in byte, word, or long word size. All but the extend flag is affected by the transfer. There are several variants of the MOVE instruction. The MOVEA instruction is used whenever the destination is an address register. The MOVEQ instruction is used when the source is an 8-bit literal.

MOVE EA,SR   (Move to the status register)
                                                        (Privileged instruction)

Operation: IF supervisor state
                THEN (SR) ← (Source)
                ELSE TRAP
Binary code: 0100 0110 11 MMMRRR
Source addressing mode: Dn, (An), (An)+, −(An), d(An), d(An,Xn), Abs.W, Abs.L, d(PC),
                d(PC,Xn), Imm
Destination addressing mode: Implied
Addressing category: Data
Size: Word
Flags: X, N, Z, V, C

This instruction is used to transfer the contents of the destination operand to the whole status register. All 16 bits are transferred. Hence this is a privileged instruction. The primary difference between the MOVE to CCR and the MOVE to SR instructions is that the former transfers only the condition codes, whereas the latter also transfers the supervisory information. The flags affected by this instruction are

Flag	Bit
C	0
V	1
Z	2
N	3
X	4
Interrupt	8−10
S	13
T	15

MOVE SR,EA

(Move from the status register)
(Privileged instruction)

Operation: IF supervisor state
          THEN (Dest) ← (SR)
          ELSE TRAP
Binary code: 0100 0000 11 MMMRRR
Source addressing mode: Implied
Destination addressing mode: Dn, (An), (An)+, −(An), d(An), d(An,Xn), Abs.W, Abs.L
Addressing category: Data alterable
Size: Word
Flags: None affected

This instruction is used to transfer the contents of the whole status register to the destination. All 16 bits are transferred. Hence this is a privileged instruction. The primary difference between the MOVE from CCR and the MOVE from SR instructions is that the former transfers only the condition codes, whereas the latter also transfers the supervisory information. The flags affect the following bits of the destination:

Flag	Bit
C	0
V	1
Z	2
N	3
X	4
Interrupt	8–10
S	13
T	15

MOVE USP,An

(Move user stack pointer)
(Privileged instruction)

Operation: IF supervisor state
          THEN (An) ← (USP)
          ELSE TRAP
Binary code: 0100 1110 0110 1 RRR
Source addressing mode: Implied
Destination addressing mode: An
Addressing category: None
Size: Long
Flags: None affected

MOVE An,USP

(Move user stack pointer)
(Privileged instruction)

Operation: IF supervisor state
          THEN (USP) ← (An)
          ELSE TRAP

Binary code: 0100 1110 0110 0 RRR
Source addressing mode: An
Destination addressing mode: Implied
Addressing category: None
Size: Long
Flags: None affected

This instruction can be used to transfer the user stack pointer to or from the specified address register. All transfers are made in long word form. This instruction is used by the supervisor program to store and restore the user stack pointer during program context changes.

MOVEA EA,An   (Move address)

Operation: (Dest) ← (Source)
Binary code: 00 SS RRX 001 MMMRRR
Source addressing mode: Dn, An, (An), (An)+, −(An), d(An), d(An,Xn), Abs.W, Abs.L,
                        d(PC), d(PC,Xn), Imm
Destination addressing mode: An
Addressing category: All
Size: Word, long
Flags: None affected

The MOVEA instruction is a variant of the MOVE instruction and is used whenever the destination is an address register. The source operand may be specified in any one of the allowable addressing modes.

MOVEC Rc,Rn   (Move from control register)
                                                     (Privileged instruction)

Operation: IF supervisor state
                THEN (Rn) ← (Rc)
                ELSE TRAP
Binary code: 0100 1110 0111 1010, Word 2
Source addressing mode: Implied
Destination addressing mode: Dn or An
Addressing category: None
Size: Long
Flags: None affected

MOVEC Rn,Rc   (Move to control register)
                                                      (Privileged instruction)

Operation: IF supervisor state
                THEN (Rc) ← (Rn)
                ELSE TRAP
Binary code: 0100 1110 0111 1011, Word 2
Source addressing mode: Dn, An
Destination addressing mode: Implied
Addressing category: None
Size: Long
Flags: None affected

This instruction provides the transfer operation between a general register, either a data register or address register, and the specified control register. The specific control register and the general register are specified in word 2.

The most significant bit of word 2 specifies whether the general register is a data or address register. The next 3 bits specify the register number. The last 12 bits specify the control register and may be one of the following:

000H	(Source) function code register
001H	(Dest) function code register
800H	User stack pointer
801H	Vector base register

All transfers are 32 bits long. If a particular source register is less than 32 bits long, then the contents of the corresponding bits in the destination register are zero-filled.

MOVEM ⟨register list⟩,EA   (Move multiple registers)
    Operation: (Dest's) ← (Rn's)
    Binary code: 0100 1 0 001 S MMMRRR, Word 2
    Source addressing mode: Implied
    Destination addressing mode: (An), −(An), d(An), d(An,Xn), Abs.W, Abs.L
    Addressing category: Control alterable or pre-decrement
    Size: Word, long
    Flags: None affected

MOVEM EA,⟨register list⟩   (Move multiple registers)
    Operation: (Rn's) ← (Source's)
    Binary code: 0100 1 1 001 S MMMRRR, Word 2
    Source addressing mode: (An), (An)+, d(An), d(An,Xn), Abs.W, Abs.L, d(PC), d(PC,Xn)
    Destination addressing mode: Implied
    Addressing category: Control or post-increment
    Size: Word, long
    Flags: None affected

The MOVEM instruction is a powerful instruction that provides support for storing or retrieving all or selected data registers and address registers. The selection of registers is specified in word 2 for control mode addressing starting with data register 0 to data register 7 and address register 0 to address register 7. Data transfer can be performed in word size or long word size. If the operand size is specified to be word size, then any transfer to data registers or address registers will be sign-extended to 32 bits.

If the registers are being stored into memory, either the control mode or the memory pre-decrement mode of addressing is allowed. If the pre-decrement mode is used, the order of transfer begins with address register 7 to address register 0 and proceeds to data register 7 to data register 0. If the registers are being retrieved from memory, then either the control mode or the post-increment mode of addressing is allowed. The order of register specification for the post-increment mode is the same as that for the control mode.

MOVEP Dx,d(Ay)   (Move peripheral data)
MOVEP d(Ay),Dx

Operation: (Dest) ← (Source)
Binary code: 0000 RRD 1 G S 001 RRA, Word 2
Source addressing mode: Dn or d(An)
Destination addressing mode: Dn or d(An)
Addressing category: None
Size: Word, long
Flags: None affected

This instruction is provided to support byte-size data transfers between a data register (RRD) and memory as specified by address register RRA. The direction of transfer is given by bit G. If the operand size is word size, then only the lower half of the data register is used, with bits 8 to 15 transferred first, followed by bits 0 to 7. If the operand size is long word, then the most significant byte is transferred first, followed by bits 16 to 23, followed by bits 8 to 15, and finally by the least significant byte.

The byte organization in memory can be stored either in the upper half or in the lower half of the memory words, depending on the initial address stored in the specified address register.

MOVEQ #Data,Dn   (Move quick)

Operation: (Dest) ← Immediate data
Binary code: 0111 RRX 0 DDDDDDDD
Source addressing mode: Imm
Destination addressing mode: Dn
Addressing category: None
Size: Long
Flags: N, Z, V = 0, C = 0

This instruction is a variant of the MOVE instruction and is used whenever the source operand is a literal that is less than 8 bits. The literal is actually stored in the OPCODE itself. This 8-bit literal is sign-extended to 32 bits after the transfer is made to the specified data register.

MOVES Rn,EA   (Move to address space)
                                                          (Privileged instruction)

Operation: IF supervisor state
                THEN (DFC) ← (Rn)
                ELSE TRAP
Binary code: 0000 1110 SS MMMRRR, Word 2
Source addressing mode: Implied
Destination addressing mode: (An), (An)+, −(An), d(An), d(An,Xn), Abs.W, Abs.L
Addressing category: Alterable memory
Size: Byte, word, long
Flags: None affected

MOVES EA,Rn   (Move from address space)
                                                               (Privileged instruction)

Operation: IF supervisor state
                    THEN (Rn) ← (SFC)
                    ELSE TRAP
Binary code: 0000 1110 SS MMMRRR, Word 2
Source addressing mode: (An), (An)+, −(An), d(An), d(An,Xn), Abs.W, Abs.L
Destination addressing mode: Implied
Addressing category: Alterable memory
Size: Byte, word, long
Flags: None affected

This instruction is used to move data from the specified general register to an address location specified by the destination function code register or from an address location specified by the source function code register to the specified general register. The transfer can be performed in byte size, in word size, or in long word size. This is a privileged instruction.

MULS EA,Dn   (Signed multiply)

Operation: (Dest) ← (Source) * (Dest)
Binary code: 1100 RRX 111 MMMRRR
Source addressing mode: Dn, (An), (An)+, −(An), d(An), d(An,Xn), Abs.W, Abs.L,
                    d(PC), d(PC,Xn), Imm
Destination addressing mode: Dn
Addressing category: Data
Size: Word
Flags: N, Z, V = 0, C = 0

This instruction performs a signed binary multiplication. The multiplicand is a word operand (16 bits) specified by the destination operand. The multiplier is also a word operand (16 bits) specified by the source operand. The result of the multiplication is a 32-bit product, which is stored in the destination data register.

MULU EA,Dn   (Unsigned multiply)

Operation: (Dest) ← (Source) * (Dest)
Binary code: 1100 RRX 011 MMMRRR
Source addressing mode: Dn, (An), (An)+, −(An), d(An), d(An,Xn), Abs.W, Abs.L,
                    d(PC), d(PC,Xn), Imm
Destination addressing mode: Dn
Addressing category: Data
Size: Word
Flags: N, Z, V = 0, C = 0

This instruction performs an unsigned binary multiplication. The multiplicand is a word operand (16 bits) specified by the destination operand. The multiplier is also a word operand (16 bits) specified by the source operand. The result of the multiplication is a 32-bit product, which is stored in the destination data register.

NBCD EA (Negate decimal with extend)

Operation: (Dest) ← 0 − (Dest)$_{10}$ − X
Binary code: 0100 1000 00 MMMRRR
Addressing mode: Dn, (An), (An)+, −(An), d(An), d(An,Xn), Abs.W, Abs.L
Addressing category: Data alterable
Size: Byte
Flags: X, N = U, Z, V = U, C

The decimal operand specified by the effective address is negated, taking into consideration the content of the extend bit. If the extend bit is set, then a 9's complement results. If the extend bit is clear, then a 10's complement results. Since the operand is in BCD representation, only byte-size operations are performed.

NEG EA (Negate)

Operation: (Dest) ← 0 − (Dest)
Binary code: 0100 0100 SS MMMRRR
Addressing mode: Dn, (An), (An)+, −(An), d(An), d(An,Xn), Abs.W, Abs.L
Addressing category: Data alterable
Size: Byte, word, long
Flags: X, N, Z, V, C

This instruction is used to negate the specified operand. The negation is done by subtracting the operand from zero. The operation is performed in binary and can be applied to byte size, to word size, or to long word size operands.

NEGX EA (Negate with extend)

Operation: (Dest) ← 0 − (Dest) − (X)
Binary code: 0100 0000 SS MMMRRR
Addressing mode: Dn, (An), (An)+, −(An), d(An), d(An,Xn), Abs.W, Abs.L
Addressing category: Data alterable
Size: Byte, word, long
Flags: X, N, Z, V, C

This instruction is used to negate the specified operand, taking into consideration the current content of the extend bit. The negation is done by subtracting the operand and the content of the extend bit from zero. The operation is performed in binary and can be applied to byte size, to word size, or to long word size operands.

NOP (No operation)

Operation: None
Binary code: 0100 1110 0111 0001
Addressing mode: None
Addressing category: None
Size: Unsized
Flags: None affected

This instruction causes no operations to be performed at all. No flags are affected at all. The processor just proceeds to the next instruction.

NOT EA   (Logical complement)

Operation: (Dest) ← $\overline{\text{(Dest)}}$
Binary code: 0100 0110 SS MMMRRR
Addressing mode: Dn, (An), (An)+, −(An), d(An), d(An,Xn), Abs.W, Abs.L
Addressing category: Data alterable
Size: Byte, word, long
Flags: N, Z, V = 0, C = 0

This instruction performs the logical complement of the specified operand. All flags except the extend flag are affected. This operation can be performed with byte size, word size, or long word size data.

OR Dn,EA   (Inclusive OR logical)

Operation: (Dest) ← (Source) ∨ (Dest)
Binary code: 1000 RRX 1 SS MMMRRR
Source addressing mode: Dn
Destination addressing mode: (An), (An)+, −(An), d(An), d(An,Xn), Abs.W, Abs.L
Addressing category: Alterable memory
Size: Byte, word, long
Flags: N, Z, V = 0, C = 0

OR EA,Dn   (Inclusive OR logical)

Operation: (Dest) ← (Source) ∨ (Dest)
Binary code: 1000 RRX 0 SS MMMRRR
Source addressing mode: Dn, (An), (An)+, −(An), d(An), d(An,Xn), Abs.W, Abs.L,
                                  d(PC), d(PC,Xn), Imm
Destination addressing mode: Dn
Addressing category: Data
Size: Byte, word, long
Flags: N, Z, V = 0, C = 0

This instruction performs the logical inclusive OR with the specified operands. One of the operands must be in the data register. A variant of the OR instruction is the ORI instruction, which is used when the source operand is a literal.

This instruction is a variant of the OR instruction and is used whenever the source operand is a literal. The ORI instruction inclusively ORs the immediate data with the destination operand. If the size of the operation is specified to be byte size, then the lower half of word 2 is taken to be the source operand. If the operation is word size, both halves of word 2 are taken to be the source operand. If the operation is long word size, then both word 2 and word 3 are taken together to be the source operand.

ORI # Data,CCR   (Inclusive OR immediate to condition codes)

Operation: (CCR) ← (Source) ∨ (CCR)
Binary code: 0000 0000 0011 1100, Word 2
Source addressing mode: Imm
Destination addressing mode: Implied
Addressing category: None
Size: Byte
Flags: X, N, Z, V, C

The condition codes are logically Inclusive ORed with the literal. The byte-size literal is taken from the lower half of word 2. All condition codes are affected. The flags affected by this instruction are

Flag	Bit
C	0
V	1
Z	2
N	3
X	4

ORI # Data,EA   (Inclusive OR immediate)

Operation: (Dest) ← Immediate data ∨ (Dest)
Binary code: 0000 0000 SS MMMRRR, Word 2, Word 3
Source addressing mode: Imm
Destination addressing mode: Dn, (An), (An)+, −(An), d(An), d(An,Xn), Abs.W, Abs.L
Addressing category: Data alterable
Size: Byte, word, long
Flags: N, Z, V = 0, C = 0

ORI # Data,SR   (Inclusive OR immediate to the status register)
                                            (Privileged instruction)

Operation: IF supervisor state
              THEN (SR) ← (SR) ∨ (Source)
              ELSE TRAP
Binary code: 0000 0000 0111 1100, Word 2
Source addressing mode: Imm
Destination addressing mode: Implied
Addressing category: None
Size: Word
Flags: X, N, Z, V, C

The contents of the complete status register is logically Inclusive ORed with the literal data taken from word 2. This is a privileged instruction because both the supervisory and user parts of the status register are affected. The flags affected by this instruction are

Flag	Bit
C	0
V	1
Z	2
N	3
X	4
Interrupt	8–10
S	13
T	15

PEA EA   (Push effective address)

    Operation: $(-(SP)) \leftarrow$ (Dest)
    Binary code: 0100 1000 01 MMMRRR
    Addressing mode: (An), d(An), d(An,Xn), Abs.W, Abs.L, d(PC), d(PC,Xn)
    Addressing category: Control
    Size: Long
    Flags: None affected

The specified effective address is computed, and the resultant address is stored in the stack. This instruction is useful for storing addresses and other address manipulation purposes.

RESET   (Reset external devices)
                                                              (Privileged instruction)

    Operation: IF supervisor state
              THEN assert RESET line
              ELSE TRAP
    Binary code: 0100 1110 0111 0000
    Addressing mode: None
    Addressing category: None
    Size: Unsized
    Flags: None affected

This instruction can be used to cause a hardware reset of all peripheral devices by asserting the hardware RESET line. No internal processor state within the microprocessor, except the program counter, is changed. Execution continues with the next instruction. This is a privileged instruction.

ROL Dx,Dy   (Rotate without extend)
ROL #Data,Dy
ROR Dx,Dy
ROR #Data,Dy

    Operation: (Dest) $\leftarrow$ (Dest) Rotated by $\langle$count$\rangle$
    Binary code: 1110 RRX D SS E 11 RRY
    Addressing mode: Dn

Addressing category: None
Size: Byte, word, long
Flags: N, Z, V = 0, C

ROL EA   (Rotate without extend)
ROR EA

> Operation: (Dest) ← (Dest) Rotated by 1 bit
> Binary code: 1110 011 D 11 MMMRRR
> Addressing mode: (An), (An)+, −(An), d(An), d(An,Xn), Abs.W, Abs.L
> Addressing category: Alterable memory
> Size: Word
> Flags: N, Z, V = 0, C

The ROL and ROR instructions can be used to rotate the operand to the left or to the right. The direction to be rotated is determined by bit D, that is, bit 8, of the OPCODE. The operand may be stored in a data register or in memory. If the operand is being rotated to the left, the last bit rotated out from the most significant side is stored in the C flag and at the same time rotated in on the right. If the operand is being rotated to the right, the last bit shifted out on the right is stored in the C flag and at the same time rotated in on the left.

If the operand is stored in a data register, any number of rotates in either direction is possible if the rotate count is stored in a data register. If the rotate count is in the range 0 to 7, the rotate count may be stored as part of the OPCODE as determined by bit E.

If the operand is stored in memory, only one bit can be rotated at a time, and the size of the operand is always taken to be word size. A number of addressing modes are available for the effective address specification.

ROXL Dx,Dy   (Rotate with extend)
ROXL #Data,Dy
ROXR Dx,Dy
ROXR #Data,Dy

> Operation: (Dest and Extend) ← (Dest and Extend) Rotated by ⟨count⟩
> Binary code: 1110 RRX D SS E 10 RRY
> Addressing mode: Dn
> Addressing category: None
> Size: Byte, word, long
> Flags: N, Z, V = 0, C

ROXL EA   (Rotate with extend)
ROXR EA

> Operation: (Dest and Extend) ← (Dest and Extend) Rotated by 1 bit
> Binary code: 1110 010 D 11 MMMRRR
> Addressing mode: (An), (An)+, −(An), d(An), d(An,Xn), Abs.W, Abs.L
> Addressing category: Alterable memory
> Size: Word
> Flags: N, Z, V = 0, C

The ROXL and ROXR instructions can be used to rotate the operand to the left or to the right. The direction to be rotated is determined by bit D, that is, bit 8, of the OPCODE. The primary difference between the ROL/ROR instruction and the ROXL/ROXR instruction is that the former rotation does not include the extend bit whereas the latter does. The operand may be stored in a data register or in memory. If the operand is being rotated to the left, the last bit rotated out from the most significant side of the specified operand is stored in both the X flag and the C flag, and, at the same time, the X flag is rotated in on the right. If the operand is being rotated to the right, the last bit shifted out on the right is stored in both the X flag and the C flag, and, at the same time, the X flag is rotated in on the left.

If the operand is stored in a data register, any number of rotates in either direction are possible if the rotate count is stored in a data register. If the rotate count is in the range 0 to 7, the rotate count may be stored as part of the OPCODE as determined by bit E.

If the operand is stored in memory, only one bit can be rotated at a time, and the size of the operand is always taken to be word size. A number of addressing modes are available for the effective address specification.

RTD #d   (Return and deallocate parameters)

    Operation: $(PC) \leftarrow ((SP)+)$
              $(SP) \leftarrow (SP) + d$
    Binary code: 0100 1110 0111 0100, Word 2
    Addressing mode: None
    Addressing category: None
    Size: Unsized
    Flags: None affected

This instruction provides a return from subroutine with the added feature that the SP is also displaced by the given sign-extended 16-bit offset. The offset allows the proper SP deallocation when the stack is also used to pass parameters of the subroutine.

RTE   (Return from exception)
                                                                           (Privileged instruction)

    Operation: IF supervisor state
              THEN BEGIN
                    $(SR) \leftarrow ((SP)+)$
                    $(PC) \leftarrow ((SP)+)$
                    IF $((SP)+)$ = long format
                        THEN full restore
              END
              ELSE TRAP
    Binary code: 0100 1110 0111 0011
    Addressing mode: None

Addressing category: None
Size: Unsized
Flags: X, N, Z, V, C

This instruction is used to return to normal program processing after an exception processing. The return may be a short restore or a long restore, depending on the type of exception as indicated by the vector offset word retrieved from the stack. A short form indicates that 4 words are to be retrieved from the stack. A long form indicates that 29 words are to be retrieved from the stack.

RTR   (Return and restore condition codes)

Operation: (CC) ← ((SP)+)
           (PC) ← ((SP)+)
Binary code: 0100 1110 0111 0111
Addressing mode: None
Addressing category: None
Size: Unsized
Flags: X, N, Z, V, C

This is another form of subroutine return where both the PC and the condition codes are updated by the stack.

RTS   (Return from subroutine)

Operation: (PC) ← ((SP)+)
Binary code: 0100 1110 0111 0101
Addressing mode: None
Addressing category: None
Size: Unsized
Flags: None affected

This is the simplest form of subroutine return. Only the PC is pulled from the stack. No other information is pulled from the stack.

SBCD Dy,Dx   (Subtract BCD)
SBCD −(Ay),−(Ax)

Operation: (Dest) ← (Dest)$_{10}$ − (Source)$_{10}$ − X
Binary code: 1000 RRX 1000 0 A RRY
Source addressing mode: Dn or −(An)
Destination addressing mode: Dn or −(An)
Addressing category: None
Size: Byte
Flags: X, Z = U, Z, V = U, C

This instruction performs BCD subtraction taking into consideration the content of the extended bit. The operands can be stored either in the data registers (RRX and RRY) or in memory as specified by address registers RRX and RRY.

The selection between data register operands or memory operands is determined by bit A. If the operands are in memory, the pre-decrement memory addressing mode is assumed.

Scc EA   (Set according to condition)

    Operation: IF condition
               THEN (Dest) ← 1's
               ELSE (Dest) ← 0's
    Binary code: 0101 CCCC 11 MMMRRR
    Addressing mode: Dn, (An), (An)+, −(An), d(An), d(An,Xn), Abs.W, Abs.L
    Addressing category: Data alterable
    Size: Byte
    Flags: None affected

This instruction sets the destination operand specified by the effective address according to the result of the specified condition. Since the destination is a logical variable, only byte-size operation is allowed. All 16 conditions can be specified by this instruction.

STOP #Data   (Load status register and stop)
                                         (Privileged instruction)

    Operation: IF supervisor state
               THEN BEGIN
                    SR ← Immediate data;
                    STOP
                    END
               ELSE TRAP
    Binary code: 0100 1110 0111 0010, Word 2
    Addressing mode: Imm
    Addressing category: None
    Size: Unsized
    Flags: None

This instruction halts the processor. Before halting, the literal stored in word 2 is fetched and updated to the full status register. The halt state can be terminated by a trace, an interrupt with a higher priority, or an external reset. All the above-mentioned conditions cause exception processing when execution is resumed.

SUB Dn,EA   (Subtract binary)
    Operation: (Dest) ← (Dest) − (Source)
    Binary code: 1001 RRX 1 SS MMMRRR
    Source addressing mode: Dn
    Destination addressing mode: (An), (An)+, −(An), d(An), d(An,Xn), Abs.W, Abs.L
    Addressing category: Alterable memory
    Size: Byte, word, long
    Flags: X, N, Z, V, C

SUB EA,Dn   (Subtract binary)

    Operation: (Dest) ← (Dest) − (Source)
    Binary code: 1001 RRX 0 SS MMMRRR
    Source addressing mode: Dn, An, (An), (An)+, −(An), d(An), d(An,Xn), Abs.W, Abs.L,
                d(PC), d(PC,Xn), Imm
    Destination addressing mode: Dn
    Addressing category: All
    Size: Byte, word, long
    Flags: X, N, Z, V, C

This instruction performs a signed 2's complement subtraction between the two specified operands. One of the operands must be stored in a data register. The other operand, which is a source or destination operand, may be stored in memory. Whether the destination operand is in the data register or in memory is determined by bit 8 in the OPCODE. The size of the operand is given by SS. All flags are affected by the SUB instruction. The SUB instruction has several variants. When the destination is an address register, SUBA is used. When the source operand is a literal, SUBI is used. When the literal is in the range 1 to 8, SUBQ is used.

SUBA EA,An   (Subtract address)

    Operation: (Dest) ← (Dest) − (Source)
    Binary code: 1001 RRX S 11 MMMRRR
    Source addressing mode: Dn, An, (An), (An)+, −(An), d(An), d(An,Xn), Abs.W, Abs.L,
                d(PC), d(PC,Xn), Imm
    Destination addressing mode: An
    Addressing category: All
    Size: Word, long
    Flags: None affected

The SUBA instruction is a variant of the SUB instruction and is used whenever the destination operand is stored in the address register. The SUBA instruction is used to perform signed 2's complement subtraction. The source operand can be anywhere. Only word and long word operations are allowed.

SUBI #Data,EA   (Subtract immediate)

    Operation: (Dest) ← (Dest) − Immediate data
    Binary code: 0000 0100 SS MMMRRR, Word 2, Word 3
    Source addressing mode: Imm
    Destination addressing mode: Dn, (An), (An)+, −(An), d(An), d(An,Xn), Abs.W, Abs.L
    Addressing category: Data alterable
    Size: Byte, word, long
    Flags, X, N, Z, V, C

This instruction is a variant of the SUB instruction and is used when one of the source operands is a literal. The SUBI instruction subtracts the immediate data from the contents of the destination operand. The result is stored in the specified destination operand. If the size of the operation is specified to be byte size, the lower half of word 2 is taken to be the source operand. If the operation is word size, both halves of word 2 are taken to be the source operand. If the operation is long word size, the immediate data are contained in word 2 and word 3. In the last case, two words are required after the OPCODE.

SUBQ #Data,EA   (Subtract quick)

    Operation: (Dest) ← (Dest) − Immediate data
    Binary code: 0101 DDD 1 SS MMMRRR
    Source addressing mode: Imm
    Destination addressing mode: Dn, An, (An), (An)+, −(An), d(An), d(An,Xn), Abs.W,
                      Abs.L
    Addressing category: Alterable
    Size: Byte, word, long
    Flags: X, N, Z, V, C

The SUBQ instruction is a variant of the SUB instruction. The SUBQ instruction is used when the source operand is a literal and is in the range 1 to 8. The immediate data are stored as part of the OPCODE and do not require word 2 or word 3. In essence, the SUBQ instruction is a decrement instruction, but it is more powerful than a usual decrement instruction in that the amount to be decremented can be specified to be in the range 1 to 8.

SUBX Dy,Dx   (Subtract with extend)
SUBX −(Ay),−(Ax)

    Operation: (Dest) ← (Dest) − (Source) − (X)
    Binary code: 1001 RRX 1 SS 00 A RRY
    Source addressing mode: Dn or −(An)
    Destination addressing mode: Dn or −(An)
    Addressing category: None
    Size: Byte, word, long
    Flags: X, N, Z, V, C

This instruction is similar to the SUB instruction except that in addition to the signed 2's complement subtraction, the content of the extend bit (X) is also subtracted from the source operand. This instruction is useful for multiple-precision subtraction. Similar to the SUB instruction, the operation can be stored either in data registers or in memory as determined by bit A. If memory operations are specified, the address register pre-decrement mode is assumed.

SWAP Dn   (Swap register halves)

    Operation: Dn(15:0) ↔ Dn(31:16)
    Binary code: 0100 1000 0100 0 RRX
    Addressing mode: Dn
    Addressing category: None
    Size: Word
    Flags: N, Z, V = 0, C = 0

In this instruction, the two halves of the specified data register are swapped. The N and Z flags are changed accordingly.

TAS EA   (Test and set an operand)

> Operation: (CC)   ← (Dest) Tested
>   (Dest(7)) ← 1
> Binary code: 0100 1010 11 MMMRRR
> Addressing mode: Dn, (An), (An)+, −(An), d(An), d(An,Xn), Abs.W, Abs.L
> Addressing category: Data alterable
> Size: Byte
> Flags: N, Z, V = 0, C = 0

In one instruction, the value of the specified operand is tested with the results stored in the N flag and the Z flag. Within the same instruction, the most significant bit of the specified destination byte is set and is written back to its destination location. This instruction facilitates the programming of semaphores.

TRAP #Vector   (Trap)

> Operation: (−(SSP)) ← (PC)
>   (−(SSP)) ← (SR)
>   (PC)   ← (Vector)
> Binary code: 0100 1110 0100 vvvv
> Addressing mode: None
> Addressing category: None
> Size: Unsized
> Flags: None affected

This instruction causes software-initiated exception processing at the specified vector address. Any one of a total of 16 vector addresses can be specified.

TRAPV   (Trap on overflow)

> Operation: IF V THEN TRAP
> Binary code: 0100 1110 0111 0110
> Addressing mode: None
> Addressing category: None
> Size: Unsized
> Flags: None affected

If the overflow condition is set, then the processor will initiate exception processing. If the overflow condition is not set, the processor continues normal processing.

TST EA   (Test an operand)

> Operation: (CC) ← (Dest) Tested
> Binary code: 0100 1010 SS MMMRRR
> Addressing mode: Dn, (An), (An)+, −(An), d(An), d(An,Xn), Abs.W, Abs.L
> Addressing category: Data alterable
> Size: Byte, word, long
> Flags: N, Z, V = 0, C = 0

The specified operand is compared with zero. The flags N and Z are then set according to the result of the subtraction.

UNLK An   (Unlink)

Operation: $(SP) \leftarrow (An)$
$(An) \leftarrow ((SP)+)$
Binary code: 0100 1110 0101 1 RRX
Addressing mode: None
Addressing category: None
Size: Unsized
Flags: None affected

This instruction restores the SP from the location pointed to by the address register. Then the address register is updated by the top of the stack as pointed to by the updated SP. This instruction works together with the LINK instruction for the maintenance of subroutines.

Index